Solid Mechanics and Its Applications

Volume 259

The fundamental questions arising in mechanics are: Why?, How?, and How much? The aim of this series is to provide lucid accounts written by authoritative researchers giving vision and insight in answering these questions on the subject of mechanics as it relates to solids. The scope of the series covers the entire spectrum of solid mechanics. Thus it includes the foundation of mechanics; variational formulations; computational mechanics; statics, kinematics and dynamics of rigid and elastic bodies; vibrations of solids and structures; dynamical systems and chaos; the theories of elasticity, plasticity and viscoelasticity; composite materials; rods, beams, shells and membranes; structural control and stability; soils, rocks and geomechanics; fracture; tribology; experimental mechanics; biomechanics and machine design. The median level of presentation is the first year graduate student. Some texts are monographs defining the current state of the field; others are accessible to final year undergraduates; but essentially the emphasis is on readability and clarity.

Springer and Professors Barber and Klarbring welcome book ideas from authors. Potential authors who wish to submit a book proposal should contact Dr. Mayra Castro, Senior Editor, Springer Heidelberg, Germany, e-mail: mayra.castro@springer.com

Indexed by SCOPUS, Ei Compendex, EBSCO Discovery Service, OCLC, ProQuest Summon, Google Scholar and SpringerLink.

More information about this series at http://www.springer.com/series/6557

Nikolay Banichuk · Alexander Barsuk ·
Juha Jeronen · Tero Tuovinen ·
Pekka Neittaanmäki

Stability of Axially Moving Materials

Springer

Nikolay Banichuk
Russian Academy of Sciences
Institute for Problems in Mechanics
Moscow, Russia

Juha Jeronen
Faculty of Information Technology
University of Jyväskylä
Jyväskylä, Finland

Pekka Neittaanmäki
Faculty of Information Technology
University of Jyväskylä
Jyväskylä, Finland

Alexander Barsuk
State University of Moldova
Chisinau, Moldova

Tero Tuovinen
Faculty of Information Technology
University of Jyväskylä
Jyväskylä, Finland

ISSN 0925-0042 ISSN 2214-7764 (electronic)
Solid Mechanics and Its Applications
ISBN 978-3-030-23802-5 ISBN 978-3-030-23803-2 (eBook)
https://doi.org/10.1007/978-3-030-23803-2

This Springer imprint is published by the registered company Springer Nature Switzerland AG
The registered company address is: Gewerbestrasse 11, 6330 Cham, Switzerland

Preface

In this book, we discuss a variety of problems involving analyses of stability in mechanics, focusing especially on the stability of axially moving materials. This is a special topic encountered, for example, in process industry applications, such as in papermaking. In theoretical terms, the field of axially moving materials is located halfway between classical solid mechanics and fluid mechanics. The object of interest is a solid, but it flows through the domain of interest such as a particular section of a paper machine, motivating the use of an Eulerian viewpoint. In vibration problems in mechanics, a loss of stability is often (but not always) accompanied by a bifurcation in the complex eigenvalue curves describing the behavior of the system under the action of a quasistatically increasing external load. Thus in many of the chapters of this book, we will consider bifurcations in some form.

In our opinion, at the present time, analytical and semianalytical approaches are undervalued. They can serve as a basis for fundamental theoretical understanding, but also importantly, as a basis for fast numerical solvers for realtime applications, such as online prediction and control of industrial processes. In this book, a focus on analytical approaches is a recurring theme, sometimes with a change in perspective, or with an unconventional application of a known general result such as the implicit function theorem. Some of the results presented in this book are new, and some have appeared only in specialized journals or in conference proceedings. Some appear now for the first time in English.

The book is organized into nine chapters. Chapters 1 through 4 discuss bifurcations in mechanics, introducing the basic ideas, approaches and methods used throughout the book. Many topics are discussed via the use of particular examples. Chapter 1 plays an introductory role, discussing prototype problems and bifurcations of different kinds. This chapter concisely summarizes the typical simplest bifurcation problems that arise in a setting of classical solid mechanics. We concentrate on finding the eigenvalues (critical stability parameters) and eigenmodes, characterizing the shape of stability loss, of the appropriate spectral problem.

Chapter 2 is devoted to the bifurcation analysis of algebraic polynomial equations. The parametric representation of the solutions of considered equations and their bifurcations are considered. Bifurcation analyses of the cubic equation and of the fourth-order polynomial equation are presented in detail. This has applications in the stability analysis of one-dimensional differential equation models whose characteristic equations reduce into polynomials.

Chapter 3 deals with non-conservative dynamic systems with a finite number of degrees of freedom. The system is investigated under small perturbations, characterized by a small parameter. The characteristic polynomial and a series expansion with respect to the small parameter are used for the eigenvalues and eigenvectors to evaluate the critical stability parameter and to study ideal and destabilizing perturbations. Sufficient conditions for stability are obtained and described. Some examples of ideal perturbations and the structure of the corresponding perturbation matrices are considered. The stability of the systems subjected to deficient perturbations is also investigated, and the determination of the deficiency index is presented. Ziegler's double pendulum is considered as a classical example of a non-conservative system. We give a detailed exposition, deriving the governing equations starting from the principle of virtual work, analyze the stability of the system, and consider some special cases. We present numerical simulations using new visualization techniques, in which the bifurcation behavior can be seen.

Chapter 4 deals with the general methods of bifurcation analysis, applied to continuous systems, and some methods of optimization of the critical stability parameter. We will briefly introduce the different types of stability loss and then look at conditions under which merging of eigenvalues may occur. We will look at a problem where applying symmetry arguments allows us to eliminate multiple (merged) eigenvalues, thus reducing the problem to determining a classical simple eigenvalue. We then discuss a general technique to look for bifurcations in problems formulated as implicit functionals. This is useful for a wide class of problems, including many problems in axially moving materials. At the end of the chapter, we will consider a variational approach to the stability analysis of an axially moving panel (a plate undergoing cylindrical deformation).

Chapters 5 through 8 form the main content of the book, concentrating specifically on axially moving materials. We start by introducing the theory of axially moving materials in a systematic manner, the aim being to give a complete overview of the fundamentals. By presenting the theory as a self-contained unit, it is our hope that this chapter may especially help the student or new researcher just entering the research field of axially moving materials. Specialists, on the other hand, can benefit from the discussion on the effects of the axial motion on the boundary conditions, a topic that has received relatively little attention. This naturally leads to a mixed formulation, which both reduces the continuity requirements on the solution and clearly shows how the boundary conditions arise, contrasting the classical treatment of the transverse deformations of axially moving elastic and viscoelastic materials using fourth- and fifth-order partial differential equations.

Chapter 5 starts by considering the general balance laws of linear and angular momentum. We derive some general equations for beams and specialize them to the small-displacement regime. We discuss linear constitutive models for elastic and viscoelastic materials, and highlight the connection between beams and panels. We then introduce axial motion in a systematic manner, via a coordinate transformation, and discuss how this affects the boundary conditions. We consider the dynamic linear stability analysis of axially moving elastic and viscoelastic materials. We numerically look at bifurcations in the stability exponents. As a result, we find that the small-viscosity case behaves radically different from the purely elastic case. Because no real material in papermaking is purely elastic, this has important practical implications for the correct qualitative understanding of real physical systems.

Chapter 6 treats bifurcations of axially moving elastic plates made of isotropic and orthotropic materials. A static stability analysis is performed to find the critical axial drive velocity and the corresponding shape in which the system loses stability. As a special topic of particular interest for process industry applications, we then look at the axially moving isotropic plate, but with an axial tension distribution that varies along the width. It is seen that as far as the critical velocity is concerned, the classical simplification assuming homogeneous tension is acceptable, but the eigenmode is highly sensitive to even minor variations in the axial tension distribution in the width direction.

Chapter 7 deals with the theoretical analysis of bifurcations of axially moving strings and beams. Critical velocities of bifurcations of the traveling material are determined for torsional, longitudinal, and transverse vibration types. The stability analysis of the axially moving string with and without damping is performed. We also comment on exact eigensolutions for axially moving beams and panels. The stability analyses of an axially moving web with elastic supports, and when subjected to a uniform gravitational field, are also considered.

Chapter 8 deals with bifurcations in fluid–structure interaction in the context of axially moving materials. This chapter includes a brief introduction to fluid mechanics, after which we look at analytical solutions for two-dimensional potential flow in fluid–structure interaction with an axially moving panel. A Green's function approach is used to analytically derive the fluid reaction pressure in terms of the panel displacement function. This simplifies the numerical problem to solving a one-dimensional integrodifferential model. An added-mass approximation of the derived solution is discussed. Numerical results based on the original exact solution are presented, for both elastic and viscoelastic axially moving panels subjected to a potential flow.

The book concludes with Chap. 9, considering optimization problems in thermoelasticity. We find the optimal thickness distribution for a beam to resist thermal buckling and the optimal material distribution for a beam of uniform thickness, with the same goal. Finally, we derive a guaranteed double-sided estimate that governs energy dissipation in heat conduction in a locally orthotropic solid body, which holds regardless of how the local material orientation is distributed in the solid body. Connecting to the main theme of this book, this has applications in the analysis of heat conduction in paper materials, which is important when considering the drying process in papermaking.

This book is addressed to researchers, specialists, and students in the fields of theoretical and applied mechanics, and of applied and computational mathematics. Considering topics related to manufacturing and processing, the book can also be applied in industrial mathematics. We hope that contents should also be of interest to applied mathematicians and mechanicians not currently in these fields, who may nonetheless be stimulated by the material presented.

It is our hope that the various solution techniques touched upon across the chapters will benefit the reader, whether in their original context or in an unexpected new application. We also hope, via detailed exposition and examples, to have made the theory of axially moving materials more accessible to a new generation of researchers. Although the field of axially moving materials was established over a century ago, new exciting applications await, for example, in printable electronics and microfluidics. This makes the field worthy of study not only from a fundamental academic viewpoint but also for those primarily interested in applications.

Moscow, Russia — Nikolay Banichuk
Chisinau, Moldova — Alexander Barsuk
Jyväskylä, Finland — Juha Jeronen
Jyväskylä, Finland — Tero Tuovinen
Jyväskylä, Finland — Pekka Neittaanmäki

Acknowledgements The research presented in this book was supported by the Academy of Finland (grant no. 140221, 301391, 297616); RFBR (grant 14-08-00016-a); Project RSF No. 17 19-01247; RAS program 12, Program of Support of Leading Scientific Schools (grant 2954.2014.1); Jenny and Antti Wihuri Foundation. The authors would like to thank Matthew Wuetrich for proofreading. For discussions and valuable input, the authors would like to thank Svetlana Ivanova, Evgeny Makeev, Reijo Kouhia, Tytti Saksa, and Tuomo Ojala.

Contents

Chapter 1
Prototype Problems: Bifurcations of Different Kinds

In this chapter, we present some prototype bifurcation problems that arise in the mechanics of rigid and deformable structural elements. These problems are typical for engineering applications and characterize the approaches that can be applied in the investigation of stability. Some methods of bifurcation theory will be presented in the context of stability studies of the considered one-dimensional mechanical problems.

1.1 Rigid Column with Elastic Clamping

Consider the problem of the stability of an absolutely rigid column of length ℓ, which is elastically clamped at one end, and assume that a compressive force P is applied to the free end, as shown in Fig. 1.1.

For sufficiently large values of P, when the rigid column loses its vertical trivial position (i.e., loses stability), the force vector retains its original direction, but its line of action is displaced in a parallel manner, and the point of application remains at the free end of the column. We adopt rectangular coordinates x and z, where the end of the column is at the origin O of this coordinate system, and the x axis coincides with the direction of the force P, see Fig. 1.1. The axis z lies in the buckling plane for the beam. Its direction is indicated in Fig. 1.1. We restrict our discussion to small deflections of the beam. The magnitude of the deflection of the axis of the beam is measured from the line of action of the force and described by the angle θ, while ℓ and C denote the length of the beam, and the rigidity of the elastic clamping, respectively. The total potential energy J of the considered mechanical system is the sum of the elastic energy U and the potential of external force Π, in other words,

$$J = U + \Pi \,. \tag{1.1.1}$$

N. Banichuk et al., *Stability of Axially Moving Materials*, Solid Mechanics and Its Applications 259, https://doi.org/10.1007/978-3-030-23803-2_1

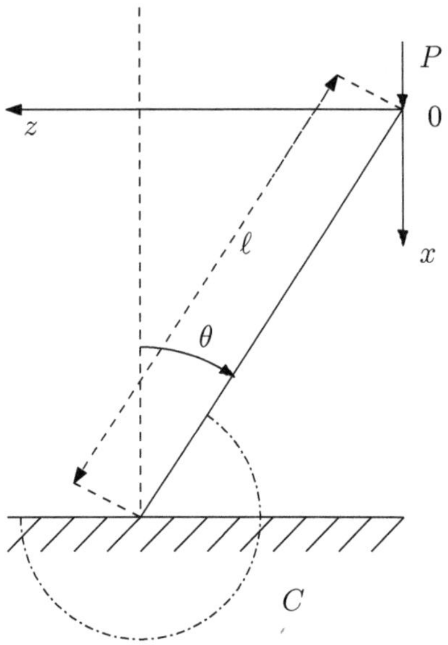

Fig. 1.1 Rigid column with elastic clamping

Taking into account the expressions for the elastic energy of a deformed hinge,

$$U = \frac{1}{2} C\theta^2 \, , \tag{1.1.2}$$

and for the potential of the external force, represented by the product of the force and the vertical displacement of the point of its application, that is,

$$\Pi = -P\ell \, (1 - \cos\theta) \ \ , \tag{1.1.3}$$

we find the expression for J, given by (1.1.1), in the form

$$J = \frac{1}{2} C\theta^2 - P\ell \, (1 - \cos\theta) \ . \tag{1.1.4}$$

In the equilibrium position the total potential energy of the considered conservative mechanical system has a stationary value. In accordance with Lagrange's theorem on the stability of an equilibrium position, if, in some position of a conservative system, the total potential energy has a strict minimum, then this position is the position of stable equilibrium of the system.

Note here that a position of a system is called a *position of equilibrium* if the system, starting from that position with zero velocity, remains in that position. The position of equilibrium is called stable if, for sufficiently small initial deviations and sufficiently small initial velocities, the system does not go beyond the limits of an arbitrary small (preassigned) neighborhood of the position of equilibrium during the entire period of motion, and similarly has arbitrary small velocities.

Note also that if the total potential energy J of a conservative system at a position of equilibrium does not have a minimum there, then the given position of equilibrium is unstable.

The system under consideration is characterized by the single coordinate θ, and to analyze the behavior of the system we will use the following expressions for the derivatives of J:

$$\frac{dJ}{d\theta} = C\theta - P\ell \sin\theta, \tag{1.1.5}$$

$$\frac{d^2J}{d\theta^2} = C - P\ell \cos\theta\ . \tag{1.1.6}$$

To find the equation of equilibrium, we equate the first derivative (1.1.5) to zero. We have

$$P\ell \sin\theta = C\theta\ . \tag{1.1.7}$$

There are several equilibrium solutions (equilibrium positions) of the Eq. (1.1.7). The equilibrium position $\theta = 0$ is stable if

$$\frac{d^2J}{d\theta^2} > 0\ . \tag{1.1.8}$$

Using (1.1.6), we have the condition

$$\frac{d^2J}{d\theta^2} = C - P\ell > 0\ . \tag{1.1.9}$$

Thus, the total potential energy is at a minimum when $P\ell < C$ and $\theta = 0$ (vertical position). If $P\ell > C$ and $\theta = 0$, then, as follows from Eq. (1.1.9), the total potential energy is at a maximum, and consequently the vertical position of the beam is unstable in this case (see Fig. 1.2).

To study the stability of an equilibrium of the system in non-trivial positions ($\theta \neq 0$), it is possible to use the Eq. (1.1.7), the expression (1.1.6), and to require

$$\frac{d^2J}{d\theta^2} = C\left(1 - \frac{\theta}{\tan\theta}\right) > 0\ . \tag{1.1.10}$$

This inequality is fulfilled when $\theta < \tan\theta$, in other words for $|\theta| < \pi$. Thus, the second derivative is positive for the deflected positions if $P\ell > k$. Consequently, the position with $\theta \neq 0$ is stable for $P\ell > C$.

In the case

$$\theta = 0 \quad \text{and} \quad P = \frac{C}{\ell}\ , \tag{1.1.11}$$

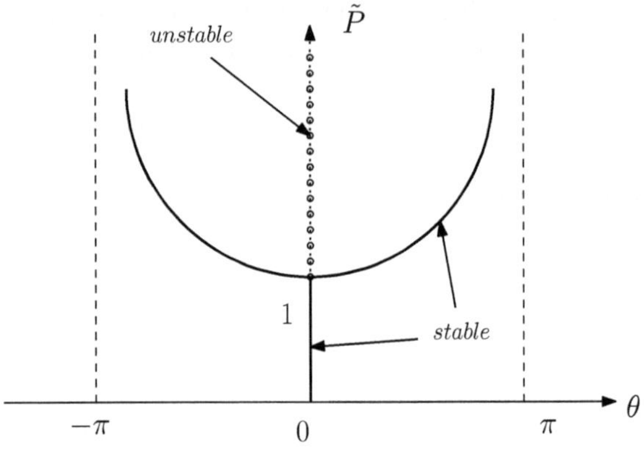

Fig. 1.2 Unstable and stable branches

the second derivative of the total potential energy is equal to zero, that is,

$$\frac{\mathrm{d}^2 J}{\mathrm{d}\theta^2} = 0\,, \tag{1.1.12}$$

and consequently it is necessary to derive and estimate the expressions for higher derivatives. We have

$$\frac{\mathrm{d}^3 J}{\mathrm{d}\theta^3} = P\ell\sin\theta, \qquad \frac{\mathrm{d}^4 J}{\mathrm{d}\theta^4} = P\ell\cos\theta. \tag{1.1.13}$$

The third derivative is equal to zero, but the fourth derivative is positive,

$$\frac{\mathrm{d}^4 J}{\mathrm{d}\theta^4} = C > 0\,. \tag{1.1.14}$$

Thus, the nondeflected position ($\theta = 0$), in this case ($P = C/\ell$), is characterized by a minimal total potential energy, and consequently this position is stable (see Fig. 1.2, where $\tilde{P} = P\ell/C$).

As it was shown previously, the vertical state of the beam is stable up to $P \le C/\ell$. Therefore, if we load the beam and $\tilde{P} \le 1$, then the beam remains in its original vertical state. If the force exceeds this value, then the vertical position becomes unstable, and the beam leaves the original position. Taking into account that, in the vicinity of the point B, there is another stable deflected position of equilibrium, the beam passes to the deflected state. The corresponding behavior is shown by the arrows in Fig. 1.3. If $\tilde{P} > 0$, then the beam cannot stay in the vertical position, and it deviates to the right or to the left.

Thus, the first stage in Fig. 1.3 is characterized by the point B, where the ordinate axis corresponding to the original vertical position crosses the curve $\bar{P} = \dfrac{\theta}{\sin\theta}$ that corresponds to the deflected position.

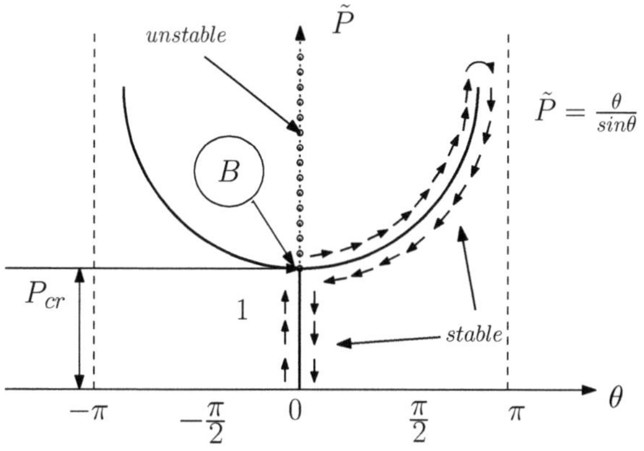

Fig. 1.3 When we load the beam and $\tilde{P} \leq 1$, then the beam remains in its original vertical state. If the force exceeds this value, then the vertical position becomes unstable, and the beam leaves the original position. Taking into account that in the vicinity of the point B there is another stable deflected position of equilibrium, the beam passes to the deflected state. The corresponding behavior is shown by the arrows

Points at which the solution splits into two (or more) branches, such as the point B, are called bifurcation points or branch points.

If we go through the bifurcation point B, then the original equilibrium state of the beam loses the stability property, and the points on the line $\theta = 0$ above the point B correspond to unstable positions.

Bifurcation Phenomenon

If the equation

$$F(u, \lambda) = 0 , \tag{1.1.15}$$

where F is a nonlinear operator, depending on a parameter λ, applied to an unknown function or vector u, has a critical value λ_0 such that the given solution of Eq. (1.1.15) splits at $\lambda = \lambda_0$, then λ_0 is called the bifurcation (or branching) value.

1.2 Elastic Column and Its Optimization

In the discussion given below, following the paper by Banichuk [1], we shall consider the problems of stability for a compressed rod (column). We shall, in the course of the investigation, generalize the well-known solution given in the article by Clausen [2], originally developed for the specific case having a rigid support.

We assume that a cantilevered column is elastically held at one end and a compressive force P is applied to the free end (as shown in Fig. 1.4).

Fig. 1.4 Elastic column with an elastic support

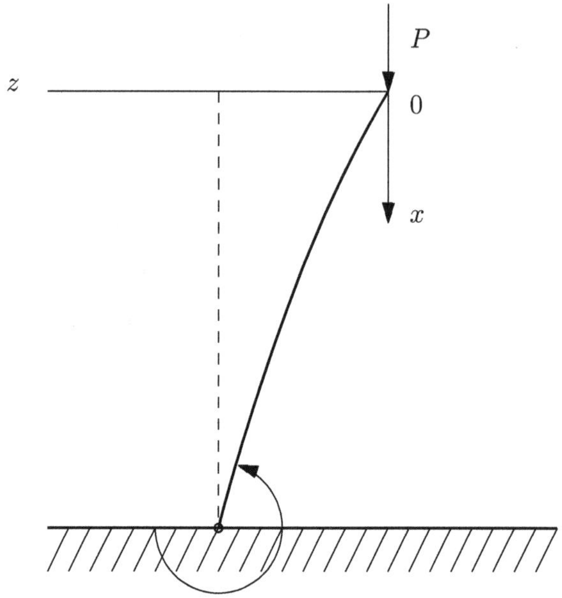

For sufficiently large values of P, while the column loses its stability and buckles, the force vector retains its original direction, but its line of action is displaced in a parallel manner, and the point of application remains at the free end of the column. We adopt rectangular coordinates x and z, where the end of the column is at the origin O of this coordinate system, and the axis x coincides with the direction of the force P, see Fig. 1.4. The axis z lies in the buckling plane of the beam. Its direction is indicated in Fig. 1.4.

We restrict our discussion to small deformations of the beam, and we study its equilibrium within the framework of the linear theory of elasticity. The magnitude of the deflection of the bent axis of the beam, denoted as $w(x)$, is measured from the line of action of the force, while ℓ denotes the length of the beam. Assuming that the shape of the cross section of the beam is uniform, we shall derive the basic relations arising in the problem of maximizing the load causing the loss of stability. We shall find the optimum (in the above sense) distribution of cross-sectional area $S = S(x)$ as a function of the distance x measured along the axis of the beam. We have

$$EI\frac{\mathrm{d}^2 w}{\mathrm{d}x^2} + Pw = 0\,, \qquad EI = C_2 S^2\,,$$

$$w(0) = 0\,, \qquad \left(c\frac{\mathrm{d}w}{\mathrm{d}x} - Pw\right)_{x=\ell} = 0\,, \tag{1.2.1}$$

$$\int_0^\ell S\,\mathrm{d}x = V\,, \qquad P \to \max_S\,,$$

where E, c, V and $I = I(x)$ are respectively Young's modulus, the coefficient of the rigidity of foundations, the given volume of the beam and moment of inertia of the cross-sectional area, relative to the axis perpendicular to the bending plane xz and intersecting the neutral axis of the beam at the point with the coordinate x. The load P denotes the eigenvalue of the boundary-value problem of Eq. (1.2.1).

A necessary condition for maximizing the critical load of compressed columns was derived in Lurie [3], Nikolai [4], Chentsov [5], Clausen [2], Keller [6]. It was obtained that

$$w^2 = S^3 \,. \tag{1.2.2}$$

This condition was originally derived for other types of boundary conditions; however, it is not hard to show that it remains valid in the case considered above. The necessary condition for maximizing the critical load relates the functions w and S and, together with the basic relations (1.2.2), leads to a solvable boundary-value problem. Using (1.2.2) to eliminate the function S from the bending equations and from the isoperimetric condition, and introducing the nondimensional variables

$$\tilde{x} = \frac{x}{\ell}\,, \quad \tilde{w} = \frac{C_2}{P\ell^2} w\,, \quad \tilde{S} = \frac{\ell}{V} S\,, \quad \tilde{P} = \frac{\ell}{C} P\,,$$

we obtain the following relations:

$$\frac{d^2 w}{dx^2} + w^{-1/3} = 0\,, \tag{1.2.3}$$

$$w(0) = 0\,, \qquad \left(\frac{dw}{dx} - Pw\right) = 0\,,$$

$$\int_0^1 w^{2/3}\, dx = \frac{\gamma}{\sqrt{P}}\,, \qquad \gamma = \frac{V}{\ell}\sqrt{\left(\frac{C_2}{c\ell}\right)}\,.$$

We have omitted the tilde from the notation.

The general solution of the differential Eq. (1.2.3) may be represented in a parametric form as

$$x = \frac{1}{2}\sqrt{3}\mu_1^{2/3}\left(\theta - \frac{1}{2}\sin 2\theta\right) + \mu_2\,, \tag{1.2.4}$$

$$w = \mu_1 \sin^3\theta, \qquad \theta_0 \le \theta \le \theta_1\,,$$

where μ_1, μ_2, θ_0 and θ_1 are constants that are to be determined. To find the values of these constants, we make use of the boundary conditions at $x = 0$ and $x = 1$, and the conditions $x(\theta_0) = 0$ and $x(\theta_1) = 1$. After substituting these values into the solution (1.2.4), we have

$$\sin \theta_0 = 0 \,, \quad \mu_2 = -\frac{1}{2}\sqrt{3}\mu_1^{2/3}\theta_0 \,, \tag{1.2.5}$$

$$\frac{1}{2}\sqrt{3}\mu_1^{2/3}\left(\theta_1 - \theta_0 - \frac{1}{2}\sin 2\theta_1\right) = 1 \,,$$

$$\mu_1^{2/3} = \sqrt{3}\frac{\cos\theta_1}{P\sin^3\theta_1} \,.$$

Equation (1.2.5) are satisfied if we substitute $\theta_0 = 0$, $\mu_2 = 0$, $\theta_1 = \theta_*$ and $\mu_1^{2/3} = \sqrt{3}\,(\cos\theta_*)\,/\,\left(P\sin^3\theta_*\right)$, where θ_* denotes a root of the equation

$$\varphi\,(\theta) = 3\cos\theta\frac{\left(\theta - \dfrac{1}{2}\sin 2\theta\right)}{\left(2\sin^3\theta\right)} = P \tag{1.2.6}$$

that satisfies the inequality $0 \le \theta_* \le \pi/2$. Equation (1.2.6) may be obtained by substituting the expression for $\mu_1^{2/3}$ into (1.2.5), satisfying (1.2.5) and setting $\theta_0 = 0$.

If we add to $\theta_0 = 0$ and to $\theta_1 = \theta_*$ a quantity that is a multiple of π (while the constants μ_1 and μ_2 are still computed by the use of (1.2.5)), this can result only in the change of sign in (1.2.4) for w. Since w is determined by (1.2.3) up to its sign, this translation along the θ axis does not result in a different solution. It is in principle possible to find the solution of (1.2.5) with $\theta_0 = 0$ and $\pi i < \theta_1 < \pi\,(i + 0.5)$, where $i = 1, 2, \ldots$. At the same time, it is easy to see by looking at (1.2.4) that the function w has additional loops. Since we are computing the eigenvalue P and the corresponding eigenfunction has a minimum number of loops, it is not necessary to consider such (more complex) solutions. The solution of the problem posed in (1.2.5) with

$$\pi(i - 0.5) < \theta_1 < \pi i, \quad i = 1, 2, \ldots$$

does not exist, since in this case the right-hand side of (1.2.5) is negative, while the left-hand side is positive.

Let us study some properties of the functions $\varphi(\theta)$ that will be needed in our future discussion. For small values of θ, we represent $\varphi(\theta)$ by a Taylor series, with accuracy up to terms of the order θ^3. Then we have $\varphi(\theta) \approx (1/5)\left(1 - \theta^2\right)$. Consequently, as $\theta \to 0$, the function $\varphi\,(\theta) \to 1$, with $\varphi(\theta) < 1$. Further, let us consider the case $\theta \to \pi/2$. For the sake of convenience, we introduce the variable $\omega = \pi/2 - \theta$, which approaches zero as $\theta \to \pi/2$. Substituting the variable ω for φ in (1.2.6) and eliminating θ and then decomposing φ in a Taylor series for small values of ω, we derive $\varphi \approx (3\omega(\pi - 4\omega))\,/4$. Therefore, φ approaches zero as $\theta \to \pi/2$. Using an asymptotic series and some simple calculations, we obtain a graph of the function $\varphi(\theta)$, which is illustrated in Fig. 1.5.

We can see from this graph that as the variable θ varies between zero and $\pi/2$, the function $\varphi(\theta)$ monotonically decreases from one to zero. Hence $P < 1$. In terms of

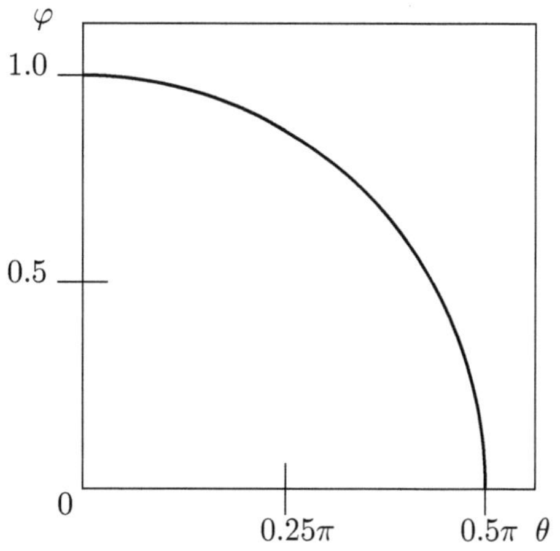

Fig. 1.5 Behaviour of function $\varphi(\theta)$

subsequently introduced dimensional quantities, this inequality can be written in the form $P < c/\ell$. It represent the condition of the stability for a perfectly rigid clamped bar.

Equation (1.2.6) gives us a relation between the unknown quantities θ_* and P. To obtain a second equation that is necessary for determining θ_* and P_*, we substitute the solution of (1.2.4), (1.2.5), into the isoperimetric condition (1.2.3) and eliminate P from the resulting equation by utilizing (1.2.6). We have

$$\Phi(\theta_1) = \gamma, \qquad \Phi(\theta_1) \equiv \Psi(\theta_1)\,\varphi^{-3/2}(\theta_1) \tag{1.2.7}$$

The graph of the function $\Phi(\theta)$ shown in Fig. 1.6 was drawn with the help of very simple calculus.

As θ decreases from $\pi/2$ to zero, the function Φ increases in a monotonic fashion from zero to infinity. As $\theta \to 0$, this function has an asymptotic representation

$$\Phi(\theta) \approx \frac{3\sqrt{3}}{5\theta} \tag{1.2.8}$$

Consequently, there is the solution for (1.2.7), which satisfies the conditions $0 < \theta_1 \leq \pi/2$ for an arbitrary nonnegative value of the parameter γ.

Thus, for given values of the constants C_2, ℓ, V, and c, finding the optimum distribution of the cross-sectional area $S(x)$ and of the corresponding value of the load P is reduced to the following sequence of computational steps.

For given values of C_2, ℓ, V, and c, we compute the value of the nondimensional parameter γ. Using this value of γ, we solve Eq. (1.2.7) and compute the value of θ_1. Next, we calculate the magnitude of the critical load P, which according to Eq. (1.2.6) is equal to $P = \varphi(\theta_1)$. The computation of the critical shape (of the deformed column) is carried out by using the formulas

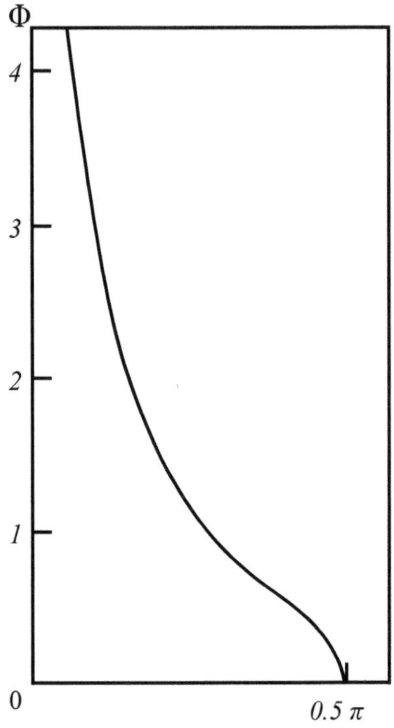

Fig. 1.6 The graph of the function $\Phi(\theta)$

Table 1.1 Parameters and values

γ	θ_1	P	γ	θ_1	P
0.5	1.283	0.488	3	0.339	0.976
1	0.882	0.811	3.5	0.292	0.983
1.5	0.640	0.909	4	0.257	0.987
2	0.496	0.948	4.5	0.229	0.989
2.5	0.404	0.966	5	0.206	0.991

$$x = \frac{\theta - \dfrac{1}{2}\sin 2\theta}{\theta_1 - \dfrac{1}{2}\sin 2\theta_1}, \qquad S = \frac{2\sqrt{P}\sin^2\theta}{\sqrt{3}\gamma\left(\theta_1 - \dfrac{1}{2}\sin 2\theta_1\right)} \tag{1.2.9}$$

where $0 \le \theta \le \theta_1$. Equation (1.2.9) follow directly from Eqs. (1.2.2), (1.2.4) and (1.2.5). The results derived for $\gamma = 0.5i,\ i = 1, 2, \dots$ and for ten values of the quantities θ_1 and P are illustrated in Table 1.1.

In Fig. 1.7, curves 1 and 2 illustrate the optimum distribution of $S = S(x)$ for $\gamma = 0.5$ and $\gamma = 5$. The dotted curve shows the distribution $S(x)$ for a rigidly held beam ($\gamma = 0$). Using (1.2.9) it is possible to show that as the rigidity of the support c

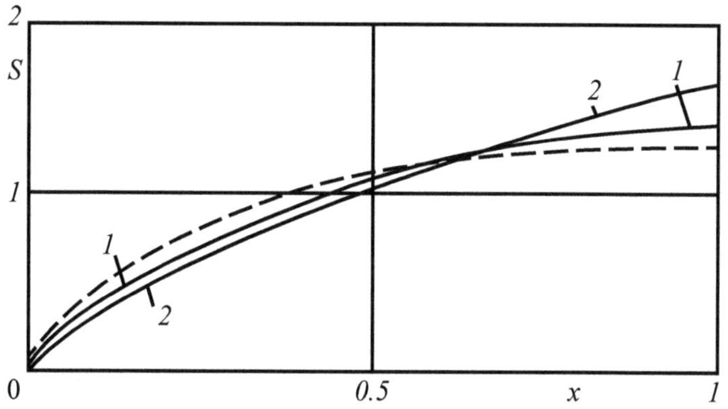

Fig. 1.7 Curves 1 and 2 illustrate the optimum distribution of $S = S(x)$ for corresponding values of $\gamma = 0.5$ and $\gamma = 5$. The dotted curve shows the distribution $S(x)$ for a rigidly held beam ($\gamma = 0$)

increases ($\gamma \to 0$), the distribution of $S(x)$ approaches the corresponding distribution for a rigidly held beam. As the rigidity c decreases (i.e. γ increases), then in the optimal shape, the material of the beam is "displaced" from the free end toward the built-in end.

1.3 Elastic Rod Under Torsion

Let a straight elastic rod of length ℓ be positioned along the x axis of a Cartesian coordinate system xyz. The ends of the rod are rigidly held at the points $x = 0$ and $x = \ell$. The rod is torsioned by a couple M applied to its end (see Fig. 1.8). It is assumed that the rod has constant rigidity in various cross-sectional planes, and therefore that

$$EI_y = EI_z = a\ ,$$

where E is Young's modulus for the material, and I_y and I_z are the moments of inertia of the transverse cross-sectional area with respect to lines passing through a point on the neutral axis of bending and parallel to the axes y and z, respectively.

In studying the stability of the rod and the critical values of the applied torque, we shall follow Euler's static approach. Let $y = y(x)$ and $z = z(x)$ be functions describing the shape of the axis of the buckled rod. We write down the corresponding equations of equilibrium and the boundary conditions:

$$\frac{d^2}{dx^2}\left(a\frac{d^2y}{dx^2}\right) = M\frac{d^3z}{dx^3}\ ,$$

Fig. 1.8 Torsioned rod

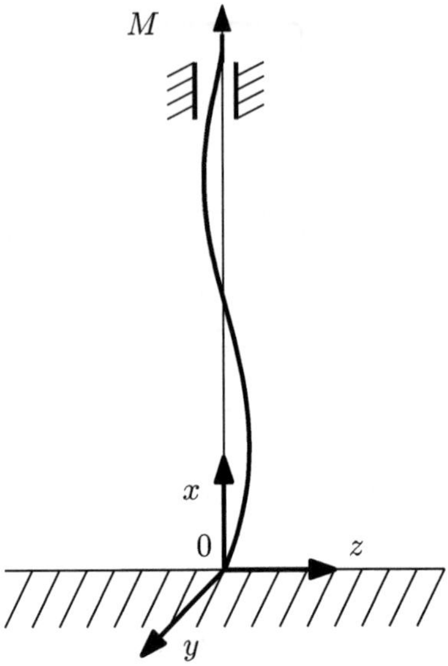

$$\frac{d^2}{dx^2}\left(a\frac{d^2z}{dx^2}\right) = M\frac{d^3y}{dx^3}\,, \tag{1.3.1}$$

$$y(0) = \left(\frac{dy}{dx}\right)_{x=0} = z(0) = \left(\frac{dz}{dx}\right)_{x=0} = 0\,,$$

$$y(\ell) = \left(\frac{dy}{dx}\right)_{x=\ell} = z(\ell) = \left(\frac{dz}{dx}\right)_{x=\ell} = 0\,,$$

We note that Eq. (1.3.1) result only if we assume that the deflections are "small". The functions $y(x) = z(x) \equiv 0$ describe the equation of the unbuckled axis satisfying the equilibrium equations and the boundary conditions (1.3.1). According to Euler's theory, the magnitude of the critical load and the loss of stability are determined by the smallest eigenvalue and the corresponding eigenfunctions $y(x) \not\equiv 0$ and $z(x) \not\equiv 0$ for the boundary-value problem (1.3.1).

For the particular case when torsional rigidity is uniformly distributed, that is $a = const$, finding the torque that causes a loss of stability is known to reduce to the computation of the smallest root for the (transcendental) equation

$$\tan\left(\frac{M\ell}{2a}\right) = \frac{M\ell}{2a}. \tag{1.3.2}$$

The magnitude of the critical torque is equal to $\pm 8.988\, a/\ell$. Following the article Banichuk and Barsuk [7], let us derive an analogue of this formula for the more general case when the distribution of torsional rigidity is a fairly arbitrary function of the variable x, i.e. $a = a(x)$. To accomplish it, we integrate twice both sides of the Eq. (1.3.1). Then we multiply the second equation in its integrated form by i (where i is the imaginary unit) and add these equations side by side. After introduction of the complex variable

$$w(x) = y(x) + iz(x)\,, \tag{1.3.3}$$

the Eq. (1.3.1) become

$$a\frac{\mathrm{d}^2 w}{\mathrm{d}x^2} = iM\frac{\mathrm{d}w}{\mathrm{d}x} + c_1 x + c_2\,. \tag{1.3.4}$$

where c_1 and c_2 are complex constants of integration, determined from the boundary conditions. By integrating the linear differential Eq. (1.3.4) twice, and taking care of the boundary conditions, we derive the formula

$$w = \int_0^x \left[e^{-iM\varphi(t)} \int_0^t \left(\frac{c_1\xi + c_2}{a(\xi)} \right) e^{iM\varphi(\xi)} \mathrm{d}\xi \right] \mathrm{d}t \tag{1.3.5}$$

and relations that determine the values of the constants

$$c_1 \int_0^\ell \frac{x}{a(x)} e^{iM\varphi(x)} \mathrm{d}x + c_2 \int_0^\ell \frac{1}{a(x)} e^{iM\varphi(x)} \mathrm{d}x = 0\,, \tag{1.3.6}$$

$$c_1 \int_0^\ell e^{-iM\varphi(x)} \int_0^x \frac{t}{a(t)} e^{iM\varphi(t)} \mathrm{d}t\mathrm{d}x + c_2 \int_0^\ell e^{-iM\varphi(x)} \int_0^x \frac{e^{iM\varphi(t)}}{a(t)} \mathrm{d}t\mathrm{d}x = 0\,, \quad \left(\varphi(x) \equiv \int_0^x \frac{\mathrm{d}t}{a(t)} \right). \tag{1.3.7}$$

The relations (1.3.6) represent a system of two homogeneous linear equations that may serve the purpose of determining values of unknown constants c_1 and c_2. A necessary condition for the existence of the nontrivial solution of (1.3.5) (i.e., when the determinant of the coefficient of (1.3.6) vanishes) can be reduced to an equation that is to be solved for determining the critical torque

$$f(M) = \int_0^\ell \frac{e^{iM\varphi(x)}}{a(x)} \mathrm{d}x \int_0^\ell e^{-iM\varphi(\xi)} \int_0^x \frac{\xi e^{iM\varphi(\xi)}}{a(\xi)} \mathrm{d}\xi \mathrm{d}x - \tag{1.3.8}$$

$$\int_0^\ell \frac{x e^{iM\varphi(x)}}{a(x)} \mathrm{d}x \int_0^\ell e^{-iM\varphi(x)} \int_0^x \frac{e^{iM\varphi(\xi)}}{a(\xi)} \mathrm{d}\xi \mathrm{d}x = 0\,.$$

When rigidity is uniformly distributed along the length of the rod, that is, if $a(x) = const$ in (1.3.8), we arrive at the well-known formula $\tan\left(M\ell/(2a)\right) = M\ell/(2a)$. We note that the complex Eq. (1.3.8) can be written as a system of two real equations

$$\mathrm{Re}\,(f(M)) = 0, \qquad \mathrm{Im}\,(f(M)) = 0\,. \tag{1.3.9}$$

In this manner a study of the stability of torsioned rods with variable rigidity may be reduced to a computation of quadratures for a given function $a(x)$ and the solution of the transcendental Eq. (1.3.8) for M.

We shall consider separately the case of the symmetric distribution of bending rigidity when $a(x) = a(\ell - x)$. In this case

$$\varphi(x) = \varphi\left(\frac{\ell}{2}\right) + \Psi(x) ,$$

where $\Psi(x)$ is an antisymmetric function, that is, $\Psi(x) = -\Psi(\ell - x)$. Using these properties of the functions $a(x)$ and $\Psi(x)$ and the rules of integration on symmetric domains of symmetric or antisymmetric functions, (1.3.8), for example, is transformed into the form

$$f(M) = \int_0^{\frac{\ell}{2}} \frac{\cos(M(\Psi(x)))}{a(x)} \mathrm{d}x \int_0^{\frac{\ell}{2}} [\chi_1 \cos(M\Psi(x)) + \chi_2 \sin(M\Psi(x))]\, \mathrm{d}x + \tag{1.3.10}$$

$$\int_0^{\frac{\ell}{2}} \frac{\left(x - \frac{\ell}{2}\right)}{a(x)} \sin(M\Psi(x))\, \mathrm{d}x \int_0^{\frac{\ell}{2}} [\chi_3 \cos(M\Psi(x)) - \chi_4 \sin(M\Psi(x))]\, \mathrm{d}x = 0 .$$

Here

$$\chi_1(x) = \int_0^x \frac{\left(\xi - \frac{\ell}{2}\right)}{a(\xi)} \cos(M\Psi(\xi))\, \mathrm{d}\xi , \qquad \chi_2(x) = \int_0^x \frac{\left(\xi - \frac{\ell}{2}\right)}{a(\xi)} \sin(M\Psi(\xi))\, \mathrm{d}\xi ,$$

$$\chi_3(x) = \int_0^x \frac{\sin(M\Psi(\xi))}{a(\xi)} \mathrm{d}\xi , \qquad \chi_4(x) = \int_0^x \frac{\cos(M\Psi(\xi))}{a(\xi)} \mathrm{d}\xi .$$

The essential difference between (1.3.8) and (1.3.10) is that (1.3.10) is a single equation in the unknown M instead of a system of equations.

Now let us consider a torsioned rod simply supported at the points $x = 0$ and $x = \ell$. Finding a static condition for the loss of stability (Greenhill's problem) leads to the boundary value problem

$$a\frac{\mathrm{d}^2 y}{\mathrm{d}x^2} = M\frac{\mathrm{d}z}{\mathrm{d}x} , \qquad a\frac{\mathrm{d}^2 z}{\mathrm{d}x^2} = -M\frac{\mathrm{d}y}{\mathrm{d}x} , \tag{1.3.11}$$

$$y(0) = y(\ell) = z(0) = z(\ell) = 0 .$$

We multiply both sides of the second equation in (1.3.11) by i and add them side by side. We introduce a function

$$w = y(x) + iz(x) \tag{1.3.12}$$

and transform the boundary-value problem (1.3.11) into the form

$$a\frac{\mathrm{d}^2 w}{\mathrm{d}x^2} = -iM\frac{\mathrm{d}w}{\mathrm{d}x}\,, \tag{1.3.13}$$

$$w(0) = w(\ell) = 0\,.$$

Integrating equation (1.3.13) twice and taking care of the assigned boundary conditions, we obtain a formula for w and a system of two simultaneous linear, homogeneous equations that can be solved to determine the constants of integration. The necessary condition on the existence of nontrivial solutions (determinant of coefficients for this system vanishing) reduces to a single complex equation in M, which may be written as a system of the two real equations

$$\int_0^\ell \cos\left(M\varphi(x)\right)\mathrm{d}x = 0\,, \qquad \int_0^\ell \sin\left(M\varphi(x)\right)\mathrm{d}x = 0\,. \tag{1.3.14}$$

The smallest common root of these equations defines a critical divergence torque. Therefore, finding critical torques amounts to finding a simultaneous solution for the system (1.3.14).

We shall show that for a symmetric distribution of rigidity, the system (1.3.14) may be replaced by a single equation. Let $a(x)$ be a quite arbitrary distribution of bending rigidity. We write $1/a(x)$ as a sum of the two terms

$$\frac{1}{a(x)} = \Omega(x) + \Psi(x)\,, \tag{1.3.15}$$

where $\Omega(x)$ and $\Psi(x)$ are, respectively, symmetric and antisymmetric functions with respect to the midpoint of the interval $[0, \ell]$. The function $\varphi(x)$ can also be represented as a sum of a symmetric and antisymmetric functions $\omega(x)$ and $\psi(x)$, respectively

$$\varphi(x) = \omega(x) + \psi(x)\,, \tag{1.3.16}$$

$$\omega(x) = \int_0^{\frac{\ell}{2}} \Omega(t)\mathrm{d}t + \int_0^x \Psi(t)\mathrm{d}t\,,$$

$$\psi(x) = \int_0^x \Omega(t)\mathrm{d}t - \int_0^{\frac{\ell}{2}} \Omega(t)\mathrm{d}t\,,$$

We note the properties of integrals in symmetric domains of symmetric and antisymmetric functions and use (1.3.16) to transform Eq. (1.3.14).

We obtain

$$\int_0^\ell \cos(M\psi(x))\cos(M\omega(x))\,\mathrm{d}x = 0\,, \tag{1.3.17}$$

$$\int_0^\ell \cos(M\psi(x))\sin(M\omega(x))\,\mathrm{d}x = 0\,.$$

For an antisymmetric distribution of rigidity (when $\Psi(x) \equiv 0$, $1/a = \Omega(x)$ and $\omega(x) \equiv \varphi(\ell/2)$), Eq. (1.3.17) assume the form

$$\cos\left(M\varphi\left(\frac{\ell}{2}\right)\right)\int_0^\ell \cos(M\psi(x))\,\mathrm{d}x = 0\,,$$

$$\sin\left(M\varphi\left(\frac{\ell}{2}\right)\right)\int_0^\ell \cos(M\psi(x))\,\mathrm{d}x = 0\,.$$

It follows directly that the system (1.3.14) can be reduced in this case to the single equation

$$\int_0^\ell \cos\left(M\psi\left(\frac{\ell}{2}\right)\right)\mathrm{d}x = 0\,,, \qquad \psi(x) = \varphi(x) - \varphi\left(\frac{\ell}{2}\right)\,. \tag{1.3.18}$$

For a constant distribution of rigidity, Eq. (1.3.18) predicts the well-known value of the critical torque

$$M = 2\pi\frac{EI}{\ell}\,.$$

The solution of Eq. (1.3.18) can be found very effectively using one of several analytic or numerical techniques, in particular, Newton's methods. Thus, equation (1.3.18) results in a real simplification particularly in solving problems, such as the optimization of rigidities, in which it is necessary to recompute the critical value of M several times for various approximate values of $a(x)$.

Let us examine some of the simplest problems of optimizing stability. Let a rod have similarly shaped cross-sections, and consequently

$$a(x) = k\,[S(x)]^2$$

where $S(x)$ is the function defining the distribution of magnitude of the area. We formulate the following problem in terms of nondimensional variables: we need to find a function of cross-sectional area $S(x)$ satisfying $S_{\min} < S(x)$, and also a constant volume condition, and assign a maximum to the critical value of the torque:

$$M_* = \max_S M\,, \tag{1.3.19}$$

where M is computed from (1.3.10) and (1.3.18), and $S_{\min}$ is a given nondimensional quantity.

For the indicated constraints on the volume of the rod, and the allowable values of its thickness, problem (1.3.19) was solved numerically in Banichuk and Barsuk [7] using an iterative algorithm. This algorithm, applied to our optimization problem, consists of finding an approximate value of $a(x)$ using Newton's method (in the process of iterating critical values of M) from the current value of $a(x)$, and then improving the variation of the control function $a(x)$ by the gradient projection technique. The computation of the critical value of M from (1.3.10) and (1.3.18) was carried out to within 10^{-5} accuracy. The computation was terminated when the magnitude of the gradient of the optimized functional (i.e. the error in satisfying the optimality conditions) was smaller than 10^{-3}.

The optimal functions $S(x)$ were found for several values of the parameter $S_{\min}$. Figure 1.9 illustrates the function $S(x)$ found as a result of these numerical computations for a rod simply supported at both ends.

Curves 1, 2, and 3 correspond to the values of the parameter $S_{\min} = 0.98,\ 0.92$ and 0.88, respectively. For these values of the parameter, the magnitude of the critical torque is $M = 6.56, 7.12, \text{and} 7.80$, respectively. The relative gain in comparison with a rod having a constant rigidity function and a unit volume amounts to 4.45%, 14.9%, and 24.2%, respectively. Numerical solutions were also carried out for rods with built-in ends. The expression for the gradient in the stability optimization problem was carried out with the help of the following representation for the eigenvalue of the boundary-value problem (1.3.1):

$$M = -\frac{\int_0^1 S^2 \mathrm{Re}\left(\dfrac{\mathrm{d}^2 w}{\mathrm{d}x^2} \cdot \dfrac{\mathrm{d}^2 w^*}{\mathrm{d}x^2}\right) \mathrm{d}x}{\int_0^1 \mathrm{Im}\left(\dfrac{\mathrm{d}w}{\mathrm{d}x} \cdot \dfrac{\mathrm{d}^2 w^*}{\mathrm{d}x^2}\right) \mathrm{d}x} \tag{1.3.20}$$

where $w(x)$ is computed from (1.3.5) and (1.3.6), and w^* is the complex conjugate function of w.

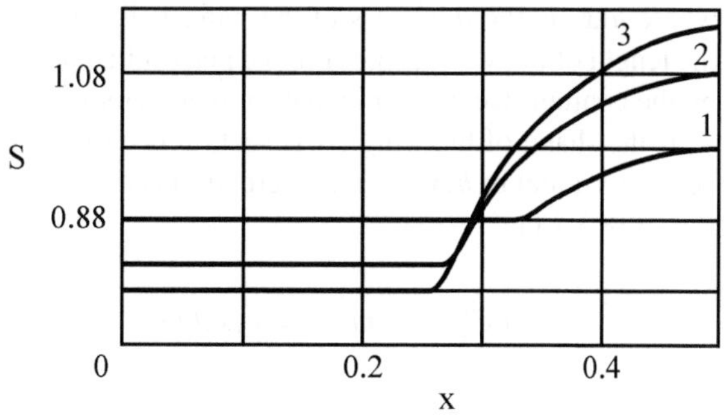

Fig. 1.9 Graph illustrates the function $S(x)$ found as a result of these numerical computations for a rod simply supported at both ends. Curves 1, 2, and 3 correspond to the values of the parameter $S_{\min} = 0.98,\ 0.92$ and 0.88

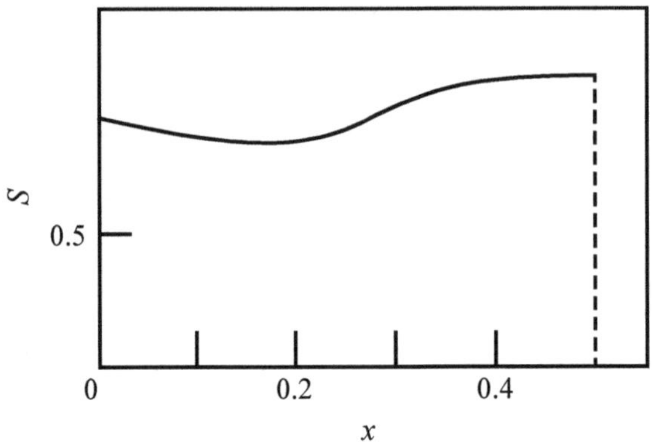

Fig. 1.10 Optimal distribution of cross-sectional area

Numerical computation indicates that for $S_{min} < 0.865$ the constraint $S_{min} < S$ becomes inactive. Figure 1.10 illustrates the optimal distribution function of the cross-sectional area for $S_{min} < 0.865$. The critical value of the torque M for an optimal rod is 9.2789. For the sake of comparison, the first four eigenvalues M for an optimal rod are 9.2789, 14.9503, 21.149 and 27.274, and the first four eigenvalues for a rod with constant cross-sectional area ($S(x) \equiv 1$) are 8.98688, 15.3506, 21.808, and 28.132.

1.4 Divergence and Optimization of Wings

Let us consider the torsional divergence of straight wings. Let the x coordinate be measured along the wing, $0 \leq x \leq \ell$. Let G denote the shear modulus, and $GI(x)$ the torsional rigidity. The torsion rigidity of a wing of large aspect ratio is determined by

$$GI = a(x)h(x) , \qquad a(x) = \frac{4G\Omega^2(x)}{s(x)} , \tag{1.4.1}$$

where $h(x)$ is the thickness of the wingshell, and $s(x)$ and $\Omega(x)$ are, respectively, the length of the wing contour in the perpendicular (yz) plane, and the area enclosed by the contour. Here and in what follows, refer to Fig. 1.11.

If the shape of the wing cross section is uniform, then the torsional rigidity will be proportional to $h(x)b^3(x)$, where $b(x)$ is the chord length (see e.g. McIntosh and Easter [8]). Let

$$\lambda^2 = \frac{1}{2}\rho v C_y^\alpha , \qquad m(x) = e(x)b(x) , \qquad GI(x) = h(x)a(x) , \tag{1.4.2}$$

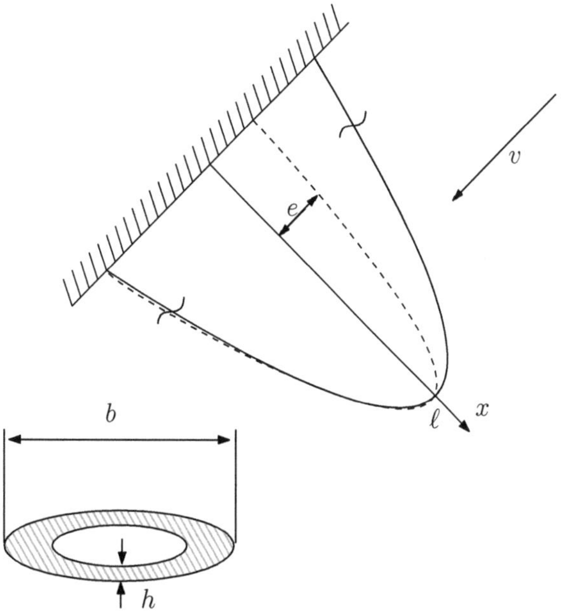

Fig. 1.11 A straight wing and its cross-section. In the cross-section, the shaded area represents the wingshell

where ρ is the density of the surrounding gas, v the free-stream velocity of the gas, C_y^{α} the aerodynamic coefficient, and $e(x)$ the distance from the aerodynamic center to the elastic axis.

Let $\theta(x)$ denote the torsion angle. The equilibrium equation of the wing is

$$\frac{\mathrm{d}}{\mathrm{d}x}\left(ah\frac{\mathrm{d}\theta}{\mathrm{d}x}\right)+\lambda^2 m\theta=0\,, \tag{1.4.3}$$

$$\left(\gamma\theta-ah\frac{\mathrm{d}\theta}{\mathrm{d}x}\right)_{x=0}=0\,,\qquad \left(ah\frac{\mathrm{d}\theta}{\mathrm{d}x}\right)_{x=\ell}=0\,. \tag{1.4.4}$$

The boundary conditions (1.4.4) correspond to elastic clamping at $x=0$ and a free end at $x=\ell$. The distribution of thickness $h(x)$ along the wing is considered an unknown design variable.

Suppose that the functions $a(x)$, $m(x)$, $s(x)$ and the parameters λ^2, γ, and ℓ are given, and consider the following wing optimization problem. The thickness distribution $h=h(x)$, which minimizes the amount of material (mass) needed for the wing, should be determined or equivalently, the volume of the wingshell,

$$V=\int_0^{\ell} sh\,\mathrm{d}x\,, \tag{1.4.5}$$

with the constraint that wing divergence occurs at the given critical velocity λ.

Note that for solution of the formulated problem it is necessary to take into account the inequality

$$\gamma - \lambda^2 \int_0^{\ell} m \, \mathrm{d}x > 0 \tag{1.4.6}$$

expressing an equilibrium condition for the wing.

The problem (1.4.3) and (1.4.4) is self-adjoint and positive definite. Consequently, the eigenvalues λ^2 are positive and one can apply the Rayleigh formula for the critical value (first eigenvalue), resulting in

$$\lambda^2 = \min_{\theta} \frac{\gamma\theta^2(0) + \displaystyle\int_0^{\ell} ah \left(\frac{\mathrm{d}\theta}{\mathrm{d}x}\right)^2 \mathrm{d}x}{\displaystyle\int_0^{\ell} m\theta^2 \mathrm{d}x} . \tag{1.4.7}$$

Using the method of Lagrange multipliers, and the expression (1.4.7), it is possible to show that the necessary optimality condition is

$$a \left(\frac{\mathrm{d}\theta}{\mathrm{d}x}\right)^2 + \mu^2 s = 0 , \tag{1.4.8}$$

where μ is a Lagrange multiplier. The condition (1.4.8) is not only necessary but also the sufficient condition for optimality.

From Eq. (1.4.8) it follows that

$$\theta = \mu\left(\Psi + \Psi_0\right) , \qquad \Psi \equiv \int_0^{x} \sqrt{\frac{s}{a}} \mathrm{d}x , \tag{1.4.9}$$

where Ψ_0 is a constant of integration. Then using (1.4.8), (1.4.9) and (1.4.3), we obtain

$$\frac{\mathrm{d}}{\mathrm{d}x}\left(h\sqrt{sa}\right) + \lambda^2 m\left(\Psi + \Psi_0\right) = 0 . \tag{1.4.10}$$

Suppose that $\sqrt{sa} \neq 0$ and determine $h(x)$ from Eq. (1.4.10),

$$h(x) = \lambda^2 \int_x^{1} m\left(\Psi + \Psi_0\right) \frac{\mathrm{d}x}{\sqrt{sa}} , \tag{1.4.11}$$

where the constant Ψ_0 is found using the first boundary condition (1.4.4), the result being

$$\Psi_0 = \frac{\lambda^2 \int_0^{\ell} m\Psi \mathrm{d}x}{\gamma - \lambda^2 \int_0^{\ell} m \mathrm{d}x} . \tag{1.4.12}$$

The value of the objective functional (wingshell volume) for the optimal distribution of thickness is

$$V_{\text{opt}} = \lambda^2 \int_0^{\ell} m\,(\Psi + \Psi_0)\,\mathrm{d}x\ . \tag{1.4.13}$$

To evaluate the gain of optimization, for comparison consider a wing with a constant distribution of thickness h, and volume $V = 1$. The critical velocity of divergence λ is determined as the smallest positive root of the equation

$$\lambda \tan \lambda = \gamma\ . \tag{1.4.14}$$

The relative gain with respect to the mass of the optimal configuration, compared with the wing of constant h and $V = 1$, having the same critical velocity, is

$$\beta = \frac{V - V_{\text{opt}}}{V} = 1 - \lambda^2 \left(\frac{1}{3} + \frac{\lambda^2}{4(\gamma - \lambda^2)} \right)\ . \tag{1.4.15}$$

Thus, as we vary the rigidity parameter γ from ∞ to 0, the gain varies from 18% up to 25%. In the case of rigid clamping ($\gamma = \infty$) we obtain the following solution:

$$h(x) = \frac{1}{2}\lambda^2 \left(1 - x^2\right)\ , \qquad \theta = \mu x\ , \qquad V = \frac{1}{3}\lambda^2\ . \tag{1.4.16}$$

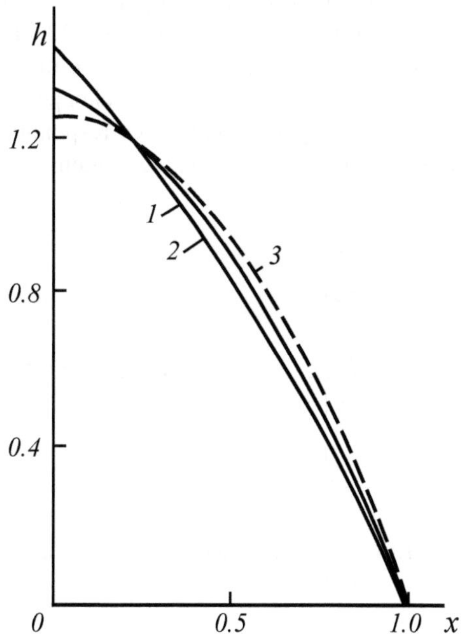

Fig. 1.12 Distribution of h as a function of x

This solution was obtained previously in Ashley and McIntosh [9] and is shown in Fig. 1.12 by curve 3. Curves 1 and 2 correspond to the cases $\gamma = 6.0,\ \lambda = 1.35$ and $\gamma = 1.3,\ \lambda = 0.94$, respectively.

1.5 Stability of Tensioned Cantilever Beam

Let us next consider an optimal design problem, using the criterion of stability, of a variable-rigidity cantilever beam. The free end of the beam is loaded in tension by a force applied to it through an attached absolutely rigid rod. The solution is sought in a closed analytical form. The bifurcation of the equilibrium form of the optimal beam is shown to take place for a certain ratio of the rigid rod and the cantilever beam lengths. The optimization of a tensioned rod has been studied in Jelicic and Atanackovic [10] and Banichuk et al. [11]. In this text we follow the paper Banichuk et al. [11].

Consider an elastic beam (rod) of variable cross-section area $S(x)$, bending rigidity $EI(x) = A_p S(x)^p$, where $p = 1, 2, 3$, and length ℓ. The volume V of the beam is given by

$$V = \int_0^{\ell} S(x)\ \mathrm{d}x\ . \tag{1.5.1}$$

Let the beam be tensioned by a force P oriented in the x direction, applied to the free end at $x = \ell$, as shown in Fig. 1.13. Here A_p is a parameter of the cross section, E is Young's modulus, and I is the moment of inertia (second moment of area). One end of the rod, at $x = 0$, is clamped, while the other end at $x = \ell$ is free. The equilibrium equation and the corresponding boundary conditions for the deflection function $w(x)$, $0 \leq x \leq \ell$, are written as

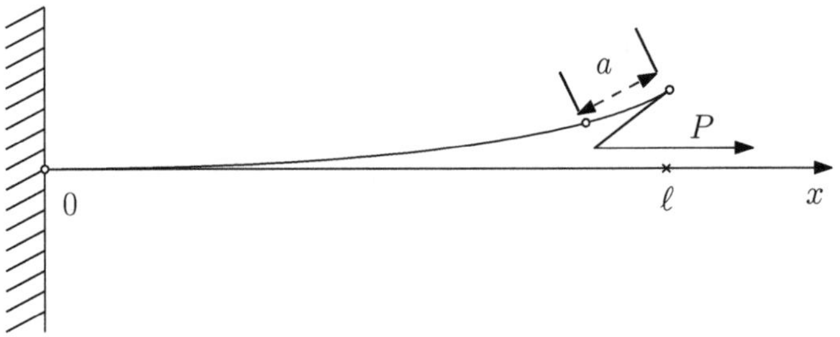

Fig. 1.13 A tensioned cantilever beam, with a rigid rod attached at the free end $x = \ell$. An external force P is applied

$$\frac{\mathrm{d}^2}{\mathrm{d}x^2}\left(EI(x)\frac{\mathrm{d}^2 w}{\mathrm{d}x^2}\right) - P\frac{\mathrm{d}^2 w}{\mathrm{d}x^2} = 0\,, \qquad 0 < x < \ell\,, \tag{1.5.2}$$

$$w(0) = \left(\frac{\mathrm{d}^2 w}{\mathrm{d}x^2}\right)_{x=0} = 0\,, \quad EI(\ell)\left(\frac{\mathrm{d}^2 w}{\mathrm{d}x^2}\right)_{x=\ell} = Pa\left(\frac{\mathrm{d}w}{\mathrm{d}x}\right)_{x=\ell}\,. \tag{1.5.3}$$

Nontrivial solutions $w(x) \not\equiv 0$ of the homogeneous boundary-value problem (1.5.2) and (1.5.3) correspond to some critical loads P_i that are known as the critical buckling values. For the fixed parameters V, ℓ, a and A_p , the critical load P depends on the thickness distribution $S(x)$, that is, $P = P(S)$.

The optimization problem consists of finding the optimal cross-section profile $S(x)$ for a given beam volume V, such that the critical load is maximized:

$$S^*(x) :\ P(S) \to \max_S \int_0^\ell S(x)\,\mathrm{d}x = V\,. \tag{1.5.4}$$

In the nondimensional variables,

$$x = \ell\tilde{x}\,, \qquad S(x) = \frac{V\tilde{S}(x)}{\ell}\,, \qquad \lambda = \frac{P\ell^{p+2}}{A_p E V^p}\,, \qquad \gamma = \frac{a}{\ell}\,, \tag{1.5.5}$$

the considered optimization problem (1.5.4) and boundary-value problem (1.5.2), (1.5.3) are represented in the following form (the tilde is omitted):

$$\frac{\mathrm{d}^2}{\mathrm{d}x^2}\left(S^p(x)\frac{\mathrm{d}^2 w}{\mathrm{d}x^2}\right) - \lambda\frac{\mathrm{d}^2 w}{\mathrm{d}x^2} = 0\,, \qquad 0 < x < 1\,, \tag{1.5.6}$$

$$w(0) = \left(\frac{\mathrm{d}w}{\mathrm{d}x}\right)_{x=0} = 0\,, \qquad S^p\,(1)\left(\frac{\mathrm{d}^2 w}{\mathrm{d}x^2}\right)_{x=1} = \lambda\gamma\left(\frac{\mathrm{d}w}{\mathrm{d}x}\right)_{x=1}\,, \tag{1.5.7}$$

$$\lambda(S) \to \max_S\,, \qquad \int_0^1 S(x)\,\mathrm{d}x = 1\,. \tag{1.5.8}$$

The bending moment is given by

$$m(x) = S^p\frac{\mathrm{d}^2 w}{\mathrm{d}x^2}\,, \tag{1.5.9}$$

and thus we can formulate the considered problem of optimal design as

$$\frac{\mathrm{d}^2 m(x)}{\mathrm{d}x^2} - \lambda\frac{m(x)}{S^p(x)} = 0\,, \qquad p = 1, 2, 3\,, \tag{1.5.10}$$

$$m(1) = \gamma \left(\frac{\mathrm{d}m}{\mathrm{d}x}\right)_{x=1}, \qquad \left(\frac{\mathrm{d}m}{\mathrm{d}x}\right)_{x=0} = 0\,, \tag{1.5.11}$$

$$\int_0^1 S(x)\,\mathrm{d}x - 1 = 0\,, \qquad \lambda(S) \to \max_S\,. \tag{1.5.12}$$

If $S(x) \equiv 1$, then the solution of the spectral boundary-value problem in (1.5.10)–(1.5.12) is reduced to the equation

$$\sqrt{\lambda} \tanh \sqrt{\lambda} = \frac{1}{\gamma} \tag{1.5.13}$$

for finding the parameter γ. Equation (1.5.13) has the following asymptotics:

$$\lambda(\gamma) = \frac{1}{\gamma^2}\,, \quad 0 < \gamma \ll 1; \qquad \lambda(\gamma) \approx \frac{1}{\gamma}\,, \quad \gamma \gg 1\,. \tag{1.5.14}$$

The spectral boundary-value problem in (1.5.10)–(1.5.12) can be presented in the form of a variational principle,

$$\lambda(S) = \min_{m(x)} \frac{-\int_0^1 \left(\frac{\mathrm{d}m(x)}{\mathrm{d}x}\right)^2 \mathrm{d}x + \frac{1}{\gamma} m^2(1)}{\int_0^1 \frac{m^2(x)}{S^p(x)} \mathrm{d}x}\,. \tag{1.5.15}$$

To exclude the integral constraint in the equation (1.5.8), we introduce a Lagrange functional

$$\mathcal{L} = \lambda(S) - \mu \left(\int_0^1 S(x)\,\mathrm{d}x - 1 \right) \tag{1.5.16}$$

and obtain the necessary optimality condition in the form

$$m^2(x) = S^{p+2}(x)\,, \qquad p = 1, 2, 3\,. \tag{1.5.17}$$

Performing differentiation of (1.5.17) we have

$$2m(x)\frac{\mathrm{d}m(x)}{\mathrm{d}x} = (p+1)S^p(x)\frac{\mathrm{d}S(x)}{\mathrm{d}x}\,. \tag{1.5.18}$$

Using necessary optimality condition (1.5.17) and relation (1.5.18) along with the boundary condition

$$\left(\frac{\mathrm{d}m}{\mathrm{d}x}\right)_{x=0} = 0\,, \tag{1.5.19}$$

we find the expression for the first integral of equations (1.5.10)–(1.5.12) as

$$\left(\frac{\mathrm{d}m(x)}{\mathrm{d}x}\right)^2 = (p+1)\lambda(S(x) - S)\,, \tag{1.5.20}$$

where $S_0 = S(0)$. From the expression (1.5.20) it follows that $S(x) \geq S(0)$. It is then possible to show that

$$\frac{\mathrm{d}m(x)}{\mathrm{d}x} = \sqrt{(\lambda(p+1)}\sqrt{S(x) - S_0}\,. \tag{1.5.21}$$

As a result of applying (1.5.17)–(1.5.21) we derive the differential equation

$$\frac{\mathrm{d}S}{\mathrm{d}x} = \sqrt{\frac{4\lambda}{p+1}}\frac{\sqrt{S(x) - S_0}}{S^{(p-1)/2}}\,, \quad p = 1, 2, 3 \tag{1.5.22}$$

for finding the optimal thickness distribution. The solution of this equation is represented in quadrature form as

$$\int_{S_0}^{S_1} \frac{\tau^{(p-1)/2}\,\mathrm{d}\tau}{\sqrt{\tau - S_0}} = x\sqrt{\frac{4\lambda}{p+1}}\,, \tag{1.5.23}$$

where $S_1 = S(1)$. It is seen from (1.5.22) that

$$\frac{\mathrm{d}S(x)}{\mathrm{d}x} > 0 \quad \text{for} \quad 0 < x < 1\,, \tag{1.5.24}$$

and consequently the optimal thickness distribution is a monotonically increasing function and $S_0 \leq S(x) \leq S_1$. Using the notation

$$I_1\,(p, \delta, s) = \int_{\delta}^{s} \frac{\tau^{(p-1)/2}\,\mathrm{d}\tau}{\sqrt{\tau - \delta}}\,, \qquad \delta = \frac{S_0}{S_1}\,, \tag{1.5.25}$$

$$I_2(p, \delta) = \int_{\delta}^{1} \frac{\tau^{(p+1)/2}\,\mathrm{d}\tau}{\sqrt{\tau - \delta}}\,,$$

and determining the values λ, S_0 and S_1, we find parametric representations for the optimal solution

$$S(x) = S_1 s(x)\,, \quad 0 \leq x \leq 1\,, \tag{1.5.26}$$

$$S_1 = \frac{I_1(p, \delta, 1)}{I_2(p, \delta)}\,, \quad 0 \le \delta \le 1\,, \tag{1.5.27}$$

$$I_1(p, \delta, s) = I_1(p, \delta, 1)x\,, \tag{1.5.28}$$

and for the critical force $\lambda(\gamma)$,

$$\gamma(\delta) = \frac{2}{(p+1)I_1(p, \delta, 1)\sqrt{1-\delta}}\,, \tag{1.5.29}$$

$$\lambda(\delta) = \frac{(p+1)I_1^{p+2}(p, \delta, 1)}{4I_2^p(p, \delta)}\,. \tag{1.5.30}$$

Thus, if we take some $\delta \in [0, 1]$, then the optimal thickness distribution $S(x)$ is found with the help of (1.5.26), while the critical buckling force λ and the problem parameter γ are obtained using (1.5.29) and (1.5.30). Note that, as it follows from the variational representation (1.5.15), the critical force decreases when the problem parameter γ increases.

Let $p = 1$. In this case

$$I_1(1, \delta, s) = s\sqrt{s-\delta}\,, \quad \delta \le s \le 1\,; \qquad I_2(1, \delta) = \frac{2}{3}\,(1+2\delta)\,\sqrt{1-\delta}\,, \tag{1.5.31}$$

$$S_1 = \frac{3}{1+2\delta}\,, \qquad S_0 = \frac{3\delta}{1+2\delta}\,,$$

$$\gamma(\delta) = \frac{1}{2(1-\delta)}\,, \qquad \lambda(\delta) = \frac{6(1-\delta)}{1+2\delta}\,,$$

$$S(x) = \frac{3}{1+2\delta}\left(\delta + (1-\delta)\,x^2\right)\,, \quad 0 \le \delta \le 1\,.$$

Note that in the considered case, $\gamma(\delta)$ is a monotonous function, and consequently there exists an inverse dependence

$$\delta(\gamma) = \frac{2\gamma - 1}{2\gamma}\,.$$

The substitution of the expression of $\delta(\gamma)$ into expressions for $\lambda(\delta)$ and $S(x)$ in (1.5.31) gives us the following representations for S and λ as a functions of the problem parameter γ:

$$S(x) = 1 + \frac{3x^2 - 1}{2(3\gamma - 1)}\,, \qquad \lambda(\gamma) = \frac{3}{3\gamma - 1}, \qquad \gamma \ge 0.5\,. \tag{1.5.32}$$

The maximum critical buckling load occurs for $\gamma_* = 0.5$ and is equal to 6, while the corresponding optimal thickness distribution is given by $S(x) = 3x^2$. A detailed analysis of the cases $p = 2$ and $p = 3$ is presented in Banichuk et al. [11].

1.6 Accelerating Motion of Rod (Rocket, Missile) Under Follower Force

The bifurcation of an elastic rod subjected to a follower force and undegoing uniformly accelerating motion was studied in Feodosjev [12]. The nonconservative form of the load causes the dynamic type of loss of stability. In this case, the critical follower force depends on both the distribution of the rigidities (as in the static problems of elastic stability) and the inertial characteristic of the rod.

In this section, we follow Banichuk and Gura [13] and consider the problem of determining the thickness distribution along a nonhomogeneous rod loaded with the maximum value of the critical follower force. Determining the optimal thickness distribution, and hence the optimal distributions of the rigidity and inertial characteristics, is performed with the help of the dynamic method Bolotin [14] and the calculus of variations.

Let the elastic rod perform a rectilinear movement at a constant acceleration. Let the rod be subjected to a follower force applied at its end (see Fig. 1.14).

We introduce a rectangular coordinate system xy, oriented along the undisturbed rod, with the origin coinciding with the end point of the rod. The differential equations and the boundary conditions of the small transverse vibration $w(x, t)$ of the rod of a

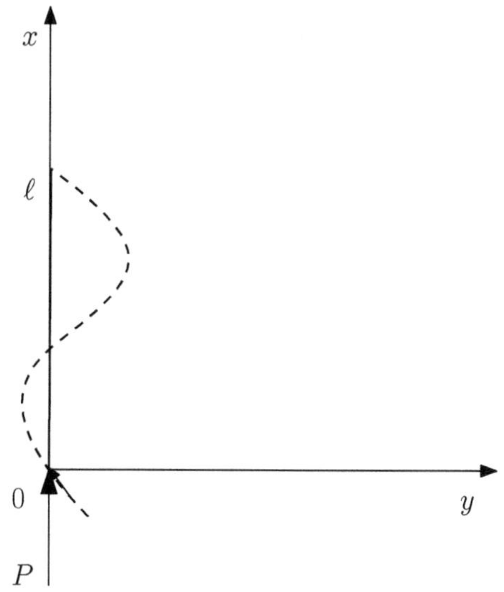

Fig. 1.14 An elastic rod performing a rectilinear motion at a constant acceleration, subjected to a follower force applied to its end

variable cross-section $S(x)$ have the following form:

$$\rho S\frac{\mathrm{d}^2w}{\mathrm{d}t^2}+\frac{\mathrm{d}^2}{\mathrm{d}x^2}\left(EI\frac{\mathrm{d}^2w}{\mathrm{d}x^2}\right)+\frac{\mathrm{d}}{\mathrm{d}x}\left(\frac{\mathrm{d}w}{\mathrm{d}x}\int_x^\ell \rho S\mathrm{d}x\right)\frac{P}{\rho V}=0\,, \tag{1.6.1}$$

$$\left(EI\frac{\mathrm{d}^2w}{\mathrm{d}x^2}\right)_{x=0,\ell}=0,\qquad \left(\frac{\mathrm{d}}{\mathrm{d}x}\left(EI\frac{\mathrm{d}^2w}{\mathrm{d}x^2}\right)\right)_{x=0,\ell}=0\,,$$

where $S = S(x)$ is the cross-sectional area, $I = I(x)$ is the moment of inertia, P is the applied follower force, ℓ is length, V is volume, ρ is density, and E is Young's modular of the rod. We will consider that the cross sections of the rod have a similar shape. In this case we have the following dependence of the moment of inertia I on the cross sectional area S:

$$I = AS^2\,. \tag{1.6.2}$$

In what follows we consider the harmonic transverse vibrations of the moving rod with frequency ω, that is,

$$w(x,t)=u(x)e^{i\omega t}\,, \tag{1.6.3}$$

and use the nondimensional variables

$$\tilde{x}=\frac{x}{\ell}\,,\qquad \tilde{u}=\frac{u}{\ell}\,,\qquad \tilde{S}=\frac{S\ell}{V}\,, \tag{1.6.4}$$

$$\tilde{\omega}^2=\frac{\omega^2\rho\ell^5}{VEA}\,,\qquad \tilde{P}=\frac{P\ell^4}{EAV^2}\,.$$

In what follows, the tilde is omitted. As a result we have the following equations and boundary conditions for the amplitude function $u(x)$:

$$\frac{\mathrm{d}^2}{\mathrm{d}x^2}\left(S^2\frac{\mathrm{d}^2u}{\mathrm{d}x^2}\right)+P\frac{\mathrm{d}}{\mathrm{d}x}\left(V_P\frac{\mathrm{d}u}{\mathrm{d}x}\right)-\omega^2Su=0\,, \tag{1.6.5}$$

$$\left(S^2\frac{\mathrm{d}^2u}{\mathrm{d}x^2}\right)_{x=0,1}=0,\qquad \left(\frac{\mathrm{d}}{\mathrm{d}x}\left(S^2\frac{\mathrm{d}^2u}{\mathrm{d}x^2}\right)\right)_{x=0,1}=0\,, \tag{1.6.6}$$

where by means of $V_P = V_P(x)$ we denote the part of the rod volume contained between the planes at $x = 1$ and at the considered point x, that is,

$$V_P(x)=\int_x^1 S(x)\mathrm{d}x\,. \tag{1.6.7}$$

We use the following constraint,

$$\int_x^1 S(x)\mathrm{d}x = 1\,, \tag{1.6.8}$$

to express the condition of a constant volume. This condition can be also written as

$$V_P(0) = 1\,.$$

The boundary value problem (1.6.5) and (1.6.6) is not a self-adjoint problem, therefore to investigate the stability of the rectilinear shape of an elastic rod we apply the following method (see, e.g., Bolotin [14], Hermann [15] and Plaut [16]). Multiply the Eq. (1.6.5) by the function $v = v(x)$ and integrate the product by parts on the interval [0, 1]. We have

$$H = \int_0^1 v\left(\frac{\mathrm{d}^2}{\mathrm{d}x^2}\left(S^2\frac{\mathrm{d}^2u}{\mathrm{d}x^2}\right) + P\frac{\mathrm{d}}{\mathrm{d}x}\left(V_P\frac{\mathrm{d}u}{\mathrm{d}x}\right) - \omega^2 Su\right)\mathrm{d}x =$$

$$= \int_0^1 u\left(\frac{\mathrm{d}^2}{\mathrm{d}x^2}\left(S^2\frac{\mathrm{d}^2v}{\mathrm{d}x^2}\right) + P\frac{\mathrm{d}}{\mathrm{d}x}\left(V_P\frac{\mathrm{d}v}{\mathrm{d}x}\right) - \omega^2 fv\right)\mathrm{d}x +$$

$$+\left[\frac{\mathrm{d}u}{\mathrm{d}x}\left(PV_Pv + S^2\frac{\mathrm{d}^2v}{\mathrm{d}x^2}\right)\right]_{x=0}^{x=1} - \left[u\left(PV_P\frac{\mathrm{d}v}{\mathrm{d}x} + \frac{\mathrm{d}}{\mathrm{d}x}\left(S^2\frac{\mathrm{d}^2v}{\mathrm{d}x^2}\right)\right)\right]_{x=0}^{x=1}.$$

The adjoint variable v we define as a solution of the following boundary-value problem

$$\frac{\mathrm{d}^2}{\mathrm{d}x^2}\left(S^2\frac{\mathrm{d}^2v}{\mathrm{d}x^2}\right) + P\frac{\mathrm{d}}{\mathrm{d}x}\left(V_P\frac{\mathrm{d}v}{\mathrm{d}x}\right) - \omega^2 Sv = 0\,, \tag{1.6.9}$$

$$Pv + S^2\frac{\mathrm{d}^2v}{\mathrm{d}x^2} = 0\,, \qquad P\frac{\mathrm{d}v}{\mathrm{d}x} + \frac{\mathrm{d}}{\mathrm{d}x}\left(S^2\frac{\mathrm{d}^2v}{\mathrm{d}x^2}\right) = 0\,, \qquad x = 0\,,$$

$$S^2\frac{\mathrm{d}^2v}{\mathrm{d}x^2} = 0\,, \qquad \frac{\mathrm{d}}{\mathrm{d}x}\left(S^2\frac{\mathrm{d}^2v}{\mathrm{d}x^2}\right) = 0\,, \qquad x = 1\,.$$

In this case, the conditions of stationary H with respect to variations u and v and the equality $H = 0$ are fulfilled. Note that for small values of P, the rectilinear rod shape is stable and all eigenvalues ω^2 are real and positive. The arising of instability corresponds to the appearance of a pair of complex conjugate frequencies. From the relations

$$H = 0\,, \qquad \frac{\mathrm{d}P}{\mathrm{d}\omega} = 0\,,$$

we obtain [16] the condition of dynamic loss of stability and the expression for the critical force

$$\int_1^0 Suv\mathrm{d}x = 0\,, \tag{1.6.10}$$

$$P = \frac{\int_0^1 S^2 \dfrac{\mathrm{d}^2u}{\mathrm{d}x^2}\dfrac{\mathrm{d}^2v}{\mathrm{d}x^2}\mathrm{d}x}{\int_0^1 V_P \dfrac{\mathrm{d}u}{\mathrm{d}x}\dfrac{\mathrm{d}v}{\mathrm{d}x}\mathrm{d}x - \dfrac{\mathrm{d}u}{\mathrm{d}x}(0)v(0)}\,. \tag{1.6.11}$$

The considered optimization problem is in finding the distribution of the sectional area, that is, the function $S = S(x)$ satisfying the isoperimetric condition (1.6.8) and maximizing the value of the critical force of the loss of stability.

To derive the necessary optimality condition we construct the Lagrange functional and derive the condition of its stationary with respect to S. We write the final result as

$$2S\frac{\mathrm{d}^2u}{\mathrm{d}x^2}\frac{\mathrm{d}^2v}{\mathrm{d}x^2} + P\int_0^x \frac{\mathrm{d}u}{\mathrm{d}x}\frac{\mathrm{d}v}{\mathrm{d}x}\mathrm{d}x - \omega^2uv = \lambda\,, \tag{1.6.12}$$

where λ is a constant. Thus, to find the unknown functions $u,\ v$, and S we have the closed system of equations and boundary conditions (1.6.5), (1.6.6), (1.6.9), and (1.6.12). The constant λ in (1.6.12) is determined with the help of the isoperimetric condition (1.6.8).

To realize the numerical solution of the optimization problem, we apply the algorithm of successive optimization Banichuk and Mironov [17], based on small variations of S and solutions of "direct" problems (1.6.9). The variations of the design functions S are performed by a gradient projection method satisfying the isoperimetric condition (1.6.8) and additional constraints on the minimal values of S introduced in the process of computations. The quantities ω^2, u and v were determinedfor fixed values of $S(x)$ using a finite-difference representation of the equation and boundary condition (1.6.9), and solving the linear algebraic eigenvalue problems that emerged.

The results for the dependences of P on ω are represented in Fig. 1.15. The broken lines correspond to the rod with constant distribution $S(x) = 1$. The continuous lines correspond to the optimal solutions.

The points $(\omega,\ P)$ of the loss of stability are indicated by the circles. At this point we have

$$\frac{\mathrm{d}P}{\mathrm{d}\omega} = 0\,.$$

For the constant distribution of thickness, we have the agreement of the computed values of P with the critical forces obtained in Feodosjev [12]. In the optimization process we obtain a meaningful increase of the critical force and the final gain is equal to 38%. The optimal distribution $S(x)$ is shown in Fig. 1.16. The function $S(x)$ has two maxima and takes the admissible minimum values at the ends and at the middle part of the rod.

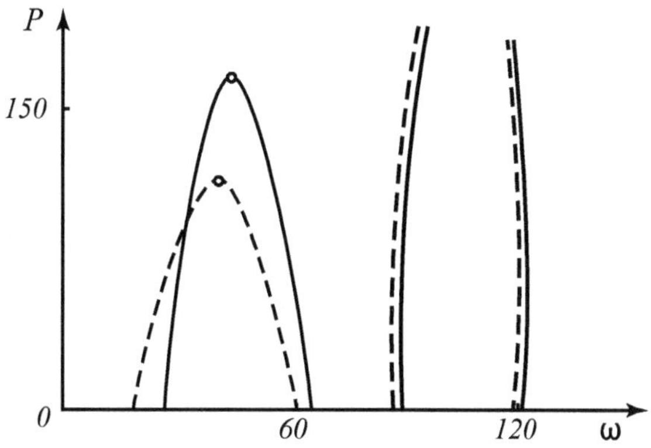

Fig. 1.15 The dependences of P on ω

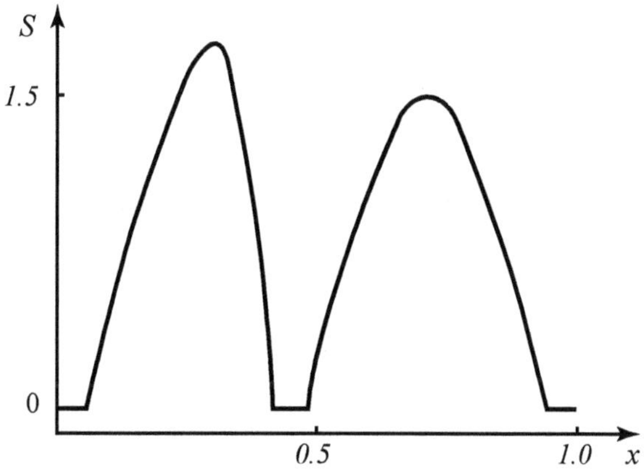

Fig. 1.16 The optimal distribution $S(x)$

In this chapter, we considered stability analysis in the mechanics of rigid and deformable structural elements. As examples, we looked at some conservative and nonconservative problems in the engineering sciences. In the next chapter, we will take a closer look at the fundamentals, considering the implicit function theorem as a tool to find bifurcations.

References

1. Banichuk NV (1978) Minimization of the weight of a wing with constraints on the velocity of divergence. Sci Notes Central Aerohydrodyn Inst (TsAGI) 9(5). in Russian
2. Clausen T (1851) About the shape of architectural columns. Bull Cl Phys-Math Acad Imp Sci St-Pétersbg 9:279–294
3. Lurie AI (1965) Application of the maximum principle to some simplest problems of mechanics. Res Works Leningr Polytechn Inst 252:34–46
4. Nikolai EL (1907) Lagrange's problem on the best shape column. Notes of the St. Petersburg Polytechnical Institute 8

5. Chentsov NG (1936) Column of least weight. Res Articles, TsA GI 265:1–48
6. Keller JB (1960) The shape of the strongest column. Arch Rat Mech Anal 5(1):275–285
7. Banichuk N, Barsuk A (192b) On stability of torsioned elastic rods. Izv Akad Nauk S.S.S.R., M.T.T. 6:148–154
8. McIntosh SC, Easter FE (1968) Design of minimum-mass structures with specificial stiffness properties. AIAA J 016(7):962–964
9. Ashley H, McIntosh SC (1968) Applications of aeroelastic constraints on structural optimization. In: Proceedings of the 12th international congress of theoretical and applied mechanics, pp 100–113, Berlin, 1969. Springer. Stanford
10. Jelicic ZD, Atanackovic TM (2006) On an optimization problem for elastic rods. Struct Multidiscip Optim 32(1):59–64
11. Banichuk NV, Makeev EV, Barsuk AA (2011g) Optimization of tension rods by the criterion of stability. Proc Nijegorod Univ 2(1):115–122 in Russian
12. Feodosjev VI (1965) On one problem of stability. Appl Math Mech 29(2). in Russian
13. Banichuk NV, Gura NM (1979b) On a dynamic problem of optimal design. Mech Deform Solid Bodies 41:20–24
14. Bolotin VV (1961) Nonconservative problems of the theory of elastic stability. Fizmatgiz, Moscow. 339 pages, in Russian
15. Prasad SN, Hermann G (1969) The usefulness of adjoint systems in solving nonconservative problems of elastic continua. Int J Solids Struct 5
16. Plaut RH (1972) Determining the nature of instability in nonconservative problems. AIAA J 10(2)
17. Banichuk NV, Mironov AA (1975) Optimization of vibration frequencies of an elastic plate in an ideal fluid. J Appl Math Mech 39(5):853–863. ISSN 0021-8928. https://doi.org/10.1016/0021-8928(75)90126-4. http://www.sciencedirect.com/science/article/pii/0021892875901264

Chapter 2
Bifurcation Analysis for Polynomial Equations

This chapter is devoted to bifurcation problems based on some models described by polynomial equations with real coefficients. Bifurcation analysis, parametric representations of solutions and their asymptotic analysis and expressions are described within a framework of analytical approaches. The results presented in this chapter can be used to help locate the bifurcation points of the solution curves. The results also allow the development of very efficient procedures for sensitivity analysis of the dependences of solutions on the problem parameters.

Many important problems of theoretical and applied mechanics lead to a need for qualitative and quantitative analysis of solutions of polynomial equations. An example is the determination of principal axes of stress and strain tensors. Despite the fact that, in some cases, there are classical formulas for equation solutions the practical, the use of these formulas encounters some difficulties, especially when there are separate continuous branches of solutions and we study the dependences of solutions on problem parameters.

As for classical approaches to bifurcation theory and its application to statical and dynamical problems arising in mechanics and engineering and in mathematical physics see, for example, Banichuk [3–5], Banichuk and Barsuk [1], Banichuk and Gura [6], Banichuk et al. [2, 7–14], Bolotin [15–20], Jeronen [21], Mote [23–26], Plaut [29], Pramila [30, 31], Saksa et al. [32–35], Wickert and Mote [37, 38], Nečas [27], Tuovinen [36], Larichev [22], Neittaanmäki and Ruotsalainen [28]. Some modern developments of bifurcation and stability analysis are known as catastrophe theory. This theory includes a variety of new problems of stability analysis and qualitative topological methods. In this chapter we address the problem of the origin of bifurcation phenomena. We present some new aspects for asymptotes as well as for parametric representations of the solutions.

N. Banichuk et al., *Stability of Axially Moving Materials*, Solid Mechanics and Its Applications 259, https://doi.org/10.1007/978-3-030-23803-2_2

2.1 Bifurcation and Parametric Representations

Very often in many problems of mechanics and physics, the relations between the state variables are given in an implicit form,

$$F(x, y) = 0 . \qquad (2.1.1)$$

In particular, this equation describes also the case of an explicit function $y = f(x)$, which corresponds to the equation

$$F(x, y) \equiv y - f(x) = 0 .$$

Let us present the following dependences in parametric form,

$$x(\tau) = \varphi(\tau), \quad y(\tau) = \psi(\tau) . \qquad (2.1.2)$$

Substituting (2.1.2) into (2.1.1) gives us an equation with respect to the parameter τ,

$$F(\varphi(\tau), \psi(\tau)) \equiv 0 . \qquad (2.1.3)$$

Thus, (2.1.3) can be considered as a parametric representation of the solution of Eq. (2.1.1). It is possible to note that the representation of the functions in parametric form includes, as a particular case, an explicit representation:

$$y = f(x), \quad x = \tau, \quad y = f(\tau) .$$

Let us perform bifurcation analysis and present basic relations for finding the bifurcation points of implicitly given functions and corresponding relations in parametric form. These relations will be used here and in what follows.

As it follows from the implicit function theorem, the uniqueness of solutions is violated when

$$F(x, y) = 0, \quad \frac{\partial F}{\partial y}(x, y) = 0 , \qquad (2.1.4)$$

where we consider the dependence $y = y(x)$. For parametrically given dependences we have

$$\frac{\partial F}{\partial y}(x(\tau), y(\tau)) = 0 . \qquad (2.1.5)$$

As an example, in a rectangular coordinate system, for a circle of unit radius with its center at (0, 0), we have

$$F(x, y) = x^2 + y^2 - 1 = 0 .$$

In parametric form we may write

$$x(\tau) = \cos\tau, \qquad y(\tau) = \sin\tau, \qquad F(x, y) = x^2 + y^2 - 1 = 0$$

and Eq. (2.1.5) takes the form

$$2y = 2\sin\tau = 0\ .$$

We have the bifurcation values of

$$\tau_1^* = 0, \qquad \tau_2^* = \pi\ ,$$

which correspond to the bifurcation values of x and y:

$$x_1^* = \cos\tau_1^* = 1 \quad y_1^* = \sin\tau_1^* = 0 \quad \text{and} \quad x_2^* = \cos\tau_2^* = -1, \quad y_2^* = \sin\tau_2^* = 0\ .$$

We obtain the same value if we use (2.1.4) on the equations

$$x^2 + y^2 - 1 = 0, \qquad 2y = 0\ .$$

Let us also consider the expression for $\mathrm{d}y/\mathrm{d}x$, which is used for the analysis of different branches of solutions for (2.1.1). Let $y = y(x)$, and differentiate (2.1.1) with respect to x. We have

$$\frac{\mathrm{d}y}{\mathrm{d}x} = \frac{-\left(\dfrac{\partial F}{\partial x}\right)}{\left(\dfrac{\partial F}{\partial y}\right)}, \qquad \frac{\partial F}{\partial y} \neq 0\ . \tag{2.1.6}$$

Using the representations for x and y in the parametric form (2.1.2) and the expressions

$$\mathrm{d}x = \dot{\varphi}(\tau)\mathrm{d}\tau, \qquad \mathrm{d}y = \dot{\psi}(\tau)\mathrm{d}\tau,$$

we obtain

$$\frac{\mathrm{d}y}{\mathrm{d}x} = \frac{\dot{\psi}}{\dot{\varphi}}\ . \tag{2.1.7}$$

The Eq. (2.1.5) and the condition

$$\dot{\varphi}(\tau_*) = 0 \quad \text{for} \quad \dot{\psi}(\tau_*) \neq 0 \tag{2.1.8}$$

are satisfied for dependences given by (2.1.1), (2.1.2) at the bifurcation points, when

$$\frac{\mathrm{d}y}{\mathrm{d}x} = \pm\infty \quad \text{for} \quad \frac{\partial F}{\partial x} \neq 0\,, \quad \frac{\partial F}{\partial y} = 0\,.$$

This is useful because (2.1.8) is simpler than the analysis of bifurcation solutions in the form (2.1.5).

Let us also present the conditions for determination of the bifurcation points when using the representation (2.1.2),

$$\dot{\varphi}(\tau_*) = 0 \quad \text{for} \quad \dot{\psi}(\tau_*) = 0\,. \tag{2.1.9}$$

Let us consider an important case, a functional dependence in the form of a polynomial $p(x)$ of order n,

$$F(x, a_1, K, a_n) = p_n(x) = x^n + a_1 x^{n-1} + K + a_{n-1}x + a_n = 0\,, \tag{2.1.10}$$

where x is a real variable, $a_1, \ldots, a_n$ are real coefficients, and K represents the terms of orders $2, \ldots, n-2$. In what follows we consider (2.1.10) as an implicit representation of the variable x as a function of the parameters $a_1, \ldots, a_n$, i.e. $x = x(a_1, \ldots, a_n)$. Note also the linear dependence of $F(x, a_1, K, a_n)$ on the parameters $a_1, \ldots, a_n$. In the considered case, the conditions of a bifurcation in the solution of Eq. (2.1.10) take the form

$$F(x, a_1, K, a_n) = 0\,, \qquad \frac{\partial F(x, a_1, K, a_n)}{\partial x} = 0\,. \tag{2.1.11}$$

Eliminating x from relations (2.1.11), we obtain the equation of a hypersurface in the space of parameters $a_1, \ldots, a_n$

$$\Phi(a_1, K, a_n) = 0\,. \tag{2.1.12}$$

This hypersurface divides the space of parameters $a_1, \ldots, a_n$ into several domains D_i, each characterized by containing some fixed number of solutions for the Eq. (2.1.10). In each internal point of the domain $D_i((a_1, K, A_n) \in D_i)$, the function $\Phi(a_1, K, a_n) \neq 0$, and the sign stays fixed for all points in the domain D_i.

To illustrate, let us present the analysis of solution of the quadratic equation

$$F(x, a_1, a_2) = x^2 + a_1 x + a_2 = 0\,.$$

If the coefficients a_1, a_2 are real, then the quadratic equation has either two real solutions or two complex-conjugate solutions.

The bifurcation points are determined with the help of relations (2.1.11). We have

$$F(x, a_1, a_2) = x^2 + a_1 x + a_2 = 0, \qquad \frac{\partial F(x, a_1, a_2)}{\partial x} = 2x + a_1 = 0\,. \tag{2.1.13}$$

Eliminating x from (2.1.13), it follows that

$$\Phi(a_1, a_2) = a_1^2 - 4a_2 = 0 . \tag{2.1.14}$$

Thus, the parabola $a_2 = a_1^2/4$ divides the plane a_1, a_2 into two domains—internal and external—with respect to the parabola. If $a_1 = 0$ and $a_2 < 0$, then the quadratic equation has two real solutions $x_{1,2} = \pm\sqrt{-a_2}$, and the point ($a_1 = 0,\ a_2 < 0$) lies in the external domain with respect to the parabola. Consequently, at all points of the external domain there are exist two real solutions (domain D_2), but there are no real solutions for points in the internal domain (domain D_0).

Returning to the polynomial equation (2.1.10), let us construct representations for its real solutions. We will vary only the parameter a_i, keeping all other parameters fixed. Thus, we consider the problem of the construction of n dependences $x = x(a_i),\quad i = 1, \ldots, n$. Let us introduce the real parameter τ and the parametric representation $x = x(\tau) = \tau$. Using (2.1.10) as well, we have

$$\begin{aligned} x(\tau) &= \tau \\ a_i(\tau) &= -\left(\tau^{n-i}\right)^{-1}\left(\tau^n + a_1\tau^{n-1} + \cdots + a_{i-1}\tau^{n-i+1} + a_{i+1}\tau^{n-i-1} + \cdots + a_n\right) , \\ &\quad i = 1, \ldots, n . \end{aligned} \tag{2.1.15}$$

The parametric representations (2.1.15) make it possible to investigate the behavior of solutions of polynomial equations (of an arbitrary order) as a function of each separate coefficient a_i. Among all parametric dependences (2.1.15), we select the dependence with $i = n$, which takes the simplest form

$$x(\tau) = \tau, \qquad a_n(\tau) = -(\tau^n + a_1\tau^{n-1} + K + a_{n-1}\tau) . \tag{2.1.16}$$

The right hand side of (2.1.16) is an analytical function of the parameter τ, and consequently the dependence of real solutions of the polynomial equation of arbitrary order on the coefficient is a continuous function (in the general case, a multifunction).

In all other dependences (2.1.15) corresponding to $i \neq n$, the right-hand sides of $a_i(\tau)$ contain singular terms. This leads to the appearance of discontinuities in the dependence $x = x(a_1)$. In the illustrative case of the quadratic equation $x^2 + a_1x + a_2 = 0$, the relations (2.1.15) are written as

$$x(\tau) = \tau, \qquad a_1(\tau) = -\tau - \frac{a_2}{\tau} \tag{2.1.17}$$

when a_2 is fixed (i.e. $x = x(a_1)$), and

$$x(\tau) = \tau, \qquad a_2(\tau) = -(\tau + a_1)\tau \tag{2.1.18}$$

when a_1 is fixed (i.e. $x = x(a_2)$).

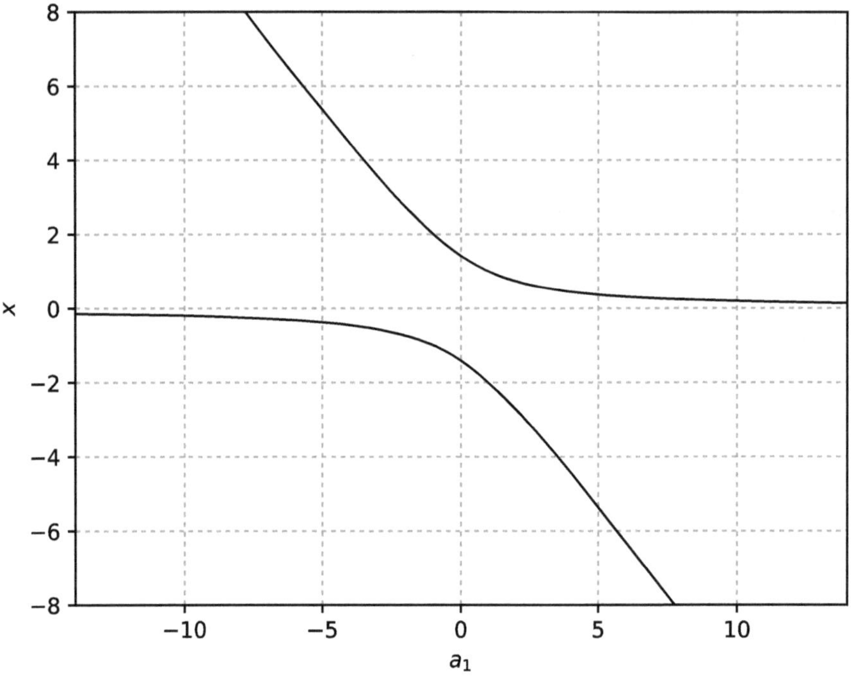

Fig. 2.1 The dependence $x = x(a_1,\ a_2 = -2)$

Examples of the dependences defined by the parametric representations (2.1.17) and (2.1.18), are respectively shown in Figs. 2.1 and 2.2

Let us formulate the conditions for determination of bifurcation points of the dependences $x = x(a_i)$. If we hold fixed the parameters $a_{1,} \ldots, a_{i-1}, a_{i+1}, \ldots a_n$, then the equations for determining the bifurcation values of the variable $x$and the parameters a_i are written in the following form:

$$F(x, a_1, K, a_{i-1}, a_i, a_{i+1}, K, a_n) = 0\,,$$
$$\frac{\partial F(x, a_1, K, a_{i-1}, a_i, a_{i+1}, K, a_n)}{\partial x} = 0\,, \tag{2.1.19}$$

where each K represents the omitted parameters. Using parametric representations in the form (2.1.15) and the general representation of implicitly given functions (2.1.2), note the correspondence between the notations of (2.1.2) and (2.1.15): $x \to a_i$ and $y \to x$. Thus, the condition of bifurcation, that is, violation of a single–valued dependence, of the parametrically given relations (2.1.15) can be written

$$\left(\frac{\partial F(x, a_1, \ldots, a_n)}{\partial x}\right)_{x=\tau} = n\tau^{n-1} + (n-1)a_1\tau^{n-2} + \cdots + a_{n-1} = 0\,. \tag{2.1.20}$$

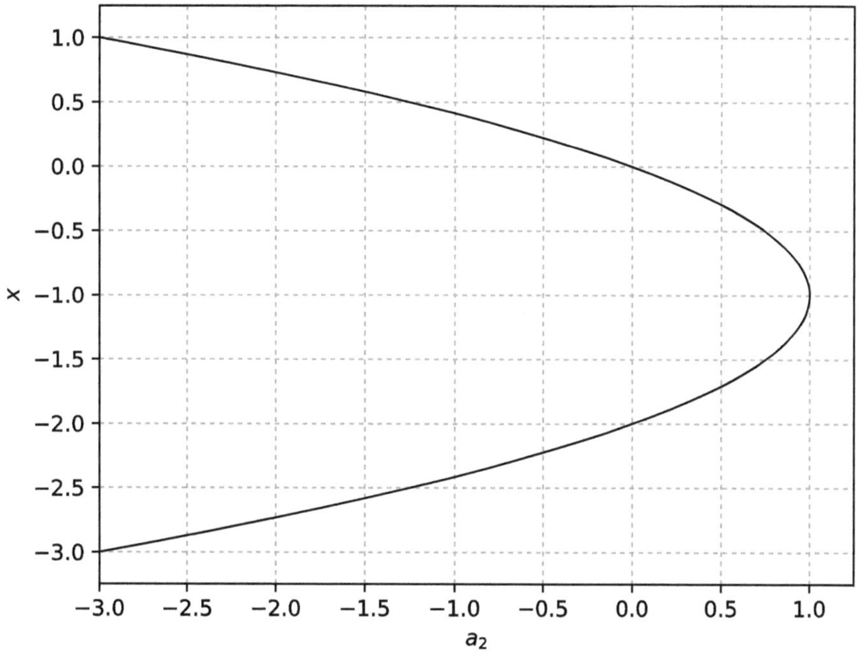

Fig. 2.2 The dependence $x = x(a_1 = 2,\ a_2)$

Using (2.1.15) and (2.1.20) yields

$$\frac{da_i(\tau)}{d\tau} \equiv -\frac{1}{\tau^{n-i}} \frac{\partial F(\tau, a_{1,} \dots, a_n)}{\partial \tau} = 0\,, \tag{2.1.21}$$

which is more convenient from the computational point of view.

In the case of the quadratic equation $x^2 + a_1 x + a_2 = 0$, we first consider the dependence $x = x(a_1)$ in accordance with Eq. (2.1.17), differentiation by τ produces

$$\dot{a}_1(\tau) = -1 + \frac{a_2}{\tau^2} = 0\,, \tag{2.1.22}$$

and consequently we obtain

$$\tau_*^2 = a_2\,, \qquad a_1^* = -\tau_* - \frac{a_2}{\tau_*} = -\frac{2a_2}{\tau_*}\,. \tag{2.1.23}$$

Thus, we find the bifurcation value of the parameter a_1, which is $(a_1^*)^2 = 4a_2$.

In accordance with (2.1.18) for the alternative version $x = x(a_2)$, we obtain

$$\dot{a}_2(\tau) = -2\tau - a_1 = 0\,, \tag{2.1.24}$$

from which it follows that

$$\tau_* = -\frac{a_1}{2}\,, \qquad a_2^* = \frac{a_1^2}{4}\,. \tag{2.1.25}$$

2.2 Analysis of a Cubic Equation

Consider the following cubic equation:

$$ey^3 + ay^2 + by + c = 0\,, \tag{2.2.1}$$

where e, a, b, c are arbitrary real coefficients, and $e > 0$. We would like to find the solution of (2.2.1) in an analytical form, and to analyze the solution as a function of the coefficients. The standard cubic equation (2.2.1) depends on the four given parameters e, a, b, c, but it can be reduced to a canonical form with one parameter. To perform this, let

$$y = \widetilde{x} - \frac{a}{3e}\,, \tag{2.2.2}$$

$$p = \frac{b}{e} - \frac{a^2}{3e^2}\,, \quad q = \frac{c}{e} - \frac{ab}{3e^2} + \frac{2}{27}\frac{a^3}{e^3}\,, \tag{2.2.3}$$

This transforms equation (2.2.1) into

$$\widetilde{x}^3 + p\widetilde{x} + q = 0\,.$$

By further transforming

$$x = \frac{1}{\sqrt[3]{q}}\widetilde{x}\,,$$

we obtain the canonical form, dependent on one parameter γ:

$$x^3 + \gamma x + 1 = 0, \qquad \gamma = \frac{p}{\sqrt[3]{q^2}}\,. \tag{2.2.4}$$

Thus, the singularities of the solutions of the Eq. (2.2.1) are determined by the unique parameter γ. The only case which does not allow defining γ is $q = 0$; in this case, the equation trivially factors into $(\widetilde{x}^2 + p)\widetilde{x} = 0$. In the following general treatment, let $q \neq 0$.

Taking into account that all cubic equations with real coefficients and $q \neq 0$ can be transformed to the canonical form (2.2.4), we conclude that all qualitative singularities of their solutions are determined by the only real parameter γ.

In what follows we perform bifurcation analysis using the canonical equation (2.2.4). Let $x_1(\gamma),\ x_2(\gamma)$ and $x_3(\gamma)$ be the solutions of the cubic equation (2.2.4), and note that these solutions represent the different branches of the function $x(\gamma)$ implicitly defined by the equation

$$F(x,\gamma) \equiv x^3 + \gamma x + 1 \,. \tag{2.2.5}$$

From the implicit function theorem, it follows that the bifurcation points of the equation $F(x, y) = 0$ are given by the conditions

$$F(x,\gamma) = x^3 + \gamma x + 1 = 0\,, \quad \frac{\partial F(x,\gamma)}{\partial x} = 3x^2 + \gamma = 0\,, \tag{2.2.6}$$

and are represented by the unique solution of the Eq. (2.2.6) as

$$x_* = \frac{1}{\sqrt[3]{2}}, \qquad \gamma_* = -\frac{3}{\sqrt[3]{2^2}}\,. \tag{2.2.7}$$

Thus, we have a unique value $\gamma = \gamma_*$ of the problem parameter for which bifurcation occurs. Consequently, we have one (or three) real solutions when $\gamma > \gamma_*$; we respectively have three (or one) real solutions when $\gamma < \gamma_*$.

Taking into account that the cubic equation (2.2.1) has a unique real solution $x = -1$ for $\gamma = 0 > \gamma_*$, we conclude that for any $\gamma > \gamma_*$, there exists only one real solution, denoted by $x_1(\gamma)$, but for $\gamma < \gamma_*$ there exist three real solutions denoted by $x_1(\gamma),\ x_2(\gamma)$ and $x_3(\gamma)$. Note also that $x = 0$ is not a solution of equations $F(x,\gamma) = 0$ in (2.2.6) for any value of γ, and consequently each of the curves $x_i(\gamma)$ $(i = 1, 2, 3)$ lies in one of the half-planes $x > 0$ or $x < 0$ of the plane (x, γ). In particular, we have $x_1(\gamma) < 0$ in the interval $-\infty < \gamma < \infty$.

At the same time,

$$x_2(\gamma_*) = x_3(\gamma_*) = \frac{1}{\sqrt[3]{2}} \tag{2.2.8}$$

for the bifurcation value $\gamma = \gamma_*$, and consequently the functions $x_2(\gamma)$ and $x_3(\gamma)$ are positive in the interval $-\infty < \gamma < \gamma_*$.

To present qualitative and asymptotic analysis of the solutions $x_i(\gamma)$, we use the derivative of an implicitly defined function:

$$\frac{\mathrm{d}x_i(\gamma)}{\mathrm{d}\gamma} = -\frac{\partial F/\partial \gamma}{\partial F/\partial x} = -\frac{x_i}{3x_i^2 + \gamma}, \qquad i = 1, 2, 3\,. \tag{2.2.9}$$

In accordance with Eqs. (2.2.7) and (2.2.8), the interval $\gamma \in (-\infty,\ \infty)$ is divided by the value $\gamma = \gamma_*$ into two parts $(-\infty,\ \gamma_*)$ and $(\gamma_*,\ \infty)$, where the sign of $\partial F/\partial x$ does not vary. Note that

$$\frac{\partial F}{\partial x} = 3x^2 > 0 \tag{2.2.10}$$

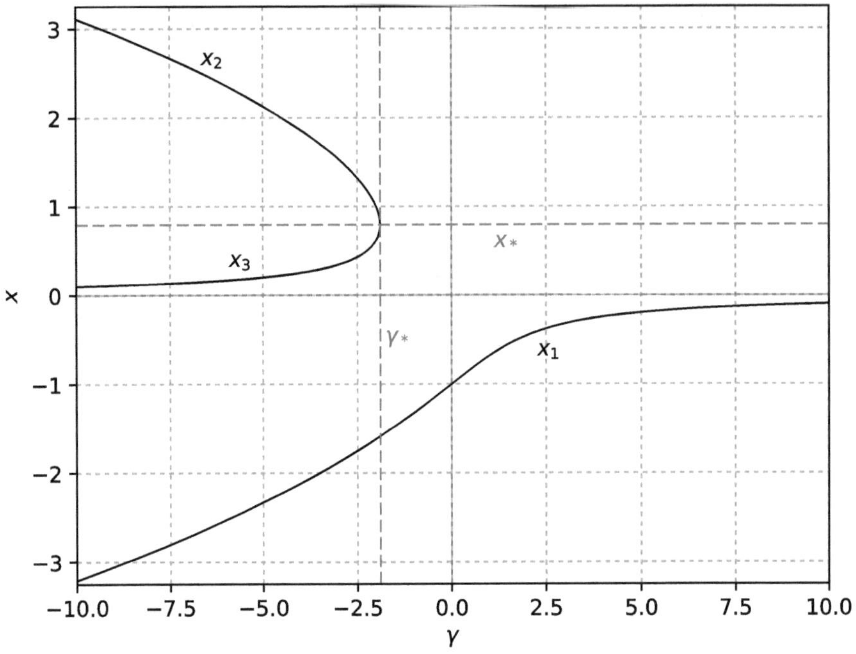

Fig. 2.3 Solutions of the canonical cubic equation. The bifurcation point is at $\gamma_* = -3/\sqrt[3]{4}$, $x_* = \sqrt[3]{4}/2$

for all real x and for $\gamma = 0 > \gamma_*$. Consequently, $\partial F/\partial x > 0$ when $\gamma \in (\gamma_*, \infty)$. But in the interval $(-\infty, \gamma_*)$, the sign of $\partial F/\partial x$ depends on the considered function $x_i(\gamma)$. In particular, the value $\partial F/\partial x$ cannot be zero for the function $x_1(\gamma)$, and the inequality $\partial F/\partial x > 0$ holds because (2.2.10) holds for $\gamma = 0 > \gamma_*$. Taking into account that $x_1(\gamma) < 0$, we arrive at the conclusion that $x_1(\gamma)$ is a monotonically increasing function of γ in the interval $-\infty < \gamma < \infty$. In the same manner, we may show that the function $x_2(\gamma)$ is a monotonically decreasing function, and $x_3(\gamma)$ is a monotonically increasing function, of the parameter γ when $-\infty < \gamma < \gamma_*$. The solutions of the canonical cubic equation are shown in Fig. 2.3. The following asymptotic expressions for $x_i(\gamma)$, when $|\gamma| \gg 1$,

$$x_1(\gamma) \approx -\frac{1}{\gamma}, \quad \gamma \gg 1,$$

$$x_1(\gamma) \approx -\sqrt{-\gamma}, \quad x_2(\gamma) \approx \sqrt{-\gamma}, \quad x_3(\gamma) \approx -\frac{1}{\gamma}, \qquad \gamma \to -\infty, \tag{2.2.11}$$

are obtained with the help of asymptotic analysis of the Eq. (2.2.7).

To obtain the parametric representations of real solutions of the canonical cubic equation (2.2.4), we solve the equation with respect to the parameter γ. We have

$$\gamma = -\frac{1+x^3}{x} = -\frac{1}{x} - x^2, \qquad x \text{ is real.} \tag{2.2.12}$$

The dependences $x(\gamma)$ have the property of having fixed signs, and consequently the set of solutions $x_1(\gamma),\ x_2(\gamma),\ x_3(\gamma)$ can be divided into two classes. The first class are the positive dependences ($x(\gamma) > 0$), and the second class are the negative dependences ($x(\gamma) < 0$).

For negative solutions $x < 0$, we introduce the positive parameter $\tau > 0$ in accordance with the relation $x = -\tau$, and as a result we obtain the parametric representation of the dependence $x(\gamma) < 0$,

$$x_1(\tau) = -\tau, \qquad \gamma(\tau) = \frac{1-\tau^3}{\tau}, \qquad \tau > 0\,, \tag{2.2.13}$$

denoted later as $x_1(\gamma)$. It follows from (2.2.13) that $\tau = 1$ for $\gamma = 0$ and $x_1(\gamma = 0) = -1$. Thus, the curve $x_1(\gamma)$ passes through the point $(0, -1)$ of the plane (x, γ). Using the definition of derivatives of parametrically defined functions, we have

$$\frac{\mathrm{d}x_1}{\mathrm{d}\gamma} = \frac{\mathrm{d}x_1/\mathrm{d}\tau}{\mathrm{d}\gamma/\mathrm{d}\tau} = \frac{\tau^2}{1+2\tau^3} > 0, \qquad \tau > 0\,, \tag{2.2.14}$$

and consequently $x_1(\gamma)$ is a monotonically increasing function of the parameter γ for $\gamma \in (-\infty, \infty)$.

Using the dependence $x_1(\gamma)$ in the parametric form (2.2.13), let us study the asymptotic behavior of the solution $x_1(\gamma)$ for asymptotic values of parameter γ, that is, $|\gamma| \gg 1$. We have $\gamma(\tau) \approx 1/\tau$ for $\tau \ll 1$, and as a result the following asymptotic behavior takes place:

$$x_1(\gamma) \approx -\frac{1}{\gamma}, \qquad \gamma \gg 1\,. \tag{2.2.15}$$

It follows also that

$$\gamma(\tau) = -\tau^2 < 0 \qquad (\tau \approx \sqrt{-\gamma}, \quad \gamma < 0)\,, \tag{2.2.16}$$

and we find asymptotic behavior for $x_1(\gamma)$ ($\gamma \to -\infty$) in the form

$$x_1(\gamma) \approx -\sqrt{-\gamma}, \qquad \gamma \to -\infty\,. \tag{2.2.17}$$

These asymptotic expressions coincide with the corresponding asymptotic expressions in (2.2.11).

Let us analyze the representation (2.2.12) in the case $x > 0$. Assuming that $x = \tau$ ($\tau > 0$) in the considered case, we obtain the following parametric representation for positive dependences $x(\gamma)$, determined by Eq. (2.2.4):

$$x(\tau) = \tau, \quad \gamma(\tau) = -\frac{1+\tau^3}{\tau}, \qquad \tau > 0\,. \tag{2.2.18}$$

Note that the dependence $\gamma(\tau)$ in (2.2.18) is not monotonic in the interval $\tau \in (0, \infty)$. In fact, using the expression

$$\frac{d\gamma}{d\tau} = \frac{1}{\tau^2} - 2\tau, \tag{2.2.19}$$

we find that the derivative vanishes for the unique value $\tau_* = 1/\sqrt[3]{2} = \sqrt[3]{4}/2$. In addition, the dependence $\gamma(\tau)$ attains its maximum value,

$$\gamma_* = \gamma(\tau_*) = -\frac{1+\tau_*^3}{\tau_*} = -3\frac{\sqrt[3]{2}}{2} = -\frac{3}{\sqrt[3]{4}}, \tag{2.2.20}$$

at the parameter value $\tau = \tau_*$. Thus, the relations (2.2.18) determine two dependences $x_2(\gamma)$ and $x_3(\gamma)$:

$$x_2(\tau) = \tau\,, \quad \gamma(\tau) = -\frac{1+\tau^3}{\tau}, \ 0 < \tau \le \tau_*\,, \tag{2.2.21}$$

$$x_3(\tau) = \tau\,, \quad \gamma(\tau) = -\frac{1+\tau^3}{\tau}, \ \tau_* \le \tau < \infty\,. \tag{2.2.22}$$

Note that

$$x_2(\tau_*) = x_3(\tau_*) = x_* = \tau_* = \frac{1}{\sqrt[3]{2}} \tag{2.2.23}$$

for $\tau = \tau_*$ and thus the obtained values γ_* and x_* correspond to the bifurcation values (2.2.7). Note also that both dependences $x_2(\gamma)$ and $x_3(\gamma)$ are monotonic in the interval $-\infty < \gamma \le \gamma_*$. Moreover, $x_2(\gamma)$ is a monotonically increasing function of parameter γ, and $x_3(\gamma)$ is a monotonically decreasing function. Concerning the asymptotic behavior of the branches $x_2(\gamma)$ and $x_3(\gamma)$ when $\gamma \to -\infty$, we have

$$\gamma(\tau) \approx -\frac{1}{\tau}\,, \quad x_2(\tau) = \tau \approx -\frac{1}{\gamma} \quad \text{for } \tau \ll 1\,, \tag{2.2.24}$$

and consequently we arrive at the following asymptotic dependence:

$$x_2(\tau) \approx \frac{1}{\gamma}\,, \quad \gamma \to -\infty\,. \tag{2.2.25}$$

Taking into account that $\gamma(\tau) \approx -\tau^2$, $x_3(\tau) = \tau \approx \sqrt{-\gamma}$ for $\tau \gg 1$, we have

$$x_3(\gamma) \approx \sqrt{-\gamma}\,, \quad \gamma \to -\infty\,. \tag{2.2.26}$$

As the final result we obtain the parametric representations of the solutions of (2.2.4):

$$\begin{aligned}
x_1(\tau) &= -\tau, \quad \gamma(\tau) = \frac{1-\tau^3}{\tau}, \quad \tau > 0\,, \\
x_2(\tau) &= \tau, \quad \gamma(\tau) = -\frac{1+\tau^3}{\tau}, \quad 0 < \tau \le \tau_*\,, \\
x_2(\tau) &= \tau, \quad \gamma(\tau) = -\frac{1+\tau^3}{\tau}, \quad \tau_* \le \tau < \infty\,,
\end{aligned} \tag{2.2.27}$$

where $\tau_* = 1/\sqrt[3]{2}$. For a visualization of $x_1(\gamma)$, $x_2(\gamma)$ and $x_3(\gamma)$, refer to Fig. 2.3.

Now let us consider a cubic equation in reduced form,

$$x^3 + px + q = 0\,, \qquad p, q \text{ are real} \tag{2.2.28}$$

and study the dependence of the solutions of the Eq. (2.2.28) on its coefficients p and q. This representation admits also $q = 0$, and is sometimes used instead of the canonical form (2.2.4).

The dependences $x(p, q)$ are represented in Figs. 2.4 and 2.5 for $p = 2$ and $p = -2$, respectively, and for $q = 2$ in Fig. 2.6.

Note that for fixed value of p the solutions of the Eq. (2.2.28) are odd functions of the parameter q, that is, $x(p, q) = -x(p, -q)$, see Figs. 2.4 and 2.5. Note also that for fixed q, the cubic equation in reduced form coincides (up to scaling) with the

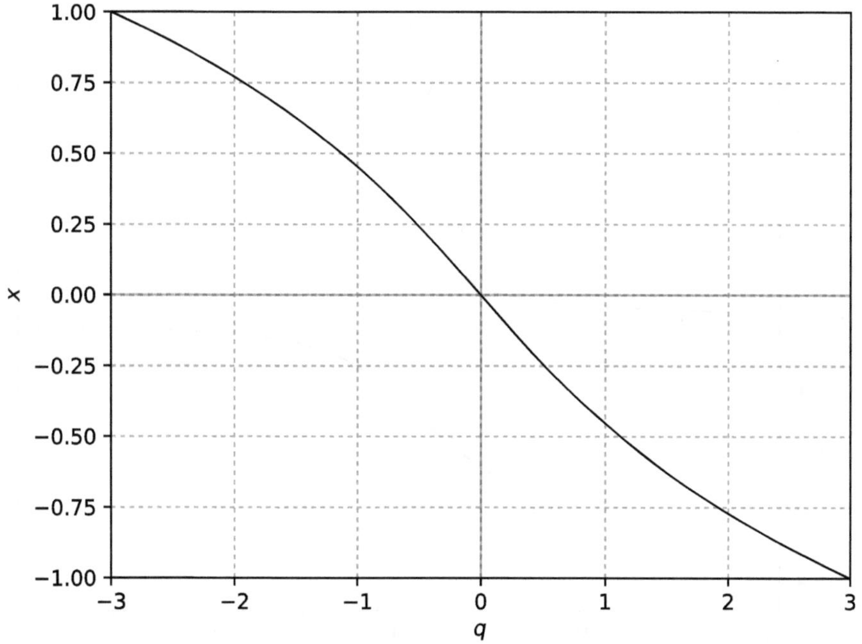

Fig. 2.4 The dependence $x(p = 2,\ q)$

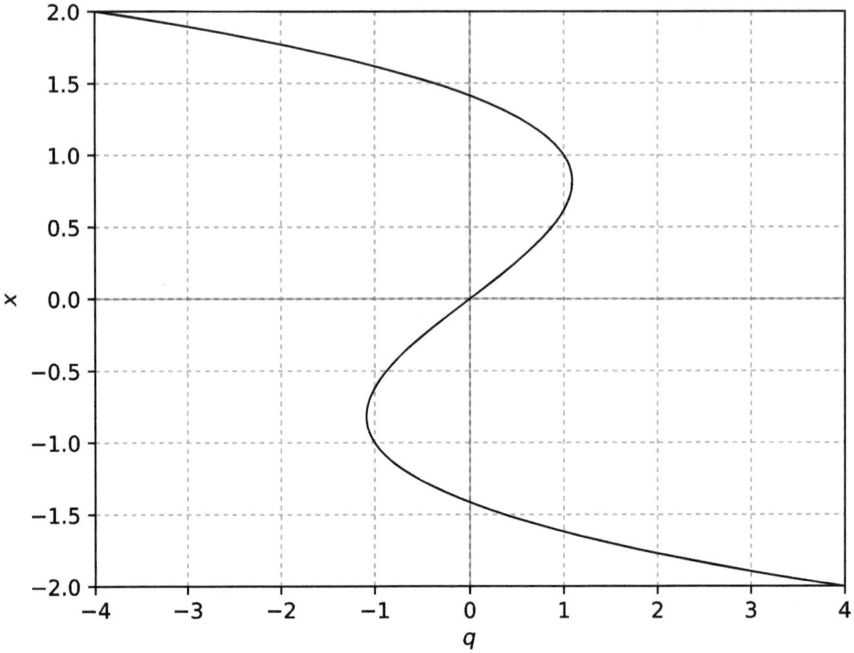

Fig. 2.5 The dependence $x(p = -2,\ q)$

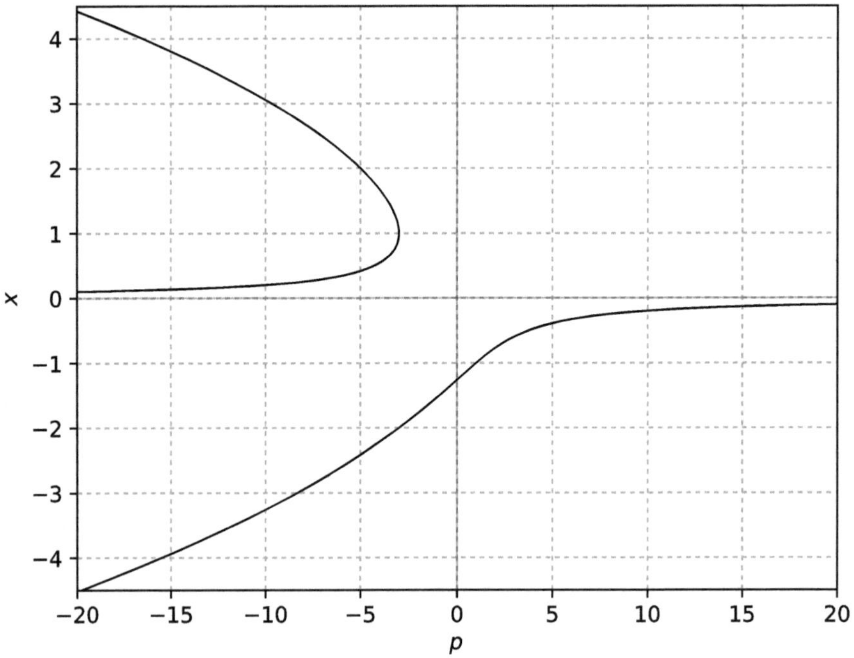

Fig. 2.6 The dependence $x(p,\ q = 2)$

equation in canonical form (2.2.4). Consequently, the behaviors of the corresponding solutions coincide, up to scaling (Figs. 2.6 and 2.3). Taking this into account, we analyze only the dependences of solutions on the coefficients.

Consider now the equation as an implicit representation for $x(p, q)$:

$$F(x, p, q) \equiv x^3 + px + q = 0 . \tag{2.2.29}$$

Using (2.2.29), let us formulate the bifurcation conditions

$$F(x, p, q) = 0 , \qquad \frac{\partial F(x, p, q)}{\partial x} = 3x^2 + p = 0 . \tag{2.2.30}$$

From the second equation in (2.2.30), it follows that for real x, a bifurcation can occur only at negative values of the coefficient p. If $p \geq 0$, then the Eq. (2.2.29) has a unique real solution and in this case $x(p, q) = -x(p, -q)$. If $p < 0$, then the solution of the system (2.2.30) gives the bifurcation values

$$x_* = \sqrt[3]{\frac{q}{2}} , \quad p_* = -3\sqrt[3]{\left(\frac{q}{2}\right)^2} . \tag{2.2.31}$$

In the case $q = 1$, the cubic equation in reduced form (2.2.28) coincides with the equation in canonical form, with the coefficient p corresponding to the parameter γ, and the bifurcation values (2.2.31) are the same as the value (2.2.7).

Solving the Eq. (2.2.29) with respect to coefficient q, and introducing the variable τ as $x = \tau$, we arrive at the parametric representation for the dependence $x(p, q)$ for fixed p:

$$x(\tau) = \tau, \quad q(\tau) = -(p + \tau^2)\tau, \qquad -\infty < \tau < \infty . \tag{2.2.32}$$

From (2.2.32) it follows that when p is fixed, negative values of $x(p, q)$ correspond to positive values of the coefficient q, and $x(p, q) \equiv -x(p, q)$. Taking into account the analyticity of the right-hand sides of the Eq. (2.2.32) with respect to the parameter τ, we conclude that the expressions (2.2.32) describe a continuous line determined by the equation $F(x, p, q) = 0$.

The graphic representations of the solutions of (2.2.32) for $x(p = 2,\ q)$ and $x(p = -2,\ q)$ are the same as those in Figs. 2.4 and 2.5, respectively.

Let us determine the ranges of the variable τ, for which the dependence $x(p, q)$ is single-valued, and for which it is many-valued. From (2.2.32), we see that for any fixed $p > 0$, we have $\mathrm{d}q(\tau)/\mathrm{d}\tau = -p - 3\tau^2 < 0$ for $\tau \in (-\infty, \infty)$. Thus, for fixed $p > 0$, the parametric representation (2.2.32) determines a single-valued dependence $x(p, q)$.

Let us analyze now the dependences $q(\tau)$ and $x(p, q)$ for $p < 0$. In this case, $q(\tau)$ vanishes at the three values

$$\tau_0 = 0 , \qquad \tau_{1,2} = \pm\sqrt{-p} . \tag{2.2.33}$$

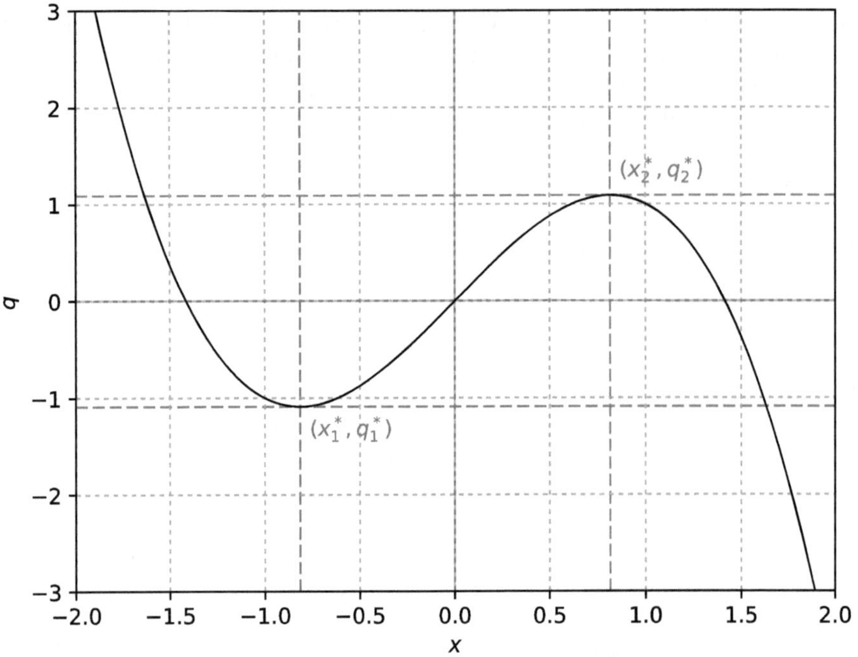

Fig. 2.7 The dependence $q(x)$ for $p = -2$, represented by $x(\tau) = \tau$, $q(\tau) = -(p + \tau^2)\tau$, $-\infty < \tau < \infty$

The local extrema of $q(\tau)$, see Fig. 2.7, are determined by the condition $\mathrm{d}q(\tau)/\mathrm{d}\tau = 0$. The local minimum is

$$\tau_1^* = -\sqrt{\frac{-p}{3}}, \quad q_{\min} = q(\tau_1^*) = \frac{2}{3}p\sqrt{\frac{-p}{3}}, \tag{2.2.34}$$

and the local maximum is

$$\tau_2^* = \sqrt{\frac{-p}{3}}, \quad q_{\max} = q(\tau_2^*) = -\frac{2}{3}p\sqrt{\frac{-p}{3}}. \tag{2.2.35}$$

These extrema characterize the bifurcation parameter values. Thus, in the considered case $p < 0$ the dependence $x(p, q)$ is single-valued in the ranges $-\infty < \tau < \tau_1^*$ and $\tau_2^* < \tau < \infty$, but this dependence is three-valued if $\tau_1^* < \tau < \tau_2^*$. Introducing the single-valued branches $x_1(p, q)$, $x_2(p, q)$ and $x_3(p, q)$, we have the following parametric representation for fixed p:

$$\begin{aligned} x_1(\tau) &= \tau, \quad q(\tau) = -(p + \tau^2)\tau, \quad -\infty < \tau < \tau_1^*, \\ x_2(\tau) &= \tau, \quad q(\tau) = -(p + \tau^2)\tau, \quad \tau_1^* < \tau < \tau_2^*, \\ x_3(\tau) &= \tau, \quad q(\tau) = -(p + \tau^2)\tau, \quad \tau_2^* < \tau < \infty. \end{aligned} \tag{2.2.36}$$

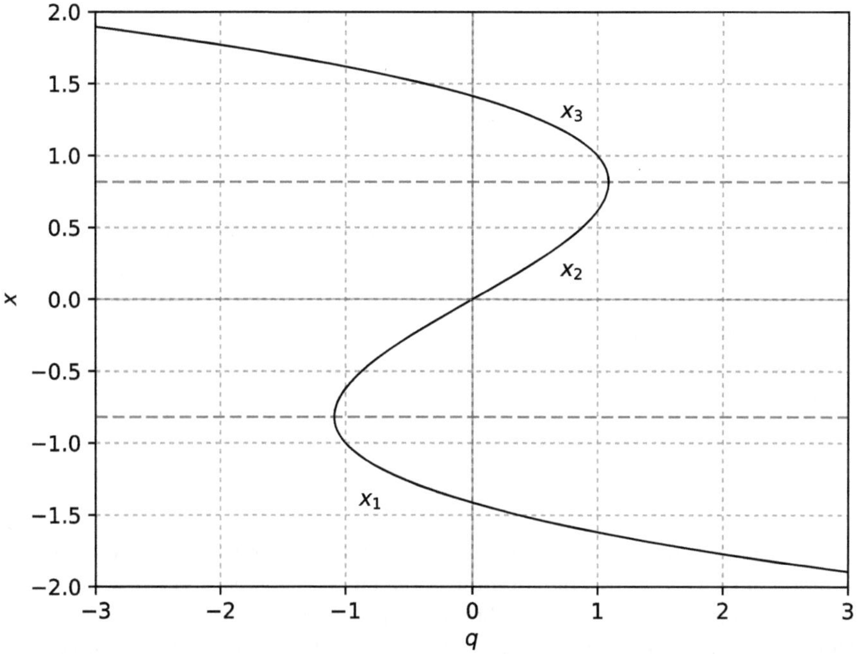

Fig. 2.8 The many-valued dependence $x(p = -2,\ q)$, determined by: $x(\tau) = \tau$, $q(\tau) = -(p + \tau^2)\tau$, $-\infty < \tau < \infty$

To present some graphs, we choose $p = -2$, leading to $\tau_1^* = -0.816494$ and $\tau_2^* = 0.816497$. The graphs of $x_1(p = -2,\ q)$, $x_2(p = -2,\ q)$ and $x_3(p = -2,\ q)$ are shown in Fig. 2.8.

2.3 Analysis of a Quartic (Fourth-Order) Polynomial Equation

Consider the quartic (fourth-order) polynomial equation

$$eu^4 + au^3 + bu^2 + cu + d = 0\ , \tag{2.3.1}$$

where $e,\ a,\ b,\ c,\ d$ are arbitrary real coefficients and $e > 0$. Let us transform (2.3.1) into canonical form. Introducing the new variable $v = u - a/4e$, we have

$$v^4 + \widetilde{p}v^2 + \widetilde{r}v + \widetilde{q} = 0\ , \tag{2.3.2}$$

where

$$\widetilde{p} = \widetilde{b} - \frac{1}{4}\widetilde{a}^2 , \quad \widetilde{r} = \frac{1}{16}\widetilde{a}^3 - \frac{1}{2}\widetilde{a}\widetilde{b} + \widetilde{c} ,$$
$$\widetilde{q} = -\frac{3}{64}\widetilde{a}^4 + \frac{1}{16}\widetilde{a}^2\widetilde{b} - \frac{1}{4}\widetilde{a}\widetilde{c} + \widetilde{d} , \tag{2.3.3}$$
$$\widetilde{a} = \frac{a}{e} , \quad \widetilde{b} = \frac{b}{e} , \quad \widetilde{c} = \frac{c}{e} , \quad \widetilde{d} = \frac{d}{e} .$$

The solutions of (2.3.2) are odd functions with respect to the transformation $v \to -v$, $\widetilde{r} \to -\widetilde{r}$, that is, $v(\widetilde{p}, \widetilde{q}, \widetilde{r}) = -v(-\widetilde{p}, \widetilde{q}, -\widetilde{r})$. Taking this into account, it is sufficient to consider only positive values of the coefficient $\widetilde{r} > 0$. If $\widetilde{r} = 0$, we have the biquadratic equation, which we analyze separately.

In Eq. (2.3.2), by introducing the new variable x defined by the substitution $v = \sqrt[3]{\widetilde{r}}x$, the quartic polynomial equation takes the canonical form

$$x^4 + px^2 + x + q = 0 , \qquad p = \frac{\widetilde{p}}{\sqrt[3]{\widetilde{r}^2}} , \qquad q = \frac{\widetilde{q}}{\sqrt[3]{\widetilde{r}^4}} , \qquad \widetilde{r} > 0 . \tag{2.3.4}$$

The canonical form (2.3.4) is characterized by the two parameters p and q, and the solution of (2.3.4) is a function of these parameters, that is, $x = x(p, q)$. According to the fundamental theorem of algebra, the Eq. (2.3.4) has four (in general complex-valued) solutions for arbitrary fixed p and q, denoted by $x_1(p, q)$, $x_2(p, q)$, $x_3(p, q)$ and $x_4(p, q)$. There are three cases:

1. All four solutions are real.
2. Two solutions are real and two are complex conjugate.
3. There are two pairs of complex conjugate solutions.

The domains in the (q, p) plane where there exist four real solutions, two real and two complex conjugate solutions, and four complex conjugate solutions are respectively denoted as D_4, D_2 and D_0.

Let us make bifurcation diagrams of the solutions. Introduce the expression $F(x, p, q)$ as the left-hand side of the Eq. (2.3.2), and consider this equation as an implicit representation for $x(p, q)$. As before, the bifurcation of the solutions of (2.3.4) is determined with the help of the conditions

$$F(x, p, q) \equiv x^4 + px^2 + x + q = 0 , \tag{2.3.5}$$

$$\frac{\partial F}{\partial x} = 4x^3 + 2px + 1 = 0 . \tag{2.3.6}$$

Solving the Eqs. (2.3.5) and (2.3.6) with respect to the parameters p and q, we have

$$p = -2x^2 - \frac{1}{2x} , \qquad q = x^4 - \frac{x}{2} , \qquad -\infty < x < \infty . \tag{2.3.7}$$

The Eq. (2.3.7) can be considered as a parametric representation of a curve separating the (q, p) plane into the domains D_0, D_2 and D_4.

For convenience we pass from the variable x to the parameter τ in (2.3.7),

$$p(\tau) = -2\tau^2 - \frac{1}{2\tau}, \qquad q(\tau) = \tau^4 - \frac{\tau}{2}, \qquad -\infty < \tau < \infty . \tag{2.3.8}$$

The parametric equations (2.3.8) describe two non-crossing contours Γ_1 and Γ_2 corresponding to the intervals $-\infty < \tau < 0$ and $0 < \tau < \infty$. We have

$$\Gamma_1 : \quad p(\tau) = -2\tau^2 - \frac{1}{2\tau}, \quad q(\tau) = \tau^4 - \frac{\tau}{2}, \quad -\infty < \tau < 0 , \tag{2.3.9}$$

$$\Gamma_2 : \quad p(\tau) = -2\tau^2 - \frac{1}{2\tau}, \quad q(\tau) = \tau^4 - \frac{\tau}{2}, \quad 0 < \tau < \infty . \tag{2.3.10}$$

It is convenient to pass from negative to positive values of the parameter ($\tau \to -\tau$) in the definition (2.3.9). We have

$$\Gamma_1 : \quad p(\tau) = -2\tau^2 + \frac{1}{2\tau}, \quad q(\tau) = \tau^4 + \frac{\tau}{2}, \quad 0 < \tau < \infty . \tag{2.3.11}$$

For Γ_1, it holds that

$$\frac{\mathrm{d}p(\tau)}{\mathrm{d}\tau} < 0 , \qquad \frac{\mathrm{d}q(\tau)}{\mathrm{d}\tau} > 0 , \qquad \frac{\mathrm{d}p}{\mathrm{d}q} = \frac{\mathrm{d}p(\tau)/\mathrm{d}\tau}{\mathrm{d}q(\tau)/\mathrm{d}\tau} < 0 . \tag{2.3.12}$$

Hence for Γ_1, $p(q)$ is a monotonically decreasing function. In accordance with (2.3.11), the parameter p vanishes when $\tau = 1/\sqrt[3]{4}$, and consequently the contour Γ_1 crosses the q axis at $q(\tau = 1/\sqrt[3]{4}) \approx 0.47247$.

The parametric representation (2.3.11) can be used to obtain an asymptotic expression for the dependence $p(q)$ on the contour Γ_1. From (2.3.11) it follows that $p \to \infty$ when $q \to 0$, and $p \to -\infty$ when $q \to \infty$. Also, for all points on Γ_1, it holds that $q > 0$. When $0 < q \ll 1$, we have $\tau \ll 1$, and hence $p(\tau) \approx 1/2\tau$ and $q(\tau) \approx \tau/2$. This gives us $p(q) \approx 1/4q$ for $0 < q \ll 1$. When $q \gg 1$, then $\tau \gg 1$, and $p(\tau) \approx -2\tau^2$, $q(\tau) \approx \tau^4$. We obtain $p(q) \approx -2\sqrt{q}$. Summarizing, the asymptotic behavior of $p(q)$ on Γ_1 is given by

$$p(q) \approx \frac{1}{4q}, \quad 0 < q \ll 1 , \qquad p(q) \approx -2\sqrt{q} , \quad q \gg 1 . \tag{2.3.13}$$

Consider now the dependence $p(q)$ determined by the parametric representation of the contour Γ_2 in (2.3.10). Note that for Γ_2, the first conditions of existence of singular points are $\mathrm{d}p(\tau)/\mathrm{d}\tau = 0$ and $\mathrm{d}q(\tau)/\mathrm{d}\tau = 0$. These conditions are satisfied for the unique value $\tau_* = 1/2$. We have $p_* = p(\tau_*) = -3/2$, $q_* = q(\tau_*) = -3/16$; geometrically, this singularity is a cusp. For any $\tau \neq \tau_*$, it holds that $\mathrm{d}p(\tau)/\mathrm{d}\tau \neq 0$

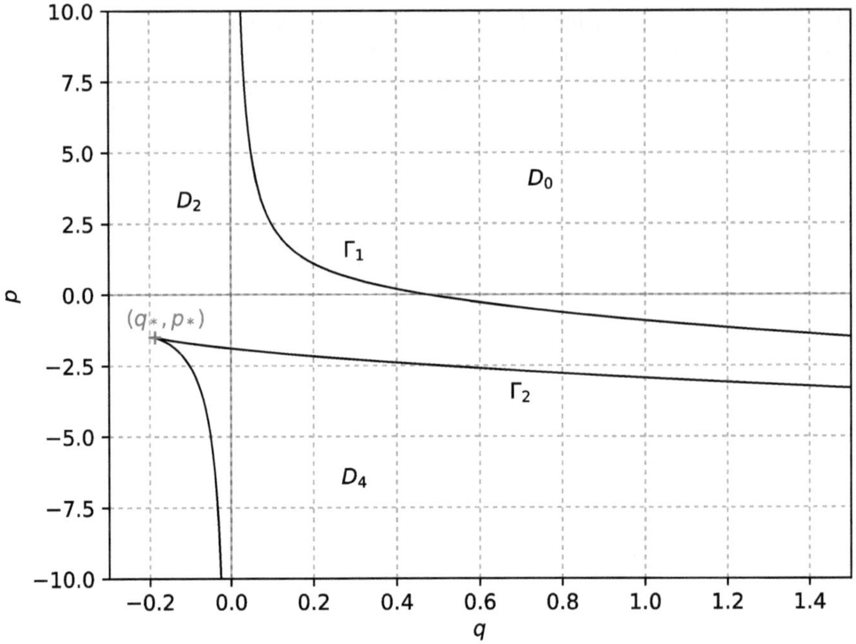

Fig. 2.9 Domains in the (q, p) parameter plane with four (D_4), two (D_2) or no real solutions (D_0) for the canonical quartic equation (2.3.4). The contour Γ_2 has a cusp at $(q_*, \ p_*) = (-3/16, \ -3/2)$

and $\mathrm{d}q(\tau)/\mathrm{d}\tau \neq 0$, and the dependences $p(q)$ are monotonic in the intervals $0 < \tau \leq \tau_*$ and $\tau_* \leq \tau < \infty$.

Next, consider the asymptotic behavior of $p(q)$ on Γ_2 for $q < 0$ with $|q| \ll 1$, and for $q \gg 1$. On one of the branches, the first case is realized for $\tau \ll 1$, and we have $q(\tau) \approx -\tau/2$, $p(\tau) \approx -1/2\tau$. We obtain $p(q) \approx 1/4q$. On the other branch, the terms τ^4 and $-\tau/2$ balance nontrivially at $\tau \approx 0.79370$, and we do not have an asymptotic relation for p. For $q \gg 1$, then $\tau \gg 1$, and we have $p(\tau) \approx -2\tau^2$ and $q(\tau) \approx \tau^4$. We obtain $p(q) \approx -2\sqrt{q}$. Summarizing, the asymptotic behavior of $p(q)$ on Γ_2 is given by

$$p(q) \approx \frac{1}{4q}\,, \quad q < 0\,, \quad |q| \ll 1\,, \qquad\qquad p(q) \approx -2\sqrt{q}\,, \quad q \gg 1\,. \tag{2.3.14}$$

The graphs of Γ_1 and Γ_2 are obtained with the help of the parametric representation (2.3.11) (contour Γ_1) and (2.3.10) (contour Γ_2). The contours Γ_1 and Γ_2 divide the parameter plane (q, p) into three subdomains D_0, D_2 and D_4, see Fig. 2.9.

Each of the domains D_4, D_2, D_0 is characterized by the number of real solutions the Eq. (2.3.4) admits in that domain. To identify which domain is which, it is necessary to find the number of real solutions at one point in each of the domains.

At the point $p = q = 0$, we have $(x^3 + 1)x = 0$, and its real solutions are $x_1 = 0$, $x_2 = -1$. This means that the domain lying between Γ_1 and Γ_2 is D_2.

Next, taking just $p = 0$, Eq. (2.3.4) becomes $F(x, q) \equiv x^4 + x + q = 0$. Let q be an arbitrary but fixed real number. Differentiating F and setting the derivative to zero, we find $\partial F/\partial x = 4x^3 + 1 = 0$, that is, $x^3 = -1/4$. The only extremum for real x occurs at $x_0 = -1/\sqrt[3]{4}$. It is a minimum, because $\partial^2 F/\partial x^2 = 12x^2 > 0$ for any real $x \neq 0$. The minimum value of $F = x_0^4 - x_0 + q \equiv q_0 + q$, where numerically $q_0 \approx -0.47247$. Therefore, if $p = 0$, and we choose $q > q_0$, Eq. (2.3.4) admits no real solutions. The set of all (q, p) lying to the right of Γ_1 forms the domain D_0 with boundary Γ_1, characterized by the absence of the real solutions to Eq. (2.3.4).

Finally, taking $q = 0$, Eq. (2.3.4) factors into $(x^3 + px + 1)x = 0$. One solution is $x_1 = 0$. For $p < 0$, $|p| \gg 1$, the canonical cubic factor admits three real solutions, recall Fig. 2.3. Therefore, there exist four real solutions to (2.3.4), and for any point on this side of the contour Γ_2, the domain is D_4. See Fig. 2.9.

Now we are ready for bifurcation analysis of the biquadratic equation. In the previous analysis it was supposed that $\widetilde{r} > 0$. Consider now the case of the biquadratic equation, with $\widetilde{r} = 0$ in Eq. (2.3.2). We have

$$v^4 + \widetilde{p}v^2 + \widetilde{q} = 0\,, \tag{2.3.15}$$

which is a quadratic equation in the variable v^2. We obtain the solutions

$$v_{1,2}^2 = -\frac{\widetilde{p}}{2} \pm \frac{1}{2}\sqrt{\widetilde{p}^2 - 4\widetilde{q}}\,. \tag{2.3.16}$$

For the solutions $v_{1,2}^2$ to be real, a necessary condition is $\widetilde{p}^2 \geq 4q$. If $\widetilde{p}^2 < 4q$, there are no real solutions of the biquadratic equation (2.3.15), i.e. this case belongs to the domain D_0.

Now let us consider the existence of real solutions when $\widetilde{p}^2 \geq 4\widetilde{q}$. First, let $\widetilde{p} = 0$. The condition $\widetilde{p}^2 \geq 4\widetilde{q}$ then requires $\widetilde{q} \leq 0$. We have $v_{1,2}^2 = \pm\sqrt{-4\widetilde{q}}$. If $\widetilde{q} < 0$, then $v_1^2 > 0$ and $v_2^2 < 0$, so there are two real solutions; this case is in D_2. If $\widetilde{q} = 0$, then $v_{1,2}^2 = 0$; zero is a quadruple root, and this case is in D_4.

Next, let $\widetilde{p} > 0$. If $0 < \widetilde{q} \leq \widetilde{p}^2/4$, then $\sqrt{\widetilde{p}^2 - 4\widetilde{q}} < \widetilde{p}$, whence $v_{1,2}^2 < 0$; this case belongs to D_0. If $\widetilde{p} > 0$ and $\widetilde{q} = 0$, then $v_1^2 = 0$, $v_2^2 < 0$; we are in D_2. If $\widetilde{p} > 0$ and $\widetilde{q} < 0$, then $\sqrt{\widetilde{p}^2 - 4\widetilde{q}} > \widetilde{p}$, which leads to $v_1^2 > 0$, $v_2^2 < 0$, so again we are in D_2.

Finally, let $\widetilde{p} < 0$. Here $0 < \widetilde{q} \leq \widetilde{p}^2/4$ yields $v_{1,2}^2 > 0$, so we have four real solutions and this case belongs to D_4. If $\widetilde{p} < 0$, $\widetilde{q} = 0$, then $v_1^2 > 0$, $v_2^2 = 0$, so we are still in D_4. Finally, if $\widetilde{p} < 0$, $\widetilde{q} < 0$, then $v_1^2 > 0$, $v_2^2 < 0$, and we are in D_2.

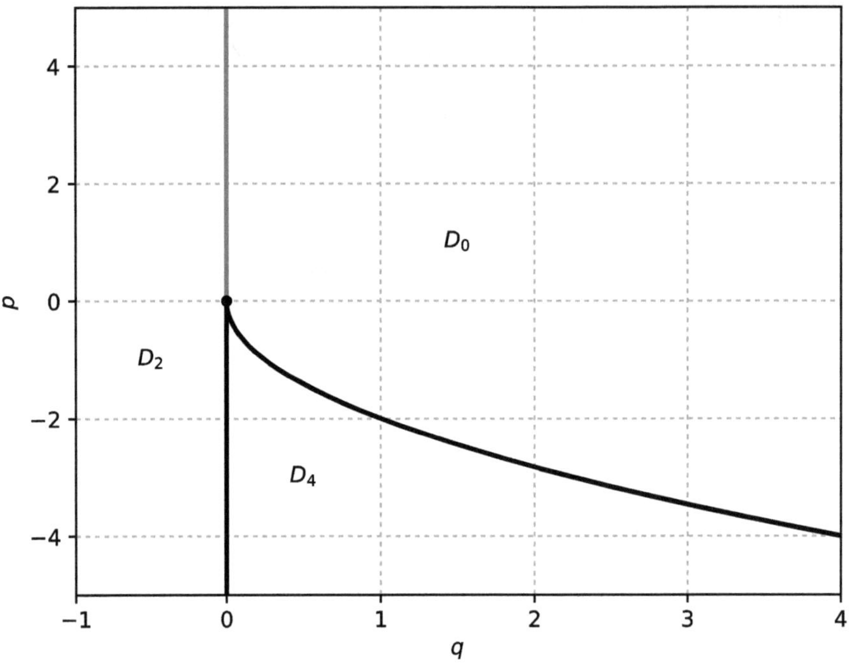

Fig. 2.10 Domains in the (q, p) parameter plane with four (D_4), two (D_2) or no real solutions (D_0) for the biquadratic equation (2.3.15). The thick black boundary and the origin belong to D_4; the $+p$ axis belongs to D_2. Compare Fig. 2.9 for the general quartic

We conclude that the domain D_0 is determined by the system of the inequalities

$$\widetilde{p}^2 - 4\widetilde{q} < 0 \quad \text{or} \quad \widetilde{p}^2 - 4\widetilde{q} \geq 0\,,\ \widetilde{q} > 0\,,\ \widetilde{p} > 0\,. \tag{2.3.17}$$

The domain D_2 is described by

$$\widetilde{p}^2 - 4\widetilde{q} \geq 0\,;\ \text{with}\ \ \widetilde{q} < 0\,,\ \widetilde{p}\ \text{arbitrary} \quad \text{or} \quad \widetilde{q} = 0\,,\ \widetilde{p} > 0\,. \tag{2.3.18}$$

Finally, the domain D_4 is given by

$$\widetilde{p}^2 - 4\widetilde{q} \geq 0\,;\ \text{with}\ \ \widetilde{q} \geq 0\,,\ \widetilde{p} < 0 \quad \text{or} \quad \widetilde{q} = 0\,,\ \widetilde{p} = 0\,. \tag{2.3.19}$$

The domains are visualized in Fig. 2.10. Some examples of $x(\widetilde{p}, \widetilde{q})$ are shown in Figs. 2.11 and 2.12, showing a transition directly from D_2 to D_0, and a transition from D_2 through D_4 to D_0.

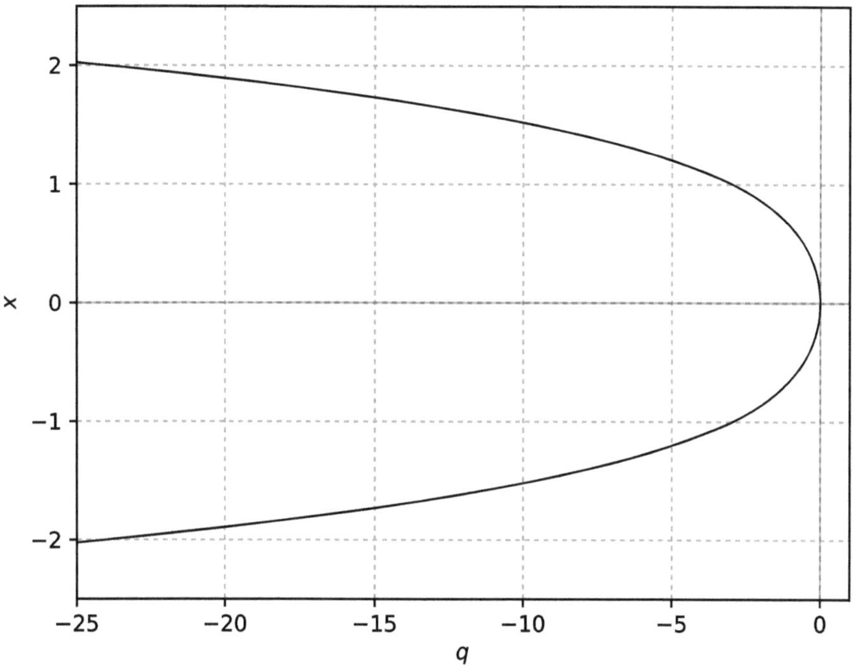

Fig. 2.11 The dependence $x(\tilde{p} = 2,\ \tilde{q})$ for the biquadratic equation (2.3.15). The system is initially in D_2 up to and including $\tilde{q} = 0$. For $\tilde{q} > 0$, the domain is D_0; there are no real solutions

It is possible to obtain the same result using the system of bifurcation equations

$$F(v, \tilde{p}, \tilde{q}) \equiv v^4 + \tilde{p}v^2 + \tilde{q} = 0\,, \tag{2.3.20}$$

$$\frac{\partial F(v, \tilde{p}, \tilde{q})}{\partial v} = 2(2v^2 + \tilde{p})v = 0\,. \tag{2.3.21}$$

From Eq. (2.3.21) we find the critical values $v_* = 0$, $\tilde{p}_* = -2v_*^2$. From (2.3.20), we have $\tilde{q} = 0$ for the critical value $v_* = 0$ and arbitrary $\tilde{p}$. Thus, the $\tilde{p}$ axis is a bifurcation line. The second solution yields that if $v_* \neq 0$, then $\tilde{p} < 0$. Inserting the bifurcation condition $v^2 = -\tilde{p}/2$ into (2.3.20), we find $\tilde{p}^2 = 4\tilde{q}$. We conclude that the plane $(\tilde{q}, \tilde{p})$ is divided by bifurcation lines into three domains. The first domain is the half-plane $\tilde{q} < 0$, and the second half-plane $\tilde{q} > 0$ is divided into two parts by the boundary $\tilde{p} = -\sqrt{4\tilde{q}}$. Each of these domains is characterized by the numbers of the real solutions of Eq. (2.3.15).

Suppose at first that $\tilde{q} < 0$ and $\tilde{p} = 0$. In this case the solutions of the biquadratic equation are represented as $\tilde{v}_{1,2} = \pm\sqrt{-\tilde{q}}$ and consequently, the considered domain is characterized by two real solutions. Next, suppose that $\tilde{q} > 0$ and $\tilde{p} = 0$. In this case there is no real solution of the biquadratic equation. Hence the domain lying to

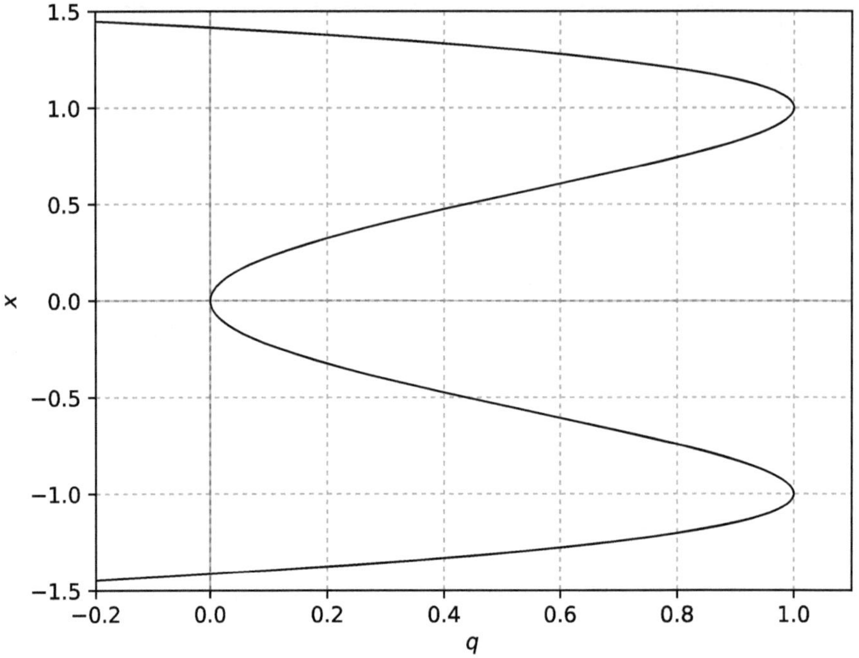

Fig. 2.12 The dependence $x(\widetilde{p} = -2,\ \widetilde{q})$ for the biquadratic equation (2.3.15). For $\widetilde{q} < 0$, the system is in D_2; then for $0 \leq \widetilde{q} \leq 1$ in D_4; and for $\widetilde{q} > 1$, in D_0

the right of the boundary $\widetilde{p} = -\sqrt{4\widetilde{q}}$ is D_0, see (2.3.17). Finally, if we take $\widetilde{p} = -2$, $\widetilde{q} = 1$, we have four real solutions, and consequently the corresponding domain is D_4.

This concludes the analysis of the biquadratic special case. Let us move forward to the last topic of this chapter, parametric representation and the analysis of solutions for a fourth order polynomial equation. Equation (2.3.4) will be considered as an implicit representation for $x(p, q)$, in other words,

$$F(x, p, q) \equiv x^4 + px^2 + x + q = 0\,. \tag{2.3.22}$$

In what follows we also use the expressions for the partial derivatives $\partial x(p, q)/\partial p$ and $\partial x(p, q)/\partial q$ in the form

$$\frac{\partial x(p, q)}{\partial p} = -\frac{x^2}{\partial F(x, p, q)/\partial x}\,, \quad \frac{\partial x(p, q)}{\partial q} = -\frac{1}{\partial F(x, p, q)/\partial x}\,. \tag{2.3.23}$$

For example, the first one follows by taking the total differential of (2.3.22), setting it to 0 (which is the condition to stay on the set where $F = \text{const.}$), then setting $q = \text{const.}$ (so $\mathrm{d}q = 0$), and rearranging. Using (2.3.22) for $F(x, p, q)$, the derivative with respect to x is

$$\frac{\partial F(x,p,q)}{\partial x} = 4x^3 + 2px + 1 \,. \tag{2.3.24}$$

Solving p from (2.3.22) and inserting the result to (2.3.24), we have

$$\frac{\partial F}{\partial x} = 2x^3 - 1 - 2\frac{q}{x} \,, \tag{2.3.25}$$

As a result we obtain the following representation:

$$\frac{\partial x}{\partial p} = -\frac{x^3}{2x^4 - x - 2q} \,, \quad \frac{\partial x}{\partial q} = -\frac{1}{4x^3 + 2px + 1} \,. \tag{2.3.26}$$

Note the right-hand side of $\partial x/\partial p$ now involves only x and q; similarly the right-hand side of $\partial x/\partial q$ now involves only x and p.

The critical values of x, that is the bifurcation values x_* for which $F = 0$, $\partial F/\partial x = 0$, satisfy the same equations for fixed values of q or p respectively. The values $\partial x(p,q)/\partial p$ and $\partial x(p,q)/\partial q$ tend to infinity at the bifurcation points, and the interval of the existence of real solutions for (2.3.22) is divided into several intervals, where $\partial F/\partial x \neq 0$. Consequently, as seen from (2.3.23), $\partial x/\partial p$ and $\partial x/\partial q$ have a constant sign in the corresponding intervals (because $\partial F/\partial x$ is continuous). We conclude that the dependences $x(q) = x(p = \widetilde{p},\ q)$ for fixed values of the parameter p and $x(p) = x(p,\ q = \widetilde{q})$ for fixed values of the parameter q are monotonic in these intervals.

Let us now construct the parametric representations for the solutions. Introducing the real variable τ, we represent the real dependences $x(p,q)$, determined by the Eq. (2.3.22) in the following parametric form:

$$x(\tau) = \tau, \quad q(\tau) = -\tau^4 - p\tau^2 - \tau, \quad \tau \text{ is real} \tag{2.3.27}$$

for fixed values of the parameter p, and

$$x(\tau) = \tau, \quad p(\tau) = -\tau^2 - \frac{1}{\tau} - \frac{q}{\tau^2} \,, \quad \tau \text{ is real} \tag{2.3.28}$$

for fixed values 0 of the parameter q. These expressions are illustrated in Figs. 2.13, 2.14 and 2.15, respectively.

Using the parametric representation (2.3.27), let us analyze $x(p,q)$ for fixed p. Consider the equation

$$\frac{\mathrm{d}q(t)}{\mathrm{d}\tau} = -(4\tau^3 + 2p\tau + 1) = 0 \,. \tag{2.3.29}$$

This is a cubic equation in τ, and depending on the value of the parameter p, it has either one or three real solutions. Using the critical parameter value (see Eq. (2.2.20)) $\gamma_* = -3/\sqrt[3]{4}$ of the canonical equation (2.2.4), and the relation

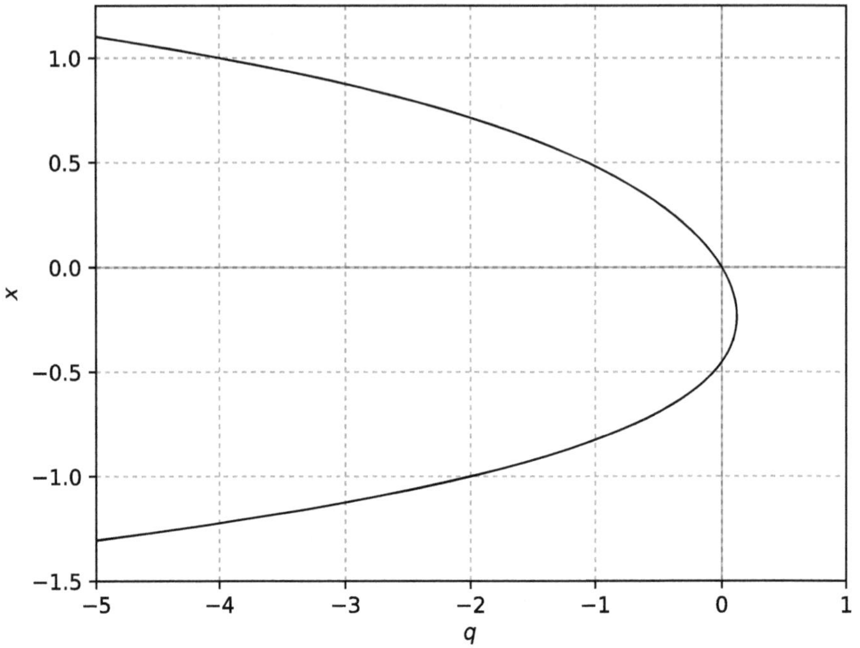

Fig. 2.13 The dependence $x(p = 2, \ q)$ for the general quartic (2.3.22). Initially there are two real solutions, until a critical value of q, after which there are no real solutions. Compare Fig. 2.11 for the biqudratic special case, where the curve becomes symmetric with regard to the q axis

$\gamma = p\sqrt[3]{2}$ (via the change of variable $\widehat{\tau} = \sqrt[3]{4}\tau$ in (2.3.29)), we find the critical value of the parameter $p_* = \gamma_*/\sqrt[3]{2} = -3/2$. Thus, there exist three real solutions, denoted as $\tau_{*1}^{(3)}(p), \ \tau_{*2}^{(3)}(p)$, and $\tau_{*3}^{(3)}(p)$, of (2.3.29) if $p < p_*$, and there exists one real solution denoted as $\tau_{*1}^{(1)}(p)$, if $p > p_*$.

When $p > p_*$, the function $q(\tau)$ achieves its maximum value at $\tau = \tau_{*1}^{(1)}(p)$, in other words, $q_{\max}(p) = q(\tau_{*1}^{(1)}(p))$. In particular, $\tau_{*1}^{(1)}(0) = -1/\sqrt[3]{4} = -0.629961$, and $q_{\max}(0) = 0.530748$. We conclude that the dependence $x(p, q)$ is two-valued for q varying in the interval $-\infty < q < q_{\max}(p)$ for any fixed $p > p_*$. For an illustration, see Fig. 2.13.

When $p < p_*$, the equation $\mathrm{d}q(\tau)/\mathrm{d}\tau = 0$ has three real solutions, denoted $\tau_{*1}^{(3)}(p), \ \tau_{*2}^{(3)}(p)$, and $\tau_{*3}^{(3)}(p)$, where $\tau_{*1}^{(3)}(p) < \tau_{*2}^{(3)}(p) < \tau_{*3}^{(3)}(p)$. (The case of coinciding roots must be considered separately). In accordance with the definition (2.3.27), we have $q(\tau) \to -\infty$ for $\tau \to \pm\infty$, and the derivative of this function vanishes at three points. Thus, the function $q(\tau)$ can have at most two local maxima and one local minimum. (If may have fewer, if some of the three points happen to be inflection points.) If the case with two maxima and one minimum is realized, the local maxima are achieved at the points $\tau_{*1}^{(3)}(p)$ and $\tau_{*3}^{(3)}(p)$, and the local minimum at the point $\tau_{*2}^{(3)}(p)$.

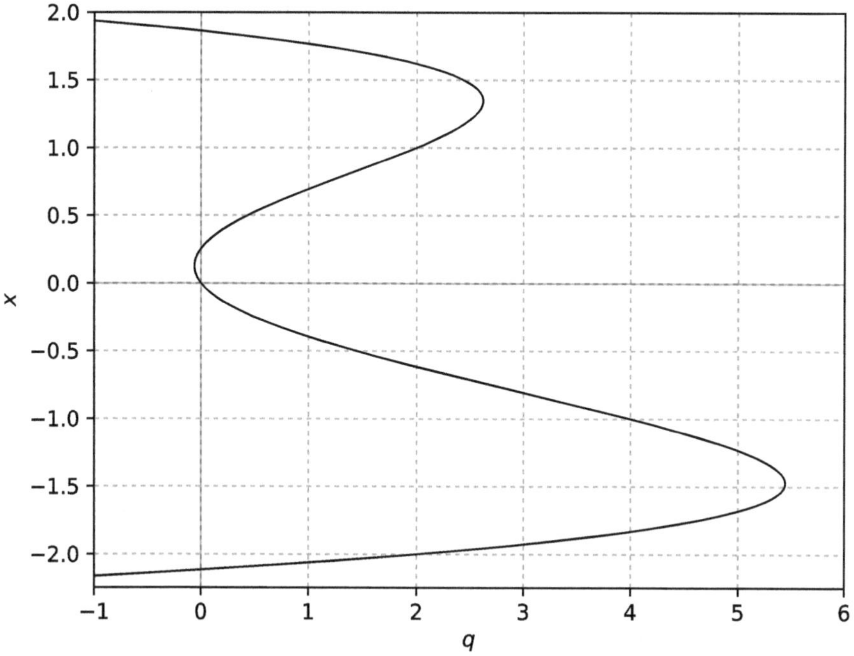

Fig. 2.14 The dependence $x(p = -4,\ q)$ for the general quartic (2.3.22). Ranges of q with two, four and no real solutions are observed. Compare Fig. 2.12 for the biquadratic special case, where the lobes become symmetric

Let us denote the local maxima as $q_{1\,\max}(p) = q(\tau_{*1}^{(3)}(p)), q_{2\,\max}(p) = q(\tau_{*3}^{(3)}(p))$, and the local minimum as $q_{\min}(p) = q(\tau_{*2}^{(3)}(p))$. The multi-valued dependence $x(p, q)$ can be described by the following conditions. When the parameter q is varied in the interval $-\infty < q < q_{\min}(p)$, then this dependence is two-valued, for the interval $q_{\min}(p) < q < \min\{q_{1\,\max}(p),\ q_{2\,\max}(p)\}$ the dependence is four-valued, and for the interval $\min\{q_{1\,\max}(p),\ q_{2\,\max}(p)\} < q < \max\{q_{1\,\max}(p),\ q_{2\max}(p)\}$ the dependence is two-valued; see Fig. 2.14.

Let us now study $x(p, q)$ for the case of varying p and fixed q. If $q \neq 0$, then $x = 0$ does not satisfy the Eq. (2.3.22) and thus the curve $x(p, q)$, in the (p, x) plane for fixed $q \neq 0$, must be completely situated in one of the half-planes $x > 0$ or $x < 0$. In the considered case, described by the relations (2.3.28), we separately write the parametric representation for positive values $x(p, q) > 0$ and negative values $x(p, q) < 0$.

As before, choosing τ to be positive, we have

$$x(p, q) > 0: \quad x(\tau) = \tau, \quad p(\tau) = -\tau^2 - \frac{1}{\tau} - \frac{q}{\tau^2}, \quad \tau > 0, \qquad (2.3.30)$$

$$x(p, q) < 0: \quad x(\tau) = -\tau, \quad p(\tau) = -\tau^2 + \frac{1}{\tau} - \frac{q}{\tau^2}, \quad \tau > 0. \qquad (2.3.31)$$

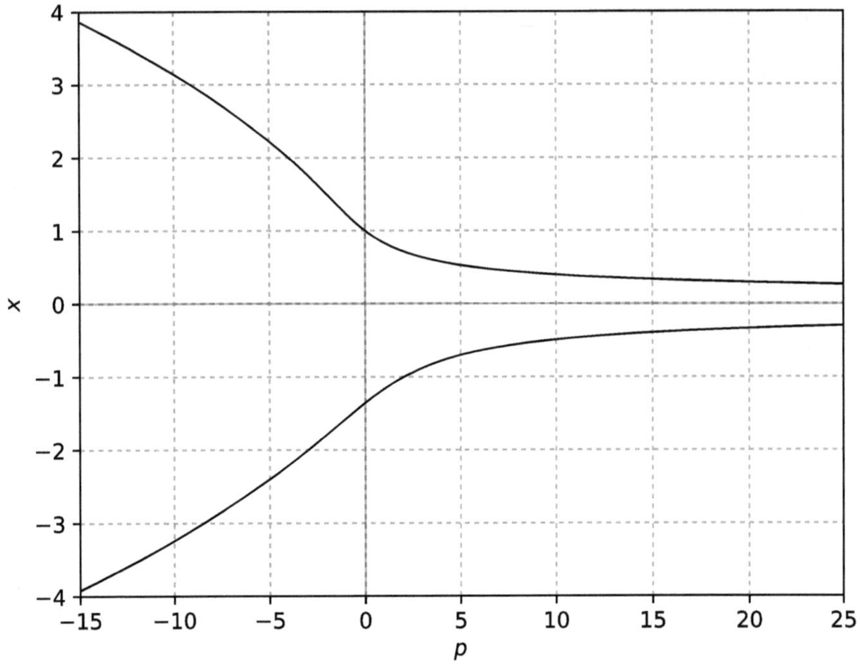

Fig. 2.15 The dependence $x(p, q = -2)$

Let us qualitatively analyze $p(\tau)$ for a fixed value of q. Let us study the solution of the equation $\partial p(\tau)/\partial\tau = 0$. From (2.3.30), we have $\partial p(\tau)/\partial\tau = -2\tau + 1/\tau^2 + 2q/\tau^3$. After multiplication of both sides of the equation $\partial p/\partial\tau = 0$ by $-\tau^3/2$, we have $P^+(q, \tau) = \tau^4 - \tau/2 - q = 0$ for positive values $x(p, q) > 0$, and (similarly from (2.3.31)) $P^-(q, \tau) = \tau^4 + \tau/2 - q = 0$ for negative values $x(p, q) < 0$.

In the analysis to follow, we must be careful. To begin with, observe that we applied the differentiation first, and only then the multiplication by $-\tau^3/2$ to simplify the expressions that we will need to manipulate in the analysis. Secondly, the zero-product property seems to imply that the equation $-(\tau^3/2)(\partial p(\tau)/\partial\tau) = 0$ should have a solution at $\tau = 0$, but writing it out explicitly, $-(\tau^3/2)(\partial p(\tau)/\partial\tau) = \tau^4 - \tau/2 - q$, we see that clearly it does not. The inconsistency is only apparent: the factor $\partial p(\tau)/\partial\tau$ is only valid for $\tau > 0$, so at the point $\tau = 0$ (which is outside its domain) we should not expect there to be a correspondence between the equation $P^+(q, \tau) = \tau^4 - \tau/2 - q = 0$ and the original equation $\partial p(\tau)/\partial\tau = 0$, because at $\tau = 0$ the factor $\partial p(\tau)/\partial\tau$ becomes undefined. The expression $P^+(q, \tau)$ gives the limit of the original product also as $\tau \to 0^+$, but we are interested in the original factor $\partial p(\tau)/\partial\tau$ directly, not in the limit. Finally, because the factor $-\tau^3/2$ is zero only at $\tau = 0$, any other zeros of P^+ (that occur for $\tau > 0$, inside the domain) coincide with the zeros of $\partial p(\tau)/\partial\tau$.

The functions $P^{\pm}(q, \tau)$ are convex with respect to τ, that is,

$$\frac{\partial^2 P^{\pm}(q, \tau)}{\partial \tau^2} = 12\tau^2 > 0 \,, \tag{2.3.32}$$

and therefore their unique extremum at $\tau_*^{\pm} = \pm 1/2$ must be a minimum. Note the values $\tau_*^{\pm}$ do not depend on q. Because $\tau > 0$, we conclude (by continuity) that $P^-(q, \tau)$ is a monotone function for $\tau \in (0, \infty)$ and

$$P^-_{\min}(q_0) \equiv P^-(q_0, 0) = -q_0 \leq P^-(q_0, \tau) < \infty \tag{2.3.33}$$

for any fixed $q = q_0$ and $\tau > 0$.

The minimum value of P^+ is achieved at $\tau = \tau_*^+ = 1/2$, and is equal to

$$P^+_{\min}(q_0) = P^+(q_0, \tau_*^+) = -\left(q_0 + \frac{3}{16}\right) \tag{2.3.34}$$

for any fixed $q = q_0$. Asymptotically, $P^{\pm}(q, \tau) \to +\infty$ as $\tau \to \pm\infty$, so the equation $P^{\pm}(q, \tau) = 0$ has no real solution if $P^{\pm}_{\min}(q) > 0$, and it has two real solutions if $P^{\pm}_{\min}(q) < 0$ (here the upper and lower signs correspond).

Recall that a real solution of $P^{\pm}(q, \tau) = 0$ for some $\tau = \tau_0 > 0$ means that $p(\tau)/\partial\tau = 0$ at τ_0, and hence $p(\tau)$ has a local extremum at τ_0. The absence of real solutions to the equation $P^+(q, \tau) = 0$ (or $P^-(q, \tau) = 0$) means that the dependence $p(\tau)$ is a monotone function for fixed $q = q_0$, and consequently the dependence $x(p, q_0) > 0$ (respectively $x(p, q_0) < 0$) is one-valued. The range of q where the dependence $x(p, q) > 0$ is one-valued is determined by the condition $P^+_{\min}(q) > 0$ (respectively $P^-_{\min}(q) > 0$). The range of q where $x(p, q) > 0$ (respectively $x(p, q) < 0$) is many-valued, is determined with the help of the condition $P^+_{\min}(q) < 0$ (respectively $P^-_{\min}(q) < 0$).

The critical values of the parameter q, where $x(p, q) > 0$ transitions between one-valued and many-valued are determined with the system of equations

$$P^+(q, \tau) = \tau^4 - \frac{\tau}{2} - q = 0 \,, \quad \frac{dP^+(q, \tau)}{d\tau} = 4\tau^3 - \frac{1}{2} = 0 \,. \tag{2.3.35}$$

By the implicit function theorem, this system represents the bifurcation conditions for $\tau(q)$, expressed with the help of $P^+(q, \tau)$. Recall from (2.3.30) that for the considered case, $x = \tau$. The unique real solution of the system (2.3.35) is $\tau_*^+ = 1/2$, $q_*^+ = -3/16$. Using the condition $P^+_{\min}(q) > 0$ for the dependence $x(p, q) > 0$ and the expression $P^+_{\min}(q) = -3/16 - q$, we conclude that $x(p, q) > 0$ is one-valued for q such that $-\infty < q < q_*^+$ and many-valued for $q_*^+ < q < \infty$.

To study the dependence $x(p, q) < 0$, we use the condition $P^-_{\min}(q) > 0$ and the expression $P^-_{\min}(q) = -q$, and find that $x(p, q) < 0$ is one-valued for $-\infty < q < 0$ and two-valued for $0 < q < \infty$.

We will next determine the range of p for which there exist the real dependences $x(p, q) > 0$ and $x(p, q) < 0$. Let q be fixed, and distinguish the cases where q is taken from the intervals for which the dependences are one-valued or many-valued.

For one-valued dependences $x(p, q)$, for fixed q we have $\mathrm{d}p(\tau)/\mathrm{d}\tau \neq 0$, and consequently

$$\frac{\mathrm{d}x}{\mathrm{d}p} = \frac{\mathrm{d}x(\tau)/\mathrm{d}\tau}{\mathrm{d}p(\tau)/\mathrm{d}\tau} \neq 0\,. \tag{2.3.36}$$

For positive dependences $\mathrm{d}x(\tau)/\mathrm{d}\tau = 1$ and $\mathrm{d}p(\tau)/\mathrm{d}\tau < 0$. To see the latter, observe that $P^+(q, 0) = -q$, and use the condition on q for the one-valuedness of $x(p, q)$, namely $-\infty < q < q_*^+ = -3/16$. We must have $q < 0$, so $P^+(q, 0) > 0$. Because P^+ is continuous in τ and has no zeros, then $P^+(q, \tau) > 0$ for all $\tau > 0$. Finally, recall the multiplier $-\tau^3/2 < 0$ (for $\tau > 0$) we used at the start of the analysis; the sign of $\partial p(\tau)/\partial\tau$ is opposite that of $P^+(q, \tau)$. Therefore, we find that in this case, $x(p, q)$ is a monotonically decreasing function of the parameter p.

For negative $x(p, q)$, for which $\mathrm{d}x/\mathrm{d}\tau = -1$, we find again $\partial p(\tau)/\partial\tau < 0$ for $\tau > 0$, and conclude that $x(p, q)$ is a monotonically increasing function of p. (Here the condition on q is $q < 0$, so $P^-(q, 0) = -q > 0$. Again $P^-(q, \tau)$ and $\partial p(\tau)/\partial\tau$ have opposite signs.)

Note the asymptotic behavior of the dependences $x(p, q)$ in the domain $q < q_*^+$ for $x > 0$, in the domain $q < 0$ for $x > 0$ and in the domain $q < 0$ for $x < 0$. Taking into account the asymptotic representation $p(\tau) \approx -q/\tau^2 \gg 1$ for $\tau \ll 1$, we find that $\tau \approx \sqrt{-q/p}$. Thus, when $p \gg 1$, we have $x = \tau \approx \sqrt{-q/p}$ for $x > 0$, and $x = -\tau \approx -\sqrt{-q/p}$ for $x < 0$. On the other hand, if $\tau \gg 1$, then $p(\tau) \approx -\tau^2$ and we have $\tau \approx \sqrt{-p}$. Consequently, we obtain the asymptotic representation $x = \tau \approx \sqrt{-p}$ for $x > 0$, and $x = -\tau \approx -\sqrt{-p}$ for $x < 0$, when$\tau \gg 1$. The graphs of $x(p, q) > 0$ and $x(p, q) < 0$ for $q = -2$, determined by the representation (2.3.28), are presented in Fig. 2.15.

Consider now the many-valued dependences $x(p, q)$ in the intervals of the parameter q where $x(p, q)$ is real and many-valued, that is, $q > q_*^+ = -3/16$ for $x(p, q) > 0$ and $q > 0$ for $x(p, q) < 0$. If the equation $P^-(\tau, q) = 0$ for $\tau > 0$ has a unique solution, and consequently the function $p(\tau)$ reaches its maximum value at $\tau = \tau_*(q)$, then $p(\tau)$ is a monotone function in the intervals $0 < \tau < \tau_*(q)$ and $\tau_*(q) < \tau < \infty$. It is follows that the dependence $x(p, q) < 0$ is two-valued and the branches of this dependence are described by the following representations:

$$x_1(\tau) = -\tau, \qquad p(\tau) = -\tau^2 + \frac{1}{\tau} - \frac{q}{\tau^2}, \quad 0 < \tau \leq \tau_*(q)\,, \tag{2.3.37}$$

$$x_2(\tau) = -\tau, \quad p(\tau) = -\tau^2 + \frac{1}{\tau} - \frac{q}{\tau^2}, \quad \tau_*(q) \leq \tau < \infty. \tag{2.3.38}$$

Let us next consider the dependence $x(p, q) > 0$. The equation $P^+(\tau, q) = 0$ has two real solutions when $q > 0$. To see this, consider the following. We have

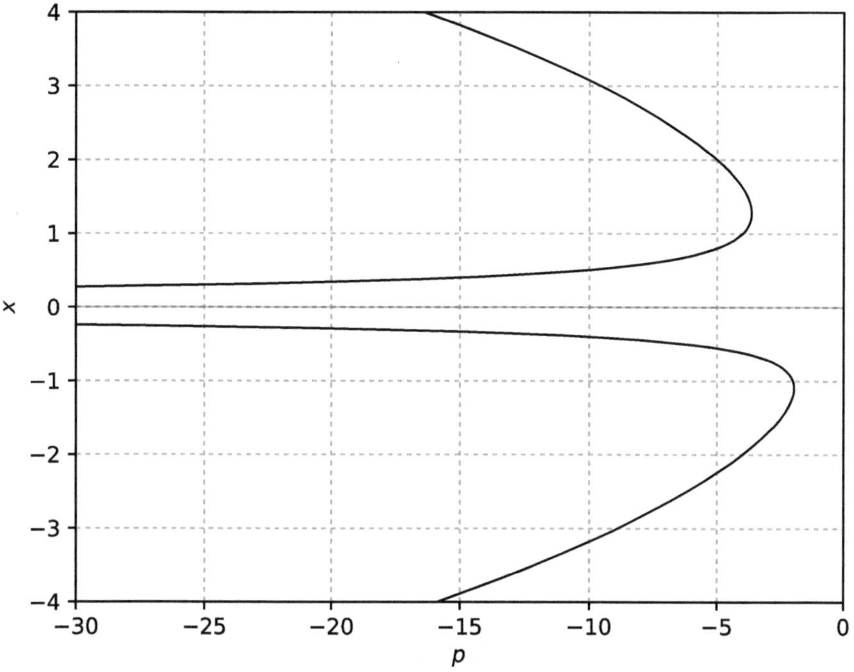

Fig. 2.16 The dependence $x(p, q = 2) > 0$ and $x(p, q = 2) < 0$

$$P^{+}(\tau, q) = \tau^4 - \frac{\tau}{2} - q = 0\,.$$

To reduce this to the canonical form, for which we know the domains D_k, we must eliminate the factor $-1/2$. Let us perform the change of variable $\tau \equiv \alpha\widehat{\tau}$, leading to

$$\alpha^4\widehat{\tau}^4 - \frac{\alpha}{2}\widehat{\tau} - q = 0\,.$$

To make the $\widehat{\tau}^4$ and $\widehat{\tau}$ terms have the same coefficient, we need

$$\alpha^4 = -\frac{\alpha}{2}\,,$$

from which

$$\alpha = \sqrt[3]{-1/2} = -\sqrt[3]{1/2}$$

We obtain the canonical form

$$\widehat{\tau}^4 + \widehat{\tau} + \widehat{q} = 0\,,$$

where

$$\widehat{q} = \frac{q}{\alpha} = -\frac{q}{\sqrt[3]{1/2}} \, .$$

Therefore, for $q > 0$, we have $\widehat{q} < 0$, and because here $\widehat{p} = 0$ (no quadratic term), by Fig. 2.9 this case is in D_2; there are two real solutions.

Furthermore, the two solutions have different signs. Now, to see this: because of the existence of only two real solutions, it is possible to represent the polynomial $P^+(\tau, q)$ as

$$P^+(\tau, q) = \tau^4 - \frac{\tau}{2} - q = (\tau - \tau_1)\left(\tau - \tau_1^*\right)(\tau - \tau_2)(\tau - \tau_3) \ , \tag{2.3.39}$$

where τ_2, τ_3 are the real solutions, and τ_1, τ_1^* are a pair of complex conjugate solutions. At $\tau = 0$ we have

$$\tau_1 \tau_1^* \tau_2 \tau_3 = -q \ . \tag{2.3.40}$$

Taking into account that $\tau_1 \tau_1^* = |\tau_1|^2 \geq 0$, we conclude that $\tau_2 \tau_3 < 0$ for $q > 0$, so τ_2 and τ_3 must have opposite signs. Therefore at any fixed $q > 0$, there is a unique $\tau > 0$ such that $P^+(\tau, q) = 0$.

Further analysis of the solution of the equation $P^+(\tau, q) = 0$ is analogous to the analysis of solutions of the equation $P^-(\tau, q)$. We conclude that the dependence $x(p, q) > 0$ is two-valued for $q > 0$, and branches of this dependence have the following representations:

$$x_1(\tau) = \tau, \quad p(\tau) = -\tau^2 - \frac{1}{\tau} - \frac{q}{\tau^2}, \quad 0 < \tau \leq \tau_*^+(q) \ ,$$

$$x_2(\tau) = \tau, \quad p(\tau) = -\tau^2 - \frac{1}{\tau} - \frac{q}{\tau^2}, \quad \tau_*^+(q) \leq \tau < \infty \ , \tag{2.3.41}$$

where $\tau_*^+(q)$ is the unique positive solution of the equation

$$\frac{\mathrm{d}p(\tau)}{\mathrm{d}\tau} = 0 \ . \tag{2.3.42}$$

As was mentioned earlier, for $x(p, q) > 0$, the positive solutions of (2.3.42) coincide with the positive solutions of $P^+(\tau, q) = 0$. Both $x_1(p, q)$ and $x_2(p, q)$ are monotone functions, similarly to the case of branches of $x(p, q) < 0$. The graphs of $x(p, q) > 0$ and $x(p, q) < 0$, with $q = 2$ and determined by the parametric representations (2.3.27) and (2.3.28), are shown in Fig. 2.16.

Let us separately consider $x(p, q) > 0$ when $q \in (q_*^+, 0)$. The graph, corresponding to $q = -3/32 \in (q_*^+, 0)$ and obtained with the help of Eq. (2.3.30) is shown in Fig. 2.17. It is clear from this graph that the interval $-\infty < p < \infty$ is divided into three subintervals, $-\infty < p < p_{\min}^+(q)$, $p_{\min}^+(q) < p < p_{\max}^+(q)$,

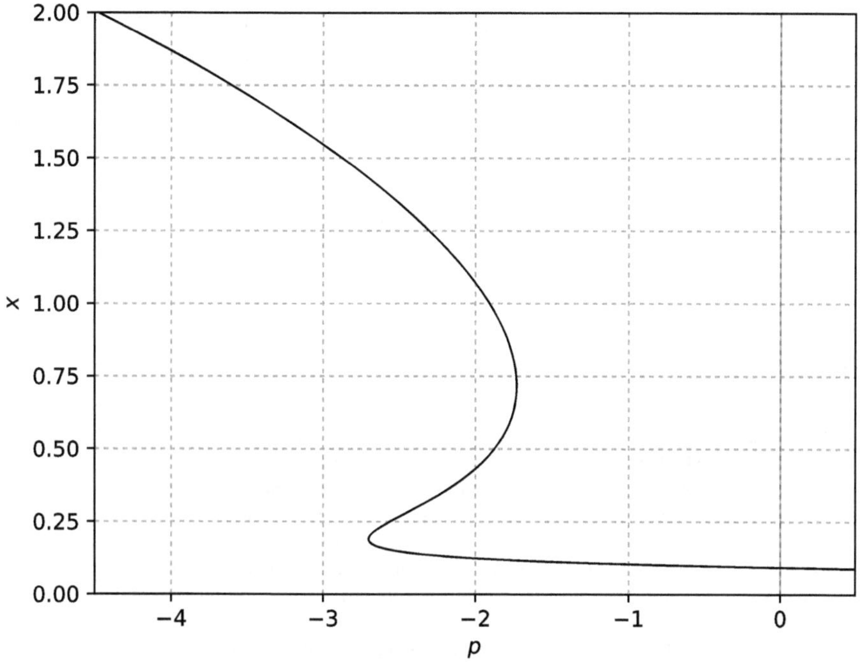

Fig. 2.17 The dependence $x(p,\ q = -3/32) > 0$

and $p^+_{\text{max}}(q) < p < \infty$, where $x(p,q)$ is one-valued in $-\infty < p < p^+_{\text{min}}(q)$ and $p^+_{\text{max}}(q) < p < \infty$, and three-valued in $p^+_{\text{min}}(q) < p < p^+_{\text{max}}(q)$.

To determine the bifurcation values $p^+_{\text{min}}(q)$ and $p^+_{\text{max}}(q)$, we analyze the positive solutions of $\mathrm{d}p(\tau)/\mathrm{d}\tau = 0$, or equivalently $P^+(q,\tau) = 0$. Denote the real solutions of this equation by $\tau^+_{\text{min}}(q) > 0$, $\tau^+_{\text{max}}(q) > 0$ for $\tau^+_{\text{min}}(q) < \tau^+_{\text{max}}(q)$. The local minimum of $p(\tau)$ is at $\tau = \tau^+_{\text{min}}(q)$ and the local maximum at $\tau = \tau^+_{\text{max}}(q)$. Let us denote by $\tau_1(q)$ the solution of the implicit equation $p(\tau, q) = p^+_{\text{min}}(q)$. We have the expressions $p^+_{\text{min}}(q) = p(\tau_1(q), q)$, $p^+_{\text{max}}(q) = p(\tau_2(q), q)$.

Thus, for example for the value $q = -3/32 \in (q^+_*, 0)$ used in Fig. 2.17, the parametric representations (2.3.30) determine the one-valued dependence $x(p,q) > 0$ for τ in $0 < \tau < \tau_1(q)$, $\tau_2(q) < \tau < \infty$, and a three-valued dependence when $\tau_1(q) < \tau < \tau_2(q)$.

The obtained conclusions concerning the one-valuedness and many-valuedness of $x(p,q)$ apply to the separate branches $x(p,q) > 0$ and $x(p,q) < 0$. Because the positive $x(p,q) > 0$ and negative $x(p,q) < 0$ dependences are mutually disjoint, we obtain the final result on the multivalence of real solutions for the fourth-order polynomial equation (2.3.22) as a simple combination of the results obtained for the sets of positive and negative solutions. For example, in the case of two-valued dependences $x(p,q) > 0$ and $x(p,q) < 0$ (for some fixed value of q), the dependence $x(p,q)$, as determined by the Eqs. (2.3.30) and (2.3.31), should be considered as four-valued.

In conclusion, in this chapter we considered the implicit function theorem as a general mathematical tool to find bifurcations, and applied it to the analysis of cubic and and quartic (fourth-order) polynomial equations. This provided, by demonstration, a practical overview of relevant techniques. The presented arguments are applicable to the stability analysis of mechanical systems, where such polynomial equations often arise as the characteristic polynomial. In the next chapter, we will focus on mechanics more directly, specifically on nonconservative problems with a finite number of degrees of freedom.

References

1. Banichuk NV, Barsuk AA (1982) On stability of torsioned elastic rods. Izvestiya Akademii Nauk SSSR, Mekhanika Tverdogo Tela (MTT), vol 6, pp 148–154 (in Russian)
2. Banichuk NV, Bratus AS, Mishkis AD (1987) Analysis of stabilizing and destabilizing effects of small damping in nonconservative systems with finite numbers of degrees of freedom. Technical report, Institute for Problems in Mechanics AN USSR. Reprint no. 312 (in Russian)
3. Banichuk NV (1978) Minimization of the weight of a wing with constraints on the velocity of divergence. Scientific Notes of the Central Aerohydrodynamic Institute (TsAGI), vol 9, no 5 (in Russian)
4. Banichuk NV (1990) Introduction to optimization of structures. Springer, New York, 300 pages
5. Banichuk VV (1983) Problems and methods of optimal structural design. Plenum Press, New York, 313 pages
6. Banichuk NV, Gura NM (1979) On a dynamic problem of optimal design. Mech Deform Solid Bodies 41:20–24
7. Banichuk NV, Bratus AS, Mishkis AD (1989) On destabilizing action on nonconservative systems of small dissipative forces. Rep USSR Acad Sci 309(6):1325–1327 (in Russian)
8. Banichuk N, Jeronen J, Saksa T, Tuovinen T (2011) Static instability analysis of an elastic band travelling in the gravitational field. Rakenteiden mekaniikka (Finn J Struct Mech) 44(3):172–185
9. Banichuk N, Jeronen J, Neittaanmäki P, Saksa T, Tuovinen T (2013) Theoretical study on travelling web dynamics and instability under non-homogeneous tension. Int J Mech Sci 66C:132–140. https://doi.org/10.1016/j.ijmecsci.2012.10.014
10. Banichuk N, Kurki M, Neittaanmäki P, Saksa T, Tirronen M, Tuovinen T (2013) Optimization and analysis of processes with moving materials subjected to fatigue fracture and instability. Mech Based Design Struct Mach Int J 41(2):146–167. https://doi.org/10.1080/15397734.2012.708630
11. Banichuk N, Jeronen J, Ivanova S, Tuovinen T (2015) Analytical approach for the problems of dynamics and stability of a moving web. Rakenteiden mekaniikka (J Struct Mech) 48(3):136–163
12. Banichuk N, Jeronen J, Kurki M, Neittaanmäki P, Saksa T, Tuovinen T (2011) On the limit velocity and buckling phenomena of axially moving orthotropic membranes and plates. Int J Solids Struct 48(13):2015–2025. ISSN 0020-7683. https://doi.org/10.1016/j.ijsolstr.2011.03.010
13. Banichuk N, Jeronen J, Neittaanmäki P, Tuovinen T (2011) Dynamic behaviour of an axially moving plate undergoing small cylindrical deformation submerged in axially flowing ideal fluid. J Fluids Struct 27(7):986–1005. ISSN 0889-9746. https://doi.org/10.1016/j.jfluidstructs.2011.07.004

14. Banichuk N, Jeronen J, Neittaanmäki P, Tuovinen T, Saksa T (2010) Theoretical study on travelling web dynamics and instability under a linear tension distribution. Technical Report B1/2010, University of Jyväskylä. Reports of the Department of Mathematical Information Technology. B: Scientific Computing
15. Bolotin VV (1961) Nonconservative problems of the theory of elastic stability. Fizmatgiz, Moscow, 339 pages (in Russian)
16. Bolotin VV (1964) The dynamic stability of elastic systems. Holden-Day, Inc. Translated from the Russian (1956) and German (1961) editions
17. Bolotin VV (1969) Statistical methods in structural mechanics. Holden-Day Inc
18. Bolotin VV (1973) Variational principles in theory of elastic stability. In Problems in mechanics of solid deformable bodies, Moscow, pp 83–88. Sudostroenie
19. Bolotin VV (1995) Dynamic stability of structures. In Kounadis AN, Krätzig WB (eds) Nonlinear stability of structures. CISM, vol 342. Springer, pp 3–72. https://doi.org/10.1007/978-3-7091-4346-9
20. Bolotin VV (1956) Dynamic stability of elastic systems. Gostekhizdat, Moscow
21. Jeronen J (2011) On the mechanical stability and out-of-plane dynamics of a travelling panel submerged in axially flowing ideal fluid: a study into paper production in mathematical terms. PhD thesis, Department of Mathematical Information Technology, University of Jyväskylä. http://urn.fi/URN:ISBN:978-951-39-4596-1. Jyväskylä studies in computing 148. ISBN 978-951-39-4595-4 (book), ISBN 978-951-39-4596-1 (PDF)
22. Larichev AD (1981) Finding a minimum volume for a beam on an elastic foundation, for a given magnitude of a critical load. In Applied and theoretical research into building structures, Moscow, pp 19–25. Kucherenko TsNIINSK (in Russian)
23. Mote CD (1968) Divergence buckling of an edge-loaded axially moving band. Int J Mech Sci 10:281–295. https://doi.org/10.1016/0020-7403(68)90013-1
24. Mote CD (1968) Dynamic stability of an axially moving band. J Frankl Inst 285(5):329–346. https://doi.org/10.1016/0016-0032(68)90482-1
25. Mote CD (1972) Dynamic stability of axially moving materials. Shock Vib Dig 4(4):2–11
26. Mote CD (1975) Stability of systems transporting accelerating axially moving materials. ASME J Dyn Syst Meas Control 97:96–98. https://doi.org/10.1115/1.3426880
27. Nečas J, Lehtonen A, Neittaanmäki P (1987) On the construction of Lusternik-Schnirelmann critical values with application to bifurcation problems. Appl Anal 25(4):253–268. https://doi.org/10.1080/00036818708839689
28. Neittaanmäki P, Ruotsalainen K (1985) On the numerical solution of the bifurcation problem for the sine-Gordon equation. Arab J Math 6(1 and 2)
29. Plaut RH (1972) Determining the nature of instability in nonconservative problems. AIAA J 10(2)
30. Pramila A (1986) Sheet flutter and the interaction between sheet and air. TAPPI J 69(7):70–74
31. Pramila A (1987) Natural frequencies of a submerged axially moving band. J Sound Vib 113(1):198–203
32. Saksa T, Banichuk N, Jeronen J, Kurki M, Tuovinen T (2012) Dynamic analysis for axially moving viscoelastic panels. Int J Solids Struct 49(23–24):3355–3366. https://doi.org/10.1016/j.ijsolstr.2012.07.017
33. Saksa T, Jeronen J, Tuovinen T (2012) Stability of moving viscoelastic panels interacting with surrounding fluid. Rakenteiden mekaniikka (Finn J Struct Mech) 45(3):88–103
34. Saksa T, Banichuk N, Jeronen J, Kurki M, Tuovinen T (2013) Dynamic behaviour of a travelling viscoelastic band in contact with rollers. In Repin S, Tiihonen T, Tuovinen T (eds) Numerical methods for differential equations, optimization, and technological problems. Dedicated to Professor P. Neittaanmäki on his 60th Birthday. Computational methods in applied sciences, vol 27. Springer Netherlands, pp 393–408. ISBN: 978-94-007-5287-0 (Print) 978-94-007-5288-7 (Online)
35. Saksa T, Jeronen J, Banichuk N, Kurki M (2013) On travelling web stability including material viscoelasticity and surrounding air. In Anson SJI' (ed) Advances in pulp and paper research, Cambridge 2013, vol 1. The Pulp & Paper Fundamental Research Society, pp 449–468. ISBN 978-0-9926163-0-4

36. Tuovinen T (2011) Analysis of stability of axially moving orthotropic membranes and plates with linear non-homogeneous tension profile. PhD thesis, Department of Mathematical Information Technology, University of Jyväskylä. http://urn.fi/URN:ISBN:978-951-39-4578-7. Jyväskylä studies in computing 147. ISBN 978-951-39-4577-0 (book), ISBN 978-951-39-4578-7 (PDF)
37. Wickert JA, Mote CD (1988) Current research on the vibration and stability of axially-moving materials. Shock Vib Dig 20:3–13
38. Wickert JA, Mote CD (1990) Classical vibration analysis of axially moving continua. ASME J Appl Mech 57:738–744. https://doi.org/10.1115/1.2897085

Chapter 3
Nonconservative Systems with a Finite Number of Degrees of Freedom

In this chapter we present some results on the stability and bifurcations of the systems with a finite number of degrees of freedom. We consider damping-induced destabilization in nonconservative systems. We start with a general theoretical treatment of the topic. As the model problem, we consider the double pendulum subject to both a follower force and gravitational loading. A special case of interest is treated with the theoretical framework.

The chapter finishes with a thorough presentation and analysis of the model problem including the nonlinear dynamics, quasistatic equilibrium paths and their stability, and special cases of interest. We first derive the governing equations, starting from the principle of virtual work. We then analyze cases both with and without damping, with and without rotational springs at the joints, with and without a point mass attached at the free end, and with and without a follower force. The mass of the rods is taken into account both in the gravitational and inertial terms. Both analytical and computational approaches are employed. In numerical examples, we plot the equilibrium paths, and show trajectory density visualizations of the time evolution of the nonlinear system. Sample-based uncertainty quantification is employed to capture both branches of a bifurcation in the same visualization.

The chapter is based on the results of the papers by Banichuk et al. [1], Vishik and Ljusternik [2], and Banichuk et al. [3], except Sect. 3.6, which is new.

3.1 Critical Parameters and Destabilizing Perturbations

Consider a linear autonomous system with a real parameter $p \in \mathbb{R}$,

$$\frac{\mathrm{d}^2 w}{\mathrm{d}t^2} = \mathcal{L}(p) w(t) \,, \qquad (3.1.1)$$

N. Banichuk et al., *Stability of Axially Moving Materials*, Solid Mechanics and Its Applications 259, https://doi.org/10.1007/978-3-030-23803-2_3

where t is time, $w(t)$ is a vector function of dimensionality $n \geq 2$, and $\mathcal{L}(p)$ is a matrix with elements, which are real analytic functions of p. Let us use the following representation,

$$w_i(t) = u_i e^{\omega t}, \qquad i = 1, 2, \dots, n\,, \tag{3.1.2}$$

and transform (3.1.1) into the following form:

$$\mathcal{L}(p)u = \omega^2(p)u, \qquad u = (u_1, u_{2,} \dots, u_n)\,. \tag{3.1.3}$$

In the general case, the set of eigenvalues $\omega_j^2(p)(j = 1, 2, \dots, n)$ may contain real values as well as complex conjugate numbers. The values $\omega_j^2(p)$ are continuously dependent on the parameter p. The components u_i of the vector u do not depend on t. Assume the following:

1. There exists a value of the parameter $p = p_0$ such that all eigenvalues $\omega_j^2(p)$ of the problem (3.1.3) are real, simple and negative ($\omega_j^2 = -\tau_j^2, \tau_j \neq 0, j = 1, 2, \dots, n$) for $p < p_0$.
2. The first two eigenvalues coincide and become double ($\omega_1^2(p_0) = \omega_2^2(p_0) = -\tau_1^2, \tau_1 \neq 0$), and the unique vector u_1^0 corresponds to these eigenvalues. The eigenvalues $\omega_1^2(p)$ and $\omega_2^2(p)$ obtain a nonzero imaginary component for $p > p_0$. The rest of the eigenvalues remain real, simple and negative ($\omega_j^2(p) = -\tau_j^2, j = 3, 4, \dots n, \tau_j \neq 0$) in some vicinity of the point p_0.

By Assumptions 1 and 2, a bifurcation takes place at $p = p_0$, and the system (3.1.1) transitions from asymptotically stable to unstable.

Let β be a real matrix of the order $n \times n$ and $\varepsilon > 0$ a small parameter. Consider the perturbed system

$$\frac{\mathrm{d}^2 w}{\mathrm{d}t^2} = \mathcal{L}(p)w(t) + \varepsilon\beta\frac{\mathrm{d}w}{\mathrm{d}t}\,. \tag{3.1.4}$$

The term $\varepsilon\beta(\mathrm{d}w/\mathrm{d}t)$ in (3.1.4) corresponds to small damping. The perturbed system (3.1.4) is transformed into the unperturbed (undamped) system (3.1.1) when $\varepsilon = 0$. The values $\omega_\varepsilon(p)$ corresponding to the perturbed system are the eigenvalues of the problem

$$\mathcal{L}(p)u^\varepsilon + \varepsilon\omega_\varepsilon(p)\beta u^\varepsilon = \omega_\varepsilon^2(p)u^\varepsilon\,. \tag{3.1.5}$$

The critical value p_ε of the parameter p is a lower bound of p for which at least one eigenvalue of the problem (3.1.5) has a positive real part. With Assumptions 1 and 2, the system (3.1.3) is stable if $p < p_\varepsilon$ and all $\omega_{j,\varepsilon}$ are simple. This system is unstable for small values $p - p_\varepsilon > 0$. Taking into account the analyticity of all considered functions, we conclude that there exists the following limit:

$$p_d = \lim_{\varepsilon \to 0} p_\varepsilon\,, \tag{3.1.6}$$

which is the critical value p_ε of the perturbed problem in the limit of vanishing damping.

If the system (3.1.1) (with its boundary conditions) is conservative, then, as it is well known, the addition of dissipative forces does not violate its stability. It is not true for nonconservative systems. In some cases, small damping has a destabilizing effect Ziegler [4], Bolotin and Zhinzher [5], that is,

$$p_d < p_0 \,. \tag{3.1.7}$$

The quantity

$$d_{\mathcal{L}}(\beta) \equiv p_0 - p_d$$

is a deficiency index of the matrix β with respect to matrix $\mathcal{L}$. If $d_{\mathcal{L}}(\beta) = 0$, then the matrix β is an ideal matrix with respect to the matrix $\mathcal{L}$ and the corresponding perturbation of the system (3.1.1) is an ideal perturbation.

A matrix β with $d_{\mathcal{L}}(\beta) > 0$ is said to be deficient and the corresponding perturbation of (3.1.1) is destabilizing. The value of the deficiency index characterizes the influence of the term $\varepsilon\beta\,(\mathrm{d}w/\mathrm{d}t)$ for small $\varepsilon > 0$. It was shown in Bolotin and Zhinzher [5] that $d_{\mathcal{L}}(\beta) \geq 0$.

In what follows, we will also use Assumption 3:

3. p_ε is an analytical function of the small parameter ε, in other words

$$p_\varepsilon = p_d + \varepsilon p_1 + \varepsilon^2 p_2 + \cdots \,, \tag{3.1.8}$$

and

$$d_{\mathcal{L}}(\beta) = p_0 - p_d \geq 0 \,. \tag{3.1.9}$$

3.2 Characteristic Polynomial and Series Expansions

The studies of stability of dynamical systems in the case of small dissipation are based on the application of the methods of the theory of perturbations. In this connection we essentially use the result of the paper Vishik and Ljusternik [2], which is devoted to the perturbation of a spectrum of matrices and non-selfadjoint operators in Hilbert spaces.

Consider the characteristic polynomial of the perturbed problem (3.1.5)

$$\det\left[\mathcal{L}\left(p_\varepsilon\right) + \varepsilon\omega_\varepsilon\left(p_\varepsilon\right)\beta - \omega_\varepsilon^2\left(p_\varepsilon\right)E\right] = 0 \,, \tag{3.2.1}$$

where E is the identity matrix. Let

$$\lim_{\varepsilon\to+0} p_\varepsilon = p_0, \quad \left(\mathrm{d}_{\mathcal{L}}\left(\beta\right) = 0\right) \,. \tag{3.2.2}$$

From Assumptions 1–3 it follows that the coefficients of (3.2.1) are analytical functions, and that the squared roots of the characteristic equation of the perturbed problem (3.2.1) are analytical functions of ε if $j = 3, 4, \ldots, n$, and the value $\omega^2_{1,\varepsilon}$ is an analytical function of $\sqrt{\varepsilon}$ (in the vicinity of the point $\varepsilon = 0$). Thus, for the squared roots $\omega^2_{j,\varepsilon} (j = 1, 2, 3, \ldots, n)$ we have

$$\omega^2_{1,\varepsilon} = -\tau_1^2 + \mu_{11}\sqrt{\varepsilon} + \mu_{12}\varepsilon + \cdots , \tag{3.2.3}$$

$$\omega^2_{j,\varepsilon} = -\tau_j^2 + \mu_{j1}\varepsilon + \mu_{j2}\varepsilon^2 + \cdots \qquad j = 3, 4, \ldots, n . \tag{3.2.4}$$

From (3.2.3) and (3.2.4), it follows that

$$\omega_{1,\varepsilon} = \pm i\tau_1 \mp \frac{i}{2\tau_1}\sqrt{\varepsilon}\mu_{11} \mp \frac{i}{2\tau_1}\varepsilon\left(\mu_{12} + \frac{\mu_{11}^2}{2\tau_1^2}\right) \mp \ldots , \tag{3.2.5}$$

$$\omega_{j,\varepsilon} = \pm i\tau_j \mp \frac{i}{2\tau_j}\varepsilon\mu_{j1} \mp \frac{i}{2\tau_j}\varepsilon^2\left(\mu_{j2} + \frac{\mu_{j1}^2}{2\tau_j^2}\right) \mp \ldots \qquad j = 3, 4, \ldots n , \tag{3.2.6}$$

where the upper sign (+ or −) corresponds to the first branch of the function and the lower sign to the the second branch. The values μ_{ji} $(j = 1, 2, \ldots, n; i = 1, 2, \ldots)$ are the perturbations to the original eigenvalues ω_j^2 and are to be determined.

As it was shown in Vishik and Ljusternik [2], the eigenvectors of the problem (3.1.5) can be found in the form of the series expansions with respect to $\sqrt{\varepsilon}$ and ε:

$$u_1^\varepsilon = u_1^0 + u_1^1\sqrt{\varepsilon} + u_1^2\varepsilon + \cdots , \tag{3.2.7}$$

$$u_j^\varepsilon = u_j^0 + u_j^1\varepsilon + u_j^2\varepsilon^2 + \cdots \qquad j = 3, 4, \ldots, n . \tag{3.2.8}$$

Here u_1^0 is the eigenvector of the problem (3.1.3), corresponding to a double eigenvalue

$$\omega_1^2(p_0) = \omega_2^2(p_0) = -\tau^2, \qquad \tau_1 \neq 0 , \tag{3.2.9}$$

and u_j^0 $(j = 3, 4, \ldots, n)$ are eigenvectors, corresponding to the eigenvalues

$$\omega_j^2(p_0) = -\tau_j^2 \qquad (\tau_j \neq 0) . \tag{3.2.10}$$

The vectors u_j^i $(j = 1, 2, \ldots, n; i = 1, 2, \ldots)$ must be determined. Without loss of generality, we consider that the eigenvectors u_j^ε are normalized, that is,

$$(u_j^\varepsilon, u_j^\varepsilon) = 1, \qquad j = 1, 2, \ldots, n , \tag{3.2.11}$$

where the parentheses mean the scalar product of the vectors in $\mathbb{R}^n$.

Let us expand the coefficients of (3.1.5) with respect to p and use the expansions (3.2.3) and (3.2.7), and collect the terms with the equal powers of ε. We obtain the following equations for determining the values $\mu_{1,i}$ and vectors u_1^i $(i = 1, 2, \ldots)$:

$$\mathcal{L}(p_0)u_1^0 = -\tau_1^2 u_1^0 \tag{3.2.12}$$

$$\mathcal{L}(p_0)u_1^1 = -\tau_1^2 u_1^1 + \mu_{11}u_1^0$$

$$p_1\mathcal{L}^1(p_0)u_1^0 + \mathcal{L}(p_0)u_1^2 \pm i\tau_1\beta u_1^0 =$$

$$-\tau_1^2 u_1^2 + \mu_{11}u_1^1 + \mu_{12}u_1^0$$

$$p_1\mathcal{L}^1(p_0)u_1^1 + \mathcal{L}(p_0)u_1^3 \pm i\tau_1\beta u_1^1 \mp \frac{i}{2\tau_1^2}\mu_{11}\beta u_1^0 =$$

$$-\tau_1^2 u_1^3 + \mu_{11}u_1^2 + \mu_{12}u_1^1 + \mu_{13}u_1^0 .$$

Here $\mathcal{L}^1(p_0)$ is a matrix with the elements $\mathrm{d}\mathcal{L}_{ij}(p)/\mathrm{d}p$ computed at $p = p_0$, and $\mathcal{L}_{ij}(p)$ are the elements of the matrix $\mathcal{L}(p)$.

For the simple eigenvalues $(j = 3, 4, \ldots, n)$, we have

$$\mathcal{L}(p_0)u_j^0 = -\tau_j^2 u_j^0 , \tag{3.2.13}$$

$$p_1\mathcal{L}^1(p_0)u_j^0 + \mathcal{L}(p_0)u_j^1 \pm i\tau_j\beta u_j^0 = -\tau_j^2 u_j^i + \mu_{ji}u_j^0 .$$

Note that the convergence of the asymptotic expansions (3.2.3), (3.2.4), (3.2.7), and (3.2.8) has been shown in Vishik and Ljusternik [2] for small ε. The equalities (3.2.12), (3.2.13) have been obtained taking into account that $p_d = p_0$, that is, under the condition that the perturbation, realized by the matrix β, is ideal: $d_{\mathcal{L}}(\beta) = 0$.

If $p_d < p_0$ $(d_{\mathcal{L}}(\beta) > 0)$, that is, if the perturbation is deficient, then only the simple eigenvalues correspond to the limiting value p_d. Therefore, the corresponding expansions with respect to ε will have the form (3.2.4), (3.2.6), (3.2.8) and the corresponding determining equations the form (3.2.13). In these relations it is necessary to replace p_0 with p_d.

3.3 Ideal Perturbations and Sufficient Conditions for Stability ($n = 2$)

In this section we formulate some statements concerning the properties of ideal perturbations and the asymptotic stability of the system (3.1.4). The proof of these statements is contained in Banichuk et al. [1].

Definition 3.1 (*Statement*) Let Assumptions 1–3 be satisfied and the matrix β realize the ideal perturbation of the system (3.1.1) in the form (3.1.4). Then

$$(\beta u_1^0, v_1^0) = 0\,, \tag{3.3.1}$$

where u_1^0 is the eigenvector of the original problem (3.1.3) with $p = p_0$, and v_1^0 is the eigenvector of the adjoint problem with respect to (3.1.3),

$$\mathcal{L}^*(p_0) = -\tau_1^2 v_1^0\,. \tag{3.3.2}$$

Without a loss of generality it is possible to use the normalizing condition

$$(v_1^0, v_1^0) = 1\,, \tag{3.3.3}$$

and as follows from Assumption 2, the vectors u_1^0 and v_1^0 are orthogonal [2]:

$$(u_1^0, v_1^0) = 0\,. \tag{3.3.4}$$

Definition 3.2 (*Statement*) If Assumptions 1 and 2 (Sect. 3.1), the condition (3.3.1) and the following representations

$$\omega_{1,\varepsilon}^2 = -\tau_1^2 + \mu_{11}\sqrt{\varepsilon} + O(\sqrt{\varepsilon})\,, \tag{3.3.5}$$

$$u_1^\varepsilon = u_1^0 + u_1^1\sqrt{\varepsilon} + O(\sqrt{\varepsilon})$$

are fulfilled, then the matrix β realizes the ideal perturbation of the system (3.1.1), and the function p_ε is differentiable with respect to ε for $\varepsilon = 0$.

Definition 3.3 (*Statement*) If Assumptions 1–3 (Sect. 3.1) and the condition (3.3.1) are fulfilled, then the equality

$$\mu_{11}\left[(\beta u_1^0, u_1^0) + (\beta v_1^0, v_1^0)\right] = 0 \tag{3.3.6}$$

is realized, and if

$$\left(\beta u_1^0, u_1^0\right) + \left(\beta v_1^0, v_1^0\right) \neq 0 \quad \text{and} \quad \left(\mathcal{L}^1(p)u_1^0, v_1^0\right) \neq 0\,, \tag{3.3.7}$$

then the value p is equal to zero in (3.1.8).

The analysis performed in Banichuk et al. [1] shows that for the asymptotic stability of the system (3.1.4), it is sufficient to fulfill the following inequalities:

$$\left(\beta u_1^0, u_1^0\right) + \left(\beta v_1^0, v_1^0\right) < 0\,, \tag{3.3.8}$$

$$\tau_1^2\left(\beta u_1^0, u_1^0\right)\left(\beta v_1^0, v_1^0\right) + p_2\left(\mathcal{L}^1(p_0)u_1^0, v_1^0\right)\left(\mathcal{L}(p_0)v_1^0, u_1^0\right) > 0\,.$$

As the final result we formulate:

Definition 3.4 (*Statement*) If Assumptions 1–3 (Sect. 3.1), the first inequality of (3.3.8) and the inequality

$$\left(\mathcal{L}^1(p_0 u_1^0, v_1^0)\right) \neq 0 \tag{3.3.9}$$

are fulfilled, then the system (3.1.4) is asymptotically stable when $p < p_\varepsilon$, where p_ε has the following representation:

$$p_\varepsilon = p_0 - \frac{\tau_1^2 \left(\beta u_1^0, u_1^0\right) \left(\beta v_1^0, v_1^0\right)}{\varkappa} \varepsilon^2 + O(\varepsilon^2), \tag{3.3.10}$$

$$\varkappa = \left(\mathcal{L}^1(p_0) u_1^0, v_1^0\right) \left(\mathcal{L}(p_0) v_1^0, v_1^0\right). \tag{3.3.11}$$

From the assumptions made above, it follows that $\varkappa \neq 0$.

3.4 Matrices and Examples of Ideal Perturbations

The condition (3.3.1) means the orthogonality of the vectors βu_1^0 and v_1^0. From the orthogonality of the vectors βu_1^0 and v_1^0, it follows that u_1^0 is an eigenvector of the matrix β, in other words,

$$\beta u_1^0 = \ell_1 u_1^0 , \tag{3.4.1}$$

where ℓ_1 is a complex value (real for particular cases). The condition (3.3.1) can be written as

$$\left(u_1^0, \beta^* v_1^0\right) = 0 , \tag{3.4.2}$$

and consequently the vectors $\beta^* v_1^0$ and u_1^0 are orthogonal. Because of (3.3.4), we have

$$\beta^* v_1^0 = \ell_2 v_1^0 , \tag{3.4.3}$$

where ℓ_2 is also some complex number (real for particular cases). The conditions (3.3.8) are represented in the form

$$\operatorname{Im}\left(\ell_1 + \bar{\ell_2}\right) = 0, \qquad \operatorname{Re}\left(\ell_1 + \bar{\ell_2}\right) < 0 \tag{3.4.4}$$

$$\ell_1 = \sigma_1 + i\theta, \qquad \ell_2 = \sigma_2 + i\theta_2$$

and also can be written as

$$\ell_1 + \ell_2 < 0 , \tag{3.4.5}$$

$$\ell_1 \ell_2 \tau_1^2 + p_2 \left(\mathcal{L}^1(p_0) u_1^0, v_1^0\right) \left(\mathcal{L}(p_0) v_1^0, u_1^0\right) > 0 . \tag{3.4.6}$$

If we take these relations and the inequality

$$\left(\mathcal{L}^1(p_0)u_1^0, v_1^0\right)\left(\mathcal{L}(p_0)v_1^0, u_1^0\right) = -\varkappa^2 < 0\,, \tag{3.4.7}$$

as well as (3.3.10) and perform the corresponding transformations, it is possible to show that

$$p_\varepsilon = p_0 + \frac{\tau_1^2}{\varkappa^2}\ell_1\ell_2\varepsilon^2 + O(\varepsilon^2)\,. \tag{3.4.8}$$

The obtained data give us the possibility to construct the structure of all 2×2 matrices realizing an ideal perturbation of the Eq. (3.1.4) such that the system (3.1.1) is asymptotically stable. Let Assumptions 1 and 2 be satisfied for $p = p_0$. The matrix $\mathcal{L} = \left\|\mathcal{L}_{ij}(p)\right\|$ $(i, j = 1, 2)$ has nonpositive double eigenvalues. It means the following:

$$(\mathcal{L}_{22} - \mathcal{L}_{11})^2 + 4\mathcal{L}_{12}\mathcal{L}_{21} = 0, \qquad \mathcal{L}_{22} + \mathcal{L}_{11} < 0, \qquad \mathcal{L}_{ij} = \mathcal{L}_{ij}(p_0)\,. \tag{3.4.9}$$

The double eigenvalue and the eigenvectors are

$$\lambda_1 = \lambda_2 = \frac{1}{2}\left(\mathcal{L}_{22} + \mathcal{L}_{11}\right), \quad u_1^0 = (k_1, 1)\,, \quad v_1^0 = (k_2, 1) \tag{3.4.10}$$

$$k_1 = \frac{\mathcal{L}_{22} - \mathcal{L}_{11}}{2\mathcal{L}_{12}}, \qquad k_2 = \frac{\mathcal{L}_{22} - \mathcal{L}_{11}}{2\mathcal{L}_{21}}\,.$$

Note that $k_1k_2 + 1 = 0$, therefore $k_1,\ k_2 \neq 0,\ k_1 \neq k_2$.

For vectors u_1^0 and v_1^0 to be (respectively) the eigenvectors of some matrix $\beta = \left\|\beta_{ij}\right\|$ and the adjoint matrix $\beta^* = \left\|\beta_{ij}^*\right\|$ with the eigenvalues ℓ_1 and ℓ_2, it is necessary to satisfy the following equations:

$$\begin{cases} \beta_{11}k_1 + \beta_{12} = \ell_1 k_1 \\ \beta_{11}k_2 + \beta_{21} = \ell_2 k_2 \\ \beta_{12}k_2 + \beta_{22} = \ell_2 \\ \beta_{21}k_2 + \beta_{22} = \ell_1\,. \end{cases} \tag{3.4.11}$$

The solvability condition $k_1k_2 + 1 = 0$ (for this system) is satisfied. The general solution has the following form:

$$\beta = \begin{bmatrix} t & k_1(\ell - t) \\ k_2(\ell_2 - t) & \ell_1 + \ell_2 - t \end{bmatrix}, \tag{3.4.12}$$

where t, ℓ_1, ℓ_2 are real numbers. To satisfy the conditions (3.4.5), it is necessary that $\ell_1 < 0$ and $\ell_2 < 0$. In particular, for

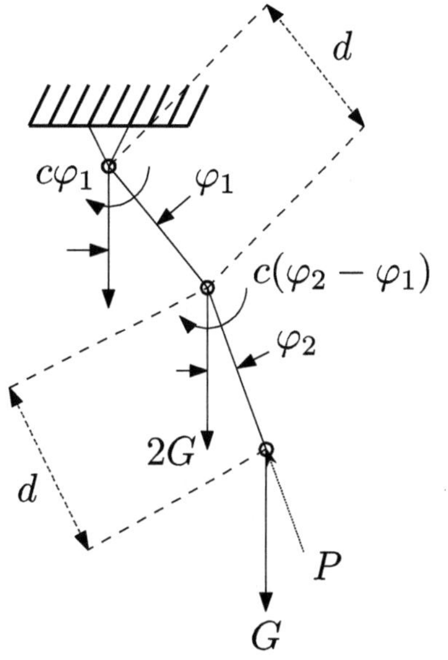

Fig. 3.1 Rod system with two degrees of freedom subjected to a follower force

$$t = \frac{k_2\ell_2 - k_1\ell_1}{k_2 - k_1} \tag{3.4.13}$$

we obtain the symmetric matrices

$$\beta = \begin{bmatrix} \dfrac{k_2\ell_2 - k_1\ell_1}{k_2 - k_1} s & \dfrac{k_2(\ell_2 - \ell_1)}{k_2 - k_1} \\ \dfrac{k_2(\ell_2 - \ell_1)}{k_2 - k_1} & \dfrac{\ell_1 k_2 - k_1\ell_1}{k_2 - k_1} \end{bmatrix}. \tag{3.4.14}$$

Thus, for a given matrix $\mathcal{L}(p)$, the set of real matrices realizing the ideal perturbation is given by a three-parameter family of the matrices (3.4.12). This set of matrices occupies a part of the three-dimensional hyperplane in the four-dimensional space of arbitrary real matrices 2×2.

As an example, consider the problem of stability of the rod system with two degrees of freedom subjected to a follower force (see Fig. 3.1) and to a gravitational force. If we exclude viscoelastic and damping properties from consideration, we may write the governing equation of the system in the form (3.1.1) with the following matrix $\mathcal{L}(p)$:

$$\mathcal{L}(p) = \frac{1}{2m^2d^5}\begin{bmatrix} p - 3k & 2k - p \\ 5k - p & p - 4k \end{bmatrix}, \qquad k = \frac{C}{d}. \tag{3.4.15}$$

The eigenvalue problem (3.1.3) has a double eigenvalue

$$\omega_1^2 = \omega_2^2 = -\frac{k}{\sqrt{2}} \qquad (\omega_1^2 = m^2 d^5 \omega_1^2)\,, \tag{3.4.16}$$

and only one eigenvector

$$u_1^0 = \left(3 - 2\sqrt{2},\, 1\right)\,. \tag{3.4.17}$$

The eigenvector of the adjoint problem (3.3.2) is

$$v_1^0 = \left(-(3 + 2\sqrt{2}),\, 1\right)\,. \tag{3.4.18}$$

Assuming $k = 1$ (for simplicity) and taking into account (3.4.14), we represent the ideal symmetric perturbations as

$$\beta = \frac{1}{6}\begin{bmatrix} 3\,(\ell_1 - \ell_2) - 2\sqrt{2}\,(\ell_1 - \ell_2) & \ell_1 - \ell_2 \\ \ell_1 - \ell_2 & 3\,(\ell_1 + \ell_2) + 2\sqrt{2}\,(\ell_1 - \ell_2) \end{bmatrix}. \tag{3.4.19}$$

Let us use the statement Definition 3.4 and the representation (3.4.8). Then

$$\left(\mathcal{L}(p_0)v_1^0, u_1^0\right) = -9, \qquad \left(\mathcal{L}^1(p_0)v_1^0, u_1^0\right) = 2\sqrt{2}\,, \tag{3.4.20}$$

$$\varkappa^2 = -\left(\mathcal{L}^1(p_0)u_1^0, v_1^0\right)\left(\mathcal{L}(p_0)v_1^0, u_1^0\right) = 18\sqrt{2}\,,$$

$$\tau_1^2 = \frac{1}{\sqrt{2}}, \qquad \left|u_1^0\right|^2 \left|v_1^0\right|^2 = 36\,.$$

As a result, we find

$$p_\varepsilon = \frac{7 - 2\sqrt{2}}{2} + \ell_1 \ell_2 \varepsilon^2 + O(\varepsilon^2)\,. \tag{3.4.21}$$

This representation can be also obtained directly from the Rauss-Hurvitz conditions.

3.5 Stability of Systems Subjected to Deficient Perturbations and Determination of the Deficiency Index

Definition 3.5 (*Statement*) Let Assumptions 1–3 (Sect. (3.1)) be satisfied, and let the matrix β realize a deficient perturbation of the system (3.1.1) in the form (3.1.4). Then, for stability of the system (3.1.4), it is necessary that the conditions

$$\left(u_i^0(p_d), v_i^0(p_d)\right)\left(\beta u_i^0(p_d), v_i(p_d)\right) \le 0 \qquad i = 1, 2 \tag{3.5.1}$$

are satisfied. If the rigorous inequality takes place in (3.5.1) for $p = p_d$, then the system is asymptotically stable.

Here $u_i^0(p_d)$ and $v_i^0(p_d)$ are the solutions of the following eigenvalue problems:

$$\mathcal{L}(p_d)u_i^0(p_d) = \omega_i^2 u_i^0(p_d) ,$$

$$\mathcal{L}^*(p_d)v_i^0(p_d) = \omega_i^2 v_i^0(p_d) , \tag{3.5.2}$$

$$\omega_i^2 = -\tau_i^2, \qquad \tau_i \ne 0 .$$

Proof The eigenvalues of (3.5.2) are simple. Therefore, it is possible to use (3.2.4), (3.2.6), (3.2.8), and (3.2.13) with p_d replacing p_0. Multiplying the second equality in 3.2.13 by the vector $v_i^0(p_d)$, which satisfies (3.5.2), we have

$$p_1\left(\mathcal{L}^1(p_d)u_i^0, v_i(p_d)\right) \pm i\tau\left(\beta u_i^0, v_i^0(p_d)\right) = \mu_{1i}\left(u_i^0(p_d), v_i^0(p_d)\right) . \tag{3.5.3}$$

Since the values $p_1, \left(\mathcal{L}^1(p_d)u_i^0, v_i(p_d)\right)$ and $\left(\beta u_i^0, v_i^0(p_d)\right)$ are real, from (3.2.6) it follows that for the fulfillment of the condition $\mathrm{Re}\,\omega_{i,\varepsilon} \le 0,\ (i = 1, 2)$, it is necessary to fulfill the inequalities (3.5.1). Since $p_d < p_0$, then $\left(u_i^0(p_d), v_i^0(p_d)\right) \ne 0$. If (3.5.1) is satisfied as a rigorous inequality, then $\mathrm{Re}\,\omega_{i,\varepsilon} < 0,\ (i = 1, 2)$ for a small enough ε. Thus, statement Definition 3.5 follows.

For a given matrix β, consider now a set $Q_\beta(p)$ of values p, such that

$$Q_\beta(p) = \left\{p : p \le p_0,\ \left(u^0(p), v^0(p)\right)\left(\beta u^0(p), v^0(p)\right) \le 0\right\} , \tag{3.5.4}$$

where $u^0(p)$ and $v^0(p)$ are the eigenvectors of the problems

$$\mathcal{L}(p)u^0(p) = \omega^2 u^0(p) \tag{3.5.5}$$

and

$$\mathcal{L}^*(p)v^0(p) = \omega^2 v^0(p) . \tag{3.5.6}$$

Let

$$p^*(\beta) = \sup_{p \in Q_\beta(p)} p . \tag{3.5.7}$$

If the set $Q_\beta(p)$ is empty, then we consider that $p^*(\beta) = -\infty$.

Definition 3.6 (*Statement*) Let Assumptions 1–3 (Sect. 3.1) besatisfied. Then the deficiency index of any matrix β is determined with the help of the formula:

$$\mathrm{d}_{\mathcal{L}}(\beta) = p_0 - p^*(\beta) \qquad \text{i.e. } p_d = p^*(\beta) . \tag{3.5.8}$$

3.6 On the Stability and Trajectories of the Double Pendulum with Linear Springs and Dampers

For a system in equilibrium under the action of potential forces, the addition of dissipative forces with complete dissipation ensures asymptotic stability of the undisturbed equilibrium, as stated in a theorem by Kelvin [6], p. 75, nowadays commonly known as the Kelvin-Tait-Chataev theorem (see, e.g., Kirillov [7], Bernstein and Bhat [8]). However, such a result does not exist for general nonconservative systems.

Indeed, in 1952, Hans Ziegler reported an anomaly observed in the stability analysis of a simple double pendulum model consisting of linear springs and dampers, and loaded by a follower force at the free end [4]. The addition of small dissipative forces in the system resulted in a destabilizing effect. This counterintuitive result has become known as Ziegler's paradox.

Bolotin attributes the discontinuity at zero damping to the inability of the linear approximation to assess the question of Lyapunov stability [6], pp. 99–100; [9]. An attempt at a physical explanation is given by Sugiyama and Langthjem [10]. For a history of damping-induced destabilization, see Kirillov and Verhulst [11].

In what follows, the behavior of Ziegler's double pendulum with vanishing dissipation is revisited in terms of the fully nonlinear model. Focus is given to a detailed presentation. The model is derived from first principles, and interesting special cases are discussed. Linearization around the trivial equilibrium is also briefly discussed.

We finish with a numerical parametric study, visualizing state space density fields for selected cases of interest. We use a new data-adaptive dynamic range compression algorithm to bring out detail while preserving visual contrast. Quasistatic equilibrium paths are also displayed, and finally we give a numerical example of Ziegler's effect, where the introduction of small but finite damping drastically reduces the critical load.

3.6.1 Problem Setup and Derivation of the Model

Consider a two-dimensional system consisting of two rigid rods and two rotational springs, see Fig. 3.2. The rods are identical; each has length L, and linear density ρ (unit $[\rho] = \mathrm{kg/m}$). The quantities k_1, k_2 represent (rotational) spring stiffnesses, and c_1, c_2 are the corresponding damping coefficients. The system is subjected to up to two external forces. The force denoted $\mathbf{P}$ is a follower force, which presses the free end inward, leading to a nonconservative problem. The force $\mathbf{P}_{\text{cons.}}$ represents a dead weight loading. Note the neutral position of the spring k_1 is with the first rod pointing upward, against gravity. This makes the neutral equilibrium position $q_1 = q_2 = 0$ unstable.

In order to develop a physically consistent dynamical model, several things must be accounted for. First, there is the inertial contribution of both the mass in the rods, and if present, of the dead weight. Secondly, unless the system is configured in a

Fig. 3.2 Problem setup

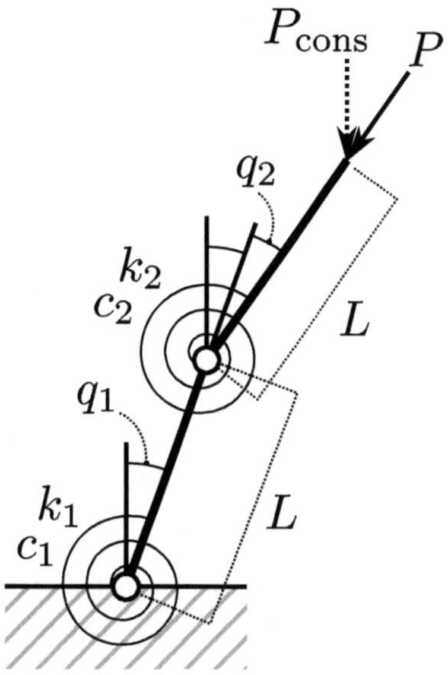

horizontal plane, there is a uniform gravitational field, the effect of which must be accounted for both for the rods and for the dead weight.

If we would like the model to be detailed, air resistance should also be modeled. However, we consider this beyond the scope of this chapter, noting only that as a first approximation, at low velocities air resistance should generate friction that depends linearly on velocity; hence, it may be possible to embed it into the values of c_j, if desired.

We start from the principle of virtual work, which for a dynamic equilibrium states that

$$\delta W_i + \delta W_e = \delta W_j \ . \tag{3.6.1}$$

Here δW_i, δW_e and δW_j are (respectively) the internal, external and inertial contributions to the virtual work:

$$\delta W_i \equiv \sum L_i \delta q_i \equiv \mathbf{L} \cdot \delta \mathbf{q} \ , \tag{3.6.2}$$

$$\delta W_e \equiv \sum Q_i \delta q_i \equiv \mathbf{Q} \cdot \delta \mathbf{q} \ , \tag{3.6.3}$$

$$\delta W_j \equiv \sum J_i \delta q_i \equiv \mathbf{J} \cdot \delta \mathbf{q} \ . \tag{3.6.4}$$

The quantities $\mathbf{L}$, $\mathbf{Q}$ and $\mathbf{J}$ are the internal, external and inertial generalized forces, and $\delta\mathbf{q}$ is the virtual generalized displacement. In our case, the generalized displacement $\mathbf{q}$ consists of the angles q_1 and q_2.

Inserting (3.6.2)–(3.6.4) into (3.6.1), we have

$$\mathbf{L} \cdot \delta\mathbf{q} + \mathbf{Q} \cdot \delta\mathbf{q} = \mathbf{J} \cdot \delta\mathbf{q} \,.$$

Rearranging, we have

$$(\mathbf{L} + \mathbf{Q} - \mathbf{J}) \cdot \delta\mathbf{q} = 0 \,.$$

Because the virtual generalized displacement $\delta\mathbf{q}$ is arbitrary, this implies that

$$\mathbf{L} + \mathbf{Q} = \mathbf{J} \,. \tag{3.6.5}$$

Equation (3.6.5) describes the dynamic equilibrium of generalized forces. In the following, we will explicitly derive **L**, **Q** and **J**.

Internal virtual work

Consider the internal virtual work δW_i. Choosing to use a linear constitutive law for our springs and dampers, we have

$$\delta W_i = -k_1 q_1 \delta q_1 - k_2 q_2 \delta q_2 - c_1 \dot{q}_1 \delta q_1 - c_2 \dot{q}_2 \delta q_2 \,. \tag{3.6.6}$$

Equating (3.6.2) and (3.6.6), and collecting in terms of δq_j, we obtain the internal generalized force,

$$\mathbf{L} = -\begin{bmatrix} k_1 q_1 + c_1 \dot{q}_1 \\ k_2 q_2 + c_2 \dot{q}_2 \end{bmatrix} \,. \tag{3.6.7}$$

External virtual work

The external virtual work can be expressed as

$$\delta W_e \;=\; \mathbf{P}_{\text{total}} \cdot \delta\mathbf{u} \;+\; \sum_{\text{rods}} \int \mathbf{a}_{\text{g}} \cdot \delta\mathbf{x} \, \mathrm{d}m \;=:\; \delta W_e' + \delta W_g \,, \tag{3.6.8}$$

where $\mathbf{P}_{\text{total}}$ is the resultant external force vector, and **u** is the displacement of the free end. The symbol $\delta\mathbf{u}$ represents the virtual displacement. The second term represents the effect of gravity on the mass in the rods; we will return to it below after we have handled the first term.

We will use y-up coordinates, setting the origin at the fixed end of the system. The displacement is taken with respect to the neutral position of the free end at $(x, y) = (0, 2L)$. From the geometry in Fig. 3.2, we have (in cartesian coordinates)

$$\mathbf{u} = L \begin{bmatrix} \sin q_1 + \sin(q_1 + q_2) \\ \cos q_1 + \cos(q_1 + q_2) - 2 \end{bmatrix} \,. \tag{3.6.9}$$

Taking the variation of (3.6.9), we obtain the virtual displacement

$$\delta \mathbf{u} = L \begin{bmatrix} (\cos q_1 + \cos(q_1 + q_2))\, \delta q_1 + \cos(q_1 + q_2)\, \delta q_2 \\ -(\sin q_1 + \sin(q_1 + q_2))\, \delta q_1 - \sin(q_1 + q_2)\, \delta q_2 \end{bmatrix} . \tag{3.6.10}$$

For our external forces $\mathbf{P}$ and $\mathbf{P}_{\text{cons.}}$, we may write (in cartesian coordinates)

$$\mathbf{P}_{\text{total}} = -P_F \begin{bmatrix} \sin(q_1 + q_2) \\ \cos(q_1 + q_2) \end{bmatrix} - P_W \begin{bmatrix} 0 \\ 1 \end{bmatrix} , \tag{3.6.11}$$

where P_F is the magnitude of the follower force, and P_W the magnitude of the dead-weight loading. Using (3.6.10)–(3.6.11), we determine that in (3.6.8),

$$\delta W_e' = \mathbf{P}_{\text{total}} \cdot \delta \mathbf{u} = \mathbf{P} \cdot \delta \mathbf{u} + \mathbf{P}_{\text{cons.}} \cdot \delta \mathbf{u} ,$$

where the contribution from the follower force is

$$\begin{aligned} \mathbf{P} \cdot \delta \mathbf{u} = -P_F L \Big[& \sin(q_1 + q_2) \left[(\cos q_1 + \cos(q_1 + q_2))\, \delta q_1 + \cos(q_1 + q_2)\, \delta q_2 \right] \\ & - \cos(q_1 + q_2) \left[(\sin q_1 + \sin(q_1 + q_2))\, \delta q_1 + \sin(q_1 + q_2)\, \delta q_2 \right] \Big] \\ = -P_F L & \Big[(\sin(q_1 + q_2) \cos q_1 - \cos(q_1 + q_2) \sin q_1)\, \delta q_1 \Big] \\ = -P_F L & \sin(q_2) \delta q_1 \\ = \mathbf{Q}_F & \cdot \delta \mathbf{q} , \end{aligned}$$

where the last step follows from (3.6.3). We obtain

$$\mathbf{Q}_F = -P_F L \begin{bmatrix} \sin q_2 \\ 0 \end{bmatrix} . \tag{3.6.12}$$

For the dead-weight contribution, we have

$$\begin{aligned} \mathbf{P}_{\text{cons.}} \cdot \delta \mathbf{u} &= -P_W L \Big[-(\sin q_1 + \sin(q_1 + q_2))\, \delta q_1 - \sin(q_1 + q_2)\, \delta q_2 \Big] \\ &= P_W L \Big[(\sin q_1 + \sin(q_1 + q_2))\, \delta q_1 + \sin(q_1 + q_2)\, \delta q_2 \Big] \\ &= \mathbf{Q}_{\text{cons.}} \cdot \delta \mathbf{q} , \end{aligned}$$

whence

$$\mathbf{Q}_{\text{cons.}} = P_W L \begin{bmatrix} \sin q_1 + \sin(q_1 + q_2) \\ \sin(q_1 + q_2) \end{bmatrix} . \tag{3.6.13}$$

We must also include the contribution of gravitational loading on the rods themselves. We have

$$\delta W_g = \sum_{\text{rods}} \int \mathbf{a}_g \cdot \delta\mathbf{x}\, dm \ , \quad \text{where} \quad dm = \rho\, d\xi \ . \tag{3.6.14}$$

Here dm is a differential mass, and ξ is the length coordinate along each rod (in the range $[0, L]$). The acceleration of the uniform gravitational field is

$$\mathbf{a}_g = -g\cos(\theta) \begin{bmatrix} 0 \\ 1 \end{bmatrix} , \tag{3.6.15}$$

where θ is the angle between the plane of the system and the vertical. In terms of Fig. 3.2, the system is considered as tilted into the page (or out of the page), keeping the whole lower edge fixed. For $\theta = 0$ we have a fully vertical configuration, and for $\theta = \pi/2$ we have a fully horizontal configuration.

For the first rod, it holds that

$$\mathbf{x} = \begin{bmatrix} \sin q_1 \\ \cos q_1 \end{bmatrix} \xi \ , \quad \delta\mathbf{x} = \begin{bmatrix} \cos q_1 \\ -\sin q_1 \end{bmatrix} \xi\, \delta q_1 \ , \tag{3.6.16}$$

and for the second rod,

$$\mathbf{x} = \begin{bmatrix} \sin q_1 \\ \cos q_1 \end{bmatrix} L + \begin{bmatrix} \sin(q_1 + q_2) \\ \cos(q_1 + q_2) \end{bmatrix} \xi \ , \tag{3.6.17}$$

$$\delta\mathbf{x} = \begin{bmatrix} \cos q_1 \\ -\sin q_1 \end{bmatrix} L\, \delta q_1 + \begin{bmatrix} \cos(q_1 + q_2) \\ -\sin(q_1 + q_2) \end{bmatrix} \xi(\delta q_1 + \delta q_2) \ . \tag{3.6.18}$$

Therefore, for the first rod,

$$\begin{aligned} \int \mathbf{a}_g \cdot \delta\mathbf{x}\, dm &= -\rho g \cos(\theta) \int_0^L \begin{bmatrix} 0 \\ 1 \end{bmatrix} \cdot \begin{bmatrix} \cos q_1 \\ -\sin q_1 \end{bmatrix} \xi\, \delta q_1\, d\xi \\ &= \rho g \cos(\theta) \sin(q_1) \delta q_1 \int_0^L \xi\, d\xi = \frac{1}{2} \rho g \cos(\theta) L^2 \sin(q_1) \delta q_1 \ . \end{aligned} \tag{3.6.19}$$

For the second rod,

$$\begin{aligned} \int \mathbf{a}_g \cdot \delta\mathbf{x}\, dm = -\rho g \cos(\theta) \int_0^L \begin{bmatrix} 0 \\ 1 \end{bmatrix} \cdot \Bigg(&\begin{bmatrix} \cos q_1 \\ -\sin q_1 \end{bmatrix} L\, \delta q_1 \\ &+ \begin{bmatrix} \cos(q_1 + q_2) \\ -\sin(q_1 + q_2) \end{bmatrix} \xi(\delta q_1 + \delta q_2) \Bigg) d\xi \end{aligned}$$

$$= \rho g \cos(\theta) \int_0^L (\sin(q_1) L\, \delta q_1 + \sin(q_1 + q_2)\xi(\delta q_1 + \delta q_2))\, d\xi$$
$$= \rho g \cos(\theta) \left[L^2 \sin(q_1)\delta q_1 + \frac{1}{2} L^2 \sin(q_1 + q_2)(\delta q_1 + \delta q_2) \right] . \tag{3.6.20}$$

Summing the contributions from the two rods and collecting in terms of δq_k, we have

$$\delta W_g = \sum_{\text{rods}} \int \mathbf{a}_g \cdot \delta \mathbf{x}\, dm = \frac{1}{2} \rho g \cos(\theta) L^2 \Big[(3 \sin(q_1) + \sin(q_1 + q_2))\, \delta q_1 + (\sin(q_1 + q_2))\, \delta q_2 \Big] . \tag{3.6.21}$$

On the other hand, with respect to the generalized coordinates q_k, we can write

$$\delta W_g = \mathbf{Q}_g \cdot \delta \mathbf{q} . \tag{3.6.22}$$

Equating the right-hand sides of (3.6.21) and (3.6.22), we identify the generalized force $\mathbf{Q}_g$ corresponding to the gravitational contribution as

$$\mathbf{Q}_g = \frac{1}{2} \rho g \cos(\theta) L^2 \begin{bmatrix} 3 \sin(q_1) + \sin(q_1 + q_2) \\ \sin(q_1 + q_2) \end{bmatrix} . \tag{3.6.23}$$

To sum up the contributions, recall that by (3.6.8) and (3.6.3),

$$\begin{aligned} \delta W_e &= \mathbf{P}_{\text{total}} \cdot \delta \mathbf{u} + \sum_{\text{rods}} \int \mathbf{a}_g \cdot \delta \mathbf{x}\, dm \\ &= \mathbf{P} \cdot \delta \mathbf{u} + \mathbf{P}_{\text{cons.}} \cdot \delta \mathbf{u} + \sum_{\text{rods}} \int \mathbf{a}_g \cdot \delta \mathbf{x}\, dm \\ &= \mathbf{Q} \cdot \delta \mathbf{q} \\ &= \left(\mathbf{Q}_F + \mathbf{Q}_{\text{cons.}} + \mathbf{Q}_g \right) \cdot \delta \mathbf{q} . \end{aligned}$$

Thus the total external generalized force becomes

$$\mathbf{Q} = -P_F L \begin{bmatrix} \sin q_2 \\ 0 \end{bmatrix} + P_W L \begin{bmatrix} \sin q_1 + \sin(q_1 + q_2) \\ \sin(q_1 + q_2) \end{bmatrix} + \frac{1}{2} \rho g \cos(\theta) L^2 \begin{bmatrix} 3 \sin(q_1) + \sin(q_1 + q_2) \\ \sin(q_1 + q_2) \end{bmatrix} . \tag{3.6.24}$$

Because the force on the dead weight is also due to gravity, its magnitude is

$$P_W = m g \cos(\theta) , \tag{3.6.25}$$

where m is the mass of the dead weight, which is modeled as a point mass attached at the free end. The dimensions of the last two terms in (3.6.24) match even though they have different powers of L, because m is a mass (SI unit $[m] = \text{kg}$), but ρ is a linear density (unit $[\rho] = \text{kg/m}$). Observe that ρL is the total mass of one rod.

The force contributions in (3.6.24) have different signs. This is explained by the physics of the situation: a dead-weight loading will tend to increase any small nonzero angles, whereas the follower force attempts to rotate the first rod in the opposite sense of the angle q_2 (think of the follower force as the reaction force from a rocket, the rod system playing the role of the spacecraft).

Inertial virtual work

To complete the condition for dynamic equilibrium, we must determine the inertial virtual work. It holds that

$$\delta W_j = \left(\sum_{\text{rods}} \int \ddot{\mathbf{x}} \cdot \delta \mathbf{x} \, \mathrm{d}m \right) + m \ddot{\mathbf{x}}_m \cdot \delta \mathbf{x}_m \,, \quad \text{where} \quad \mathrm{d}m = \rho \, \mathrm{d}\xi \,, \tag{3.6.26}$$

and ξ is the rod length coordinate. The second term is the contribution of the point mass representing the dead weight; $\mathbf{x}_m$ denotes its position.

Let us deal with the rods first. For the first (lower) rod, we have

$$\mathbf{x} = \begin{bmatrix} \sin q_1 \\ \cos q_1 \end{bmatrix} \xi \,, \tag{3.6.27}$$

from which

$$\dot{\mathbf{x}} = \begin{bmatrix} \cos q_1 \\ -\sin q_1 \end{bmatrix} \xi \dot{q}_1 \,, \quad \delta \mathbf{x} = \begin{bmatrix} \cos q_1 \\ -\sin q_1 \end{bmatrix} \xi \, \delta q_1 \,, \tag{3.6.28}$$

and

$$\ddot{\mathbf{x}} = \begin{bmatrix} \cos q_1 \\ -\sin q_1 \end{bmatrix} \xi \ddot{q}_1 - \begin{bmatrix} \sin q_1 \\ \cos q_1 \end{bmatrix} \xi \, (\dot{q}_1)^2 \,. \tag{3.6.29}$$

With constant linear density ρ, the contribution of the first rod to the inertial virtual work thus becomes

$$\int_0^L \ddot{\mathbf{x}} \cdot \delta \mathbf{x} \, \rho \, \mathrm{d}\xi = \int_0^L \xi^2 \ddot{q}_1 \delta q_1 \, \rho \, \mathrm{d}\xi = \rho \ddot{q}_1 \delta q_1 \int_0^L \xi^2 \, \mathrm{d}\xi = \frac{1}{3} \rho L^3 \ddot{q}_1 \delta q_1 \,. \tag{3.6.30}$$

For the second rod,

$$\mathbf{x} = \begin{bmatrix} \sin q_1 \\ \cos q_1 \end{bmatrix} L + \begin{bmatrix} \sin(q_1 + q_2) \\ \cos(q_1 + q_2) \end{bmatrix} \xi \,, \tag{3.6.31}$$

from which

$$\dot{\mathbf{x}} = \begin{bmatrix} \cos q_1 \\ -\sin q_1 \end{bmatrix} L\dot{q}_1 + \begin{bmatrix} \cos(q_1 + q_2) \\ -\sin(q_1 + q_2) \end{bmatrix} \xi(\dot{q}_1 + \dot{q}_2) , \tag{3.6.32}$$

$$\delta\mathbf{x} = \begin{bmatrix} \cos q_1 \\ -\sin q_1 \end{bmatrix} L\,\delta q_1 + \begin{bmatrix} \cos(q_1 + q_2) \\ -\sin(q_1 + q_2) \end{bmatrix} \xi(\delta q_1 + \delta q_2) , \tag{3.6.33}$$

and

$$\begin{aligned} \ddot{\mathbf{x}} = &\begin{bmatrix} \cos q_1 \\ -\sin q_1 \end{bmatrix} L\ddot{q}_1 - \begin{bmatrix} \sin q_1 \\ \cos q_1 \end{bmatrix} L\,(\dot{q}_1)^2 \\ &+ \begin{bmatrix} \cos(q_1 + q_2) \\ -\sin(q_1 + q_2) \end{bmatrix} \xi(\ddot{q}_1 + \ddot{q}_2) - \begin{bmatrix} \sin(q_1 + q_2) \\ \cos(q_1 + q_2) \end{bmatrix} \xi\,(\dot{q}_1 + \dot{q}_2)^2 . \end{aligned} \tag{3.6.34}$$

Using (3.6.33) and (3.6.34), we obtain

$$\begin{aligned} \int_0^L \ddot{\mathbf{x}} \cdot \delta\mathbf{x}\,\rho\,\mathrm{d}\xi = &\int_0^L \Big[L^2\ddot{q}_1\delta q_1 + \cos(q_2)L\delta q_1\xi(\ddot{q}_1 + \ddot{q}_2) - \sin(q_2)L\delta q_1\xi\,(\dot{q}_1 + \dot{q}_2)^2 \\ &+ \cos(q_2)\xi(\delta q_1 + \delta q_2)L\ddot{q}_1 + \sin(q_2)\xi(\delta q_1 + \delta q_2)L(\dot{q}_1)^2 \\ &+ \xi(\delta q_1 + \delta q_2)\xi(\ddot{q}_1 + \ddot{q}_2)\Big]\rho\,\mathrm{d}\xi \\ = &\,\rho L^3\ddot{q}_1\delta q_1 + \frac{1}{2}\rho L^3\cos(q_2)\delta q_1(\ddot{q}_1 + \ddot{q}_2) - \frac{1}{2}\rho L^3\sin(q_2)\delta q_1\,(\dot{q}_1 + \dot{q}_2)^2 \\ &+ \frac{1}{2}\rho L^3\cos(q_2)(\delta q_1 + \delta q_2)\ddot{q}_1 + \frac{1}{2}\rho L^3\sin(q_2)(\delta q_1 + \delta q_2)(\dot{q}_1)^2 \\ &+ \frac{1}{3}\rho L^3(\delta q_1 + \delta q_2)(\ddot{q}_1 + \ddot{q}_2) \\ = &\,\rho L^3\Big[\ddot{q}_1\delta q_1 + \frac{1}{2}\cos(q_2)(\ddot{q}_1 + \ddot{q}_2)\delta q_1 - \frac{1}{2}\sin(q_2)\,(\dot{q}_1 + \dot{q}_2)^2\,\delta q_1 \\ &+ \frac{1}{2}\cos(q_2)\ddot{q}_1(\delta q_1 + \delta q_2) + \frac{1}{2}\sin(q_2)(\dot{q}_1)^2(\delta q_1 + \delta q_2) \\ &+ \frac{1}{3}(\ddot{q}_1 + \ddot{q}_2)(\delta q_1 + \delta q_2)\Big] , \end{aligned} \tag{3.6.35}$$

where we have used the trigonometric identities $\cos(\alpha \pm \beta) = \cos\alpha\cos\beta \mp \sin\alpha \sin\beta$ and $\cos(-\alpha) = \cos(\alpha)$.

Let us split the generalized inertial force into contributions from the rods and from the dead weight:

$$\mathbf{J} = \mathbf{J}_{\text{rods}} + \mathbf{J}_m . \tag{3.6.36}$$

Noting (3.6.4), summing the contributions from (3.6.30) and (3.6.35) and collecting terms with respect to δq_i, we obtain

$$\mathbf{J}_{\text{rods}} = \rho L^3 \begin{bmatrix} \frac{4}{3}\ddot{q}_1 + \frac{1}{2}\cos q_2(\ddot{q}_1 + \ddot{q}_2) - \frac{1}{2}\sin q_2\,(\dot{q}_1 + \dot{q}_2)^2 \\ +\frac{1}{2}\cos(q_2)\ddot{q}_1 + \frac{1}{2}\sin q_2\,(\dot{q}_1)^2 + \frac{1}{3}(\ddot{q}_1 + \ddot{q}_2) \\ \frac{1}{2}\cos(q_2)\ddot{q}_1 + \frac{1}{2}\sin q_2\,(\dot{q}_1)^2 + \frac{1}{3}(\ddot{q}_1 + \ddot{q}_2) \end{bmatrix} . \tag{3.6.37}$$

Finally, we determine the contribution of the dead weight. We have

$$\mathbf{x}_m = \begin{bmatrix} \sin q_1 \\ \cos q_1 \end{bmatrix} L + \begin{bmatrix} \sin(q_1 + q_2) \\ \cos(q_1 + q_2) \end{bmatrix} L ,$$

$$\delta\mathbf{x}_m = \begin{bmatrix} \cos q_1 \\ -\sin q_1 \end{bmatrix} L\,\delta q_1 + \begin{bmatrix} \cos(q_1 + q_2) \\ -\sin(q_1 + q_2) \end{bmatrix} L(\delta q_1 + \delta q_2) ,$$

$$\dot{\mathbf{x}}_m = \begin{bmatrix} \cos q_1 \\ -\sin q_1 \end{bmatrix} L\dot{q}_1 + \begin{bmatrix} \cos(q_1 + q_2) \\ -\sin(q_1 + q_2) \end{bmatrix} L(\dot{q}_1 + \dot{q}_2) ,$$

$$\begin{aligned} \ddot{\mathbf{x}}_m = &\begin{bmatrix} \cos q_1 \\ -\sin q_1 \end{bmatrix} L\ddot{q}_1 - \begin{bmatrix} \sin q_1 \\ \cos q_1 \end{bmatrix} L\,(\dot{q}_1)^2 \\ &+ \begin{bmatrix} \cos(q_1 + q_2) \\ -\sin(q_1 + q_2) \end{bmatrix} L(\ddot{q}_1 + \ddot{q}_2) - \begin{bmatrix} \sin(q_1 + q_2) \\ \cos(q_1 + q_2) \end{bmatrix} L\,(\dot{q}_1 + \dot{q}_2)^2 . \end{aligned}$$

This leads to

$$\begin{aligned} m\ddot{\mathbf{x}}_m \cdot \delta\mathbf{x}_m = m\Big(&L^2\ddot{q}_1\delta q_1 + \cos(q_2)L^2(\ddot{q}_1 + \ddot{q}_2)\delta q_1 - \sin(q_2)L^2\,(\dot{q}_1 + \dot{q}_2)^2\,\delta q_1 \\ &+ \cos(q_2)L^2(\delta q_1 + \delta q_2)\ddot{q}_1 + \sin(q_2)L^2(\delta q_1 + \delta q_2)\,(\dot{q}_1)^2 \\ &+ L^2(\delta q_1 + \delta q_2)(\ddot{q}_1 + \ddot{q}_2)\Big) \\ = mL^2\Big(&\ddot{q}_1\delta q_1 + \cos(q_2)(\ddot{q}_1 + \ddot{q}_2)\delta q_1 - \sin(q_2)\,(\dot{q}_1 + \dot{q}_2)^2\,\delta q_1 \\ &+ \cos(q_2)\ddot{q}_1(\delta q_1 + \delta q_2) + \sin(q_2)\,(\dot{q}_1)^2\,(\delta q_1 + \delta q_2) \\ &+ (\ddot{q}_1 + \ddot{q}_2)(\delta q_1 + \delta q_2)\Big) . \end{aligned}$$

In terms of the generalized inertial force $\mathbf{J}$, we have the result

$$\mathbf{J}_m = mL^2 \begin{bmatrix} \ddot{q}_1 + \cos q_2(\ddot{q}_1 + \ddot{q}_2) - \sin q_2\,(\dot{q}_1 + \dot{q}_2)^2 \\ +\cos(q_2)\ddot{q}_1 + \sin q_2\,(\dot{q}_1)^2 + (\ddot{q}_1 + \ddot{q}_2) \\ \cos(q_2)\ddot{q}_1 + \sin q_2\,(\dot{q}_1)^2 + (\ddot{q}_1 + \ddot{q}_2) \end{bmatrix} . \tag{3.6.38}$$

Finally, from (3.6.36), (3.6.37) and (3.6.38), the total inertial generalized force is

$$\mathbf{J} = \mathbf{J}_{\text{rods}} + \mathbf{J}_m$$
$$= \rho L^3 \begin{bmatrix} \frac{4}{3}\ddot{q}_1 + \frac{1}{2}\cos q_2(\ddot{q}_1 + \ddot{q}_2) - \frac{1}{2}\sin q_2\,(\dot{q}_1 + \dot{q}_2)^2 \\ +\frac{1}{2}\cos(q_2)\ddot{q}_1 + \frac{1}{2}\sin q_2\,(\dot{q}_1)^2 + \frac{1}{3}(\ddot{q}_1 + \ddot{q}_2) \\ \frac{1}{2}\cos(q_2)\ddot{q}_1 + \frac{1}{2}\sin q_2\,(\dot{q}_1)^2 + \frac{1}{3}(\ddot{q}_1 + \ddot{q}_2) \end{bmatrix}$$
$$+ mL^2 \begin{bmatrix} \ddot{q}_1 + \cos q_2(\ddot{q}_1 + \ddot{q}_2) - \sin q_2\,(\dot{q}_1 + \dot{q}_2)^2 \\ +\cos(q_2)\ddot{q}_1 + \sin q_2\,(\dot{q}_1)^2 + (\ddot{q}_1 + \ddot{q}_2) \\ \cos(q_2)\ddot{q}_1 + \sin q_2\,(\dot{q}_1)^2 + (\ddot{q}_1 + \ddot{q}_2) \end{bmatrix} . \tag{3.6.39}$$

Observe that ρ is a linear density, whereas m is a mass; hence $[\rho L^3] = (\text{kg/m}) \cdot \text{m}^3 = \text{kg m}^2$ and $[mL^2] = \text{kg m}^2$; both have the same unit as expected. The factor $1/\text{s}^2$, required to produce N m so that the units match in the balance for generalized forces, Eq. (3.6.5), is supplied by the time differentiations performed in the vector components.

A variant with point-mass rods

In the original variant of the model that was explored by H. Ziegler, instead of taking the rod material to be uniformly distributed along the length of the rod, it was approximated by point masses having mass ρL (each) placed at the midpoint of each rod.

Obtaining this version of the model requires re-deriving the external and inertial contributions to virtual work, for those terms that are related to the rods. For the gravitational loading on the rods, instead of (3.6.14), we now have

$$\delta W_{\text{g}} = \sum_{\text{rods}} \rho L \mathbf{a}_{\text{g}} \cdot \delta\mathbf{x} . \tag{3.6.40}$$

For the first rod,

$$\mathbf{x} = \begin{bmatrix} \sin q_1 \\ \cos q_1 \end{bmatrix} \frac{L}{2} , \quad \delta\mathbf{x} = \begin{bmatrix} \cos q_1 \\ -\sin q_1 \end{bmatrix} \frac{L}{2}\,\delta q_1 ,$$

whence

$$\rho L \mathbf{a}_{\text{g}} \cdot \delta\mathbf{x} = -\rho g \cos(\theta) L \begin{bmatrix} 0 \\ 1 \end{bmatrix} \cdot \begin{bmatrix} \cos q_1 \\ -\sin q_1 \end{bmatrix} \frac{L}{2}\,\delta q_1 = \frac{1}{2}\rho g \cos(\theta) L^2 \sin(q_1)\delta q_1 . \tag{3.6.41}$$

This is exactly the same result that was obtained for a distributed mass. The reason is that the distributed mass was uniform (constant density along rod length), and the force was also a constant. Hence, both of these could be taken outside the integral, and the remaining integrand was just ξ (producing $L^2/2$). Here, the mass point contributes one "L", and the $\delta\mathbf{x}$ contributes "$L/2$", for a total of $L^2/2$, which coincides with the distributed-mass case.

By the same argument, this conclusion holds also for the second rod. Thus, it happens that (3.6.19)–(3.6.20), and hence the result (3.6.23), can be used for the point-mass variant as-is.

However, the inertial virtual work turns out to require changes. We have

$$\delta W_j = \left(\sum_{\text{rods}} \rho L \ddot{\mathbf{x}} \cdot \delta\mathbf{x}\right) + m\ddot{\mathbf{x}}_m \cdot \delta\mathbf{x}_m \,, \tag{3.6.42}$$

where $\mathbf{x}$ is the position of the relevant point mass. The second term represents the dead weight and requires no modification.

For the first rod, $\mathbf{x}$ and $\delta\mathbf{x}$ are as above, whence

$$\dot{\mathbf{x}} = \begin{bmatrix} \cos q_1 \\ -\sin q_1 \end{bmatrix} \frac{L}{2}\dot{q}_1 \,, \quad \ddot{\mathbf{x}} = \begin{bmatrix} \cos q_1 \\ -\sin q_1 \end{bmatrix} \frac{L}{2}\ddot{q}_1 - \begin{bmatrix} \sin q_1 \\ \cos q_1 \end{bmatrix} \frac{L}{2} (\dot{q}_1)^2 \,.$$

For the second rod,

$$\mathbf{x} = \begin{bmatrix} \sin q_1 \\ \cos q_1 \end{bmatrix} L + \begin{bmatrix} \sin(q_1 + q_2) \\ \cos(q_1 + q_2) \end{bmatrix} \frac{L}{2} \,,$$

$$\delta\mathbf{x} = \begin{bmatrix} \cos q_1 \\ -\sin q_1 \end{bmatrix} L\,\delta q_1 + \begin{bmatrix} \cos(q_1 + q_2) \\ -\sin(q_1 + q_2) \end{bmatrix} \frac{L}{2}(\delta q_1 + \delta q_2) \,,$$

$$\dot{\mathbf{x}} = \begin{bmatrix} \cos q_1 \\ -\sin q_1 \end{bmatrix} L\dot{q}_1 + \begin{bmatrix} \cos(q_1 + q_2) \\ -\sin(q_1 + q_2) \end{bmatrix} \frac{L}{2}(\dot{q}_1 + \dot{q}_2) \,,$$

$$\begin{aligned} \ddot{\mathbf{x}} &= \begin{bmatrix} \cos q_1 \\ -\sin q_1 \end{bmatrix} L\ddot{q}_1 - \begin{bmatrix} \sin q_1 \\ \cos q_1 \end{bmatrix} L\,(\dot{q}_1)^2 \\ &\quad + \begin{bmatrix} \cos(q_1 + q_2) \\ -\sin(q_1 + q_2) \end{bmatrix} \frac{L}{2}(\ddot{q}_1 + \ddot{q}_2) - \begin{bmatrix} \sin(q_1 + q_2) \\ \cos(q_1 + q_2) \end{bmatrix} \frac{L}{2}\,(\dot{q}_1 + \dot{q}_2)^2 \,. \end{aligned}$$

The contribution of the first rod to the inertial virtual work thus becomes

$$\rho L \ddot{\mathbf{x}} \cdot \delta\mathbf{x} = \rho L \left(\frac{L^2}{4}\ddot{q}_1 \delta q_1\right) = \frac{1}{4}\rho L^3 \ddot{q}_1 \delta q_1 \,. \tag{3.6.43}$$

Recall that for the distributed mass case, we obtained the same term, but with coefficient $1/3$ instead of $1/4$.

For the second rod,

$$\begin{aligned}\rho L\ddot{\mathbf{x}}\cdot\delta\mathbf{x} &= \rho L\Big(L^2\ddot{q}_1\delta q_1 + \cos(q_2)\frac{L^2}{2}(\ddot{q}_1+\ddot{q}_2)\delta q_1 - \sin(q_2)\frac{L^2}{2}\,(\dot{q}_1+\dot{q}_2)^2\,\delta q_1\\ &+\cos(q_2)\frac{L^2}{2}(\delta q_1+\delta q_2)\ddot{q}_1 + \sin(q_2)\frac{L^2}{2}(\delta q_1+\delta q_2)\,(\dot{q}_1)^2\\ &+\frac{L^2}{4}(\delta q_1+\delta q_2)(\ddot{q}_1+\ddot{q}_2)\Big)\\ &= \rho L^3\Big(\ddot{q}_1\delta q_1 + \frac{1}{2}\cos(q_2)(\ddot{q}_1+\ddot{q}_2)\delta q_1 - \frac{1}{2}\sin(q_2)\,(\dot{q}_1+\dot{q}_2)^2\,\delta q_1\\ &+\frac{1}{2}\cos(q_2)\ddot{q}_1(\delta q_1+\delta q_2) + \frac{1}{2}\sin(q_2)\,(\dot{q}_1)^2\,(\delta q_1+\delta q_2)\\ &+\frac{1}{4}(\ddot{q}_1+\ddot{q}_2)(\delta q_1+\delta q_2)\Big)\,. \end{aligned}\tag{3.6.44}$$

We obtain, by summing the contributions from the two rods, and collecting in terms of δq_j, that the contribution of the point-mass rods to the generalized inertial force is

$$\mathbf{J}_{\text{rods}} = \rho L^3\begin{bmatrix} \frac{5}{4}\ddot{q}_1 + \frac{1}{2}\cos q_2(\ddot{q}_1+\ddot{q}_2) - \frac{1}{2}\sin q_2\,(\dot{q}_1+\dot{q}_2)^2\\ +\frac{1}{2}\cos(q_2)\ddot{q}_1 + \frac{1}{2}\sin q_2\,(\dot{q}_1)^2 + \frac{1}{4}(\ddot{q}_1+\ddot{q}_2)\\ \frac{1}{2}\cos(q_2)\ddot{q}_1 + \frac{1}{2}\sin q_2\,(\dot{q}_1)^2 + \frac{1}{4}(\ddot{q}_1+\ddot{q}_2)\end{bmatrix}\,. \tag{3.6.45}$$

We see that changing the rods to use point masses has replaced the coefficient $4/3$ by $5/4$, and each $1/3$ by $1/4$, but there are no other changes. By replacing the original $\mathbf{J}_{\text{rods}}$ in (3.6.39) by (3.6.45), the inertial virtual work is changed to treat the variant of the model with point masses located at the midpoints of the rods.

3.6.2 Governing Equations

Summing up the developments so far, the dynamic equilibrium of generalized forces is given by Eq. (3.6.5)

$$\mathbf{L}+\mathbf{Q}=\mathbf{J}\,.$$

Here **L**, **Q** and **J** are, respectively, the internal, external and inertial generalized forces (expressed in terms of the angle variables q_1 and q_2):

$$\mathbf{L} = -\begin{bmatrix} k_1q_1 + c_1\dot{q}_1\\ k_2q_2 + c_2\dot{q}_2\end{bmatrix}\,, \tag{3.6.46}$$

$$\mathbf{Q} = -P_F L \begin{bmatrix} \sin q_2 \\ 0 \end{bmatrix} + P_W L \begin{bmatrix} \sin q_1 + \sin(q_1 + q_2) \\ \sin(q_1 + q_2) \end{bmatrix} + \frac{1}{2} \rho g \cos(\theta) L^2 \begin{bmatrix} 3\sin(q_1) + \sin(q_1 + q_2) \\ \sin(q_1 + q_2) \end{bmatrix}, \tag{3.6.47}$$

$$\mathbf{J} = \rho L^3 \begin{bmatrix} \frac{4}{3}\ddot{q}_1 + \frac{1}{2}\cos q_2(\ddot{q}_1 + \ddot{q}_2) - \frac{1}{2}\sin q_2\,(\dot{q}_1 + \dot{q}_2)^2 \\ \quad + \frac{1}{2}\cos(q_2)\ddot{q}_1 + \frac{1}{2}\sin q_2\,(\dot{q}_1)^2 + \frac{1}{3}(\ddot{q}_1 + \ddot{q}_2) \\ \frac{1}{2}\cos(q_2)\ddot{q}_1 + \frac{1}{2}\sin q_2\,(\dot{q}_1)^2 + \frac{1}{3}(\ddot{q}_1 + \ddot{q}_2) \end{bmatrix} + mL^2 \begin{bmatrix} \ddot{q}_1 + \cos q_2(\ddot{q}_1 + \ddot{q}_2) - \sin q_2\,(\dot{q}_1 + \dot{q}_2)^2 \\ \quad + \cos(q_2)\ddot{q}_1 + \sin q_2\,(\dot{q}_1)^2 + (\ddot{q}_1 + \ddot{q}_2) \\ \cos(q_2)\ddot{q}_1 + \sin q_2\,(\dot{q}_1)^2 + (\ddot{q}_1 + \ddot{q}_2) \end{bmatrix} \quad \text{(distributed-mass rods)}\,. \tag{3.6.48}$$

In (3.6.47), P_F is the magnitude of the follower force, while $P_W = mg\cos(\theta)$ (Eq. (3.6.25)) is the magnitude of dead-weight loading. In (3.6.48), "distributed mass" refers to having constant linear density ρ.

If the distributed rod masses are replaced by point masses of magnitude ρL located at the midpoint of each rod, as in H. Ziegler's original version, the inertial generalized force (3.6.48) is replaced by

$$\mathbf{J}_{\text{pm}} = \rho L^3 \begin{bmatrix} \frac{5}{4}\ddot{q}_1 + \frac{1}{2}\cos q_2(\ddot{q}_1 + \ddot{q}_2) - \frac{1}{2}\sin q_2\,(\dot{q}_1 + \dot{q}_2)^2 \\ \quad + \frac{1}{2}\cos(q_2)\ddot{q}_1 + \frac{1}{2}\sin q_2\,(\dot{q}_1)^2 + \frac{1}{4}(\ddot{q}_1 + \ddot{q}_2) \\ \frac{1}{2}\cos(q_2)\ddot{q}_1 + \frac{1}{2}\sin q_2\,(\dot{q}_1)^2 + \frac{1}{4}(\ddot{q}_1 + \ddot{q}_2) \end{bmatrix} + mL^2 \begin{bmatrix} \ddot{q}_1 + \cos q_2(\ddot{q}_1 + \ddot{q}_2) - \sin q_2\,(\dot{q}_1 + \dot{q}_2)^2 \\ \quad + \cos(q_2)\ddot{q}_1 + \sin q_2\,(\dot{q}_1)^2 + (\ddot{q}_1 + \ddot{q}_2) \\ \cos(q_2)\ddot{q}_1 + \sin q_2\,(\dot{q}_1)^2 + (\ddot{q}_1 + \ddot{q}_2) \end{bmatrix} \quad \text{(point-mass rods)}\,. \tag{3.6.49}$$

Note the two instances of $5/4$ and $1/4$; everything else remains the same. Equations (3.6.5)–(3.6.47) require no modification.

Equations (3.6.5), (3.6.46)–(3.6.48) (or (3.6.5), (3.6.46)–(3.6.47) and (3.6.49) for the point-mass variant) determine the exact (nonlinear) dynamics of the considered model. To compute a trajectory, in addition an initial condition for the state vector $\mathbf{u} := (q_1, q_2, \dot{q}_1, \dot{q}_2)$ must be supplied.[1]

3.6.3 Nondimensional Form

For the rest of the discussion, let us restrict our consideration to the case where the springs are identical; $k_1 = k_2 =: k$ and $c_1 = c_2 =: c$. We will work in nondimensional time

[1]This $\mathbf{u}$ is unrelated to the displacement that was used when deriving the model; we will not need the displacement any more.

$$t' := \frac{t}{\tau} , \tag{3.6.50}$$

where τ (unit $[\tau] = \mathrm{s}$) is an arbitrary constant (scaling factor; sometimes known as the characteristic time or reference time). We will consider the choice of τ further below.

In the main part of the text to follow, the usual shortcuts will be taken, and the primes related to definitions of nondimensional quantities will be immediately omitted from the notation. But to make the treatment rigorous, we first note the following.

Each time differentiation by the dimensional time t (in the original equations) will introduce a factor of $1/\tau$ when the problem is expressed in terms of the nondimensional time t'. To see why, let us denote

$$q'(t/\tau) = q'(t') \equiv q(\tau t') = q(t) , \tag{3.6.51}$$

where the first and last equalities follow from (3.6.50), and the middle equivalence is a definition. By (3.6.51) and the chain rule, we have

$$\dot{q} \equiv \frac{\mathrm{d}q(t)}{\mathrm{d}t} = \frac{\mathrm{d}q'(t')}{\mathrm{d}t} = \frac{\mathrm{d}q'}{\mathrm{d}t'}\frac{\mathrm{d}t'}{\mathrm{d}t} = \frac{1}{\tau}\frac{\mathrm{d}q'}{\mathrm{d}t'} .$$

At this point, it is conventional to omit the primes from the notation. As a commonly used shortcut, in the original equations, one just replaces $\dot{q} \to (1/\tau)\dot{q}$ (and respectively, $\ddot{q} \to (1/\tau^2)\ddot{q}$), and then treats the equations as having been converted into nondimensional time.

In our problem, q happens to be an angle variable, which is already nondimensional; this simplifies the treatment slightly. If we think of the quantity of interest as a function, we have so far only nondimensionalized its input (arguments). To generalize this treatment to dimensional quantities, such as a linear displacement u, we define another scaling factor to make also the output nondimensional:

$$u''(t') \equiv \frac{1}{u_0} u'(t') .$$

Here u' is defined in the same manner as q', Eq. (3.6.51). The constant u_0 (unit $[u_0] = \mathrm{m}$) is an arbitrary scaling factor. Then we insert $u' = u_0 u''$ — or as a shortcut, just replace $u \to u_0 u$ (primes omitted), and then treat the equations as having been converted into nondimensional displacement.

Details of the treatment change slightly also if the function has several arguments. This will be of interest later for the main topic of the book, axially moving materials. Each argument is treated using the approach of Eq. (3.6.51). As an example of the two-argument case, let $w = w(x, t)$. Using nondimensional variables $t' = t/\tau$ and $x' = x/\ell$ (for arbitrary constants τ and ℓ), we define

$$w'(x/\ell\,,\ t/\tau\,) = w'(\,x',\ t'\,) \equiv w(\,\ell x',\ \tau t'\,) = w(x,t)\ .$$

Then

$$\dot{w} \equiv \frac{\partial w(x,t)}{\partial t} = \frac{\partial w'(x',t')}{\partial t} = \frac{\partial w'}{\partial x'}\frac{\partial x'}{\partial t} + \frac{\partial w'}{\partial t'}\frac{\partial t'}{\partial t} = \frac{1}{\tau}\frac{\partial w'}{\partial t'}\ .$$

The first term in the sum vanishes, since x' does not depend on t. Similarly,

$$\frac{\partial w}{\partial x} \equiv \frac{\partial w(x,t)}{\partial x} = \frac{\partial w'(x',t')}{\partial x} = \frac{\partial w'}{\partial x'}\frac{\partial x'}{\partial x} + \frac{\partial w'}{\partial t'}\frac{\partial t'}{\partial x} = \frac{1}{\ell}\frac{\partial w'}{\partial x'}\ .$$

This works only for partial derivatives. If any total (material, Lagrange) derivatives appear in the problem, the conventional approach is to first express these in terms of partial derivatives, and only then perform the nondimensionalization.

Performing the nondimensionalization

Beside the reference time τ, we similarly define a reference length ℓ (unit $[\ell] = \mathrm{m}$) and a reference mass μ (unit $[\mu] = \mathrm{kg}$). As implied by dimensional analysis, this set of reference quantities defines a reference acceleration ℓ/τ^2, a reference force $\mu\ell/\tau^2$, a reference moment $\mu\ell^2/\tau^2$, and a reference kinetic energy $(1/2)\mu(\ell/\tau)^2$.

Thus, fixing the values of τ, ℓ and μ fixes the scaling for all mechanical quantities encountered in the system under study. It is convenient at first to keep these scaling factors arbitrary, and only later decide what would be convenient *natural units* for the system under study.

This motivates the following definitions:

$$\alpha := \frac{L}{\ell}\,,\quad \beta := \frac{c\,\tau}{\mu\ell^2}\,,\quad \gamma := \frac{k\,\tau^2}{\mu\ell^2}\,,\quad \mu_m := \frac{m}{\mu}\,,\quad \mu_\rho := \frac{\rho L}{\mu}\,,\quad G := \frac{g\,\tau^2}{\ell}\,,\quad \Pi_j := \frac{P_j\,\tau^2}{\mu\ell}$$
$$\text{(where } j = F, W\text{)}\,. \tag{3.6.52}$$

The parameter α is a length scale, β characterizes the ratio of spring damping to rod inertia, γ the ratio of spring stiffness to rod inertia, μ_m (respectively μ_ρ) the mass of the dead weight (respectively one rod) in relation to the reference mass, and G the gravitational acceleration. The parameters Π_j characterize the magnitudes of the applied forces; Π_W for the dead weight, and Π_F for the follower force.

Observe that (3.6.52) is obtained trivially in a systematic way: we simply cancel out the dimension of each physical parameter by using appropriate powers of our reference values τ, ℓ and μ. The only noteworthy thing here is that it is convenient to talk of only masses, instead of both masses and densities, taking ρL instead of just ρ. We will leave the geometric factor $\cos(\theta)$ in the dynamical equation, to keep the definitions (3.6.52) based on dimensional analysis only.

There are now eight parameters, seven of which are independent. Observe that

$$\cos(\theta)\mu_m G = \cos(\theta)\frac{m}{\mu}\frac{g\,\tau^2}{\ell} = \frac{\cos(\theta) m g\,\tau^2}{\mu\ell} = \frac{P_W\,\tau^2}{\mu\ell} = \Pi_W\,, \tag{3.6.53}$$

obtained via (3.6.52) and (3.6.25). Thus one of μ_m, G and Π_W can be eliminated.

If we have a horizontal system ($\theta = \pi/2$), then (3.6.53) says only that $\Pi_W = 0$, making no claim regarding μ_m or G. Indeed, for a horizontal system, the gravitational external force is zero.

To nondimensionalize the generalized forces, we will multiply by $\tau^2/\mu\ell^2$ (unit: 1/Nm), which contains only the reference constants. All of them can be chosen to be nonzero; how to do this in a meaningful way, we will explore below. Starting from (3.6.46)–(3.6.48), letting $k_1 = k_2 = k$ and $c_1 = c_2 = c$, multiplying by $\tau^2/\mu\ell^2$, and applying (3.6.52), we obtain

$$\frac{\tau^2}{\mu\ell^2}\mathbf{L} = -\begin{bmatrix} \gamma q_1 + \beta\dot{q}_1 \\ \gamma q_2 + \beta\dot{q}_2 \end{bmatrix}, \tag{3.6.54}$$

$$\frac{\tau^2}{\mu\ell^2}\mathbf{Q} = \alpha\left(-\Pi_F\begin{bmatrix} \sin q_2 \\ 0 \end{bmatrix} + \cos(\theta)G\left(\mu_m\begin{bmatrix} \sin q_1 + \sin(q_1+q_2) \\ \sin(q_1+q_2) \end{bmatrix} + \frac{1}{2}\mu_\rho\begin{bmatrix} 3\sin(q_1) + \sin(q_1+q_2) \\ \sin(q_1+q_2) \end{bmatrix}\right)\right), \tag{3.6.55}$$

$$\begin{aligned}\frac{\tau^2}{\mu\ell^2}\mathbf{J} = \alpha^2\mu_\rho &\begin{bmatrix} \frac{4}{3}\ddot{q}_1 + \frac{1}{2}\cos q_2(\ddot{q}_1+\ddot{q}_2) - \frac{1}{2}\sin q_2\,(\dot{q}_1+\dot{q}_2)^2 \\ +\frac{1}{2}\cos(q_2)\ddot{q}_1 + \frac{1}{2}\sin q_2\,(\dot{q}_1)^2 + \frac{1}{3}(\ddot{q}_1+\ddot{q}_2) \\ \frac{1}{2}\cos(q_2)\ddot{q}_1 + \frac{1}{2}\sin q_2\,(\dot{q}_1)^2 + \frac{1}{3}(\ddot{q}_1+\ddot{q}_2) \end{bmatrix} \\ +\alpha^2\mu_m &\begin{bmatrix} \ddot{q}_1 + \cos q_2(\ddot{q}_1+\ddot{q}_2) - \sin q_2\,(\dot{q}_1+\dot{q}_2)^2 \\ +\cos(q_2)\ddot{q}_1 + \sin q_2\,(\dot{q}_1)^2 + (\ddot{q}_1+\ddot{q}_2) \\ \cos(q_2)\ddot{q}_1 + \sin q_2\,(\dot{q}_1)^2 + (\ddot{q}_1+\ddot{q}_2) \end{bmatrix}.\end{aligned} \tag{3.6.56}$$

In the case of point-mass rods, (3.6.56) is replaced by

$$\begin{aligned}\frac{\tau^2}{\mu\ell^2}\mathbf{J}_{\text{pm}} = \alpha^2\mu_\rho &\begin{bmatrix} \frac{5}{4}\ddot{q}_1 + \frac{1}{2}\cos q_2(\ddot{q}_1+\ddot{q}_2) - \frac{1}{2}\sin q_2\,(\dot{q}_1+\dot{q}_2)^2 \\ +\frac{1}{2}\cos(q_2)\ddot{q}_1 + \frac{1}{2}\sin q_2\,(\dot{q}_1)^2 + \frac{1}{4}(\ddot{q}_1+\ddot{q}_2) \\ \frac{1}{2}\cos(q_2)\ddot{q}_1 + \frac{1}{2}\sin q_2\,(\dot{q}_1)^2 + \frac{1}{4}(\ddot{q}_1+\ddot{q}_2) \end{bmatrix} \\ +\alpha^2\mu_m &\begin{bmatrix} \ddot{q}_1 + \cos q_2(\ddot{q}_1+\ddot{q}_2) - \sin q_2\,(\dot{q}_1+\dot{q}_2)^2 \\ +\cos(q_2)\ddot{q}_1 + \sin q_2\,(\dot{q}_1)^2 + (\ddot{q}_1+\ddot{q}_2) \\ \cos(q_2)\ddot{q}_1 + \sin q_2\,(\dot{q}_1)^2 + (\ddot{q}_1+\ddot{q}_2) \end{bmatrix},\end{aligned} \tag{3.6.57}$$

and the other equations remain the same.

The next task is to fix the reference values τ, ℓ and μ. Consider a related system: a simple pendulum consisting of a massless string of length $\ell = L$, and a bob of arbitrary nonzero mass, the value of which will not matter here. Let it hang vertically, subjected to a uniform gravitational field having acceleration g. As usual, we take one end of the pendulum to be fixed, and one end free.

In accordance with this idea, we choose $\ell = L$, leading to $\alpha = 1$, and $\tau = \sqrt{\ell/g}$, leading to $G = 1$. Note $\tau = t_0/2\pi$, where t_0 is the period of oscillation of the simple pendulum undergoing oscillations of a small amplitude. This eliminates two parameters, namely α and G, by making them equal unity, suggesting that these values of ℓ and τ are the natural units for length and time in the system under study.

This still leaves open the choice of the reference mass. All the remaining parameters have a factor of $1/\mu$; hence the relative scale of these parameters to α and G, which do not involve μ, will be fixed by the choice of μ. If we desire to explore all possible special cases of the system under study, any of the remaining six parameters, of which five are independent, may be reduced to zero.

One possibility for fixing μ is to return to the system under study, and note that it makes no physical sense to set both ρ and m simultaneously to zero. A completely massless system lacks any inertia, whence its physical behavior is qualitatively so radically different from a system with any nonzero mass, that no meaningful direct comparison can be made.[2] This suggests the choice $\mu = \rho L + m$, which is always nonzero. However, the mass of the dead weight (m) is essentially a loading parameter (via (3.6.53)), and hence in a parametric study, one must be able to vary it without affecting the values of the other parameters. This is not possible if $\mu = \rho L + m$, because the scaling

$$\ell = L\ , \quad \tau = \sqrt{\ell/g}\ , \quad \mu = \rho L + m$$

leads to the parameterization

$$\alpha = 1\ , \quad \beta = \frac{c}{(\rho L + m)\sqrt{g}\,L^{3/2}}\ , \quad \gamma = \frac{k}{(\rho L + m)gL}\ ,$$

$$\mu_m = \frac{m}{\rho L + m}\ , \quad \mu_\rho = \frac{\rho L}{\rho L + m}\ , \quad G = 1\ , \quad \Pi_j = \frac{P_j}{(\rho L + m)g} \quad (\text{where } j = F, W)\ ,$$

where varying m will affect the values of six parameters.

Hence, we will choose a different value for μ. There are only two sources of mass in the system, namely the rods (ρL each) and the dead weight (m). If ρ is allowed to be zero, and m is a loading parameter, neither of these can be utilized to fix μ.

An indirect approach, utilizing the other remaining parameters, is also possible. If we require one of them (any parameter that references μ) to equal unity, we can use that to define a natural reference mass, but there is a cost. When a parameter is made equal to unity, it can no longer be varied in a parametric study, and especially, it cannot be made to tend to zero. Observing that α and G were already set to unity

---

[2] As is seen from (3.6.54)–(3.6.57), setting $\mu_\rho = \mu_m = 0$ completely eliminates the inertial generalized force, and thus removes all occurrences of second time derivatives from the dynamical equation. It also eliminates the external force in the dead-weight case. In such a system, time evolution is driven by the damping terms in (3.6.54). If there is no damping, then there are no time derivatives anywhere in the dynamical equation, and it becomes simply a set of constraints for admissible combinations of q_1 and q_2. In the follower-force case, for example, the admissible positions form a continuous path, by the continuity of (3.6.54) and (3.6.55), possibly branching at certain points as q_2 increases, due to the sine in (3.6.55).

(and in any case, did not reference μ), and μ_m, Π_W and Π_F are related to loading which we must be able to vary, we are left with the candidates β, γ, and μ_ρ.

However, we would like to be able to vary each of these three parameters, and to zero out some of them for special cases. Hence, there are no parameters left that we are at liberty to fix to unity, that could be used to define a reference mass μ.

The conclusion is that it is not possible for this system to choose a natural mass while keeping the nondimensional representation fully configurable. We may either take an arbitrary reference, such as $\mu = 1\,\text{kg}$ lacking any physical motivation, or choose $\mu = \rho L$, and accept the fact that the special case of massless rods will be inaccessible to the parameterization. Here we make the latter choice.

We conclude that the natural units of length, time and mass for the system under study are

$$\ell = L \;, \quad \tau = \sqrt{\ell/g} \;, \quad \mu = \rho L \;. \tag{3.6.58}$$

Inserting (3.6.58) into (3.6.52), we obtain the final nondimensional parameters as

$$\alpha := 1 \;, \quad \beta := \frac{c}{\rho\sqrt{g}\,L^{5/2}} \;, \quad \gamma := \frac{k}{\rho g L^2} \;, \quad \mu_m := \frac{m}{\rho L} \;, \quad \mu_\rho := 1 \;, \quad G := 1 \;, \quad \Pi_j := \frac{P_j}{\rho L g}$$
$$\text{(where } j = F, W) \;. \tag{3.6.59}$$

Equation (3.6.53) still holds (*a fortiori*). We are left with five parameters that are not identically equal to unity, four of which are independent.

Consider the range of validity of the definitions (3.6.59). The quantities appearing in the denominators are ρ, g, and L. Arguably, the requirement $L > 0$ is reasonable in order for the problem setup to make any geometric sense. If we exclude the case of massless rods, both ρ and L can be reasonably considered nonzero. As for g, if the system is configured vertically, g is a natural choice of reference. However, for a horizontally configured system, upon which gravity has no effect, using the acceleration of gravity as part of a reference constant makes no physical sense. It seems unreasonable to require specifying g in this case, just to be able to compute numerical values for (3.6.59). It would make more physical sense to calibrate a horizontally configured system using elastic vibrations; but this is possible only if the spring stiffness k is nonzero.

This issue cannot be worked around completely, because when gravity, spring stiffness and damping are all absent, no process exists that would conveniently define a physically meaningful characteristic timescale for the system. Also, generally speaking, a unified framework in which to make direct comparisons only makes sense for a set of situations which are in some sense directly comparable.

Hence, we may just as well simply use g, regardless of whether the system is configured vertically or horizontally, or in an inclined configuration in between, allowing to use the same scaling for all situations.

Beside massless rods, there is one notable special case where the parameterization (3.6.59) is not meaningful: a zero-gravity environment. This is because due to our definition of τ, a nonzero g is needed. The characteristic time is calibrated to the period of a vertical reference pendulum *in the same environment* as the system under study. Obviously, in a zero-gravity environment, the reference must be chosen in some other way. (Equation (3.6.52) remain valid, with $g = 0$.) In this chapter we will only consider environments with nonzero g.

In practical numerical work with the nondimensional form (3.6.5), (3.6.54)–(3.6.57), one takes the left-hand sides of (3.6.59) as the primary parameters of interest, and does not care about the right-hand sides. The definitions (3.6.59) simply provide mappings, from the set of physical quantities appearing on the right-hand sides (c, k, ρ, g, L, m, P_W and P_F), to the abstract mathematical quantities $\beta, \gamma, \mu_m, \Pi_W$ and Π_F. These mathematical quantities are then those that are varied in a parametric study. It is then easy to determine, using these mappings, which point of the parametric study any given physical instance of the system corresponds to.

3.6.4 Energy Considerations

Considering the total energy of the system under study is not only interesting from a physical viewpoint, but it can also act as an indicator to help determine the validity of computed numerical solutions. The total energy can be written as the sum of kinetic, gravitational potential and elastic potential contributions:

$$\begin{aligned} E &= K + V_{\mathrm{g}} + V_{\mathrm{elastic}} \\ &= \left(\frac{1}{2}\int \dot{\mathbf{x}}^2\,\mathrm{d}m + \frac{1}{2}m\dot{\mathbf{x}}_{\mathrm{m}}^2\right) + \cos(\theta)\left(g\int h(\xi)\,\mathrm{d}m + mgh(\mathbf{x}_{\mathrm{m}})\right) + \left(\frac{1}{2}kq_1^2 + \frac{1}{2}kq_2^2\right). \end{aligned} \tag{3.6.60}$$

The integrals are taken over both rods. As before, the dead weight at the free end of the system has mass m. The location of the free end is denoted by $\mathbf{x}_{\mathrm{m}}$. The quantity $h(\xi)$ is the height of the point ξ, as measured from a reference level. The reference level is arbitrary; gravitational potential energy is only unique up to an additive constant. We fix the reference at the level of the fixed end of the system, $y = 0$.

The mass differential is

$$\mathrm{d}m = \rho\,\mathrm{d}\xi\,, \tag{3.6.61}$$

and thus we have

$$E = \frac{1}{2}\rho\int \dot{\mathbf{x}}^2\,\mathrm{d}\xi + \frac{1}{2}m\dot{\mathbf{x}}_{\mathrm{m}}^2 + \cos(\theta)\rho g\int h(\xi)\,\mathrm{d}\xi + \cos(\theta)mgh(\mathbf{x}_{\mathrm{m}}) + \frac{1}{2}kq_1^2 + \frac{1}{2}kq_2^2\,. \tag{3.6.62}$$

As was determined earlier (Eqs. (3.6.27)–(3.6.32)), for the first rod,

$$\mathbf{x} = \begin{bmatrix} \sin q_1 \\ \cos q_1 \end{bmatrix} \xi \,, \tag{3.6.63}$$

$$\dot{\mathbf{x}} = \begin{bmatrix} \cos q_1 \\ -\sin q_1 \end{bmatrix} \xi \dot{q}_1 \,, \tag{3.6.64}$$

and for the second rod,

$$\mathbf{x} = \begin{bmatrix} \sin q_1 \\ \cos q_1 \end{bmatrix} L + \begin{bmatrix} \sin(q_1 + q_2) \\ \cos(q_1 + q_2) \end{bmatrix} \xi \,, \tag{3.6.65}$$

$$\dot{\mathbf{x}} = \begin{bmatrix} \cos q_1 \\ -\sin q_1 \end{bmatrix} L\dot{q}_1 + \begin{bmatrix} \cos(q_1 + q_2) \\ -\sin(q_1 + q_2) \end{bmatrix} \xi(\dot{q}_1 + \dot{q}_2) \,. \tag{3.6.66}$$

For the dead-weight load,

$$\mathbf{x}_\mathrm{m} = \begin{bmatrix} \sin q_1 \\ \cos q_1 \end{bmatrix} L + \begin{bmatrix} \sin(q_1 + q_2) \\ \cos(q_1 + q_2) \end{bmatrix} L \,, \tag{3.6.67}$$

and

$$\dot{\mathbf{x}}_\mathrm{m} = \begin{bmatrix} \cos q_1 \\ -\sin q_1 \end{bmatrix} L\dot{q}_1 + \begin{bmatrix} \cos(q_1 + q_2) \\ -\sin(q_1 + q_2) \end{bmatrix} L(\dot{q}_1 + \dot{q}_2) \,. \tag{3.6.68}$$

Inserting (3.6.64) and (3.6.66) to the kinetic energy of the rods in (3.6.62), we obtain

$$\begin{aligned} E = {} & \frac{1}{2}\rho\Big(\dot{q}_1^2 \int_0^L \xi^2 \,\mathrm{d}\xi + L^2 \dot{q}_1^2 \int_0^L 1 \,\mathrm{d}\xi \\ & + 2L\dot{q}_1(\dot{q}_1 + \dot{q}_2) \left[\cos(q_1)\cos(q_1 + q_2) + \sin(q_1)\sin(q_1 + q_2)\right] \int_0^L \xi \,\mathrm{d}\xi \\ & + (\dot{q}_1 + \dot{q}_2)^2 \int_0^L \xi^2 \,\mathrm{d}\xi \Big) \\ & + \frac{1}{2} m \dot{\mathbf{x}}_\mathrm{m}^2 + \cos(\theta)\rho g \int h(\xi) \,\mathrm{d}\xi + \cos(\theta) m g h(\mathbf{x}_\mathrm{m}) + \frac{1}{2} k q_1^2 + \frac{1}{2} k q_2^2 \,. \end{aligned}$$

Using the trigonometric identity $\cos(a)\cos(b) + \sin(a)\sin(b) = \cos(a - b) = \cos(b - a)$, evaluating the integrals, and collecting terms, we obtain

$$\begin{aligned} E = {} & \frac{1}{2}\rho L^3 \Big(\frac{4}{3}\dot{q}_1^2 + \cos(q_2)\dot{q}_1(\dot{q}_1 + \dot{q}_2) + \frac{1}{3}(\dot{q}_1 + \dot{q}_2)^2 \Big) \\ & + \frac{1}{2} m \dot{\mathbf{x}}_\mathrm{m}^2 + \cos(\theta)\rho g \int h(\xi) \,\mathrm{d}\xi + \cos(\theta) m g h(\mathbf{x}_\mathrm{m}) + \frac{1}{2} k q_1^2 + \frac{1}{2} k q_2^2 \,. \end{aligned} \tag{3.6.69}$$

Next, we deal with the kinetic energy of the dead weight by (3.6.68). We have

$$\frac{1}{2}m\dot{\mathbf{x}}_{\mathrm{m}}^2 = \frac{1}{2}mL^2\left(\dot{q}_1^2 + 2\cos(q_2)\dot{q}_1(\dot{q}_1 + \dot{q}_2) + (\dot{q}_1 + \dot{q}_2)^2\right) , \tag{3.6.70}$$

and thus the total energy becomes

$$\begin{aligned} E = {} & \frac{1}{2}\rho L^3\left(\frac{4}{3}\dot{q}_1^2 + \cos(q_2)\dot{q}_1(\dot{q}_1 + \dot{q}_2) + \frac{1}{3}(\dot{q}_1 + \dot{q}_2)^2\right) \\ & + \frac{1}{2}mL^2\left(\dot{q}_1^2 + 2\cos(q_2)\dot{q}_1(\dot{q}_1 + \dot{q}_2) + (\dot{q}_1 + \dot{q}_2)^2\right) \\ & + \cos(\theta)\rho g\int h(\xi)\,\mathrm{d}\xi + \cos(\theta)mgh(\mathbf{x}_{\mathrm{m}}) + \frac{1}{2}kq_1^2 + \frac{1}{2}kq_2^2 . \end{aligned} \tag{3.6.71}$$

The products ρL^3 and mL^2 have the same dimension, because m is mass, while ρ is mass per unit length.

To finish, we compute the potential energy contributions. For our choice of the reference level, the function h simply picks the y component (second component) of the position vectors (3.6.63), (3.6.65) and (3.6.67). We have

$$\begin{aligned} \cos(\theta)\rho g\int h(\xi)\,\mathrm{d}\xi &= \cos(\theta)\rho g\left(\cos(q_1)\int_0^L \xi\,\mathrm{d}\xi + L\cos(q_1)\int_0^L 1\,\mathrm{d}\xi + \cos(q_1 + q_2)\int_0^L \xi\,\mathrm{d}\xi\right) \\ &= \cos(\theta)\rho g L^2\left(\frac{3}{2}\cos(q_1) + \frac{1}{2}\cos(q_1 + q_2)\right) . \end{aligned}$$

For the dead weight, we obtain

$$\cos(\theta)mgh(\mathbf{x}_{\mathrm{m}}) = \cos(\theta)mgL\left(\cos(q_1) + \cos(q_1 + q_2)\right) .$$

Therefore, the final result for the total energy is

$$\begin{aligned} E = {} & \frac{1}{2}\rho L^3\left(\frac{4}{3}\dot{q}_1^2 + \cos(q_2)\dot{q}_1(\dot{q}_1 + \dot{q}_2) + \frac{1}{3}(\dot{q}_1 + \dot{q}_2)^2\right) \\ & + \frac{1}{2}mL^2\left(\dot{q}_1^2 + 2\cos(q_2)\dot{q}_1(\dot{q}_1 + \dot{q}_2) + (\dot{q}_1 + \dot{q}_2)^2\right) \\ & + \cos(\theta)\rho g L^2\left(\frac{3}{2}\cos(q_1) + \frac{1}{2}\cos(q_1 + q_2)\right) \\ & + \cos(\theta)mgL\left(\cos(q_1) + \cos(q_1 + q_2)\right) \\ & + \frac{1}{2}kq_1^2 + \frac{1}{2}kq_2^2 . \end{aligned} \tag{3.6.72}$$

For the model variant with point-mass rods, the mass distribution is

$$\mathrm{d}m = \rho L\,\delta(\xi - \frac{1}{2}L)\,\mathrm{d}\xi , \tag{3.6.73}$$

where $\delta(\dots)$ denotes the Dirac delta. Equation (3.6.62) is thus replaced by

$$E_{\text{pm}} = \frac{1}{2}\rho L \int \dot{\mathbf{x}}^2\, \delta(\xi - \frac{1}{2}L)\,\mathrm{d}\xi + \frac{1}{2}m\dot{\mathbf{x}}_{\text{m}}^2 + \cos(\theta)\rho L g \int h(\xi)\, \delta(\xi - \frac{1}{2}L)\,\mathrm{d}\xi + \cos(\theta) m g h(\mathbf{x}_{\text{m}}) + \frac{1}{2}kq_1^2 + \frac{1}{2}kq_2^2\,.$$

We obtain

$$\begin{aligned} E_{\text{pm}} = \frac{1}{2}\rho L\Big(&\dot{q}_1^2 \int_0^L \xi^2\, \delta(\xi - \frac{1}{2}L)\,\mathrm{d}\xi + L^2\dot{q}_1^2 \int_0^L \delta(\xi - \frac{1}{2}L)\,\mathrm{d}\xi \\ &+ 2L\dot{q}_1(\dot{q}_1 + \dot{q}_2)\,[\cos(q_1)\cos(q_1+q_2) + \sin(q_1)\sin(q_1+q_2)] \int_0^L \xi\, \delta(\xi - \frac{1}{2}L)\,\mathrm{d}\xi \\ &+ (\dot{q}_1 + \dot{q}_2)^2 \int_0^L \xi^2\, \delta(\xi - \frac{1}{2}L)\,\mathrm{d}\xi\Big) \\ &+ \frac{1}{2}m\dot{\mathbf{x}}_{\text{m}}^2 + \cos(\theta)\rho L g \int h(\xi)\, \delta(\xi - \frac{1}{2}L)\,\mathrm{d}\xi + \cos(\theta) m g h(\mathbf{x}_{\text{m}}) + \frac{1}{2}kq_1^2 + \frac{1}{2}kq_2^2\,. \end{aligned}$$

Evaluating the integrals,

$$\begin{aligned} E_{\text{pm}} = \frac{1}{2}\rho L\Big(&\dot{q}_1^2 \cdot \frac{1}{4}L^2 + L^2\dot{q}_1^2 \cdot 1 \\ &+ 2L\dot{q}_1(\dot{q}_1 + \dot{q}_2)\,[\cos(q_1)\cos(q_1+q_2) + \sin(q_1)\sin(q_1+q_2)] \cdot \frac{1}{2}L \\ &+ (\dot{q}_1 + \dot{q}_2)^2 \cdot \frac{1}{4}L^2\Big) \\ &+ \frac{1}{2}m\dot{\mathbf{x}}_{\text{m}}^2 + \cos(\theta)\rho L g \int h(\xi)\, \delta(\xi - \frac{1}{2}L)\,\mathrm{d}\xi + \cos(\theta) m g h(\mathbf{x}_{\text{m}}) + \frac{1}{2}kq_1^2 + \frac{1}{2}kq_2^2\,. \end{aligned}$$

Again, collecting terms and using the trigonometric identity $\cos(a)\cos(b) + \sin(a)\sin(b) = \cos(a-b) = \cos(b-a)$, we have

$$\begin{aligned} E_{\text{pm}} = &\frac{1}{2}\rho L^3\Big(\frac{5}{4}\dot{q}_1^2 + \cos(q_2)\dot{q}_1(\dot{q}_1 + \dot{q}_2) + \frac{1}{4}(\dot{q}_1 + \dot{q}_2)^2\Big) \\ &+ \frac{1}{2}m\dot{\mathbf{x}}_{\text{m}}^2 + \cos(\theta)\rho L g \int h(\xi)\, \delta(\xi - \frac{1}{2}L)\,\mathrm{d}\xi + \cos(\theta) m g h(\mathbf{x}_{\text{m}}) + \frac{1}{2}kq_1^2 + \frac{1}{2}kq_2^2\,. \end{aligned}$$

As for the gravitational contribution of the rods, we have

$$\begin{aligned} \cos(\theta)\rho L g \int h(\xi)\, \delta(\xi - \frac{1}{2}L)\,\mathrm{d}\xi &= \cos(\theta)\rho L g\Big(\cos(q_1) \int_0^L \xi\, \delta(\xi - \frac{1}{2}L)\,\mathrm{d}\xi \\ &\quad + L\cos(q_1) \int_0^L \delta(\xi - \frac{1}{2}L)\,\mathrm{d}\xi + \cos(q_1+q_2) \int_0^L \xi\, \delta(\xi - \frac{1}{2}L)\,\mathrm{d}\xi\Big) \\ &= \cos(\theta)\rho L g\Big(\cos(q_1) \cdot \frac{1}{2}L + L\cos(q_1) \cdot 1 + \cos(q_1+q_2) \cdot \frac{1}{2}L\Big) \\ &= \cos(\theta)\rho g L^2\Big(\frac{3}{2}\cos(q_1) + \frac{1}{2}\cos(q_1+q_2)\Big)\,. \end{aligned}$$

We see that the gravitational contribution remains the same.

The final total energy for model variant with point-mass rods is thus

$$\begin{aligned} E_{\text{pm}} &= \frac{1}{2}\rho L^3\left(\frac{5}{4}\dot{q}_1^2 + \cos(q_2)\dot{q}_1(\dot{q}_1+\dot{q}_2) + \frac{1}{4}(\dot{q}_1+\dot{q}_2)^2\right) \\ &+ \frac{1}{2}mL^2\left(\dot{q}_1^2 + 2\cos(q_2)\dot{q}_1(\dot{q}_1+\dot{q}_2) + (\dot{q}_1+\dot{q}_2)^2\right) \\ &+ \cos(\theta)\rho g L^2\left(\frac{3}{2}\cos(q_1) + \frac{1}{2}\cos(q_1+q_2)\right) \\ &+ \cos(\theta) m g L\left(\cos(q_1) + \cos(q_1+q_2)\right) \\ &+ \frac{1}{2}kq_1^2 + \frac{1}{2}kq_2^2\,. \end{aligned} \tag{3.6.74}$$

Note the 5/4 and the 1/4, both on the first line; everything else remains the same.

Nondimensional total energy

Let us express the total energy in nondimensional form. Multiplying both sides of Eq. (3.6.72) by $\tau^2/\mu\ell^2$, we obtain

$$\begin{aligned} \widetilde{E} := \frac{\tau^2}{\mu\ell^2}E &= \frac{1}{2}\frac{\tau^2\rho L^3}{\mu\ell^2}\left(\frac{4}{3}\dot{q}_1^2 + \cos(q_2)\dot{q}_1(\dot{q}_1+\dot{q}_2) + \frac{1}{3}(\dot{q}_1+\dot{q}_2)^2\right) \\ &+ \frac{1}{2}\frac{\tau^2 mL^2}{\mu\ell^2}\left(\dot{q}_1^2 + 2\cos(q_2)\dot{q}_1(\dot{q}_1+\dot{q}_2) + (\dot{q}_1+\dot{q}_2)^2\right) \\ &+ \cos(\theta)\frac{\tau^2\rho g L^2}{\mu\ell^2}\left(\frac{3}{2}\cos(q_1) + \frac{1}{2}\cos(q_1+q_2)\right) \\ &+ \cos(\theta)\frac{\tau^2 m g L}{\mu\ell^2}\left(\cos(q_1) + \cos(q_1+q_2)\right) \\ &+ \frac{1}{2}\frac{\tau^2 k}{\mu\ell^2}q_1^2 + \frac{1}{2}\frac{\tau^2 k}{\mu\ell^2}q_2^2\,. \end{aligned}$$

After applying $\gamma = \tau^2 k/\mu\ell^2$ and reorganizing,

$$\begin{aligned} \widetilde{E} &= \frac{1}{2}\tau^2\frac{\rho L}{\mu}\frac{L^2}{\ell^2}\left(\frac{4}{3}\dot{q}_1^2 + \cos(q_2)\dot{q}_1(\dot{q}_1+\dot{q}_2) + \frac{1}{3}(\dot{q}_1+\dot{q}_2)^2\right) \\ &+ \frac{1}{2}\tau^2\frac{m}{\mu}\frac{L^2}{\ell^2}\left(\dot{q}_1^2 + 2\cos(q_2)\dot{q}_1(\dot{q}_1+\dot{q}_2) + (\dot{q}_1+\dot{q}_2)^2\right) \\ &+ \cos(\theta)\frac{g\tau^2}{\ell}\frac{\rho L}{\mu}\frac{L}{\ell}\left(\frac{3}{2}\cos(q_1) + \frac{1}{2}\cos(q_1+q_2)\right) \\ &+ \cos(\theta)\frac{g\tau^2}{\ell}\frac{m}{\mu}\frac{L}{\ell}\left(\cos(q_1) + \cos(q_1+q_2)\right) \\ &+ \frac{1}{2}\gamma q_1^2 + \frac{1}{2}\gamma q_2^2\,. \end{aligned}$$

Using the definitions of the nondimensional quantities in (3.6.52), and combining some terms, the nondimensional total energy becomes

$$\widetilde{E} = \frac{1}{2}\tau^2\alpha^2\left(\left[\frac{4}{3}\mu_\rho + \mu_m\right]\dot{q}_1^2 + \left[\mu_\rho + 2\mu_m\right]\cos(q_2)\dot{q}_1(\dot{q}_1 + \dot{q}_2) + \left[\frac{1}{3}\mu_\rho + \mu_m\right](\dot{q}_1 + \dot{q}_2)^2\right)$$
$$+ \cos(\theta)G\alpha\left(\left[\frac{3}{2}\mu_\rho + \mu_m\right]\cos(q_1) + \left[\frac{1}{2}\mu_\rho + \mu_m\right]\cos(q_1 + q_2)\right)$$
$$+ \frac{1}{2}\gamma\left(q_1^2 + q_2^2\right), \tag{3.6.75}$$

which clearly shows the split to kinetic, gravitational potential and elastic potential contributions.

Finally, note that $[\tau] = \mathrm{s}$, and the time differentiations in the $\dot{q}_j$ are still performed in dimensional time. Performing the same time scaling as before, each time differentiation yields a factor of $1/\tau$, yielding the final result:

$$\widetilde{E} = \frac{1}{2}\alpha^2\left(\left[\frac{4}{3}\mu_\rho + \mu_m\right]\dot{q}_1^2 + \left[\mu_\rho + 2\mu_m\right]\cos(q_2)\dot{q}_1(\dot{q}_1 + \dot{q}_2) + \left[\frac{1}{3}\mu_\rho + \mu_m\right](\dot{q}_1 + \dot{q}_2)^2\right)$$
$$+ \cos(\theta)G\alpha\left(\left[\frac{3}{2}\mu_\rho + \mu_m\right]\cos(q_1) + \left[\frac{1}{2}\mu_\rho + \mu_m\right]\cos(q_1 + q_2)\right)$$
$$+ \frac{1}{2}\gamma\left(q_1^2 + q_2^2\right), \tag{3.6.76}$$

where also the time is now dimensionless. The result (3.6.76) is written in the general parametrization (3.6.52).

In the natural parametrization (3.6.59), which uses the specific choices

$$\ell = L\,, \quad \tau = \sqrt{\ell/g}\,, \quad \mu = \rho L\,, \tag{3.6.77}$$

implying, as we recall, $\alpha = 1$, $G = 1$, $\mu_\rho = 1$, the nondimensional total energy (3.6.76) becomes

$$\widetilde{E} = \frac{1}{2}\left(\left[\frac{4}{3} + \mu_m\right]\dot{q}_1^2 + [1 + 2\mu_m]\cos(q_2)\dot{q}_1(\dot{q}_1 + \dot{q}_2) + \left[\frac{1}{3} + \mu_m\right](\dot{q}_1 + \dot{q}_2)^2\right)$$
$$+ \cos(\theta)\left(\left[\frac{3}{2} + \mu_m\right]\cos(q_1) + \left[\frac{1}{2} + \mu_m\right]\cos(q_1 + q_2)\right)$$
$$+ \frac{1}{2}\gamma\left(q_1^2 + q_2^2\right). \tag{3.6.78}$$

For the model variant with point-mass rods, Eq. (3.6.76) and (3.6.78) are replaced, respectively, by

$$\widetilde{E}_{\rm pm} = \frac{1}{2}\alpha^2\left(\left[\frac{5}{4}\mu_\rho + \mu_m\right]\dot{q}_1^2 + \left[\mu_\rho + 2\mu_m\right]\cos(q_2)\dot{q}_1(\dot{q}_1 + \dot{q}_2) + \left[\frac{1}{4}\mu_\rho + \mu_m\right](\dot{q}_1 + \dot{q}_2)^2\right)$$
$$+ \cos(\theta)G\alpha\left(\left[\frac{3}{2}\mu_\rho + \mu_m\right]\cos(q_1) + \left[\frac{1}{2}\mu_\rho + \mu_m\right]\cos(q_1 + q_2)\right)$$
$$+ \frac{1}{2}\gamma\left(q_1^2 + q_2^2\right) \tag{3.6.79}$$

and

$$\widetilde{E}_{\rm pm} = \frac{1}{2}\left(\left[\frac{5}{4} + \mu_m\right]\dot{q}_1^2 + [1 + 2\mu_m]\cos(q_2)\dot{q}_1(\dot{q}_1 + \dot{q}_2) + \left[\frac{1}{4} + \mu_m\right](\dot{q}_1 + \dot{q}_2)^2\right)$$
$$+ \cos(\theta)\left(\left[\frac{3}{2} + \mu_m\right]\cos(q_1) + \left[\frac{1}{2} + \mu_m\right]\cos(q_1 + q_2)\right)$$
$$+ \frac{1}{2}\gamma\left(q_1^2 + q_2^2\right) . \tag{3.6.80}$$

Internal mechanical power

The instantaneous power input to the system is

$$p = \frac{\mathrm{d}E}{\mathrm{d}t} . \tag{3.6.81}$$

We denote the power by a lowercase p, because uppercase P denotes the external forces.

We start from the nondimensional form. By differentiating the nondimensional total energy (3.6.76) with respect to the dimensionless time, and noting that the state variables $q_1, \dot{q}_1, q_2, \dot{q}_2$ are functions of time (hence we apply the chain rule), we have

$$\widetilde{p} = \alpha^2\left(\left[\frac{4}{3}\mu_\rho + \mu_m\right]\dot{q}_1\ddot{q}_1\right.$$
$$+ \frac{1}{2}\left[\mu_\rho + 2\mu_m\right]\left[-\sin(q_2)\dot{q}_1\dot{q}_2(\dot{q}_1 + \dot{q}_2) + \cos(q_2)\big(\ddot{q}_1(\dot{q}_1 + \dot{q}_2) + \dot{q}_1(\ddot{q}_1 + \ddot{q}_2)\big)\right]$$
$$\left.+ \left[\frac{1}{3}\mu_\rho + \mu_m\right](\dot{q}_1 + \dot{q}_2)(\ddot{q}_1 + \ddot{q}_2)\right)$$
$$- \cos(\theta)G\alpha\left(\left[\frac{3}{2}\mu_\rho + \mu_m\right]\sin(q_1)\dot{q}_1 + \left[\frac{1}{2}\mu_\rho + \mu_m\right]\sin(q_1 + q_2)(\dot{q}_1 + \dot{q}_2)\right)$$
$$+ \gamma\left(q_1\dot{q}_1 + q_2\dot{q}_2\right) . \tag{3.6.82}$$

For the model variant with point-mass rods, Eq. (3.6.82) is replaced by

$$
\begin{aligned}
\widetilde{p}_{\mathrm{pm}} = \alpha^2 \Bigg(& \left[\frac{5}{4}\mu_\rho + \mu_m \right] \dot{q}_1 \ddot{q}_1 \\
& + \frac{1}{2} \left[\mu_\rho + 2\mu_m \right] \left[-\sin(q_2)\dot{q}_1\dot{q}_2(\dot{q}_1 + \dot{q}_2) + \cos(q_2)\big(\ddot{q}_1(\dot{q}_1 + \dot{q}_2) + \dot{q}_1(\ddot{q}_1 + \ddot{q}_2)\big) \right] \\
& + \left[\frac{1}{4}\mu_\rho + \mu_m \right] (\dot{q}_1 + \dot{q}_2)(\ddot{q}_1 + \ddot{q}_2) \Bigg) \\
& - \cos(\theta) G \alpha \left(\left[\frac{3}{2}\mu_\rho + \mu_m \right] \sin(q_1)\dot{q}_1 + \left[\frac{1}{2}\mu_\rho + \mu_m \right] \sin(q_1 + q_2)(\dot{q}_1 + \dot{q}_2) \right) \\
& + \gamma \left(q_1 \dot{q}_1 + q_2 \dot{q}_2 \right) . \qquad (3.6.83)
\end{aligned}
$$

Note the $5/4$ and the $1/4$; everything else remains the same.

Observe that while the gravitational and elastic power can be evaluated using only the state vector, the kinetic power requires access to $\ddot{q}_1$ and $\ddot{q}_2$. In a practical numerical implementation, this can be obtained by evaluating the right-hand side in the ODE system solver for the dynamics of the system, using the state vector data.

Depending on numerical methods used, one may need to be very careful here. For example, in the time-discontinuous Galerkin (dG) method, the state vector $(q_1, \dot{q}_1, q_2, \dot{q}_2)$, as a function of time, belongs to the class C^{-1} of finitely discontinuous functions. Thus all the above energy expressions, and the parts of the power expression that depend only on $(q_1, \dot{q}_1, q_2, \dot{q}_2)$, also belong to this class.

Therefore, the time derivative of the state vector—which is needed by the kinetic power contribution in (3.6.82)—has Dirac deltas at the timestep boundaries. The $\ddot{q}_j$ terms appear only as linear factors, so the kinetic power is indeed integrable (recalling that the Dirac delta is in L^1, but not in L^2 or higher). A better approach is to observe that there is no need to numerically integrate the kinetic power, because we already have an expression in (3.6.76) for the exact value of its integral, i.e., the kinetic energy.

External mechanical power

There are two external sources of mechanical power to consider: the dampers, which dissipate energy, and the follower force, which may either add or remove energy depending on the state of the system.

Each damper dissipates energy at the power

$$
p_{\mathrm{d}} = M \, \dot{q} = (c \, \dot{q})\dot{q} = c \, \dot{q}^2 , \qquad (3.6.84)
$$

where M is the moment imposed by the damper, and c is the damping coefficient. Here dissipation is taken as positive.

Thus, for the system with two dampers the dissipation power is

$$
p_{\mathrm{d}} = c \, (\dot{q}_1^2 + \dot{q}_2^2) . \qquad (3.6.85)
$$

To nondimensionalize, we multiply by $\tau^3 / \mu \ell^2$:

$$\widetilde{p}_{\mathrm{d}} := \frac{\tau^3}{\mu\ell^2} p_{\mathrm{d}} = \frac{c\tau}{\mu\ell^2}\tau^2(\dot{q}_1^2 + \dot{q}_2^2) = \beta\tau^2(\dot{q}_1^2 + \dot{q}_2^2)\ .$$

Converting the $\dot{q}$ to dimensionless time, we have the nondimensional dissipation power

$$\widetilde{p}_{\mathrm{d}} = \beta\,(\dot{q}_1^2 + \dot{q}_2^2)\ . \tag{3.6.86}$$

Let us next consider the power input by the follower force. In general, mechanical power in linear motion is

$$p = \mathbf{F}\cdot\mathbf{v}\ . \tag{3.6.87}$$

In our case, $\mathbf{F}$ is the follower force, and $\mathbf{v}$ is the velocity vector of the point where the force is applied, i.e. the free end of the system. The position of the free end, $\mathbf{x}_{\mathrm{m}}$, and its velocity vector, $\dot{\mathbf{x}}_{\mathrm{m}}$, were already determined for the dead weight load, so we may simply re-use the results, Eqs. (3.6.67)–(3.6.68).

The follower force points inward along the second rod, which is at an angle $q_1 + q_2$ to the $+y$ axis. This was already given in Eq. (3.6.11) at the beginning, repeated here for convenience:

$$\mathbf{P} = -P\begin{bmatrix}\sin(q_1+q_2)\\ \cos(q_1+q_2)\end{bmatrix}\ . \tag{3.6.88}$$

Here P is the magnitude of the follower force, positive taken inward. Hence

$$\begin{aligned} p_{\mathrm{F}} &= \mathbf{P}\cdot\dot{\mathbf{x}}_{\mathrm{m}} \\ &= -P\begin{bmatrix}\sin(q_1+q_2)\\ \cos(q_1+q_2)\end{bmatrix}\cdot\left(\begin{bmatrix}\cos q_1\\ -\sin q_1\end{bmatrix}L\dot{q}_1 + \begin{bmatrix}\cos(q_1+q_2)\\ -\sin(q_1+q_2)\end{bmatrix}L(\dot{q}_1+\dot{q}_2)\right) \\ &= -PL\dot{q}_1\,[\sin(q_1+q_2)\cos(q_1) - \cos(q_1+q_2)\sin(q_1))] \\ &\quad - PL(\dot{q}_1+\dot{q}_2)\,[\sin(q_1+q_2)\cos(q_1+q_2) - \cos(q_1+q_2)\sin(q_1+q_2)] \\ &= -PL\dot{q}_1\,[\sin(q_1+q_2)\cos(q_1) - \cos(q_1+q_2)\sin(q_1))] \\ &= -PL\dot{q}_1\sin(q_2)\ . \end{aligned} \tag{3.6.89}$$

To nondimensionalize, we again multiply by $\tau^3/\mu\ell^2$:

$$\begin{aligned} \widetilde{p}_{\mathrm{F}} :&= \frac{\tau^3}{\mu\ell^2}p_{\mathrm{F}} = -\frac{\tau^3}{\mu\ell^2}PL\dot{q}_1\sin(q_2) \\ &= -\frac{\tau^2 P}{\mu\ell}\frac{L}{\ell}\tau\dot{q}_1\sin(q_2) = -\Pi_F\alpha\tau\dot{q}_1\sin(q_2)\ . \end{aligned} \tag{3.6.90}$$

Using nondimensional time, we have the nondimensional power input by the follower force:

$$\widetilde{p}_{\mathrm{F}} = -\Pi_F\alpha\dot{q}_1\sin(q_2)\ . \tag{3.6.91}$$

The power input by the follower force is positive when the force is pushing the *first* rod in the same direction it is already turning. For small q_2, if $\dot{q}_1 > 0$ and $q_2 < 0$, $\widetilde{P} > 0$; likewise, if $\dot{q}_1 < 0$ and $q_2 > 0$, $\widetilde{P} > 0$. This matches the physical understanding.

3.6.5 Static Equilibrium Paths

From the viewpoint of fundamental behavior of the system under study, let us consider the question of finding all static equilibria. At a static equilibrium, $\dot{\mathbf{q}} = \ddot{\mathbf{q}} = 0$. Hence, by (3.6.56) (or (3.6.57)), also $\mathbf{J} = 0$. The equilibrium equation (3.6.5) thus becomes

$$-\mathbf{L} - \mathbf{Q} = 0\,, \tag{3.6.92}$$

where the minus sign has been taken for convenience in what follows. In the nondimensional form, we will use the general parameterization (3.6.52). Inserting (3.6.54) and (3.6.55) into (3.6.92), and setting $\dot{\mathbf{q}} = 0$, we obtain

$$\gamma \begin{bmatrix} q_1 \\ q_2 \end{bmatrix} + \alpha \left(\Pi_F \begin{bmatrix} \sin q_2 \\ 0 \end{bmatrix} - \cos(\theta) G \left(\mu_m \begin{bmatrix} \sin q_1 + \sin(q_1 + q_2) \\ \sin(q_1 + q_2) \end{bmatrix} + \frac{1}{2}\mu_\rho \begin{bmatrix} 3\sin(q_1) + \sin(q_1 + q_2) \\ \sin(q_1 + q_2) \end{bmatrix} \right) \right) = 0\,, \tag{3.6.93}$$

or in component form,

$$\gamma q_1 + \alpha \Pi_F \sin(q_2) - \alpha \cos(\theta) G \left[\mu_m \left(\sin q_1 + \sin(q_1 + q_2)\right) + \frac{1}{2}\mu_\rho \left(3\sin(q_1) + \sin(q_1 + q_2)\right) \right] = 0\,, \tag{3.6.94}$$

$$\gamma q_2 - \alpha \cos(\theta) G (\mu_m + \frac{1}{2}\mu_\rho) \sin(q_1 + q_2) = 0\,. \tag{3.6.95}$$

In the parameterization (3.6.59), $\alpha = G = \mu_\rho = 1$, and we have the somewhat simpler (but not as symmetric) form

$$\gamma q_1 + \Pi_F \sin(q_2) - \cos(\theta) \left[\mu_m \left(\sin q_1 + \sin(q_1 + q_2)\right) + \frac{1}{2} \left(3\sin(q_1) + \sin(q_1 + q_2)\right) \right] = 0\,, \tag{3.6.96}$$

$$\gamma q_2 - \cos(\theta) (\mu_m + \frac{1}{2}) \sin(q_1 + q_2) = 0\,. \tag{3.6.97}$$

which has three remaining parameters: the spring stiffness γ, the follower force magnitude Π_F, and the dead weight mass μ_m. Equivalently to μ_m, we may consider the dead weight gravitational load Π_W; when $G = 1$, by (3.6.53) it holds that $\cos(\theta)\mu_m = \Pi_W$. For any given values of the parameters, this system can be solved numerically to obtain the states (q_1, q_2) that correspond to static equilibria.

Observe that because in any static equilibrium $\dot{\mathbf{q}} = 0$, the damping factor β has no effect on static solutions of the system. Thus, regardless of the presence or absence of damping or its magnitude, a quasistatic analysis results in the same equilibrium paths.

On dynamical stability however, damping does have a dramatic effect. As is well-known, in the follower-force case, any nonzero damping, no matter how small, will drastically decrease the critical load where the trivial equilibrium becomes unstable. This damping-induced destabilization is known as *Ziegler's paradox*. Although, following in the spirit of Lighthill,[3] it would perhaps be more constructive to term this *Ziegler's effect*; Ziegler having presented a mathematical counterexample to the physically intuitive notion of damping always being a stabilizing property. It is clear that when a mathematical formalization and intuition disagree, intuition is simply in error; there is no paradox.

Ziegler's effect only arises in nonconservative systems. As Bolotin [6], p. 99 remarks, a theorem by Kelvin (nowadays known as the Kelvin-Tait-Chataev theorem; see, e.g., Kirillov [7, 8]) guarantees that for a conservative system this cannot occur. For more on damping-induced destabilization, see the paper by Kirillov and Verhulst [11].

Aside from the trivial solution $q_1 = q_2 = 0$, the general case of (3.6.94)–(3.6.95) has no analytical solution. Below, we will examine some special cases, and finish with a discussion of the bifurcation of the trivial equilibrium in the general case.

Limit of infinite stiffness

Dividing both sides of (3.6.94)–(3.6.95) by γ, and then taking the limit $\gamma \to +\infty$, we obtain the equations $q_1 = q_2 = 0$. A system with infinite stiffness has its only static equilibrium in the neutral position.

Classical double pendulum

As a special case, let us consider the double pendulum with zero stiffness, that is without the springs. Setting $\gamma = 0$ in (3.6.94)–(3.6.95), and noting that this allows us to cancel the α, we have

$$\Pi_F \sin(q_2) - \cos(\theta) G \left[\mu_m \left(\sin q_1 + \sin(q_1 + q_2)\right) + \frac{1}{2}\mu_\rho \left(3\sin(q_1) + \sin(q_1 + q_2)\right) \right] = 0 \,,$$

$$\cos(\theta) G (\mu_m + \frac{1}{2}\mu_\rho) \sin(q_1 + q_2) = 0 \,.$$

Subtracting the second equation from the first one, the terms involving $\sin(q_1 + q_2)$ vanish, resulting in

$$\Pi_F \sin(q_2) - \cos(\theta) G \left[\mu_m \sin(q_1) + \frac{3}{2}\mu_\rho \sin(q_1) \right] = 0 \,.$$

After collecting terms,

---

[3] Lighthill [12], pp. 113–114 suggested, in the name of being constructive, renaming *d'Alembert's paradox* to *d'Alembert's theorem*: a mathematical demonstration showing that if one were able to physically construct shapes which, when subjected to an external flow, have the flow fulfill the very stringent conditions of potential flow, then very low (essentially zero) drag could be observed.

$$\Pi_F \sin(q_2) - \cos(\theta) G \left[\mu_m + \frac{3}{2}\mu_\rho \right] \sin(q_1) = 0 \,. \tag{3.6.98}$$

If $\Pi_F = 0$ (dead weight only), then $q_1 = j_1\pi$ with j_1 any integer, and q_2 is free. If $\cos(\theta) = 0$ (follower force only), then $q_2 = j_2\pi$ with j_2 any integer, and q_1 is free.

Now let $\Pi_F \neq 0$, $\cos(\theta) \neq 0$. In this general case, a family of trivial solutions is obtained with $q_1 = j_1\pi$ and $q_2 = j_2\pi$, where j_1 and j_2 are any integers.

There is also a family of nontrivial solutions where the forces balance statically, satisfying

$$\sin(q_2) = \frac{\cos(\theta) G \left[\mu_m + \frac{3}{2}\mu_\rho \right]}{\Pi_F} \sin(q_1) \,. \tag{3.6.99}$$

Both sides of (3.6.99) must remain in the range $[-1, 1]$. Even if the coefficient on the right-hand side is large, this is always possible, if $|\sin(q_1)|$ is small enough to balance out the large coefficient. Hence, if the coefficient is large, $q_1 \approx j_1\pi$ with j_1 any integer, and $|\sin(q_2)|$ can still be near unity, that is, $q_2 \approx \pi/2 + j_2\pi$ with j_2 any integer.

When the coefficient is near zero, $q_2 \approx j_2\pi$ with j_2 any integer, and q_1 is practically free.

For intermediate values of the coefficient, a nontrivial connection between q_1 and q_2 appears, depending on the numerical value. Note the periodicity under shifting both angles simultaneously by π, or either one independently by 2π.

Follower force only

The only case where all gravity effects, also those pertaining to the rods, vanish, is that of a horizontal inclination. We obtain this from (3.6.94)–(3.6.95) by setting $\theta = \pi/2$.

We have

$$\gamma q_1 + \alpha \Pi_F \sin(q_2) = 0 \,,$$
$$\gamma q_2 = 0 \,.$$

The only solutions are $q_1 = q_2 = 0$ (regardless of the magnitude of the follower force Π_F); or alternatively, $\gamma = 0$ (no stiffness), q_1 free, and $q_2 = j\pi$ for any integer j. The latter agrees with the solution of the classical double pendulum.

Dead weight only

This is perhaps the most complex, and the most interesting, of the special cases.

We have $\Pi_F = 0$ (no follower force). Additionally, let $\gamma \neq 0$; the case $\gamma = 0$ was already covered above. We have

$$\gamma q_1 - \alpha \cos(\theta) G \left[\mu_m \left(\sin q_1 + \sin(q_1 + q_2)\right) + \frac{1}{2}\mu_\rho \left(3\sin(q_1) + \sin(q_1 + q_2)\right) \right] = 0 \,,$$
$$\gamma q_2 - \alpha \cos(\theta) G (\mu_m + \frac{1}{2}\mu_\rho) \sin(q_1 + q_2) = 0 \,.$$

If the inclination $\theta = \pi/2$ ($+j\pi$ for any integer j), this reduces to the previous case with $\Pi_F = 0$. Thus, let us restrict our consideration to the case $\theta \neq \pi/2$, whence $\cos(\theta) \neq 0$.

Again, it is useful to subtract the second component equation from the first one, cancelling out terms involving $\sin(q_1 + q_2)$. We have

$$\gamma(q_1 - q_2) - \alpha\cos(\theta)G\left(\mu_m + \frac{3}{2}\mu_\rho\right)\sin(q_1) = 0 ,$$

which, because now $\gamma \neq 0$, can be solved explicitly for q_2:

$$q_2 = q_1 - \kappa\sin(q_1) , \tag{3.6.100}$$

where we have shortened the notation by defining

$$\kappa := \lambda\left(\mu_m + \frac{3}{2}\mu_\rho\right) , \quad \text{where } \lambda := \frac{1}{\gamma}\alpha\cos(\theta)G . \tag{3.6.101}$$

This will be convenient below. Note $0 < \kappa < \infty$, because in the considered case, $\cos(\theta) \neq 0$ and $\gamma \neq 0$.

Recall that μ_m is the loading parameter for the dead-weight case. Thus, κ is an effective loading. It is a linear function of μ_m; the offset is generated by the fact that also the rods have mass, and scaling comes from the parameterization.

Inserting (3.6.100) into the first component equation, and dividing both sides of the equation by γ, we obtain a transcendental equation for q_1:

$$q_1 - \lambda\left[\mu_m\left[\sin q_1 + \sin\big(2q_1 - \kappa\sin(q_1)\big)\right] + \frac{1}{2}\mu_\rho\left[3\sin(q_1) + \sin\big(2q_1 - \kappa\sin(q_1)\big)\right]\right] = 0 ,$$

where we have used the definition of λ from (3.6.101). Note the nested sines.

Collecting terms and using (3.6.101) in the coefficients of the $\sin q_1$ terms, we are left with

$$q_1 - \kappa\sin q_1 - \lambda\left(\mu_m + \frac{1}{2}\mu_\rho\right)\sin\big(2q_1 - \kappa\sin(q_1)\big) = 0 .$$

In order to eliminate the remaining μ_m, and thus write all instances of the load in terms of the effective load κ, we add a suitably formatted zero:

$$q_1 - \kappa\sin q_1 - \lambda\left[\left(\mu_m + \frac{3}{2}\mu_\rho\right) - \mu_\rho\right]\sin\big(2q_1 - \kappa\sin(q_1)\big) = 0 .$$

Now using (3.6.101) again, we have the final form

$$q_1 - \kappa\sin q_1 - \left[\kappa - \kappa_\rho\right]\sin\big(2q_1 - \kappa\sin(q_1)\big) = 0 , \tag{3.6.102}$$

where we have defined

$$\kappa_\rho := \lambda\mu_\rho \,. \tag{3.6.103}$$

Equation (3.6.102) is exact, because it follows directly from the original nonlinear model. It can be solved numerically to obtain admissible q_1, which allows us to draw a bifurcation diagram. This only gives us the equilibrium states; their stability must be determined separately. For each value of q_1 that is a solution of (3.6.102), the corresponding q_2 can be obtained from (3.6.100).

Bifurcation of the trivial equilibrium with dead weight only

For small values of q_1, approximations for the critical loads, where the trivial equilibrium undergoes bifurcation, can be obtained analytically.

Taking the Taylor series up to the linear term, it holds that $\sin(q_1) \approx q_1$ for $|q_1| \ll 1$. Hence, we have the following $O(\,(q_1)^2\,)$ approximation of (3.6.102):

$$(1-\kappa)q_1 - \left[\kappa - \kappa_\rho\right]\sin\bigl((2-\kappa)q_1\bigr) = 0 \quad \text{for}\, |q_1| \ll 1 \,.$$

Provided that also $|(2-\kappa)q_1| \ll 1$, which can be always made to hold by taking $|q_1|$ small enough, this further transforms into

$$\Bigl((1-\kappa) - \left[\kappa - \kappa_\rho\right](2-\kappa)\Bigr)q_1 = 0 \quad \text{for } |q_1| \ll 1 \,, \tag{3.6.104}$$

which can be rearranged into

$$\left(\left[1+2\kappa_\rho\right] - \left[3+\kappa_\rho\right]\kappa + \kappa^2\right)q_1 = 0 \quad \text{for } |q_1| \ll 1 \,. \tag{3.6.105}$$

We have arrived at (3.6.105) by first determining the condition for static equilibrium states for the original nonlinear model, and then linearizing the result around the trivial equilibrium. The other approach is to first linearize the dynamical equation around the trivial equilibrium (applying a Taylor series already at that point), and then find conditions that give rise to nontrivial static equilibrium states in a small neighborhood of the trivial equilibrium, this neighbourhood being what the linearized model is able to represent.

Aside from the trivial solution $q_1 = 0$, let us find the other solutions of (3.6.105). Let $q_1 \neq 0$, $|q_1| \ll 1$. The parameter κ plays the role of an effective load, so we solve (3.6.105) for κ, denoting this critical value of κ by κ^*:

$$\begin{aligned}\kappa^* &= \frac{1}{2}\left(3+\kappa_\rho \pm \sqrt{9+6\kappa_\rho+\kappa_\rho^2 - 4(1+2\kappa_\rho)}\right)\\ &= \frac{1}{2}\left(3+\kappa_\rho \pm \sqrt{5-2\kappa_\rho+\kappa_\rho^2}\right) \,.\end{aligned} \tag{3.6.106}$$

Equation (3.6.106) is written in terms of the generic parameterization (3.6.52). In the specific parameterization (3.6.59), $\alpha = G = \mu_\rho = 1$, and $\gamma = k/\rho g L^2$. Then

$$\lambda = \frac{1}{\gamma}\alpha\cos(\theta)G = \frac{1}{\gamma}\cos(\theta) = \frac{\rho g L^2}{k}\cos(\theta) = \frac{(\rho L\, g\, \cos(\theta))\, L}{k} =: \frac{P_\rho L}{k} \quad (3.6.107)$$

and $\kappa_\rho = \lambda\mu_\rho = \lambda$.

In (3.6.107), P_ρ is the magnitude of the gravitational force felt by one rod. Note that this is different from $P_W = mg\cos(\theta)$, which is the gravitational force felt by the dead weight attached at the end. The choice of the symbol λ mimics the notation used in Jeronen and Kouhia [13], where $\lambda = PL/k$. We conclude that λ is the ratio of a reference gravitational force to a reference elastic force, characteristic to the system under study.

At the critical load, any small value of q_1 is at static equilibrium,[4] that is, the trivial static equilibrium at $q_1 = 0$ undergoes bifurcation. From (3.6.100) we have the corresponding q_2 at the critical load, for any arbitrary small q_1:

$$q_2 = (1 - \kappa)q_1 \quad \text{for } |q_1| \ll 1\ .$$

When interpreting the results, recall the definition of κ, Eq. (3.6.101). In the parameterization (3.6.59), because $\mu_\rho = 1$, it becomes

$$\kappa = \lambda\left(\mu_m + \frac{3}{2}\right), \quad (3.6.108)$$

whence

$$\mu_m = \frac{\kappa}{\lambda} - \frac{3}{2} = \Pi_W\ , \quad (3.6.109)$$

where the second equality follows from the relation connecting μ_m, G and Π_W, Eq. (3.6.53), after applying $G = 1$ from the parameterization (3.6.59). Finally, the corresponding dimensional value is

$$P_W = mg = \mu_m\rho L g = (\frac{\kappa}{\lambda} - \frac{3}{2})\rho L g\ . \quad (3.6.110)$$

Classically, the effect of gravity on the rods has been neglected. To obtain this case, we set $\mu_\rho = 0$. By Eq. (3.6.103), then $\kappa_\rho = 0$. In the inertial terms, this would not be allowed, because the rods' inertia is included in the classical model; but those terms have already vanished, because we are considering static equilibrium positions only. Hence the only remaining factors of μ_ρ originate from the external load **Q**, corresponding to gravitational loading.

We have

---

[4]Strictly speaking, at the critical load, Eq. (3.6.105) itself—when considering only the expressions on both sides of the equals sign—admits *any* value of q_1, no matter how large. Thus we may say that according to the linearized model, at the bifurcation point the angular displacement q_1 shoots off to infinity. However, to remain rigorous, we must consider the assumptions made when deriving (3.6.105). We have used a Taylor series, up to the linear term, requiring $|q_1| \ll 1$ for validity.

$$\kappa^*_{\text{class.}} = \frac{1}{2}\left(3 \pm \sqrt{5}\right) \quad \text{(classical case)}\,, \tag{3.6.111}$$

which agrees with the well-known classical result. Numerical evaluation obtains $\kappa^*_{\text{class.}} \approx \{0.382,\ 2.618\}$. Due to our approach in deriving (3.6.105), there is no reason to reject either solution; both are consistent with our only requirement, $|q_1| \ll 1$. In all other respects (3.6.105), being based on (3.6.102), is a consequence of the original nonlinear model. Obviously, the smaller value corresponds to the first critical load; but another bifurcation point also exists around the trivial equilibrium, for a higher value of the load.

The definition of κ corresponding to (3.6.111), from Eq. (3.6.101), is

$$\kappa_{\text{class.}} = \frac{\alpha \cos(\theta) G \mu \ell^2}{k \tau^2} \mu_m \quad \text{(classical case)}\,. \tag{3.6.112}$$

If we use the parameterization (3.6.59), we have

$$\kappa_{\text{class.}} = \frac{\cos(\theta)}{\gamma} \mu_m = \lambda \mu_m\,. \tag{3.6.113}$$

Compare with (3.6.108). The value of κ corresponding to a given value of μ_m is smaller; vice versa, the value of μ_m, the loading parameter, corresponding to a given value of κ is larger. However, this does not say anything about the critical load, because the expressions for κ^* for the two cases are different.

In order to go to dimensional values from (3.6.112), we must use the generic parameterization (3.6.52). Solving (3.6.112) for μ_m, we have

$$\mu_m = \frac{k \tau^2}{\alpha \cos(\theta) G \mu \ell^2} \kappa_{\text{class.}}\,. \tag{3.6.114}$$

Now, from (3.6.52), recall $\mu_m = m/\mu$. Hence

$$P_W = mg = \mu \mu_m g = \frac{\mu k \tau^2 g}{\alpha \cos(\theta) G \mu \ell^2} \kappa_{\text{class.}} = \frac{k \tau^2 g}{\alpha \cos(\theta) G \ell^2} \kappa_{\text{class.}} \tag{3.6.115}$$

in the system of units defined by τ, ℓ and μ.

Bifurcation of the trivial equilibrium in the general case

Let us perform the above bifurcation analysis also for the general case, where a follower force may be present in addition to any gravitational effects.

Provided that $\left|q_j\right| \ll 1$, $j = 1, 2$, we may apply a Taylor series, up to the linear term, directly to the nonlinear equilibrium equations (3.6.94)–(3.6.95), effectively replacing all sines by their arguments. Thus we obtain a linear analysis of the general case. Above, it was more convenient to first eliminate q_2, because that gave us an exact nonlinear result for the case with dead weight only.

Consider the general case where $\gamma \neq 0, \cos(\theta) \neq 0$; the special cases have already been analyzed above. Now we optionally allow a follower force; no assumption is made of Π_F.

Applying the Taylor series in (3.6.94)–(3.6.95) and dividing by γ, we have

$$q_1 + \frac{\alpha \Pi_F}{\gamma} q_2 - \lambda \left[\mu_m (2q_1 + q_2) + \frac{1}{2} \mu_\rho (4q_1 + q_2) \right] = 0 \,,$$

$$q_2 - \lambda(\mu_m + \frac{1}{2}\mu_\rho)(q_1 + q_2) = 0 \,,$$

with λ as defined in (3.6.101). This is a linear equation system, which can be written as

$$\mathbf{A}\mathbf{q} = 0 \,, \tag{3.6.116}$$

where $\mathbf{q} = (q_1, q_2)$ and

$$\mathbf{A} = \begin{bmatrix} 1 - 2\lambda(\mu_m + \mu_\rho) & \frac{\alpha \Pi_F}{\gamma} - \lambda(\mu_m + \frac{1}{2}\mu_\rho) \\ -\lambda(\mu_m + \frac{1}{2}\mu_\rho) & 1 - \lambda(\mu_m + \frac{1}{2}\mu_\rho) \end{bmatrix} . \tag{3.6.117}$$

Note that the load is denoted by μ_m, not λ. We see that the follower force introduces an asymmetry to the matrix.

It is convenient to define the effective load

$$\varkappa := \lambda(\mu_m + \frac{1}{2}\mu_\rho) \,, \quad \text{where } \lambda := \frac{1}{\gamma}\alpha \cos(\theta) G \,. \tag{3.6.118}$$

Note the 1/2 instead of 3/2 in (3.6.101) earlier. The definition of λ is the same as before. Let us define also

$$\varkappa_\rho := \lambda \mu_\rho \,, \quad \varkappa_F := \frac{\alpha \Pi_F}{\gamma} \,, \tag{3.6.119}$$

The term in the matrix element A_{11} becomes

$$2\lambda(\mu_m + \mu_\rho) = 2\left[\lambda(\mu_m + \mu_\rho)\right] = 2\left[\lambda(\mu_m + \frac{1}{2}\mu_\rho) + \lambda\frac{1}{2}\mu_\rho\right] = 2\varkappa + \lambda\mu_\rho = 2\varkappa + \varkappa_\rho \,. \tag{3.6.120}$$

Equations (3.6.118)–(3.6.120) allow us to rewrite (3.6.117) as

$$\mathbf{A} = \begin{bmatrix} 1 - 2\varkappa - \varkappa_\rho & \varkappa_F - \varkappa \\ -\varkappa & 1 - \varkappa \end{bmatrix} . \tag{3.6.121}$$

Nontrivial solutions exist if and only if $\det \mathbf{A} = 0$. Similarly to before, this leads to a quadratic polynomial, now for $\varkappa$, which gives the critical loads:

$$\left[1 - 2\varkappa - \varkappa_\rho\right][1 - \varkappa] - [\varkappa_F - \varkappa][-\varkappa] = 0\,. \tag{3.6.122}$$

Expanding, we obtain

$$1 - 2\varkappa - \varkappa_\rho - \varkappa + 2\varkappa^2 + \varkappa_\rho \varkappa + \varkappa_F \varkappa - \varkappa^2 = 0\,.$$

Collecting in powers of $\varkappa$, the final result is

$$(1 - \varkappa_\rho) - (3 - (\varkappa_\rho + \varkappa_F))\varkappa + \varkappa^2 = 0\,. \tag{3.6.123}$$

The solution of the general case, that is, the critical value $\varkappa^*$, is

$$\begin{aligned}
\varkappa^* &= \frac{1}{2}\left(3 - (\varkappa_\rho + \varkappa_F) \pm \sqrt{9 - 6(\varkappa_\rho + \varkappa_F) + (\varkappa_\rho + \varkappa_F)^2 - 4(1 - \varkappa_\rho)}\right) \\
&= \frac{1}{2}\left(3 - (\varkappa_\rho + \varkappa_F) \pm \sqrt{5 - 2\varkappa_\rho - 6\varkappa_F + (\varkappa_\rho + \varkappa_F)^2}\right)\,.
\end{aligned} \tag{3.6.124}$$

Equation (3.6.124) gives the values of the effective load that correspond to the two bifurcation points that exist around the trivial equilibrium. Again, we have no reason to reject either of the solutions, as long as $|q_j| \ll 1$, $j = 1, 2$, which can always be made to hold.

The classical solution, with $\mu_\rho = \Pi_F = 0$, whence $\varkappa_\rho = \varkappa_F = 0$, remains the same:

$$\varkappa^*_{\text{class.}} = \frac{1}{2}\left(3 \pm \sqrt{5}\right) \quad \text{(classical case)}\,,$$

but now the correction terms (involving $\varkappa_\rho$, existing in both formulations) have a different sign due to the difference in the definitions of the effective load, (3.6.118) versus (3.6.101).

When $\Pi_F = 0$, the physical critical load μ_m must be the same as before. Hence, inevitably $\varkappa^* \neq \kappa^*$ whenever the correction is nonzero, and thus the coefficients of the polynomials determining $\varkappa^*$ and κ^* indeed must be different.

3.6.6 *Linearization*

From this point on, let us return to the full dynamical model. Although our main focus is on the nonlinear model, we first consider the linearization of the dynamical model around points of static equilibrium. This can be used to determine their stability.

Let us choose the point of linearization $(\mathbf{q}_e, \dot{\mathbf{q}}_e, \ddot{\mathbf{q}}_e)$ as an arbitrary static equilibrium point, $(\mathbf{q}_e, \dot{\mathbf{q}}_e, \ddot{\mathbf{q}}_e) = (\mathbf{q}_e, 0, 0)$, and define the perturbation

$$(\mathbf{q}^*, \dot{\mathbf{q}}^*, \ddot{\mathbf{q}}^*) \equiv (\mathbf{q} - \mathbf{q}_e\,,\ \dot{\mathbf{q}} - \dot{\mathbf{q}}_e\,,\ \ddot{\mathbf{q}} - \ddot{\mathbf{q}}_e)\,. \tag{3.6.125}$$

Developing $\mathbf{L}$, $\mathbf{Q}$ and $\mathbf{J}$ in a multivariate Taylor series in $(\mathbf{q}^*, \dot{\mathbf{q}}^*, \ddot{\mathbf{q}}^*)$ around the point of linearization, up to first order, we have

$$\widehat{\mathbf{L}}(\mathbf{q}^*, \dot{\mathbf{q}}^*, \ddot{\mathbf{q}}^*) \equiv \mathbf{L}(\mathbf{q}_e, 0, 0) + \frac{\partial \mathbf{L}}{\partial \mathbf{q}}(\mathbf{q}_e, 0, 0)\, \mathbf{q}^* + \frac{\partial \mathbf{L}}{\partial \dot{\mathbf{q}}}(\mathbf{q}_e, 0, 0)\, \dot{\mathbf{q}}^* + \frac{\partial \mathbf{L}}{\partial \ddot{\mathbf{q}}}(\mathbf{q}_e, 0, 0)\, \ddot{\mathbf{q}}^* \,, \tag{3.6.126}$$

$$\widehat{\mathbf{Q}}(\mathbf{q}^*, \dot{\mathbf{q}}^*, \ddot{\mathbf{q}}^*) \equiv \mathbf{Q}(\mathbf{q}_e, 0, 0) + \frac{\partial \mathbf{Q}}{\partial \mathbf{q}}(\mathbf{q}_e, 0, 0)\, \mathbf{q}^* + \frac{\partial \mathbf{Q}}{\partial \dot{\mathbf{q}}}(\mathbf{q}_e, 0, 0)\, \dot{\mathbf{q}}^* + \frac{\partial \mathbf{Q}}{\partial \ddot{\mathbf{q}}}(\mathbf{q}_e, 0, 0)\, \ddot{\mathbf{q}}^* \,, \tag{3.6.127}$$

$$\widehat{\mathbf{J}}(\mathbf{q}^*, \dot{\mathbf{q}}^*, \ddot{\mathbf{q}}^*) \equiv \mathbf{J}(\mathbf{q}_e, 0, 0) + \frac{\partial \mathbf{J}}{\partial \mathbf{q}}(\mathbf{q}_e, 0, 0)\, \mathbf{q}^* + \frac{\partial \mathbf{J}}{\partial \dot{\mathbf{q}}}(\mathbf{q}_e, 0, 0)\, \dot{\mathbf{q}}^* + \frac{\partial \mathbf{J}}{\partial \ddot{\mathbf{q}}}(\mathbf{q}_e, 0, 0)\, \ddot{\mathbf{q}}^* \,. \tag{3.6.128}$$

The hat denotes a first-order Taylor approximation. For interpreting the derivatives, we use the standard Jacobian

$$\left(\frac{\partial \mathbf{F}}{\partial \mathbf{x}}\right)_{ij} \equiv (\nabla \mathbf{F})_{ij} = \frac{\partial F_i}{\partial x_j} \tag{3.6.129}$$

for $\mathbf{F}$ any differentiable vector quantity and $\mathbf{x}$ any vector-valued independent variable. With this convention, the product with the $\mathbf{q}^*$ variables is then a standard matrix multiplication.

Five of the nine perturbation terms vanish. First, from (3.6.54) and (3.6.55) we see that $\mathbf{L}$ and $\mathbf{Q}$ are both identically constant with respect to $\ddot{\mathbf{q}}$, so the term involving the second time derivative is needed only for $\mathbf{J}$. Similarly, $\mathbf{Q}$ is identically constant with respect to $\dot{\mathbf{q}}$. Then, from (3.6.56) we observe that at our particular choice for the point of linearization, that is, any static equilibrium $(\mathbf{q}_e, 0, 0)$, the derivatives of $\mathbf{J}$ with respect to $\mathbf{q}$ and $\dot{\mathbf{q}}$ happen to be zero. Thus, we are left with

$$\widehat{\mathbf{L}} = \mathbf{L}(\mathbf{q}_e, 0, 0) + \frac{\partial \mathbf{L}}{\partial \mathbf{q}}(\mathbf{q}_e, 0, 0)\, \mathbf{q}^* + \frac{\partial \mathbf{L}}{\partial \dot{\mathbf{q}}}(\mathbf{q}_e, 0, 0)\, \dot{\mathbf{q}}^* \,, \tag{3.6.130}$$

$$\widehat{\mathbf{Q}} = \mathbf{Q}(\mathbf{q}_e, 0, 0) + \frac{\partial \mathbf{Q}}{\partial \mathbf{q}}(\mathbf{q}_e, 0, 0)\, \mathbf{q}^* \,, \tag{3.6.131}$$

$$\widehat{\mathbf{J}} = \mathbf{J}(\mathbf{q}_e, 0, 0) + \frac{\partial \mathbf{J}}{\partial \ddot{\mathbf{q}}}(\mathbf{q}_e, 0, 0)\, \ddot{\mathbf{q}}^* \,. \tag{3.6.132}$$

Now, consider the expression

$$\begin{aligned}
\widehat{\mathbf{J}} - \widehat{\mathbf{L}} - \widehat{\mathbf{Q}} = {} & \mathbf{J}(\mathbf{q}_e, 0, 0) + \frac{\partial \mathbf{J}}{\partial \ddot{\mathbf{q}}}(\mathbf{q}_e, 0, 0)\, \ddot{\mathbf{q}}^* \\
& - \mathbf{L}(\mathbf{q}_e, 0, 0) - \frac{\partial \mathbf{L}}{\partial \mathbf{q}}(\mathbf{q}_e, 0, 0)\, \mathbf{q}^* - \frac{\partial \mathbf{L}}{\partial \dot{\mathbf{q}}}(\mathbf{q}_e, 0, 0)\, \dot{\mathbf{q}}^* \\
& - \mathbf{Q}(\mathbf{q}_e, 0, 0) - \frac{\partial \mathbf{Q}}{\partial \mathbf{q}}(\mathbf{q}_e, 0, 0)\, \mathbf{q}^* \,.
\end{aligned}$$

Because the dynamical equation (3.6.5) holds anywhere, it must hold also at the specific point $(\mathbf{q}_e, 0, 0)$. Hence also at this point, $\mathbf{J} - \mathbf{L} - \mathbf{Q} = 0$, and we are left with

$$\frac{\partial \mathbf{J}}{\partial \ddot{\mathbf{q}}}(\mathbf{q}_e, 0, 0)\, \ddot{\mathbf{q}}^* - \frac{\partial \mathbf{L}}{\partial \dot{\mathbf{q}}}(\mathbf{q}_e, 0, 0)\, \dot{\mathbf{q}}^* - \left(\frac{\partial \mathbf{L}}{\partial \mathbf{q}}(\mathbf{q}_e, 0, 0) + \frac{\partial \mathbf{Q}}{\partial \mathbf{q}}(\mathbf{q}_e, 0, 0) \right) \mathbf{q}^* = 0 \,, \tag{3.6.133}$$

Equation (3.6.133) is the linearized dynamical equation, which is valid for small perturbations (3.6.125). To justify the statement, observe that by their construction, $\widehat{\mathbf{L}}$, $\widehat{\mathbf{Q}}$ and $\widehat{\mathbf{J}}$ are linear approximations of $\mathbf{L}$, $\mathbf{Q}$ and $\mathbf{J}$, respectively:

$$\begin{aligned}
\widehat{\mathbf{L}} &= \mathbf{L} + O(\mathbf{q}^*)^2 + O(\dot{\mathbf{q}}^*)^2 + O(\ddot{\mathbf{q}}^*)^2 \,, \\
\widehat{\mathbf{Q}} &= \mathbf{Q} + O(\mathbf{q}^*)^2 + O(\dot{\mathbf{q}}^*)^2 + O(\ddot{\mathbf{q}}^*)^2 \,, \\
\widehat{\mathbf{J}} &= \mathbf{J} + O(\mathbf{q}^*)^2 + O(\dot{\mathbf{q}}^*)^2 + O(\ddot{\mathbf{q}}^*)^2 \,.
\end{aligned}$$

Hence the left-hand side of (3.6.133) becomes

$$\begin{aligned}
\widehat{\mathbf{J}} - \widehat{\mathbf{L}} - \widehat{\mathbf{Q}} &= \mathbf{J} - \mathbf{L} - \mathbf{Q} + O(\mathbf{q}^*)^2 + O(\dot{\mathbf{q}}^*)^2 + O(\ddot{\mathbf{q}}^*)^2 \\
&= 0 + O(\mathbf{q}^*)^2 + O(\dot{\mathbf{q}}^*)^2 + O(\ddot{\mathbf{q}}^*)^2 \,,
\end{aligned} \tag{3.6.134}$$

where we have lumped all error terms of the same order, and again applied (3.6.5). In other words, $\widehat{\mathbf{J}}$, $\widehat{\mathbf{L}}$ and $\widehat{\mathbf{Q}}$ obey the dynamic Eq. (3.6.5), up to first order, around the chosen point of linearization.

Finally, we define the stiffness, damping and mass matrices

$$\mathbf{K} \equiv - \left(\frac{\partial \mathbf{L}}{\partial \mathbf{q}}(\mathbf{q}_e, 0, 0) + \frac{\partial \mathbf{Q}}{\partial \mathbf{q}}(\mathbf{q}_e, 0, 0) \right) \,, \tag{3.6.135}$$

$$\mathbf{C} \equiv - \frac{\partial \mathbf{L}}{\partial \dot{\mathbf{q}}}(\mathbf{q}_e, 0, 0) \,, \tag{3.6.136}$$

$$\mathbf{M} \equiv \frac{\partial \mathbf{J}}{\partial \ddot{\mathbf{q}}}(\mathbf{q}_e, 0, 0) \,. \tag{3.6.137}$$

transforming Eq. (3.6.133) into the standard form

$$\mathbf{M}\ddot{\mathbf{q}}^* + \mathbf{C}\dot{\mathbf{q}}^* + \mathbf{K}\mathbf{q}^* = 0 \,. \tag{3.6.138}$$

Equation (3.6.138) is a system of second-order ordinary differential equations with constant coefficients.

Provided that $\mathbf{M}$ is invertible, we may apply the standard trick of order reduction via auxiliary variables. This will reduce (3.6.138) to a twice larger system of first-order ordinary differential equations. Beside $\mathbf{q}^*$ itself, let us include $\dot{\mathbf{q}}^*$ in the vector of unknowns. After reordering terms and introducing the trivial equation $(\partial/\partial t)\mathbf{q}^* = \dot{\mathbf{q}}^*$, we have

$$\frac{\partial}{\partial t}\begin{bmatrix}\dot{\mathbf{q}}^* \\ \mathbf{q}^*\end{bmatrix} = \begin{bmatrix}-\mathbf{M}^{-1}\mathbf{C} & -\mathbf{M}^{-1}\mathbf{K} \\ \mathbf{I} & \end{bmatrix}\begin{bmatrix}\dot{\mathbf{q}}^* \\ \mathbf{q}^*\end{bmatrix} . \tag{3.6.139}$$

Empty space denotes a block of zeroes, and $\mathbf{I}$ is the identity matrix. Equation (3.6.139) is of the standard form

$$\dot{\mathbf{w}} = \mathbf{J}\mathbf{w} , \tag{3.6.140}$$

where

$$\mathbf{w} := \begin{bmatrix}\dot{\mathbf{q}}^* \\ \mathbf{q}^*\end{bmatrix} , \quad \mathbf{J} := \begin{bmatrix}-\mathbf{M}^{-1}\mathbf{C} & -\mathbf{M}^{-1}\mathbf{K} \\ \mathbf{I} & \end{bmatrix} . \tag{3.6.141}$$

We now have a system of first-order ordinary differential equations with constant coefficients. By its construction above, the matrix $\mathbf{J}$ is essentially the Jacobian of the original nonlinear problem, evaluated at the static equilibrium point $\mathbf{q}_e$, that was chosen as the reference point for the perturbation variable $\mathbf{q}^*$.

The matrices

Although the nondimensional parameterization (3.6.59) is the one we will use, we will here keep the equations in a general form to allow for any parameterization based on (3.6.52), using any admissible choices for τ, ℓ and μ.

Below, the symbols q_1 and q_2 refer to the values of these angles at the point of linearization, $\mathbf{q}_e$. Once the desired static equilibrium point has been chosen, the corresponding values should be inserted in order to obtain numerical representations for the matrices. The matrices are given in this general format, because for the dead-weight case, the trivial equilibrium position is not the only possible static equilibrium.

Using (3.6.54)–(3.6.56) in (3.6.135)–(3.6.137), we obtain the matrices

$$\mathbf{K} = \gamma\begin{bmatrix}1 & 0 \\ 0 & 1\end{bmatrix} + \alpha\Bigg(\Pi_F\begin{bmatrix}0 & \cos q_2 \\ 0 & 0\end{bmatrix} - \cos(\theta)G\Bigg(\mu_m\begin{bmatrix}\cos(q_1)+\cos(q_1+q_2) & \cos(q_1+q_2) \\ \cos(q_1+q_2) & \cos(q_1+q_2)\end{bmatrix}$$
$$+\frac{1}{2}\mu_p\begin{bmatrix}3\cos(q_1)+\cos(q_1+q_2) & \cos(q_1+q_2) \\ \cos(q_1+q_2) & \cos(q_1+q_2)\end{bmatrix}\Bigg)\Bigg) , \tag{3.6.142}$$

$$\mathbf{C} = \beta\begin{bmatrix}1 & 0 \\ 0 & 1\end{bmatrix} , \tag{3.6.143}$$

$$\mathbf{M} = \alpha^2\left(\mu_p\begin{bmatrix}\frac{5}{3}+\cos q_2 & \frac{1}{3}+\frac{1}{2}\cos q_2 \\ \frac{1}{3}+\frac{1}{2}\cos q_2 & \frac{1}{3}\end{bmatrix} + \mu_m\begin{bmatrix}2+2\cos q_2 & 1+\cos q_2 \\ 1+\cos q_2 & 1\end{bmatrix}\right) . \tag{3.6.144}$$

In the case with a follower force only, set $\mu_m = 0$. In the case with a dead-weight loading only, set $\Pi_F = 0$.

For the case of point-mass rods, from (3.6.57) we see that (3.6.144) is replaced by

$$\mathbf{M}_{\mathrm{pm}} = \alpha^2 \left(\mu_p \begin{bmatrix} \frac{3}{2} + \cos q_2 & \frac{1}{4} + \frac{1}{2}\cos q_2 \\ \frac{1}{4} + \frac{1}{2}\cos q_2 & \frac{1}{4} \end{bmatrix} + \mu_m \begin{bmatrix} 2 + 2\cos q_2 & 1 + \cos q_2 \\ 1 + \cos q_2 & 1 \end{bmatrix} \right) . \tag{3.6.145}$$

The other equations remain the same and all the above notes apply.

Linear dynamic stability analysis

Equation (3.6.138) is a system of ordinary differential equations with constant coefficients. We will apply the standard technique of the time-harmonic trial function to determine its harmonic vibrations. This method was extensively used by Bolotin [6, 14] for dynamic stability analysis of elastic systems.

The trial function is

$$\mathbf{q}^*(t) = e^{s\,t}\,\mathbf{q}_0^* , \tag{3.6.146}$$

where $\mathbf{q}_0^*$ is an unknown eigenstate vector, and s is the stability exponent. We insert (3.6.146) into (3.6.138) and discard the common factor $e^{s\,t}$, obtaining

$$\mathbf{L}(s)\,\mathbf{q}_0^* := \big(\mathbf{K} + s\mathbf{C} + s^2\mathbf{M}\big)\,\mathbf{q}_0^* = 0 , \tag{3.6.147}$$

which has nontrivial solutions if and only if the determinant of the matrix $\mathbf{L}(s)$ vanishes. Using (3.6.142)–(3.6.144), we construct the characteristic polynomial, which is of the fourth order in the stability exponent s.

To shorten the notation, let us denote

$$c_1 := \cos(q_1) , \quad c_2 := \cos(q_2) , \quad c_p := \cos(q_1 + q_2) , \tag{3.6.148}$$

which are to be evaluated at the point of linearization. For the trivial equilibrium $q_1 = q_2 = 0$, we have $c_1 = c_2 = c_p = 1$.

The characteristic polynomial is

$$a_{11}a_{22} - a_{12}a_{21} = 0 , \tag{3.6.149}$$

where the a_{ij} $(i, j = 1, 2)$ are the following quadratic polynomials of s, which can be read off Eqs. (3.6.142)–(3.6.144):

$$a_{11}(s) := \gamma - \alpha\cos(\theta)G\left[\mu_m(c_1 + c_p) + \frac{1}{2}\mu_p(3c_1 + c_p)\right] + \beta s + \alpha^2\left[\mu_p(\frac{5}{3} + c_2) + 2\mu_m(1 + c_2)\right]s^2 , \tag{3.6.150}$$

$$a_{21}(s) := -\alpha\cos(\theta)G(\mu_m + \frac{1}{2}\mu_p)c_p + \alpha^2\left[\mu_p(\frac{1}{3} + \frac{1}{2}c_2) + \mu_m(1 + c_2)\right]s^2 , \tag{3.6.151}$$

$$a_{12}(s) := a_{21}(s) + \alpha\Pi_F c_2 , \tag{3.6.152}$$

$$a_{22}(s) := \gamma - \alpha\cos(\theta)G(\mu_m + \frac{1}{2}\mu_p)c_p + \beta s + \alpha^2\left[\frac{1}{3}\mu_p + \mu_m\right]s^2 . \tag{3.6.153}$$

For the case of point-mass rods, from (3.6.142), (3.6.143) and (3.6.145), these are replaced by

$$a_{11}^{\mathrm{pm}}(s) := \gamma - \alpha\cos(\theta)G\left[\mu_m(c_1+c_p)+\frac{1}{2}\mu_\rho(3c_1+c_p)\right]+\beta s+\alpha^2\left[\mu_\rho(\frac{3}{2}+c_2)+2\mu_m(1+c_2)\right]s^2\,, \tag{3.6.154}$$

$$a_{21}^{\mathrm{pm}}(s) := -\alpha\cos(\theta)G(\mu_m+\frac{1}{2}\mu_\rho)c_p+\alpha^2\left[\mu_\rho(\frac{1}{4}+\frac{1}{2}c_2)+\mu_m(1+c_2)\right]s^2\,, \tag{3.6.155}$$

$$a_{12}^{\mathrm{pm}}(s) := a_{21}(s)+\alpha\Pi_F c_2\,, \tag{3.6.156}$$

$$a_{22}^{\mathrm{pm}}(s) := \gamma - \alpha\cos(\theta)G(\mu_m+\frac{1}{2}\mu_\rho)c_p+\beta s+\alpha^2\left[\frac{1}{4}\mu_\rho+\mu_m\right]s^2\,. \tag{3.6.157}$$

The only changes are in the terms involving s^2; the $5/3$ has become $3/2$, and each $1/3$ has become $1/4$.

Due to the generic form of (3.6.147)–(3.6.157), we may use the same equations for any specific nondimensional parameterization based on (3.6.52), such as (3.6.59), around any static equilibrium state (q_1, q_2). Such states, beside the trivial one, can be obtained numerically from (3.6.94)–(3.6.95).

Together, Eqs. (3.6.94)–(3.6.95) and (3.6.147)–(3.6.157) provide the means to find all static equilibrium states, and characterize the local stability of each one, for each desired value of the loading parameters μ_m and Π_F.

Relation between Bolotin's and Lyapunov's methods of stability analysis

Bolotin's method of stability analysis is based on Lyapunov's first, that is linearization-based, method. The first-order reduced Eq. (3.6.140) involves the Jacobian of the system under study, evaluated at the static equilibrium point $\mathbf{q}_e$ that was chosen as reference. Determining the eigenvalues of the Jacobian, we obtain Lyapunov's local stability exponents at the reference point; these are exactly the stability exponents s obtained by Bolotin's method.

To justify the statement, consider Eq. (3.6.147), which is a quadratic eigenvalue problem for the eigenvalue-eigenvector pair $(s, \mathbf{q}_0^*)$. It can be reduced to a twice larger linear one by the companion method [15]. Provided that $\mathbf{M}$ is invertible, (3.6.147) can be cast into block-matrix form as

$$\begin{bmatrix} -\mathbf{M}^{-1}\mathbf{C} & -\mathbf{M}^{-1}\mathbf{K} \\ \mathbf{I} & \end{bmatrix}\begin{bmatrix} s\mathbf{q}_0^* \\ \mathbf{q}_0^* \end{bmatrix} = s\begin{bmatrix} s\mathbf{q}_0^* \\ \mathbf{q}_0^* \end{bmatrix}\,, \tag{3.6.158}$$

where empty space denotes a block of zeroes, and $\mathbf{I}$ is the identity matrix. Defining

$$\mathbf{z} := \begin{bmatrix} s\mathbf{q}_0^* \\ \mathbf{q}_0^* \end{bmatrix}\,, \tag{3.6.159}$$

Equation (3.6.158) becomes

$$\mathbf{J}\mathbf{z} = s\mathbf{z}\,, \tag{3.6.160}$$

where $\mathbf{J}$ is as defined in (3.6.141). The matrix is the same as in the reduction of the original linearized problem to its first-order form.

Observe that by (3.6.141) and (3.6.146), we have

$$\mathbf{w} = \begin{bmatrix} \dot{\mathbf{q}}^* \\ \mathbf{q}^* \end{bmatrix} = \begin{bmatrix} s\, e^{s\,t}\, \mathbf{q}_0^* \\ e^{s\,t}\, \mathbf{q}_0^* \end{bmatrix} = e^{s\,t} \begin{bmatrix} s\mathbf{q}_0^* \\ \mathbf{q}_0^* \end{bmatrix} = e^{s\,t} \mathbf{z}\,. \tag{3.6.161}$$

The extended state variables $\mathbf{w}$ and $\mathbf{z}$ are essentially the same; in $\mathbf{z}$, we have simply discarded the common factor $e^{s\,t}$, which was done when deriving (3.6.147).

Equation (3.6.160) is a standard linear eigenvalue problem for the eigenvalue-eigenvector pair $(s, \mathbf{z})$. It is straightforward to show that the values of s which allow nontrivial solutions of (3.6.147) (i.e., s such that $\det(\mathbf{L}(s)) = 0$) are exactly the eigenvalues of $\mathbf{J}$. For details, see, e.g., Jeronen [16].

Hence Bolotin's method, at least when applied to second-order systems with constant coefficients, is equivalent with Lyapunov's first method. This implies that all of Lyapunov's original caveats also apply; especially, that of the case of purely imaginary stability exponents. Discussing Ziegler's problem, Bolotin writes:

> Since the present physical problem is non-linear, and since according to Liapunov the case of purely imaginary characteristic exponents is one of the doubtful cases (when the linear approximation is insufficient for an assessment of stability), there is every reason to check the solution obtained taking damping into account. This whole problem requires further investigation, and in particular, investigation based on non-linear equations.
>
> Bolotin [6], pp. 99–100

In another monograph of his, Bolotin clarifies this statement, discussing the dynamic stability of straight rods:

> If it is proved that the introduction of an arbitrarily small damping transforms the questionable case to asymptotic stability, then it is not necessary to make a strict analysis where only an approximate solution is required. Here, generally, this does not take place. Thus, Ziegler [1] has shown that in the case of a simple system subjected to the action of a 'following' force, the introduction of damping lowers the critical value in comparison with the value which is obtained from the simplified analysis.
>
> [1] H. Ziegler. Die Stabilitatskriterien der Elastomechanik. Ing.-Arch. 20, 49–56 (1952).
>
> Bolotin [14], §72, pp. 290–291

3.6.7 Numerical Considerations

In the case with a dead-weight loading only, in the absence of damping, the system reduces to a conservative one. The other cases can be thought of as different generalizations or modifications of this case.

Thus, the implicit midpoint rule (IMR) is a reasonable choice for the time integrator, since it approximately conserves energy over very long time intervals; the error in the energy oscillates around zero instead of accumulating. For Hamiltonian systems, IMR has the additional advantage of exactly preserving symplecticity.

Numerical validity checking can be performed by comparing the results against those using other integrators, such as RK4. Although the error in RK4 accumulates, rendering it useless for accurate simulation over very long time intervals, for shorter simulations it is useful to have as a point of comparison a method that comes from a different family. If the numerical scheme introduces spurious solutions (i.e., solutions that do not correspond to a solution of the original continuum problem), using integrators from different method families makes it more unlikely that both methods would produce the same spurious solutions. Hence, if the methods produce the same solution, it is likely, although of course not proven, that it actually corresponds to a solution of the original continuum problem.

Symplectic Euler (SE), although only $O(\Delta t)$, may also be useful as a check against programming mistakes, as it is very simple, but behaves qualitatively similarly to IMR.

Another possible time integrator is the time-discontinuous Galerkin method dG(q), which is an implicit method that often produces accurate results. In the examples we will use dG(2); for details on the method, see Appendix B.

Transformation of the nonlinear system into standard form

Standard time integrator methods for systems of autonomous differential equations accept equations of the form

$$\dot{\mathbf{u}} = f(\mathbf{u}) \,. \tag{3.6.162}$$

Let us transform the nonlinear dynamical equations into the form (3.6.162). Define the state vector

$$\mathbf{u} := (q_1, q_2, \dot{q}_1, \dot{q}_2) \,. \tag{3.6.163}$$

Expanding (3.6.162) into component form, we have

$$\frac{\partial}{\partial t}\begin{bmatrix} q_1 \\ q_2 \\ \dot{q}_1 \\ \dot{q}_2 \end{bmatrix} = \begin{bmatrix} f_1(q_1, q_2, \dot{q}_1, \dot{q}_2) \\ f_2(q_1, q_2, \dot{q}_1, \dot{q}_2) \\ f_3(q_1, q_2, \dot{q}_1, \dot{q}_2) \\ f_4(q_1, q_2, \dot{q}_1, \dot{q}_2) \end{bmatrix} \,. \tag{3.6.164}$$

The left-hand side is written this way, instead of directly as $(\dot{q}_1, \dot{q}_2, \ddot{q}_1, \ddot{q}_2)$, in order to avoid obscuring what the equation says: the time derivative of the state vector is a function of the state vector. This function is algebraic in the sense that it does not involve differentiation or integration of its arguments. At each timestep, we readily have access to the value of the state vector at the start of the timestep. Hence, we have values not only for q_1 and q_2, but also for $\dot{q}_1$ and $\dot{q}_2$.

Two of the component functions in (3.6.164) are trivial; $f_1 = f_1(\dot{q}_1) = \dot{q}_1$ and $f_2 = f_2(\dot{q}_2) = \dot{q}_2$. In effect, it is sufficient to just copy data from the last two components of the state vector; they are exactly the time derivatives of the first two components.

The other two component functions, f_3 and f_4, are determined from the dynamical equation, by solving it for $\ddot{q}_1$ and $\ddot{q}_2$.[5] We start from the nondimensional dynamical

[5] Solving at least in principle. What matters is that, given a state vector $(q_1, q_2, \dot{q}_1, \dot{q}_2)$, we somehow extract the corresponding values for $(\ddot{q}_1, \ddot{q}_2)$ from the dynamical equation. It does not matter whether

equation, that is, (3.6.5) multiplied by $-\tau^2/\mu\ell^2$, taking a minus sign for convenience:

$$\frac{\tau^2}{\mu\ell^2}(-\mathbf{L}-\mathbf{Q}+\mathbf{J})=0\,. \tag{3.6.165}$$

The nondimensional generalized forces are given by (3.6.54)–(3.6.56),

$$\frac{\tau^2}{\mu\ell^2}\mathbf{L}=-\begin{bmatrix}\gamma q_1+\beta\dot{q}_1\\ \gamma q_2+\beta\dot{q}_2\end{bmatrix},$$

$$\frac{\tau^2}{\mu\ell^2}\mathbf{Q}=\alpha\left(-\Pi_F\begin{bmatrix}\sin q_2\\ 0\end{bmatrix}+\cos(\theta)G\left(\mu_m\begin{bmatrix}\sin q_1+\sin(q_1+q_2)\\ \sin(q_1+q_2)\end{bmatrix}+\frac{1}{2}\mu_\rho\begin{bmatrix}3\sin(q_1)+\sin(q_1+q_2)\\ \sin(q_1+q_2)\end{bmatrix}\right)\right),$$

$$\frac{\tau^2}{\mu\ell^2}\mathbf{J}=\alpha^2\mu_\rho\begin{bmatrix}\frac{4}{3}\ddot{q}_1+\frac{1}{2}\cos q_2(\ddot{q}_1+\ddot{q}_2)-\frac{1}{2}\sin q_2\,(\dot{q}_1+\dot{q}_2)^2\\ +\frac{1}{2}\cos(q_2)\ddot{q}_1+\frac{1}{2}\sin q_2\,(\dot{q}_1)^2+\frac{1}{3}(\ddot{q}_1+\ddot{q}_2)\\ \frac{1}{2}\cos(q_2)\ddot{q}_1+\frac{1}{2}\sin q_2\,(\dot{q}_1)^2+\frac{1}{3}(\ddot{q}_1+\ddot{q}_2)\end{bmatrix}$$
$$+\alpha^2\mu_m\begin{bmatrix}\ddot{q}_1+\cos q_2(\ddot{q}_1+\ddot{q}_2)-\sin q_2\,(\dot{q}_1+\dot{q}_2)^2\\ +\cos(q_2)\ddot{q}_1+\sin q_2\,(\dot{q}_1)^2+(\ddot{q}_1+\ddot{q}_2)\\ \cos(q_2)\ddot{q}_1+\sin q_2\,(\dot{q}_1)^2+(\ddot{q}_1+\ddot{q}_2)\end{bmatrix}.$$

In the case of point-mass rods, (3.6.56) is replaced by (3.6.57),

$$\frac{\tau^2}{\mu\ell^2}\mathbf{J}_{\mathrm{pm}}=\alpha^2\mu_\rho\begin{bmatrix}\frac{5}{4}\ddot{q}_1+\frac{1}{2}\cos q_2(\ddot{q}_1+\ddot{q}_2)-\frac{1}{2}\sin q_2\,(\dot{q}_1+\dot{q}_2)^2\\ +\frac{1}{2}\cos(q_2)\ddot{q}_1+\frac{1}{2}\sin q_2\,(\dot{q}_1)^2+\frac{1}{4}(\ddot{q}_1+\ddot{q}_2)\\ \frac{1}{2}\cos(q_2)\ddot{q}_1+\frac{1}{2}\sin q_2\,(\dot{q}_1)^2+\frac{1}{4}(\ddot{q}_1+\ddot{q}_2)\end{bmatrix}$$
$$+\alpha^2\mu_m\begin{bmatrix}\ddot{q}_1+\cos q_2(\ddot{q}_1+\ddot{q}_2)-\sin q_2\,(\dot{q}_1+\dot{q}_2)^2\\ +\cos(q_2)\ddot{q}_1+\sin q_2\,(\dot{q}_1)^2+(\ddot{q}_1+\ddot{q}_2)\\ \cos(q_2)\ddot{q}_1+\sin q_2\,(\dot{q}_1)^2+(\ddot{q}_1+\ddot{q}_2)\end{bmatrix},$$

In component form, we have

the solution process is an explicit formula, is based on iterating an implicit system of equations, or is based on some entirely different approach.

$$\gamma q_1 + \beta \dot{q}_1 + \alpha \Pi_F \sin(q_2) - \alpha \cos(\theta) G \left[\mu_m (\sin q_1 + \sin(q_1 + q_2)) + \frac{1}{2}\mu_\rho (3 \sin(q_1) + \sin(q_1 + q_2)) \right]$$
$$+ \alpha^2 \mu_\rho \left[\frac{4}{3}\ddot{q}_1 + \frac{1}{2}\cos q_2 (\ddot{q}_1 + \ddot{q}_2) - \frac{1}{2}\sin q_2 (\dot{q}_1 + \dot{q}_2)^2 + \frac{1}{2}\cos(q_2)\ddot{q}_1 + \frac{1}{2}\sin q_2 (\dot{q}_1)^2 + \frac{1}{3}(\ddot{q}_1 + \ddot{q}_2) \right]$$
$$+ \alpha^2 \mu_m \left[\ddot{q}_1 + \cos q_2 (\ddot{q}_1 + \ddot{q}_2) - \sin q_2 (\dot{q}_1 + \dot{q}_2)^2 + \cos(q_2)\ddot{q}_1 + \sin q_2 (\dot{q}_1)^2 + (\ddot{q}_1 + \ddot{q}_2) \right] = 0 , \tag{3.6.166}$$

$$\gamma q_2 + \beta \dot{q}_2 - \alpha \cos(\theta) G (\mu_m + \frac{1}{2}\mu_\rho) \sin(q_1 + q_2)$$
$$+ \alpha^2 \mu_\rho \left[\frac{1}{2}\cos(q_2)\ddot{q}_1 + \frac{1}{2}\sin q_2 (\dot{q}_1)^2 + \frac{1}{3}(\ddot{q}_1 + \ddot{q}_2) \right]$$
$$+ \alpha^2 \mu_m \left[\cos(q_2)\ddot{q}_1 + \sin q_2 (\dot{q}_1)^2 + (\ddot{q}_1 + \ddot{q}_2) \right] = 0 . \tag{3.6.167}$$

For point-mass rods, we replace each instance of $4/3$ by $5/4$, and each $1/3$ by $1/4$; no other changes are needed.

Let us isolate the terms related to $\ddot{q}_1$ and $\ddot{q}_2$ in the second component equation, (3.6.167):

$$\gamma q_2 + \beta \dot{q}_2 - \alpha \cos(\theta) G (\mu_m + \frac{1}{2}\mu_\rho) \sin(q_1 + q_2)$$
$$+ \alpha^2 \left[(\frac{1}{2}\mu_\rho + \mu_m) \right] \sin(q_2)(\dot{q}_1)^2$$
$$+ \alpha^2 \left[(\frac{1}{2}\mu_\rho + \mu_m)\cos(q_2) + (\frac{1}{3}\mu_\rho + \mu_m) \right] \ddot{q}_1$$
$$+ \alpha^2 (\frac{1}{3}\mu_\rho + \mu_m)\ddot{q}_2 = 0 . \tag{3.6.168}$$

If we wish, we may explicitly solve this for $\ddot{q}_2$:

$$\alpha^2 (\frac{1}{3}\mu_\rho + \mu_m)\ddot{q}_2 = - \gamma q_2 - \beta \dot{q}_2 + \alpha \cos(\theta) G (\mu_m + \frac{1}{2}\mu_\rho) \sin(q_1 + q_2)$$
$$- \alpha^2 \left[(\frac{1}{2}\mu_\rho + \mu_m) \right] \sin(q_2)(\dot{q}_1)^2$$
$$- \alpha^2 \left[(\frac{1}{2}\mu_\rho + \mu_m)\cos(q_2) + (\frac{1}{3}\mu_\rho + \mu_m) \right] \ddot{q}_1 . \tag{3.6.169}$$

For point-mass rods, replace each instance of $1/3$ by $1/4$ in (3.6.168) and (3.6.169).

The coefficient of $\ddot{q}_2$ is always nonzero, because α is the length scale and at least one of the masses is always nonzero. Recall that we have the state vector $(q_1, q_2, \dot{q}_1, \dot{q}_2)$ explicitly available at each timestep. Thus, if we can obtain a value for $\ddot{q}_1$, we obtain the corresponding value of $\ddot{q}_2$ from (3.6.169).

To do this, we obviously need to use the other component equation. As before, subtracting the second equation, (3.6.167), from the first one, (3.6.166), eliminates some terms. We have

$$
\begin{aligned}
\gamma(q_1 - q_2) + \beta(\dot{q}_1 - \dot{q}_2) + \alpha \Pi_F \sin(q_2) - \alpha \cos(\theta) G \left[\mu_m + \frac{3}{2}\mu_\rho\right] \sin(q_1) \\
+ \alpha^2 \mu_\rho \left[\frac{4}{3}\ddot{q}_1 + \frac{1}{2}\cos q_2 (\ddot{q}_1 + \ddot{q}_2) - \frac{1}{2}\sin q_2 \, (\dot{q}_1 + \dot{q}_2)^2\right] \\
+ \alpha^2 \mu_m \left[\ddot{q}_1 + \cos q_2 (\ddot{q}_1 + \ddot{q}_2) - \sin q_2 \, (\dot{q}_1 + \dot{q}_2)^2\right] = 0 \, .
\end{aligned}
$$

For point-mass rods, replace the $4/3$ by $5/4$.

Similarly to above, we reorganize the result to isolate the terms related to $\ddot{q}_1$ and $\ddot{q}_2$:

$$
\begin{aligned}
\gamma(q_1 - q_2) + \beta(\dot{q}_1 - \dot{q}_2) + \alpha \Pi_F \sin(q_2) - \alpha \cos(\theta) G \left[\mu_m + \frac{3}{2}\mu_\rho\right] \sin(q_1) \\
- \alpha^2 \left[\frac{1}{2}\mu_\rho + \mu_m\right] \sin(q_2) \, (\dot{q}_1 + \dot{q}_2)^2 \\
+ \alpha^2 \left[\mu_\rho (\frac{4}{3} + \frac{1}{2}\cos q_2) + \mu_m (1 + \cos q_2)\right] \ddot{q}_1 \\
+ \alpha^2 \left[\frac{1}{2}\mu_\rho + \mu_m\right] \cos(q_2) \ddot{q}_2 = 0 \, .
\end{aligned}
\tag{3.6.170}
$$

At this point, we observe that (3.6.168) and (3.6.170) depend linearly on $\ddot{q}_1$ and $\ddot{q}_2$, and set up a linear equation system.

We have

$$
\mathbf{A}\mathbf{a} = \mathbf{b} \, , \tag{3.6.171}
$$

where the unknown generalized acceleration $\mathbf{a} := (\ddot{q}_1, \ddot{q}_2)$, and $\mathbf{A}$ and $\mathbf{b}$ depend only on the explicitly known state vector $(q_1, q_2, \dot{q}_1, \dot{q}_2)$:

$$
\mathbf{A} := \alpha^2 \begin{bmatrix} (\frac{1}{2}\mu_\rho + \mu_m)\cos(q_2) + (\frac{1}{3}\mu_\rho + \mu_m) & (\frac{1}{3}\mu_\rho + \mu_m) \\ \mu_\rho (\frac{4}{3} + \frac{1}{2}\cos q_2) + \mu_m (1 + \cos q_2) & (\frac{1}{2}\mu_\rho + \mu_m)\cos(q_2) \end{bmatrix} , \tag{3.6.172}
$$

$$
\mathbf{b} := \begin{bmatrix} -\gamma q_2 - \beta \dot{q}_2 + \alpha \cos(\theta) G (\mu_m + \frac{1}{2}\mu_\rho) \sin(q_1 + q_2) - \alpha^2 \left[(\frac{1}{2}\mu_\rho + \mu_m)\right] \sin(q_2) (\dot{q}_1)^2 \\ -\gamma(q_1 - q_2) - \beta(\dot{q}_1 - \dot{q}_2) - \alpha \Pi_F \sin(q_2) + \alpha \cos(\theta) G \left[\mu_m + \frac{3}{2}\mu_\rho\right] \sin(q_1) + \alpha^2 \left[\frac{1}{2}\mu_\rho + \mu_m\right] \sin(q_2) \, (\dot{q}_1 + \dot{q}_2)^2 \end{bmatrix} . \tag{3.6.173}
$$

Equation (3.6.171) can then be solved via any method for solving linear equation systems.

The matrix $\mathbf{A}$ is always invertible. To show this, let us assume the opposite, and determine the conditions under which the rows become linearly dependent. In order for the second row to become some constant times the first one, it must hold that

$$\frac{\mu_\rho(\frac{4}{3}+\frac{1}{2}\cos q_2)+\mu_m(1+\cos q_2)}{(\frac{1}{2}\mu_\rho+\mu_m)\cos(q_2)+(\frac{1}{3}\mu_\rho+\mu_m)}=\frac{(\frac{1}{2}\mu_\rho+\mu_m)\cos(q_2)}{(\frac{1}{3}\mu_\rho+\mu_m)},$$

which can be reorganized as

$$\frac{(\frac{1}{2}\mu_\rho+\mu_m)\cos q_2+(\frac{4}{3}\mu_\rho+\mu_m)}{(\frac{1}{2}\mu_\rho+\mu_m)\cos(q_2)+(\frac{1}{3}\mu_\rho+\mu_m)}=\frac{(\frac{1}{2}\mu_\rho+\mu_m)\cos(q_2)}{(\frac{1}{3}\mu_\rho+\mu_m)}.$$

To shorten the notation, let

$$x := (\frac{1}{2}\mu_\rho+\mu_m)\cos q_2\ , \quad r := (\frac{1}{3}\mu_\rho+\mu_m)\ .$$

We have

$$\frac{x+r+\mu_\rho}{x+r}=\frac{x}{r},$$

which yields

$$rx+r^2+r\mu_\rho=x^2+rx\ ,$$

in other words,

$$x^2=r(r+\mu_\rho)\ ,$$

or in the original notation,

$$(\frac{1}{2}\mu_\rho+\mu_m)^2(\cos q_2)^2=(\frac{1}{3}\mu_\rho+\mu_m)(\frac{4}{3}\mu_\rho+\mu_m)\ .$$

Considering that μ_ρ and μ_m remain constant during any given simulation, and q_2 varies, it makes the most sense to solve for q_2. We have

$$(\cos q_2)^2=\frac{(\frac{1}{3}\mu_\rho+\mu_m)(\frac{4}{3}\mu_\rho+\mu_m)}{(\frac{1}{2}\mu_\rho+\mu_m)^2}\ . \tag{3.6.174}$$

The left-hand side is always in the range [0, 1]. Let us determine when the right-hand side is in this same range. Denoting

$$M := \mu_m/\mu_\rho\ , \quad \text{i.e.}\quad \mu_m = M\mu_\rho\ ,$$

we proceed by rewriting

$$(\cos q_2)^2 = \frac{(\frac{1}{3}\mu_\rho + \mu_m)(\frac{4}{3}\mu_\rho + \mu_m)}{(\frac{1}{2}\mu_\rho + \mu_m)^2} = \frac{\frac{4}{9}\mu_\rho^2 + \frac{5}{3}\mu_\rho\mu_m + \mu_m^2}{\frac{1}{4}\mu_\rho^2 + \mu_\rho\mu_m + \mu_m^2} = \frac{\frac{4}{9} + \frac{5}{3}M + M^2}{\frac{1}{4} + M + M^2} =: Q\,,$$

which now depends on the nondimensional mass ratio M only. At $M = 0$, which is the smallest admissible value of M, corresponding to no dead weight loading, we have $Q = 16/9 > 1$. As $M \to +\infty$, $Q \to 1_+$. The derivative is easily obtained by the quotient rule:

$$\frac{\partial Q}{\partial M} = \frac{(\frac{5}{3} + 2M)(\frac{1}{4} + M + M^2) - (\frac{4}{9} + \frac{5}{3}M + M^2)(1 + 2M)}{(\frac{1}{4} + M + M^2)^2} = -\frac{4}{9} \cdot \frac{12M + 1}{(2M + 1)^3}\,.$$

This is negative for all $M \geq 0$. Since $Q(M = 0) > 1$, and Q is continuous and strictly decreasing for all $M \geq 0$, and is bounded from below by the limiting value 1 as $M \to +\infty$, it follows that $Q > 1$ for all $M \geq 0$.

Hence however we choose the values of μ_ρ and μ_m, the condition (3.6.174) is never satisfied, in other words, the rows of **A** cannot become linearly dependent. Therefore the matrix **A** is always invertible.

Observe that the diagonal elements of **A** are never simultaneously zero, but each of them becomes zero for some values of q_2. Hence, standard LU decomposition based methods will require pivoting. This leads to a relatively complicated algorithm (see, e.g., Golub and Loan [17]), especially considering the trivial size of the problem.

Thus, for simplicity it is thus actually preferable to explicitly set up the inverse matrix:

$$\mathbf{A} =: \begin{bmatrix} a_{11} & a_{12} \\ a_{21} & a_{22} \end{bmatrix}, \quad \mathbf{A}^{-1} = (a_{11}a_{22} - a_{12}a_{21})^{-1} \begin{bmatrix} a_{22} & -a_{12} \\ -a_{21} & a_{11} \end{bmatrix}. \tag{3.6.175}$$

Computing $\mathbf{A}^{-1}$ requires only four floating point operations to get the reciprocal of the determinant, which is always finite because **A** is invertible, and then two negations and four multiplications to obtain the elements of $\mathbf{A}^{-1}$, bringing the operation count of the inversion to ten. To obtain the generalized acceleration **a**, additionally three operations per component are needed, in the matrix-vector product $\mathbf{A}^{-1}\mathbf{b}$. At a total of sixteen operations for obtaining **a** (given **A** and **b**), this approach is still relatively cheap. Also importantly, it does not matter if a_{11} or a_{22} is zero; we can always run the same operations without branching, making the algorithm very simple.

Equations (3.6.171)–(3.6.173) can be used to obtain $(\ddot{q}_1, \ddot{q}_2)$, which are then fed into (3.6.164) (providing values for f_3 and f_4) in the time integrator. This concludes the numerical solution process for the nonlinear system.

3.6.8 Results

To illustrate the dynamical behavior of the double pendulum, we have computed some trajectories using dG(2), the second-order time-discontinuous Galerkin method (an exposition of the method is provided in Appendix B). When performing Banach fixed point iteration for a timestep, we iterate until the 64-bit floating-point representation converges down to its last bit, or up to 100 iterations, whichever occurs first. (It was observed that for a majority of timesteps, convergence occurs first.)

We have performed a parametric study with regard to damping (β) and loading parameters (μ_m and Π_F), to illustrate how the system behaves in different regions of the (nondimensional) parameter space. We use the parameterization (3.6.59). For the most part of the results, we have used a fixed value for the spring stiffness, $\gamma = 0.5$, as this was observed to yield the most interesting dynamics.

We visualize the trajectories of the state vector $(q_1, \dot{q}_1, q_2, \dot{q}_2)$ as density fields of its two-dimensional projections $(q_1, \dot{q}_1)$, $(q_2, \dot{q}_2)$, (q_1, q_2) and $(\dot{q}_1, \dot{q}_2)$, gathered over the history of a given simulation. The density field is constructed from the discrete state samples (value of the state vector at each timestep) using kernel density estimation. Density estimation, generally speaking, is a family of techniques to extract an approximation of a continuous density function, when given only a set of samples drawn from it. This family of methods is most commonly used in statistics, but can be applied to the approximation any density field for which samples are available. The core idea of kernel density estimation, specifically, is to place a copy of a given function (called the kernel; typically, a unimodal distribution) on each of the samples, and then sum the contributions. The method is fairly insensitive to the shape of the kernel, but can be sensitive to the bandwidth (the width of the peak of the kernel). For discussion on the effects of kernel bandwidth and how to choose it, see e.g. Sheather and Jones [18], Duong and Hazelton [19], Chacón and Duong [20], Botev et al. citeBotevEtAl2010.

We use density fields, because they give useful information also when the figure is so densely populated that individual trajectory lines can no longer be seen. This will occur when a simulation runs long enough, as the system revisits parts of the state space it has already been in. Secondly, this occurs when several simulations are plotted on top of each other, which is useful for uncertainty quantification (UQ). Sample-based UQ is reviewed in Helton et al. [21]; see also Helton and Davis [22].

The idealized special case with no damping ($\beta = 0$) is especially interesting, to see qualitative changes in the dynamics when compared to the same case with small but finite damping (i.e., in the limit of vanishing damping). When there is no damping, the investigated system propagates with no decay not only the solution, but also all numerical error (such as roundoff, cancellation, and truncation), so error may in principle accumulate exponentially. Whether and how much this matters depends on how sensitive the limiting motion in a long simulation, considered as a whole (as a subset of state space), is to small perturbations to the state-space position anywhere along the trajectory. Here sample-based UQ can be seen as a tool to capture not only one arbitrary trajectory (with an arbitrary history of accumulated numerical error, depending on implementation details), but also some nearby trajectories. It stands to reason that a set of such computed trajectories, considered as a whole, should be much more stable against small perturbations than any individual member of the set.

The trajectory density exhibits very large variations. To produce a clear, readable visualization, we use a new method based on data-adaptive dynamic range compression. The color scale is neither linear nor logarithmic, but computationally adapted to the data specifically displayed in each figure, in order to bring out as much detail as possible. The algorithm was introduced by Ward-Larson et al. [23] in the context of tone-mapping high dynamic range (HDR) 3D computer graphics for visualization on regular (low dynamic range) computer displays. Although the original study was published over 20 years ago, which in information technology may seem a long time, the issue remains as relevant as ever. Computer displays are still limited to a dynamic range of approximately 45 dB, and print media even less than that. The algorithm was first used for postprocessing of high dynamic range mathematical functions, for their visualization on regular computer displays, in Jeronen [24]. Even in the last decade, tone mapping has remained mainly a special-interest topic in computer graphics and HDR photography, despite being very promising for the visualization of density fields (which typically are HDR) in the computational sciences.

Because in our examples here the trajectory density is only a visualization aid, the actual values of the density are not important. Therefore, we have used a qualitative shading, with darker shades corresponding to higher densities. The computational steps in the tone mapper remain the same as in Jeronen [24, 25]; what this means in practice is only that no color bar is shown. What is quantitatively important is the shape of the limiting motion; this can be read off the figures by observing where trajectories are the densest. The density field visualization allows us to distinguish, at a glance, between attractive steady states (appearing in the visualization as points of high density), limit cycles (curve-like regions of high density), and chaotic motion (filled regions with no obvious pattern).

Finally, it is physically impossible to set up a system precisely at an unstable equilibrium point. To simulate this uncertainty of the initial position, we have performed sample-based uncertainty quantification, to obtain a sample of trajectories that start near an unstable equilibrium point. For an equilibrium point at $(q_1^0, 0, q_2^0, 0)$, we have sampled $(q_1^0 \pm \Delta q, 0, q_2^0 \pm \Delta q, 0)$ on a uniform grid, and performed an ensemble simulation. Gathering the data from all the individual simulations into the same trajectory density field causes both branches of a bifurcation to contribute to the trajectory density, allowing us to see both solutions simultaneously. In all the direct time simulations shown, we have used $q_1^0 = q_2^0 = 0$, and $\Delta q = 10^{-2}$, with a grid size of $4 \times 4 = 16$ samples.

Because the setup of the ensemble is symmetric with respect to the origin, it also provides a small but important visual aid to help judge the validity of the numerical results. We know that the system is symmetric with regard to the trivial equilibrium position, so any significant asymmetry in any density field produced by an UQ simulation (provided the simulation is long enough to exhibit the limiting motion) indicates that accumulated error has caused at least some of the individual trajectories to diverge from the correct solution. During the study, when comparing results obtained with different time integration algorithms, it was this feature that hinted that e.g. with RK4, using some form of compensated summation (e.g. Kahan [26]) in the state vector update was critical in order to produce correct results in cases with no damping, and that dG(2) seemed reliable even at relatively large timestep sizes. Note that if maximal accuracy is desired, Kahan's algorithm has been superseded by the binary floating-point summation accurate to full precision introduced by Shewchuk [27].

Results are collected into Figs. 3.3 and 3.14. First, Figs. 3.3 and 3.4 display a state space atlas in a small-multiples format for the cases with dead-weight loading and follower-force loading, respectively. Note the mass of the rods themselves always contributes some dead-weight loading (whenever the system inclination $\theta \neq 0$, as is the case for our parameterization). The timestep used in Figs. 3.3 and 3.4 is $\Delta t = 5 \cdot 10^{-2}$, with $n_t = 2000$ timesteps taken, yielding a simulation end time of $t_f = 100$.

In each subfigure, the top-left projection (q_1, q_2) is a goniometer-like representation of the angular position of both springs; the axes represent the angles q_k. Technically the data is not a Lissajous curve, because the problem in nonlinear, so large motions are not harmonic. The benefit of this projection is that it allows to see at a glance whether the springs bend in the same (upper right and lower left quadrants) or opposite (the other two quadrants) directions. The top-center projection $(\dot{q}_1, \dot{q}_2)$ is similar, but for the angular velocities.

The bottom-left and bottom-center projections display $(q_1, \dot{q}_1)$ and $(q_2, \dot{q}_2)$, respectively. These correspond to the phase spaces of the first and second rotational spring. Because in these projections, the vertical axis represents the time derivative $\dot{q}_k$, if a given point on the trajectory is above the horizontal axis, q_k must be increasing, so the trajectory is to be followed toward the right. Similarly, if a point is below the horizontal axis, the trajectory is to be followed left. Hence these projections are read in a clockwise fashion around the origin.

Finally, the top-right plot directly visualizes the spatial position of the free end of the system, and the bottom-right plot visualizes the spatial position of the joint between the rods. Because the rods are rigid and the fixed end of the system is located at the origin, the joint is constrained to move on the unit circle, $(x_1, y_1) = (\sin q_1, \cos q_1)$. Note q_1 is measured from the positive vertical direction (up). The free end is mechanically able to reach positions $(x_2, y_2) = (x_1, y_1) + (\sin(q_1 + q_2), \cos(q_1 + q_2))$, which amounts to a circle of radius 2 centered on the origin. Note q_2 is measured from the current direction of the first rod, so the angle of the second rod as measured from the positive vertical is $q_1 + q_2$.

From Figs. 3.3 and 3.4, we observe that in the unloaded case, with no damping the system fills a region of the state space with chaotic motion. With dead-weight loading, Fig. 3.3, as damping is increased, first limit cycles appear, and once the damping is high enough, the limit cycles become replaced with a pair of stable equilibrium positions. As loading is increased, the vertical spikes in the $(q_2, \dot{q}_2)$ plots hint at the existence of further equilibrium positions.

Figure 3.4 considers the case with follower-force loading. We see that unlike in the dead-weight loading case, limit cycles exist also for low values of damping (and

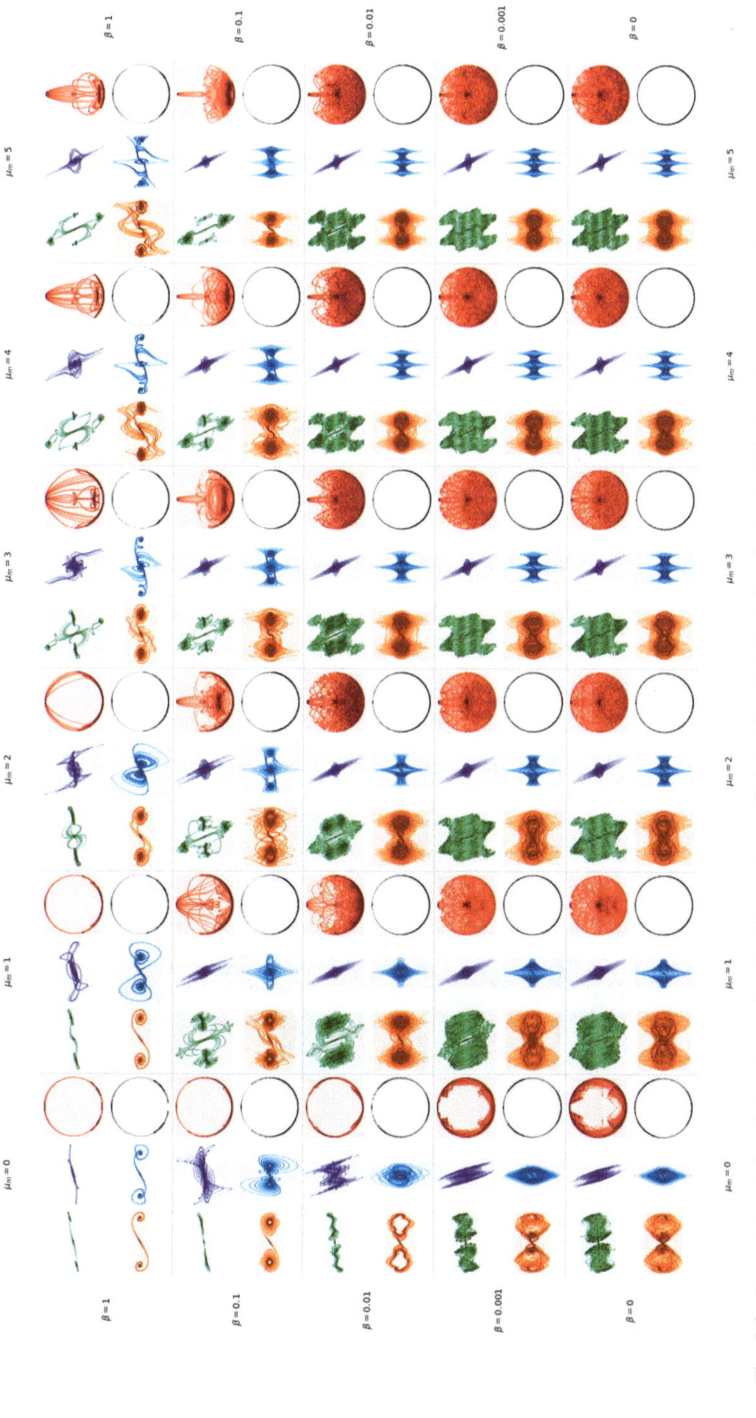

Fig. 3.3 Parametric study of state space density. Stiffness $\gamma = 0.5$, varying damping β and dead-weight loading μ_m. No follower-force loading, $\Pi_F = 0$. In each subfigure, *Top left:* (q_1, q_2); *Top center:* $(\dot{q}_1, \dot{q}_2)$; *Top right:* position of free end; *Bottom left:* $(q_1, \dot{q}_1)$; *Bottom center:* $(q_2, \dot{q}_2)$; *Bottom right:* position of the joint between the rods

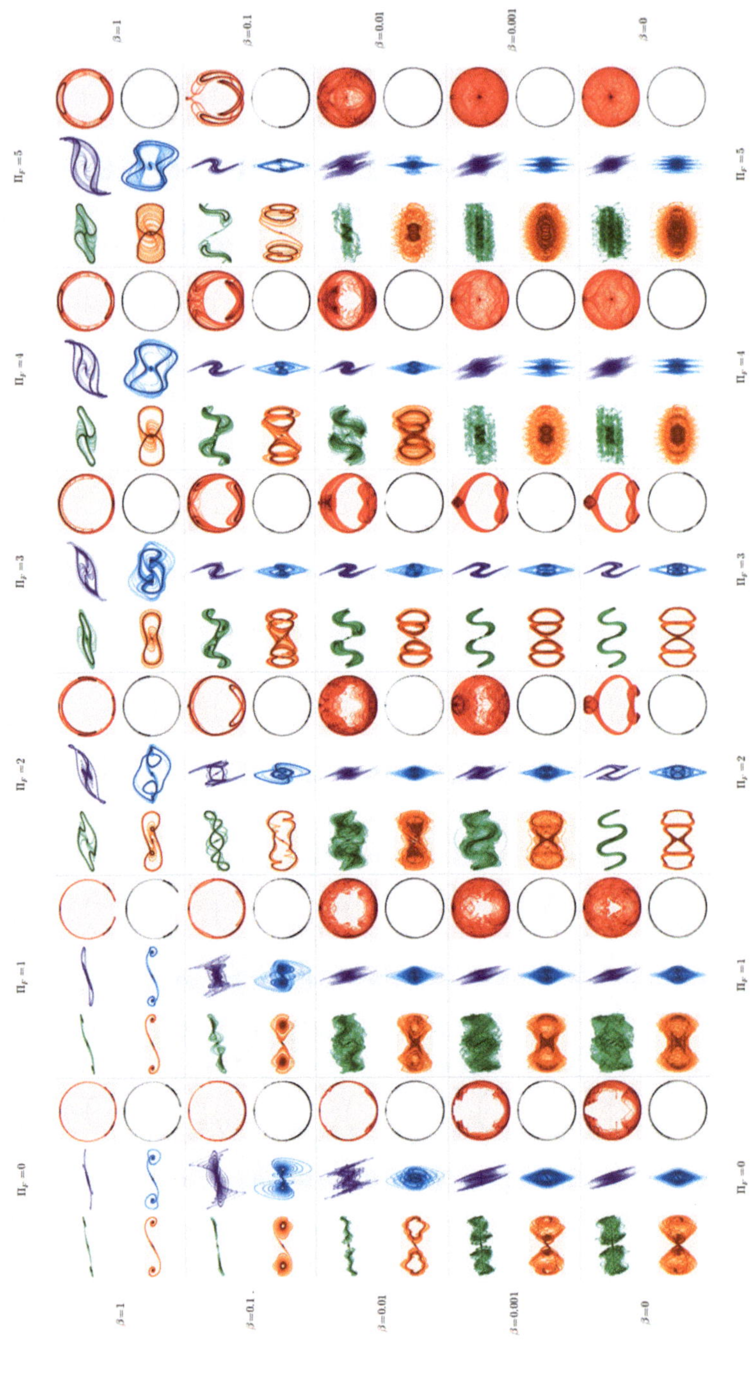

Fig. 3.4 Parametric study of state space density. Stiffness $\gamma = 0.5$, varying damping β and follower-force loading Π_F. No dead-weight loading, $\mu_m = 0$. In each subfigure, *Top left*: (q_1, q_2); *Top center*: $(\dot{q}_1, \dot{q}_2)$; *Top right*: position of free end; *Bottom left*: $(q_1, \dot{q}_1)$; *Bottom center*: $(q_2, \dot{q}_2)$; *Bottom right*: position of the joint between the rods

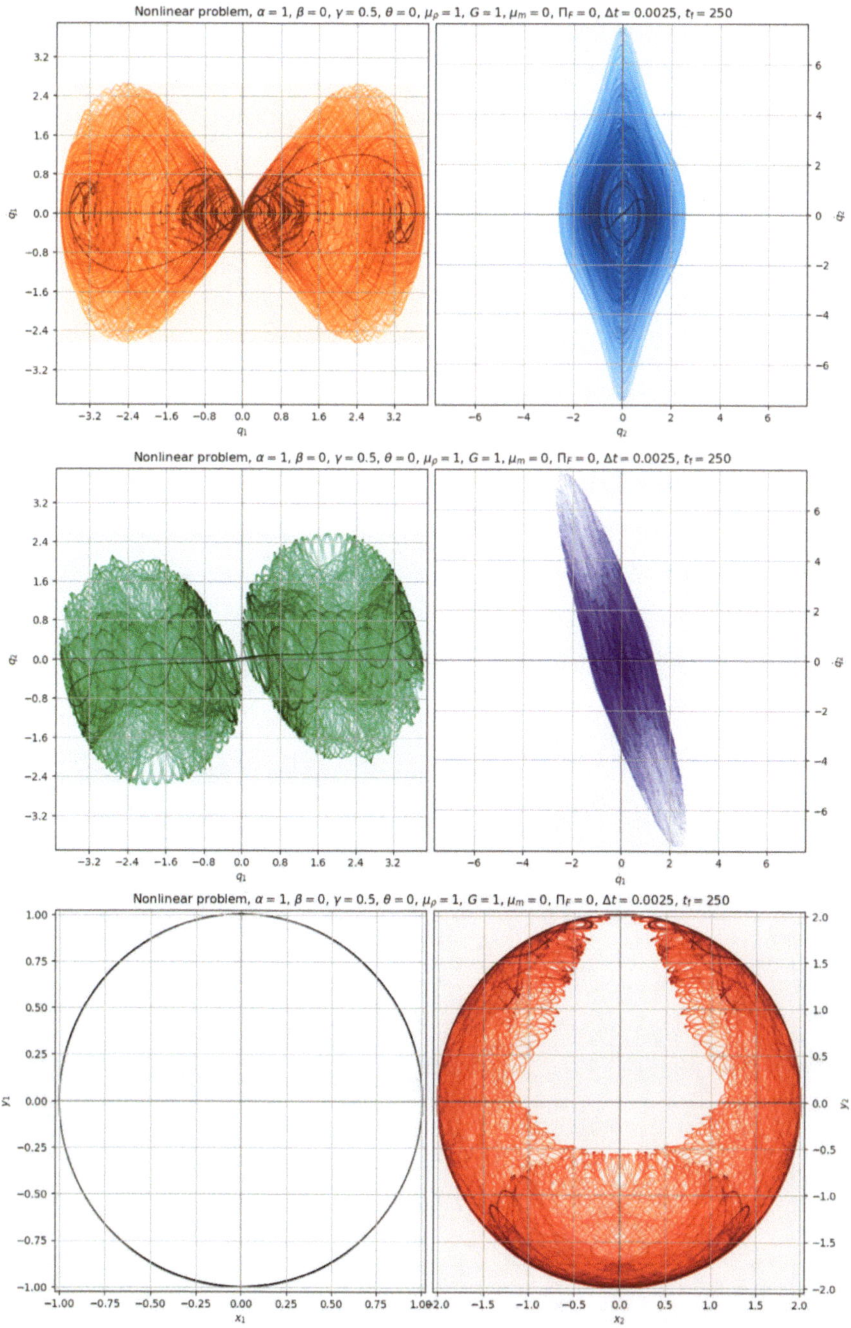

Fig. 3.5 State space density. Stiffness $\gamma = 0.5$, damping $\beta = 0$. No loading, $\mu_m = 0$, $\Pi_F = 0$. *Top*: $(q_1, \dot{q}_1)$ and $(q_2, \dot{q}_2)$, *Middle*: (q_1, q_2) and $(\dot{q}_1, \dot{q}_2)$. *Bottom*: positions of the joint between the rods and the free end

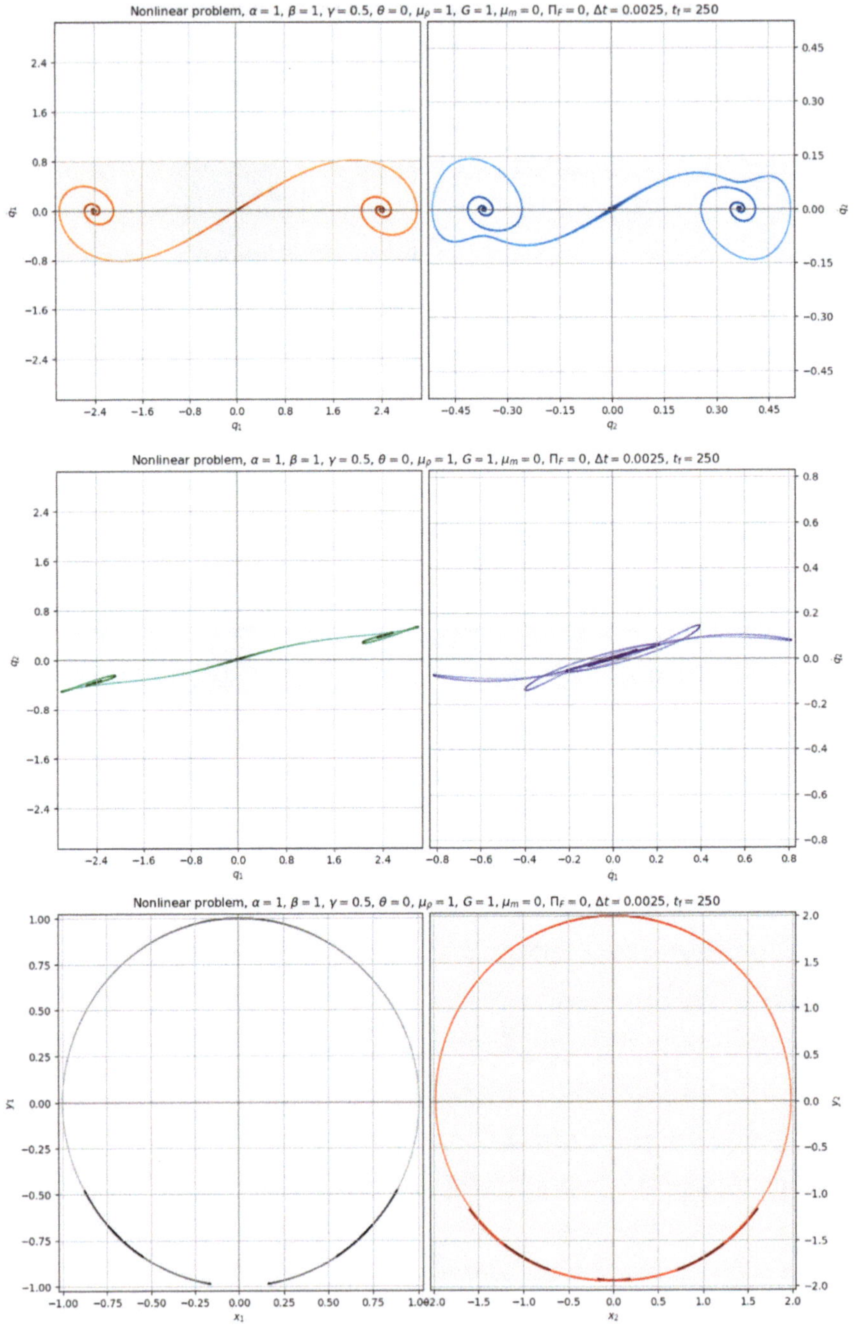

Fig. 3.6 State space density. Stiffness $\gamma = 0.5$, damping $\beta = 1$. No loading, $\mu_m = 0$, $\Pi_F = 0$. *Top*: $(q_1, \dot{q}_1)$ and $(q_2, \dot{q}_2)$, *Middle*: (q_1, q_2) and $(\dot{q}_1, \dot{q}_2)$. *Bottom*: positions of the joint between the rods and the free end

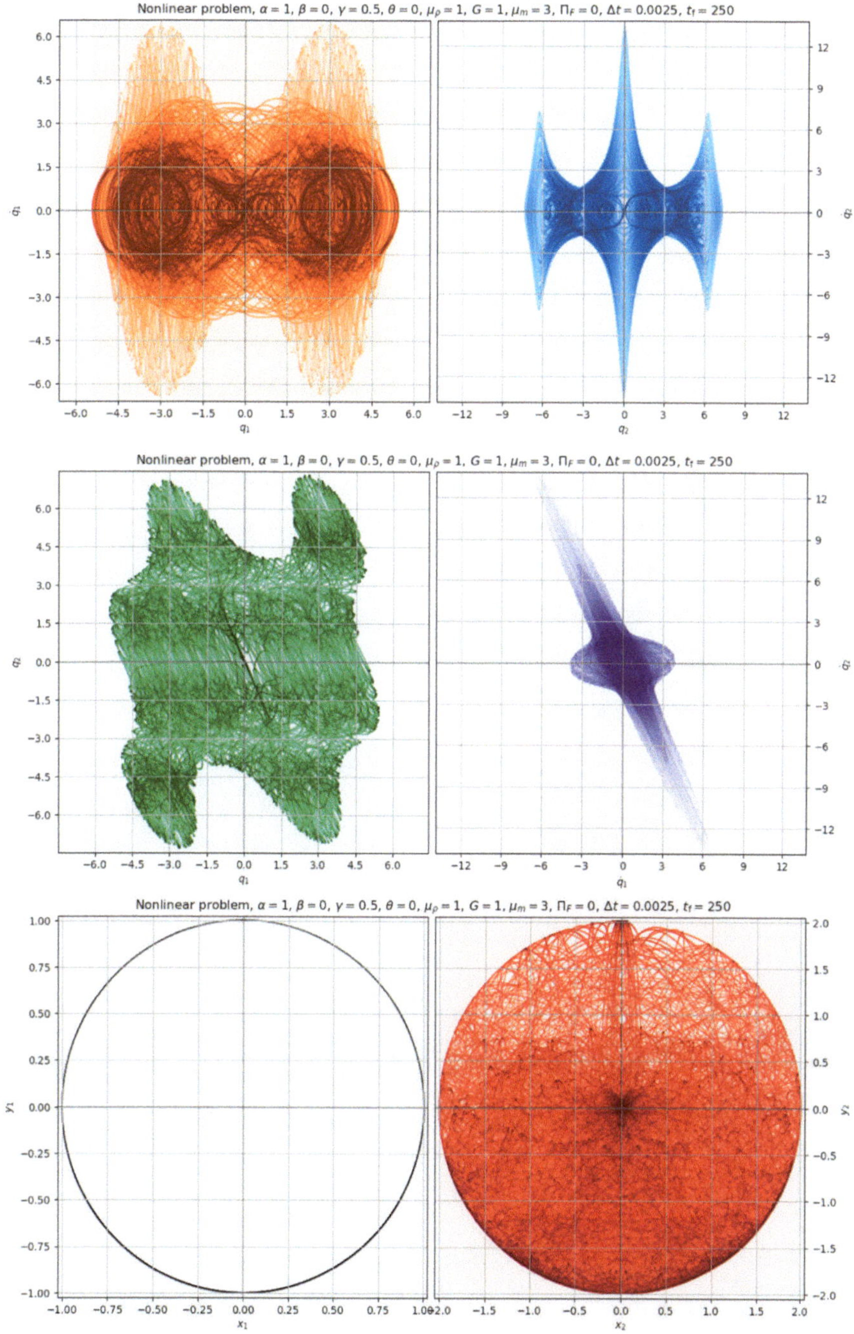

Fig. 3.7 State space density. Stiffness $\gamma = 0.5$, damping $\beta = 0$. Dead-weight loading, $\mu_m = 3$. No follower-force loading, $\Pi_F = 0$. *Top*: $(q_1, \dot{q}_1)$ and $(q_2, \dot{q}_2)$, *Middle*: (q_1, q_2) and $(\dot{q}_1, \dot{q}_2)$. *Bottom*: positions of the joint between the rods and the free end

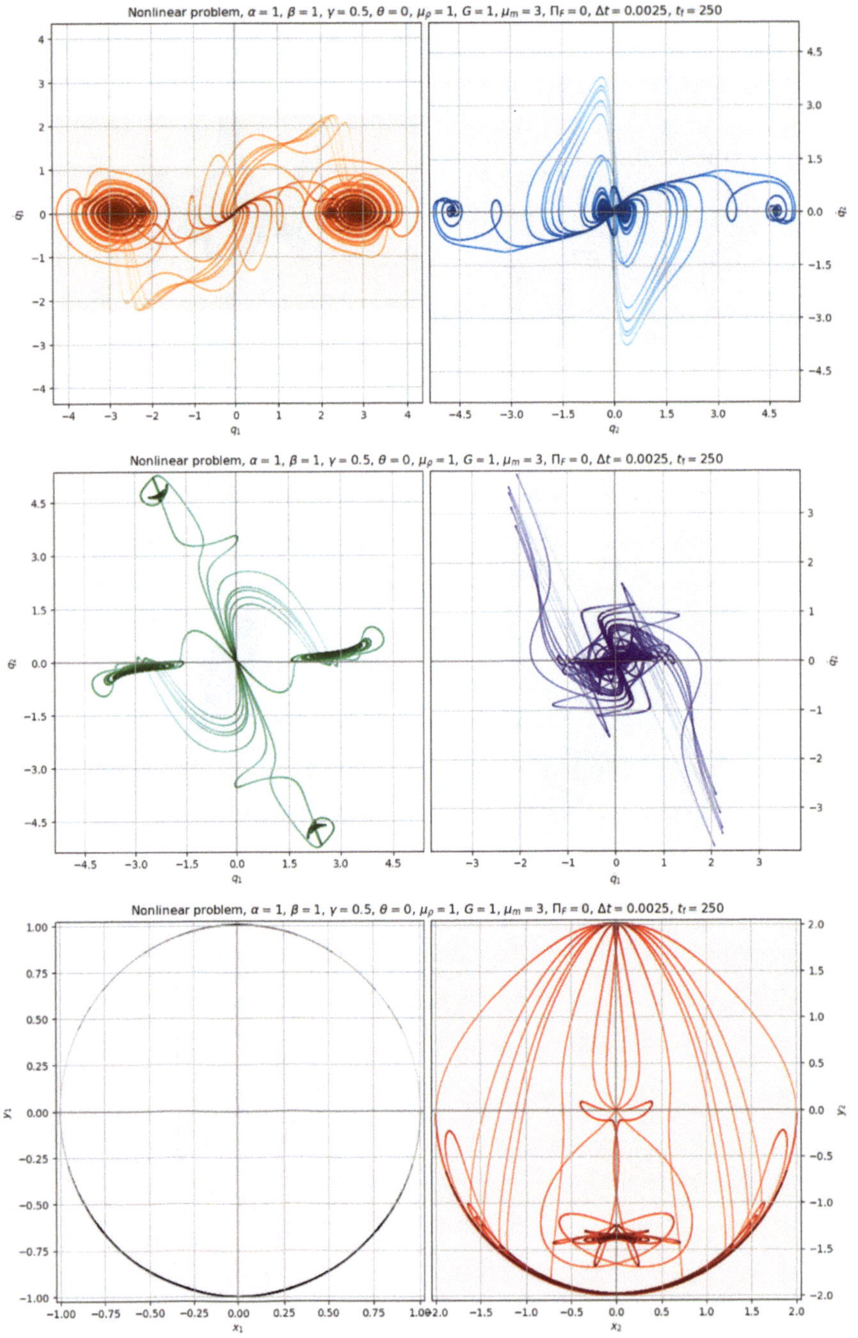

Fig. 3.8 State space density. Stiffness $\gamma = 0.5$, damping $\beta = 1$. Dead-weight loading, $\mu_m = 3$. No follower-force loading, $\Pi_F = 0$. *Top*: $(q_1, \dot{q}_1)$ and $(q_2, \dot{q}_2)$, *Middle*: (q_1, q_2) and $(\dot{q}_1, \dot{q}_2)$. *Bottom*: positions of the joint between the rods and the free end

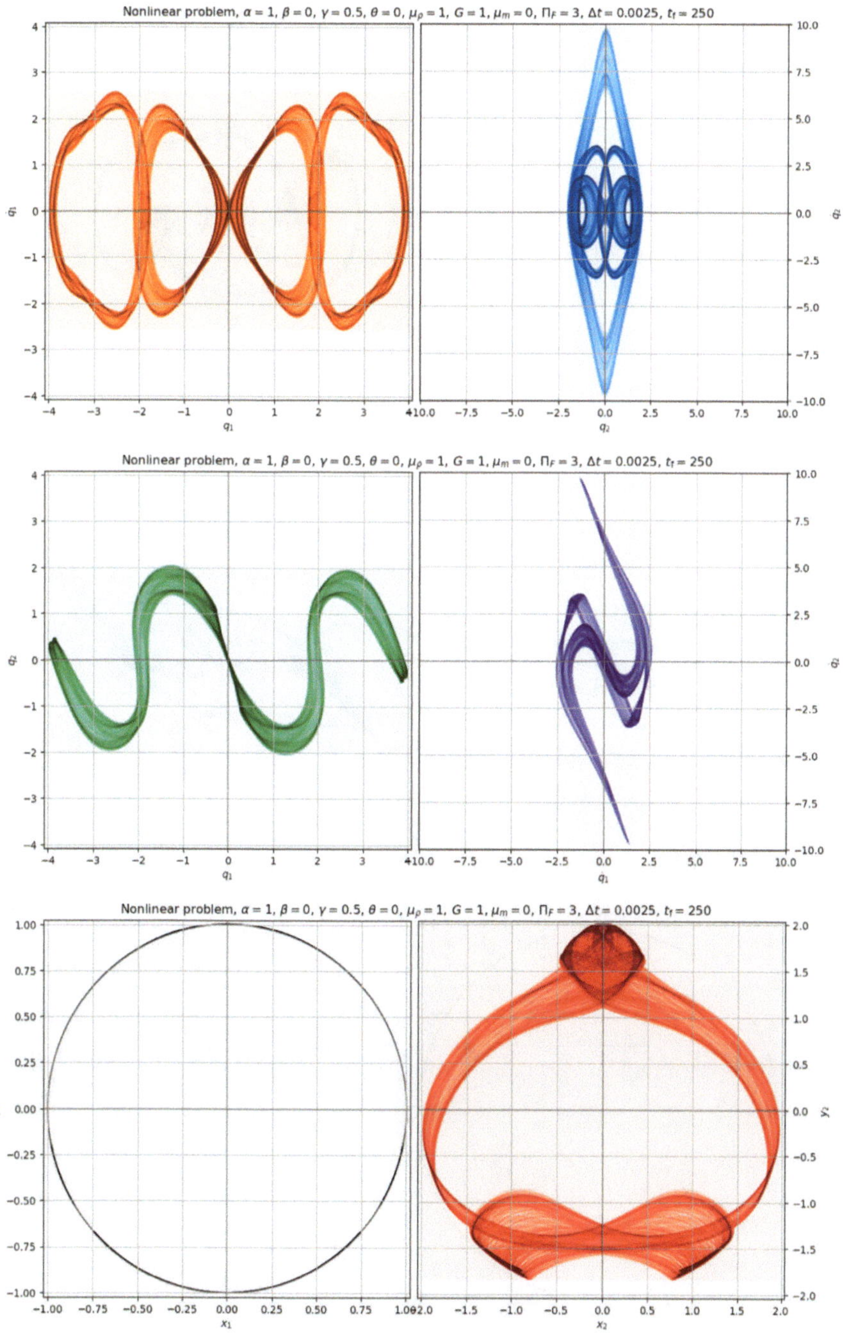

Fig. 3.9 State space density. Stiffness $\gamma = 0.5$, damping $\beta = 0$. No dead-weight loading, $\mu_m = 0$. Follower-force loading, $\Pi_F = 3$. *Top*: $(q_1, \dot{q}_1)$ and $(q_2, \dot{q}_2)$, *Middle*: (q_1, q_2) and $(\dot{q}_1, \dot{q}_2)$. *Bottom*: positions of the joint between the rods and the free end

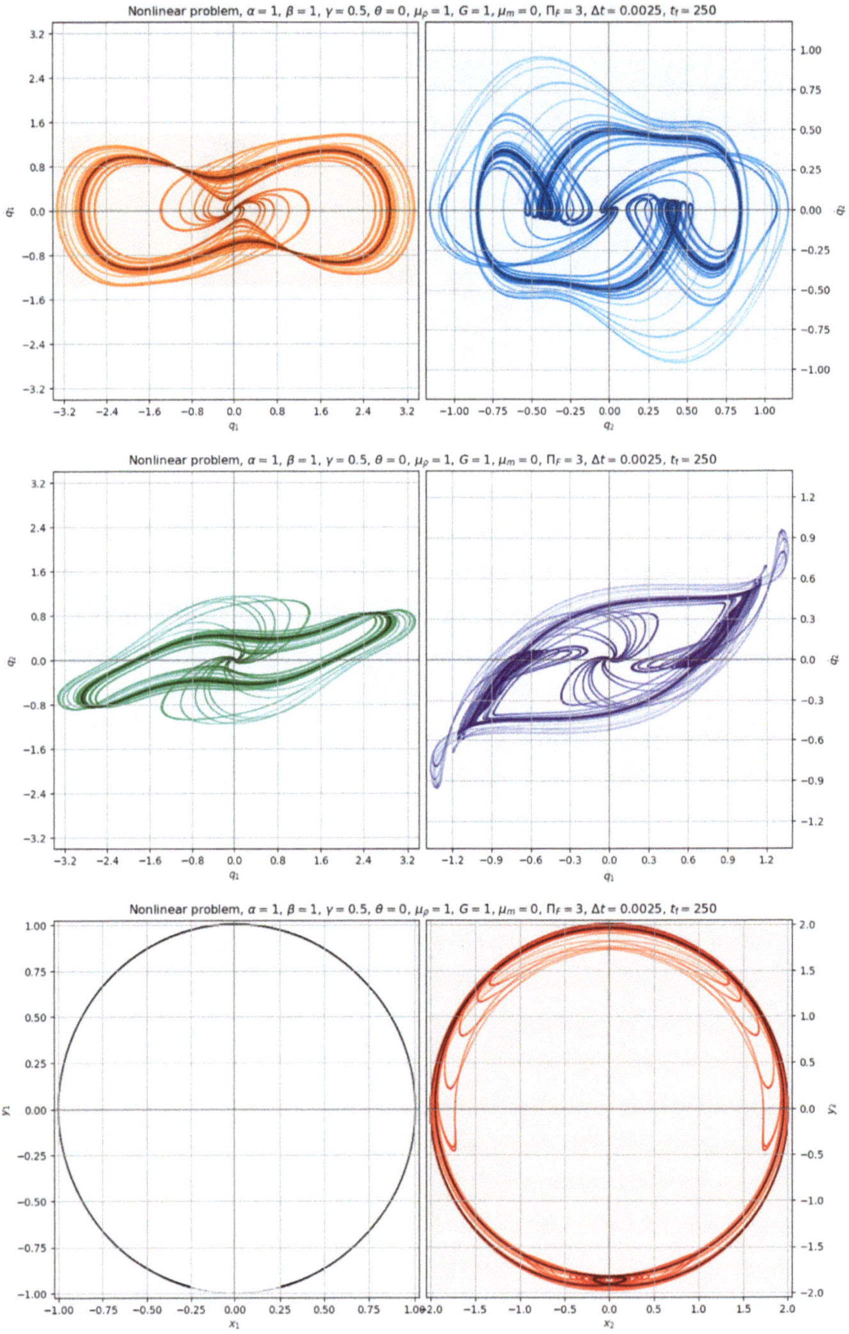

Fig. 3.10 State space density. Stiffness $\gamma = 0.5$, damping $\beta = 1$. No dead-weight loading, $\mu_m = 0$. Follower-force loading, $\Pi_F = 3$. *Top*: $(q_1, \dot{q}_1)$ and $(q_2, \dot{q}_2)$, *Middle*: (q_1, q_2) and $(\dot{q}_1, \dot{q}_2)$. *Bottom*: positions of the joint between the rods and the free end

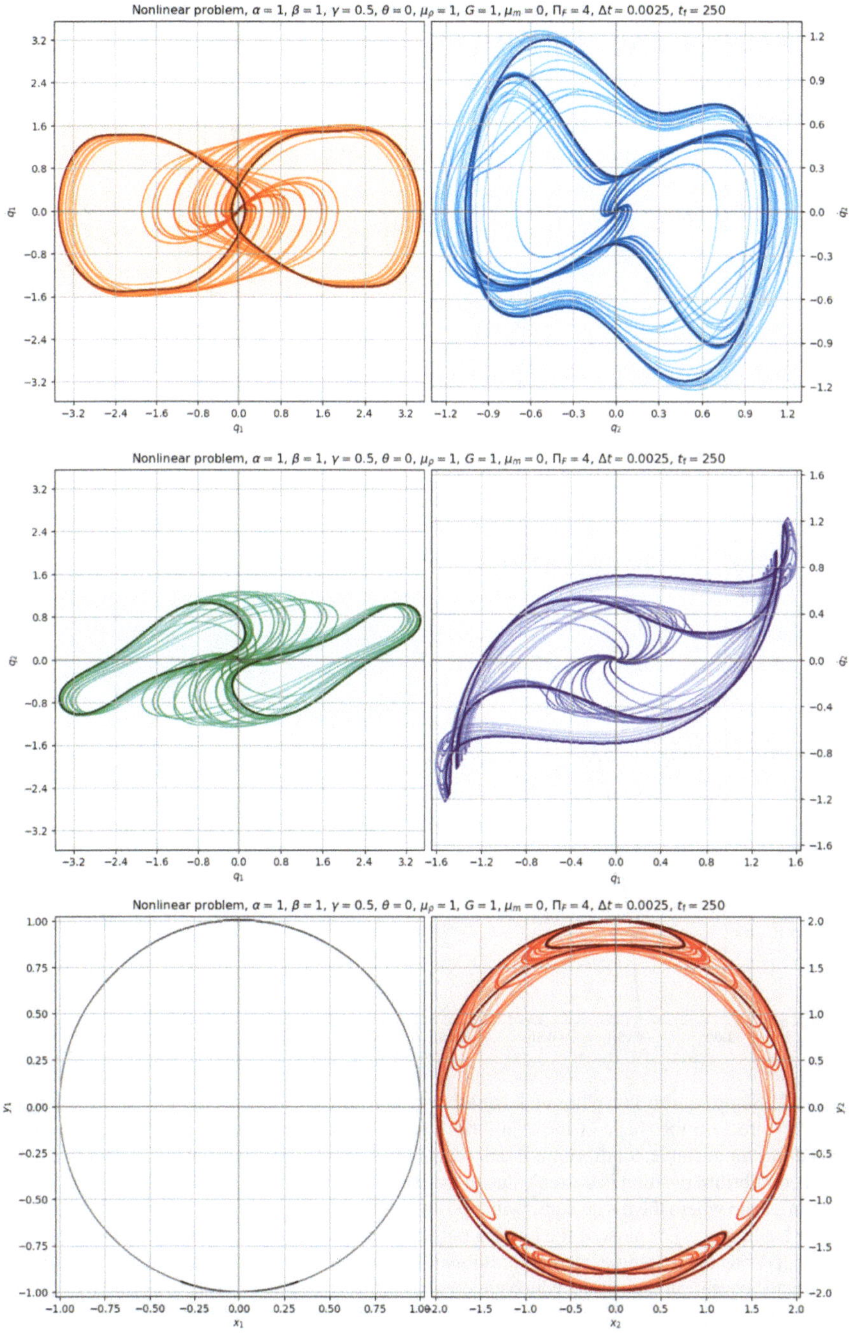

Fig. 3.11 State space density. Stiffness $\gamma = 0.5$, damping $\beta = 1$. No dead-weight loading, $\mu_m = 0$. Follower-force loading, $\Pi_F = 4$. *Top*: $(q_1, \dot{q}_1)$ and $(q_2, \dot{q}_2)$, *Middle*: (q_1, q_2) and $(\dot{q}_1, \dot{q}_2)$. *Bottom*: positions of the joint between the rods and the free end

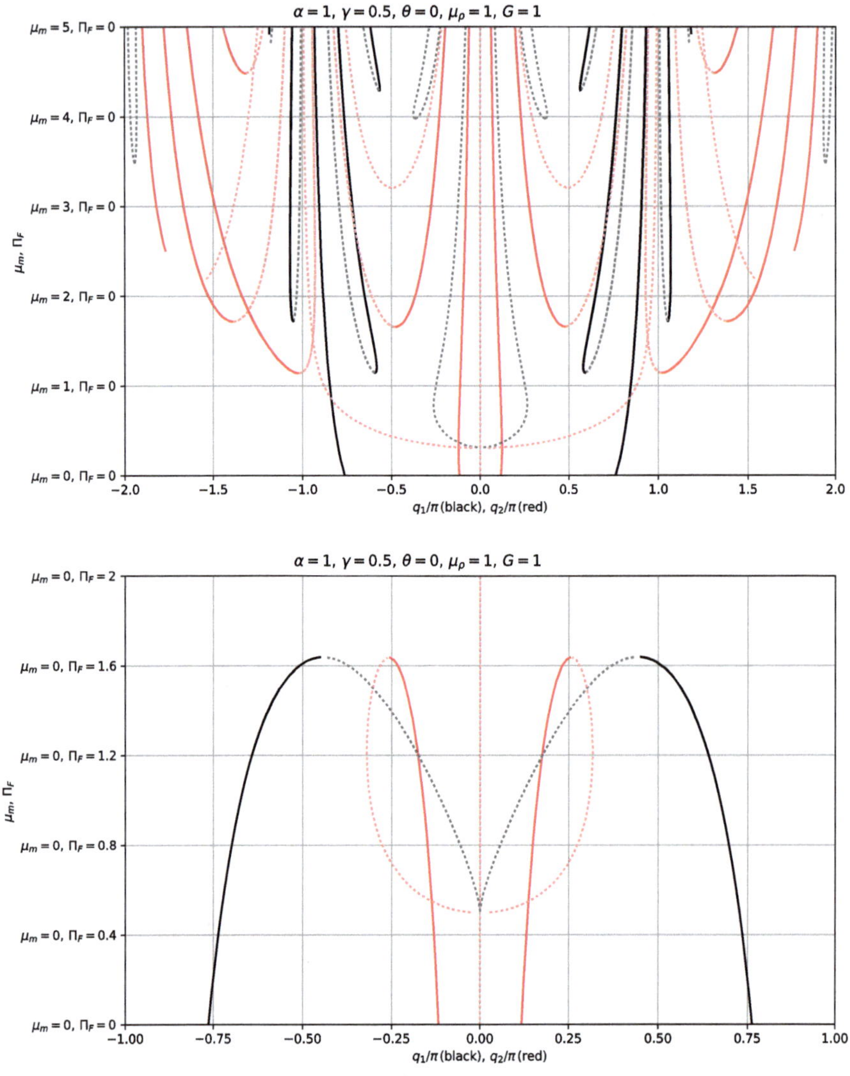

Fig. 3.12 Static equilibrium paths under quasistatically varied loading. Stiffness $\gamma = 0.5$. Damping β has no effect on the shape of the equilibrium paths. Solid line indicates a stable equilibrium, a dashed line an unstable equilibrium. Black (dark) lines denote q_1; red (light) lines denote q_2. The trivial equilibrium position is unstable due to the mass of the rods themselves; the unloaded system starts in a state where the trivial equilibrium position has already undergone bifurcation. *Top*: dead-weight loading μ_m. For most of the loading range, several stable static equilibrium positions exist. More appear as the loading is increased. *Bottom*: follower-force loading Π_F. There is initially one stable static equilibrium position, which undergoes a bifurcation and vanishes at a critical load value slightly above $\Pi_F = 1.6$. Above the critical load, no stable static equilibrium positions exist

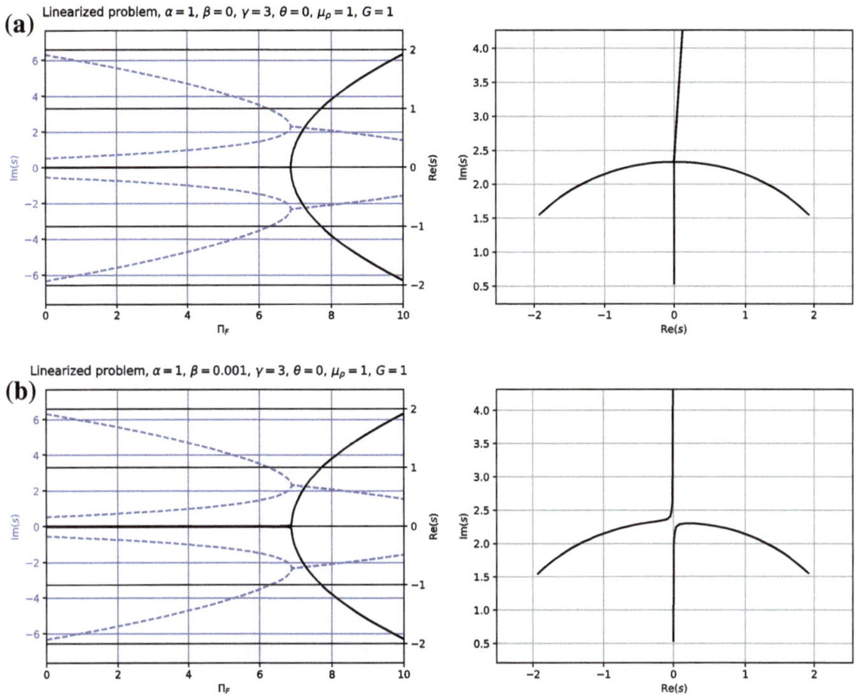

Fig. 3.13 Ziegler's effect, overview. Stiffness $\gamma = 3$. **a** No damping, $\beta = 0$. **b** Small but finite damping, $\beta = 0.001$. The bifurcation has vanished. Although in other respects the overall appearance of the solution remains similar, the stability behavior is very different; see the detail plots in Fig. 3.14

even at no damping). With high damping, $\beta = 1$, a limit cycle at $\Pi_F = 3$ splits into two when the loading is increased to $\Pi_F = 4$.

Figures 3.5, 3.6, 3.7, 3.8, 3.9, 3.10 and 3.11 display a closer look at the state space density in individual selected cases, picked from Figs. 3.3 and 3.4 and re-computed at a higher resolution. For all of these simulations the timestep is $\Delta t = 2.5 \cdot 10^{-3}$, with $n_t = 10^5$, yielding a simulation end time of $t_f = 250$. The setup for the UQ grid is the same as above. The smaller timestep was used to obtain a higher number of samples for approximating the density function, as well as to improve accuracy. It was observed that specifically for the case with $\beta = 0$, $\mu_m = 0$, $\Pi_F = 3$, the previous timestep size was fine for up to $t_f = 100$, but for larger values of the end time the error build-up became significant, causing the system to escape to a different, higher-energy region of the state space. With the smaller timestep and $t_f = 250$ this did not occur.

Figure 3.12 shows paths of static equilibrium, and their stability (for $\beta = 0$), under quasistatically varied loading. We consider the dead-weight loading and follower-force loading cases separately. Damping does not contribute to the shape of the equilibrium paths, but that it has a dramatic effect on the dynamics, and it may also affect the stability of equilibrium positions. In the equilibrium paths, bifurcations

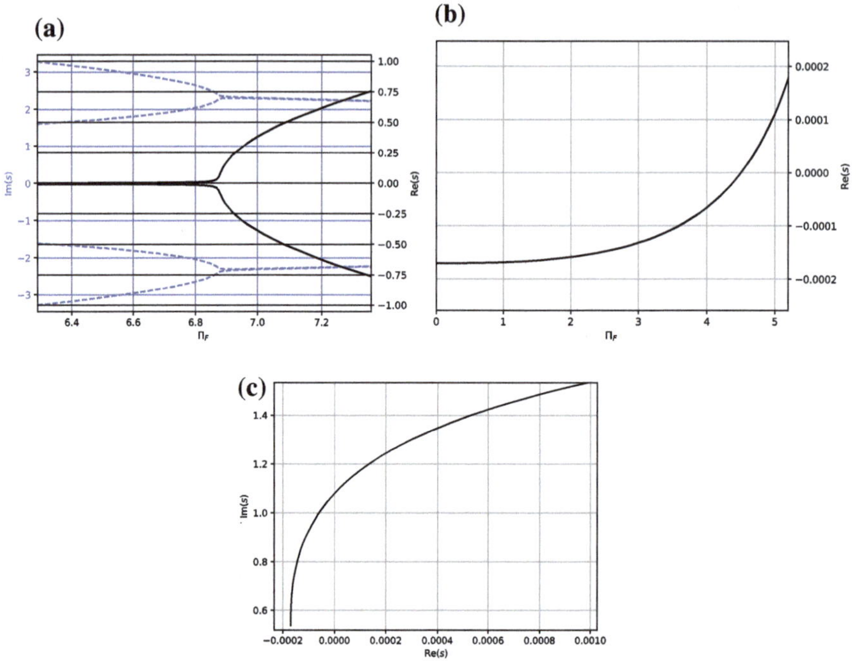

Fig. 3.14 Ziegler's effect. Stiffness $\gamma = 3$. Small but finite damping, $\beta = 0.001$. **a** Close-up of the region that used to contain the bifurcation; the solution curves no longer touch. **b** Stability is lost much earlier, at $\Pi_F \approx 4.500$, as one of the stability exponents crosses over to the half-plane with $\text{Re}\, s > 0$. **c** Parametric ($\text{Re}\, s$, $\text{Im}\, s$) plot near the instability. Note the scales; in the overview in Fig. 3.13b, this loss of stability occurs in the lower solution branch in the part that appears purely vertical

can be observed. At zero loading, the equilibrium paths start in a bifurcated state, which is due to the always present dead-weight loading of the rods themselves. Since the trivial equilibrium position corresponds to the situation where both rods point upward, we indeed physically expect it to be unstable.

In the case with dead-weight loading only, more nontrivial equilibrium positions appear as the load is increased. In the case where the system is loaded (beside the mass of the rods) with only a follower force, there is initially one nontrivial equilibrium state, which undergoes bifurcation and vanishes at a critical load value slightly above $\Pi_F = 1.6$. For higher values of follower-force loading in the range investigated, this system admits no static equilibrium solutions.

Finally, Figs. 3.13 and 3.14 demonstrate Ziegler's effect. We plot the complex-valued stability exponents s of the problem linearized around the trivial equilibrium position. For these examples, we use a relatively high spring stiffness, $\gamma = 3$, to stabilize the trivial equilibrium position in the no-loading state. The dead-weight load is set to zero, $\mu_m = 0$, and the follower-force load Π_F is varied quasistatically. When damping is absent ($\beta = 0$), the trivial equilibrium position is seen to lose stability at the bifurcation point at $\Pi_F \approx 6.876$. When small but finite damping is introduced ($\beta = 0.001$), the bifurcation vanishes, and stability is lost much earlier, at $\Pi_F \approx 4.500$, as one of the stability exponents crosses over to the half-plane with $\mathrm{Re}\, s > 0$.

To sum up, in this chapter, we looked at the stability analysis of nonconservative systems with a finite number of degrees of freedom. In this final section, attention was focused on Ziegler's double pendulum, which is a classical example of a system in the considered class. The next chapter finishes the first part of the book on prototype problems and techniques. We will consider continuous systems in classical solid mechanics. We will present a general technique for finding bifurcations in problems described by implicit functionals, and start the treatment of the main topic of this book, axially moving materials, as the example target of a variational solution technique.

References

1. Banichuk NV, Bratus AS, Mishkis AD (1987) Analysis of stabilizing and destabilizing effects of small damping in nonconservative systems with finite numbers of degrees of freedom. Technical report, Institute for Problems in Mechanics AN USSR. Reprint no. 312, in Russian
2. Vishik MI, Ljusternik LA (1960) Solution of some perturbation problems in the case of matrices and selfadjoint and nonselfadjoint differential equations. I. Adv Math Sci 15(3(93)): 30–80
3. Banichuk NV, Bratus AS, Mishkis AD (1989a) On the effects of stabilization and destabilization in nonconservative systems. Appl Math Mech (PMM) 53(2):206–214 in Russian
4. Ziegler H (1952) Die stabilitätskriterien der elastomechanik. Ingenieur Archiv 20:49–56
5. Bolotin VV, Zhinzher NI (1969) Effects of damping on stability of elastic systems subjected to nonconservative forces. Int J Solids Struct 5(9):965–989
6. Bolotin VV (1963) Nonconservative problems of the theory of elastic stability. Pergamon Press, New York
7. Kirillov ON (2013) Nonconservative stability problems of modern physics. de Gruyter. ISBN 978-3-11-027043-3
8. Bernstein DS, Bhat SP (1995) Lyapunov stability, semistability, and asymptotic stability of matrix second-order systems. Trans ASME 117:145–153
9. Bolotin VV (1995) Dynamic stability of structures. In: Kounadis AN, Krätzig WB (eds), Nonlinear stability of structures, volume 342 of CISM, pp 3–72. Springer. https://doi.org/10.1007/978-3-7091-4346-9
10. Sugiyama Y, Langthjem MA (2007) Physical mechanism of the destabilizing effect of damping in continuous non-conservative dissipative systems. Int J Nonlin Mech 42(1):132–145
11. Kirillov ON, Verhulst F (2010) Paradoxes of dissipation-induced destabilization or who opened Whitney's umbrella? Zeitschrift für Angewandte Mathematik und Mechanik 90(6):462–488. https://doi.org/10.1002/zamm.200900315
12. Lighthill J (1986) An informal introduction to theoretical fluid mechanics. Oxford Science Publications. ISBN 0-19-853630-5
13. Jeronen J, Kouhia R (2015) On the effect of damping on stability of non-conservative systems. In: Kouhia R, Mäkinen J, Pajunen S, Saksala T (eds) Proceedings of the XII finnish mechanics days, pp 77–82, 2015. http://rmseura.tkk.fi/smp_proceedings/SMP12_Proceedings.pdf. ISBN 978-952-93-5608-9 (printed), ISBN 978-952-93-5609-6 (electronic)
14. Bolotin VV (1964) The dynamic stability of elastic systems. Holden–Day, Inc. Translated from the Russian (1956) and German (1961) editions

15. Tisseur F, Meerbergen K (2001) The quadratic eigenvalue problem. SIAM Rev 43:235–286
16. Jeronen J (2011) On the mechanical stability and out-of-plane dynamics of a travelling panel submerged in axially flowing ideal fluid: a study into paper production in mathematical terms. PhD thesis, Department of Mathematical Information Technology, University of Jyväskylä. http://urn.fi/URN:ISBN:978-951-39-4596-1. Jyväskylä studies in computing 148. ISBN 978-951-39-4595-4 (book), ISBN 978-951-39-4596-1 (PDF)
17. Golub GH, van Loan CF (1996) Matrix computations. Johns Hopkins, 3rd edition. ISBN 0-8018-5414-8
18. Sheather SJ, Jones MC (1991) A reliable data-based bandwidth selection method for kernel density estimation. J Royal Stat Soc Ser B (Methodological) 53(3):683–690
19. Duong T, Hazelton ML (2003) Plug-in bandwidth selectors for bivariate kernel density estimation. J Nonparametric Stat 15:17–30
20. Chacón JE, Duong T (2010) Multivariate plug-in bandwidth selection with unconstrained pilot bandwidth matrices. Test 19:375–398
21. Helton JC, Johnson JD, Sallaberry CJ, Storlie CB (2006) Survey of sampling-based methods for uncertainty and sensitivity analysis. Technical report, Sandia National Laboratories. Report SAND2006-2901
22. Helton JC, Davis FJ (2002a) Illustration of sampling-based methods for uncertainty and sensitivity analysis. Risk Anal 22(3):591–622. https://doi.org/10.1111/0272-4332.00041. A preprint exists as 10.1.1.119.1645
23. Ward-Larson G, Rushmeier H, Piatko C (1997) Visibility matching tone reproduction operator for high dynamic range scenes. IEEE Trans Visu Comput Gr 3:291–306
24. Jeronen J (2013a) Visual contrast preserving representation of high dynamic range mathematical functions. In: Repin S, Tiihonen T, Tuovinen T (eds) Numerical methods for differential equations, optimization, and technological problems. Dedicated to Professor P. Neittaanmäki on his 60th Birthday., vol 27 of Computational Methods in Applied Sciences, pp 409–429. Springer Netherlands. ISBN: 978-94-007-5287-0 (Print) 978-94-007-5288-7 (Online)
25. Jeronen J (2011a) SAVU: sample-based analysis and visualization of uncertainty. In: Proceedings of CAO2011 – ECCOMAS thematic conference on computational analysis and optimization, pp 165–168. University of Jyväskylä, 2011a. ISBN 978-951-39-4331-8
26. William K (1965) Further remarks on reducing truncation errors. Commun ACM 8(1):40. https://doi.org/10.1145/363707.363723
27. Shewchuk JR (1997) Adaptive precision floating-point arithmetic and fast robust geometric predicates. Discret Comput Geom 18(3):305–363

Chapter 4
Some General Methods

In this chapter, we take a brief general look into elastic stability in the setting of classical solid mechanics. We will briefly introduce the different types of stability loss, and then look at conditions under which merging of eigenvalues may occur. We will look at a problem where applying symmetry arguments allows us to eliminate multiple (merged) eigenvalues, thus reducing the problem to determining a classical simple eigenvalue. We then discuss a general technique to look for bifurcations in problems formulated as implicit functionals. This is useful for a wide class of problems, including many problems in axially moving materials. At the end of the chapter, we will consider a variational approach to the stability analysis of an axially moving panel (a plate undergoing cylindrical deformation).

4.1 Criteria of Elastic Stability

Stability constitutes one of the basic demands that must be satisfied in designing elastic structures. It is particularly important in the design of slender structures or structures made of high-strength materials.

Elastic stability analysis comes with a long tradition. The present form of static stability analysis was originally developed by Euler [14, 15], for a differential equation describing the bending of a beam or column. Dynamic stability analysis for linear elastic systems, extending Euler's method, is due to Bolotin [12] following the pioneering work by Lyapunov. The stability behavior of some axially moving materials is mathematically analogous to the buckling of a compressed column, enabling the use of these techniques also in the context of axially moving materials [30].

The concept of elastic stability is closely related to the response of a structure to applied forces. A study of the change in geometry as a function of time is the primary purpose of any investigation of structural stability. Therefore, a dynamic approach to problems of stability is the most general. The idea of a static approach in the study of problems of elastic stability, however, dates back to Euler. In this chapter, we offer a

N. Banichuk et al., *Stability of Axially Moving Materials*, Solid Mechanics and Its Applications 259, https://doi.org/10.1007/978-3-030-23803-2_4

brief outline of some approaches and of criteria that may be applied to evaluate the elastic stability of structural elements. For a more detailed exposition, see Bolotin [9–11].

Let the structure be subjected to a static loads that is proportional to a parameter p. Let the structure perform small elastic vibrations around an equilibrium point. The function $w(x, t)$ describing the displacement of the structure in the statically loaded but unbuckled state satisfies the differential equation

$$A\frac{\partial^2 w}{\partial t^2} + B\frac{\partial w}{\partial t} + Cw - pKw = 0\,. \tag{4.1.1}$$

Here A, B, C and K are linear differential operators, p is the loading parameter, and t is time. The inertial operator A is represented by a symmetric matrix of mass coefficients. The stiffness operator C characterizes the static response of the elastic structure whose application to (vector) w reduces to a differential of that function with respect to space variables. The operator B arises in two different contexts: it may represent dissipation, or gyroscopic behavior, depending on its form.

Let us represent the state function $w(x, t)$ as

$$w(x, t) = u(x)e^{i\omega t}\,, \tag{4.1.2}$$

and substitute (4.1.2) in (4.1.1). As a result, we obtain an equation for the amplitude function $u(x)$:

$$-\omega^2 Au + i\omega Bu + Cu - pKu = 0\,. \tag{4.1.3}$$

The stability properties of the structure are determined by the magnitude of the eigenvalues $\omega = \omega(p)$ of the Eq. (4.1.3). The structure will remain stable if all eigenvalues $\omega(p)$ have positive imaginary parts or, in the limiting case, if they are all real. A loss of stability occurs when at least one eigenvalue obtains a negative imaginary part. Figure 4.1 shows a schematic representation of the static and dynamic types of stability loss.

One of the eigenvalues $\omega(p)$ represented in subfigure (a) enters the lower half-plane as the value of the loading parameter p increases and the curve passes through the origin. The loss of stability occurs in the form of a monotone increase in the amplitude of the disturbance, and has an essentially static character. Subfigures (b) and (c) correspond to the dynamic loss of stability, which is characterized by the initiation of vibrations at an increasing amplitude.

Let us consider the case where $B = 0$, and where external forces can be derived from a potential. Then the system is conservative. The operator K (as shown in Bolotin [9]; in English, see Bolotin [12]) is then independent of ω, and Eq. (4.1.3) reduces to

$$-\omega^2 Au + Cu - pKu = 0\,. \tag{4.1.4}$$

The eigenvalue ω^2 is a function of parameter p, that is, $\omega^2 = \omega^2(p)$. Equation (4.1.4) with the appropriate boundary conditions constitutes a self-adjoint boundary-value

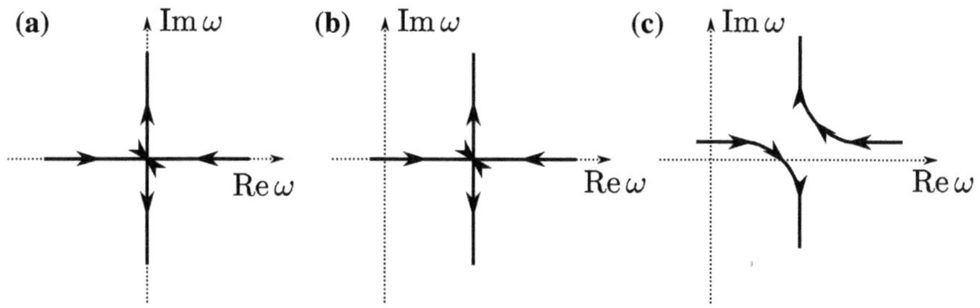

Fig. 4.1 Schematic diagram of the static and dynamic types of stability loss. One pair of eigenvalues is shown. **a** Static loss of stability. A bifurcation occurs at the origin. **b** Dynamic loss of stability in a system with no dissipation. A bifurcation occurs on the real axis at some $\mathrm{Re}\,\omega \neq 0$. Stability is lost for the eigenvalue that obtains $\mathrm{Im}\,\omega < 0$. **c** Dynamic loss of stability in a system with dissipation. Compared to the corresponding ideal system, dissipation tends to shift all eigenvalues (in the unloaded state) toward positive $\mathrm{Im}\,\omega$. Loss of stability takes place as $\mathrm{Im}\,\omega$ becomes negative for one of the eigenvalues. No bifurcation occurs

problem. Therefore, the eigenvalues ω^2 are real. A loss of stability occurs when ω^2 changes from positive to negative real values. The critical value of the parameter p corresponds to the equality $\omega^2 = 0$. A transition into the unstable regime, which occurs when the square of the natural frequency vanishes, turns out to be a consequence of the continuous dependence of ω^2 on the loading parameter p.

Consequently, for any system under the above conditions, the loss of stability must occur when $\omega^2 = 0$. That is, we have a static type of instability. For this reason, research into the elastic stability of conservative systems with no gyroscopic termsgenerally follows the static techniques originated by Leonhard Euler.

For the above conclusion, it is important that not only is dissipation absent, but so are gyroscopic terms, because in the general case, these together form the operator B. A system that has no dissipation but has gyroscopic terms, such as an axially moving structure made of linear elastic material, may still be conservative (if the external forces can be derived form a potential), but it may have values of the loading parameter p where the system undergoes a dynamic stability loss.

Concentrating now on a static loss of stability, let us set $\omega^2 = 0$, then

$$Cu - pKu = 0\,. \tag{4.1.5}$$

The problem of elastic stability now consists in finding the minimal value of the parameter p (i.e., the first eigenvalue) and the corresponding deflection function u (the eigenfunction) that solves the boundary-value problem. As we have stated above, the self-adjoint property of a boundary-value operator is a sufficient condition for all eigenvalues p to be real-valued.

Static techniques for investigating stability can be reduced to a formulation and solution of a bifurcation problem. From a mathematical point of view, this depends on the self-adjoint and positive-definite properties of the eigenvalue problem. This means that for fairly arbitrary functions u and v,

$$\int_\Omega uCv\,\mathrm{d}\Omega = \int_\Omega vCu\,\mathrm{d}\Omega\,, \qquad \int_\Omega uCu\,\mathrm{d}\Omega > 0\,, \tag{4.1.6}$$

$$\int_\Omega uKv\,\mathrm{d}\Omega = \int_\Omega vKu\,\mathrm{d}\Omega\,, \qquad \int_\Omega uKu\,\mathrm{d}\Omega > 0\,.$$

We note that a self-adjoint positive-definite operator (for a boundary-value problem of classical elasticity) that does not contain a parameter p always has an infinite spectrum comprised of positive eigenvalues (see Alfutov [1] and Naimark [31]).

In structural stability problems it generally suffices to find the smallest eigenvalue, which determines the critical load. For a self-adjoint, positive-definite eigenvalue problem, the solution minimizes the Rayleigh quotient (see Mikhlin [27]). Using energy methods, the stability problem reduces to a variational problem of minimizing the nonadditive functional

$$p = \min_u \frac{\int_\Omega uCu\,\mathrm{d}\Omega}{\int_\Omega uKu\,\mathrm{d}\Omega}\,. \tag{4.1.7}$$

The minimum with respect to u in (4.1.7) is sought in the class of sufficiently smooth functions that satisfy the assigned boundary conditions.

To this class of static research techniques in elastic stability also belongs the *nonideal* approach. This technique consists of introducing into the homogeneous equations of equilibrium a small additional term that makes these equations inhomogeneous. These additional terms may, for example, represent a small eccentricity in the application of the compressive load to a column, the action of a small transverse force in the compression of a plate by forces acting in the plane of the plate, or the initial curvature of a panel that is being compressed. Taking into account these nonideal corrections described by a function h, the basic equilibrium equations assume the form

$$Cu - pKu + h = 0\,. \tag{4.1.8}$$

Unlike the bifurcation equations (4.1.5), (4.1.8) has a unique solution u (with nonzero norm, $\|u\| \neq 0$) even for an arbitrarily small value of the loading parameter p. As the value of p increases, the maximal value (amplitude) of the deflection function increases. This maximum approaches infinity as the loading parameter approaches a critical value. This value coincides with Euler's value for the critical load, which can be found by computing the first eigenvalue for the bifurcation problem (Eq. 4.1.5). Therefore, the study of elastic stability may also be carried out using this nonideal approach, which leads to the problem of finding the magnitude of a load (i.e., of the parameter p) for which the deflections of the nonideal system become infinitely large.

4.2 Bifurcations and Multiplicity of Critical Loads

In many problems of buckling analysis and optimal design, it has been observed that the points of the discrete spectrum frequently come close to each other and multiple critical loads appear. This phenomenon creates real difficulties in designing structures with optimal stability properties. Many studies on the subject of structural optimization aim to examine different aspects of cases with multiple eigenvalues.

Let us first discuss the possibility of the occurrence of double eigenvalues. We formulate a general optimization problem depending on a single parameter, and study the behavior of the eigenvalues as we change this parameter. Let $\mathcal{H}_\alpha$ denote the set of admissible design parameters, that is, the design vector $h \in \mathcal{H}_\alpha$. The subscript signifies the dependence of admissible values on the parameter α. For example, in many problems of optimizing stability, the design variable is taken as the thickness function $h(x)$ for a structural part, which must satisfy a constant volume condition and a constraint on the minimal value of the thickness:

$$\int_\Omega h \, d\Omega = 1 \,, \qquad h \gg h_{min} = \alpha \,. \tag{4.2.1}$$

Here we have selected the quantity h_{min} as the parameter that defines the admissible set. Note that we use nondimensional variables.

Let us examine the following optimization problem. We wish to find a function $h(x)$ that assigns a maximum to the first eigenvalue

$$p_1 = \max_h p_1(h) \,, \tag{4.2.2}$$

$$h \in \mathcal{H}_\alpha \,, \tag{4.2.3}$$

for the boundary-value problem

$$\mathcal{L}(h)u = pu \,. \tag{4.2.4}$$

We assume that the optimization problem (4.2.2)–(4.2.4) can be solved for values of α filling some interval and that the exact dependence of the design function $h(x, \alpha)$ on the parameter α is known. The corresponding eigenfunctions u_i and eigenvalues p_i of the boundary-value problem (4.2.4) also depend on the parameter α, in other words, $u_i = u_i(x, \alpha)$ and $p_i = p_i(\alpha)$.

The behavior of eigenvalues that depend on a design parameter is of special interest in the theory of optimal design. This problem has been widely discussed in relation to other important problems (see Arnol'd [3, 4]). For example, for systems that have only a finite number of degrees of freedom and do not have any symmetry (i.e., systems in general position), a general result given in Arnol'd [3] is widely known. According to this result, in general a change in the value of a single parameter will not make it possible to merge two natural frequencies. An analogous assertion can be made for the vibrations of a continuous medium (see Arnol'd [3]).

Let $h^0(x)$ be a solution of the optimal design problem (4.2.2) and (4.2.3) for some fixed value α^0 of the parameter α. For a given design variable $h^0(x)$, there exists a corresponding spectrum of eigenvalues and eigenfunctions for the boundary-value problem (4.2.4). Let us look at the first two eigenvalues p_1 and p_2 of the problem (4.2.4). Aside from the fixed value α^0, we consider small perturbations of the parameter $\alpha = \alpha^0 + \delta\alpha$, where we assume that $\delta\alpha$ is small. To this (perturbed) value of α corresponds an optimal solution $h(x) = h^0(x) + \delta h$. For this changed value of the parameter, the eigenvalue problem may be written as

$$\left(\mathcal{L}^0 + \delta\mathcal{L}\right) u = pu \ , \tag{4.2.5}$$

$$\mathcal{L}^0 = \mathcal{L}\left(h^0(x)\right) \ ,$$

$$\delta\mathcal{L} = \mathcal{L}\left(h^0(x) + \delta h(x)\right) - \mathcal{L}\left(h^0(x)\right) \ .$$

We determine the magnitude of the perturbations for which the eigenvalues are equal to

$$p_1^0 = p_1\left(\alpha^0\right), \qquad p_2^0 = p_2\left(\alpha^0\right) \tag{4.2.6}$$

as we vary the parameter. The solution of the problem (4.2.5) is written as a sum

$$u = c_1 u_1^0(x) + c_2 u_2^0(x) \ , \tag{4.2.7}$$

where $u_i^0(x)$ with $i = 1, 2$ are the eigenfunctions corresponding to the eigenvalues p_i. After some substitutions, we have

$$p_{1,2} = \frac{1}{2}\left(p_1^0 + p_2^0 + A_{11} + A_{22}\right) \pm \sqrt{\frac{1}{4}\left(p_1^0 - p_2^0 + A_{11} - A_{22}\right)^2 + A_{12}^2} \ , \tag{4.2.8}$$

$$A_{ij} = \int_0^\ell u_i^0(x)\delta\mathcal{L}u_j^0(x)\,\mathrm{d}x \ .$$

It is clear from (4.2.8) that merging the first two eigenvalues is possible only if the expression under the square root vanishes. Since the radical is a sum of two squares, it can be equal to 0 only if the two conditions

$$p_1 - p_2 + A_{11} - A_{22} = 0 \ , \tag{4.2.9}$$

$$A_{12} = 0 \tag{4.2.10}$$

hold simultaneously.

The quantities appearing on the left-hand side of (4.2.9) and (4.2.10) are functions of α. Therefore, in general, it is impossible to cause the two eigenvalues to merge by altering the value of α. (The curves $p_1 = p_1(\alpha)$ and $p_2 = p_2(\alpha)$ do not, in general, intersect each other.) An exception to this rule is a rare case, the probability of which is equal to the probability of finding equal roots of independent algebraic equations. However, for problems with symmetries such an occurrence is possible, because then the condition $A_{12} = 0$ may turn out as an identity that is satisfied for any choice of α. For example, if $u_1^0(x)$ and $\delta\mathcal{L}$ are both symmetric with respect to the midpoint $x = \ell/2$ of the interval $(0, \ell)$, and $u_2^0(x)$ is antisymmetric with respect to the midpoint, then $A_{12} \equiv 0$. This occurs in the study of optimization of the stability of compressed columns that have symmetric support conditions at both ends. In design problems for compressed columns, the oscillation theorem of Sturm can also be useful.

4.3 Decomposition Method for Bimodal Solutions

Symmetry arguments can sometimes be used to eliminate multiple eigenvalues. As an example, we shall consider a problem of optimal design of an elastic straight column with built-in ends at the points $x = -\ell$ and $x = \ell$ subjected to a compressive load p. Many research papers on this problem have provided both analytical and numerical solutions (see, e.g., Banichuk and Barsuk [6]). In this section we introduce a solution based on symmetry properties and the decomposition of the spectrum consisting of eigenvalues for the boundary-value problem of elastic stability. At the same time, we eliminate properties connected with the multiplicities of the critical loads, so that the original optimization problem reduces to the classical problem of maximizing a simple eigenvalue.

Let us write down the basic formulas for maximizing a load that causes a loss of stability by finding the best possible distribution of the cross-sectional area. We may use a Rayleigh quotient approach similar to Eq. (4.1.7):

$$p = \max_S \min_u \Phi(u, S) , \tag{4.3.1}$$

$$\Phi(u, S) = \frac{\int_{-\ell}^{\ell} EI \left(\frac{d^2u}{dx^2}\right)^2 dx}{\int_{-\ell}^{\ell} \left(\frac{du}{dx}\right)^2 dx} , \tag{4.3.2}$$

$$u(\ell) = \left(\frac{du}{dx}\right)_{x=\ell} = u(-\ell) = \left(\frac{du}{dx}\right)_{x=-\ell} = 0 , \tag{4.3.3}$$

$$\int_{-\ell}^{\ell} S(x)dx = V , \qquad S(x) \geq S_{\min} , \tag{4.3.4}$$

where

$$I = A_\alpha S^\alpha, \qquad \alpha = 1, 2, 3\,.$$

We seek the maximum in (4.3.1) in the class of symmetric smooth functions $S(x)$, denoting the distribution of the cross-sectional area, that satisfy the constraints (4.3.4), on the constant total volume of the material and the minimum admissible value of thickness. We observe that for a symmetric distribution of rigidity $EI(x)$, the Euler-Lagrange differential equation for a minimum of the functional (4.3.1), i.e., the equation

$$\frac{\mathrm{d}^2}{\mathrm{d}x^2}\left(EI\frac{\mathrm{d}^2u}{\mathrm{d}x^2}\right) + p\frac{\mathrm{d}^2u}{\mathrm{d}x^2} = 0 \tag{4.3.5}$$

with the homogeneous boundary conditions (4.3.3), remains invariant under the transformation $x \to -x$. This implies that our previous arguments from Sect. 4.2 on page 123 are applicable. We arrive at the following formulas for computing the first eigenvalues for symmetric and antisymmetric solutions:

$$p_1 = \min_{u^i} \Phi\left(u^i, S\right), \qquad i = \mathrm{s}, \mathrm{a}\,, \tag{4.3.6}$$

$$\left(\frac{\mathrm{d}u^{\mathrm{s}}}{\mathrm{d}x}\right)_{x=0} = \left(\frac{\mathrm{d}^3u^{\mathrm{s}}}{\mathrm{d}x^3}\right)_{x=0} = 0\,, \qquad u^{\mathrm{s}}(\ell) = \left(\frac{\mathrm{d}u^{\mathrm{s}}}{\mathrm{d}x}\right)_{x=\ell} = 0\,, \tag{4.3.7}$$

$$\left(\frac{\mathrm{d}u^{\mathrm{a}}}{\mathrm{d}x}\right)_{x=0} = \left(\frac{\mathrm{d}^3u^{\mathrm{a}}}{\mathrm{d}x^3}\right)_{x=0} = 0\,, \qquad u^{\mathrm{a}}(\ell) = \left(\frac{\mathrm{d}u^{\mathrm{a}}}{\mathrm{d}x}\right)_{x=\ell} = 0\,. \tag{4.3.8}$$

In computing scalar products in (4.3.1) for the functional Φ, integration is now carried out on the interval $[0, \ell]$. The optimization problem can be solved numerically. In carrying out numerical computations, we use the nondimensional variables

$$x' = \frac{x}{2\ell}, \qquad S' = \frac{2\ell S}{V}, \qquad p' = \frac{(2\ell)^{\alpha+2}\,p}{A_\alpha E V^\alpha}\,. \tag{4.3.9}$$

In nondimensional variables (with the prime omitted), the search for an optimal thickness function is carried out on the interval $0 \le x \le 1/2$, while the total volume is equal to 1.

We shall investigate the auxiliary problem of finding the maximum of the first eigenvalue of a symmetric mode,

$$\max_S p_1^{\mathrm{s}}\,,$$

with constraints (4.3.4)–(4.3.7) and (4.3.8), corresponding to a symmetric shape $u^{\mathrm{s}}(x)$. The solution to this optimization problem was derived by a numerical technique, using iterative optimization for different values of the parameter $S_{\min}$, with $0 < S_{\min} < 1$, and for different types of cross-sections, $\alpha = 1, 2, 3$ (see Banichuk and Barsuk [5]).

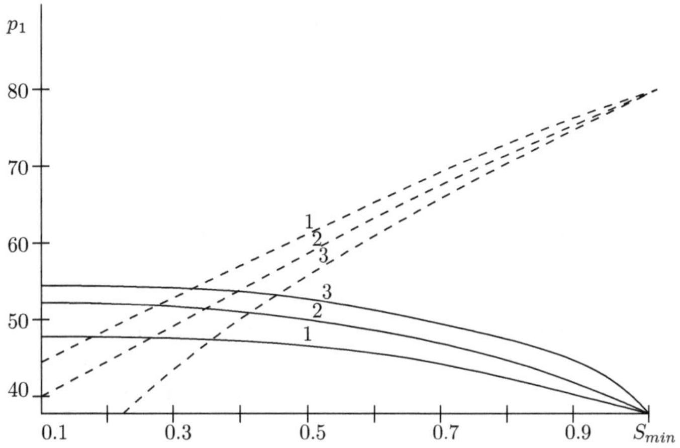

Fig. 4.2 The magnitude of the first eigenvalue as a function of $S_{\min}$

Aside from $p_1^s\left(S_0^s\right)$, computed for the solution function S_0^s, the values $p_1^a\left(S_0^s\right)$, $p_2^a\left(S_0^s\right)$ and $p_2^a\left(S_0^s\right)$ were also computed. In Fig. 4.2, the solid lines 1, 2, and 3 indicate cases corresponding to values of $\alpha = 1, 2$ and 3, respectively. The magnitude of $p_1^a(S_0^s)$ as a function of the parameter $S_{\min}$ is shown in the Fig. 4.2 by the broken lines 1, 2 and 3 again corresponding to the cases $\alpha = 1, 2$ and 3, respectively. Looking at the curves displayed in Fig. 4.2, it is clear that there exist intervals in which the parameter $S_{\min}$ varies between $S_{\min}^{\alpha} \leq S_{\min} \leq 1$, on which it is true that

$$p_1^s\left(S_0^s\right) \leq p_1^a\left(S_0^s\right) . \tag{4.3.10}$$

Making use of the inequality (4.3.10), it is easy to show that the solution of the auxiliary problem $\max_S p_1$ of maximizing the first eigenvalue p^a corresponding to an antisymmetric case with constraints (4.3.4), (4.3.6) and (4.3.8) must satisfy the inequality

$$p_1^s\left(S_0^a\right) < p_1^a\left(S_0^a\right) . \tag{4.3.11}$$

To prove this inequality, we introduce the chain of inequalities

$$p_1^s\left(S_0^a\right) < p_1^s\left(S_0^s\right) < p_1^a\left(S_0^s\right) < p_1^a\left(S_0^a\right) . \tag{4.3.12}$$

The inequality between the first and the last item in this chain expresses the statement that for a beam with non-optimal distribution of thickness, the force causing a loss of stability is smaller than the critical force for an optimal beam. The second inequality in the chain (4.3.12) is true on any interval $S_{\min}^a \leq S_{\min} \leq 1$ (this follows from computations). Therefore, for $S_{\min}^a \leq S_{\min} \leq 1$, a solution S_0^s of the auxiliary problem determines the optimal design for a beam that is to be subjected to compressive loading.

4.4 Bifurcation and Analysis of Implicitly Given Functionals

In this section, we will analyze the stability and bifurcation of elastic systems using a general scheme developed for problems with implicitly given functionals. An asymptotic property for the behavior of the natural frequency curves in a small vicinity of each bifurcation point is obtained for the considered class of systems. Two examples are given. The first is the stability analysis of an axially moving elastic panel, with no external applied tension, performing transverse vibrations. The second is the free vibration problem of a stationary compressed panel. The approach is applicable to a class of problems in mechanics, for example, in elasticity, aeroelasticity and axially moving materials (such as in papermaking or bandsaw blades).

The most often used models for an axially moving material have been travelling flexible strings, membranes, beams, and plates. The research field of axially moving materials can be traced back to Skutch [40]. Among the first English-language papers on moving materials were [28, 36]. All these studies considered axially moving ideal strings. The analytical solution describing the free vibrations of the axially moving ideal string was derived by Swope and Ames [41]. Dynamics and stability considerations were first reviewed in the article by Mote [29].

The effects of axial motion of the material on its frequency spectrum and eigenfunctions were investigated in the classic papers by Archibald and Emslie [2] and by Simpson [39]. It was shown that the natural frequency of each mode decreases when the transport speed increases, and that the travelling string and beam both experience divergence instability at a sufficiently high speed. However, in the case of the string, this result was contrasted by Wang and Huang [44], who used Hamiltonian mechanics to show that the ideal string remains stable at any speed.

Wickert [45] studied the loss of stability with an application of dynamic and static approaches It was shown by means of numerical analysis that in all cases instability occurs when the frequency is zero and the critical velocity coincides with the corresponding velocity obtained from static analysis. Similar results were obtained for travelling plates by Lin [23].

The dynamical properties of axially moving plates have been studied by Shen et al. [37] and by Shen et al. [38], and the properties of a moving paper web have been studied in the two-part article by Kulachenko et al. [20, 21]. Critical regimes and other problems of stability analysis have been studied, for example, by Wang [43] and Sygulski [42]. Moreover, [24–26] discusses widely the dynamical aspects of axially moving webs. Yang and Chen [46] considered transverse vibrations of the axially accelerating viscoelastic beam, and in Pellicano and Vestroni [34] the dynamic behavior of a simply supported beam subjected to an axial transport of mass was studied. An extensive literature review can be found, for example, in Mergen et al. [16]. Some approaches to bifurcation problems and estimation of critical parameters were also presented by Nečas et al. [32] and Neittaanmäki and Ruotsalainen [33].

Let us consider the spectral boundary value problem described by the equation

$$\mathcal{L}(u(x), \lambda, \gamma) = \sum_{k=0}^{m} \sum_{\ell=0}^{n} \lambda^k \gamma^\ell \mathcal{L}_{k\ell}(u(x)) = 0 , \tag{4.4.1}$$

where γ is a real-valued loading parameter, characterizing the interaction of the structure and an external medium, λ is a spectral parameter, and $\mathcal{L}_{k\ell}(u(x))$ are given differential operators applied to the behavior function $u(x)$, defined in the domain Ω, $x \in \Omega$. Boundary conditions are considered as included in the differential operator $\mathcal{L}(u(x))$.

The problems of free harmonic vibrations, and the stability of elastic systems interacting with an external fluid (liquid or gas) can be reduced to the formulation (4.4.1).

Let the function $v(x)$ be the solution of the spectral problem

$$\mathcal{L}^*(v(x), \lambda, \gamma) = 0 , \tag{4.4.2}$$

which is adjoint to the problem (4.4.1). In the case of a self-adjoint problem, $v(x)$ coincides with $u(x)$.

If we multiply Eq. (4.4.1) by $v(x)$ and integrate over the domain, we have

$$\Phi(\lambda, J_{00}, \ldots, J_{mn}, \gamma) = \sum_{k=0}^{m} \sum_{\ell=0}^{n} \lambda^k \gamma^\ell J_{k\ell} = 0 , \tag{4.4.3}$$

where the functionals $J_{k\ell}$, $k = 1, 2, \ldots m$; $\ell = 1, 2, \ldots, n$ are defined as

$$J_{k\ell} = (v, \mathcal{L}_{k\ell} u) = \int_\Omega v(x) \mathcal{L}_{k\ell} u(x)\, d\Omega . \tag{4.4.4}$$

Here and in the following, the relation (4.4.3) is considered an implicit expression for the spectral parameter λ, which is also considered as a functional. The function $\Phi(\lambda, J_{00}, \ldots, J_{mn}, \gamma)$ is a polynomial of degree m with respect to λ, and the solutions of the Eq. (4.4.3) are

$$\lambda_1 = \varphi_1(J_{00}, \ldots, J_{mn}, \gamma) , \quad \ldots , \quad \lambda_m = \varphi_m(J_{00}, \ldots, J_{mn}, \gamma) . \tag{4.4.5}$$

Keep in mind the functionals $J_{k\ell}$ depend on the functions $u(x)$ and $v(x)$.

In the general case, the values $\lambda_1, \ldots, \lambda_m$ have a local extremum at the solutions $u(x)$ and $v(x)$ of the direct and adjoint spectral problems (4.4.1), (4.4.2), that is,

$$\lambda_k(u', v') = \varphi_k\left(J_{00}(u', v'), \ldots, J_{mn}(u', v'), \gamma\right) \overset{u'=u, v'=v}{\rightarrow} \text{extr} . \tag{4.4.6}$$

To show this, suppose that the solutions $u(x)$, $v(x)$ of the problems (4.4.1), (4.4.2), and the functionals $J_{00}, \ldots, J_{mn}$ defined in accordance with (4.4.4), correspond to a fixed value of the parameter γ. Suppose also that the variations

$$u(x) \to u(x) + \delta u(x)\ , \quad v(x) \to v(x) + \delta v(x) \tag{4.4.7}$$

of the solutions of (4.4.1) and (4.4.2) correspond to the variation

$$\lambda \to \lambda + \delta\lambda \tag{4.4.8}$$

of the spectral parameter. Consider the expression for the perturbed value $\widetilde{\Phi}$:

$$\begin{aligned}\widetilde{\Phi} &= \Phi(\lambda + \delta\lambda,\ J_{00} + \delta J_{00},\ \ldots,\ J_{mn} + \delta J_{mn},\ \gamma)\\ &= \sum_{k=0}^{m}\sum_{\ell=0}^{n}(\lambda + \delta\lambda)^k \gamma^\ell (J_{k\ell} + \delta J_{k\ell})\ .\end{aligned} \tag{4.4.9}$$

Using Eq. (4.4.1) for $u(x)$ and adjoint Eq. (4.4.2) for $v(x)$, noting definition (4.4.4), and performing elementary operations, we have the following perturbation of the Eq. (4.4.3):

$$\begin{aligned}\widetilde{\Phi} &= \Phi(\lambda, J_{00}, \ldots, J_{mn}, \gamma) + \frac{\partial\Phi}{\partial\lambda}\delta\lambda + \sum_{k=0}^{m}\sum_{\ell=0}^{n}\lambda^k\gamma^\ell\delta J_{k\ell}\\ &= \Phi(\lambda, J_{00}, \ldots, J_{mn}, \gamma) + \frac{\partial\Phi}{\partial\lambda}\delta\lambda + (\delta v,\ \mathcal{L}u) + (v,\ \delta\mathcal{L}u)\\ &= \Phi(\lambda, J_{00}, \ldots, J_{mn}, \gamma) + \frac{\partial\Phi}{\partial\lambda}\delta\lambda + (\delta v,\ \mathcal{L}u) + \big(\mathcal{L}^*v,\ \delta u\big) = 0\ .\end{aligned} \tag{4.4.10}$$

The first (unperturbed) term is zero because of (4.4.3). The third and fourth terms are also equal to zero because $u(x)$ and $v(x)$ satisfy, respectively, the equations $\mathcal{L}u = 0$, $\mathcal{L}^*v = 0$. Thus, it follows from (4.4.10) that

$$\frac{\partial\Phi}{\partial\lambda}\delta\lambda = 0\ , \tag{4.4.11}$$

and if $\partial\Phi/\partial\lambda \neq 0$, then

$$\delta\lambda = 0\ , \quad \lambda = \lambda_1, \ldots, \lambda_m\ , \tag{4.4.12}$$

thus establishing the extremal property in the general case. Only in the special case $\partial\Phi/\partial\lambda = 0$ (at the solution point, where the perturbation was made) it may occur that an extremum does not appear.

Let us study the dependencies of λ_k, $k = 1, 2, \ldots, m$, on the parameter γ in more detail. There is an important peculiarity in Eq. (4.4.3): when $u(x)$ and

$v(x)$ are the solution functions of the original and adjoint problems, respectively, the functionals $J_{00}, \ldots, J_{mn}$ can be considered as constant when the function $\Phi(\lambda,\ J_{00},\ \ldots,\ J_{mn},\ \gamma)$ is differentiated with respect to γ. To show this, let us write the total derivative of Φ,

$$\frac{\mathrm{d}\Phi}{\mathrm{d}\gamma} = \frac{\mathrm{d}\Phi}{\mathrm{d}\lambda}\frac{\mathrm{d}\lambda}{\mathrm{d}\gamma} + \frac{\partial\Phi}{\partial\gamma} + \sum_{k=0}^{m}\sum_{\ell=0}^{n}\frac{\partial\Phi}{\partial J_{k\ell}}\frac{\mathrm{d}J_{k\ell}}{\mathrm{d}\gamma}\,. \tag{4.4.13}$$

The double sum in (4.4.13) is evaluated as

$$\begin{aligned}
&\sum_{k=0}^{m}\sum_{\ell=0}^{n}\frac{\partial\Phi}{\partial J_{k\ell}}\frac{\mathrm{d}J_{k\ell}}{\mathrm{d}\gamma}\\
=&\sum_{k=0}^{m}\sum_{\ell=0}^{n}\lambda^{k}\gamma^{\ell}\left[\left(\frac{\mathrm{d}v}{\mathrm{d}\gamma},\ \mathcal{L}_{k\ell}u\right)+\left(v,\ \mathcal{L}_{k\ell}\frac{\mathrm{d}u}{\mathrm{d}\gamma}\right)\right]\\
=&\left[\left(\frac{\mathrm{d}v}{\mathrm{d}\gamma},\ \mathcal{L}u\right)+\left(v,\ \mathcal{L}\frac{\mathrm{d}u}{\mathrm{d}\gamma}\right)\right]\\
=&\left[\left(\frac{\mathrm{d}v}{\mathrm{d}\gamma},\ \mathcal{L}u\right)+\left(\mathcal{L}^{*}v,\ \frac{\mathrm{d}u}{\mathrm{d}\gamma}\right)\right]=0\,,
\end{aligned} \tag{4.4.14}$$

by taking into account the equalities $\mathcal{L}u = 0$ and $\mathcal{L}^{*}v = 0$. Thus the function $\Phi = \Phi(\lambda,\ J_{00},\ \ldots,\ J_{mn},\ \gamma)$ can be considered as a function of just two variables λ and γ, and denoted as $F(\lambda, \gamma)$, that is,

$$F(\lambda,\gamma) = \sum_{k=0}^{m}\sum_{\ell=0}^{n}\lambda^{k}\gamma^{n}J_{k\ell} = 0\,. \tag{4.4.15}$$

This equation can be taken to determine the dependence $\lambda = \lambda(\gamma)$. To be exact, it determines a set of functions $\lambda_1(\gamma), \ldots, \lambda_m(\gamma)$.

In correspondence with the fundamental theorem on implicit functions (see, e.g., Rektorys [35]), a unique solution of (4.4.15) exists in a small vicinity of the fixed values $\lambda = \widetilde{\lambda}, \gamma = \widetilde{\gamma}$, if $\partial F/\partial\lambda \neq 0$.

Thus the nonuniqueness of the solution of (4.4.15), or in other words, a bifurcation of the considered system, may occur for some values $\lambda = \lambda^{*}$, $\gamma = \gamma^{*}$ when the condition of the theorem on implicit functions is violated. Hence the bifurcation values λ^{*} and γ^{*} are found with the help of the equations

$$F(\lambda^{*},\gamma^{*}) = 0\,,\quad \frac{\partial F(\lambda^{*},\gamma^{*})}{\partial\lambda} = 0\,. \tag{4.4.16}$$

Let us denote by $(\lambda_1^{*}, \gamma_1^{*}), (\lambda_2^{*}, \gamma_2^{*}), \ldots$ the solutions of the nonlinear system of Eqs. (4.4.16), representing points on the λ, γ plane. Let us investigate the behavior of functions $\lambda_i = \lambda_i(\gamma)$ in a small vicinity of the bifurcation points $(\lambda_k^{*}, \gamma_k^{*})$.

For simplicity, the subscript indices of the considered functions and points will be omitted.

Let us represent the function $F(\lambda, \gamma)$ in a small vicinity of the point (λ^*, γ^*) as a series expansion,

$$F(\lambda,\gamma) = F(\lambda^*,\gamma^*) + \frac{\partial F(\lambda^*,\gamma^*)}{\partial \lambda}[\lambda-\lambda^*] + \frac{\partial F(\lambda^*,\gamma^*)}{\partial \gamma}[\gamma-\gamma^*]$$
$$+\frac{1}{2}\frac{\partial^2 F(\lambda^*,\gamma^*)}{\partial \lambda^2}[\lambda-\lambda^*]^2 + \frac{\partial^2 F(\lambda^*,\gamma^*)}{\partial \lambda \partial \gamma}[\lambda-\lambda^*][\gamma-\gamma^*] + \frac{1}{2}\frac{\partial^2 F(\lambda^*,\gamma^*)}{\partial \gamma^2}[\gamma-\gamma^*]^2 .$$
$$+\ldots \tag{4.4.17}$$

Taking into account the bifurcation conditions (4.4.16), $F(\lambda^*, \gamma^*) = 0$ and $\partial F/\partial\lambda = 0$.

For simplicity, in the following let us consider only the general case, for which at the bifurcation point, it holds that

$$\frac{\partial F(\lambda^*,\gamma^*)}{\partial \gamma} \neq 0\,, \quad \frac{\partial^2 F(\lambda^*,\gamma^*)}{\partial \lambda^2} \neq 0\,. \tag{4.4.18}$$

This means that the remaining terms displayed in Eq. (4.4.17) include the lowest-order nonzero ones. We may omit the term quadratic in γ, because (4.4.17) includes also a term linear in γ, which is nonzero by (4.4.18). Hence, under the conditions (4.4.18), we have

$$F(\lambda,\gamma) = \frac{\partial F(\lambda^*,\gamma^*)}{\partial \gamma}[\gamma-\gamma^*] + \frac{\partial^2 F(\lambda^*,\gamma^*)}{\partial \lambda \partial \gamma}[\lambda-\lambda^*][\gamma-\gamma^*] + \frac{1}{2}\frac{\partial^2 F(\lambda^*,\gamma^*)}{\partial \lambda^2}[\lambda-\lambda^*]^2 + \ldots . \tag{4.4.19}$$

Provided only sufficient continuity, we may represent the behavior of the function $\lambda = \lambda(\gamma)$ in the vicinity of the bifurcation point (λ^*, γ^*) as a power series

$$\lambda(\gamma) = \lambda^* + \alpha_1[\gamma-\gamma^*]^{\varepsilon_1} + \alpha_2[\gamma-\gamma^*]^{\varepsilon_2} + \ldots\,, \quad \text{where } 0 < \varepsilon_1 < \varepsilon_2 < \ldots \tag{4.4.20}$$

where $\alpha_1, \alpha_2, \ldots$ and $\varepsilon_1, \varepsilon_2, \ldots$ are to be determined with the help of the condition $F(\lambda, \gamma) = 0$. For simplicity, we will approximate, taking only one pair of unknowns α and ε:

$$\lambda(\gamma) = \lambda^* + \alpha\left[\gamma-\gamma^*\right]^{\varepsilon} + \ldots\,, \tag{4.4.21}$$

By substituting (4.4.21) into (4.4.19), the Eq. (4.4.19) is transformed into

$$\widetilde{F} = \widetilde{F}(\gamma-\gamma^*) \equiv 0\,,$$

which must be satisfied identically, that is, the coefficient of each power of γ must vanish separately. Here $\widetilde{F}(\gamma - \gamma^*)$ is a series expansion with respect to the perturbation $(\gamma - \gamma^*)$. We have

$$F(\lambda,\gamma) = \frac{\partial F(\lambda^*,\gamma^*)}{\partial\gamma}\left[\gamma-\gamma^*\right] + \alpha\frac{\partial^2 F(\lambda^*,\gamma^*)}{\partial\lambda\partial\gamma}\left[\gamma-\gamma^*\right]^{1+\varepsilon} + \frac{\alpha^2}{2}\frac{\partial^2 F(\lambda^*,\gamma^*)}{\partial\lambda^2}\left[\gamma-\gamma^*\right]^{2\varepsilon} + \cdots \equiv 0\,. \tag{4.4.22}$$

For any $\varepsilon > 0$, we have $1+\varepsilon > 1$. Therefore, because a term linear in γ is present in (4.4.22), the middle term can never be of leading order. We may omit it, obtaining

$$F(\lambda,\gamma) = \frac{\partial F(\lambda^*,\gamma^*)}{\partial\gamma}\left[\gamma-\gamma^*\right] + \frac{\alpha^2}{2}\frac{\partial^2 F(\lambda^*,\gamma^*)}{\partial\lambda^2}\left[\gamma-\gamma^*\right]^{2\varepsilon} + \cdots \equiv 0\,. \tag{4.4.23}$$

To find the leading term, there are now three possibilities:

$$2\varepsilon < 1\,, \quad 2\varepsilon = 1\,, \quad 2\varepsilon > 1\,.$$

If $0 < 2\varepsilon < 1$, the first term in (4.4.23) is of a higher order in γ than the second one, and hence can be omitted. The equality (4.4.23) is then satisfied if and only if $\partial^2 F(\lambda^*,\gamma^*)/\partial\lambda^2 = 0$, which contradicts the second of our general-case assumptions in (4.4.18). Hence in the general case being considered, it must hold that $2\varepsilon \geq 1$.

Similarly, if $2\varepsilon > 1$, we are left with the first term only, contradicting the first of our general-case assumptions in (4.4.18). The only remaining possibility is $2\varepsilon = 1$, leading to

$$\alpha^2 = -2\left(\frac{\partial F(\lambda^*,\gamma^*)}{\partial\gamma}\right)\left(\frac{\partial^2 F(\lambda^*,\gamma^*)}{\partial\lambda^2}\right)^{-1}\,. \tag{4.4.24}$$

Using (4.4.21), our final result is

$$\lambda(\gamma) = \lambda^* + \alpha\sqrt{\gamma-\gamma^*}\,, \quad \left|\gamma-\gamma^*\right| \ll 1\,, \tag{4.4.25}$$

provided that the general-case conditions (4.4.18) are satisfied. If the conditions (4.4.18) are violated, then the terms displayed in Eq. (4.4.22) are not sufficient to perform the analysis. Such cases can be analyzed by a minor modification to the above derivation. One starts from Eq. (4.4.17), accounting for more terms in the multivariate Taylor series, and under the particular conditions being investigated, again keeps the lowest-order nonzero terms.

Note that by (4.4.24), for a real-valued function F, the quantity α is either real or pure imaginary. This means that for any real F, as the real-valued load γ approaches any bifurcation value γ^*, in the $(\mathrm{Re}\,\lambda,\ \mathrm{Im}\,\lambda)$ plane the two eigenvalues λ forming a pair (corresponding to different choices of sign for the square root in (4.4.25)) must approach the bifurcation point $\lambda^* = (\mathrm{Re}\,\lambda^*,\ \mathrm{Im}\,\lambda^*)$ from opposite directions aligned with either the real or the imaginary axis. After the bifurcation point, they must leave in opposite directions aligned with the other axis.

Speaking even more exactly, at the bifurcation point, each solution curve turns counterclockwise by $\pi/2$. For example, an eigenvalue whose imaginary component decreases before the bifurcation point (as γ is increased), becomes an eigenvalue whose real component increases after the bifurcation point. This identification, and the fact that this is a general property of the class of problems investigated, is something that would be impossible to ascertain by a numerical approach. (See Jeronen [17] for a discussion concerning the identification of points belonging on the same curve in numerical parametric studies of eigenvalue problems.)

Furthermore, Eq. (4.4.25) tells us that in the class of problems analyzed, if we plot $(\mathrm{Re}\,\lambda,\ \mathrm{Im}\,\lambda)$ as a function γ, then in a small vicinity of any bifurcation point, the solution curves always take on a locally square-root shape. This is essentially a direct consequence of the implicit function theorem. The bifurcation condition $\partial F/\partial\lambda = 0$ eliminates the term linear in λ from (4.4.17), which then leads to $2\varepsilon = 1$, producing the square root.

As an example of using the presented bifurcation analysis, let us consider the stability problem of an axially moving elastic panel, with no external applied tension, performing transverse vibrations. The only restoring force arises from the bending rigidity of the panel. In the fixed (laboratory, Euler) coordinate system, the equation of small transverse vibrations and the corresponding boundary conditions can be written as

$$\begin{aligned} &\frac{\partial^2 w}{\partial t^2} + 2V_0\frac{\partial^2 w}{\partial x \partial t} + V_0^2\frac{\partial^2 w}{\partial x^2} + \frac{D}{\rho S}\frac{\partial^4 w}{\partial x^4} = 0\,, \\ &w(0,t) = w(\ell,t) = 0\,, \quad D\frac{\partial^2 w(0,t)}{\partial x^2} = D\frac{\partial^2 w(\ell,t)}{\partial x^2} = 0\,, \end{aligned} \tag{4.4.26}$$

where $w = w(x,t)$ describes the transverse displacement, ρ is the density of the material, S the cross-sectional area of the panel, t time and $x \in [0, \ell]$.

Time-harmonic transverse vibrations of the panel are represented as

$$w(x,t) = e^{i\omega t}u(x)\,, \tag{4.4.27}$$

and we will use the nondimensional variables

$$x = \ell\widetilde{x}\,, \quad \widetilde{\omega}^2 = \frac{\rho S\omega^2\ell^4}{D}\,, \quad \widetilde{V_0}^2 = \frac{\rho S\ell^2}{D}V_0^2\,. \tag{4.4.28}$$

The tilde will be omitted.

We obtain

$$\begin{aligned} &\omega^2 u - 2i\omega V_0\frac{\mathrm{d}u}{\mathrm{d}x} - V_0^2\frac{\mathrm{d}^2 u}{\mathrm{d}x^2} - \frac{\mathrm{d}^4 u}{\mathrm{d}x^4} = 0\,, \\ &u(0) = u(1) = 0\,, \quad \left(\frac{\mathrm{d}^2 u}{\mathrm{d}x^2}\right)_{x=0} = \left(\frac{\mathrm{d}^2 u}{\mathrm{d}x^2}\right)_{x=1} = 0\,. \end{aligned} \tag{4.4.29}$$

In (4.4.27)–(4.4.29), ω is a complex-valued frequency, $u = u(x)$ is the amplitude function, and i the imaginary unit.

After multiplication of the Eq. (4.4.29) by the complex conjugate amplitude function $u^*(x)$ and performing integration, taking into account the boundary conditions (4.4.29), we obtain

$$\Phi = a\omega^2 + 2bV_0\omega + V_0^2 c - d = 0 \,, \tag{4.4.30}$$

where

$$\begin{aligned}
a &= \int_0^1 uu^* \,\mathrm{d}x > 0 \,, \\
ib &= \int_0^1 u^* \frac{\mathrm{d}u}{\mathrm{d}x} \,\mathrm{d}x = -\int_0^1 u \frac{\mathrm{d}u^*}{\mathrm{d}x} \,\mathrm{d}x > 0 \,, \quad (b \text{ real}) \\
c &= -\int_0^1 u^* \frac{\mathrm{d}^2 u}{\mathrm{d}x^2} \,\mathrm{d}x = \int_0^1 \frac{\mathrm{d}u}{\mathrm{d}x}\frac{\mathrm{d}u^*}{\mathrm{d}x} \,\mathrm{d}x > 0 \,, \\
d &= \int_0^1 u^* \frac{\mathrm{d}^4 u}{\mathrm{d}x^4} \,\mathrm{d}x = \int_0^1 \frac{\mathrm{d}^2 u}{\mathrm{d}x^2}\frac{\mathrm{d}^2 u^*}{\mathrm{d}x^2} \,\mathrm{d}x > 0 \,,
\end{aligned} \tag{4.4.31}$$

Using the definitions (4.4.31), it is possible to find the coefficient α in the asymptotic representation of the function $\lambda(\gamma)$. Denoting $\Phi = F$, we have

$$\frac{\partial F}{\partial V_0} = 2\,(b\omega + cV_0) \;, \qquad \frac{\partial^2 F}{\partial \omega^2} = 2a \;, \tag{4.4.32}$$

and consequently,

$$\alpha^2 = -2\frac{b\omega + cV_0}{a} \,. \tag{4.4.33}$$

Thus we find the following asymptotic representation for the dependence $\omega(V_0)$ in the vicinity of the bifurcation point (ω_k^*, V_0^*):

$$\begin{aligned}
\omega(V_0) &\approx \omega^* \pm \sqrt{-2\frac{b\omega^* + cV_0^*}{a}}\sqrt{V_0 - V_0^*} \\
&= \omega^* \pm \sqrt{2\frac{b^2 - ac}{a^2} V_0^*}\sqrt{V_0 - V_0^*} \,.
\end{aligned} \tag{4.4.34}$$

Note that for this particular problem, the equation $\Phi(\omega, a, b, c, d, V_0)$ can be solved with respect to the variable ω. As a result, we have

$$\omega_{1,2}(V_0) = \frac{-bV_0 \pm \sqrt{(b^2 - ac)V_0^2 + ad}}{a} \,. \tag{4.4.35}$$

Let us analyze the dependence $\omega(V_0)$, determined by the expression (4.4.35), in the small vicinity of the bifurcation point (ω^*, V_0^*). At the bifurcation point, the square root term vanishes. Hence at that point, the harmonic vibration frequency ω^* and the velocity of axial motion V_0^* must satisfy

$$\omega^* = -\frac{b}{a}V_0^* , \quad (ac - b^2)(V_0^*)^2 = ad . \tag{4.4.36}$$

On the other hand, by Taylor series expansion,

$$V_0^2 \approx (V_0^*)^2 + 2V_0^* \left[V - V_0^*\right] , \quad \left|V - V_0^*\right| \ll 1 . \tag{4.4.37}$$

From (4.4.35)–(4.4.37), we obtain the asymptotic result

$$\omega_{1,2} \approx \omega^* \pm \sqrt{2\frac{b^2 - ac}{a^2} V_0^*} \sqrt{V_0 - V_0^*} , \quad \left|V_0 - V_0^*\right| \ll 1 , \tag{4.4.38}$$

which coincides with the asymptotic representation (4.4.34).

As a second example of application of the bifurcation method, we consider the problem of harmonic vibrations of a (stationary, not axially travelling) panel compressed by the force γ, with $\gamma > 0$. The governing equation and boundary conditions are

$$\begin{aligned} &\frac{\mathrm{d}^4 u}{\mathrm{d}x^4} + \gamma\frac{\mathrm{d}^2 u}{\mathrm{d}x^2} - \omega^2 u = 0 , \qquad x \in (0, 1) , \\ &u(0) = u(1) = 0 , \quad \left(\frac{\mathrm{d}^2 u}{\mathrm{d}x^2}\right)_{x=0} = \left(\frac{\mathrm{d}^2 u}{\mathrm{d}x^2}\right)_{x=1} = 0 . \end{aligned} \tag{4.4.39}$$

Let us investigate the asymptotic behavior of the frequency ω as a function of the loading parameter γ, that is, $\omega = \omega(\gamma)$, using the discussed perturbation method. To do this, we multiply the Eq. (4.4.39) by the function $u(x)$, which coincides in the considered case with $u^*(x)$ (because the problem (4.4.39) is self-adjoint), and perform integration.

We find the following expression for Φ as a function of the functionals a, c and d (as defined in (4.4.31)). We have

$$\Phi(\omega, a, c, d, \gamma) = -a\omega^2 - \gamma c + d = 0 . \tag{4.4.40}$$

The functionals a, c and d can be expressed with the help of eigenmodes of vibrations

$$u_k(x) = B_k \sin(k\pi x) .$$

We find

$$\begin{aligned} a_k &= \int_0^1 (u_k(x))^2 \,\mathrm{d}x = \frac{B_k^2}{2} , \\ c_k &= \int_0^1 \left(\frac{\mathrm{d}u_k}{\mathrm{d}x}\right)^2 \mathrm{d}x = \frac{k^2\pi^2}{2} B_k^2 , \\ d_k &= \int_0^1 \left(\frac{\mathrm{d}^2 u_k}{\mathrm{d}x^2}\right)^2 \mathrm{d}x = \frac{k^4\pi^4}{2} B_k^2 . \end{aligned} \tag{4.4.41}$$

In correspondence with the general formulas (4.4.24)–(4.4.25), the asymptotic behavior of the frequencies in the vicinity of the bifurcation points

$$\omega_k^* = 0\,, \quad \gamma_k^* = k^2\pi^2 \tag{4.4.42}$$

is described by the expressions

$$\omega_k = \omega_k(\gamma) = \pm\alpha\sqrt{\gamma - k^2\pi^2}\,, \quad \left|\gamma - k^2\pi^2\right| \ll 1\,, \tag{4.4.43}$$

and the value of the coefficient α is given by

$$\alpha^2 = -2\left(\frac{\partial F(\omega_k^*, \gamma_k^*)}{\partial \gamma}\right)\left(\frac{\partial^2 F(\omega_k^*, \gamma_k^*)}{\partial \omega^2}\right)^{-1} = -k^2\pi^2\,. \tag{4.4.44}$$

In this section we presented a method for bifurcation analysis, based on the introduction of the adjoint spectral problem and implicitly given functionals. The general analysis provided us insight into the general properties of bifurcation points for the class of systems considered. Namely, in a small vicinity of each bifurcation point, when viewed in the (Re λ, Im λ) plane, the two eigenvalues that form a pair approach and then depart in directions aligned with the real and imaginary axes. The approach occurs from opposite directions, and each eigenvalue turns counterclockwise by $\pi/2$ at the bifurcation point. Furthermore, if we plot λ versus γ, the shape of the solution curve near each bifurcation point is always that of a square root function. This was seen to arise essentially directly from the implicit function theorem.

Finally, we applied the presented method to some example problems in stability analysis.

4.5 Variational Principle and Bifurcation Analysis

The focus of this section is the stability of a simply supported axially moving elastic panel. By "panel" we refer to a plate undergoing cylindrical deformation. The governing equation for transverse displacement is of the same form as for an Euler–Bernoulli beam, but its interpretation is different, and each term in the governing equation represents a pressure (force per unit area). Details regarding this will be explained later, in Chap. 5.

We will use a complex variable technique and bifurcation theory. Our main task is the derivation of variational equations and a variational principle. Moreover, we will perform an analysis of the variational principle, which allows the study of qualitative properties of the bifurcation points. Furthermore, asymptotic behavior around an arbitrary bifurcation point will be analyzed and presented. As a result, we will show analytically that for this particular problem, the eigenvalue curves in the (ω, V_0) plane cross both the ω and V_0 axes perpendicularly. It is also seen that near each bifurcation point, the dependence $\omega(V_0)$ for each mode approximately follows the shape of a square root function. As this problem belongs to the class we analyzed using the implicit function theorem, this exemplifies the general result obtained from that analysis.

The results presented in this section complement existing numerical studies on the stability of axially moving materials, and especially materials with finite bending rigidity. From a rigorous mathematical viewpoint, the presence of bending rigidity is essential, because the presence of the fourth-order term in the model changes the qualitative behavior of the bifurcation points. Wang et al. [44] have used Hamiltonian mechanics to show that, unlike the panel analyzed here, an axially moving ideal string (exactly zero bending rigidity) remains stable at any axial velocity. This is not physically realizable; even paper or thin plastic films have a small but finite bending rigidity. For correct fundamental understanding it is critically important to acknowledge the role of bending rigidity as a factor that not only quantitatively, but also qualitatively contributes to the stability behavior of panels. This is akin to the well-known boundary layer effect that arises in the dynamics of fluids with small but finite viscosity (see e.g. Batchelor [7]); both effects involve a singular perturbation of the governing equation (see Lagerstrom and Casten [22], Bender and Orszag [8], Chen et al. [13]) that leads to a qualitative change in behavior.

Finally, keep in mind that singular perturbations are not the only way qualitative surprises may arise in mechanics, when new terms are added to a model. Recall our discussion in Chap. 3. As Ziegler [47] discovered, the introduction of small but finite damping in a nonconservative system may cause a dramatic decrease in the critical value of the loading where the system loses its stability (see Kirillov and Verhulst [19], Kirillov [18]).

Let us consider the problem of free harmonic vibrations of an elastic panel, moving axially at a constant velocity V_0. In a stationary orthogonal coordinate system, the transverse vibrations are characterized by the function $w = w(x, t)$, which is determined by the following partial differential equation:

$$\frac{\partial^2 w}{\partial t^2} + 2V_0 \frac{\partial^2 w}{\partial x \partial t} + (V_0^2 - C^2)\frac{\partial^2 w}{\partial x^2} + \frac{D}{\rho S}\frac{\partial^4 w}{\partial x^4} = 0\,, \quad 0 < x < \ell\,, \quad C = \sqrt{\frac{T_0}{\rho S}}\,. \tag{4.5.1}$$

At the ends of the considered interval $x \in [0, \ell]$, we apply the simply supported boundary conditions

$$w(0,t) = w(\ell,t) = D\frac{\partial^2 w}{\partial x^2}(0,t) = D\frac{\partial^2 w}{\partial x^2}(\ell,t) = 0\,. \tag{4.5.2}$$

Here x is the axial space coordinate, t time, T_0 tension, ρ material density, S the area of the panel cross section, ℓ the length of the free span, and D the bending rigidity of the panel.

For harmonic vibrations at frequency ω, the transverse displacement can be represented in the form

$$w(x,t) = e^{i\omega t} u(x) \,, \quad i = \sqrt{-1} \,, \tag{4.5.3}$$

where $u(x)$ is an amplitude function that satisfies the following boundary value problem:

$$\frac{\partial^4 u}{\partial x^4} + (V_0^2 - C^2)\frac{\partial^2 u}{\partial x^2} + 2i\omega V_0 \frac{\partial u}{\partial x} - \omega^2 u = 0 \,, \tag{4.5.4}$$

$$u(0) = u(1) = \frac{\partial^2 u}{\partial x^2}(0) = \frac{\partial^2 u}{\partial x^2}(1) = 0 \,, \tag{4.5.5}$$

which is written in the nondimensional variables

$$x = \ell \tilde{x} \,, \quad \frac{\rho S \omega^2 \ell^4}{D} = \tilde{\omega}^2 \,, \quad \frac{\rho S \ell^2}{D} V_0^2 = \tilde{V}_0^2 \,, \quad \frac{\rho S \ell^2}{D} C^2 = \tilde{C}^2 \,. \tag{4.5.6}$$

In the following, the tilde will be omitted. Note that (4.5.4) belongs to the class of problems that was analyzed via the implicit function theorem in the previous section.

The amplitude function $u(x)$ determined from the boundary value problem (4.5.4)–(4.5.5) is a complex-valued function, that is,

$$u(x) = u^{(1)}(x) + i u^{(2)}(x) \,, \quad \hat{u}(x) = u^{(1)}(x) - i u^{(2)}(x) \,, \tag{4.5.7}$$

where $u^{(1)}(x)$ and $u^{(2)}(x)$ are real-valued functions and $\hat{u}(x)$ is the complex conjugate of $u(x)$. Because the original problem (4.5.1) is linear in w, the real and imaginary parts of $w(x)$ constructed with the help of (4.5.3) are its real-valued solutions.

Let us consider a variational formulation of the spectral problem (4.5.4)–(4.5.5). To derive it, we multiply the differential equation by the complex conjugate (adjoint) amplitude function $\hat{u}(x)$ and integrate the result on the interval (0, 1). We also take into account the boundary conditions

$$\begin{gathered} u^{(1)}(0) = u^{(2)}(0) = u^{(1)}(1) = u^{(2)}(1) = 0 \,, \\ \frac{\partial^2 u^{(1)}}{\partial x^2}(0) = \frac{\partial^2 u^{(2)}}{\partial x^2}(0) = \frac{\partial^2 u^{(1)}}{\partial x^2}(1) = \frac{\partial^2 u^{(2)}}{\partial x^2}(1) = 0 \,, \end{gathered} \tag{4.5.8}$$

which follow from the boundary conditions (4.5.5). We obtain the functional equation

$$a\omega^2 + 2bV_0\omega + (V_0^2 - C^2)c - d = 0 \,, \tag{4.5.9}$$

where a, b, c and d are integral functionals depending on the problem (4.5.4)–(4.5.5). The functional a is given by

$$a = \int_0^1 u\hat{u} \, \mathrm{d}x = \int_0^1 \left((u^{(1)})^2 + (u^{(2)})^2\right) \mathrm{d}x > 0 \,. \tag{4.5.10}$$

Using the boundary conditions (4.5.8), we can write the functional b as

$$\int_0^1 \frac{\partial u}{\partial x}\hat{u}\,\mathrm{d}x = i\int_0^1 \left(\frac{\partial u^{(2)}}{\partial x}u^{(1)} - \frac{\partial u^{(1)}}{\partial x}u^{(2)}\right)\mathrm{d}x = ib\,, \tag{4.5.11}$$

where b is real-valued. The functionals c and d are obtained by integrating by parts (once in the case of c and twice for d), and taking into account the corresponding boundary conditions in (4.5.8). We have

$$c = -\int_0^1 \frac{\partial^2 u}{\partial x^2}\hat{u}\,\mathrm{d}x = \int_0^1 \left(\left[\frac{\partial u^{(1)}}{\partial x}\right]^2 + \left[\frac{\partial u^{(2)}}{\partial x}\right]^2\right)\mathrm{d}x > 0\,, \tag{4.5.12}$$

$$d = \int_0^1 \frac{\partial^4 u}{\partial x^4}\hat{u}\,\mathrm{d}x = \int_0^1 \left(\left[\frac{\partial^2 u^{(1)}}{\partial x^2}\right]^2 + \left[\frac{\partial^2 u^{(2)}}{\partial x^2}\right]^2\right)\mathrm{d}x > 0\,. \tag{4.5.13}$$

To show that b is indeed real-valued, we may proceed as follows. Consider the integrand in (4.5.11). The functions u and $\partial u/\partial x$ are linearly independent, so at each fixed x, they can be considered as two independent complex numbers of the form $A + Bi$, $C + Di$. Their product is of the form

$$(A + Bi)(C + Di) = (AC - BD) + i(BC + AD)\,. \tag{4.5.14}$$

The condition for (4.5.14) to be real-valued is

$$(BC + AD) = 0\,. \tag{4.5.15}$$

This holds identically if $C = A$, $D = -B$, that is, if $C + Di = \mathrm{conj}(A + Bi)$. This is indeed the case with the functionals a and c.

At first glance, it would appear that in the case of b, we must deal with the case where $C + Di$ is chosen independently of $A + Bi$. However, it turns out that this is not needed. Consider the functional b as defined in Eq. (4.5.11),

$$ib = \int_0^1 \frac{\partial u}{\partial x}\hat{u}\,\mathrm{d}x = -\int_0^1 u\frac{\partial \hat{u}}{\partial x}\,\mathrm{d}x\,,$$

where the second equality follows from integration by parts and the boundary conditions (4.5.5). Utilizing both forms, it follows that

$$\int_0^1 \frac{\partial u}{\partial x}\hat{u}\,\mathrm{d}x - \int_0^1 u\frac{\partial \hat{u}}{\partial x}\,\mathrm{d}x = \int_0^1 \left[\frac{\partial u}{\partial x}\hat{u} - u\frac{\partial \hat{u}}{\partial x}\right]\mathrm{d}x = 2ib\,, \tag{4.5.16}$$

Because conjugation distributes, that is,

$$[(\alpha+\beta i)(\gamma+\delta i)]^* = [(\alpha\gamma-\beta\delta)+i(\alpha\delta+\beta\gamma)]^* = (\alpha\gamma-\beta\delta)-i(\alpha\delta+\beta\gamma)$$
$$= (\alpha-\beta i)(\gamma-\delta i) = (\alpha+\beta i)^*(\gamma+\delta i)^* \,,$$

we see that the integrand in (4.5.16) is of the form $z - \hat{z}$ with $z = u \cdot \partial\hat{u}/\partial x$. Hence the real part cancels, $2ib$ is always pure imaginary, and b is always real-valued, as claimed.

Note that the boundary conditions that allowed us to write (4.5.16) are important; in the general case, integrands of the form (4.5.11) are not necessarily real-valued. Setting $A = \operatorname{Re} u$, $B = \operatorname{Im} u$, $C = \operatorname{Re} \partial\hat{u}/\partial x = \operatorname{Re} \partial u/\partial x$, $D = \operatorname{Im} \partial\hat{u}/\partial x = -\operatorname{Im} \partial u/\partial x$, and considering that it is not necessary to satisfy the condition (4.5.15) pointwise, as long as the imaginary part cancels out in the integration in (4.5.11), for the general case the condition for the functional to be real-valued becomes

$$\int_0^1 \left[\operatorname{Im}(u)\operatorname{Re}\left(\frac{\partial u}{\partial x}\right) - \operatorname{Re}(u)\operatorname{Im}\left(\frac{\partial u}{\partial x}\right)\right] \mathrm{d}x = 0 \,. \tag{4.5.17}$$

In the following, we will use the method of variational analysis and variational principle in complex variables. Let us write the variation of the functional equation (4.5.9). To do this, we take the variations of the considered functionals,

$$\begin{aligned}
\delta a &= \int_0^1 \left(\hat{u}\delta u + u\delta\hat{u}\right) \mathrm{d}x \,, \\
i\delta b &= \int_0^1 \left(\hat{u}\delta\frac{\partial u}{\partial x} + \frac{\partial u}{\partial x}\delta\hat{u}\right) \mathrm{d}x \,, \\
\delta c &= \int_0^1 \left(\frac{\partial u}{\partial x}\delta\frac{\partial\hat{u}}{\partial x} + \frac{\partial\hat{u}}{\partial x}\delta\frac{\partial u}{\partial x}\right) \mathrm{d}x \,, \\
\delta d &= \int_0^1 \left(\frac{\partial^2 u}{\partial x^2}\delta\frac{\partial^2\hat{u}}{\partial x^2} + \hat{u}\delta\frac{\partial^2 u}{\partial x^2}\right) \mathrm{d}x \,,
\end{aligned} \tag{4.5.18}$$

and perform standard transformations in (4.5.9), replacing u, $\hat{u}$ and ω with $u + \delta u$, $\hat{u} + \delta\hat{u}$ and $\omega + \delta\omega$, respectively. We have the variation

$$\begin{aligned}
2(a\omega + bV_0)\delta\omega &+ \int_0^1 \left[-\omega^2 u + 2i\omega V_0\frac{\partial u}{\partial x} + (V_0^2 - C^2)\frac{\partial^2 u}{\partial x^2} + \frac{\partial^4 u}{\partial x^4}\right]\delta\hat{u}\,\mathrm{d}x \\
&+ \int_0^1 \left[-\omega^2\hat{u} - 2i\omega V_0\frac{\partial\hat{u}}{\partial x} + (V_0^2 - C^2)\frac{\partial^2\hat{u}}{\partial x^2} + \frac{\partial^4\hat{u}}{\partial x^4}\right]\delta u\,\mathrm{d}x = 0 \,.
\end{aligned} \tag{4.5.19}$$

For $u(x)$ and $\hat{u}(x)$, which are solutions of the spectral boundary value problem (4.5.4)–(4.5.5) and its complex conjugate, the integral expressions in (4.5.19) are identically zero. Taking this into account, we are left with

$$2(a\omega + bV_0)\,\delta\omega = 0 \,. \tag{4.5.20}$$

By the rule of zero product, there are two possibilities. If $a\omega + bV_0 \neq 0$ for the spectral problem (4.5.4)–(4.5.5), then the frequency variation for free vibrations $\delta\omega$ must be zero. That is, the first possibility is

$$a\omega + bV_0 \neq 0 \,, \quad \delta\omega = 0 \,. \tag{4.5.21}$$

Solving (4.5.9) for ω, we arrive at the variational representation for harmonic vibrations, corresponding to each of the two solution branches of Eq. (4.5.9):

$$\omega_{\pm}(V_0) = \frac{1}{a}\left(-bV_0 \pm \sqrt{(b^2 - ac)V_0^2 + acC^2 + ad}\right) . \tag{4.5.22}$$

As was observed in the previous section, the functional (4.5.22) has a local extremum at the solution functions $u(x)$, $\hat{u}(x)$.

The other possibility in Eq. (4.5.20) is

$$a\omega + bV_0 = 0 \tag{4.5.23}$$

with $\delta\omega$ free. To analyze this case, we consider Eq. (4.5.9) as an implicit function $F(\omega, V_0)$:

$$F(\omega, V_0) = 0 \,, \quad F(\omega, V_0) = a\omega^2 + 2bV_0\omega + (V_0^2 - C^2)c - d \,. \tag{4.5.24}$$

Again, we may solve (4.5.9) for ω, obtaining the following two solution branches:

$$\omega_{\pm}(V_0) = \frac{1}{a}\left(-bV_0 \pm \sqrt{(b^2 - ac)V_0^2 + acC^2 + ad}\right) . \tag{4.5.25}$$

This is the same expression as (4.5.22).

Let (ω^*, V_0^*) denote a bifurcation point, that is, the values of ω and V_0 at which the solution of (4.5.24) branches. At a bifurcation point, the conditions of the implicit function theorem must be violated, in other words, we will have

$$F(\omega, V_0) = 0 \,, \quad \frac{\partial F(\omega, V_0)}{\partial \omega} = 0 \,. \tag{4.5.26}$$

Using (4.5.24), these conditions become

$$a\omega^2 + 2bV_0\omega + (V_0^2 - C^2)c - d = 0 \,, \quad a\omega + bV_0 = 0 \,. \tag{4.5.27}$$

As a result, we find the following representation for bifurcation values of the frequency and panel velocity:

$$\omega^* = -\frac{b}{a}V_0^* \,, \quad (ac - b^2)(V_0^*)^2 = acC^2 + ad \,. \tag{4.5.28}$$

Alternatively, these values can be obtained from the condition $\omega_+(V_0) = \omega_-(V_0)$ that holds at a bifurcation point, and the representation (4.5.25) for $\omega_\pm(V_0)$. Note also that if some solutions have $b = 0$, the corresponding bifurcation points are distributed along the V_0 axis in the (V_0, ω) plane, that is,

$$\omega^* = 0\,, \quad (V_0^*)^2 = C^2 + \frac{d}{c} \qquad \text{for solutions with } b = 0\,. \tag{4.5.29}$$

Let us differentiate $\omega(V_0)$ with respect to the parameter V_0. To do this, we perform a variation. In (4.5.24) we replace V_0, u and ω with $V_0 + \delta V_0$, $u + \delta u$ and $\omega + \delta\omega$, respectively. Using the standard transformations (as was done in (4.5.20)) we obtain

$$2(a\omega + bV_0)\,\delta\omega + 2(b\omega + cV_0)\,\delta V_0 = 0\,. \tag{4.5.30}$$

Consequently,

$$\frac{\mathrm{d}\omega}{\mathrm{d}V_0} = -\frac{b\omega + cV_0}{a\omega + bV_0}\,. \tag{4.5.31}$$

In particular, it follows from (4.5.31) that for all bifurcation points (ω^*, V_0^*) we have the limit

$$\lim_{V_0 \to V_0^*} \frac{\mathrm{d}\omega_\pm(V_0)}{\mathrm{d}V_0} = \pm\infty\,. \tag{4.5.32}$$

In the case $V_0 = 0$, we have $b = 0$, and find that

$$\frac{\mathrm{d}\omega_\pm(V_0{=}0)}{\mathrm{d}V_0} = 0\,. \tag{4.5.33}$$

It follows from (4.5.32) and (4.5.33) that the curves $\omega_\pm(V_0)$ cross the ω and V_0 axes at right angles; see Fig. 4.3.

Let us now perform a nonlinear analysis of asymptotic behavior of the frequencies in the vicinity of bifurcation points. Let (ω_1^*, V_{01}^*), (ω_2^*, V_{02}^*), ... be solutions of the system of nonlinear equations (4.5.24). Consider the behavior of the functions $\omega_i(V_0)$, $i = 1, 2, \ldots$, determined in a small neighborhood of the bifurcation point (ω_k^*, V_{0k}^*), in implicit form, by the equation $F(\omega, V_0) = 0$. For brevity, we will omit the indices of the functions $\omega_i(V_0)$ and the bifurcation points (ω_k^*, V_{0k}^*).

To study the behavior of the function $F(\omega, V_0)$, we expand it in series around the bifurcation point (ω^*, V_0^*). We have

$$F(\omega, V_0) = F(\omega^*, V_0^*) + \frac{\partial F(\omega^*, V_0^*)}{\partial\omega}(\omega - \omega^*) + \frac{\partial F(\omega^*, V_0^*)}{\partial V_0}(V_0 - V_0^*) + \frac{1}{2}\frac{\partial^2 F(\omega^*, V_0^*)}{\partial\omega^2}(\omega - \omega^*)^2 + \ldots \tag{4.5.34}$$

Because at each bifurcation point (ω^*, V_0^*), the bifurcation conditions (4.5.26) hold, the first two terms in (4.5.34) vanish, yielding

$$F(\omega, V_0) = \frac{\partial F(\omega^*, V_0^*)}{\partial V_0}(V_0 - V_0^*) + \frac{1}{2}\frac{\partial^2 F(\omega^*, V_0^*)}{\partial\omega^2}(\omega - \omega^*)^2 + \ldots \tag{4.5.35}$$

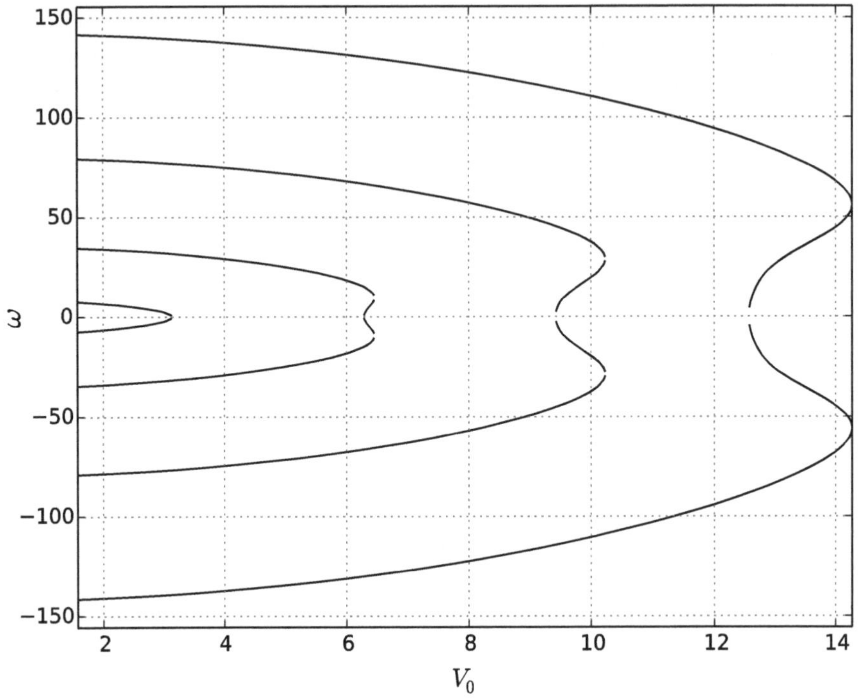

Fig. 4.3 Behavior of the natural frequencies ω as a function of the panel axial velocity V_0. No axial tension. Numerical solution using finite elements

Observe that all terms that have been omitted in (4.5.34) have a higher order of smallness. The expression (4.5.35) thus contains all leading-order terms, and describes completely general behavior of $F(\omega, V_0)$ in a small neighborhood of a given bifurcation point (ω^*, V_0^*). This is the general case; as before, the special cases where one or both of $\partial F(\omega^*, V_0^*)/\partial V_0$ and $\partial^2 F(\omega^*, V_0^*)/\partial \omega^2$ are zero must be studied separately.

Again, we may represent the function $\omega = \omega(V_0)$ in a small neighborhood of the bifurcation point (ω^*, V_0^*) as a power series:

$$\omega(V_0) = \omega^* + \alpha_1 (V_0 - V_0^*)^{\varepsilon_1} + \alpha_2 (V_0 - V_0^*)^{\varepsilon_2} + \dots , \quad \text{where } 0 < \varepsilon_1 < \varepsilon_2 < \dots \tag{4.5.36}$$

The values of the constants $\alpha_1, \alpha_2, \dots$ and $\varepsilon_1, \varepsilon_2, \dots$ are to be determined with the help of the condition $F(\omega, V_0) = 0$. After substitution of (4.5.36) into (4.5.35), the equation $F(\omega, V_0) = 0$ reduces to the corresponding equation

$$\Psi(V - V_0^*) = 0 , \tag{4.5.37}$$

where Ψ is a function of one variable and $V - V_0^*$ is the perturbation.

In order for (4.5.37) to hold, the coefficient of each power of $(V - V_0^*)$ in the expression of Ψ must be equal to zero. This requirement allows us to determine the values of $\alpha_1, \alpha_2, \ldots$ and $\varepsilon_1, \varepsilon_2, \ldots$ in the power series (4.5.36). In the following, for simplicity we consider only the determination of α_1 and ε_1, that is, we approximate $\omega(V_0)$ as

$$\omega(V_0) \approx \omega^* + \alpha_1 (V_0 - V_0^*)^{\varepsilon_1} . \tag{4.5.38}$$

After substitution of (4.5.38) into (4.5.35), we obtain

$$\Psi(V_0 - V_0^*) = \frac{\partial F(\omega^*, V_0^*)}{\partial V_0}(V_0 - V_0^*) + \frac{\alpha_1^2}{2}\frac{\partial F(\omega^*, V_0^*)}{\partial \omega^2}(V_0 - V_0^*)^{2\varepsilon_1} + \cdots \equiv 0 . \tag{4.5.39}$$

The expression (4.5.39) contains the leading-order terms; all omitted terms are of a higher order of smallness.

We will analyze the general case where

$$\frac{\partial F(\omega^*, V_0^*)}{\partial V_0} \neq 0 , \quad \frac{\partial^2 F(\omega^*, V_0^*)}{\partial \omega^2} \neq 0 . \tag{4.5.40}$$

As before, a separate analysis is needed if one or both values in (4.5.40) are zero.

Consider now the possibilities $2\varepsilon_1 < 1$, $2\varepsilon_1 = 1$ and $2\varepsilon_1 > 1$, which together cover all choices of ε_1. If $2\varepsilon_1 < 1$, then the first term in (4.5.39) is of a higher order of smallness with respect to the second term, and consequently in order for (4.5.39) to hold, $\partial^2 F(\omega^*, V_0^*)/\partial \omega^2$ must be zero, which contradicts the second condition in (4.5.40). Similarly, if $2\varepsilon_1 > 1$, then in order for (4.5.39) to hold, $\partial F(\omega^*, V_0^*)/\partial V_0^2$ must be zero, which contradicts the first condition in (4.5.40). Again the only possible value is thus $2\varepsilon_1 = 1$, which transforms (4.5.39) into

$$\Psi(V_0 - V_0^*) = \left[\frac{\partial F(\omega^*, V_0^*)}{\partial V_0} + \frac{\alpha_1^2}{2}\frac{\partial F(\omega^*, V_0^*)}{\partial \omega^2}\right](V_0 - V_0^*) + \cdots \equiv 0 . \tag{4.5.41}$$

The value of α_1 is then found from the condition that the coefficient of $(V - V_0^*)$ is zero, so that (4.5.37) holds identically. We have

$$\alpha_1^2 = -2\frac{\partial F(\omega^*, V_0^*)/\partial V_0}{\partial^2 F(\omega^*, V_0^*)/\partial \omega^2} . \tag{4.5.42}$$

Because the functionals (4.5.10)–(4.5.13) are all real-valued, and thus $F(\omega, V_0)$ is real-valued, it follows from (4.5.42) that α_1^2 is real-valued, and thus α_1 is either purely real or purely imaginary.

We find the asymptotic dependence $\omega(V_0)$ in the small neighborhood of the bifurcation point as

$$\omega(V_0) \approx \omega^* \pm \alpha_1 \sqrt{V_0 - V_0^*} , \quad |V_0 - V_0^*| \ll 1 . \tag{4.5.43}$$

From (4.5.43) it follows that in the small neighborhood around each bifurcation point (ω^*, V_0^*), the frequency of harmonic vibrations ω obtains complex values. If the coefficient α_1 is real, then the frequency becomes complex for $V_0 < V_0^*$; otherwise (α_1 imaginary) the frequency becomes complex for $V_0 > V_0^*$.

The appearance of complex frequencies and their complex conjugates means that according to the model considered, the displacement will grow exponentially, which corresponds to instability in the Lyapunov sense. Thus, the considered elastic system exhibits elastic instability at the bifurcation points, and in the Bolotin classification, the bifurcation points correspond to static (divergence, buckling, $\omega^* = 0$) and dynamic (flutter, $\omega^* \neq 0$) kinds of instability. Both kinds of instabilities are caught by the present analysis, because in both cases ($\omega^* = 0$ and $\omega^* \neq 0$) we have instability in the Lyapunov sense.

Finally, consider the case with zero axial tension ($C = 0$). In the nondimensional parameters (4.5.6), let $C = 0$, $\rho S/D = 1$ and $\ell = 1$. In this case, some of the bifurcation points lie on the V_0 axis ($\omega = 0$), corresponding to static instabilities (divergence). For this set of points, the bifurcation values of the velocities for these points are

$$V_{0k}^* = k\pi\ , \quad k = 1, 2, 3, \dots \quad \text{(static instabilities)}$$

and the dependences $\omega_k(V_0)$ in a small neighborhood of each point $(0,\ V_{0k}^*)$ are given by

$$\omega_k(V_0) \approx \pm\alpha_{1k}\sqrt{V_0 - k\pi} + \dots\ , \quad |V_0 - k\pi| \ll 1\ , \quad k = 1, 2, \dots \tag{4.5.44}$$

where we have used $\varepsilon_1 = 1/2$. Taking into account that $\alpha_{11}^2 < 0$, the first branch $\omega_1(V_0)$ is complex for $V_0 > \pi$, and consequently we will have a loss of stability at $V_0 = V_{01}^* = \pi$. It can be shown that the values α_{1k}^2 are positive for all $k \geq 2$, and consequently each branch $\omega_k(V_0)$ takes complex values at $V_0 < k\pi$; that branch of the solution regains stability at $V_0 = k\pi$, where $k \geq 2$. The system may still remain unstable above $V_0 = k\pi$, if some other branch of the solution has $\operatorname{Im}\omega < 0$.

The above results of asymptotic analysis of $\omega_k(V_0)$ agree with the numerical solution presented in Fig. 4.3, which was obtained by solving the spectral boundary value problem (4.5.4)–(4.5.5) as an eigenvalue problem for $(\omega,\ u(x))$ using finite elements of the Hermite type. Note that at least C^1 continuity is required at the element boundaries due to the term with the fourth derivative in the strong form.

In Fig. 4.3, the first two bifurcation values for critical points outside the axis $V_0(\omega{=}0)$, denoted by two lower indices, are

$$(V_{021}^* = 6.45\ ,\ \omega_{21}^* = \pm 10.58)\ ,$$
$$(V_{031}^* = 10.23\ ,\ \omega_{31}^* = \pm 32.01)\ .$$

In the numerical results shown here, 40 elements were used, with C^2 continuity across element boundaries (see Appendix C). Each element had six local degrees of freedom, leading to three global degrees of freedom $(u,\ u',\ u'')$ per node. With

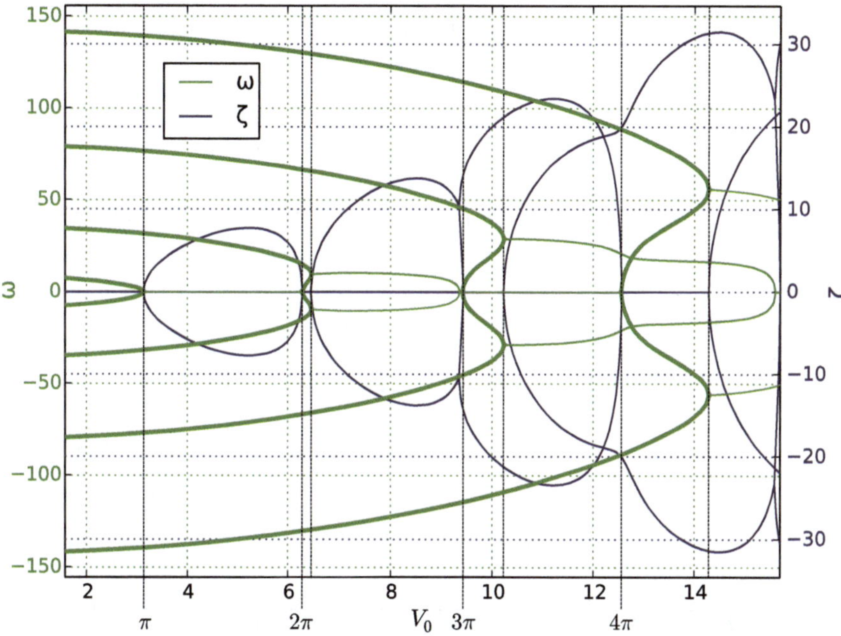

Fig. 4.4 Behavior of the complex natural frequencies $s = \zeta + i\omega$ as a function of the panel axial velocity V_0, drawn as projections onto the (V_0, ζ) and (V_0, ω) planes, overlaid into the same picture. No axial tension. Numerical solution using finite elements. Parts of the solution corresponding to Fig. 4.3 have been **bolded**

N_e elements, the total number of global nodes was $N_e + 1$. For these elements, the boundary conditions (4.5.5) are of Dirichlet (essential) type; they were applied to eliminate four degrees of freedom. The total number of degrees of freedom in the discretization was thus $(N_e + 1) * 3 - 4 = 119$.

Figures 4.4 and 4.5 show the full complex-valued numerical solution, of which Fig. 4.3 shows only those solutions for which the real part of the Lyapunov exponent is zero. We replace (4.5.3) with

$$w(x, t) = e^{st}\, u(x) \,, \quad s = \zeta + i\omega \,,$$

in other words, we now allow complex-valued frequencies to appear. Here the real part ζ describes the stability behavior in the Bolotin sense. If, at any fixed value of V_0, one or more of the natural frequencies has $\zeta > 0$, the system is unstable for that value of V_0. The imaginary part ω is the same as above.

From Fig. 4.4, we see that the model predicts a small supercritical stable range of V_0 (marked by the vertical lines near $V_0 = 2\pi$), beginning after the divergence gap spanning $(\pi,\ 2\pi)$. In this numerical example, no further stable ranges are seen; at any value of V_0 after this second stable range, there is always at least one solution with a positive real part.

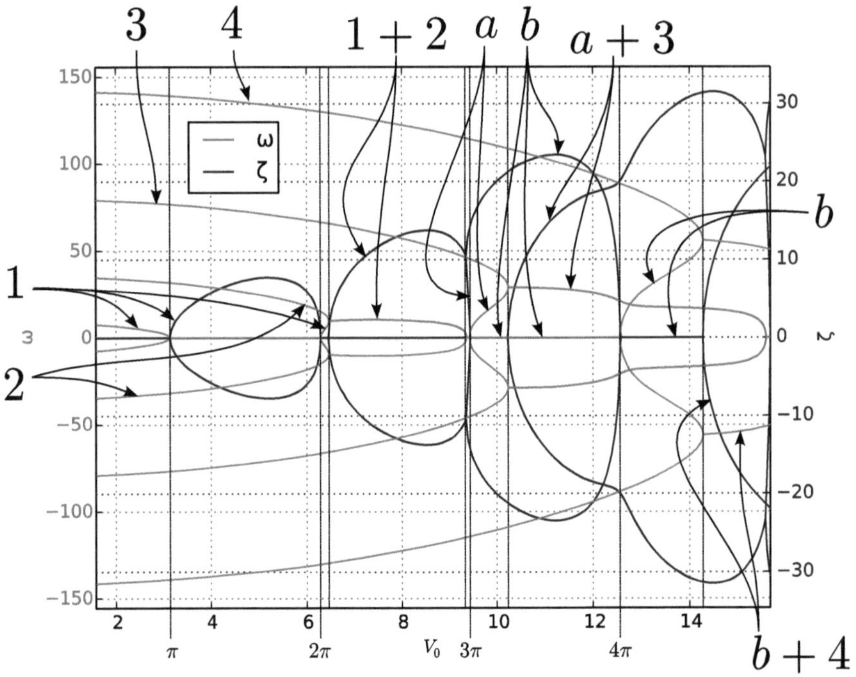

Fig. 4.5 Behavior of the complex natural frequencies $s = \zeta + i\omega$ as a function of the panel axial velocity V_0, drawn as projections onto the (V_0, ζ) and (V_0, ω) planes, overlaid into the same picture. No axial tension. Numerical solution using finite elements. Same data as in Fig. 4.4, distinct modes labelled. If no real part ζ is labelled for a given mode, its real part is zero. For the mode "b", zero imaginary and real parts have been indicated where appropriate, to emphasize that at first it is a divergence mode, later becoming a stable vibration mode

The overlaid projection plot is highly useful for an overview at a glance, but it has the drawback that it may become difficult to identify which real and imaginary parts belong to the same mode. Figure 4.5 shows the same data as Fig. 4.4, but with distinct modes labeled to aid identification.

We see that after the short supercritical stable range above the divergence gap, the system again loses stability, this time by coupled-mode flutter. This is marked by the vertical line just after $V_0 = 2\pi$.

Shortly before $V_0 = 3\pi$, the imaginary part of this coupled mode (labelled as "$1 + 2$" in the Figure) vanishes, and the mode splits into two new modes (labelled "a" and "b"). This bifurcation point is marked by the vertical line just before $V_0 = 3\pi$.

Both "a" and "b" are initially divergence modes (nonzero real part, zero imaginary part). The real part of mode "a" reaches zero at $V_0 = 3\pi$, and at this critical point, mode "a" stabilizes. However, mode "b" remains a divergence mode until $V_0 = 4\pi$, where it stabilizes.

Slightly above $V_0 = 10$, mode "a" combines with mode "3", producing a new flutter mode of the coupled-mode type (labelled "$a + 3$"). Later, slightly above $V_0 = 14$, the mode "b" combines with mode "4", producing another coupled-mode flutter mode (labelled "$b + 4$"). Near the right edge of the Figure, the imaginary part of mode "$a + 3$" becomes zero, and it splits into two new divergence modes (not labelled), similar to the earlier split of the coupled mode "$1 + 2$".

In conclusion, useful information about this system can be derived even from just the imaginary parts of the complex-valued natural frequencies. However, when making stability conclusions, it is useful to look also at general complex-valued solutions.

As a conclusion, in this section, the stability of an axially moving elastic panel was considered. The panel was travelling at constant velocity, with simply supported boundary conditions. Small transverse elastic displacements of the panel were described by a fourth-order differential equation that included the centrifugal and Coriolis effects (induced by the axial motion), axial tension, and bending resistance. The same formulation directly applies also to the small out-of-plane elastic displacements of an axially travelling beam.

To study the stability of the system, a complex variable technique and bifurcation theory were applied. As a result, variational equations and a variational principle were derived. Analysis of the variational principle allowed the study of qualitative properties of the bifurcation points. Asymptotic behavior in a small neighborhood around an arbitrary bifurcation point was analyzed and presented. The bifurcation points were found by determining where the conditions of the implicit function theorem are violated.

It was shown analytically that the eigenvalue curves in the (ω, V_0) plane cross both the ω and V_0 axes perpendicularly. It was also shown that near each bifurcation point, the dependence $\omega_k(V_0)$, for each mode k, approximately follows the shape of a square root function. From this analysis it was also seen that the eigenvalues appear in conjugate pairs.

The obtained results complement existing numerical studies on the stability of axially moving materials, especially those with finite bending rigidity. From a rigorous mathematical viewpoint, the presence of bending rigidity is essential, because the presence of the fourth-order term in the model changes the qualitative behavior of the bifurcation points.

In the next chapter, we will lay out the fundamentals of the theory of axially moving materials in a systematic manner. We will especially concentrate on axially moving strings, panels and beams, and look in detail at the connection between beams and panels. We will also consider in detail the question of boundary conditions when a uniform axial motion in present. An essential difference from classical solid mechanics is that the axial motion turns the problem essentially into flow problem involving a solid material. Thus the inflow and outflow ends of the domain must be considered differently. Finally, we will give some numerical results concerning stability behavior of axially moving elastic and viscoelastic panels that provide a point of comparison for the complex eigenfrequency curves presented above.

References

1. Alfutov NA (1978) Foundations for computation of stability of elastic systems. Mashinostroienie, Moscow, 312 pages (in Russian)
2. Archibald FR, Emslie AG (1958) The vibration of a string having a uniform motion along its length. ASME J Appl Mech 25:347–348
3. Arnol'd VI (1978) Methods mathematical, of classical mechanics. Springer, Berlin, 464 pages (translated from Russian by Vogtman K, Weinstein A)
4. Arnol'd VI (1972) Modes and quasi-modes. Funct Anal Appl 6(2):12–20
5. Banichuk NV, Barsuk AA (1982) On stability of torsioned elastic rods. In: Izvestiya Akademii Nauk SSSR, Mekhanika Tverdogo Tela (MTT) (6):148–154 (in Russian)
6. Banichuk N, Barsuk A (1983) Application of decomposition of the spectrum of eigenvalues to problems of optimal design. In: Problems of stability and critical load carrying capacity of structures, pp 17–24
7. Batchelor GK (1967) An introduction to fluid dynamics. Cambridge University Press
8. Bender CM, Orszag SA (1978) Advanced mathematical methods for scientists and engineers I: asymptotic methods and perturbation theory. Springer (1999 reprint: ISBN 978-0-387-98931-0)
9. Bolotin VV (1961) Nonconservative problems of the theory of elastic stability. Fizmatgiz, Moscow, 339 pages (in Russian)
10. Bolotin VV (1973) Variational principles in theory of elastic stability. In: Problems in mechanics of solid deformable bodies. Moscow, pp 83–88 (Sudostroenie)
11. Bolotin VV (1956) Dyn Stab Elast Syst. Gostekhizdat, Moscow
12. Bolotin VV (1963) Nonconservative problems of the theory of elastic stability. Pergamon Press, New York
13. Chen L-Y, Goldenfeld N, Oono Y (1996) Renormalization group and singular perturbations: Multiple scales, boundary layers, and reductive perturbation theory. Phys Rev E 54(1):376–394
14. Euler L (1744) A method for finding curves possessing a maximal or minimal property, or solving an isoperimetric problem in the broadest possible sense. Gostekhizdat, Moscow-Leningrad, 1934, 569 pages (originally published in 1744, Methodus inveniendi lineas curvas)
15. Euler L (1766) De motu vibratorio tympanorum. Novi Commentarii academiae scientiarum imperialis Petropolitanae, 10:243–260. http://eulerarchive.maa.org/pages/E302.html
16. Ghayesh MH, Amabili M, Michael M, Païdoussis P (2013) Nonlinear dynamics of axially moving plates. J Sound Vib 332(2):391–406. https://doi.org/10.1016/j.jsv.2012.08.013. ISSN 0022-460X
17. Jeronen J (2011) On the mechanical stability and out-of-plane dynamics of a travelling panel submerged in axially flowing ideal fluid: a study into paper production in mathematical terms. PhD thesis, Department of Mathematical Information Technology, University of Jyväskylä. http://urn.fi/URN:ISBN:978-951-39-4596-1 (Jyväskylä studies in computing 148. ISBN 978-951-39-4595-4 (book), ISBN 978-951-39-4596-1 (PDF))
18. Kirillov ON (2013) Nonconservative stability problems of modern physics. de Gruyter. ISBN 978-3-11-027043-3
19. Kirillov ON, Verhulst F (2010) Paradoxes of dissipation-induced destabilization or who opened Whitney's umbrella? Zeitschrift für Angewandte Mathematik und Mechanik 90(6):462–488. https://doi.org/10.1002/zamm.200900315
20. Kulachenko A, Gradin P, Koivurova H (2007a) Modelling the dynamical behaviour of a paper web. Part I. Comput Struct 85:131–147. https://doi.org/10.1016/j.compstruc.2006.09.006
21. Kulachenko A, Gradin P, Koivurova H (2007b) Modelling the dynamical behaviour of a paper web. Part II. Comput Struct 85:148–157. https://doi.org/10.1016/j.compstruc.2006.09.007
22. Lagerstrom PA, Casten RG (1972) Basic concepts underlying singular perturbation techniques. SIAM Rev 14(1):63–120
23. Lin CC (1997) Stability and vibration characteristics of axially moving plates. Int J Solids Struct 34(24):3179–3190. https://doi.org/10.1016/S0020-7683(96)00181-3

24. Marynowski K (2002) Non-linear dynamic analysis of an axially moving viscoelastic beam. J Theor Appl Mech 2(40):465–482. http://ptmts.org.pl/jtam/index.php/jtam/article/view/v40n2p465
25. Marynowski K (2004) Non-linear vibrations of an axially moving viscoelastic web with time-dependent tension. Chaos, Solitons & Fractals 21(2):481–490. https://doi.org/10.1016/j.chaos.2003.12.020. ISSN 0960-0779
26. Marynowski K (2008) Non-linear vibrations of the axially moving paper web. J Theor Appl Mech 46(3):565–580
27. Mikhlin SG (1970) Variational methods of mathematical physics. Nauka, Moscow, 512 p (in Russian)
28. Miranker WL (1960) The wave equation in a medium in motion. IBM J Res Dev 4:36–42. https://doi.org/10.1147/rd.41.0036
29. Mote CD (1972) Dynamic stability of axially moving materials. Shock Vib Dig 4(4):2–11
30. Mote CD Jr, Wickert JA (1991) Response and discretization methods for axially moving materials. Appl Mech Rev 44(11):S279–S284. https://doi.org/10.1115/1.3121365
31. Naimark MA (1969) Linear differential equations. Nauka, Moscow, 526 p (in Russian)
32. Nečas J, Lehtonen A, Neittaanmäki P (1987) On the construction of Lusternik-Schnirelmann critical values with application to bifurcation problems. Appl Anal 25(4):253–268. https://doi.org/10.1080/00036818708839689
33. Neittaanmäki P, Ruotsalainen K (1985) On the numerical solution of the bifurcation problem for the sine-Gordon equation. Arab J Math 6(1 and 2)
34. Pellicano F, Vestroni F (2000) Nonlinear dynamics and bifurcations of an axially moving beam. J Vib Acoust 122(1):21–30
35. Rektorys K (1968) Survey of applicable mathematics. Iliffe Books London Ltd
36. Sack RA (1954) Transverse oscillations in traveling strings. Br J Appl Phys 5:224–226. http://stacks.iop.org/0508-3443/5/i=6/a=307
37. Shen JY, Sharpe L, McGinley WM (1995) Identification of dynamic properties of plate-like structures by using a continuum model. Mech Res Commun 22(1):67–78. https://doi.org/10.1016/0093-6413(94)00043-D
38. Shin Changho, Chung Jintai, Kim Wonsuk (2005) Dynamic characteristics of the out-of-plane vibration for an axially moving membrane. J Sound Vib 286(4–5):1019–1031. https://doi.org/10.1016/j.jsv.2005.01.013
39. Simpson A (1973) Transverse modes and frequencies of beams translating between fixed end supports. J Mech Eng Sci 15:159–164. https://doi.org/10.1243/JMES_JOUR_1973_015_031_02
40. Skutch R (1897) Uber die Bewegung eines gespannten Fadens, weicher gezwungen ist durch zwei feste Punkte, mit einer constanten Geschwindigkeit zu gehen, und zwischen denselben in Transversal-schwingungen von gerlinger Amplitude versetzt wird. Annalen der Physik und Chemie 61:190–195
41. Swope RD, Ames WF (1963) Vibrations of a moving threadline. J Frankl Inst 275:36–55. https://doi.org/10.1016/0016-0032(63)90619-7
42. Sygulski R (2007) Stability of membrane in low subsonic flow. Int J Non-Linear Mech 42(1):196–202. https://doi.org/10.1016/j.ijnonlinmec.2006.11.012
43. Wang X (2003) Instability analysis of some fluid-structure interaction problems. Comput Fluids 32(1):121–138. https://doi.org/10.1016/S0045-7930(01)00103-7
44. Wang Y, Huang L, Liu X (2005) Eigenvalue and stability analysis for transverse vibrations of axially moving strings based on Hamiltonian dynamics. Acta Mech Sin 21:485–494. https://doi.org/10.1007/s10409-005-0066-2
45. Wickert JA (1992) Non-linear vibration of a traveling tensioned beam. Int J Non-Linear Mech 27(3):503–517. https://doi.org/10.1016/0020-7462(92)90016-Z
46. Yang X-D, Chen L-Q (2005) Bifurcation and chaos of an axially accelerating viscoelastic beam. Chaos, Solitons & Fractals 23(1):249–258. https://doi.org/10.1016/j.chaos.2004.04.008
47. Ziegler H (1952) Die stabilitätskriterien der elastomechanik. Ingenieur Archiv 20:49–56

Chapter 5
Modeling and Stability Analysis of Axially Moving Materials

Bifurcations are often encountered in stability analysis, whether one considers quasi-static loading or the natural frequencies of a system. Axially moving materials, also known as (axially) traveling materials, have many applications in the process industry, such as in the making of steel, textiles and paper. In this chapter, we combine these two topics. We systematically develop and solve a simplified model for the small-amplitude free vibrations of a one-dimensional axially moving structure. The aim of the systematic presentation is to clearly expose the construction of the model, from physical principles through to the final linearized equations, which are then used to determine the stability of the physical system by linear stability analysis.

The field of study of *axially moving materials* was founded by Skutch [127]. In the paper, written and published in German, he considered the vibrations of an axially traveling ideal string constrained by two stationary pinholes. English-language literature on axially moving materials began to appear half a century later, Among the first were Sack [118], Archibald and Emslie [6]. The modern mixed Eulerian–Lagrangean approach was established in Koivurova and Salonen [63], laying out the foundation for the rigorous derivation of the equations of motion for axially moving materials.

The natural frequencies and modes of the axially traveling string were obtained analytically by Swope and Ames [131], essentially by diagonalizing the principal part of the differential operator with the help of the characteristic equation and an appropriate coordinate transformation. Simpson [126] obtained an analytical solution for the transverse free vibrations of an axially traveling Euler–Bernoulli beam, but without axial tension. Much later, Kong and Parker [65] determined an analytical approximation for the natural frequencies of a tensioned traveling beam in the case where the bending rigidity is small. The result of Swope and Ames [131] was extended in Jeronen [54] to the case of a traveling string with damping and an elastic foundation.

The early study by Miranker [98] spawned the field of *axially accelerating materials*, where the axial drive velocity is considered uniform (constant in space) but is allowed to vary in time. It was also one of the first papers to apply Hamiltonian

N. Banichuk et al., *Stability of Axially Moving Materials*, Solid Mechanics and Its Applications 259, https://doi.org/10.1007/978-3-030-23803-2_5

mechanics to axially moving materials. More recent studies on accelerating materials include Chen [24], Chen et al. [28], Chen and Wang [25], Chen et al. [26, 27], Ding and Chen [32], Wang and Chen [144], Yang and Chen [155], Zhang et al. [158], Zhou and Wang [160].

Among the first to study the mechanical stability of axially moving materials was Mote [100, 101]. The investigation was extended to accelerating materials in Mote [102]. The stability of gyroscopic continua, which include (beside rotating systems) axially moving materials, was studied by Parker [107]. Supercritical behavior was investigated in Parker [108].

The classical wisdom on stability is summarized by Wickert and Mote [153], which concludes that axially traveling tensioned strings and Euler–Bernoulli beams, with symmetric boundary conditions, both experience divergence instability at a sufficiently high drive velocity. This phenomenon is analogous to the buckling of the classical compressed Euler column.

In contrast to this classical wisdom, using Hamiltonian mechanics, Wang et al. [145] showed that the axially traveling string remains (neutrally) stable at any value of the axial drive velocity. This agrees with the immediate conclusion (initial post-buckling analysis) from the analytical free-vibration solution of Swope and Ames [131]. This result is extremely important for mathematical correctness. However, in the context of practical applications, it must be kept in mind that all real strings have nonzero bending rigidity, no matter how small. Thus the beam model, which does experience instability at its first critical velocity, is qualitatively more appropriate.

The axially traveling string with damping, analyzed in Jeronen [54], remains stable at any drive velocity if the damping is external with constant coefficients. However, if the damping coefficients are of the form that is generated by internal damping (inside the string material), then the traveling string becomes unstable at the first critical velocity.

When making conclusions about stability, one must be careful. Indeed, in the context of linear stability analysis, Bolotin [19, pp. 99–100], [18, pp. 290–291] cautions that in cases where the Lyapunov exponents initially reside on the imaginary axis (i.e., neutral stability), the idealized model where damping is not taken into account is not sufficient to perform a reliable analysis. As we see by comparing the damped and undamped cases of the traveling ideal string, this indeed occurs here, too. Depending on the form of the damping, its introduction may either stabilize or destabilize the system at the first critical point.

When any mechanical system is loaded by conservative forces only, the Kelvin–Tait–Chataev theorem [61, p. 163], [14] guarantees that the introduction of damping will move the system, from the neutral stability of the undamped case, into asymptotic stability. This is the result [19, pp. 75, 99] refers to as a theorem by Kelvin.

However, as these authors point out, for general nonconservative systems, no similar result can exist. The classical counter example that proves the claim is the double pendulum subjected to a follower force [161]. Introducing dissipation makes the system lose stability at a significantly lower value of the loading parameter than the idealized system [62].

In our context, one must additionally take into account that introducing a small bending rigidity (such as for paper materials) causes a singular perturbation to the partial differential equation (PDE), as the new term that appears is of the highest order in $\partial/\partial x$. This causes the qualitative behavior of the equation to change, introducing boundary layers. Theory and analytical solution techniques for singularly perturbed PDEs are provided in the book by Carl et al. [12]; see also Lagerstrom and Casten [73].

As can be seen, although a more marginal topic than general solid mechanics, axially moving materials are still a rather wide field. Spinning discs, that is, axially moving materials in polar coordinates, have been considered by Hosaka and Crandall [48]. Traveling elastic plates have been studied by Lin [80], membranes by Shin et al. [125], and viscoelastic plates by Zhou and Wang [159, 160], Tang and Chen [132]. Axially tensioned traveling materials have been investigated in Lee and Oh [77], Wickert [151], Ghayesh et al. [44], Kim et al. [60], Kurki et al. [71]. Then there is the extremely extensive body of research by the group of C. D. Mote spanning several decades, including Mote [99, 100], Thurman and Mote [134], Mote [101, 102], Ulsoy et al. [142], Ulsoy and Mote [140, 141], Wickert and Mote [152], Mote and Wickert [103], Yang and Mote [156], Lin and Mote [81, 82], Lee and Mote [76], Renshaw and Mote [116], Luo and Mote [84].

A review focusing specifically on axially traveling strings can be found in Chen [23]. General recent developments on axially moving materials are reported in Marynowski and Kapitaniak [95].

We will consider papermaking as our practical example, and giving, therefore, special emphasis to paper materials. Structurally speaking, paper materials are essentially a random network of fibers. We refer the interested reader to the excellent overview by Alava and Niskanen [3]. The study covers the physics of paper materials from several viewpoints, including a qualitative overview, the structure of paper as a random-fiber network, paper as a solid from the viewpoint of elasticity and viscoelasticity, and finally the behavior of liquids in fiber networks in terms of mass flow and two-phase flows, the last topic being important for understanding the drying process.

In the context of continuum mechanics, paper can be approximated by an orthotropic model, originally introduced for steel-reinforced concrete by Huber [49–51]. For orthotropicity in the context of paper materials, see Mann et al. [86], Baum et al. [10], Thorpe [133], Yokoyama and Nakai [157], Erkkilä et al. [35], Kurki et al. [69].

The dominant fiber direction, along or nearly along which most of the fibers are aligned, is taken as material axis 1. This is also called the machine direction (MD), because it coincides with the direction of travel of the paper web as it runs through the paper machine. Material axis 2 is taken to be the perpendicular in-plane direction, called the cross direction (CD), since it is aligned with the width direction of the paper machine. Material axis 3 is thickness, which is also the direction of out-of-plane transverse motion. From the viewpoint of mechanical stability, the transverse direction is the most interesting one.

In some applications, solids exhibit not only an elastic response, but also significant viscous behavior. This is the case for paper in a paper machine at the wet end, which includes the press nip and the start of the dryer section. At this stage of the production process, still one half of the total mass of the paper web is in the water contained in it. When the paper web travels through the dryer section, its dry solids content (DSC) rises from 50% to (typically) approximately 90%. Constitutively speaking, in the drying process the material transforms from viscoelastic to nearly fully elastic. For the reader interested in exploring this particular topic further, an overview of drying in the papermaking process can be found in the book edited by Karlsson [57].

Further studies relevant to the physics of drying, that is, of heat and mass transport, for the specific case of paper materials include DeCrosta and Vennos [31], Lampinen and Toivonen [74], Karlsson [56], Lepomäki [78], Lu and Shen [83]. Heat and mass transport in porous materials in general has been studied by Whitaker [148–150], Kaviany and Mittal [59], Gibson [45], Quintard and Whitaker [111–115], Gibson and Charmchi [46], Berryman and Pride [15], Souza and Whitaker [139]. Of the mentioned general studies, Gibson [45] is especially relevant for paper materials, because paper fibers have hygroscopic properties [74]. Mechanical properties of paper as it is subjected to drying have been considered in Wahlström and Fellers [143], Erkkilä et al. [36].

The dynamics of a viscoelastic traveling paper web have been considered in the books by Marynowski [88], Banichuk et al. [8], and in studies by, among others, Kurki et al. [71], Marynowski and Kapitaniak [89, 93, 94], Kurki [68], Marynowski [90–92], Kurki and Lehtinen [72], Kurki et al. [69, 70], Saksa et al. [120–122, 124], Saksa [119], Saksa and Jeronen [123]. In the context of papermaking, the Kelvin–Voigt constitutive model has been considered in, for example, [69, 70, 119, 121, 122, 124]. The Poynting–Thomson model has been considered in Saksa and Jeronen [123].

An important point, when choosing the material model, is that with paper, one cannot make the standard modeling assumption that in a viscoelastic material the volumetric response is fully elastic, and viscous response occurs only for shear deformation. This assumption is valid for polymers and for Newtonian fluids, but it is not true for paper. Uesaka et al. [138], Thorpe [133] note this, and discuss an orthotropic two-dimensional Kelvin–Voigt solid for modeling the in-plane behavior of paper. This model is also discussed in the book by Sobotka [128], and has been used in studies by Kurki et al. [69–71], Kurki [68], Kurki and Lehtinen [72], Tang and Chen [132], Saksa [119].

Paper is also auxetic in the 1–3 material plane. In other words, paper thickens when stretched axially; the Poisson effect is reversed. This is a metamaterial effect, due to the geometric structure of the fiber network, where straightening the fibers (by tensioning them) reduces their vertical overlap. Stenberg and Fellers [130] note that for paper materials, $\nu_{13} < 0$, and $|\nu_{13}|$ may be as large as 3.0. For more on the material properties of paper, see also Wahlström and Fellers [143], Lif et al. [79].

In this text, we will ignore the microstructure of the fiber network, and will only consider the macroscopic scale using a classical homogenized model. Any continuum-mechanical material constants should thus be understood as average values, which arise from volume averaging of the microstructure consisting of fibers and voids.

We will eventually work in the small-displacement regime only, simplifying the theory somewhat especially in the case of axially moving materials. For readers interested in large deformations in solid mechanics, we recommend the book by Holzapfel [47].

Finally, for a general presentation of the continuum mechanics of solids, see Holzapfel [47], Malvern [85], Fung [43]. For a handbook, see Bower [20]. Specifically for elasticity, see Timoshenko and Goodier [136], Landau and Lifshitz [75], Feng and Shi [38], Antman [5]. A fundamental mathematical presentation of elasticity is provided in Marsden and Hughes [87]. For plates and shells, see Timoshenko and Woinowsky-Krieger [135]. Some of the mentioned works make use of tensor calculus, which is presented in books by, among others, Flügge [40], Sokolnikoff [129].

The outline of this chapter is as follows. We start by considering a classical beam, and its dynamic balance of forces. We then introduce small-displacement kinematics, and simplify the equations into the small-displacement regime. Following that, we discuss constitutive models, systematically developing spring–dashpot models of linear viscoelasticity. Then the connection between beams and *panels*, that is, plates undergoing cylindrical deformation, is considered. The fundamental theory of axially moving materials is introduced. The weak form of the beam problem, and the boundary conditions, are considered for both the stationary and axially moving cases. Finally, we give numerical examples concerning small vibrations and the stability of axially moving viscoelastic panels.

5.1 General Dynamics and Geometric Considerations

Our task in this chapter is to develop a model for the axially traveling panel, starting from physical principles. Although the physical interpretation is different, the equation of motion of a plate undergoing cylindrical deformation coincides with that of a beam. Therefore, it will be convenient to first derive the equation of motion for the beam, and only then introduce the appropriate modifications.

We will start with a fairly general two-dimensional problem of the dynamics of a curved beam. We will work in the plane only, ignoring torsion. We will first consider the force and moment balance, deferring kinematic (strain–displacement) and constitutive (stress–strain) considerations until later. To derive the general equations of motion, we will use the d'Alembert principle, which states that the internal, external and inertial forces must be in dynamic equilibrium.

For the sake of simplicity, we will limit our consideration to classical beams of the Euler–Bernoulli type. For beams of the Timoshenko type, the direction of the shear force depends on the shear angle, which must also be taken into account in the problem setup. In the following exposition, we assume that the shear force is perpendicular to the mid-line, which holds for Euler–Bernoulli beams, and for Kirchhoff–Love plates undergoing cylindrical deformation.

At first, we will ignore the axial motion of the beam. For axially moving materials, we will later interpret the classical equations as describing the physics in the axially

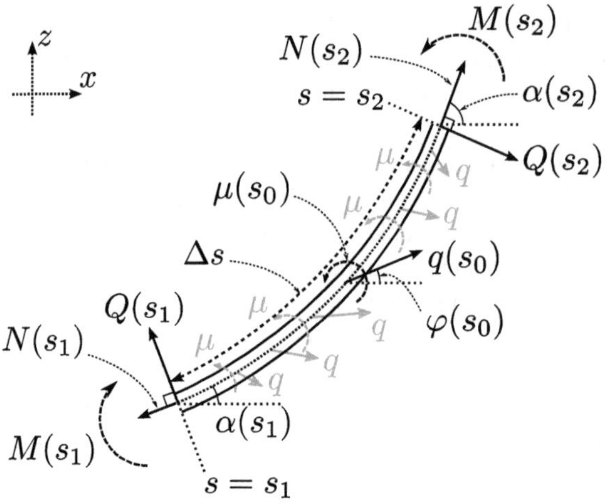

Fig. 5.1 Forces and moments acting on an arbitrary interval (s_1, s_2) of a curved beam of the Euler–Bernoulli type. The curve length coordinate s is measured along the mid-line (dotted) of the beam. Angles are measured counterclockwise from $+x$. Here N is an axial force, Q is a shear force, $q = q(s)$ is a distributed, vector-valued load, M is a bending moment, and $\mu = \mu(s)$ is a distributed moment load. The pairs of N, Q and M at the ends (at points s_1 and s_2) correspond to the action of the parts of the beam outside the considered section. Only true forces are pictured; the inertial force that arises in dynamics will be accounted for in the calculation to follow. The curved state $\alpha(s_2) \neq \alpha(s_1)$ is required in order to make the straightening effect of the axial tension N to appear in the transverse component of the final result

co-moving coordinate system, and transform them into Euler (laboratory, stationary) coordinates. In that context, any partial time derivatives $\partial/\partial t$ taken here must be understood as taken in the axially co-moving coordinate system.

We will elaborate on axially moving materials later, in Sect. 5.5. Briefly stated, when the equations are transformed into Euler coordinates, the coordinate transformation will have the effect of replacing each axially co-moving $\partial/\partial t$ by the material (Lagrange) derivative $\mathrm{d}/\mathrm{d}t \equiv \partial/\partial t + V_0\partial/\partial x$, where on the right-hand side the $\partial/\partial t$ is taken in the Euler coordinate system. This is a general observation that applies to all time derivatives in this derivation, including those that will appear when we later consider viscoelastic constitutive relations.

Consider a segment of the curved beam, as shown in Fig. 5.1. We will work in the Cartesian xz coordinate system. Let us define the vector-valued function

$$\mathbf{x}(s,t) \equiv (x(s,t),\ z(s,t))\ , \tag{5.1.1}$$

which expresses the position at time t, in the ambient Euclidean space $\mathbb{R}^2$, of the point s on the beam. For the purposes of deriving the force and moment balance equations, the beam is identified with its one-dimensional mid-line (dotted in Fig. 5.1), where the coordinate s measures the curve length.

For brevity of notation, in the following we will often omit the t in the function arguments. The dynamic equilibrium of forces in the considered section of the beam reads

$$-\mathbf{t}(s_1)\,N(s_1) + \mathbf{t}(s_2)\,N(s_2) + \mathbf{n}(s_1)\,Q(s_1) - \mathbf{n}(s_2)\,Q(s_2) + \int_{s_1}^{s_2} \mathbf{q}(s)\,\mathrm{d}s - \int_{s_1}^{s_2} m\frac{\partial^2 \mathbf{x}}{\partial t^2}\,\mathrm{d}s = 0\,, \tag{5.1.2}$$

where

$$\mathbf{t} \equiv (\cos(\alpha),\ \sin(\alpha))\,, \quad \mathbf{n} \equiv (-\sin(\alpha),\ \cos(\alpha)) \tag{5.1.3}$$

are the nondimensional local unit tangent and unit normal vectors of the mid-line. The normal is oriented $\pi/2$ counterclockwise from the tangent. Observe that $\mathbf{t} = \mathbf{t}(s,t)$ and $\mathbf{n} = \mathbf{n}(s,t)$, because $\alpha = \alpha(s,t)$.

The last term in (5.1.2) represents the inertial loading that appears in the dynamic case. The correct sign is readily determined by a physical consideration. First, in order for the resultant force to sum to zero, if the resultant of the true forces pushes the system along $+x$, the inertial force must push along $-x$. On the other hand, by Newton's second law, a surplus of force in the $+x$ direction causes the system to accelerate toward $+x$, so in this situation we must have $\partial^2 x/\partial t^2 > 0$ (because $m > 0$). Thus the x component of the inertial term must have a minus sign. The same argument can be repeated along the z axis, leading to the conclusion that the inertial term comes with the minus sign.

From the viewpoint of dimensional analysis, each term in (5.1.2) represents a force (SI unit N). The first four terms represent forces as-is ($[N] = [Q] = \mathrm{N}$). In the distributed load term, dimensional analysis requires $[\mathbf{q}] = \mathrm{N/m}$. Because the integral is taken along the curve, the function $\mathbf{q}(s)$ represents *external load per arc length*, or put more simply, *external load per unit length* of the beam. If we would like $\mathbf{q}$ to represent gravity, this is achieved by taking $\mathbf{q} = m\mathbf{g}$, since $[m] = \mathrm{kg/m}$ and $[\mathbf{g}] = \mathrm{m/s^2}$. Recall that in the SI system, $\mathrm{N} = \mathrm{kg\,m/s^2}$.

In the inertial term, dimensional analysis requires that $[m] = \mathrm{kg/m}$. Because this integral too is taken along the curve, the m in the integrand represents *mass per arc length*, i.e., *mass per unit length* of the beam, which is also known as the *linear density*.

We follow the approach of Paavola and Salonen [106]. Let us regroup (5.1.2) as

$$(\mathbf{t}(s_2)\,N(s_2) - \mathbf{t}(s_1)\,N(s_1)) - (\mathbf{n}(s_2)\,Q(s_2) - \mathbf{n}(s_1)\,Q(s_1)) + \int_{s_1}^{s_2} \mathbf{q}(s)\,\mathrm{d}s - \int_{s_1}^{s_2} m\frac{\partial^2 \mathbf{x}}{\partial t^2}\,\mathrm{d}s = 0\,. \tag{5.1.4}$$

This allows us to use the fundamental theorem of calculus (see e.g. [117, pp. 126–127]),

$$\int_a^b \frac{\partial f}{\partial x}(x)\,\mathrm{d}x = f(b) - f(a)\,, \tag{5.1.5}$$

in the "backward" direction to transform the grouped pairs of terms into integrals over the interval (s_1, s_2):

$$\int_{s_1}^{s_2} \frac{\partial}{\partial s} [\mathbf{t}(s)\, N(s)]\, \mathrm{d}s - \int_{s_1}^{s_2} \frac{\partial}{\partial s} [\mathbf{n}(s)\, Q(s)]\, \mathrm{d}s + \int_{s_1}^{s_2} \mathbf{q}(s)\, \mathrm{d}s - \int_{s_1}^{s_2} m \frac{\partial^2 \mathbf{x}}{\partial t^2}\, \mathrm{d}s = 0\,. \quad (5.1.6)$$

Gathering all integrals, we have

$$\int_{s_1}^{s_2} \left(\frac{\partial}{\partial s} [\mathbf{t}(s)\, N(s)] - \frac{\partial}{\partial s} [\mathbf{n}(s)\, Q(s)] + \mathbf{q}(s) - m \frac{\partial^2 \mathbf{x}}{\partial t^2} \right) \mathrm{d}s = 0\,. \quad (5.1.7)$$

Performing the differentiations yields

$$\int_{s_1}^{s_2} \left(\frac{\partial \mathbf{t}(s)}{\partial s} N(s) + \mathbf{t}(s) \frac{\partial N(s)}{\partial s} - \frac{\partial \mathbf{n}(s)}{\partial s} Q(s) - \mathbf{n}(s) \frac{\partial Q(s)}{\partial s} + \mathbf{q}(s) - m \frac{\partial^2 \mathbf{x}}{\partial t^2} \right) \mathrm{d}s = 0\,. \quad (5.1.8)$$

By the chain rule and Eq. (5.1.3),

$$\frac{\partial \mathbf{t}}{\partial s} = \frac{\partial \mathbf{t}}{\partial \alpha} \frac{\partial \alpha}{\partial s} = \frac{\partial}{\partial \alpha} (+\cos(\alpha),\ +\sin(\alpha)) \cdot \frac{\partial \alpha}{\partial s} = (-\sin(\alpha),\ +\cos(\alpha)) \cdot \frac{\partial \alpha}{\partial s} = \mathbf{n} \frac{\partial \alpha}{\partial s}\,, \quad (5.1.9)$$

$$\frac{\partial \mathbf{n}}{\partial s} = \frac{\partial \mathbf{n}}{\partial \alpha} \frac{\partial \alpha}{\partial s} = \frac{\partial}{\partial \alpha} (-\sin(\alpha),\ +\cos(\alpha)) \cdot \frac{\partial \alpha}{\partial s} = (-\cos(\alpha),\ -\sin(\alpha)) \cdot \frac{\partial \alpha}{\partial s} = -\mathbf{t} \frac{\partial \alpha}{\partial s}\,. \quad (5.1.10)$$

We thus obtain the result

$$\int_{s_1}^{s_2} \left(\mathbf{n}(s) \frac{\partial \alpha(s)}{\partial s} N(s) + \mathbf{t}(s) \frac{\partial N(s)}{\partial s} + \mathbf{t}(s) \frac{\partial \alpha(s)}{\partial s} Q(s) - \mathbf{n}(s) \frac{\partial Q(s)}{\partial s} + \mathbf{q}(s) - m \frac{\partial^2 \mathbf{x}}{\partial t^2} \right) \mathrm{d}s = 0\,. \quad (5.1.11)$$

Because s_1 and s_2 are arbitrary, the integrand must vanish pointwise for all s. Grouping terms that have the same unit vector, we obtain the following pointwise equation for the force balance:

$$\mathbf{t} \left[\frac{\partial N}{\partial s} + \frac{\partial \alpha}{\partial s} Q \right] + \mathbf{n} \left[\frac{\partial \alpha}{\partial s} N - \frac{\partial Q}{\partial s} \right] + \mathbf{q} - m \frac{\partial^2 \mathbf{x}}{\partial t^2} = 0\,. \quad (5.1.12)$$

Note that in (5.1.12), the bracketed expressions multiplied by **t** and **n** are the local tangential and normal components of the force balance without the distributed and inertial loads; they are not the global x and z components. To obtain those, **t** and **n** must be represented in global (x, z) components by inserting their expressions from Eq. (5.1.3).

Each quantity in (5.1.12), including the vectors **t** and **n**, is a function of the curve length coordinate s. The equation is general and exact for an arbitrary curved Euler–Bernoulli beam. So far, the only assumption we have made concerning the beam geometry is that the shear force is perpendicular to the mid-line.

Let us next consider the dynamic balance of moments. Recall that torque is defined as

$$\boldsymbol{\tau} = \mathbf{r} \times \mathbf{F}\,, \quad (5.1.13)$$

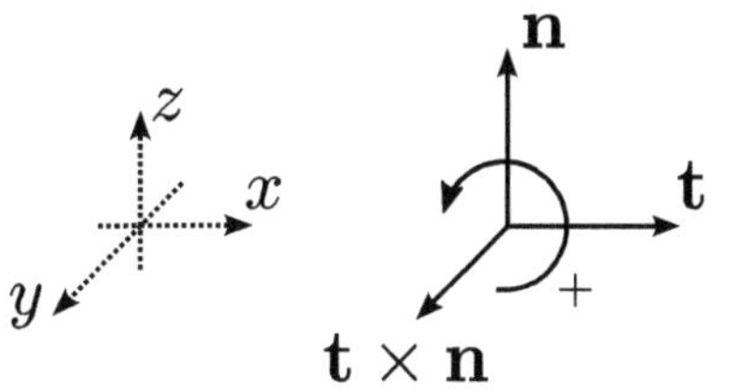

Fig. 5.2 The right-handed xzy coordinate system. Note the ordering of the axes; xz is the plane where we consider the dynamics, and y is the beam width coordinate. Local unit tangent vector $\mathbf{t}(s)$, local unit normal vector $\mathbf{n}(s)$, and their cross product are shown. Positive direction of bending moment is indicated by the plus sign

where $\mathbf{r}$ is the lever arm and $\mathbf{F}$ is the force vector. The lever arm is a vector, pointing from a reference point (the fulcrum) to the point at which the force $\mathbf{F}$ is applied.

Since we work in the xz plane, all moments are aligned with the y axis, which is perpendicular to the page. We may thus identify the moment vectors with their y components, since their x and z components are always zero.

Let us define

$$\mathbf{r}(s) \equiv \mathbf{x}(s) - \mathbf{x}(s_0) \equiv (x(s) - x(s_0),\ z(s) - z(s_0)) \,, \tag{5.1.14}$$

where s_0 is an arbitrary reference point on the considered segment of the beam.

Taking the direction of $\mathbf{t} \times \mathbf{n}$ as positive—refer to Fig. 5.2—the dynamical balance of moments reads

$$\begin{aligned} &M(s_2) - M(s_1) + \mathbf{r}(s_2) \times [-\mathbf{n}(s_2)Q(s_2) + \mathbf{t}(s_2)N(s_2)] + \mathbf{r}(s_1) \times [\mathbf{n}(s_1)Q(s_1) - \mathbf{t}(s_1)N(s_1)] \\ &\quad + \int_{s_1}^{s_2} \mu(s)\,\mathrm{d}s + \int_{s_1}^{s_2} \mathbf{r}(s) \times \mathbf{q}(s)\,\mathrm{d}s - \int_{s_1}^{s_2} \mathbf{r}(s) \times m\frac{\partial^2 \mathbf{x}}{\partial t^2}\,\mathrm{d}s = 0 \,. \end{aligned} \tag{5.1.15}$$

The last term is the contribution of the inertial force. Note that the dimension of the integrands, such as $[\mu] = \mathrm{Nm/m} = \mathrm{N}$, so that after integration the dimension will match $[M] = \mathrm{Nm}$. Each integrand represents a distributed moment, that is, a moment per unit length of the beam.

We must account for the axial forces $N(s_1)$ and $N(s_2)$ during the derivation, because $\mathbf{r}$ is measured in a straight line, but the beam is curved. Hence, in general, the local tangent of the beam $\mathbf{t}$ and the lever arm $\mathbf{r}$ point in different directions. Thus the axial forces have a component perpendicular to $\mathbf{r}$, leading to a contribution to the torque. Refer to Fig. 5.3.

Using the fundamental theorem of calculus, we convert expressions of the form $f(s_2) - f(s_1)$ into the corresponding integrals of $\partial f/\partial s$:

$$\begin{aligned} &\int_{s_1}^{s_2} \frac{\partial M}{\partial s}(s)\,\mathrm{d}s + \int_{s_1}^{s_2} \frac{\partial}{\partial s}\Big(\mathbf{r}(s) \times [-\mathbf{n}(s)Q(s) + \mathbf{t}(s)N(s)]\Big)\,\mathrm{d}s \\ &\quad + \int_{s_1}^{s_2} \mu(s)\,\mathrm{d}s + \int_{s_1}^{s_2} \mathbf{r}(s) \times \mathbf{q}(s)\,\mathrm{d}s - \int_{s_1}^{s_2} \mathbf{r}(s) \times m\frac{\partial^2 \mathbf{x}}{\partial t^2}\,\mathrm{d}s = 0 \,. \end{aligned} \tag{5.1.16}$$

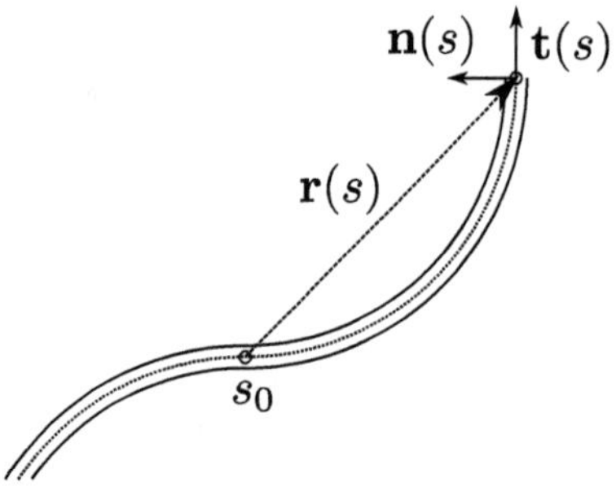

Fig. 5.3 The local unit tangent vector $\mathbf{t}(s)$, local unit normal vector $\mathbf{n}(s)$, and the lever arm vector $\mathbf{r}(s)$. The reference point s_0 is arbitrary

Performing the differentiation, we obtain

$$\begin{aligned}\int_{s_1}^{s_2} \frac{\partial M}{\partial s}(s)\,\mathrm{d}s + \int_{s_1}^{s_2} \biggl(\frac{\partial \mathbf{r}}{\partial s}(s) \times [-\mathbf{n}(s)Q(s) + \mathbf{t}(s)N(s)] \\ + \mathbf{r}(s) \times \left[-\frac{\partial \mathbf{n}}{\partial s}(s)Q(s) + \frac{\partial \mathbf{t}}{\partial s}(s)N(s) - \mathbf{n}(s)\frac{\partial Q}{\partial s}(s) + \mathbf{t}(s)\frac{\partial N}{\partial s}(s) \right] \biggr) \mathrm{d}s \\ + \int_{s_1}^{s_2} \mu(s)\,\mathrm{d}s + \int_{s_1}^{s_2} \mathbf{r}(s) \times \mathbf{q}(s)\,\mathrm{d}s - \int_{s_1}^{s_2} \mathbf{r}(s) \times m\frac{\partial^2 \mathbf{x}}{\partial t^2}\,\mathrm{d}s = 0 \,. \end{aligned} \tag{5.1.17}$$

Combining integrals, we then obtain

$$\begin{aligned}\int_{s_1}^{s_2} \biggl(\frac{\partial M}{\partial s}(s) + \frac{\partial \mathbf{r}}{\partial s}(s) \times [-\mathbf{n}(s)Q(s) + \mathbf{t}(s)N(s)] + \mu(s) \\ + \mathbf{r}(s) \times \left[-\frac{\partial \mathbf{n}}{\partial s}(s)Q(s) + \frac{\partial \mathbf{t}}{\partial s}(s)N(s) - \mathbf{n}(s)\frac{\partial Q}{\partial s}(s) + \mathbf{t}(s)\frac{\partial N}{\partial s}(s) + \mathbf{q}(s) - m\frac{\partial^2 \mathbf{x}}{\partial t^2} \right] \biggr) \mathrm{d}s = 0 \,. \end{aligned} \tag{5.1.18}$$

Again, because s_1 and s_2 are arbitrary, the integrand must vanish identically for all s. We obtain the pointwise equation

$$\begin{aligned}\frac{\partial M}{\partial s}(s) + \frac{\partial \mathbf{r}}{\partial s} \times [-\mathbf{n}(s)Q(s) + \mathbf{t}(s)N(s)] + \mu(s) \\ + \mathbf{r}(s) \times \left[-\frac{\partial \mathbf{n}}{\partial s}(s)Q(s) + \frac{\partial \mathbf{t}}{\partial s}(s)N(s) - \mathbf{n}(s)\frac{\partial Q}{\partial s}(s) + \mathbf{t}(s)\frac{\partial N}{\partial s}(s) + \mathbf{q}(s) - m\frac{\partial^2 \mathbf{x}}{\partial t^2} \right] = 0 \,. \end{aligned} \tag{5.1.19}$$

Using Eqs. (5.1.9) and (5.1.10) to express the derivatives of the unit vectors, Eq. (5.1.19) becomes

$$\begin{aligned}\frac{\partial M}{\partial s}(s) + \frac{\partial \mathbf{r}}{\partial s} \times [-\mathbf{n}(s)Q(s) + \mathbf{t}(s)N(s)] + \mu(s) \\ + \mathbf{r}(s) \times \left[\mathbf{t}(s)\frac{\partial \alpha(s)}{\partial s}Q(s) + \mathbf{n}(s)\frac{\partial \alpha(s)}{\partial s}N(s) - \mathbf{n}(s)\frac{\partial Q}{\partial s}(s) + \mathbf{t}(s)\frac{\partial N}{\partial s}(s) + \mathbf{q}(s) - m\frac{\partial^2 \mathbf{x}}{\partial t^2} \right] = 0 \,. \end{aligned} \tag{5.1.20}$$

Now, recall the force balance from Eq. (5.1.12):

$$\mathbf{t}\left[\frac{\partial N}{\partial s}+\frac{\partial \alpha}{\partial s}Q\right]+\mathbf{n}\left[\frac{\partial \alpha}{\partial s}N-\frac{\partial Q}{\partial s}\right]+\mathbf{q}-m\frac{\partial^2 \mathbf{x}}{\partial t^2}=0\,.$$

We see that the last expression in the brackets in (5.1.20) is exactly the force balance equation. Hence, it vanishes identically, leaving us with

$$\frac{\partial M}{\partial s}(s)+\frac{\partial \mathbf{r}}{\partial s}\times[-\mathbf{n}Q(s)+\mathbf{t}N(s)]+\mu(s)=0\,. \tag{5.1.21}$$

Observe that

$$\frac{\partial \mathbf{r}}{\partial s}=\frac{\partial}{\partial s}[\mathbf{x}(s)-\mathbf{x}(s_0)]=\frac{\partial \mathbf{x}}{\partial s}\,, \tag{5.1.22}$$

because s_0, and hence also $\mathbf{x}(s_0)$, is a constant. Therefore, we may write

$$\frac{\partial M}{\partial s}(s)+\frac{\partial \mathbf{x}}{\partial s}(s)\times[-\mathbf{n}Q(s)+\mathbf{t}N(s)]+\mu(s)=0\,. \tag{5.1.23}$$

The final task in deriving the moment balance equation is to express $\partial\mathbf{x}/\partial s$ in terms of the local unit vectors $\mathbf{t}(s)$ and $\mathbf{n}(s)$, and a magnitude $\|\partial\mathbf{x}/\partial s\|$. Let us start with a finite vector-valued increment $\Delta\mathbf{x}$; refer to Fig. 5.4. By the definition of the derivative,

$$\lim_{\Delta s\to 0}\frac{\Delta\mathbf{x}}{\Delta s}=\lim_{\Delta s\to 0}\frac{\mathbf{x}(s_0+\Delta s)-\mathbf{x}(s_0)}{\Delta s}\equiv\frac{\partial \mathbf{x}}{\partial s}(s_0)\,. \tag{5.1.24}$$

Geometrically speaking, $\Delta\mathbf{x}$ is the secant vector pointing from $\mathbf{x}(s_0)$ to $\mathbf{x}(s_0+\Delta s)$, denoted as $\sec(\mathbf{x}(s))$ in Fig. 5.4.

Consider the magnitude $\|\partial\mathbf{x}/\partial s\|$ at the point s_0. Because s was defined as the curve length coordinate, the arc length of $\mathbf{x}(s)$ from s_0 to $s_0+\Delta s$ is precisely the difference in the argument, namely Δs. Intuitively, it must approach the length of the secant $\Delta\mathbf{x}$ in the limit as $\Delta s\to 0$.

To ensure that this claim is accurate, we follow the study by Kasner [58], where it is observed that this property holds only for regular curves. Beside explaining the regular case, the study presents a counterexample from complex analysis, where

Fig. 5.4 A segment of the curve $\mathbf{x}(s)$ from s_0 to $s_0+\Delta s$, its local tangent $\mathbf{t}(s_0)$ and secant vector $\Delta\mathbf{x}$

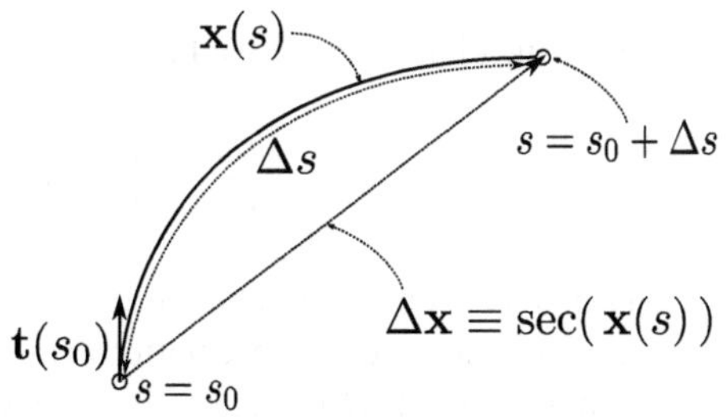

this intuitive property fails even at some points on an ellipse, when complex-valued coordinates are allowed.

In this chapter, we limit our consideration to regular curves with real-valued coordinates. Consider a local coordinate system with rectangular axes x and z. Without a loss of generality, we may place the origin of the local coordinates at $\mathbf{x}(s_0)$. We choose an arbitrary orientation for the axes, with only the requirement that $\mathbf{t}(s_0)$ must not point along the $\pm z$ direction. An arbitrary regular plane curve can then be locally represented by a power series

$$z(x) = c_1 x + c_2 x^2 + c_3 x^3 + \ldots \tag{5.1.25}$$

Let us denote by γ the length of the chord from the origin of the local coordinates to a nearby point (x, z) on the curve. In terms of Fig. 5.4, this is the length of the secant $\Delta\mathbf{x}$. By taking the Euclidean distance, and inserting the series representation (5.1.25) for z, we have

$$\begin{aligned}\gamma &= \sqrt{x^2 + z^2} \\ &= x\sqrt{1 + c_1^2 + 2c_1c_2x + (2c_1c_3 + c_2^2)x^2 + \ldots}\ .\end{aligned} \tag{5.1.26}$$

As is known from calculus (see e.g. [2]), the corresponding arc length Δs along the curve is

$$\begin{aligned}\alpha &= \int_0^x \sqrt{1 + \left(\frac{\partial z}{\partial x}\right)^2}\, \mathrm{d}x \\ &= \int_0^x \sqrt{1 + c_1^2 + 4c_1c_2x + (6c_1c_3 + 4c_2^2)x^2 + \ldots}\ \mathrm{d}x\ ,\end{aligned} \tag{5.1.27}$$

where in the last form we have differentiated the series (5.1.25) with respect to x and inserted the result.

Provided that $1 + c_1^2 \neq 0$—which is the case for all plane curves for real-valued coordinates—γ and α can be developed into integral power series starting with the first power of x, the coefficient of x in both being $(1 + c_1^2)^{1/2}$. It follows that

$$\lim_{x\to 0} \frac{\gamma}{\alpha} = 1\ , \tag{5.1.28}$$

justifying the claim. Hence we have

$$\left\| \frac{\partial \mathbf{x}}{\partial s}(s_0) \right\| = \lim_{\Delta s \to 0} \left\| \frac{\Delta \mathbf{x}}{\Delta s} \right\| = 1\ . \tag{5.1.29}$$

What remains to consider is the direction of the vector $\Delta\mathbf{x}/\Delta s$ in the limit $\Delta s \to 0$. We again use the series representation (5.1.25). The secant of $z(x)$, with the reference point on the curve taken as $z(x_0)$, is the line segment

$$\sec(x;\ x_0) \equiv C(x_0)\,x\ , \quad 0 \le x \le x_0\ , \tag{5.1.30}$$

where the slope is

$$C(x_0) \equiv \frac{\Delta z}{\Delta x} \equiv \frac{z(x_0) - z(0)}{x_0 - 0} = \frac{1}{x_0}\left(c_1 x_0 + x_2 x_0^2 + c_3 x_0^3 + \dots\right) = c_1 + x_2 x_0 + c_3 x_0^2 + \dots \tag{5.1.31}$$

As the reference point $z(x_0)$ tends to the origin of the local xz coordinates, we have

$$\lim_{x_0 \to 0} C(x_0) = c_1\ . \tag{5.1.32}$$

The slope of the curve itself, which is also the direction of the tangent, is the derivative of $z(x)$. From the series representation (5.1.25),

$$\frac{\partial z}{\partial x}(x) = c_1 + 2c_2 x + 2c_3 x^2 + \dots\ , \tag{5.1.33}$$

whence immediately

$$\frac{\partial z}{\partial x}(0) = c_1\ . \tag{5.1.34}$$

Therefore, as $\Delta s \to 0$, the slopes (5.1.32) and (5.1.34) coincide, so in this limit $\Delta\mathbf{x}/\Delta s$ becomes parallel to $\mathbf{t}(s_0)$. By combining this result with Eqs. (5.1.29) and (5.1.24), we obtain

$$\frac{\partial \mathbf{x}}{\partial s}(s_0) = \mathbf{t}(s_0)\ . \tag{5.1.35}$$

Finally, since s_0 was arbitrary, we have the result

$$\frac{\partial \mathbf{x}}{\partial s}(s) \equiv \mathbf{t}(s)\ . \tag{5.1.36}$$

Observe that by the chain rule, we may write

$$\frac{\partial}{\partial s}(\dots) = \sum_{j=1}^{3} \frac{\partial x_j}{\partial s}\frac{\partial}{\partial x_j}(\dots) = \frac{\partial \mathbf{x}}{\partial s} \cdot \nabla(\dots) = \mathbf{t} \cdot \nabla(\dots)\ , \tag{5.1.37}$$

in other words, $\partial/\partial s$ is the derivative along the local tangential direction.

Using the result (5.1.36) in the moment balance Eq. (5.1.23), and noting that $\mathbf{t} \times (-\mathbf{n}) = -1$ and $\mathbf{t} \times \mathbf{t} = 0$, the final result for the moment balance is

$$\frac{\partial M}{\partial s}(s) - Q(s) + \mu(s) = 0\ . \tag{5.1.38}$$

Again, note the dimension: Eq. (5.1.38) represents distributed moments, with each term having SI unit $\mathrm{N} = \mathrm{Nm/m}$.

Effects such as gravity or an (visco-)elastic foundation can be included in the model by simply defining the distributed load $\mathbf{q}$ appropriately; obviously, several effects can be combined by simply summing them. For example, for a uniform gravitational field in the xz plane,

$$\mathbf{q} = mg\,(\cos\varphi,\ \sin\varphi)\ , \tag{5.1.39}$$

where $g\ \left[\mathrm{m/s^2}\right]$ is the gravitational acceleration and φ is the angle of the gravity vector, measured counterclockwise from the $+x$ axis (recall Fig. 5.1). Later in this chapter, we will consider examples including elastic and viscoelastic foundations.

It is possible to eliminate the shear force Q by solving (5.1.38) for Q (and also, by differentiation, for $\partial Q/\partial s$) and then inserting the results into (5.1.12), leaving just one equation. We have $Q = \partial M/\partial s + \mu$, whence we obtain

$$\mathbf{t}\left[\frac{\partial N}{\partial s} + \frac{\partial\alpha}{\partial s}\frac{\partial M}{\partial s} + \frac{\partial\alpha}{\partial s}\mu\right] + \mathbf{n}\left[\frac{\partial\alpha}{\partial s}N - \frac{\partial^2 M}{\partial s^2} - \frac{\partial\mu}{\partial s}\right] + \mathbf{q} - m\frac{\partial^2\mathbf{x}}{\partial t^2} = 0\ . \tag{5.1.40}$$

Equation (5.1.40) represents the force balance, with the moment balance automatically accounted for. We will call (5.1.40) the *combined force balance* equation.

As we will see below, in the case of the classical beam with small displacements, (5.1.40) leads to the standard beam equation for the dynamic case, accounting also for axial loading.

Although our problem of plane dynamics is two-dimensional—and the functions appearing in the force and moment balance Eqs. (5.1.12) and (5.1.38) are one-dimensional—for some purposes it is useful to consider the three-dimensional object that is represented by the beam model. For this we need the resultant forces and moments.

In the force and moment balance equations, we will use the following expressions [106] for the resultant axial and shear forces and the bending moment, for each fixed value of s:

$$N(s) = \int_A \sigma_{ss}\,\mathrm{d}A\ ,\quad Q(s) = \int_A \sigma_{sz}\,\mathrm{d}A\ ,\quad M(s) = -\int_A \sigma_{ss}\,z\,\mathrm{d}A\ . \tag{5.1.41}$$

Because the moment balance equation can be used to eliminate the shear force $Q(s)$—keep in mind the combined force balance Eq. (5.1.40)—we actually only need the axial stress field σ_{ss}.

For a rectangular cross-section of width b and height h, we may decompose the integral over the cross-section area as

$$\int_A \ldots\,\mathrm{d}A = \int_{-b/2}^{+b/2}\int_{-h/2}^{+h/2} \ldots\,\mathrm{d}z\,\mathrm{d}y = b\int_{-h/2}^{+h/2} \ldots\,\mathrm{d}z\ , \tag{5.1.42}$$

where the last form holds because we are considering planar dynamics, where all quantities are independent of the beam width coordinate y.

The minus sign in the expression for $M(s)$ in (5.1.41) comes from the vector nature of the bending moment. Consider the bending moment in a cross section at a fixed value of s. In local xzy coordinates (recall Fig. 5.2), where $+x$ coincides with the tangential direction at s, the resultant bending moment vector $\mathbf{M}$ is given by

$$\mathbf{M} = \int_A \mathbf{r} \times (\sigma_{xx}\mathbf{e}_x + \sigma_{xz}\mathbf{e}_z + \sigma_{xy}\mathbf{e}_y)\,\mathrm{d}A\,, \tag{5.1.43}$$

where σ_{xx} is the axial stress, σ_{xz} and σ_{xy} are the shear stresses along the plane of the constant-x cross-section, and the lever arm is

$$\mathbf{r} = z\mathbf{e}_z + y\mathbf{e}_y\,.$$

The reference point $\mathbf{x}(s_0)$ is placed on the mid-line ($y = 0$, $z = 0$) of the beam at the considered position s (where $x = 0$), so the x component of $\mathbf{r}$ is zero.

Expanding the cross product, the integrand becomes

$$\mathbf{r} \times (\sigma_{xx}\mathbf{e}_x + \sigma_{xz}\mathbf{e}_z + \sigma_{xy}\mathbf{e}_y) = \begin{vmatrix} \mathbf{e}_x & \mathbf{e}_z & \mathbf{e}_y \\ 0 & z & y \\ \sigma_{xx} & \sigma_{xz} & \sigma_{xy} \end{vmatrix} = \begin{bmatrix} z\sigma_{xy} - y\sigma_{xz} \\ y\sigma_{xx} \\ -z\sigma_{xx} \end{bmatrix}. \tag{5.1.44}$$

Because we are considering plane dynamics without torsion, the x component of $\mathbf{M}$ (twisting around the x axis) is zero. At first glance, Eqs. (5.1.43) and (5.1.44) seem to tell us something more; for this to happen, the stresses must satisfy the relation

$$M_x = \int_A (z\sigma_{xy} - y\sigma_{xz})\,\mathrm{d}A = 0\,. \tag{5.1.45}$$

The problem setup implies that $\sigma_{xy} \equiv 0$; there are no forces shearing a constant-x cross section in the y direction (out of the xz plane). Furthermore, because in our setup σ_{xz} is independent of y, the second term in the integral (5.1.45) becomes antisymmetric in y, and hence vanishes (considering (5.1.42)). Hence (5.1.45) is satisfied, but it does not actually provide us with any new information.

Because in our problem setup any rotation is known to occur in the xz plane only, the z (second) component of $\mathbf{M}$ must be zero:

$$M_z = \int_A y\sigma_{xx}\,\mathrm{d}A = 0\,.$$

This is satisfied because in our problem setup the axial stress σ_{xx} does not depend on y.

We are left with the y (third) component of $\mathbf{M}$, namely

$$M_y = -\int_A z\sigma_{xx}\,\mathrm{d}A\,,$$

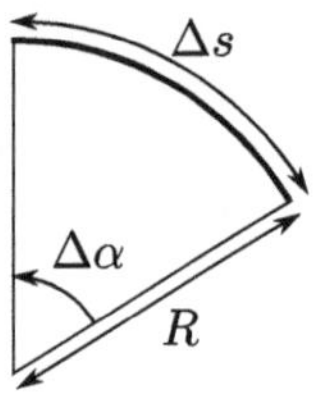

Fig. 5.5 A circular arc of length Δs, spanning the angle $\Delta\alpha$, having radius R

which is exactly $M(s)$ in (5.1.41), once we account for the fact that the local x axis coincides with the tangent of the curve $\mathbf{x}(s)$ at the considered point. This expression for $M(s)$, with the minus sign, is also given e.g. in Feng and Shi [38, p. 75]; our xzy coordinate system corresponds to the xyz coordinate system used in the reference.

The quantity $\partial\alpha/\partial s$ that appears in the force balance Eq. (5.1.12) is known as the curvature. The reason is as follows. Consider a circular arc spanning an angle $\Delta\alpha$, having arc length Δs; see Fig. 5.5. By definition of the radian,

$$R\,\Delta\alpha = \Delta s\;,$$

whence immediately

$$\frac{\Delta\alpha}{\Delta s} = \frac{1}{R} \qquad \text{for any } \Delta s \neq 0. \tag{5.1.46}$$

Hence, finding a circular arc that matches a given ratio of $\Delta\alpha/\Delta s$ is always possible by choosing the radius R according to (5.1.46). In the limit $\Delta s \to 0$, (5.1.46) becomes

$$\frac{\partial\alpha}{\partial s} = \frac{1}{R}\;. \tag{5.1.47}$$

This reciprocal of the radius, $1/R$, is defined as the *curvature*. For a general regular curve, the radius R itself is known as the *radius of curvature*, which is a local quantity: $R = R(s)$. The sign of R just indicates the direction of the curvature, positive corresponding to a counterclockwise turn.

In geometry, $\partial\alpha/\partial s$ is called the *extrinsic curvature*, because while it measures the rate of change of direction of the plane curve, it does this with reference to the ambient Euclidean space in which the curve is embedded.

This raises the following question: how well does the circular arc approximate a general regular plane curve $\mathbf{x}(s)$ locally? Without loss of generality, let us choose a Cartesian coordinate system (x, z) such that the considered point $\mathbf{x}(s_0)$ is at the origin, and $\mathbf{t}(s_0) = (1, 0)$. We expand $\mathbf{x}(s)$ in a power series, locally representing z as a function of x:

$$z = c_0 + c_1 x + c_2 x^2 + c_3 x^3 + c_4 x^4 \ldots\;. \tag{5.1.48}$$

Because the coordinate system was chosen such that the function $z(x)$ goes through the origin, $c_0 = 0$.

The equation of a circle with radius R, centered at (x_0, z_0) is

$$(x - x_0)^2 + (z - z_0)^2 = R^2 ,$$

whence

$$z = z_0 \pm \sqrt{R^2 - (x - x_0)^2} .$$

Let $(x_0, z_0) = (0, R)$. Then the lower edge of the circle touches the origin horizontally. On the lower branch of this circle, we can write

$$z = R\left(1 - \sqrt{1 - \left(\frac{x}{R}\right)^2}\right) .$$

Expanding the square root term as a Taylor series, we obtain

$$z = R\left(1 - \left[1 - \frac{1}{2}\left(\frac{x}{R}\right)^2 - \frac{1}{8}\left(\frac{x}{R}\right)^4 - \frac{1}{16}\left(\frac{x}{R}\right)^6 - \dots\right]\right) ,$$

which is valid for $|x/R| \ll 1$. This immediately simplifies to

$$z = R\left[\frac{1}{2}\left(\frac{x}{R}\right)^2 + \frac{1}{8}\left(\frac{x}{R}\right)^4 + \frac{1}{16}\left(\frac{x}{R}\right)^6 + \dots\right] . \tag{5.1.49}$$

If the circle exactly coincided with the curve $\mathbf{x}(s)$ around the origin, the coefficients of equal powers of x in (5.1.48) and (5.1.49) would be equal, and thus we would have

$$c_1 = 0 , \quad c_2 = \frac{1}{2R} , \quad c_3 = 0 , \quad c_4 = \frac{1}{8R^3} , \quad c_5 = 0 , \quad c_6 = \frac{1}{16R^5} , \quad \dots . \tag{5.1.50}$$

Considering (5.1.48) as a Taylor series for $z(x) - z(0)$, we identify that the values of c_j are actually

$$c_j = \frac{1}{j!}\frac{\partial^j z}{\partial x^j}(0) . \tag{5.1.51}$$

Due to our choice of coordinate system, we have $\mathbf{t}(s_0) = (1, 0)$. For $z(x)$ to be horizontal at the origin, the slope must be zero. By (5.1.48), the slope of $z(x)$ is

$$\frac{\partial z}{\partial x} = c_1 + 2c_2 x + 3c_3 x^2 + 4c_4 x^3 + \dots ,$$

and especially $(\partial z/\partial x)(0) = c_1$, whence we have $c_1 = 0$, as required for (5.1.50).

The first nonzero coefficient is

$$c_2 = \frac{1}{2}\frac{\partial^2 z}{\partial x^2}(0) , \tag{5.1.52}$$

which may have any value depending on the particular curve $\mathbf{x}(s)$. Comparing (5.1.52) to (5.1.50), we have

$$\frac{\partial^2 z}{\partial x^2}(0) = \frac{1}{R} \,. \tag{5.1.53}$$

So far, the radius R is free; we may thus choose it to satisfy (5.1.53). This gives us the *osculating circle*: the circle that touches the curve at the point $\mathbf{x}(s_0)$, has the same tangent there, and the same curvature.

For a general regular curve, there is no requirement for $c_3, c_5, \ldots$ to be zero, or the actual values of $c_4, c_6, \ldots$ to agree with (5.1.50) once the value of R has been chosen by (5.1.53).

Therefore, the local approximation as a circle is valid up to second order in x, in the Cartesian coordinate system with origin at $\mathbf{x}(s_0)$ and the $+x$ axis aligned with $\mathbf{t}(s_0)$. Hence, in a Cartesian coordinate system in general orientation and position, the approximation is valid, around the point $\mathbf{x}(s_0)$, up to second order in Δs along the line $\mathbf{t}(s_0)$.

To work with the force balance equation, it is convenient to have an expression for the curvature $\partial\alpha/\partial s$ in terms of the Cartesian components of $\mathbf{x}(s)$. By the chain rule, at an arbitrary point s_0 on the curve, we may write

$$\frac{\partial \alpha}{\partial s} = \frac{\partial \alpha}{\partial x}\frac{\partial x}{\partial s} \,, \tag{5.1.54}$$

provided that $\partial x/\partial s \neq 0$ at the considered point s_0, i.e. $\mathbf{t}(s_0) \neq (0, \pm 1)$. We can assume this without loss of generality by orienting a local xz coordinate system appropriately.

Because the tangent is not vertical, for the points on the curve $\mathbf{x}(s) = (x(s),\ z(s))$ we may consider z as a function of x in a small neighborhood of s_0. If $z(x)$ happens to be multi-valued, we pick the branch that goes through $\mathbf{x}(s_0)$. Refer to Fig. 5.6.

The slope of the function $z(x)$ is $\partial z/\partial x$. On the other hand, the slope is the tangent of the angle $\alpha(s)$:

$$\tan\alpha = \frac{\partial z}{\partial x} \,, \tag{5.1.55}$$

where $z = z(x)$. Differentiating both sides of (5.1.55) with respect to x,

$$\frac{\partial}{\partial x}(\tan\alpha) = \frac{1}{(\cos\alpha)^2}\frac{\partial \alpha}{\partial x} = \frac{\partial^2 z}{\partial x^2} \,,$$

whence

$$\frac{\partial \alpha}{\partial x} = (\cos\alpha)^2 \frac{\partial^2 z}{\partial x^2} \,. \tag{5.1.56}$$

On the other hand, from the triangle in Fig. 5.7, we have

$$\cos(\alpha(s_0)) = \frac{\Delta x}{\Delta s} \,,$$

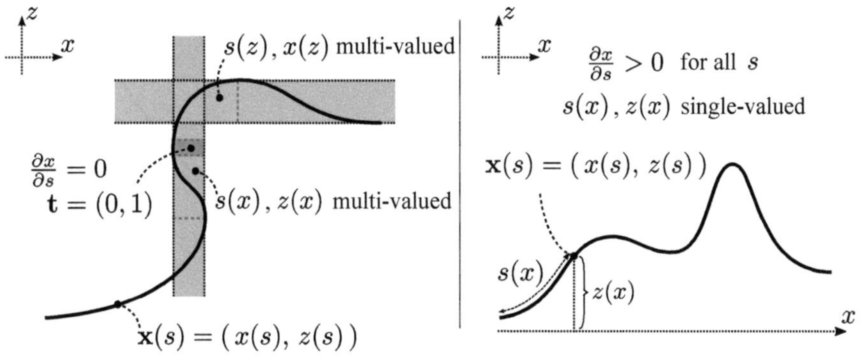

Fig. 5.6 The parametric curve $\mathbf{x} = \mathbf{x}(s)$ in the xz plane. *Left*: General case in global xz coordinates. It is always possible to write $\mathbf{x}(s) = (x(s), z(s))$, whereas in the shaded regions, single-valued inverse functions $s(x)$ and $s(z)$ do not exist. Points dividing the inverse functions into branches are marked with dashed lines. Furthermore, there may be regions where the curve is fully vertical ($\partial x/\partial s = 0$; an example is indicated) or fully horizontal ($\partial z/\partial s = 0$). In these regions, the "value" of s that corresponds to the given x (respectively z) is the whole region. *Right*: A local xz coordinate system. If $\partial x/\partial s > 0$ for all $s \in [s_1, s_2]$ for some $s_1 \leq s_2$ (allowing the special case $s_1 \to -\infty$, $s_2 \to +\infty$), then $x = x(s)$ can be inverted in this interval (resp. globally) to obtain a single-valued function $s = s(x)$. We may also consider $z(x)$ as a function that gives the z coordinate of the point(s) on the curve for a given x

where Δs denotes the hypotenuse parallel to $\mathbf{t}(s_0)$. It will coincide with the actual curve length increment in the limit as the triangle is uniformly shrunk to a point at s_0. Because s_0 is arbitrary, in general we have

$$\cos\alpha = \frac{\partial x}{\partial s} . \tag{5.1.57}$$

By the Pythagorean theorem,

$$\Delta s = \sqrt{(\Delta x)^2 + (\Delta z)^2} = \sqrt{1 + \left(\frac{\Delta z}{\Delta x}\right)^2}\, \Delta x , \tag{5.1.58}$$

whence immediately

$$\frac{\Delta x}{\Delta s} = \frac{1}{\sqrt{1 + \left(\frac{\Delta z}{\Delta x}\right)^2}} .$$

Again, shrinking the triangle uniformly to a point at s_0 and noting that s_0 was arbitrary, we have

$$\frac{\partial x}{\partial s} = \frac{1}{\sqrt{1 + \left(\frac{\partial z}{\partial x}\right)^2}} . \tag{5.1.59}$$

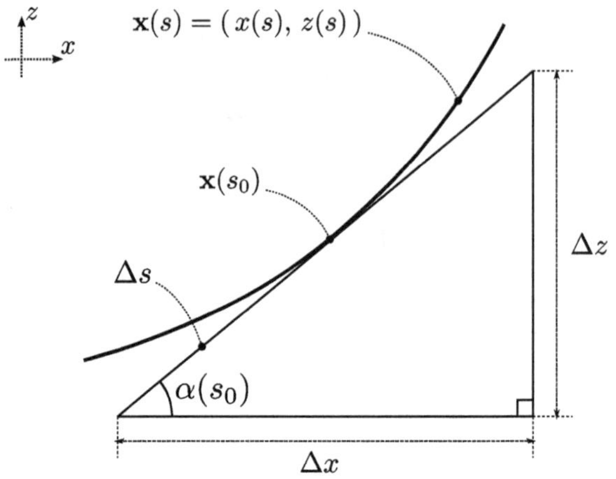

Fig. 5.7 A small segment of the curve $\mathbf{x}(s)$. The hypotenuse Δs is parallel to $\mathbf{t}(s_0)$; hence, the angle at the lower left corner is $\alpha(s_0)$

For more rigor, it is possible to instead consider the ratio $\Delta s/\Delta x$, making all Δx appear in denominators, allowing one to define Δs and Δz as functions of Δx. Then at the limit $\Delta x \to 0$, we obtain $\partial s/\partial x$. Finally, invoking the chain rule,

$$1 = \frac{\partial x}{\partial x} = \frac{\partial x}{\partial s}\frac{\partial s}{\partial x} ,$$

gives the expression for the derivative of the inverse function,

$$\frac{\partial x}{\partial s} = \frac{1}{\dfrac{\partial s}{\partial x}} ,$$

which holds at an arbitrary point (x_0, s_0) such that $x(s_0) = x_0$ and $s(x_0) = s_0$. The result is (5.1.59).

Using (5.1.56), (5.1.57) and (5.1.59) in Eq. (5.1.54), we obtain the standard expression of the extrinsic curvature,

$$\frac{\partial \alpha}{\partial s} = \frac{\dfrac{\partial^2 z}{\partial x^2}}{\left[1 + \left(\dfrac{\partial z}{\partial x}\right)^2\right]^{3/2}} . \tag{5.1.60}$$

Equation (5.1.60) is valid in any xz coordinate system where the tangent is not exactly vertical. However, in each instance, the particular function $z(x)$ will depend on the orientation of the coordinate system.

If the slope $|\partial z/\partial x| \ll 1$, we may approximate (5.1.60) as

$$\frac{\partial \alpha}{\partial s} \approx \frac{\partial^2 z}{\partial x^2}, \tag{5.1.61}$$

which is commonly used in cases where $\|\partial z/\partial x\| \ll 1$. The expressions (5.1.60) and (5.1.61) will be important for the kinematical considerations that will follow later.

By Eq. (5.1.47), the expressions (5.1.60) and (5.1.61) represent $1/R$, the reciprocal of the local radius of curvature. One may interpret R as the radius of the local second-order approximation of the general curve as a circle, or as the radius of the circle whose constant curvature matches the locally observed value of $\partial \alpha/\partial s$ on the curve.

In summary to this section, the fundamental physical laws describing the balance of forces and moments in the plane dynamics for a curved beam are (5.1.12) and (5.1.38):

$$\mathbf{t}\left[\frac{\partial N}{\partial s} + \frac{\partial \alpha}{\partial s} Q\right] + \mathbf{n}\left[\frac{\partial \alpha}{\partial s} N - \frac{\partial Q}{\partial s}\right] + \mathbf{q} - m\frac{\partial^2 \mathbf{x}}{\partial t^2} = 0\,, \tag{5.1.62}$$

$$\frac{\partial M}{\partial s} - Q + \mu = 0\,. \tag{5.1.63}$$

In terms of the stress field inside the beam, the resultant forces and moments are given by (5.1.41):

$$N(s) = \int_A \sigma_{ss}\, \mathrm{d}A\,, \quad Q(s) = \int_A \sigma_{sz}\, \mathrm{d}A\,, \quad M(s) = -\int_A \sigma_{ss}\, z\, \mathrm{d}A\,. \tag{5.1.64}$$

Using (5.1.63) in (5.1.62), it is possible to eliminate the shear force Q. The combined result reads

$$\mathbf{t}\left[\frac{\partial N}{\partial s} + \frac{\partial \alpha}{\partial s}\frac{\partial M}{\partial s} + \frac{\partial \alpha}{\partial s}\mu\right] + \mathbf{n}\left[\frac{\partial \alpha}{\partial s} N - \frac{\partial^2 M}{\partial s^2} - \frac{\partial \mu}{\partial s}\right] + \mathbf{q} - m\frac{\partial^2 \mathbf{x}}{\partial t^2} = 0\,. \tag{5.1.65}$$

Either (5.1.62) and (5.1.63) or (5.1.65) may be more convenient to use, depending on the case studied and the solution approach chosen.

Each quantity in the Eqs. (5.1.62)–(5.1.65), including the vectors **t** and **n**, is a function of the curve length coordinate s; the equations are written in a local tangent-normal coordinate system.

In deriving these fundamental equations, the only assumption we have made concerning the problem geometry is that the shear force is perpendicular to the mid-line. Therefore, Eqs. (5.1.62)–(5.1.65) are general and exact for an arbitrary curved Euler–Bernoulli beam.

This is the appropriate point to remark on the ideal string, which is a classical related model. By definition, an ideal string is a thin stringlike structure that can only support tensile loads, i.e. $N(s) > 0$; the bending moment $M(s)$ and the shear force $Q(s)$ are both identically zero.

In what limit does the beam model produce a string? Consider the resultants, Eq. (5.1.64). The resultant shear force $Q(s)$ in a beam vanishes if $\sigma_{sz} \equiv 0$, i.e. if the beam does not resist shearing. In other words, it does not produce shear stress when subjected to a shear strain in the sz plane. The resultant moment $M(s)$ in a beam vanishes only if the axial stress field σ_{ss} is symmetric with respect to z, which is not realistic; in bending, one side of the mid-line always experiences compression while the other side experiences expansion. Thus, in this sense the string model is not a consistent reduction of the beam model (at any finite thickness).

On the other hand, we are free to consider a model that has $M(s) \equiv 0$ and $\mu \equiv 0$ already at the outset, before starting the derivation. The force and moment balance equations for the ideal string are then derived similarly to how we derived them for the beam; in fact, the result will be the same in other respects, so we may take a shortcut, setting $M(s) \equiv 0$ and $\mu \equiv 0$ in the result that we already obtained.

With $M(s) \equiv 0$ and $\mu \equiv 0$, the moment balance Eq. (5.1.63) yields $Q(s) \equiv 0$. This simplifies the force balance Eq. (5.1.62) to

$$\mathbf{t}\frac{\partial N}{\partial s} + \mathbf{n}\frac{\partial \alpha}{\partial s}N + \mathbf{q} - m\frac{\partial^2 \mathbf{x}}{\partial t^2} = 0 \,. \tag{5.1.66}$$

Equation (5.1.66) describes the balance of forces for an ideal string in any arbitrary configuration. For the ideal string, no equation is needed for the moment balance, due to the modeling assumption $M(s) \equiv 0$.

5.2 Kinematic Relations of Small Deformations

The stress field, whose components appear in the resultant forces and moments (5.1.41) in Sect. 5.1, is connected to the strain field via the constitutive relations. These relations are specific to each material model, such as the linear elastic or Kelvin–Voigt viscoelastic materials that will be discussed later in this chapter. The strain field, in turn, is connected to the displacement field by the kinematic relations, which arise from geometry. This section is devoted to a more detailed exposition of the kinematic relations for small deformations.

In order to be able to determine strains, we define a reference state $\mathbf{x}_0(s, t)$, which is usually undeformed, and consider how the positions of the points on the beam in its current state $\mathbf{x}(s, t)$ differ from the reference state. We define the displacement as follows:

$$\mathbf{u}(s, t) \equiv \mathbf{x}(s, t) - \mathbf{x}_0(s, t) \,. \tag{5.2.1}$$

In the xz plane, it is customary to denote the components of $\mathbf{u}$ by u and w:

$$\mathbf{u}(s, t) \equiv (u(s, t),\ w(s, t)) \,. \tag{5.2.2}$$

Using (5.2.1), the local acceleration $\partial^2 \mathbf{x}/\partial t^2$ that appears in the equations of motion becomes

$$\frac{\partial^2 \mathbf{x}}{\partial t^2} = \frac{\partial^2 \mathbf{x}_0}{\partial t^2} + \frac{\partial^2 \mathbf{u}}{\partial t^2} \,. \tag{5.2.3}$$

Typically the reference state is constant in time, $\mathbf{x}_0(s, t) \equiv \mathbf{x}_0(s)$, and even for axially moving materials, it travels at a constant velocity. The first term in (5.2.3) vanishes and we have

$$\frac{\partial^2 \mathbf{x}}{\partial t^2} = \frac{\partial^2 \mathbf{u}}{\partial t^2} \,. \tag{5.2.4}$$

Note, however, that for axially moving materials, each $\partial/\partial t$ in (5.2.4) is taken in the axially co-moving frame. Refer to Koivurova and Salonen [63] and Sect. 5.5.

The small-displacement regime, where the displacement and its space derivatives are considered small, is a classical simplification. This is often sufficient for the analysis of stability of the trivial equilibrium position, which in many practical cases is straight, in other words, it can be taken to lie along the x axis. Furthermore, if a linear constitutive model is used, the equations of motion will be linear. This allows us to split the strain into a linear superposition of contributions due to different effects. Let us now look at how these different contributions arise.

In local xz coordinates, let us denote a small undeformed segment of the (midline of the) beam by Δx. To begin with, we will allow for moderate displacements, where we require only that the x component $x = x(s)$ of the parametric curve $\mathbf{x}(s) = (x(s), z(s))$ is invertible to obtain $s = s(x)$; recall Fig. 5.6. ! Consider the deformation by (u, w), as shown in Fig. 5.8. We will look at a linear displacement only and disregard the curvature, as its effect will vanish in the final result in the limit $\Delta x \to 0$.

To justify this claim, consider that in order to determine the axial strain due to a displacement by (u, w), we are ultimately interested in $\partial s/\partial x$, $\partial u/\partial x$ and $\partial w/\partial x$. Let f denote any of s, u and w. Expanding $\Delta f \equiv f(x) - f(x_0)$ as a Taylor series in $\Delta x \equiv x - x_0$, then dividing by Δx to obtain $\Delta f/\Delta x$, and finally letting $\Delta x \to 0$ to obtain $\partial f/\partial x$, we see that terms originally of second order or higher in Δx vanish from the result.

Using the Pythagorean theorem, from Fig. 5.8 we have

$$\begin{aligned}(\Delta s)^2 &= (\Delta x + \Delta u)^2 + (\Delta w)^2 \\ &= \left[\left(1 + \frac{\Delta u}{\Delta x}\right)^2 + \left(\frac{\Delta w}{\Delta x}\right)^2\right](\Delta x)^2 \\ &= \left[1 + 2\frac{\Delta u}{\Delta x} + \left(\frac{\Delta u}{\Delta x}\right)^2 + \left(\frac{\Delta w}{\Delta x}\right)^2\right](\Delta x)^2 \,.\end{aligned} \tag{5.2.5}$$

Thus we may express the ratio of deformed and undeformed lengths as

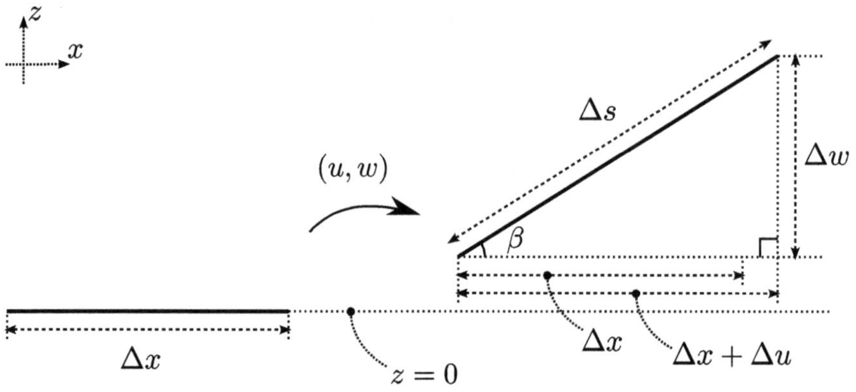

Fig. 5.8 Deformation of a small segment of the beam mid-line in local xz coordinates, based on a first-order Taylor expansion. The undeformed position is on the x axis ($z = 0$). Note axial change in length by Δu, and nonzero z coordinate also at the left endpoint after deformation by (u, w)

$$\frac{\Delta s}{\Delta x} = \sqrt{1 + 2\frac{\Delta u}{\Delta x} + \left(\frac{\Delta u}{\Delta x}\right)^2 + \left(\frac{\Delta w}{\Delta x}\right)^2} \; . \tag{5.2.6}$$

The axial strain ε_{xx} is defined as the relative change in length in the deformed versus the undeformed state:

$$\varepsilon_{xx} = \frac{\Delta s - \Delta x}{\Delta x} = \frac{\Delta s}{\Delta x} - 1 \; . \tag{5.2.7}$$

In the limit $\Delta x \to 0$, (5.2.7) becomes

$$\varepsilon_{xx} = \frac{\partial s}{\partial x} - 1 \; . \tag{5.2.8}$$

Inserting (5.2.6) at the limit $\Delta x \to 0$, we have

$$\varepsilon_{xx} = \sqrt{1 + 2\frac{\partial u}{\partial x} + \left(\frac{\partial u}{\partial x}\right)^2 + \left(\frac{\partial w}{\partial x}\right)^2} - 1 \; . \tag{5.2.9}$$

Keep in mind that by definition, $\Delta s = s(x_0 + \Delta x) - s(x_0)$; hence, taking the limits produces (5.2.8) and (5.2.9) rigorously.

Equations (5.2.8) and (5.2.9) are exact as long as $s = s(x)$ is single-valued. Restricting the result to the small-displacement regime, we may neglect the higher-order terms in (5.2.6), obtaining the approximation

$$\frac{\Delta s}{\Delta x} \approx \sqrt{1 + 2\frac{\Delta u}{\Delta x}} \; . \tag{5.2.10}$$

Equation (5.2.10) is valid up to the first order in the small quantities. Considering furthermore that in the small-displacement regime, $|\Delta u/\Delta x| \ll 1$, we may use the following series expansion, developed around $x = 0$:

$$\sqrt{1+x} = 1 + \frac{1}{2}x - \frac{1}{8}x^2 + \frac{1}{16}x^3 - \ldots \ . \tag{5.2.11}$$

Hence, the ratio of the deformed and undeformed lengths is approximately

$$\frac{\Delta s}{\Delta x} \approx 1 + \frac{\Delta u}{\Delta x}\ , \tag{5.2.12}$$

and at the limit $\Delta x \to 0$,

$$\frac{\partial s}{\partial x} \approx 1 + \frac{\partial u}{\partial x}\ . \tag{5.2.13}$$

Thus by (5.2.8) and (5.2.13), in the small-displacement regime the axial strain is

$$\varepsilon_{xx} \approx \frac{\partial u}{\partial x}\ . \tag{5.2.14}$$

Equation (5.2.14) is valid up to first order in the small quantities. In the strain contribution from a linear displacement, up to first order, only the axial displacement u matters.

The angle β in Fig. 5.8 is related to the shear strain. But to determine the shear strain, we must consider more than just the mid-line, because the shear strain measures how adjacent layers of the continuous material slide with respect to each other. Consider the deformation of a small rectangular piece of the beam, as shown in Fig. 5.9. The same remark about first-order Taylor expansions also applies to this figure. The *engineering shear strain* γ_{xz} is defined as

$$\gamma_{xz} = \beta + \delta\ . \tag{5.2.15}$$

As can be seen in Fig. 5.9, it describes the total change of angle, which the originally straight angle at the lower left corner of the rectangular piece experiences under the applied deformation.

By the standard indexing convention for strains (and stresses), the quantity γ_{xz} is the strain in the z direction, experienced along the plane of constant x. The engineering shear strain is symmetric in the sense that $\gamma_{zx} = \gamma_{xz}$, which is readily seen by redrawing Fig. 5.9 with the roles of the coordinate axes swapped.

An expression for γ_{xz} is obtained from the geometry of Fig. 5.9. In terms of the first-order Taylor expansions shown in the figure, we have

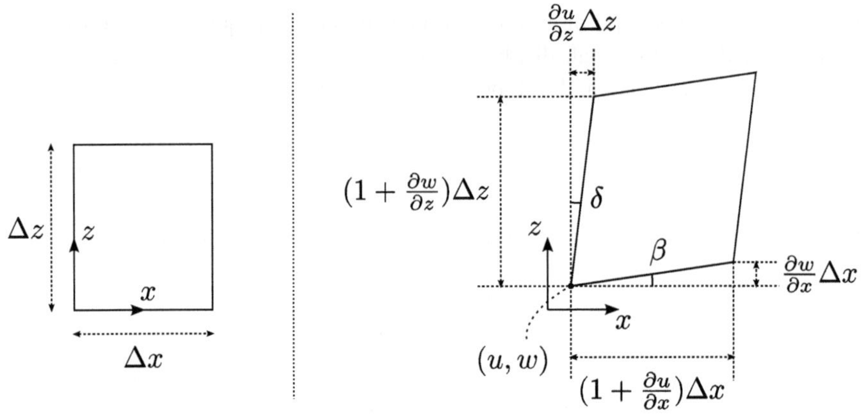

Fig. 5.9 Deformation of a small rectangular piece of the beam in local xz coordinates, using a Taylor expansion up to first order. *Left*: Initial undeformed state. Before the deformation, the origin is on the mid-line of the beam. *Right*: State after a small deformation by (u, w). Note that the lower left corner is no longer at the origin

$$\tan\beta = \frac{\dfrac{\partial w}{\partial x}\Delta x}{(1+\dfrac{\partial u}{\partial x})\Delta x} = \frac{\dfrac{\partial w}{\partial x}}{(1+\dfrac{\partial u}{\partial x})}, \tag{5.2.16}$$

$$\tan\delta = \frac{\dfrac{\partial u}{\partial z}\Delta z}{(1+\dfrac{\partial w}{\partial z})\Delta z} = \frac{\dfrac{\partial u}{\partial z}}{(1+\dfrac{\partial w}{\partial z})}. \tag{5.2.17}$$

Note that in the last forms, we have cancelled one Δx in (5.2.16), and respectively one Δz in (5.2.17). Taking the limit $\Delta x \to 0$ in (5.2.16) and $\Delta z \to 0$ in (5.2.17), the omitted higher-order terms vanish, and these expressions become exact.

In the small-displacement regime, where $\partial u/\partial x$ and $\partial w/\partial z$ are assumed small, we may use the following series expansion of $a/(b+x)$ for $|x| \ll 1$:

$$\frac{a}{b+x} = \frac{a}{b}\left(1 - \frac{x}{b} + \frac{x^2}{b^2} - \frac{x^3}{b^3} + \dots\right). \tag{5.2.18}$$

We have

$$\tan\beta = \frac{\partial w}{\partial x}\left(1 - \frac{\partial u}{\partial x} + \left(\frac{\partial u}{\partial x}\right)^2 - \left(\frac{\partial u}{\partial x}\right)^3 + \dots\right),$$

$$\tan\delta = \frac{\partial u}{\partial z}\left(1 - \frac{\partial w}{\partial z} + \left(\frac{\partial w}{\partial z}\right)^2 - \left(\frac{\partial w}{\partial z}\right)^3 + \dots\right).$$

Discarding small quantities of second and higher orders on the right-hand side, we obtain

$$\tan\beta \approx \frac{\partial w}{\partial x}\,, \quad \tan\delta \approx \frac{\partial u}{\partial z}\,. \tag{5.2.19}$$

We then develop $\tan\alpha$ into a Taylor series around $\alpha = 0$:

$$\tan\alpha = \alpha + \frac{1}{3}\alpha^3 + \frac{2}{15}\alpha^5 + \dots\,. \tag{5.2.20}$$

Using (5.2.20) on the left-hand sides of (5.2.19), and keeping just the leading term, we obtain that up to first order in the small quantities,

$$\beta \approx \frac{\partial w}{\partial x}\,, \quad \delta \approx \frac{\partial u}{\partial z}\,. \tag{5.2.21}$$

Inserting (5.2.21) into (5.2.15), the engineering shear strain is thus, up to the first order,

$$\gamma_{xz} \approx \frac{\partial w}{\partial x} + \frac{\partial u}{\partial z}\,.$$

The component ε_{xz} of the small-strain tensor is defined as

$$\varepsilon_{xz} \equiv \frac{1}{2}\gamma_{xz} = \frac{1}{2}\left(\frac{\partial w}{\partial x} + \frac{\partial u}{\partial z}\right)\,. \tag{5.2.22}$$

This definition makes the strain tensor symmetric, while also attributing the change of angle to contributions from both $\partial w/\partial x$ and $\partial u/\partial z$.

Equations (5.2.14) and (5.2.22), describing the first-order effects of a linear displacement in the small-displacement regime, are sufficient for our purposes. For readers interested in a general treatment of strain in solid mechanics, including finite strains in curvilinear coordinate systems, we refer to Holzapfel [47].

However, in addition to a linear displacement, we also need to consider bending. Because we are operating in the framework of beams of the Euler–Bernoulli type, we may use the Kirchhoff hypotheses, which are as follows:

1. Surface normals remain perpendicular to the surface after deformation.
2. In-plane displacements are negligible.
3. Changes in thickness are negligible.

These are the classical modeling assumptions for the pure bending of Kirchhoff–Love plates, but they also apply to the classical Euler–Bernoulli beam. See Fig. 5.10.

The length of a small curved ($0 < R < \infty$) segment of the beam, at perpendicular signed distance z from the mid-line, is

$$\ell(z) = (R - z)\,\Delta\alpha\,. \tag{5.2.23}$$

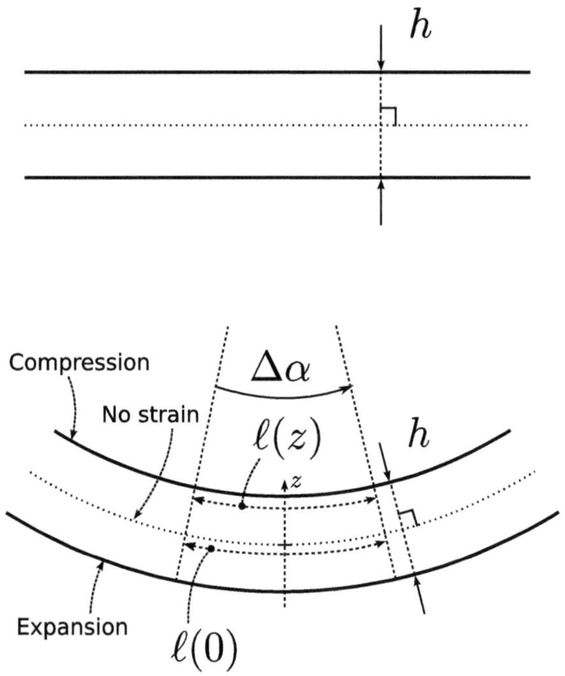

Fig. 5.10 Pure bending in the plane under the Kirchhoff hypotheses

Observe that Eq. (5.2.23) implies that the curved segment is locally approximated by a circle; as was already considered, this is valid up to second order along the local tangential direction.

For determining the correct signs for (5.2.23), recall Eq. (5.1.47) for the curvature $1/R$,

$$\frac{1}{R} = \frac{\partial \alpha}{\partial s} .$$

A positive $\partial\alpha/\partial s$, that is, a counterclockwise turn, corresponds to a positive radius of curvature R. The signs of $\Delta\alpha$ and R always match, so the product $R\,\Delta\alpha > 0$ regardless of the direction of the turn. In a counterclockwise turn in z-up coordinates, the upper surface becomes shorter, hence the minus sign in the $-z\,\Delta\alpha$ term.

By hypothesis, the mid-line experiences no axial strain, so the rest length at any z is the length of the corresponding segment of the mid-line, $\ell(0) = R\,\Delta\alpha$. Hence, the axial strain is

$$\varepsilon_{xx} \equiv \frac{\ell(z) - \ell(0)}{\ell(0)} = \frac{\ell(z)}{\ell(0)} - 1 = \frac{(R - z)\,\Delta\alpha}{R\,\Delta\alpha} - 1 = -\frac{z}{R} . \tag{5.2.24}$$

Taking the limit $\Delta\alpha \to 0$, Eq. (5.2.24) becomes exact.

If $|\partial w/\partial x| \ll 1$, we may approximate the curvature by Eq. (5.1.61),

$$\frac{\partial \alpha}{\partial s} \approx \frac{\partial^2 w}{\partial x^2} .$$

Combining (5.2.24), (5.1.47) and (5.1.61) in that order, we have

$$\varepsilon_{xx} = -\frac{z}{R} = -z\frac{\partial \alpha}{\partial s} \approx -z\frac{\partial^2 w}{\partial x^2} .$$

Thus, we have obtained the classical kinematic relation

$$\varepsilon_{xx}(x, z) = -z\,\frac{\partial^2 w}{\partial x^2}(x) , \tag{5.2.25}$$

which describes, in the small-displacement regime, the axial strain contribution due to pure bending.

The first condition that normals remain perpendicular implies that the deformation causes no shear, so the contribution of this type of deformation to the shear strain ε_{xz} is zero; hence, the name *pure bending*.

Before moving on, let us consider force balance in the small-displacement regime. As above, let the neutral position of the beam coincide with the x axis. Recall Eq. (5.1.3) in Sect. 5.1 for the unit vectors $\mathbf{t}$ and $\mathbf{n}$:

$$\mathbf{t} \equiv (\cos(\alpha),\ \sin(\alpha)) , \quad \mathbf{n} \equiv (-\sin(\alpha),\ \cos(\alpha)) .$$

In the small-displacement regime, it is assumed that $|\alpha| \ll 1$, allowing us to use only the leading terms of the Taylor series of the trigonometric functions, developed around $\alpha = 0$:

$$\cos\alpha = 1 - \frac{1}{2}\alpha^2 + \frac{1}{24}\alpha^4 - \dots , \quad \sin\alpha = \alpha - \frac{1}{6}\alpha^3 + \frac{1}{120}\alpha^5 - \dots . \tag{5.2.26}$$

Hence, for $|\alpha| \ll 1$,

$$\mathbf{t} \approx (1,\ \alpha) , \quad \mathbf{n} \approx (-\alpha,\ 1) . \tag{5.2.27}$$

Furthermore, to express α in terms of the displacement, we use the leading term of the Taylor expansion of $\tan\alpha$ from Eq. (5.2.20)Thus, for $|\alpha| \ll 1$, we may write

$$\alpha \approx \tan\alpha = \frac{\partial w}{\partial x} . \tag{5.2.28}$$

Hence in the small-displacement regime, we have

$$\mathbf{t} \approx (1,\ \partial w/\partial x) , \quad \mathbf{n} \approx (-\partial w/\partial x,\ 1) . \tag{5.2.29}$$

The combined force balance equation for a beam is (5.1.40), viz.

$$\mathbf{t}\left[\frac{\partial N}{\partial s}+\frac{\partial \alpha}{\partial s}\frac{\partial M}{\partial s}+\frac{\partial \alpha}{\partial s}\mu\right]+\mathbf{n}\left[\frac{\partial \alpha}{\partial s}N-\frac{\partial^2 M}{\partial s^2}-\frac{\partial \mu}{\partial s}\right]+\mathbf{q}-m\frac{\partial^2 \mathbf{x}}{\partial t^2}=0\,.$$

We may use (5.2.29) to represent **t** and **n** in the small-displacement regime. Using Eq. (5.2.4) to replace $\partial^2\mathbf{x}/\partial t^2$, and replacing $\partial\alpha/\partial s$ by its small-displacement expression from (5.1.61), in component form we obtain the equations

$$\left[\frac{\partial N}{\partial s}+\frac{\partial^2 w}{\partial x^2}\frac{\partial M}{\partial s}+\frac{\partial^2 w}{\partial x^2}\mu\right]-\frac{\partial w}{\partial x}\left[\frac{\partial^2 w}{\partial x^2}N-\frac{\partial^2 M}{\partial s^2}-\frac{\partial \mu}{\partial s}\right]+q_x-m\frac{\partial^2 u}{\partial t^2}=0\,,$$
$$\frac{\partial w}{\partial x}\left[\frac{\partial N}{\partial s}+\frac{\partial^2 w}{\partial x^2}\frac{\partial M}{\partial s}+\frac{\partial^2 w}{\partial x^2}\mu\right]+\left[\frac{\partial^2 w}{\partial x^2}N-\frac{\partial^2 M}{\partial s^2}-\frac{\partial \mu}{\partial s}\right]+q_z-m\frac{\partial^2 w}{\partial t^2}=0\,.$$

Because in the small-displacement regime $\partial/\partial s\approx\partial/\partial x$, we have

$$\left[\frac{\partial N}{\partial x}+\frac{\partial^2 w}{\partial x^2}\frac{\partial M}{\partial x}+\frac{\partial^2 w}{\partial x^2}\mu\right]-\frac{\partial w}{\partial x}\left[\frac{\partial^2 w}{\partial x^2}N-\frac{\partial^2 M}{\partial x^2}-\frac{\partial \mu}{\partial x}\right]+q_x-m\frac{\partial^2 u}{\partial t^2}=0\,,$$
$$\frac{\partial w}{\partial x}\left[\frac{\partial N}{\partial x}+\frac{\partial^2 w}{\partial x^2}\frac{\partial M}{\partial x}+\frac{\partial^2 w}{\partial x^2}\mu\right]+\left[\frac{\partial^2 w}{\partial x^2}N-\frac{\partial^2 M}{\partial x^2}-\frac{\partial \mu}{\partial x}\right]+q_z-m\frac{\partial^2 w}{\partial t^2}=0\,.$$

Finally, because $\partial w/\partial x$ and $\partial^2 w/\partial x^2$ are both assumed small, their product is a second-order small quantity and hence negligible. Expanding the brackets, and dropping the second-order small terms, the result is

$$\frac{\partial N}{\partial x}+\frac{\partial^2 w}{\partial x^2}\frac{\partial M}{\partial x}+\frac{\partial^2 w}{\partial x^2}\mu+\frac{\partial w}{\partial x}\frac{\partial^2 M}{\partial x^2}+\frac{\partial w}{\partial x}\frac{\partial \mu}{\partial x}+q_x-m\frac{\partial^2 u}{\partial t^2}=0\,,\quad(5.2.30)$$
$$\frac{\partial w}{\partial x}\frac{\partial N}{\partial x}+\frac{\partial^2 w}{\partial x^2}N-\frac{\partial^2 M}{\partial x^2}-\frac{\partial \mu}{\partial x}+q_z-m\frac{\partial^2 w}{\partial t^2}=0\,.\quad(5.2.31)$$

Equations (5.2.30) and (5.2.31) represent the general force balance of a beam in the small-displacement regime, with the moment balance already accounted for.

Now consider the ideal string. It is common wisdom (e.g., [30]) that in the small-displacement regime, the force balance of the ideal string, Eq. (5.1.66), leads to the classical wave equation. Less widely known is the fact that under the assumption of a purely transverse displacement ($\Delta u\equiv 0$ in terms of Fig. 5.8), the same classical wave equation is obtained for arbitrarily large transverse displacements (Δw in Fig. 5.8). In other words, if axial displacements are not allowed, the small-displacement assumption in the transverse direction is unnecessary [17].

Clelland and Vassiliou [29] provide a short, simple proof of this and generalize to vibrations of a string on Riemannian manifolds. This observation, however, is mainly of importance for mathematical correctness, because in physical systems that can be modelled as ideal strings, typically nothing prevents axial displacement (or indeed deformation) from occurring.

For purposes of comparison with (5.2.30) and (5.2.31), consider the string in the small-displacement regime. Inserting (5.2.29) for **t** and **n**, (5.2.4) for $\partial^2\mathbf{x}/\partial t^2$, and (5.1.61) for $\partial\alpha/\partial s$ into the general force balance of the ideal string, Eq. (5.1.66), and finally noting that in the small-displacement regime $\partial/\partial s \approx \partial/\partial x$, in component form we have the small-displacement ideal string force balance equations

$$\begin{aligned}
\frac{\partial N}{\partial x} - \frac{\partial w}{\partial x}\frac{\partial^2 w}{\partial x^2} N + q_x - m\frac{\partial^2 u}{\partial t^2} &= 0 \,, \\
\frac{\partial w}{\partial x}\frac{\partial N}{\partial x} + \frac{\partial^2 w}{\partial x^2} N + q_z - m\frac{\partial^2 w}{\partial t^2} &= 0 \,.
\end{aligned}$$

Dropping the second-order small term, the final result is

$$\frac{\partial N}{\partial x} + q_x - m\frac{\partial^2 u}{\partial t^2} = 0 \,, \tag{5.2.32}$$

$$\frac{\partial w}{\partial x}\frac{\partial N}{\partial x} + \frac{\partial^2 w}{\partial x^2} N + q_z - m\frac{\partial^2 w}{\partial t^2} = 0 \,. \tag{5.2.33}$$

Compare (5.2.30)–(5.2.31) for the beam. Because here $M(x) \equiv 0$, the axial equation has become much simpler.

Let us now consider an ideal string lying on a viscoelastic foundation. In this context, a *foundation* represents an external continuous medium on which the string lies (e.g., an elastic solid), or with which the string interacts (e.g., a highly viscous fluid). Flügge [41, p. 93] notes that the most classical case, a linear elastic foundation, was first studied by Winkler [154, p. 182].

In the discussion for Eqs. (5.1.62) and (5.1.63), it was hinted that an elastic or viscoelastic foundation can be included in the model by choosing the body force **q** appropriately. As an example, let us consider a Kelvin–Voigt viscoelastic foundation in the small-displacement regime. The Kelvin–Voigt solid will be introduced in the next section.

As the string deflects in the transverse direction from its neutral position, it experiences a restoring force in the z direction. From the viewpoint of the string, this can be treated as a body force **q**. In component form, we have

$$q_x = 0 \,, \tag{5.2.34}$$

$$q_z = -k_1 w - k_2 \frac{\partial w}{\partial t}\Big|_{\text{lab.}} \,. \tag{5.2.35}$$

The coefficients k_1 and k_2 describe the elastic and viscous reactions of the foundation, respectively, and their SI units are $[k_1] = 1/s^2$ and $[k_2] = 1/s$. The coefficient k_1 gives the acceleration of the reactive body force per unit displacement of the foundation from its rest position at $z = 0$, while k_2 gives the acceleration per unit deformation velocity of the foundation.

In the case of an axially traveling string or beam, to be considered later, the $\partial/\partial t$ in Eq. (5.2.35) must be understood as written in laboratory coordinates at constant x, because (5.2.35)—although it refers to the string or beam displacement w—is actually related to the displacement of the *foundation*, which is not traveling axially. We have indicated this explicitly by the notation $\partial/\partial t|_{\text{lab.}}$.

If the foundation is also moving axially, the axial velocity of the foundation (not that of the axially traveling string!) must be used when converting the local $\partial/\partial t$ to the laboratory coordinates.

Inserting the body force (5.2.34) and (5.2.35) into the small-displacement force balance equations for an ideal string, (5.2.32) and (5.2.33), we obtain

$$\frac{\partial N}{\partial x} + m\frac{\partial^2 u}{\partial t^2} = 0\,, \tag{5.2.36}$$

$$\frac{\partial w}{\partial x}\frac{\partial N}{\partial x} + \frac{\partial^2 w}{\partial x^2}N - k_1 w - k_2\frac{\partial w}{\partial t}\Big|_{\text{lab.}} - m\frac{\partial^2 w}{\partial t^2} = 0\,. \tag{5.2.37}$$

Equations (5.2.36) and (5.2.37) describe the force balance of a general ideal string in the small-displacement regime, lying on a Kelvin–Voigt viscoelastic foundation. Concerning the string itself, no constitutive assumptions have been made.

Finally, let us briefly consider mass balance in the small-displacement regime. For a classical beam or string that does not move axially, mass conservation in the considered domain is trivially satisfied, but such considerations will become extremely important when later in this chapter we move onto axially moving materials.

Perhaps the only interesting question here is, how much does the local density change as the beam undergoes a small strain? Let m_0 denote the linear density, $[m_0] = \text{kg/m}$, of the beam or string in the neutral state, where $u \equiv w \equiv 0$. The total mass of a segment of length Δx is $M = m_0\,\Delta x$. In the deformed state, the length has become $\Delta s = (1 + \varepsilon_{xx})\Delta x$. See Fig. 5.8 in Sect. 5.2. The total mass must remain the same, i.e. $M = m\,\Delta s$.

By equating the two expressions for the total mass of the segment, we have

$$m_0\Delta x = m\Delta s = m(1 + \varepsilon_{xx})\Delta x\,, \tag{5.2.38}$$

whence it follows for the linear density m that

$$m = m_0\,\frac{1}{1 + \varepsilon_{xx}}\,. \tag{5.2.39}$$

In the small-displacement regime, where $|\varepsilon_{xx}| \ll 1$, we may use the series expansion for $a/(b + x)$ around $x = 0$, which was already given in Eq. (5.2.18) in Sect. 5.2. Up to first order in ε_{xx}, we thus have

$$m \approx m_0\,[1 - \varepsilon_{xx}]\,. \tag{5.2.40}$$

In the small-displacement regime, we have $\varepsilon_{xx} = \partial u/\partial x$, transforming (5.2.40) into

$$m \approx m_0 \left[1 - \frac{\partial u}{\partial x}\right] . \tag{5.2.41}$$

Consider the term $m\ddot{\mathbf{u}}$ in the force balance Eq. (5.1.62). We have

$$m\ddot{\mathbf{u}} = m_0 \left[1 - \frac{\partial u}{\partial x}\right] \ddot{\mathbf{u}} \approx m_0 \ddot{\mathbf{u}} , \tag{5.2.42}$$

because the quantity $\partial u/\partial x\, \ddot{\mathbf{u}}$ is second-order small, provided that m_0 is of the order of unity.

Thus in a small-displacement analysis, this change in density can be neglected. In Eq. (5.1.62), m appears only in the term $m\ddot{\mathbf{u}}$, so a local change in density caused by the axial straining has no other effects. However, be aware that we will need to be careful when we later consider axially moving materials undergoing a small strain.

In summary to this section, in the small-displacement regime, where the equations are linear, the total axial strain is the sum of the contributions due to the linear displacement and pure bending. Summing the axial strain contributions (5.2.14) and (5.2.25), we have the result

$$\varepsilon_{xx}(x, z) \approx \frac{\partial u}{\partial x}(x) - z\frac{\partial^2 w}{\partial x^2}(x) . \tag{5.2.43}$$

Note that u and w are the mid-line displacements, which do not depend on z. However, ε_{xx} is defined for any point (x, z) in the region occupied by the beam.

The pure bending contribution to the deformation, as was noted, produces no shear strain. The shear strain thus comes only from the contribution of the linear displacement, namely (5.2.22):

$$\varepsilon_{xz}(x, z) \approx \frac{1}{2}\left(\frac{\partial w}{\partial x}(x) + \frac{\partial u}{\partial z}(x)\right) . \tag{5.2.44}$$

This shear strain is constant across z.

The force balance equations of a beam in the small-displacement regime are (5.2.30) and (5.2.31), that is,

$$\frac{\partial N}{\partial x} + \frac{\partial^2 w}{\partial x^2}\frac{\partial M}{\partial x} + \frac{\partial^2 w}{\partial x^2}\mu + \frac{\partial w}{\partial x}\frac{\partial^2 M}{\partial x^2} + \frac{\partial w}{\partial x}\frac{\partial \mu}{\partial x} + q_x - m\frac{\partial^2 u}{\partial t^2} = 0 ,$$
$$\frac{\partial w}{\partial x}\frac{\partial N}{\partial x} + \frac{\partial^2 w}{\partial x^2}N - \frac{\partial^2 M}{\partial x^2} - \frac{\partial \mu}{\partial x} + q_z - m\frac{\partial^2 w}{\partial t^2} = 0 .$$

Moment balance is already accounted for. The corresponding equations for the ideal string are (5.2.32) and (5.2.33), namely,

$$\frac{\partial N}{\partial x} + q_x - m\frac{\partial^2 u}{\partial t^2} = 0 ,$$

$$\frac{\partial w}{\partial x}\frac{\partial N}{\partial x} + \frac{\partial^2 w}{\partial x^2} N + q_z - m\frac{\partial^2 w}{\partial t^2} = 0 .$$

5.3 Constitutive Linear Elastic and Visco-Elastic Relations

In continuum mechanics, a constitutive relation describes the behavior of a specific class of materials, by specifying the correspondence between the strain and stress fields inside the material. Allen et al. [4] point out that in continuum mechanics, the form of the constitutive relation is the critical difference between solids and fluids. Solids respond to strain, whereas fluids respond to strain rate only.

There are several types of constitutive response. An elastic constitutive response is time-independent in the sense that the elastic stress depends only on the instantaneous value of the elastic strain. Time derivatives or integration across the time variable appear, respectively, in the cases of viscous and plastic responses.

Combinations are also possible. For example, *viscoelastic* materials exhibit both elastic and viscous behavior, and this can have interesting consequences. Viscoelastic materials exhibit the phenomena of *stress relaxation* and *viscoelastic creep*. Stress relaxation is a time-dependent reduction in stress when the material is held at a constant strain. Viscoelastic creep refers to a time-dependent increase in strain when held at a constant stress. For more on viscoelasticity, including a variety of constitutive models, see the books by Flügge [41], Sobotka [128], Ottosen and Ristinmaa [105].

Constitutive relations cannot be completely arbitrary, because they must fulfill certain physical constraints, namely thermodynamic admissibility (the Clausius–Duhem inequality) and objectivity. See Allen et al. [4] for a clear presentation of both topics in this context. Objectivity in the context of solid mechanics is discussed in Holzapfel [47], and briefly in Bower [20]. We restrict our exposition to simple linear constitutive models that are known to fulfill these constraints.

In this chapter we will concentrate on solids only, and exclude plastic behavior from consideration. Excluding plasticity gives us the memory-free property: with respect to time, the stress-strain relations can be worked with in a pointwise manner, without considering history, that is, in this context, the path taken in the (σ, ε) plane from the $t = 0$ initial state to the $t = t_0$ current state.

For simplicity, we will concentrate on linear constitutive relations only. By linearity, it is meant that the constitutive law is of the form

$$\mathcal{L}_1(\sigma) = \mathcal{L}_2(\varepsilon) , \tag{5.3.1}$$

where $\mathcal{L}_1(\dots)$ and $\mathcal{L}_2(\dots)$ are linear integrodifferential operators.

We will work in the small-displacement regime, using the Cauchy stress. Recall that the resultant axial force $N(s)$ and the resultant moment $M(s)$, given by

Eq. (5.1.41), depend only on the axial stress σ_{ss}. Recall also that we are working with planar dynamics, where all quantities are independent of y.

In the small-displacement regime, we may approximate $s \approx x$. Let us now restrict the discussion to beams having a rectangular cross-section at each fixed value of x, which lets us use (5.1.42) to split each area integral in (5.1.41) into one-dimensional integrals along the y and z directions. We may thus rewrite $N(s)$ and $M(s)$ as

$$N(x) = b \int_{-h/2}^{+h/2} \sigma_{xx} \, \mathrm{d}z \ , \quad M(x) = -b \int_{-h/2}^{+h/2} \sigma_{xx} \, z \, \mathrm{d}z \ . \tag{5.3.2}$$

The uniaxial constitutive relations, to be considered below, will relate the axial stress field σ_{xx} to the axial strain field ε_{xx}. We may then apply Eq. (5.2.43) to express the axial strain ε_{xx} in terms of the displacements. The result is an equation of motion, describing everything in terms of displacements, for a given combination of kinematic and constitutive models.

Uniaxial linear viscoelastic constitutive models are often represented as mechanical networks constructed from linear (Hookean) springs and linear (Newtonian) dashpots. Let us consider how to obtain the stress–strain relations for some of the simplest models of viscoelastic materials.

First, we will consider the building blocks and how to connect them. A Hookean spring follows the constitutive law

$$\sigma = E \, \varepsilon \ , \tag{5.3.3}$$

where E, a constant, is the Young's modulus of the material. A Newtonian dashpot follows the law

$$\sigma = \eta \frac{\partial}{\partial t} \varepsilon \ , \tag{5.3.4}$$

where η is the viscous modulus. Note that the dashpot responds to the strain rate.

In a network, any two components connected in series follow the rule

$$\varepsilon_{\text{total}} = \varepsilon_1 + \varepsilon_2 \ , \quad \sigma_{\text{total}} = \sigma_1 = \sigma_2 \qquad \text{(serial)} \ , \tag{5.3.5}$$

whereas any two components connected in parallel follow the rule

$$\varepsilon_{\text{total}} = \varepsilon_1 = \varepsilon_2 \ , \quad \sigma_{\text{total}} = \sigma_1 + \sigma_2 \qquad \text{(parallel)} \ . \tag{5.3.6}$$

A subnetwork with one attachment node at each end counts as one component. Together with (5.3.5) and (5.3.6), this is the key to building networks.

The fact that the strains must follow the rules (5.3.5) and (5.3.6) is a simple consequence of geometry, due to the behavior of displacements. We have assumed identical rest lengths for elements connected in parallel. We may make this assumption, because we are working in the context of continuum models. The component size is Δx, which at the end is taken to the limit $\Delta x \to 0$. The components are

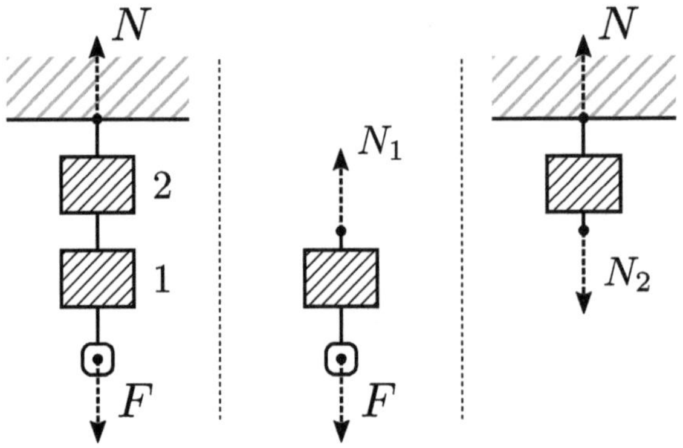

Fig. 5.11 Two structural components connected in series and subjected to an external force F. The force N represents the mechanical support of the fixed end, which keeps the system in static equilibrium. Each black box represents a generic structural component: a spring, a dashpot, or a subnetwork with one input and one output constructed out of springs and dashpots. Both components (1 and 2) are seen to experience the same force F. *Left*: The free-body diagram for the full system, which shows that $N = F$. *Center and right*: A sectioning of the free-body diagram. *Center*: Forces experienced by component 1. We see that $N_1 = F$. *Right*: Forces experienced by component 2. This reveals that $N_2 = N$, so $N_2 = F$. Hence, with regard to the applied force, each component behaves as if it was alone in the chain; each component is stressed by the same force F

abstract mathematical entities representing certain types of stress–strain response; no generality is lost by assuming their rest lengths to be identical.

For the serial connection, we may justify the total stress rule in (5.3.5) by a physical argument. Consider one end of the system fixed, and one end attached to a movable block that is pulled by a prescribed external force F, as shown in Fig. 5.11. In a vertically hanging configuration such as that shown in the figure, we have $F = mg$, where m is the mass of the block; the masses of the components are assumed negligible. Let the system be in static equilibrium. From the free-body diagrams in Fig. 5.11, we see that the internal force at the midpoint between the components must have the same magnitude as F. Hence, each component experiences the same force, namely F. By extension, the same property holds for stresses, because stress is simply force per unit area.

In the parallel case, the sum rule for stresses in (5.3.6) arises as follows. Again, consider one end of the system fixed, with the other end attached to a movable block, which is pulled by a prescribed external force F, as shown in Fig. 5.12. By considering the free-body diagrams in Fig. 5.12, the result is that $N_{11} + N_{21} = F$, and $N_{12} + N_{22} = F$. Again, by extension, this property holds also for stresses.

Perhaps the most classical constitutive model is the linear elastic solid. The linear elastic model is the small-strain limit characterizing many elastic materials. In this case we have just one spring, as in Fig. 5.13, and the stress field is given by Hooke's law,

$$\sigma = E\,\varepsilon \,. \tag{5.3.7}$$

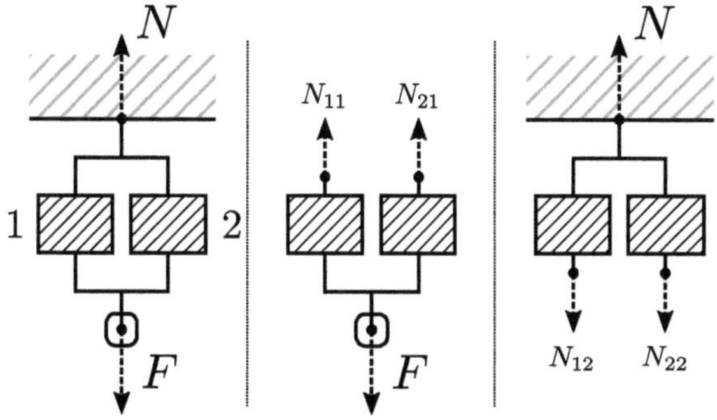

Fig. 5.12 Two structural components connected in parallel and subjected to an external force F. The force N represents the mechanical support of the fixed end, which keeps the system in static equilibrium. Each black box represents a generic structural component; a spring, a dashpot, or a subnetwork with one input and one output constructed out of springs and dashpots. *Left*: From the free-body diagram for the full system, $N = F$. *Center*: This half of a sectioned free-body diagram reveals that $N_{11} + N_{21} = F$. *Right*: This half of an alternative sectioning shows that $N_{12} + N_{22} = N$, so $N_{12} + N_{22} = F$. Hence, at both ends of the black-box components, the internal forces (representing the effect of the rest of the system) pulling on the components sum to F

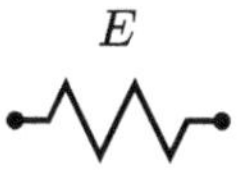

Fig. 5.13 Schematic representation of the linear elastic constitutive model

Interpreting the constitutive law (5.3.7) to represent the axial stress σ_{xx} in terms of the axial strain ε_{xx}, the integrals (5.3.2) become

$$N(x) = bE \int_{-h/2}^{+h/2} \varepsilon_{xx}\, \mathrm{d}z \ , \quad M(x) = -bE \int_{-h/2}^{+h/2} \varepsilon_{xx}\, z\, \mathrm{d}z \ . \tag{5.3.8}$$

Now inserting (5.2.43) into (5.3.8), for the resultant axial force $N(x)$ we have

$$\begin{aligned} N(x) &= bE \int_{-h/2}^{+h/2} \frac{\partial u}{\partial x} - z\frac{\partial^2 w}{\partial x^2}\, \mathrm{d}z \\ &= bhE\frac{\partial u}{\partial x} - bE\frac{\partial^2 w}{\partial x^2} \int_{-h/2}^{+h/2} z\, \mathrm{d}z \\ &= bhE\frac{\partial u}{\partial x} \\ &= EA\frac{\partial u}{\partial x} \ , \end{aligned} \tag{5.3.9}$$

where $A = bh$ is the area of the beam cross-section. Similarly, for the resultant moment $M(x)$,

$$\begin{aligned} M(x) &= -bE \int_{-h/2}^{+h/2} \left(\frac{\partial u}{\partial x} - z \frac{\partial^2 w}{\partial x^2} \right) z \, \mathrm{d}z \\ &= -bE \frac{\partial u}{\partial x} \int_{-h/2}^{+h/2} z \, \mathrm{d}z + bE \frac{\partial^2 w}{\partial x^2} \int_{-h/2}^{+h/2} z^2 \, \mathrm{d}z \\ &= bE \frac{\partial^2 w}{\partial x^2} \left[\frac{1}{3} z^3 \right]_{z=-h/2}^{h/2} \\ &= \frac{bh^3}{12} E \frac{\partial^2 w}{\partial x^2} \\ &= EI \frac{\partial^2 w}{\partial x^2} \,, \end{aligned} \tag{5.3.10}$$

where

$$I \equiv \frac{bh^3}{12} \tag{5.3.11}$$

is the moment of inertia (also known as the second moment of area) of the rectangular cross-section of the beam. Because the first term in the total axial strain (5.2.43) does not depend on z, while the second one depends on z linearly, the effect of the strain on the resultants $N(x)$ and $M(x)$ splits into two parts, the axial displacement affecting only the axial force $N(x)$, and the curvature affecting only the bending moment $M(x)$.

We next explore the Kelvin–Voigt viscoelastic solid, which is the simplest constitutive model for a viscoelastic solid. It consists of a spring and a dashpot connected in parallel, as shown in Fig. 5.14.

Using the rule (5.3.6) for the parallel connection, we have

$$\varepsilon = \varepsilon_E = \varepsilon_\eta \,, \tag{5.3.12}$$

$$\sigma = \sigma_E + \sigma_\eta \,, \tag{5.3.13}$$

where the subscripts indicate the network component the strain or stress belongs to. We simply insert the strain ε, which by (5.3.12) is the same for both network components, into the constitutive laws (5.3.3) and (5.3.4),

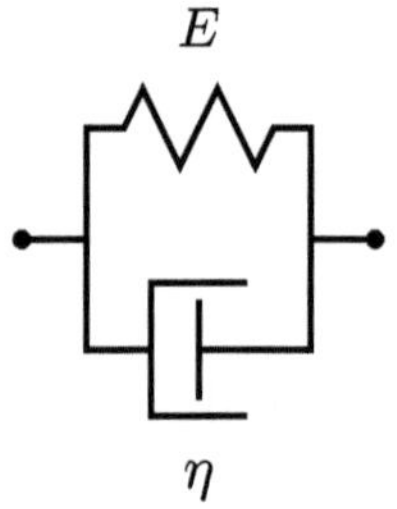

Fig. 5.14 Schematic representation of the Kelvin–Voigt constitutive model

$$\sigma_E = E\,\varepsilon_E = E\,\varepsilon\,,$$
$$\sigma_\eta = \eta\frac{\partial}{\partial t}\varepsilon_\eta = \eta\frac{\partial}{\partial t}\varepsilon\,,$$

obtaining the stresses σ_E and σ_η in terms of the total strain ε. Inserting the results into (5.3.13) yields the stress-strain relation for a Kelvin–Voigt viscoelastic material,

$$\sigma = E\,\varepsilon + \eta\frac{\partial}{\partial t}\varepsilon\,, \tag{5.3.14}$$

where E is the Young's modulus, and η is the viscous modulus. Some authors, such as Sobotka [128], define the retardation time τ (SI unit $[\tau] = \mathrm{s}$),

$$\tau \equiv \frac{\eta}{E}\,, \tag{5.3.15}$$

which characterizes the relative strength of the viscous response to the elastic one. Saksa [119] points out that (5.3.15) is sometimes called a creep time constant [88] or a delay time [159]. Holzapfel [47, p. 280] notes that if the parameter τ relates to the decay of stress and strain in a viscoelastic process, then it is called the relaxation time, and if it is associated with a creeping process, then τ is referred to as the retardation time. Since the Kelvin–Voigt model exhibits creep but not relaxation, we will call τ the retardation time.

Using (5.3.15), we have $\eta = E\tau$, and the constitutive law (5.3.14) becomes

$$\sigma = E\left[1 + \tau\frac{\partial}{\partial t}\right]\varepsilon\,, \tag{5.3.16}$$

which highlights the formal similarity to the linear elastic law (5.3.7).

The simplicity of the Kelvin–Voigt model has its drawbacks. Particularly, as can be seen from Eq. (5.3.16), a Kelvin–Voigt material responds to a sudden change in strain—a Heaviside step function with respect to time—by an infinite (Dirac delta) stress, which is of course unphysical. Also, (5.3.16) shows that the Kelvin–Voigt model responds to a constant strain by a constant stress, that is, it does not model stress relaxation.

However, the model can be useful e.g. for analyzing an idealized steady state of industrial processes, where no sudden changes occur in strains in the material being produced. This helps in understanding the fundamental physics in a simplified context, before all effects from a realistic environment are added in.

Similarly to before, by interpreting the constitutive law (5.3.14) to represent the axial stress σ_{xx} in terms of the axial strain ε_{xx}, the integrals (5.3.2) become

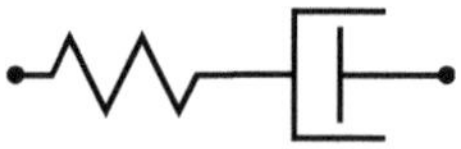

Fig. 5.15 Schematic representation of the Maxwell constitutive model

$$N(x) = bE \int_{-h/2}^{+h/2} \varepsilon_{xx} \, \mathrm{d}z + b\eta \int_{-h/2}^{+h/2} \frac{\partial}{\partial t}(\varepsilon_{xx}) \, \mathrm{d}z \,, \tag{5.3.17}$$

$$M(x) = -bE \int_{-h/2}^{+h/2} \varepsilon_{xx} \, z \, \mathrm{d}z - b\eta \int_{-h/2}^{+h/2} \frac{\partial}{\partial t}(\varepsilon_{xx}) \, z \, \mathrm{d}z \,. \tag{5.3.18}$$

Inserting (5.2.43) we obtain, by performing the exact same sequence of steps as earlier,

$$N(x) = EA\frac{\partial u}{\partial x} + \eta A \frac{\partial^2 u}{\partial x \partial t} = EA\left[1 + \tau \frac{\partial}{\partial t}\right]\frac{\partial u}{\partial x} \,, \tag{5.3.19}$$

$$M(x) = EI\frac{\partial^2 w}{\partial x^2} + \eta I \frac{\partial^3 w}{\partial x^2 \partial t} = EI\left[1 + \tau \frac{\partial}{\partial t}\right]\frac{\partial^2 w}{\partial x^2} \,. \tag{5.3.20}$$

Again, $A = bh$, and I is given by (5.3.11). We assume sufficient regularity such that the indicated differentiations can be performed.

Be aware that the regularity requirements will be very high if (5.3.19) and (5.3.20) are inserted into the combined force balance Eq. (5.1.40), instead of keeping N and M as auxiliary variables. Because Eq. (5.1.40) contains the quantity $\partial^2 M / \partial s^2$, the resulting behavioral equation for w, describing small out-of-plane for vibrations of a Kelvin–Voigt beam, will be a fifth-order partial differential equation.

The other classical two-element configuration is the one where the spring and the dashpot are connected in series. This is known as the Maxwell viscoelastic fluid (see Fig. 5.15). Using the rule (5.3.5) for the serial connection, we have

$$\varepsilon = \varepsilon_E + \varepsilon_\eta \,, \tag{5.3.21}$$

$$\sigma = \sigma_E = \sigma_\eta \,. \tag{5.3.22}$$

Again, the individual constitutive laws for the components are (5.3.3) for the spring, and (5.3.4) for the dashpot:

$$\sigma_E = E \, \varepsilon_E \,, \tag{5.3.23}$$

$$\sigma_\eta = \eta \frac{\partial}{\partial t} \varepsilon_\eta \,. \tag{5.3.24}$$

Differentiating (5.3.21) with respect to time gives us the total strain rate:

$$\frac{\partial}{\partial t}\varepsilon = \frac{\partial}{\partial t}\varepsilon_E + \frac{\partial}{\partial t}\varepsilon_\eta \,. \tag{5.3.25}$$

By differentiating the elastic Eq. (5.3.23) with respect to time and rearranging, we have

$$\frac{\partial}{\partial t}\varepsilon_E = \frac{1}{E}\frac{\partial}{\partial t}\sigma_E \,. \tag{5.3.26}$$

Dividing both sides of the viscous Eq. (5.3.24) by η, we obtain the other term for (5.3.25):

$$\frac{\partial}{\partial t}\varepsilon_\eta = \frac{1}{\eta}\sigma_\eta \,. \tag{5.3.27}$$

Inserting (5.3.26) and (5.3.27) into the total strain rate (5.3.25) yields

$$\frac{\partial}{\partial t}\varepsilon = \frac{1}{E}\frac{\partial}{\partial t}\sigma_E + \frac{1}{\eta}\sigma_\eta \,.$$

Using (5.3.22) and collecting terms, we have the stress–strain relation for a Maxwell viscoelastic fluid:

$$\frac{\partial}{\partial t}\varepsilon = \left[\frac{1}{\eta} + \frac{1}{E}\frac{\partial}{\partial t}\right]\sigma \,.$$

Finally, let us get rid of the reciprocals by multiplying both sides by $E\,\eta$, and swap sides to have σ on the left-hand side and ε on the right-hand side, as for the previously considered models:

$$\left[E + \eta\frac{\partial}{\partial t}\right]\sigma = E\,\eta\frac{\partial}{\partial t}\varepsilon \,. \tag{5.3.28}$$

Equation (5.3.28) is a first-order ordinary differential equation for σ. Instead of having an explicit expression for stress, like in the linear elastic and Kelvin–Voigt models, in the Maxwell model the stress is the solution of this first-order ordinary differential equation.

From (5.3.28) we see that the Maxwell model essentially represents a fluid, because when the material is held at a constant stress σ, then $\partial\varepsilon/\partial t$ is a constant, and thus the strain will creep up indefinitely.

Holding a Maxwell material at a constant strain ε leads to

$$\frac{\partial}{\partial t}\sigma + \tau\sigma = 0 \,,$$

where we have used $\partial\varepsilon/\partial t = 0$ for constant strain, divided both sides by E, and inserted the definition (5.3.15) for τ. The solution is $\sigma(t) = \sigma(0)\exp(-\tau t)$, which shows that the Maxwell model incorporates stress relaxation.

Connecting either a spring and a Kelvin–Voigt component in series, or a spring and a Maxwell component in parallel, we obtain a three-parameter model that describes a solid (see Fig. 5.16). The first variant is the Poynting–Thomson model, and the second variant is the Zener model. These are collectively known as the standard linear solid (SLS), or the 3-parameter solid. The models have equivalent behavior,

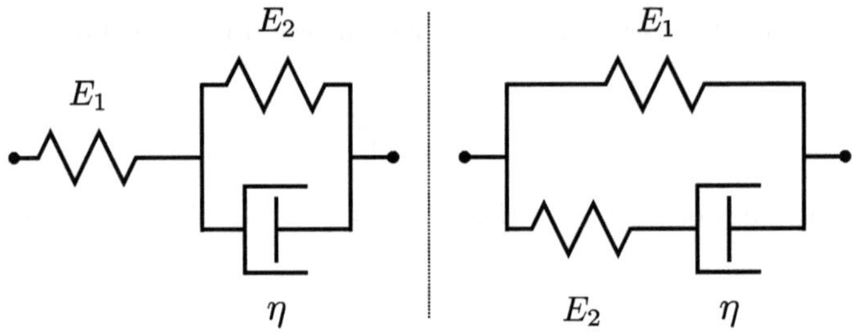

Fig. 5.16 Schematic representation of the standard linear solid (SLS), also known as the 3-parameter solid, constitutive model. *Left*: The Poynting–Thomson variant. *Right*: The Zener variant

but the coefficients in the stress–strain relation are different because the elements are positioned differently in the network.

Let us start with the Poynting–Thomson variant, the left subfigure in Fig. 5.16. Using Eq. (5.3.5) for the serial connection, and the constitutive laws, namely (5.3.3) for the Hookean spring and (5.3.14) for the Kelvin–Voigt component, we have

$$\varepsilon = \varepsilon_E + \varepsilon_{\mathrm{KV}} \,, \tag{5.3.29}$$

$$\sigma = \sigma_E = \sigma_{\mathrm{KV}} \,, \tag{5.3.30}$$

$$\sigma_E = E_1 \, \varepsilon_E \,, \tag{5.3.31}$$

$$\sigma_{\mathrm{KV}} = E_2 \, \varepsilon_{\mathrm{KV}} + \eta \frac{\partial}{\partial t} \varepsilon_{\mathrm{KV}} \,. \tag{5.3.32}$$

The subscripts E and KV refer to the spring and to the Kelvin–Voigt component, respectively. First, we use (5.3.29) and (5.3.31) to express $\varepsilon_{\mathrm{KV}}$ in terms of the total strain ε and total stress σ:

$$\varepsilon_{\mathrm{KV}} = \varepsilon - \varepsilon_E = \varepsilon - \frac{1}{E_1}\sigma_E = \varepsilon - \frac{1}{E_1}\sigma \,. \tag{5.3.33}$$

Then, from (5.3.30), (5.3.32) and (5.3.33), we may express the total stress σ as

$$\sigma = \sigma_{\mathrm{KV}} = E_2 \left(\varepsilon - \frac{1}{E_1}\sigma \right) + \eta \frac{\partial}{\partial t} \left(\varepsilon - \frac{1}{E_1}\sigma \right) \,. \tag{5.3.34}$$

Multiplying both sides by E_1 and expanding the parentheses, we have

$$E_1 \sigma = E_1 E_2 \varepsilon - E_2 \sigma + E_1 \eta \frac{\partial}{\partial t} \varepsilon - \eta \frac{\partial}{\partial t} \sigma \,.$$

Rearranging to place all terms involving σ to the left-hand side and all terms involving ε to the right-hand side, we obtain the stress–strain relation for the Poynting–Thomson variant of a standard linear solid:

$$(E_1 + E_2)\sigma + \eta \frac{\partial}{\partial t}\sigma = E_1 E_2 \varepsilon + E_1 \eta \frac{\partial}{\partial t}\varepsilon \,. \tag{5.3.35}$$

Collecting terms, we may also express this as

$$\left[(E_1 + E_2) + \eta \frac{\partial}{\partial t}\right]\sigma = E_1 \left[E_2 + \eta \frac{\partial}{\partial t}\right]\varepsilon \,. \tag{5.3.36}$$

As in the Maxwell model, this is a first-order ordinary differential equation for σ. However, now the right-hand side contains both the strain ε and the strain rate $\partial\varepsilon/\partial t$.

Next, consider the Zener variant, the right subfigure in Fig. 5.16. With (5.3.6) for the parallel connection and the constitutive laws, (5.3.3) for the spring and (5.3.28) for the Maxwell component, we have

$$\varepsilon = \varepsilon_E = \varepsilon_{\mathrm{M}} \,, \tag{5.3.37}$$

$$\sigma = \sigma_E + \sigma_{\mathrm{M}} \,, \tag{5.3.38}$$

$$\sigma_E = E_1 \, \varepsilon_E \,, \tag{5.3.39}$$

$$\frac{\partial}{\partial t}\varepsilon_{\mathrm{M}} = \left(\frac{1}{\eta} + \frac{1}{E_2}\frac{\partial}{\partial t}\right)\sigma_{\mathrm{M}} \,, \tag{5.3.40}$$

where the subscripts E and M refer to the spring and the Maxwell component, respectively.

Due to (5.3.37), Eq. (5.3.40) represents the total strain rate. We eliminate σ_{M} by solving (5.3.38) for it, inserting (5.3.39) to the result and finally using (5.3.37):

$$\sigma_{\mathrm{M}} = \sigma - \sigma_E = \sigma - E_1 \, \varepsilon_E = \sigma - E_1 \, \varepsilon \,. \tag{5.3.41}$$

Now starting with (5.3.40), using (5.3.37) on the left-hand side and inserting (5.3.41) to the right-hand side yields

$$\frac{\partial}{\partial t}\varepsilon = \left(\frac{1}{\eta} + \frac{1}{E_2}\frac{\partial}{\partial t}\right)(\sigma - E_1 \, \varepsilon) \,.$$

Multiplying both sides by $E_2\eta$, we have

$$E_2\eta \frac{\partial}{\partial t}\varepsilon = \left(E_2 + \eta \frac{\partial}{\partial t}\right)(\sigma - E_1 \, \varepsilon) \,.$$

Moving the terms involving ε from the right-hand side to the left-hand side and then swapping sides, we obtain the stress–strain relation for the *Zener* variant of a standard linear solid:

$$E_2\sigma + \eta \frac{\partial}{\partial t}\sigma = E_1 E_2 \, \varepsilon + (E_1 + E_2)\eta \frac{\partial}{\partial t}\varepsilon \,. \tag{5.3.42}$$

Collecting terms, we obtain

$$\left[E_2 + \eta \frac{\partial}{\partial t}\right] \sigma = \left[E_1 E_2 + (E_1 + E_2)\eta \frac{\partial}{\partial t}\right] \varepsilon \,. \tag{5.3.43}$$

Again, this is a first-order ordinary differential equation for σ. Note that the Poynting–Thomson and Zener variants of the SLS model, represented by Eqs. (5.3.36) and (5.3.43), respectively, produce equations of the same form but with different values for the coefficients. As was noted, this is due to the different placement of the three elementary components (two springs and one dashpot) in these two variants of the SLS model.

Next, let us consider an example of the force balance in a Kelvin–Voigt solid. Often when considering small transverse deformations, especially in the context of ideal strings, the axial resultant force $N(x)$ is simply prescribed as some function that makes physical sense for the situation under consideration. For example, $N(x)$ is taken equal to a prescribed constant, or in the case of a vertical configuration, as a linear function of the axial coordinate to account for a uniform gravitational field in the axial direction.

Indeed, in the case of a string, from (5.2.32) in Sect. 5.2 it is obvious that if we wish to have $\ddot{u} = 0$—i.e. static equilibrium in the axial direction—then in the absence of external axial loading q_x, the axial force must be a constant, $N(x) \equiv N_0$.

Similarly, if there is a constant external axial loading $q_x(x) \equiv g$, e.g. a uniform gravitational field for a string hanging vertically with $+x$ pointing downward, then in order to have $\ddot{u} = 0$, we must have $N(x) = N_0 - gx$. Physically, this means that the differentially small segment of the string at x must support the weight of the part of the string below it.

However, using the kinematic and constitutive relations discussed above, it is possible to link $N(x)$ to other mechanical quantities in the system in a rigorous manner in a fairly general case. This approach is valid also for situations where the form of the axial force field is not known beforehand. Let us illustrate this for a Kelvin–Voigt string in the small-displacement regime.

The stress–strain relation of a Kelvin–Voigt beam or string is given by Eq. (5.3.14). We have the axial stress

$$\sigma_{xx} = E\,\varepsilon_{xx} + \eta \frac{\partial}{\partial t} \varepsilon_{xx} \,.$$

The total axial strain for small displacements, under combined action of a linear displacement and bending in the xz plane, is given by Eq. (5.2.43) in Sect. 5.2, namely,

$$\varepsilon_{xx}(x, z) \approx \frac{\partial u}{\partial x}(x) - z \frac{\partial^2 w}{\partial x^2}(x) \,.$$

Recall that by using the constitutive relation (5.3.14) and the kinematic relation (5.2.43) in the expression for the resultant axial force in (5.1.41) in Sect. 5.1, the result is (5.3.19), as in,

$$N(x) = EA\frac{\partial u}{\partial x} + \eta A\frac{\partial^2 u}{\partial x \partial t} ,$$

where $A = bh$ is the area of the cross-section of the beam or string (at a fixed value of x). Equation (5.3.19) represents the internal axial reaction force of a Kelvin–Voigt material as it experiences an axial displacement field u.

Inserting (5.3.19) into the small-displacement force balance equations for a general ideal string, (5.2.32) and (5.2.33) in Sect. 5.2, we specialize them for a viscoelastic ideal string made out of Kelvin–Voigt material:

$$EA\frac{\partial^2 u}{\partial x^2} + \eta A\frac{\partial^3 u}{\partial x^2 \partial t} + q_x - m\frac{\partial^2 u}{\partial t^2} = 0 , \tag{5.3.44}$$

$$\frac{\partial w}{\partial x}\left(EA\frac{\partial^2 u}{\partial x^2} + \eta A\frac{\partial^3 u}{\partial x^2 \partial t}\right) + \frac{\partial^2 w}{\partial x^2}\left(EA\frac{\partial u}{\partial x} + \eta A\frac{\partial^2 u}{\partial x \partial t}\right) + q_z - m\frac{\partial^2 w}{\partial t^2} = 0 . \tag{5.3.45}$$

Here some care must be taken handling small quantities. The displacements u and w and their derivatives are assumed small. Hence most terms on the left-hand side of the transverse small displacement Eq. (5.3.45) appear to be second-order small. Naively, this leads us to expect that (5.3.45) should reduce to

$$q_z - m\frac{\partial^2 w}{\partial t^2} = 0 ,$$

but this equation describes a free particle subjected to a force. What went wrong?

The key observation is that the axial force N *is not a small quantity*; this property obviously does not change, even if we replace N by its expression (5.3.19). The coefficients E and η in (5.3.19) typically have very large numeric values (e.g., $E = 10^9 \dots 10^{11}$ Pa). Even if $\partial u/\partial x$ by itself is considered a small quantity, the numeric value of the product $EA\,\partial u/\partial x$ is at least of the order of unity, and typically more.

A straightforward application of the standard rules of small quantities thus fails here, because the rules are based on the assumption that constants are of the order of unity, whence multiplication by a constant does not change the order of smallness. The coefficients in the stress-strain relation have such a large magnitude that this classical assumption is violated.

Now we are ready to analyze coupling of displacement components. From (5.3.44) to (5.3.45), we observe that the displacement equations are coupled even for small displacements. Furthermore, (5.3.45) appears nonlinear in the displacements, involving products of derivatives of u and w. However, the coupling is one way only, allowing a single-pass solution process, where furthermore only linear partial differential equations need to be solved.

Equation (5.3.44) can be solved first, obtaining the axial displacement u. Its solution gives access to the axial force N, which can then be used in (5.3.45), allowing its solution for w. Thus it is useful to think of the solution process of (5.3.44) and (5.3.45) as the following sequence of equations:

$$EA\frac{\partial^2 u}{\partial x^2} + \eta A\frac{\partial^3 u}{\partial x^2 \partial t} + q_x - m\frac{\partial^2 u}{\partial t^2} = 0 \,,$$

$$N(x,t) = EA\frac{\partial u}{\partial x} + \eta A\frac{\partial^2 u}{\partial x \partial t} \,,$$

$$\frac{\partial w}{\partial x}\frac{\partial N}{\partial x} + \frac{\partial^2 w}{\partial x^2}N + q_z - m\frac{\partial^2 w}{\partial t^2} = 0 \,.$$

The first equation connects the axial displacement u to the internal axial reaction force N, for this particular combination of kinematic and constitutive models. Recall the general small-displacement force balance of an ideal string, Eq. (5.2.32), which states that $\partial N/\partial x + q_x - m\partial^2 u/\partial t^2 = 0$. The second equation is just (5.3.19), explicitly representing $N = N(u)$. The third equation is the force balance of an ideal string in the transverse direction, Eq. (5.2.33), relating the transverse displacement w and the axial force N.

One first solves u, then determines N based on the obtained u, and finally solves w. The last step for solving w is identical with the usual solution of ideal string problems with a prescribed axial force $N(x)$. The difference is that now the force is determined from the longitudinal (x direction) force balance, utilizing the constitutive and kinematic models.

In summary to this section, following Sobotka [128], we may condense our discussion of uniaxial spring–dashpot linear constitutive models into one equation of the form

$$\Gamma\sigma = \Xi\varepsilon \,, \qquad (5.3.46)$$

where $\Gamma(\dots)$ and $\Xi(\dots)$ are up to first-order linear differential operators, depending on the constitutive model used. Explicitly, we have

$$\left[a_0 + a_1\frac{\partial}{\partial t}\right]\sigma = \left[b_0 + b_1\frac{\partial}{\partial t}\right]\varepsilon \,. \qquad (5.3.47)$$

The coefficients a_0, a_1, b_0 and b_1 for the considered models are given in Table 5.1.

Flügge [41] tabulates linear viscoelastic models for solids and fluids, see Tables 1.2 (pp. 22–23) and 1.3 (p. 25) in the reference. Both of the SLS variants are there listed under the 3-parameter solid.

Table 5.1 Coefficients for the general stress–strain relation for linear constitutive models. Blank entries indicate zeroes. Note placement of springs E_1 and E_2 in the Poynting–Thomson and Zener variants of the standard linear solid (SLS) model, as shown in Fig. 5.16

# comp.	Model	Type	a_0	a_1	b_0	b_1
1	Hookean spring; linear elastic	Solid, elastic	1		E	
1	Newtonian dashpot	Fluid, viscous	1			η
2	Kelvin–Voigt	Solid, viscoelastic	1		E	η
2	Maxwell	Fluid, viscoelastic	E	η		$E\,\eta$
3	Poynting–Thomson (variant of SLS)	Solid, viscoelastic	$E_1 + E_2$	η	$E_1 E_2$	$E_1\eta$
3	Zener (variant of SLS)	Solid, viscoelastic	E_2	η	$E_1 E_2$	$(E_1 + E_2)\eta$

5.4 Modeling of Beams and Panels

Consider a panel, which is defined as a plate undergoing cylindrical deformation. The term panel (or flat panel) has been used by, among others, Bisplinghoff and Ashley [16]. In cylindrical deformation in the xz plane, there is no variation in the displacement in the y direction, and no loading in the y direction. This allows us to re-use the equations of the beam model, requiring just a few appropriate changes, which will be introduced in this section.

Concerning the balance of forces, the planar dynamics work in exactly the same manner as in the case of a beam, so Eqs. (5.1.62) and (5.1.63), and consequently (5.1.65), are valid as they are. Also the kinematic relations, being based on pure geometry, remain valid. We may thus use (5.2.43) and (5.2.44) to express the strains in the small-displacement range.

What must be modified for the panel are the expressions for the force and moment resultants (5.1.41), and the constitutive laws. In the panel case, the relevant quantities are force per unit width, and moment per unit width. Because all our quantities are independent of the width b, we may simply divide the right-hand sides of (5.1.41) by b and define these quantities as the resultants for the panel. Recalling Eq. (5.1.42) for the area integrals for a beam of rectangular cross-section, we obtained a factor of b in front; now this factor simply cancels. We obtain

$$\widehat{N}(s) = \int_{-h/2}^{+h/2} \sigma_{ss}\,\mathrm{d}z\,, \quad \widehat{Q}(s) = \int_{-h/2}^{+h/2} \sigma_{sz}\,\mathrm{d}z\,, \quad \widehat{M}(s) = -\int_{-h/2}^{+h/2} \sigma_{ss}\,z\,\mathrm{d}z\,, \tag{5.4.1}$$

where the hats indicate that these resultants are defined per unit width, and are to be used with the panel model.

Recall the discussion on dimensional analysis at the very beginning. For the beam, the distributed external load $\mathbf{q}(s)$ represented load per unit length. Dividing by the width b, we have

$$\widehat{\mathbf{q}}(s) \equiv \frac{1}{b}\mathbf{q}(s) , \tag{5.4.2}$$

where the left-hand side is now the prescribed distributed external load. The dimension of $\widehat{\mathbf{q}}$ is $[\widehat{\mathbf{q}}] = [1/b][\mathbf{q}] = (1/\mathrm{m})(\mathrm{N/m}) = \mathrm{N/m^2} = \mathrm{Pa}$. Because one of the dividing m represents length and one width, physically $\widehat{\mathbf{q}}$ represents force per unit area, that is, pressure.

Similarly, by dimensional analysis, for the panel we define m as mass per unit area (or area density), $[m] = \mathrm{kg/m^2}$. In the context of the paper industry, this is often expressed in $\mathrm{g/m^2}$ and called the grammage.

If we wish $\widehat{\mathbf{q}}$ to represent gravity, observe that similarly to the original case, we may again choose $\widehat{\mathbf{q}} = m\mathbf{g}$. Because now $[m] = \mathrm{kg/m^2}$ and the vector $\mathbf{g}$ still represents gravitational acceleration, $[\mathbf{g}] = \mathrm{m/s^2}$, we have $[\widehat{\mathbf{q}}] = [m][\mathbf{g}] = (\mathrm{kg/m^2})(\mathrm{m/s^2}) = \mathrm{N/m^2}$, which is the correct dimension for $\widehat{\mathbf{q}}$.

The final conversions to be performed are the constitutive relations connecting strains and stresses. Again, we are only interested in the axial stress, but to do this rigorously, we must take into account the three-dimensional nature of the object being modelled, and use the appropriate three-dimensional constitutive relations.

In the following we will consider some constitutive laws for the panel model. For stable operation in paper industry applications, the variants of the Kelvin–Voigt model will be the most interesting due to their combination of simplicity and just enough detail to capture some viscoelastic behavior.

At first, we will consider isotropic linear elastic panel. As can be found in textbooks on elasticity (e.g., [136, p. 8 ff.], [75, p. 11 ff.], [38, p. 60 ff.]), the three-dimensional stress–strain response of a linear elastic material with Young's modulus E and Poisson ratio ν is, in compliance form,

$$\begin{bmatrix} \varepsilon_{xx} \\ \varepsilon_{yy} \\ \varepsilon_{zz} \\ \varepsilon_{yz} \\ \varepsilon_{zx} \\ \varepsilon_{xy} \end{bmatrix} = \frac{1}{E} \begin{bmatrix} 1 & -\nu & -\nu & & & \\ -\nu & 1 & -\nu & & & \\ -\nu & -\nu & 1 & & & \\ & & & 1+\nu & & \\ & & & & 1+\nu & \\ & & & & & 1+\nu \end{bmatrix} \begin{bmatrix} \sigma_{xx} \\ \sigma_{yy} \\ \sigma_{zz} \\ \sigma_{yz} \\ \sigma_{zx} \\ \sigma_{xy} \end{bmatrix} , \tag{5.4.3}$$

which is, essentially, a formalization of the elastic Poisson effect. For example, the normal strain along the x axis, ε_{xx}, is affected by a stretching force along the x axis (which causes a stress $\sigma_{xx} > 0$), but due to the Poisson effect, also by stretching forces along the y and z axes. We then require that the total strain is a linear superposition of these strains. Because most materials contract in the perpendicular directions when stretched along an axis, stretching along y (i.e., $\sigma_{yy} > 0$) causes compression, that is, a negative strain, along x, hence the minus signs in the nondiagonal terms. In this interpretation, the Poisson ratio ν can be understood as just an arbitrary coefficient characterizing the strength of the Poisson effect, while Young's modulus characterizes the overall stiffness. The $1 + \nu$ in the shear terms is obtained by considering a pure shear, where $\sigma_{xx} + \sigma_{yy} + \sigma_{zz} = 0$, and then using the stress-strain relation in the axial direction. For details, see the aforementioned textbooks.

We note in passing that for the isotropic elastic solid, the incompressibility condition $\nu = 1/2$ is easily derived from (5.4.3). For orthotropic materials, to be discussed later, the issue is much more complex. Readers interested in the conditions for incompressibility in orthotropic elastic materials are advised to look at the study by Itskov and Aksel [53], which concentrates exclusively on this particular issue.

Consider now the isotropic case. The volume of a small rectangular piece of the material, in the neutral state, is $V_1 = \Delta x\, \Delta y\, \Delta z$. In a state with nonzero normal strains, the volume becomes $V_2 = (1 + \varepsilon_{xx})\Delta x\, (1 + \varepsilon_{yy})\Delta y\, (1 + \varepsilon_{zz})\Delta z$. Applying a uniaxial stress, with only $\sigma_{xx} \neq 0$, by (5.4.3) we have $\varepsilon_{xx} = \sigma_{xx}/E$, $\varepsilon_{yy} = -\nu\sigma_{xx}/E = -\nu\varepsilon_{xx}$ and $\varepsilon_{zz} = -\nu\varepsilon_{xx}$. Thus the volume ratio V_2/V_1 is

$$\begin{aligned} \frac{V_2}{V_1} &= (1 + \varepsilon_{xx})(1 - \nu\varepsilon_{xx})^2 \\ &= (1 + \varepsilon_{xx})(1 - 2\nu\varepsilon_{xx} + \nu^2\varepsilon_{xx}^2) \\ &= 1 - 2\nu\varepsilon_{xx} + \nu^2\varepsilon_{xx}^2 + x - 2\nu\varepsilon_{xx}^2 + \nu^2\varepsilon_{xx}^3 \\ &= 1 + (1 - 2\nu)\varepsilon_{xx} + \nu(1 - 2\nu)\varepsilon_{xx}^2 + \nu^2\varepsilon_{xx}^3 . \end{aligned} \tag{5.4.4}$$

We obtain the relative volume change per unit strain as

$$\frac{1}{\varepsilon_{xx}}\left[\frac{V_2}{V_1} - 1\right] = (1 - 2\nu) + \nu(1 - 2\nu)\varepsilon_{xx} + \nu^2\varepsilon_{xx}^2 . \tag{5.4.5}$$

Taking the small-strain limit $\varepsilon_{xx} \to 0$, the right-hand side vanishes if and only if the parameter $\nu = 1/2$, which was to be shown.

For an isotropic elastic material, the shear modulus is

$$G = \frac{E}{2(1 + \nu)} , \tag{5.4.6}$$

so in the entries corresponding to shear in (5.4.3), we may write $(1 + \nu)/E = 1/2G$. The resulting scaling of G by a factor of 2 accounts for the fact that the left-hand side of (5.4.3) is given in terms of the strain tensor components ε_{jk} ($j = x, y, z$ and $k = x, y, z$) instead of engineering shear strains $\gamma_{jk} = 2\varepsilon_{jk}$ (where $k \neq j$). Note that the normal (xx, yy, zz) and shear (yz, zx and xy) effects are decoupled in (5.4.3).

In our case of a thin Kirchhoff–Love plate occupying the xy plane, we require $\sigma_{zz} = \sigma_{yz} = \sigma_{zx} = 0$. Note that σ_{zz}, σ_{yz} and σ_{zx} do not appear in our force and moment balance equations. In linear elasticity, this is known as a plane stress configuration, and the typical application is a thin slab that is stressed only in its plane. From (5.4.3), we see that $\sigma_{yz} = \sigma_{zx} = 0$ immediately implies also $\varepsilon_{yz} = \varepsilon_{zx} = 0$.

From the Kirchhoff hypotheses, we have $\varepsilon_{zz} = 0$, since thickness changes were assumed negligible. Strictly speaking, this is an approximation not rigorously compatible with the three-dimensional stress–strain relation (5.4.3). Under the requirement $\sigma_{zz} = 0$, the third equation becomes $\varepsilon_{zz} = -(E/\nu)\,(\sigma_{xx} + \sigma_{yy})$. Thus, $\varepsilon_{zz} = 0$ implies $\sigma_{xx} + \sigma_{yy} = 0$; however, Kirchhoff–Love plate theory makes no such requirement on the axial stresses.

We then drop the equations concerning $\varepsilon_{zz}, \varepsilon_{yz}, \varepsilon_{zx}$ and the corresponding stresses, considering the in-plane quantities only. We are left with the equations

$$\begin{bmatrix} \varepsilon_{xx} \\ \varepsilon_{yy} \\ \varepsilon_{xy} \end{bmatrix} = \frac{1}{E} \begin{bmatrix} 1 & -\nu & \\ -\nu & 1 & \\ & & 1+\nu \end{bmatrix} \begin{bmatrix} \sigma_{xx} \\ \sigma_{yy} \\ \sigma_{xy} \end{bmatrix} . \tag{5.4.7}$$

Note that the shear strain ε_{zx} depends only on σ_{zx}, and both the strain and stress tensors are symmetric, $\varepsilon_{xz} = \varepsilon_{zx}, \sigma_{xz} = \sigma_{zx}$. Thus, if we wish, we may later modify our assumptions to include the shear effect of linear displacement, that is, shear strain (5.2.44), without affecting the present consideration.

The stiffness form of (5.4.7) is the inverse,

$$\begin{bmatrix} \sigma_{xx} \\ \sigma_{yy} \\ \sigma_{xy} \end{bmatrix} = E \begin{bmatrix} \dfrac{1}{1-\nu^2} & \dfrac{\nu}{1-\nu^2} & \\ \dfrac{\nu}{1-\nu^2} & \dfrac{1}{1-\nu^2} & \\ & & \dfrac{1}{1+\nu} \end{bmatrix} \begin{bmatrix} \varepsilon_{xx} \\ \varepsilon_{yy} \\ \varepsilon_{xy} \end{bmatrix} , \tag{5.4.8}$$

obtained simply by inverting the matrix. In a cylindrical deformation, only ε_{xx} is nonzero. The axial stress is then

$$\sigma_{xx} = \frac{E}{1-\nu^2}\,\varepsilon_{xx} . \tag{5.4.9}$$

Equation (5.4.9) is the axial stress–strain response of a linear elastic panel. Compared to the linear elastic beam, Eq. (5.3.7), we see that a factor of $1/(1-\nu^2)$ has been introduced, due to the fact that the object being modelled is a plate instead of a beam.

We may now insert the constitutive law (5.4.9), and the kinematic expression for the axial strain, Eq. (5.2.43), into the resultant force and moment integrals (5.3.8), and divide the equations by the width b. We obtain the resultant axial force and resultant moment, per unit width of the linear elastic panel, as

$$\widehat{N}(x) = \frac{Eh}{1-\nu^2}\frac{\partial u}{\partial x} , \tag{5.4.10}$$

$$\widehat{M}(x) = \frac{E\widehat{I}}{1-\nu^2}\frac{\partial^2 w}{\partial x^2} , \tag{5.4.11}$$

where we have defined

$$\widehat{I} \equiv \frac{I}{b} \equiv \frac{1}{b}\cdot\frac{bh^3}{12} = \frac{h^3}{12} , \tag{5.4.12}$$

and I is given by (5.3.11). The quantity $\widehat{I}$ represents the second moment of area per unit width of the panel.

The expression

$$D \equiv \frac{E\widehat{I}}{1-\nu^2} = \frac{Eh^3}{12(1-\nu^2)} , \tag{5.4.13}$$

which appears in (5.4.11), is called the flexural rigidity [135], cylindrical rigidity, or bending rigidity.

For orthotropic elastic materials, or in other words, materials having three orthogonal principal axes, each with its own values for the elastic parameters, the relation (5.4.3) is replaced by

$$\begin{bmatrix} \varepsilon_{xx} \\ \varepsilon_{yy} \\ \varepsilon_{zz} \\ \varepsilon_{yz} \\ \varepsilon_{zx} \\ \varepsilon_{xy} \end{bmatrix} = \begin{bmatrix} \frac{1}{E_x} & -\frac{\nu_{yx}}{E_y} & -\frac{\nu_{zx}}{E_z} & & & \\ -\frac{\nu_{xy}}{E_x} & \frac{1}{E_y} & -\frac{\nu_{zy}}{E_z} & & & \\ -\frac{\nu_{xz}}{E_x} & -\frac{\nu_{yz}}{E_y} & \frac{1}{E_z} & & & \\ & & & \frac{1}{2G_{yz}} & & \\ & & & & \frac{1}{2G_{zx}} & \\ & & & & & \frac{1}{2G_{xy}} \end{bmatrix} \begin{bmatrix} \sigma_{xx} \\ \sigma_{yy} \\ \sigma_{zz} \\ \sigma_{yz} \\ \sigma_{zx} \\ \sigma_{xy} \end{bmatrix} . \tag{5.4.14}$$

The indexing convention for the Poisson ratios is that ν_{jk} represents the effect that stretching applied on axis j has in the direction of axis k. For example, ν_{yx} represents the response that stretching applied on the y axis causes in the x direction.

Elastic compatibility (essentially, symmetry of the compliance matrix) requires that

$$E_j \nu_{kj} = E_k \nu_{jk} , \tag{5.4.15}$$

where $j = x, y, z$ and $k = x, y, z$, and $k \neq j$; no summation over repeated indices. Equation (5.4.15) is sometimes called the Maxwell relation (unrelated to the Maxwell material model).

The shear compliances $1/2G_{jk}$ account for the fact that the left-hand side is given in terms of the strain tensor components ε_{jk} instead of engineering shear strains $\gamma_{jk} = 2\varepsilon_{jk}$ (where $k \neq j$).

When data is not available, or if one wishes to simplify the model, the shear modulus G_{jk} is sometimes approximated by using geometric averages of the appropriate Young moduli and Poisson ratios (no summation over repeated indices):

$$G_{jk} \approx \frac{\sqrt{E_j E_k}}{2(1+\sqrt{\nu_{jk}\nu_{kj}})} . \tag{5.4.16}$$

Equation (5.4.16) is called the Huber value for the shear modulus, after M. T. Huber, who originally suggested this approximation in Huber [50]. If $E_k = E_j$ and $\nu_{kj} = \nu_{jk}$, Eq. (5.4.16) reduces to the isotropic case (5.4.6).

In the bending of Kirchhoff–Love plates, including the case of axially moving plates, approximating the shear modulus by (5.4.16) causes (after an appropriate scaling of coordinate axes) the behavioral equation of plate bending for the orthotropic model to reduce to the same form as in the isotropic case; see Timoshenko and Woinowsky-Krieger [135], Banichuk et al. [7].

For an orthotropic linear elastic Kirchhoff–Love plate, the compliance Eq. (5.4.14) becomes

$$\begin{bmatrix} \varepsilon_{xx} \\ \varepsilon_{yy} \\ \varepsilon_{xy} \end{bmatrix} = \begin{bmatrix} \dfrac{1}{E_x} & -\dfrac{\nu_{yx}}{E_y} & \\ -\dfrac{\nu_{xy}}{E_x} & \dfrac{1}{E_y} & \\ & & \dfrac{1}{2G_{xy}} \end{bmatrix} \begin{bmatrix} \sigma_{xx} \\ \sigma_{yy} \\ \sigma_{xy} \end{bmatrix} . \tag{5.4.17}$$

Inverting (5.4.17), we have the stiffness form

$$\begin{bmatrix} \sigma_{xx} \\ \sigma_{yy} \\ \sigma_{xy} \end{bmatrix} = \begin{bmatrix} \dfrac{E_x}{1-\nu_{xy}\nu_{yx}} & \dfrac{E_x \nu_{yx}}{1-\nu_{xy}\nu_{yx}} & \\ \dfrac{E_y \nu_{xy}}{1-\nu_{xy}\nu_{yx}} & \dfrac{E_y}{1-\nu_{xy}\nu_{yx}} & \\ & & 2G_{xy} \end{bmatrix} \begin{bmatrix} \varepsilon_{xx} \\ \varepsilon_{yy} \\ \varepsilon_{xy} \end{bmatrix} . \tag{5.4.18}$$

Therefore, in a cylindrical deformation, where only ε_{xx} is nonzero, the axial stress–strain relation for an *orthotropic linear elastic panel* is

$$\sigma_{xx} = \frac{E_x}{1-\nu_{xy}\nu_{yx}} \varepsilon_{xx} . \tag{5.4.19}$$

Note the factor of $1/(1-\nu_{xy}\nu_{yx})$ when compared to a beam.

For the resultant forces and moments per unit width, we have

$$\widehat{N}(x) = \frac{E_x h}{1-\nu_{xy}\nu_{yx}} \frac{\partial u}{\partial x} , \tag{5.4.20}$$

$$\widehat{M}(x) = \frac{E_x \widehat{I}}{1-\nu_{xy}\nu_{yx}} \frac{\partial^2 w}{\partial x^2} , \tag{5.4.21}$$

where $\widehat{I}$ is given by (5.4.12).

By analogy with the isotropic linear elastic case (5.4.9), for a Kelvin–Voigt panel we have

$$\sigma_{xx} = \frac{E}{1-\nu^2} \varepsilon_{xx} + \frac{\eta}{1-\mu^2} \frac{\partial}{\partial t} \varepsilon_{xx} , \tag{5.4.22}$$

where μ is the viscous Poisson ratio, a viscous analogue of the classical elastic Poisson ratio. Again, (5.4.22) is of the same form as (5.3.14), but corrective factors have appeared due to the fact that the object modelled is a plate instead of a beam.

The Kelvin–Voigt model is mainly relevant for analyzing the steady-state operation of industrial processes, such as paper machines, where the singular stress response of the model to fast transient strains does not present problems.

In the Kelvin–Voigt solid, viscous effects affect the volumetric response. As was noted, this is not the case for polymers or Newtonian fluids. The constitutive law (5.4.22) is valid for paper materials; see Uesaka et al. [138], Thorpe [133] for an orthotropic two-dimensional in-plane version, also discussed in Sobotka [128]. In addition, this model has also been used in Kurki et al. [69–71], Kurki [68], Kurki and Lehtinen [72], Tang and Chen [132], Saksa [119].

The resultants per unit width become

$$\widehat{N}(x) = \frac{Eh}{1-\nu^2}\frac{\partial u}{\partial x} + \frac{\eta h}{1-\mu^2}\frac{\partial^2 u}{\partial x \partial t} = h\left[\frac{E}{1-\nu^2} + \frac{\eta}{1-\mu^2}\frac{\partial}{\partial t}\right]\frac{\partial u}{\partial x}\,, \tag{5.4.23}$$

$$\widehat{M}(x) = \frac{E\widehat{I}}{1-\nu^2}\frac{\partial^2 w}{\partial x^2} + \frac{\eta \widehat{I}}{1-\mu^2}\frac{\partial^3 w}{\partial x^2 \partial t} = \widehat{I}\left[\frac{E}{1-\nu^2} + \frac{\eta}{1-\mu^2}\frac{\partial}{\partial t}\right]\frac{\partial^2 w}{\partial x^2} \equiv \left[D + D_{\mathrm{VE}}\frac{\partial}{\partial t}\right]\frac{\partial^2 w}{\partial x^2}\,, \tag{5.4.24}$$

where $\widehat{I}$ is given by (5.4.12). In analogy with the (elastic) flexural rigidity D given by Eq. (5.4.13), in Eq. (5.4.24) we have defined the viscoelastic flexural rigidity as

$$D_{\mathrm{VE}} \equiv \frac{\eta \widehat{I}}{1-\mu^2} = \frac{\eta h^3}{12(1-\mu^2)}\,. \tag{5.4.25}$$

Instead of relying on similarity of form with the isotropic case, the constitutive law for the Kelvin–Voigt panel, Eq. (5.4.22), can be explicitly derived as follows. Generalizing the one-dimensional case, we may think of a three-dimensional Kelvin–Voigt material as an abstract parallel network. In analogy with the three-dimensional stress–strain response of a linear elastic material, Eq. (5.4.3), we introduce a three-dimensional Newtonian fluid that follows a relation of the same form between its strain rate and stress:

$$\frac{\partial}{\partial t}\begin{bmatrix} \varepsilon_{xx} \\ \varepsilon_{yy} \\ \varepsilon_{zz} \\ \varepsilon_{yz} \\ \varepsilon_{zx} \\ \varepsilon_{xy} \end{bmatrix} = \frac{1}{\eta}\begin{bmatrix} 1 & -\mu & -\mu & & & \\ -\mu & 1 & -\mu & & & \\ -\mu & -\mu & 1 & & & \\ & & & 1+\mu & & \\ & & & & 1+\mu & \\ & & & & & 1+\mu \end{bmatrix}\begin{bmatrix} \sigma_{xx} \\ \sigma_{yy} \\ \sigma_{zz} \\ \sigma_{yz} \\ \sigma_{zx} \\ \sigma_{xy} \end{bmatrix}. \tag{5.4.26}$$

This fluid will play the role of a three-dimensional Newtonian dashpot, while the linear elastic material plays the role of a three-dimensional spring.

The fluid (5.4.26) responds to volumetric stresses (corresponding to normal strain rates $\partial\varepsilon_{xx}/\partial t$, $\partial\varepsilon_{yy}/\partial t$ and $\partial\varepsilon_{zz}/\partial t$) in a viscous manner, so as was pointed out above, it is appropriate for modeling paper materials, but not for polymers, among others. Although this is not the standard Newtonian fluid that is used in, for example, the Navier–Stokes equations, the relation between strain rate and stress is linear, hence the label.

Following the exact same steps as in the linear elastic case, that is, first reducing (5.4.26) to the plane case for a Kirchhoff–Love plate, and then inverting to obtain an explicit expression for the stress, we have

$$\begin{bmatrix} \sigma_{xx} \\ \sigma_{yy} \\ \sigma_{xy} \end{bmatrix} = \eta \begin{bmatrix} \dfrac{1}{1-\mu^2} & \dfrac{\mu}{1-\mu^2} & \\ \dfrac{\mu}{1-\mu^2} & \dfrac{1}{1-\mu^2} & \\ & & \dfrac{1}{1+\mu} \end{bmatrix} \frac{\partial}{\partial t} \begin{bmatrix} \varepsilon_{xx} \\ \varepsilon_{yy} \\ \varepsilon_{xy} \end{bmatrix} . \tag{5.4.27}$$

Equation (5.4.27) describes the stress–strain response of a Newtonian viscous plate, which is a two-dimensional analogue of a Newtonian dashpot. Compare with the one-dimensional strain–stress response of a dashpot, Eq. (5.3.4).

Now that we have the stress–strain responses of the elastic and viscous plate components (5.4.8) and (5.4.27), we may use rules (5.3.5) and (5.3.6) to represent the connections of such components in a network. Essentially, we apply the one-dimensional rules (5.3.5) and (5.3.6) componentwise in the stress and strain tensors, thus justifying their generalization into several space dimensions. Note that we may still use one-dimensional schematic illustrations such as Figs. 5.13, 5.14, 5.15 and 5.16 to represent the structure of the material; the picture is then understood as referring to each tensor component independently.

In a parallel connection, the strains are the same, while the stresses sum. Explicitly, (5.3.6) says that

$$\varepsilon_E = \varepsilon_\eta = \varepsilon ,$$

which in several space dimensions actually means

$$(\varepsilon_E)_{jk} = (\varepsilon_\eta)_{jk} = (\varepsilon)_{jk} ,$$

where $j = x, y, z$ and $k = x, y, z$. Similarly, for the stresses (5.3.6) states

$$\sigma = \sigma_E + \sigma_\eta ,$$

which means

$$(\sigma)_{jk} = (\sigma_E)_{jk} + (\sigma_\eta)_{jk} ,$$

where $j = x, y, z$ and $k = x, y, z$. Working in any appropriate notation (e.g. tensor notation or vector-matrix notation), the componentwise relations are enforced automatically when performing algebraic manipulation on the whole entities. Hence, in practice, we may use the original forms of the rules (5.3.5) and (5.3.6) directly.

Summing the stiffness forms of the stress–strain responses, Eqs. (5.4.8) and (5.4.27), each representing the stress in one component of the parallel network, we obtain

$$\begin{bmatrix} \sigma_{xx} \\ \sigma_{yy} \\ \sigma_{xy} \end{bmatrix} = E \begin{bmatrix} \frac{1}{1-\nu^2} & \frac{\nu}{1-\nu^2} & \\ \frac{\nu}{1-\nu^2} & \frac{1}{1-\nu^2} & \\ & & \frac{1}{1+\nu} \end{bmatrix} \begin{bmatrix} \varepsilon_{xx} \\ \varepsilon_{yy} \\ \varepsilon_{xy} \end{bmatrix} + \eta \begin{bmatrix} \frac{1}{1-\mu^2} & \frac{\mu}{1-\mu^2} & \\ \frac{\mu}{1-\mu^2} & \frac{1}{1-\mu^2} & \\ & & \frac{1}{1+\mu} \end{bmatrix} \frac{\partial}{\partial t} \begin{bmatrix} \varepsilon_{xx} \\ \varepsilon_{yy} \\ \varepsilon_{xy} \end{bmatrix} . \tag{5.4.28}$$

Therefore, in a cylindrical deformation, where only ε_{xx} is nonzero, we have

$$\sigma_{xx} = \frac{E}{1-\nu^2}\varepsilon_{xx} + \frac{\eta}{1-\mu^2}\frac{\partial}{\partial t}\varepsilon_{xx} ,$$

which is exactly (5.4.22), the stress–strain relation of a Kelvin–Voigt panel.

By analogy with the orthotropic linear elastic case (5.4.19), or by repeating the above derivation using orthotropic materials, we have

$$\sigma_{xx} = \frac{E_x}{1-\nu_{xy}\nu_{yx}}\varepsilon_{xx} + \frac{\eta_x}{1-\mu_{xy}\mu_{yx}}\frac{\partial}{\partial t}\varepsilon_{xx} , \tag{5.4.29}$$

where now also the viscous material parameters are orthotropic (see [128, 133, 138]).

Equation (5.4.29) leads to the resultants

$$\widehat{N}(x) = h\left[\frac{E_x}{1-\nu_{xy}\nu_{yx}} + \frac{\eta_x}{1-\mu_{xy}\mu_{yx}}\frac{\partial}{\partial t}\right]\frac{\partial u}{\partial x} , \tag{5.4.30}$$

$$\widehat{M}(x) = \widehat{I}\left[\frac{E_x}{1-\nu_{xy}\nu_{yx}} + \frac{\eta_x}{1-\mu_{xy}\mu_{yx}}\frac{\partial}{\partial t}\right]\frac{\partial^2 w}{\partial x^2} , \tag{5.4.31}$$

where $\widehat{I}$ is given by (5.4.12).

Finally, considering that all the considered models are linear in ε and σ, and noting that in the previous cases when going from the beam to the panel, the process has essentially resulted in the following scalings for the Young and viscous moduli:

$$E \mapsto \frac{E}{1-\nu^2} , \quad \eta \mapsto \frac{\eta}{1-\mu^2} , \tag{5.4.32}$$

we will take a shortcut, and proceed by analogy when considering the standard linear solid panel.

Rigorously, the pattern (5.4.32) is a consequence of the two-dimensional stress–strain relations of the linear elastic material, Eq. (5.4.8), and that of the Newtonian fluid, Eq. (5.4.27). Each term except shear comes with these scaling factors, and because in the panel only ε_{xx} and $\partial\varepsilon_{xx}/\partial t$ are nonzero, the rest of the terms are effectively discarded, and we are left with simple scalings for ε_{xx} and $\partial\varepsilon_{xx}/\partial t$.

Be aware that the scaling (5.4.32) is specific to the two-dimensional case. For example, for a linear elastic material filling a general three-dimensional region of space, inverting the compliance form (5.4.3) produces the stiffness form

$$\begin{bmatrix} \sigma_{xx} \\ \sigma_{yy} \\ \sigma_{zz} \\ \sigma_{yz} \\ \sigma_{zx} \\ \sigma_{xy} \end{bmatrix} = \frac{E}{(1+\nu)(1-2\nu)} \begin{bmatrix} 1-\nu & \nu & \nu & & & \\ \nu & 1-\nu & \nu & & & \\ \nu & \nu & 1-\nu & & & \\ & & & 1-2\nu & & \\ & & & & 1-2\nu & \\ & & & & & 1-2\nu \end{bmatrix} \begin{bmatrix} \varepsilon_{xx} \\ \varepsilon_{yy} \\ \varepsilon_{zz} \\ \varepsilon_{yz} \\ \varepsilon_{zx} \\ \varepsilon_{xy} \end{bmatrix},$$

suggesting a scaling factor of $(1-\nu)/[(1+\nu)(1-2\nu)]$ for the corresponding case where only ε_{xx} is nonzero. The reduction to two dimensions is thus essential to produce correct stress–strain responses for the panel and plate models in the plane stress configuration.

The final linear model to be treated is the standard linear solid panel. Applying the scalings (5.4.32) to (5.3.36), for an isotropic Poynting–Thomson panel, we have the axial stress–strain relation

$$\left[\frac{E_1+E_2}{1-\nu^2} + \frac{\eta}{1-\mu^2}\frac{\partial}{\partial t}\right]\sigma_{xx} = \frac{E_1}{1-\nu^2}\left[\frac{E_2}{1-\nu^2} + \frac{\eta}{1-\mu^2}\frac{\partial}{\partial t}\right]\varepsilon_{xx}\,, \tag{5.4.33}$$

and from (5.4.32) to (5.3.43), for an isotropic Zener panel, we have

$$\left[\frac{E_2}{1-\nu^2} + \frac{\eta}{1-\mu^2}\frac{\partial}{\partial t}\right]\sigma_{xx} = \left[\frac{E_1E_2}{(1-\nu^2)^2} + \frac{E_1+E_2}{1-\nu^2}\frac{\eta}{1-\mu^2}\frac{\partial}{\partial t}\right]\varepsilon_{xx}\,. \tag{5.4.34}$$

Note that the subscripts 1 and 2 simply label the springs in the material element (no relation to material axes).

For an orthotropic Poynting–Thomson panel, we have

$$\left[\frac{E_{x1}+E_{x2}}{1-\nu_{xy}\nu_{yx}} + \frac{\eta_x}{1-\mu_{xy}\mu_{yx}}\frac{\partial}{\partial t}\right]\sigma_{xx} = \frac{E_{x1}}{1-\nu_{xy}\nu_{yx}}\left[\frac{E_{x2}}{1-\nu_{xy}\nu_{yx}} + \frac{\eta_x}{1-\mu_{xy}\mu_{yx}}\frac{\partial}{\partial t}\right]\varepsilon_{xx}\,, \tag{5.4.35}$$

and for an orthotropic Zener panel,

$$\left[\frac{E_{x2}}{1-\nu_{xy}\nu_{yx}} + \frac{\eta_x}{1-\mu_{xy}\mu_{yx}}\frac{\partial}{\partial t}\right]\sigma_{xx} = \left[\frac{E_{x1}E_{x2}}{(1-\nu_{xy}\nu_{yx})^2} + \frac{E_{x1}+E_{x2}}{1-\nu_{xy}\nu_{yx}}\frac{\eta_x}{1-\mu_{xy}\mu_{yx}}\frac{\partial}{\partial t}\right]\varepsilon_{xx}\,. \tag{5.4.36}$$

The Young moduli E_{x1} and E_{x2}, similarly to the earlier orthotropic x-directional Young modulus E_x, refer to springs 1 and 2 related to the x material axis of the orthotropic material.

5.5 Modeling of Axially Moving Materials

When working with axially moving materials, we interpret the force and moment balance Eqs. (5.1.62) and (5.1.63) as describing the physics in the axially co-moving frame, where the constant-velocity axial motion is cancelled out. The co-moving frame is an inertial reference frame, so the principle of Galilean relativity applies. In the co-moving frame, the physics of the axially moving beam is identical to that of a classical stationary beam.

However, because in applications of axially moving materials we are mainly interested in an Eulerian description—that is, what happens in a stationary control volume as the solid material flows through it—we must transform our equations into the Eulerian frame.

To do this, it is convenient to use a moving reference state that follows the axial motion. After the terminology of Koivurova and Salonen [63], this is called the mixed Eulerian–Lagrangean approach. The coordinate parameters to the functions describing the displacement are Eulerian, while the value of the displacement is given in terms of the axially co-moving description.

The problems of axially moving materials share some similarities with fluid mechanics. The Eulerian governing equations are transport-dominated, which implies a preferred direction of information flow. In simple cases such as axially moving strings (with or without damping), the equations are easily shown to be hyperbolic (see e.g. [54, pp. 58, 69]). The mixed Eulerian–Lagrangean approach, having a second reference configuration beside the material one, is related to the arbitrary Lagrangean–Eulerian (ALE) description, which is often used in finite element treatment of problems with a moving mesh. See the book by Donea and Huerta [33] for a compact description of ALE, and a detailed exposition of numerical stabilization techniques for transport-dominated problems.

Previously in this chapter, the symbol $\mathbf{x}$ was used to denote the Cartesian position vector of points on the beam. We will now need to consider two Cartesian coordinate systems filling all of space, and for this section, will re-use the symbol $\mathbf{x}$ to denote a general position vector in the Eulerian frame.

We define the coordinate transformation for axially moving materials as

$$\mathbf{x}(\boldsymbol{\xi}, t) \equiv \boldsymbol{\xi} + (V_0 t,\ 0,\ 0)\ , \tag{5.5.1}$$

where

$$\boldsymbol{\xi} \equiv (\xi, y, z) \tag{5.5.2}$$

is a position vector expressed in the axially co-moving coordinates, V_0 is the axial drive velocity (taken constant), and $\mathbf{x}$ is the corresponding position vector in laboratory (stationary, Eulerian) coordinates. Without loss of generality, the axial motion can be taken to occur along the x axis. See Fig. 5.17.

Here $\boldsymbol{\xi}$ and t are the independent variables; $\mathbf{x}$ is taken as a dependent variable. With respect to the axial motion, in the co-moving system, the structure (such as a

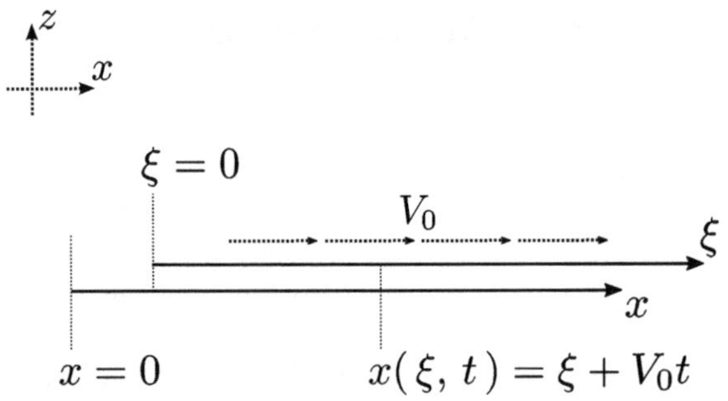

Fig. 5.17 Coordinate systems for an axially moving material. The Eulerian (laboratory, stationary) axial coordinate is denoted by x and the axially co-moving axial coordinate is denoted by ξ. The ξ axis is in axial motion toward $+x$ at a constant velocity V_0. The y and z coordinates are not affected by the axial motion. In the figure, the axes are offset vertically for legibility of the illustration

beam) remains stationary, whereas in the Eulerian system, the structure flows past the observer toward $+x$ at velocity V_0.

The transformation (5.5.1) is one-to-one, hence invertible. As an alternative formulation, we could invert (5.5.1) by moving the $V_0 t$ term, taking $\mathbf{x}$ and t as the independent variables, and $\boldsymbol{\xi}$ as the dependent variable. But from the viewpoint of re-using derivations for equations of motion written in a co-moving frame, the first choice is the appropriate one.

Frame invariance, i.e. the principle that regardless of the chosen coordinate system, the described physical situation is the same, requires that

$$f(\mathbf{x}(\boldsymbol{\xi}, t),\ t) \equiv \widetilde{f}(\boldsymbol{\xi}(\mathbf{x}, t),\ t)\ , \tag{5.5.3}$$

where $f = f(\mathbf{x}, t)$ is any scalar function defined in terms of the Eulerian coordinates, and $\widetilde{f} = \widetilde{f}(\boldsymbol{\xi}, t)$ is the corresponding function defined in terms of the co-moving coordinates.

In order to be able to work with partial differential equations describing axially moving materials, we need a way to convert differentiations in one coordinate system to differentiations in the other. Consider transforming a local time derivative of a differentiable scalar function $\widetilde{f} = \widetilde{f}(\boldsymbol{\xi}, t)$, taken in the co-moving $(\boldsymbol{\xi}, t)$ coordinate system, into how it appears in the Eulerian $(\mathbf{x}, t)$ coordinate system, see Fig. 5.18.

By the frame invariance principle (5.5.3), the chain rule, and the coordinate transformation $\mathbf{x}(\boldsymbol{\xi}, t)$ given by Eq. (5.5.1), we have

$$\begin{aligned} \frac{\partial \widetilde{f}(\boldsymbol{\xi}, t)}{\partial t}\Big|_{\boldsymbol{\xi}=\text{const.}} &= \frac{\partial f(\mathbf{x}(\boldsymbol{\xi}, t),\ t)}{\partial t}\Big|_{\boldsymbol{\xi}=\text{const.}} \\ &= \sum_{j=1}^{3} \frac{\partial x_i}{\partial t}\frac{\partial f}{\partial x_i} + \frac{\partial f}{\partial t}\Big|_{\mathbf{x}=\text{const.}} = V_0 \frac{\partial f}{\partial x} + \frac{\partial f}{\partial t}\Big|_{\mathbf{x}=\text{const.}}\ . \end{aligned} \tag{5.5.4}$$

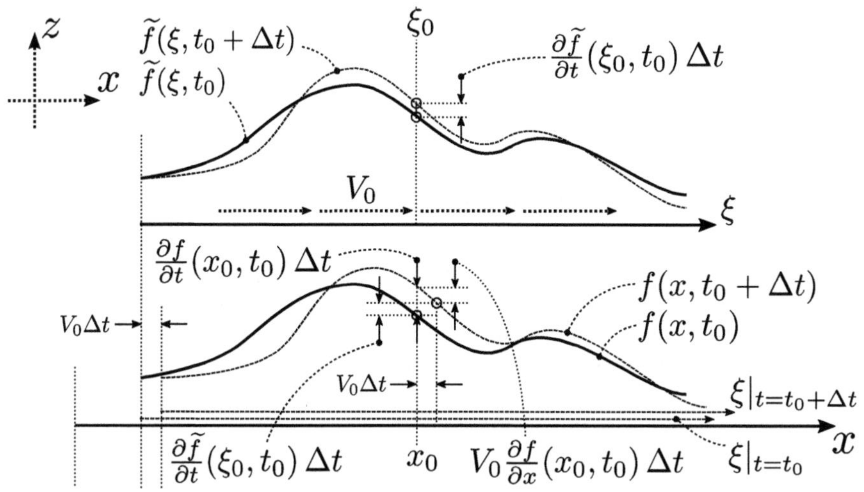

Fig. 5.18 The co-moving derivative, i.e. the material derivative with respect to the co-moving frame. The co-moving ξ axis is in axial motion toward $+x$ at a constant velocity V_0. The Eulerian reference point is taken as $x_0 = x(\xi_0, t_0)$. Both co-moving and Eulerian descriptions are illustrated at time instants $t = t_0$ and $t = t_0 + \Delta t$. In each instance, the point $(\xi_0, \widetilde{f}(\xi_0, t))$ is marked with a circle. At x_0, the Eulerian function $f(x_0, t)$ changes in time not only due to $\partial \widetilde{f}/\partial t$, but also due to the axial motion of the ξ coordinate axis. The situation is illustrated over a small finite time interval Δt, using a Taylor expansion up to first order. Recall the rigorous result $\partial \widetilde{f}/\partial t = \partial f/\partial t + V_0 \partial f/\partial x$, and note the signs at the considered point in the illustration: $V_0 > 0$, $\partial \widetilde{f}/\partial t > 0$, $\partial f/\partial x < 0$

Motivated by (5.5.4), the material derivative (also known as Lagrange derivative or total derivative) is defined as

$$\frac{\mathrm{d}f}{\mathrm{d}t} \equiv \frac{\partial \widetilde{f}}{\partial t}\Big|_{\xi=\text{const.}} = \frac{\partial f}{\partial t}\Big|_{\mathbf{x}=\text{const.}} + V_0 \frac{\partial f}{\partial x} \tag{5.5.5}$$

for any differentiable function $f = f(\mathbf{x}, t)$.

It must be emphasized that in the context of axially moving materials, to avoid confusion (5.5.5) should be properly called the *(axially) co-moving derivative*, because the axial drive is not the only motion that the material particles experience. The velocity V_0, appearing in (5.5.5), is not the velocity of the material particles, but the axial drive velocity. Contrast this with the usage of the term *material derivative* in fluid mechanics, where it is indeed the *velocity of the material particles* that is used as the velocity field $\partial x_i/\partial t$ in Eq. (5.5.4).

Applying the operator $\mathrm{d}/\mathrm{d}t$ to (5.5.5) again gives the second material derivative (second co-moving derivative),

$$\frac{\mathrm{d}^2 f}{\mathrm{d}t^2} \equiv \left(\frac{\partial}{\partial t} + V_0 \frac{\partial}{\partial x}\right)\left(\frac{\partial f}{\partial t} + V_0 \frac{\partial f}{\partial x}\right) = \frac{\partial^2 f}{\partial t^2} + 2V_0 \frac{\partial^2 f}{\partial x \partial t} + V_0^2 \frac{\partial^2 f}{\partial x^2}\,, \tag{5.5.6}$$

where on the right-hand side we have used the fact that V_0 = constant. In (5.5.6), we have denoted

$$\frac{\partial}{\partial t}(\ldots) \equiv \frac{\partial}{\partial t}(\ldots)|_{\mathrm{x=const.}} \tag{5.5.7}$$

for brevity. The expression (5.5.6) represents the local acceleration in the co-moving (Lagrange) coordinates, as it appears when viewed in the laboratory (Euler) coordinates.

It is illustrative to observe that for the z component of the displacement of the beam, $f = w$, in the small-displacement regime, the three terms on the right-hand side of (5.5.6) physically represent the accelerations of local inertia, the Coriolis effect, and the centrifugal effect, respectively. We have

$$\frac{\partial^2 w}{\partial x \partial t} = \frac{\partial}{\partial t}\frac{\partial w}{\partial x} = \frac{\partial}{\partial t}(\tan\alpha) \approx \frac{\partial}{\partial t}\alpha$$

for small values of α. Hence, $\partial^2 w/\partial x \partial t$ is a linear approximation of local rotation speed. The presence of time-dependent rotation leads to the Coriolis effect.

For the local radius of curvature R, recall Eqs. (5.1.47) and (5.1.60), which yield

$$\frac{1}{R} = \frac{\dfrac{\partial^2 w}{\partial x^2}}{[\,1 + (\dfrac{\partial w}{\partial x})^2\,]^{3/2}}\,.$$

In the small displacement regime, $\partial w/\partial x$ is assumed small, and thus $1/R \approx \partial^2 w/\partial x^2$. Therefore, we have an acceleration that is inversely proportional to the local radius of curvature; hence this is a centrifugal effect. These effects appear although the overall axial motion is purely a constant-velocity linear translation, because the material may undergo local rotation.

In the general case (for any f), we observe the very important fact that the right-hand side of (5.5.6) contains a $V_0^2 \partial^2 f/\partial x^2$ term, which does not explicitly depend on t. When choosing the equations to transform from the co-moving to the Eulerian frame—even if the goal is to analyze a steady state—the original equation must be dynamic, so that it will generate this term. The steady state of a moving material is different from a classical static equilibrium due to the presence of the centrifugal effect, arising from the steady axial motion combined with local rotation.

After applying the coordinate transformation, and only then, it is possible to look for a steady state, where the motion of the material appears stationary in the Eulerian frame despite the fact that the material flows through the control volume. Such a situation is analogous to a steady-state flow in fluid mechanics, where the Eulerian velocity field remains constant in time.

By Eq. (5.5.5), the same caveat applies to first-order time derivatives. For example, with time-dependent constitutive laws such as Kelvin–Voigt viscoelasticity, one must begin with the dynamic formulation, and only after the coordinate transformation, drop the time-dependent terms if the goal is to find a steady state.

The most important points to remember are that when deriving the behavioral equations, before the coordinate transformation is applied, everything is expressed in the co-moving frame, and that the coordinate transformation will generate additional space derivatives out of time derivatives taken in the co-moving frame.

The final task is to transform the axial space derivatives. In the same manner as above, we write

$$\frac{\partial \tilde{f}(\boldsymbol{\xi}, t)}{\partial \xi} = \frac{\partial f(\mathbf{x}(\boldsymbol{\xi}, t), t)}{\partial \xi} = \sum_{j=1}^{3} \frac{\partial x_i}{\partial \xi} \frac{\partial f}{\partial x_i} = \frac{\partial f}{\partial x} , \tag{5.5.8}$$

where, using (5.5.1) and (5.5.2), we have simplified

$$\frac{\partial \mathbf{x}}{\partial \xi} = \frac{\partial}{\partial \xi} [\, \boldsymbol{\xi} + V_0 t \,] = \frac{\partial}{\partial \xi} [\, (\xi, y, z) + V_0 t \,] = (1, 0, 0) \; .$$

Note that regardless of the coordinate system, t is an independent variable; $\partial t / \partial \xi_i = \partial t / \partial x_i = 0$ (where $i = 1, 2, 3$).

Thus, for the coordinate transformation (5.5.1), by (5.5.8) the appropriate transformation for the axial space derivative is the identity function. In other words, we may simply replace

$$\frac{\partial \tilde{f}}{\partial \xi} = \frac{\partial f}{\partial x} \; . \tag{5.5.9}$$

By repeated application, also $\partial^2 \tilde{f} / \partial \xi^2 = \partial^2 f / \partial x^2$, and likewise for higher space derivatives.

Partial derivatives with respect to the y and z coordinates also remain unaffected by the coordinate transformation (5.5.1). The fact that all space derivatives are simply passed through, highlights the property that (5.5.1) is a rigid-body translation; there is no rotation, and no expansion or compression of coordinates.

Let us now consider the Eulerian displacement. As was mentioned, problems of axially moving materials are commonly solved using an Eulerian description. How can one define a sensible Eulerian displacement, especially in the axial direction where an overall axial motion occurs? Furthermore, how can this be done in a way that allows for a steady state from the laboratory viewpoint although the material is traveling? In the following exposition, we will answer these questions.

In general, in the co-moving frame, displacement is defined as in the classical case (5.2.1),

$$\tilde{\mathbf{u}}(\boldsymbol{\xi}, t) \equiv \tilde{\mathbf{r}}(\boldsymbol{\xi}, t) - \boldsymbol{\xi} \; , \tag{5.5.10}$$

where the function $\tilde{\mathbf{r}}(\boldsymbol{\xi}, t)$ gives the position in (ξ, y, z) coordinates, at time t, of the material particle originally, in an initial or undeformed state, located at $\boldsymbol{\xi}$. The displacement (5.5.10) is a Lagrangean quantity; it measures the change in position in terms of the (ξ, y, z) coordinate system. In the co-moving frame, the Lagrangean description of solid mechanics works as usual.

Recall the coordinate transformation (5.5.1),

$$\mathbf{x}(\boldsymbol{\xi}, t) \equiv \boldsymbol{\xi} + (V_0 t,\ 0,\ 0)\ .$$

Noting that this transformation is invertible, we have

$$\boldsymbol{\xi}(\mathbf{x}, t) = \mathbf{x} - (V_0 t,\ 0,\ 0)\ . \tag{5.5.11}$$

Substituting (5.5.11) into (5.5.10), we may represent the displacement $\widetilde{\mathbf{u}}(\boldsymbol{\xi}, t)$ for given $\mathbf{x}$ and t as

$$\widetilde{\mathbf{u}}(\boldsymbol{\xi}(\mathbf{x}, t),\ t) \ = \ \widetilde{\mathbf{r}}(\boldsymbol{\xi}(\mathbf{x}, t),\ t) - [\, \mathbf{x} - (V_0 t,\ 0,\ 0)\,]\ . \tag{5.5.12}$$

The expression in brackets is expressed in the (ξ, y, z) coordinate system, due to (5.5.11), so the subtraction on the right-hand side of (5.5.12) is valid (i.e., both operands are expressed in the same coordinate system).

Recall the frame invariance principle (5.5.3). Obviously, we may apply it componentwise to vector quantities. Thus, we may define the mixed Eulerian–Lagrangean functions

$$\mathbf{u}(\mathbf{x}, t) \ \equiv \ \widetilde{\mathbf{u}}(\boldsymbol{\xi}(\mathbf{x}, t),\ t)\ , \tag{5.5.13}$$

$$\mathbf{r}(\mathbf{x}, t) \ \equiv \ \widetilde{\mathbf{r}}(\boldsymbol{\xi}(\mathbf{x}, t),\ t)\ . \tag{5.5.14}$$

In terms of (5.5.13) and (5.5.14), Eq. (5.5.12) becomes

$$\mathbf{u}(\mathbf{x}, t) \ = \ \mathbf{r}(\mathbf{x}, t) - \mathbf{x} + (V_0 t,\ 0,\ 0)\ . \tag{5.5.15}$$

It is extremely important to keep in mind that, although the input to the expression (5.5.15) is given in terms of (x, y, z) coordinates, its output has not changed. Equation (5.5.15) gives the Lagrangean displacement at the Eulerian position $\mathbf{x}$ at time t.

It is possible for the traveling material to flow through the domain of interest (*control interval*) in such a way that its displacement profile, as viewed in the laboratory, stays unchanged with respect to time. In other words, the process being modelled may allow a steady state in terms of the laboratory coordinates. This is analogous to steady flows in fluid mechanics, where the Eulerian velocity stays constant, although the fluid flows through the domain.

The known overall axial motion represented by the $V_0 t$ term is obviously of no interest for the analysis of the displacement. Hence, to get rid of it, and to allow for steady states in the laboratory, the final step is to subtract the overall axial motion, defining the *laboratory displacement*

$$\mathbf{u}_{\text{lab}}(\mathbf{x}, t) \ \equiv \ \mathbf{u}(\mathbf{x}, t) - (V_0 t,\ 0,\ 0) \ = \ \mathbf{r}(\mathbf{x}, t) - \mathbf{x}\ , \tag{5.5.16}$$

which describes the *Lagrangean displacement, with respect to a reference state uniformly traveling toward positive x at constant velocity* V_0, *at the Eulerian position* $\mathbf{x}$ *at time* t. This is the displacement that is actually used in the analysis of problems of axially moving materials. This is called the *mixed Eulerian–Lagrangean* formulation [63].

From the viewpoint of small-displacement theory, $\mathbf{u}(\mathbf{x}, t)$ is not a small quantity, but $\mathbf{u}_{\text{lab}}(\mathbf{x}, t)$ is. The same observation applies to $\partial\mathbf{u}(\mathbf{x}, t)/\partial t$. However, all other derivatives of $\mathbf{u}(\mathbf{x}, t)$ are small, because the additional term is a constant-velocity rigid-body translation.

In a sense, (5.5.16) can be thought of as an *Eulerian displacement*, but this is only because the units of length and the directions of the axes coincide in the (x, y, z) and (ξ, y, z) coordinate systems. For this reason, although the right-hand side of Eq. (5.5.16) gives its output in units of $\boldsymbol{\xi}$, we may treat it as if it were in units of $\mathbf{x}$. Thus we may interpret (5.5.16) also as the *Eulerian displacement, with respect to a material uniformly traveling toward* $+x$ *at constant velocity* V_0, *at the Eulerian position* $\mathbf{x}$ *at time* t.

Summarizing the above: the original dynamical equations, written in the co-moving frame, use the description $\widetilde{\mathbf{u}}(\boldsymbol{\xi}, t)$ that is defined by Eq. (5.5.10). The coordinate transformation (5.5.1) gets us from there to a description in terms of $\mathbf{u}(\mathbf{x}, t)$, as defined by (5.5.13) and (5.5.15).

Since the overall axial motion is a rigid-body translation, it introduces no strains or stresses. Because V_0 is constant, it neither introduces any acceleration. Hence the co-moving frame is an inertial one, and thus by the principle of Galilean relativity, the dynamical equations for $\mathbf{u}(\mathbf{x}, t)$ work just as well for $\mathbf{u}_{\text{lab}}(\mathbf{x}, t)$; this justifies the transition from (5.5.15) to (5.5.16).

In practice, after the coordinate transformation has been applied as discussed above, we simply replace $\mathbf{u}(\mathbf{x}, t)$ by $\mathbf{u}_{\text{lab}}(\mathbf{x}, t)$ in the result, and then rename this quantity back to $\mathbf{u}(\mathbf{x}, t)$, omitting the label lab, to obtain the final equations.

To instead perform this final step rigorously, we may invert (5.5.16) to represent $\mathbf{u}(\mathbf{x}, t)$, obtaining

$$\mathbf{u}(\mathbf{x}, t) = \mathbf{u}_{\text{lab}}(\mathbf{x}, t) + (V_0 t,\ 0,\ 0)\ , \tag{5.5.17}$$

and then insert (5.5.17) into the dynamical equations. In most cases this results in no changes to the equations, because any $\partial/\partial x$ makes the $V_0 t$ term vanish, and the time differentiation in the dynamical equations is of second order,

$$\mathrm{d}^2/\mathrm{d}t^2 = \partial^2/\partial t^2 + 2V_0\partial^2/\partial x\partial t + V_0^2\partial^2/\partial x^2,$$

also making the $V_0 t$ term vanish. The same observation applies to the representations of $M(x)$ and $N(x)$ in the case of linear elastic and Kelvin–Voigt materials, where each term contains at least one $\partial/\partial x$.

Analogously to the laboratory displacement defined in Eq. (5.5.16), we may define the laboratory velocity by

$$\mathbf{v}_{\text{lab}}(\mathbf{x}, t) \equiv \mathbf{v}(\mathbf{x}, t) - (V_0,\ 0,\ 0)\ , \tag{5.5.18}$$

which gives the velocity relative to the known constant-velocity contribution. To express $\mathbf{v}(\mathbf{x}, t)$ in terms of $\mathbf{u}(\mathbf{x}, t)$, we must consider the velocity of a material point. In the co-moving frame,

$$\widetilde{\mathbf{v}}(\boldsymbol{\xi}, t) = \frac{\partial \widetilde{\mathbf{u}}(\boldsymbol{\xi}, t)}{\partial t} , \tag{5.5.19}$$

where the $\partial/\partial t$ is taken at constant $\boldsymbol{\xi}$. The initial position of the material point (labeled $\boldsymbol{\xi}$) does not change in time, so only the displacement contributes to the velocity; recall Eq. (5.2.1) in Sect. 5.2.

Transforming both sides of (5.5.19) to the laboratory frame, and expanding the co-moving derivative,

$$\mathbf{v}(\mathbf{x}, t) \equiv \frac{\mathrm{d}\mathbf{u}(\mathbf{x}, t)}{\mathrm{d}t} = \left(\frac{\partial}{\partial t} + V_0 \frac{\partial}{\partial x} \right) \mathbf{u}(\mathbf{x}, t) , \tag{5.5.20}$$

where the $\partial/\partial t$ is now taken at constant $\mathbf{x}$. Thus,

$$\mathbf{v}_{\mathrm{lab}}(\mathbf{x}, t) = \left(\frac{\partial}{\partial t} + V_0 \frac{\partial}{\partial x} \right) \mathbf{u}(\mathbf{x}, t) - (V_0,\ 0,\ 0) . \tag{5.5.21}$$

As was discussed, in the typical case, the final equations are formulated for $\mathbf{u}_{\mathrm{lab}}$ instead of $\mathbf{u}$. In this case we may write

$$\begin{aligned}
\mathbf{v}_{\mathrm{lab}}(\mathbf{x}, t) &= \left(\frac{\partial}{\partial t} + V_0 \frac{\partial}{\partial x} \right) [\mathbf{u}_{\mathrm{lab}}(\mathbf{x}, t) + (V_0 t,\ 0,\ 0)] - (V_0,\ 0,\ 0) \\
&= \left(\frac{\partial}{\partial t} + V_0 \frac{\partial}{\partial x} \right) \mathbf{u}_{\mathrm{lab}}(\mathbf{x}, t) \\
&= \frac{\mathrm{d}}{\mathrm{d}t} \mathbf{u}_{\mathrm{lab}}(\mathbf{x}, t) ,
\end{aligned} \tag{5.5.22}$$

i.e. the laboratory velocity is simply the co-moving derivative of the laboratory displacement.

As an example, consider now the modeling of the Eulerian balance of forces of a traveling beam. This also brings forth an important issue that requires us to modify our approach slightly.

We start with the combined force balance Eq. (5.1.40), where momentum balance has already been accounted for. We interpret the force balance as written in the co-moving frame (using tilded quantities):

$$\mathbf{t} \left[\frac{\partial \widetilde{N}}{\partial s} + \frac{\partial \alpha}{\partial s} \frac{\partial \widetilde{M}}{\partial s} + \frac{\partial \alpha}{\partial s} \widetilde{\mu} \right] + \mathbf{n} \left[\frac{\partial \alpha}{\partial s} \widetilde{N} - \frac{\partial^2 \widetilde{M}}{\partial s^2} - \frac{\partial \widetilde{\mu}}{\partial s} \right] + \widetilde{\mathbf{q}} - m \frac{\partial^2 \widetilde{\mathbf{u}}}{\partial t^2} = 0 . \tag{5.5.23}$$

In the inertial term, we have used (5.2.4) to replace $\ddot{\mathbf{x}} = \ddot{\mathbf{u}}$; this is valid also for axially moving materials when the drive velocity is a constant. Because there is no rotation between the co-moving and laboratory frames, the angle $\widetilde{\alpha} = \alpha$. In addition, the curvature $\partial\alpha/\partial s$ obviously has the same value in both frames.

For any constant drive velocity field $\mathbf{V}$, the inertial term transforms into the laboratory frame using the second co-moving derivative,

$$\frac{\partial^2 \widetilde{\mathbf{u}}}{\partial t^2} = \left(\frac{\partial}{\partial t} + \mathbf{V}\cdot\nabla\right)\left(\frac{\partial}{\partial t} + \mathbf{V}\cdot\nabla\right)\mathbf{u}(\mathbf{x},t) = \frac{\partial^2 \mathbf{u}}{\partial t^2} + 2\mathbf{V}\cdot\nabla\frac{\partial \mathbf{u}}{\partial t} + (\mathbf{V}\cdot\nabla)(\mathbf{V}\cdot\nabla)\mathbf{u}\,. \tag{5.5.24}$$

The drive velocity $\mathbf{V}$ is the velocity of the co-moving frame, from the viewpoint of the laboratory frame.

Observe that the beam (or panel) is essentially a lower-dimensional manifold embedded in a higher-dimensional ambient space. Specifically, in the considered case of plane dynamics, the beam is a one-dimensional manifold embedded in a two-dimensional Euclidean space. This implies, among other things, that some care must be taken when dealing with directional derivatives.

An important implication is as follows. In addition to the coordinate transformation from the co-moving frame to the laboratory frame, $\mathbf{x}(\boldsymbol{\xi}, t)$, we must consider the mapping $\boldsymbol{\xi} = \boldsymbol{\xi}(s, t)$, which was earlier in this section called $\mathbf{x}(s)$. This mapping, describing the position of the beam particles in the co-moving frame, in plane dynamics maps from one space dimension, the curve length coordinate s, into two (the position $\boldsymbol{\xi}$ in the ambient space). Indeed, the quantities $\mathbf{u}$, α, $\widetilde{N}$, $\widetilde{M}$ and $\widetilde{\mu}$ are functions of s and t only, and do not explicitly refer to $\boldsymbol{\xi}$. They are only defined for points on the mid-line of the beam.

Obviously, most points $\boldsymbol{\xi}$ of the ambient space at any given t do not lie on the mid-line. There is no meaningful inverse transformation from $\boldsymbol{\xi}$ to s, and hence, neither from $\mathbf{x}$ to s, since for most inputs there is no meaningful value of s. Because the above discussion defining the Eulerian displacement relies on invertibility, it does not apply in this context, and another approach must be sought.

Consider now the directional derivatives. Provided enough continuity, the quantities $\mathbf{u}$, α, $\widetilde{N}$, $\widetilde{M}$ and $\widetilde{\mu}$ may be differentiated along s. Geometrically speaking, as was observed in Eq. (5.1.37) in Sect. 5.1, $\partial/\partial s$ is the directional derivative in the tangential direction:

$$\frac{\partial}{\partial s}(\ldots) = \mathbf{t}\cdot\nabla(\ldots)\,. \tag{5.5.25}$$

However, the derivative in the normal direction,

$$\frac{\partial}{\partial n}(\ldots) \equiv \mathbf{n}\cdot\nabla(\ldots)\,, \tag{5.5.26}$$

is not defined for these quantities, as they are defined on the mid-line only, and $\mathbf{n}$ is perpendicular to the mid-line. Writing the drive velocity field $\mathbf{V}$ in (t, n) coordinates,

$$\mathbf{V} = v_t\mathbf{t} + v_n\mathbf{n}\,, \tag{5.5.27}$$

the directional derivative $\mathbf{V} \cdot \nabla$ appearing in the co-moving derivative can be expressed as

$$\mathbf{V} \cdot \nabla(\ldots) = v_t \mathbf{t} \cdot \nabla(\ldots) + v_n \mathbf{n} \cdot \nabla(\ldots) \,. \tag{5.5.28}$$

Thus, if time derivatives appear, for example, in $\widetilde{M}$, such as for a Kelvin–Voigt material, we may not apply the co-moving derivative in the usual manner, because in the general case $v_n \neq 0$ in (5.5.27), and thus $\mathbf{V} \cdot \nabla$ is not defined for the considered quantities. For the displacement $\mathbf{u}$, it is possible to model also the behavior of material points off the mid-line of the beam, thus making $\mathbf{n} \cdot \nabla \mathbf{u}$ meaningful, but this does not fix the issue of invertibility for points of the ambient space outside the region occupied by the beam.

Both of the above issues—the absence of a meaningful Eulerian description with respect to the ambient space, and nondifferentiability along the normal direction—occur because the object of interest is a lower-dimensional manifold embedded in a higher-dimensional space. For other types of problems involving moving materials, where the dimensionalities of the object of interest and the ambient space match, these issues do not arise. Examples include the purely axial vibrations of an axially traveling rod, and the in-plane deformation of a traveling sheet.

We will step around both of the mentioned issues as follows. In the small-displacement regime—which is useful for stability analysis—we may consider both the beam and the ambient space as one-dimensional. We take the axial drive motion to occur along the $+x$ direction (for $V_0 > 0$), which gives us

$$\mathbf{V} \cdot \nabla(\ldots) = V_0 \frac{\partial}{\partial x}(\ldots) \,. \tag{5.5.29}$$

Approximating $s \approx x$, Eq. (5.5.29) is valid for quantities defined on the mid-line, which now approximately coincides with the x axis. We can now work with the force balance Eqs. (5.2.30) and (5.2.31), written in the co-moving frame using tilded quantities.

For the sake of demonstration, let us consider a panel. We replace the resultant axial force N and resultant moment M by the corresponding resultants per unit width, $\widehat{N}$ and $\widehat{M}$. We have

$$\frac{\partial \widetilde{\widehat{N}}}{\partial \xi} + \frac{\partial^2 \widetilde{w}}{\partial \xi^2} \frac{\partial \widetilde{\widehat{M}}}{\partial \xi} + \frac{\partial^2 \widetilde{w}}{\partial \xi^2} \widetilde{\mu} + \frac{\partial \widetilde{w}}{\partial \xi} \frac{\partial^2 \widetilde{\widehat{M}}}{\partial \xi^2} + \frac{\partial \widetilde{w}}{\partial \xi} \frac{\partial \widetilde{\mu}}{\partial \xi} + \widetilde{q}_\xi - m \frac{\partial^2 \widetilde{u}}{\partial t^2} = 0 \,, \tag{5.5.30}$$

$$\frac{\partial \widetilde{w}}{\partial \xi} \frac{\partial \widetilde{\widehat{N}}}{\partial \xi} + \frac{\partial^2 \widetilde{w}}{\partial \xi^2} \widetilde{\widehat{N}} - \frac{\partial^2 \widetilde{\widehat{M}}}{\partial \xi^2} - \frac{\partial \widetilde{\mu}}{\partial \xi} + \widetilde{q}_z - m \frac{\partial^2 \widetilde{w}}{\partial t^2} = 0 \,. \tag{5.5.31}$$

In the inertial terms, each $\partial/\partial t$ is taken in the co-moving frame, that is, keeping ξ fixed. Transforming to the laboratory frame using the second co-moving derivative (5.5.6), we have

$$\frac{\partial \widehat{N}}{\partial x} + \frac{\partial^2 w}{\partial x^2}\frac{\partial \widehat{M}}{\partial x} + \frac{\partial^2 w}{\partial x^2}\mu + \frac{\partial w}{\partial x}\frac{\partial^2 \widehat{M}}{\partial x^2} + \frac{\partial w}{\partial x}\frac{\partial \mu}{\partial x} + q_x - m\left[\frac{\partial^2}{\partial t^2} + 2V_0\frac{\partial^2}{\partial x \partial t} + V_0^2\frac{\partial^2}{\partial x^2}\right]u = 0\,, \tag{5.5.32}$$

$$\frac{\partial w}{\partial x}\frac{\partial \widehat{N}}{\partial x} + \frac{\partial^2 w}{\partial x^2}\widehat{N} - \frac{\partial^2 \widehat{M}}{\partial x^2} - \frac{\partial \mu}{\partial x} + q_z - m\left[\frac{\partial^2}{\partial t^2} + 2V_0\frac{\partial^2}{\partial x \partial t} + V_0^2\frac{\partial^2}{\partial x^2}\right]w = 0\,. \tag{5.5.33}$$

Now each $\partial/\partial t$ is taken in the laboratory frame, that is, keeping x fixed. Inserting (5.5.17) to rewrite (5.5.32) and (5.5.33) in terms of the Eulerian displacement components u_{lab} and w_{lab}, we obtain the exact same equations, so we rename these quantities back to u and w, omitting the label lab.

In order to perform this step rigorously, we should make this observation only after inserting the representations of $\widehat{N}$ and $\widehat{M}$ in terms of the displacements. However, they contain at least one $\partial/\partial x$ in the case of all constitutive models considered in this chapter, hence eliminating the overall axial motion, as was mentioned above in the discussion for Eq. (5.5.17).

Be aware that when inserting $\widehat{N}$ and $\widehat{M}$ later, any time derivatives in their expressions must be replaced by the co-moving derivative (5.5.5).

In the context of the linear models considered here, the sole exception to the simple replacement of $\mathbf{u}(\mathbf{x}, t)$ by $\mathbf{u}_{\text{lab}}(\mathbf{x}, t)$ is the case of an elastic or viscoelastic foundation, where the response depends on $\mathbf{u}$ and possibly on $\partial \mathbf{u}/\partial t$.

In the transverse direction, $w = w_{\text{lab}}$, so no changes are introduced, but in the axial direction, $u = u_{\text{lab}} + V_0 t$. This implies that the response of a classical elastic (Winkler type) foundation that reacts directly to u (in the axial direction) increases without bound as time t increases, which suggests that in our context, such a setup makes no physical sense.

The conclusion is of course correct. For an axially moving material traveling over a stationary foundation, such a result is to be expected. A stationary foundation with an elastic response in the x direction essentially models a situation where the material is elastically attached to the foundation, with no strain in the initial state. As the axially moving material travels over such a stationary elastic foundation, the foundation becomes increasingly strained, and thus its elastic response—within the limitations of the linear model—increases without bound.

To remedy this, an axially moving foundation, traveling at the same velocity as the material, i.e. V_0, can be used to eliminate the diverging part of the response. Another possibility is to make the foundation react only in the z and possibly y directions, making it slip axially with no resistance (no response to u); then the axial velocity of the foundation does not matter.

A third possibility, which perhaps makes the most physical sense, is to prescribe a viscous-only response in the x direction, modeling kinetic dry friction between the surfaces of the traveling material and the foundation. The derivative $\partial u/\partial t = \partial u_{\text{lab}}/\partial t + V_0$ stays bounded under the same conditions u_{lab} itself does.

Thus, one physically interesting type of response is

$$\mathbf{q}(u, v, w) = -(c\,\partial u/\partial t,\ c\,\partial v/\partial t,\ kw)\,, \tag{5.5.34}$$

where the $\partial/\partial t$ are taken in the laboratory frame. Equation (5.5.34) describes a stationary foundation that resists tangential movement in the x and y directions by kinetic dry friction, and transverse compression and expansion (in the z direction) elastically. The constant c is the coefficient of friction (SI unit Ns/m), while k is the spring coefficient (SI unit N/m).

In this extremely simplified abstract mathematical model, the foundation remains in contact with the object under study at all $\mathbf{x}$ and all t regardless, for example, of the sign of the transverse displacement w.

In terms of the laboratory displacement (5.5.17), we may rewrite Eq. (5.5.34) as

$$\mathbf{q}(u_{\text{lab}},\ v_{\text{lab}},\ w_{\text{lab}}) = -(c\,[\partial u_{\text{lab}}/\partial t + V_0],\ c\,\partial v_{\text{lab}}/\partial t,\ k w_{\text{lab}})\ . \tag{5.5.35}$$

The $-cV_0$ term in the axial component represents the kinetic dry friction resistance to the axial motion of the traveling material.

On the left-hand side, we have taken a shortcut in the notation: the $\mathbf{q}$ in the original equation (5.5.34) takes the total displacement $\mathbf{u}$ as its input, whereas the $\mathbf{q}$ in (5.5.35) takes the corresponding value of the laboratory displacement $\mathbf{u}_{\text{lab}}$. The output has not, of course, changed.

If we wish to generalize the model to incorporate an axially moving foundation, the setup becomes slightly more complicated. Let V_f denote the axial velocity of the foundation, as seen in the laboratory frame. There are two main points to consider. First, the $\partial/\partial t$ in the response of the foundation must be considered as taken in the co-moving frame of the foundation, as that is the frame where the response is most naturally defined. This is easily accounted for by using V_f instead of V_0 in the conversion of the $\partial/\partial t$ into laboratory coordinates.

Secondly, when defining the response of the foundation, instead of just $\mathbf{u}$, one must consider the displacement of the traveling material as seen in the co-moving frame of the traveling foundation. By analogy with (5.5.16) in Sect. 5.5, we define

$$\mathbf{u}_{\text{rel}}(\mathbf{x}, t) \ \equiv\ \mathbf{u}(\mathbf{x}, t) - (V_f\, t,\ 0,\ 0)\ . \tag{5.5.36}$$

Keep in mind that the total displacement $\mathbf{u}(\mathbf{x}, t)$ includes the contribution to the displacement from the overall axial motion of the traveling material. To remove this contribution, we use (5.5.17) to represent $\mathbf{u}$ in terms of $\mathbf{u}_{\text{lab}}$:

$$\begin{aligned}\mathbf{u}_{\text{rel}}(\mathbf{x}, t) &= [\ \mathbf{u}_{\text{lab}}(\mathbf{x}, t) + (V_0 t,\ 0,\ 0)\ ] - (V_f\, t,\ 0,\ 0) \\ &= \mathbf{u}_{\text{lab}}(\mathbf{x}, t) + ([V_0 - V_f]t,\ 0,\ 0)\ .\end{aligned} \tag{5.5.37}$$

Analogously to (5.5.34), let us define the response of the traveling foundation as

$$\mathbf{q}(u_{\text{rel}},\ v_{\text{rel}},\ w_{\text{rel}}) = -(c\,\partial u_{\text{rel}}/\partial t,\ c\,\partial v_{\text{rel}}/\partial t,\ k w_{\text{rel}})\ , \tag{5.5.38}$$

where the $\partial/\partial t$ are taken in the co-moving frame of the foundation. Transforming to laboratory coordinates, we have

$$\mathbf{q}(u_{\text{rel}}, v_{\text{rel}}, w_{\text{rel}}) = -\Big(c\,[\partial u_{\text{rel}}/\partial t + V_{\text{f}}\,\partial u_{\text{rel}}/\partial x],\; c\,[\partial v_{\text{rel}}/\partial t + V_{\text{f}}\,\partial v_{\text{rel}}/\partial x],\; k w_{\text{rel}}\Big)\,, \tag{5.5.39}$$

where the $\partial/\partial t$ are now taken in the laboratory frame. Finally, inserting (5.5.37) gives

$$\begin{aligned} &\mathbf{q}(u_{\text{lab}}, v_{\text{lab}}, w_{\text{lab}}) \\ &\quad = -\Big(c\,[\,\partial u_{\text{lab}}/\partial t + V_{\text{f}}\,\partial u_{\text{lab}}/\partial x + (V_0 - V_{\text{f}})\,],\; c\,[\,\partial v_{\text{lab}}/\partial t + V_{\text{f}}\,\partial v_{\text{lab}}/\partial x\,],\; k w_{\text{lab}}\Big)\,, \end{aligned} \tag{5.5.40}$$

where the $\partial/\partial t$ are taken in the laboratory frame. In the special case $V_{\text{f}} = 0$, Eq. (5.5.40) reduces to (5.5.35).

We see that in addition to resistance to vibrations (represented by $\mathbf{u}_{\text{lab}}$), in the axial direction there is a friction contribution depending on the axial drive velocity difference $V_0 - V_{\text{f}}$.

In continuum mechanics, there are four fundamental balance laws, namely those of linear momentum, angular momentum (i.e., moment of linear momentum), mass, and internal energy. Our discussion of axially moving materials has thus far concentrated on the first two. The balance of internal energy is mainly of interest for thermoelastic applications, and will not be discussed in this chapter. What remains to consider is the mass balance for the axially moving material.

For a stationary beam, mass balance across the whole domain is trivially satisfied, but in an axially moving configuration, material particles constantly enter and leave the Eulerian control volume. A pure constant-velocity translation has no effect on the mass balance, because then the mass flow rates in and out of the control volume are obviously equal. However, small deviations from such a state may have interesting consequences.

Following Kurki et al. [69], let us derive a practically useful consequence of the mass balance. In a papermaking environment, the axial tension that mechanically stabilizes the paper web is generated by a small velocity difference between subsequent rollers. It is physically obvious that this induces an axial strain in the traveling material. By considering the mass balance, it is possible to determine this strain as a function of the axial drive velocities at the rollers.

To begin, we define a control volume, in the laboratory frame, as

$$\Omega = \{\,(x, y, z) : 0 < x < \ell,\; 0 < y < b,\; 0 < z < h\,\}\,, \tag{5.5.41}$$

where ℓ is the length of the span between the rollers, b is the width of the paper web, and h its thickness (see Fig. 5.19). Consider a steady state, where the general Eulerian mass balance equation (see e.g. [4] for a derivation),

$$\frac{\partial \rho}{\partial t} + \nabla \cdot (\rho \mathbf{v}) = 0\,, \tag{5.5.42}$$

reduces to

$$\nabla \cdot (\rho \mathbf{v}) = 0\,. \tag{5.5.43}$$

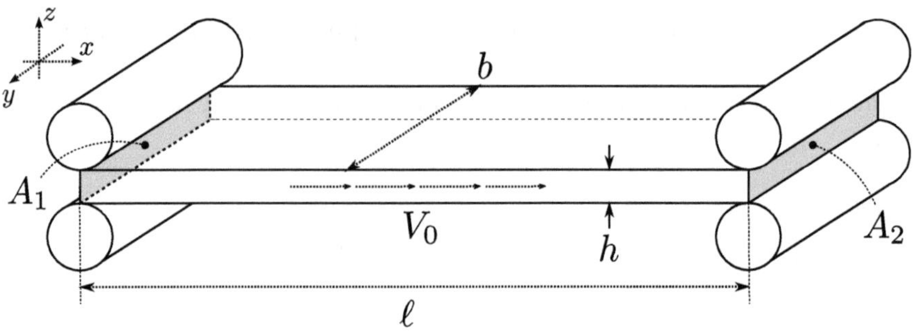

Fig. 5.19 A simplified control volume $\Omega = \{ (x, y, z) : 0 < x < \ell,\ 0 < y < b,\ 0 < z < h \}$, for the mass balance of an axially moving material that is traveling toward $+x$ at constant axial drive velocity V_0. Inflow and outflow surfaces are indicated by shading. Inflow occurs at the surface A_1 at $x = 0$, outflow at the surface A_2 at $x = \ell$. The illustration shows the neutral state before considering the deformation due to axial tension

Here $\mathbf{v}$ is the velocity field of the material particles, as seen in the Eulerian frame. In practice, (5.5.43) means that the mass flow rates at the inflow and outflow surfaces of the control volume must be equal. Integrating (5.5.43) over the control volume Ω, applying the Gauss–Green–Ostrogradsky divergence theorem, and noting that ρ is a scalar, we have

$$\int_\Omega \nabla \cdot (\rho \mathbf{v})\, \mathrm{d}\Omega = \int_{\partial\Omega} \rho\, (\mathbf{n} \cdot \mathbf{v})\, \mathrm{d}\Gamma = 0\,, \tag{5.5.44}$$

where $\partial\Omega$ is the closed surface bounded the control volume Ω, and $\mathbf{n}$ is the outer unit normal vector.

On the Gauss–Green–Ostrogradsky divergence theorem, see calculus textbooks, such as Adams and Essex [2, pp. 906–911], which includes a proof of the two- and three-dimensional cases, and also lists the less known variants for the gradient and curl operators. The theorem has many names, including Gauss's theorem, and the divergence theorem. Some authors, such as Evans [37], call it the Gauss–Green theorem.

In our setup, flows in and out of the control volume occur only at $x = 0$ and $x = \ell$. Let us make the simplifying approximation that ρ and $\mathbf{v}$ are constant across the inflow and outflow surfaces, but that their values may arbitrarily change between these surfaces. In practice, there may exist small variations in the velocity at the outflow surface due to material straining (that will generally depend on position), but we will neglect that for simplicity. Further, let us approximate $\mathbf{v}$ as the axial drive velocity only, ignoring small axial vibrations of the material. so we set $\mathbf{v} = (v_x, 0, 0)$.

Under these approximations, Eq. (5.5.44) becomes

$$-\rho_1 A_1 v_1 + \rho_2 A_2 v_2 = 0\,. \tag{5.5.45}$$

The subscripts on ρ and v refer to the values of these quantities on the surfaces A_1 and A_2 (see Fig. 5.19).

When a small deformation $\mathbf{u}$ is applied, the volume of a small piece (*differential element*) of the material initially having volume V_0 becomes

$$V = V_0[1 + \nabla \cdot \mathbf{u}] = V_0[1 + \varepsilon_{xx} + \varepsilon_{yy} + \varepsilon_{zz}] \,, \tag{5.5.46}$$

which is familiar e.g. from the theory of elasticity. If necessary, Eq. (5.5.46) can be proved via a Taylor expansion of the displacement $\mathbf{u}$, truncated to first order. We write the volume of the differential element, originally a rectangular box having volume $V_0 = \mathrm{d}x\,\mathrm{d}y\,\mathrm{d}z$, subjected to the small deformation $\mathbf{u}$; refer to the corresponding treatment of a plane deformation shown earlier in Fig. 5.9 in Sect. 5.2. An arbitrary small deformation in three dimensions, up to first order, transforms the box into a parallelepiped. The volume of the parallelepiped is obtained as $V = |\mathbf{a} \cdot (\mathbf{b} \times \mathbf{c})|$, where $\mathbf{a}$, $\mathbf{b}$ and $\mathbf{c}$ are vectors representing the three edges that meet at one vertex. Dropping small terms of second order and higher, we obtain (5.5.46). To make the result waterproof, we may write $\mathrm{d}y = \alpha_1 \mathrm{d}x$, $\mathrm{d}z = \alpha_2 \mathrm{d}x$ (allowing us to shrink the box uniformly), divide both sides of (5.5.46) by V_0, and then take the limit $\mathrm{d}x \to 0$, making the higher-order terms vanish exactly.

Because the total mass M of the considered piece is conserved, from (5.5.46) and the definition of the density in the neutral state,

$$\rho_0 \equiv \frac{M}{V_0} \,, \tag{5.5.47}$$

where $[\rho_0] = \mathrm{kg/m^3}$ (not normalized for area or length), it follows that

$$\rho \equiv \frac{M}{V} = \frac{1}{V_0} \frac{M}{V/V_0} = \frac{M}{V_0} \frac{1}{V/V_0} = \rho_0[1 + \varepsilon_{xx} + \varepsilon_{yy} + \varepsilon_{zz}]^{-1} \,. \tag{5.5.48}$$

Let us now consider applying a pure axial stress to the material. This induces an axial strain ε_{xx}, and via the Poisson effect, also the strains ε_{yy} and ε_{zz} in the two perpendicular directions:

$$\varepsilon_{yy} = -\nu_{12}\varepsilon_{xx} \,, \quad \varepsilon_{zz} = -\nu_{13}\varepsilon_{xx} \,. \tag{5.5.49}$$

We consider the material to be orthotropic. The purely elastic approximation made here neglects all viscous effects.

The cross-sectional area of the paper web, at constant x, is

$$A = (1 + \varepsilon_{zz})h(1 + \varepsilon_{yy})b \approx bh(1 + \varepsilon_{yy} + \varepsilon_{zz}) \,, \tag{5.5.50}$$

up to first order in the small quantities. Combining (5.5.49) and (5.5.50) yields

$$A = bh(1 - (\nu_{12} + \nu_{13})\varepsilon_{xx}) \equiv bh(1 - \nu_{1A}\varepsilon_{xx}) \,, \tag{5.5.51}$$

where we have defined the *effective Poisson ratio* for the change in cross-sectional area, when the material is stretched along material axis 1:

$$\nu_{1A} \equiv \nu_{12} + \nu_{13} \ . \tag{5.5.52}$$

From (5.5.46), (5.5.49) and (5.5.52), for the volume it follows that

$$V = V_0(1 + (1 - \nu_{1A})\varepsilon_{xx}) \ . \tag{5.5.53}$$

For $\nu_{1A} = 1$, Eq. (5.5.53) reduces to $V = V_0$, and such material behaves incompressibly when stretched along material axis 1.

The classical isotropic incompressible case with $\nu_{ij} = 1/2$ for all $i, j = 1, 2, 3$ leads to $\nu_{1A} = 1$, but this is by far not the only possibility. For an orthotropic material, when ν_{12} and ν_{13} are fixed, the values of ν_{23}, ν_{21}, ν_{31} and ν_{32} remain free. Elastic compatibility requires $E_i \nu_{ji} = E_j \nu_{ij}$ (no summation), but this brings in additional free parameters in the form of the Young moduli. Thus, it is possible for an orthotropic material to behave incompressibly in axial stretching only when the deformation is applied along some particular axis. We refer the reader interested in incompressibility conditions for orthotropic materials to the study by Itskov and Aksel [53].

Combining equations (5.5.48), (5.5.49) and (5.5.52), we have

$$\rho = \frac{\rho_0}{1 + (1 - \nu_{1A})\varepsilon_{xx}} \ . \tag{5.5.54}$$

Let us now subject the material to constant axial tension along the x axis at $x = 0$ and $x = \ell$. We assume that the material has zero strain at $x = 0$, and experiences some nonzero axial strain ε_{xx} at $x = \ell$ due to the applied axial stress. This is known to be admissible at least for an axially moving Kelvin–Voigt material.

Be aware that both axial motion and nonzero material viscosity are required. Otherwise, ε_{xx} will be constant with respect to x, and the state assumed here will not be admissible in light of the axial small-displacement force balance, Eq. (5.5.32). Details are given in Kurki et al. [69], with an analytical solution.

Under the present assumptions, by Eq. (5.5.51), the cross-sectional areas A_1 and A_2 become

$$A_1 = bh \ , \quad A_2 = bh(1 - \nu_{1A}\varepsilon_{xx}) \ . \tag{5.5.55}$$

By Eq. (5.5.54), the material density at these surfaces is

$$\rho_1 = \rho_0 \ , \quad \rho_2 = \frac{\rho_0}{1 + (1 - \nu_{1A})\varepsilon_{xx}} \ . \tag{5.5.56}$$

Inserting (5.5.55) and (5.5.56) into the mass balance (5.5.45) and solving for ε_{xx}, we obtain the result

$$\varepsilon_{xx}(\ell) = \frac{\dfrac{v_2}{v_1} - 1}{1 + \left[\dfrac{v_2}{v_1} - 1\right] \nu_{1A}} \, . \tag{5.5.57}$$

Equation (5.5.57) gives the axial strain of the viscoelastic material at the end of the span, induced by the velocity difference of the rollers, when the axial drive velocity of the material changes from v_1 at $x = 0$ to v_2 at $x = \ell$.

The validity requirements of the result (5.5.57) are as follows. The material is viscoelastic and in axial motion. The drive velocity ratio v_2/v_1 is such that $\varepsilon_{xx}(\ell)$ remains in the small-deformation range. The transport velocity is controlled only in the x direction, with all deformations in the y (width) and z (transverse) directions determined by the material response. The viscous component of the material response is small. Finally, strictly speaking, the result only applies to a steady-state case, where the material flows smoothly without time-dependent disturbances.

To conclude, let us consider some important special cases of (5.5.57). For a material that behaves incompressibly in stretching along material axis 1, $\nu_{1A} = 1$. This gives the special case

$$\varepsilon_{xx}(\ell) = 1 - \frac{v_1}{v_2} \, . \tag{5.5.58}$$

In the limit of materials that do not exhibit the Poisson effect (such as cork), we have $\nu_{1A} \to 0$. In this case, Eq. (5.5.57) simplifies to

$$\varepsilon_{xx}(\ell) = \frac{v_2}{v_1} - 1 \, . \tag{5.5.59}$$

The case (5.5.59) also arises for a one-dimensional compressible traveling rod. Assuming that the cross-sectional area of the rod remains constant, the mass balance (5.5.45) becomes

$$\rho_1 v_1 - \rho_2 v_2 = 0 \, , \tag{5.5.60}$$

where now $[\rho] = \mathrm{kg/m}$. Under stretching or compression, the linear density changes as

$$\rho = \frac{\rho_0}{1 + \varepsilon_{xx}} \, . \tag{5.5.61}$$

As before, let $\varepsilon_{xx}(0) = 0$ and $\varepsilon_{xx}(\ell) \neq 0$, as is consistent for a traveling rod made of Kelvin–Voigt material. We have $\rho_1 = \rho_0$, and ρ_2 is given by Eq. (5.5.61). Combining (5.5.60) and (5.5.61) yields the result (5.5.59).

The general case (5.5.57) is plotted in Fig. 5.20 for various values of the effective Poisson ratio ν_{1A}, including auxetic materials such as paper ($\nu_{1A} < 0$), and the corklike ($\nu_{1A} = 0$) and incompressible ($\nu_{1A} = 1$) special cases.

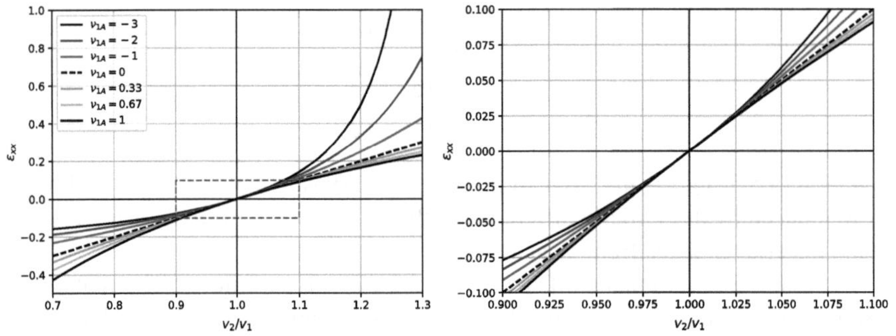

Fig. 5.20 Result (5.5.57). Axial strain ε_{xx} at the end of the span, for an axially moving viscoelastic material, as a function of the axial drive velocity ratio v_2/v_1. *Left*: overview. *Right*: detail of the region indicated on the left. Adapted from Kurki et al. [69]

5.6 Transformation to Weak Form

For the numerical solution of partial differential equations, it is preferable to consider the weak form of the problem, because this mitigates the regularity (continuity) requirements that must be imposed on the solution. Indeed, finite element methods, which operate on the weak form, have become the standard tool for numerically solving partial differential equations. Finite elements also offer geometric flexibility, simplifying the practical work needed for spatially nonuniform discretizations. A further advantage of finite elements, against the classical background of collocation methods (e.g. finite differences), is that because finite element methods represent the solution as a linear combination of known basis functions, the numerical solution is obtained as a function that can be evaluated anywhere in the domain.

A standard comprehensive reference work on finite elements is Zienkiewicz et al. [162–164]; see also e.g. Bathe [9], Belytschko et al. [11]. For a mathematical focus, see Brenner and Scott [21], for which background can be found in Adams [1], Evans [37], Brezis [22]. Flow problems require special treatment; we refer the interested reader to Donea and Huerta [33]. The reader new to finite elements may be interested in Johnson [55], Eriksson et al. [34], Hughes [52], Fish and Belytschko [39]. The book [4] especially discusses both finite elements and finite differences.

The weak form is closely related to the calculus of variations, which is treated, e.g. in the books by Mikhlin [97], Washizu [146], Komkov [64], Krizek and Neittaanmäki [66], Eriksson et al. [34], Weinstock [147], Berdichevsky [13]. In the calculus of variations, in the prototypical case, one seeks to minimize a functional representing some appropriate energy. Generally however, allowing for non-conservative forces and damping, such an energy to be minimized does not necessarily exist.

A classical approach for the general case is to start with the strong form of the problem, and transform it into a weak form. In mechanics, this transformation is embodied in the *principle of virtual work*. The weak form is then taken as the new *definition* of the problem, the motivation being that under conditions where both forms are valid, the weak form implies the strong form.

The transformation process will also generate boundary conditions, in mechanics often ones that are physically appropriate for the problem under consideration. Since the question of Eulerian boundary conditions for an axially moving material requires some care, the present section concentrates on transforming the problem to the weak form. Boundary conditions will be discussed later.

Aiming at a numerical solution, let us derive the weak form of the combined force balance Eq. (5.1.40), namely,

$$\mathbf{t}\left[\frac{\partial N}{\partial s}+\frac{\partial \alpha}{\partial s}\frac{\partial M}{\partial s}+\frac{\partial \alpha}{\partial s}\mu\right]+\mathbf{n}\left[\frac{\partial \alpha}{\partial s}N-\frac{\partial^2 M}{\partial s^2}-\frac{\partial \mu}{\partial s}\right]+\mathbf{q}-m\frac{\partial^2 \mathbf{x}}{\partial t^2}=0\,.$$

Using the principle of virtual work, we take the dot product of both sides of the equation with a vector-valued test function that represents an arbitrary virtual displacement,

$$\boldsymbol{\psi}(s)\equiv(\phi(s),\,\psi(s))\,. \tag{5.6.1}$$

Engineering texts often denote this quantity by $\delta\mathbf{u}$. We then integrate over an arbitrary segment (s_1, s_2) of the beam. We have

$$\int_{s_1}^{s_2}\mathbf{t}\left[\frac{\partial N}{\partial s}+\frac{\partial \alpha}{\partial s}\frac{\partial M}{\partial s}+\frac{\partial \alpha}{\partial s}\mu\right]\cdot\boldsymbol{\psi}\,\mathrm{d}s+\int_{s_1}^{s_2}\mathbf{n}\left[\frac{\partial \alpha}{\partial s}N-\frac{\partial^2 M}{\partial s^2}-\frac{\partial \mu}{\partial s}\right]\cdot\boldsymbol{\psi}\,\mathrm{d}s$$
$$+\int_{s_1}^{s_2}\mathbf{q}\cdot\boldsymbol{\psi}\,\mathrm{d}s-\int_{s_1}^{s_2}m\frac{\partial^2 \mathbf{x}}{\partial t^2}\cdot\boldsymbol{\psi}\,\mathrm{d}s=0\,. \tag{5.6.2}$$

To make (5.6.2) into a weak form, we would like to get rid of differentiations of $N(s)$ and $M(s)$. To do this, we will apply integration by parts. We must keep in mind that $\mathbf{t}=\mathbf{t}(s)$ and $\mathbf{n}=\mathbf{n}(s)$. Recall the result from Eqs. (5.1.9) to (5.1.10) in Sect. 5.1:

$$\frac{\partial \mathbf{t}}{\partial s}=\mathbf{n}\frac{\partial \alpha}{\partial s}\,,$$
$$\frac{\partial \mathbf{n}}{\partial s}=-\mathbf{t}\frac{\partial \alpha}{\partial s}\,.$$

The first term in the first integrand in (5.6.2) is

$$\left(\mathbf{t}\frac{\partial N}{\partial s}\right)\cdot\boldsymbol{\psi}=\frac{\partial N}{\partial s}\mathbf{t}\cdot\boldsymbol{\psi}\,, \tag{5.6.3}$$

because $\partial N/\partial s$ is scalar. For any vector $\mathbf{v}=\mathbf{v}(s)$, we observe that

$$\frac{\partial}{\partial s}\left(N\mathbf{v}\cdot\boldsymbol{\psi}\right)=\frac{\partial N}{\partial s}\mathbf{v}\cdot\boldsymbol{\psi}+N\frac{\partial \mathbf{v}}{\partial s}\cdot\boldsymbol{\psi}+N\mathbf{v}\cdot\frac{\partial \boldsymbol{\psi}}{\partial s}\,. \tag{5.6.4}$$

We see that choosing $\mathbf{v} = \mathbf{t}$ gives us a term of the form (5.6.3) on the right-hand side of (5.6.4). Making this choice and rearranging terms, we have

$$\begin{aligned} \frac{\partial N}{\partial s}\mathbf{t} \cdot \boldsymbol{\psi} &= \frac{\partial}{\partial s}(N\mathbf{t} \cdot \boldsymbol{\psi}) - N\frac{\partial \mathbf{t}}{\partial s} \cdot \boldsymbol{\psi} - N\mathbf{t} \cdot \frac{\partial \boldsymbol{\psi}}{\partial s} \\ &= \frac{\partial}{\partial s}(N\mathbf{t} \cdot \boldsymbol{\psi}) - N\frac{\partial \alpha}{\partial s}\mathbf{n} \cdot \boldsymbol{\psi} - N\mathbf{t} \cdot \frac{\partial \boldsymbol{\psi}}{\partial s}, \end{aligned} \tag{5.6.5}$$

where on the second line we have used (5.1.9). Integrating (5.6.5) over (s_1, s_2), we have

$$\int_{s_1}^{s_2} \frac{\partial N}{\partial s}\mathbf{t} \cdot \boldsymbol{\psi}\, \mathrm{d}s = \int_{s_1}^{s_2} \frac{\partial}{\partial s}(N\mathbf{t} \cdot \boldsymbol{\psi})\, \mathrm{d}s - \int_{s_1}^{s_2} N\frac{\partial \alpha}{\partial s}\mathbf{n} \cdot \boldsymbol{\psi}\, \mathrm{d}s - \int_{s_1}^{s_2} N\mathbf{t} \cdot \frac{\partial \boldsymbol{\psi}}{\partial s}\, \mathrm{d}s\,. \tag{5.6.6}$$

Now we invoke the fundamental theorem of calculus (Eq. (5.1.5) in Sect. 5.1), obtaining an integration by parts formula for the left-hand side of (5.6.6), conveniently split into contributions along the local directions of $\mathbf{t}$ and $\mathbf{n}$:

$$\int_{s_1}^{s_2} \frac{\partial N}{\partial s}\mathbf{t} \cdot \boldsymbol{\psi}\, \mathrm{d}s = [\, N\mathbf{t} \cdot \boldsymbol{\psi}\,]_{s=s_1}^{s_2} - \int_{s_1}^{s_2} N\frac{\partial \alpha}{\partial s}\mathbf{n} \cdot \boldsymbol{\psi}\, \mathrm{d}s - \int_{s_1}^{s_2} N\mathbf{t} \cdot \frac{\partial \boldsymbol{\psi}}{\partial s}\, \mathrm{d}s\,, \tag{5.6.7}$$

where we have used the common shorthand notation for substitution resulting from a definite integral,

$$[\ldots]_{s=s_1}^{s_2} \equiv (\ldots)|_{s=s_2} - (\ldots)|_{s=s_1}\,. \tag{5.6.8}$$

The second term in the first integrand in (5.6.2) is

$$\left(\mathbf{t}\frac{\partial \alpha}{\partial s}\frac{\partial M}{\partial s}\right) \cdot \boldsymbol{\psi} = \frac{\partial \alpha}{\partial s}\frac{\partial M}{\partial s}\mathbf{t} \cdot \boldsymbol{\psi}\,. \tag{5.6.9}$$

We see that the following expression produces this term:

$$\begin{aligned} \frac{\partial}{\partial s}\left(\frac{\partial \alpha}{\partial s}M\mathbf{t} \cdot \boldsymbol{\psi}\right) &= \frac{\partial^2 \alpha}{\partial s^2}M\mathbf{t} \cdot \boldsymbol{\psi} + \frac{\partial \alpha}{\partial s}\frac{\partial M}{\partial s}\mathbf{t} \cdot \boldsymbol{\psi} + \frac{\partial \alpha}{\partial s}M\frac{\partial \mathbf{t}}{\partial s} \cdot \boldsymbol{\psi} + \frac{\partial \alpha}{\partial s}M\mathbf{t} \cdot \frac{\partial \boldsymbol{\psi}}{\partial s} \\ &= \frac{\partial^2 \alpha}{\partial s^2}M\mathbf{t} \cdot \boldsymbol{\psi} + \frac{\partial \alpha}{\partial s}\frac{\partial M}{\partial s}\mathbf{t} \cdot \boldsymbol{\psi} + \left(\frac{\partial \alpha}{\partial s}\right)^2 M\mathbf{n} \cdot \boldsymbol{\psi} + \frac{\partial \alpha}{\partial s}M\mathbf{t} \cdot \frac{\partial \boldsymbol{\psi}}{\partial s}\,, \end{aligned}$$

where we have used (5.1.9). Reorganizing terms, we have

$$\frac{\partial \alpha}{\partial s}\frac{\partial M}{\partial s}\mathbf{t} \cdot \boldsymbol{\psi} = \frac{\partial}{\partial s}\left(\frac{\partial \alpha}{\partial s}M\mathbf{t} \cdot \boldsymbol{\psi}\right) - \frac{\partial^2 \alpha}{\partial s^2}M\mathbf{t} \cdot \boldsymbol{\psi} - \left(\frac{\partial \alpha}{\partial s}\right)^2 M\mathbf{n} \cdot \boldsymbol{\psi} - \frac{\partial \alpha}{\partial s}M\mathbf{t} \cdot \frac{\partial \boldsymbol{\psi}}{\partial s}\,. \tag{5.6.10}$$

Integrating (5.6.10) over (s_1, s_2) and again applying the fundamental theorem of calculus, we obtain the integration by parts formula

$$\int_{s_1}^{s_2} \frac{\partial \alpha}{\partial s}\frac{\partial M}{\partial s}\mathbf{t}\cdot\boldsymbol{\psi}\,\mathrm{d}s = \left[\frac{\partial \alpha}{\partial s}M\mathbf{t}\cdot\boldsymbol{\psi}\right]_{s=s_1}^{s_2} - \int_{s_1}^{s_2}\frac{\partial^2\alpha}{\partial s^2}M\mathbf{t}\cdot\boldsymbol{\psi}\,\mathrm{d}s$$
$$- \int_{s_1}^{s_2}\left(\frac{\partial \alpha}{\partial s}\right)^2 M\mathbf{n}\cdot\boldsymbol{\psi}\,\mathrm{d}s - \int_{s_1}^{s_2}\frac{\partial \alpha}{\partial s}M\mathbf{t}\cdot\frac{\partial \boldsymbol{\psi}}{\partial s}\,\mathrm{d}s\,. \quad (5.6.11)$$

The last term to be integrated by parts is the second one in the second integral in (5.6.2), namely,

$$-\left(\mathbf{n}\frac{\partial^2 M}{\partial s^2}\right)\cdot\boldsymbol{\psi} = -\frac{\partial^2 M}{\partial s^2}\mathbf{n}\cdot\boldsymbol{\psi}\,. \quad (5.6.12)$$

This term is generated by the expression

$$\frac{\partial}{\partial s}\left(-\frac{\partial M}{\partial s}\mathbf{n}\cdot\boldsymbol{\psi}\right) = -\frac{\partial^2 M}{\partial s^2}\mathbf{n}\cdot\boldsymbol{\psi} - \frac{\partial M}{\partial s}\frac{\partial \mathbf{n}}{\partial s}\cdot\boldsymbol{\psi} - \frac{\partial M}{\partial s}\mathbf{n}\cdot\frac{\partial \boldsymbol{\psi}}{\partial s}$$
$$= -\frac{\partial^2 M}{\partial s^2}\mathbf{n}\cdot\boldsymbol{\psi} + \frac{\partial \alpha}{\partial s}\frac{\partial M}{\partial s}\mathbf{t}\cdot\boldsymbol{\psi} - \frac{\partial M}{\partial s}\mathbf{n}\cdot\frac{\partial \boldsymbol{\psi}}{\partial s}\,.$$

Rearranging, we have

$$-\frac{\partial^2 M}{\partial s^2}\mathbf{n}\cdot\boldsymbol{\psi} = -\frac{\partial}{\partial s}\left(\frac{\partial M}{\partial s}\mathbf{n}\cdot\boldsymbol{\psi}\right) - \frac{\partial \alpha}{\partial s}\frac{\partial M}{\partial s}\mathbf{t}\cdot\boldsymbol{\psi} + \frac{\partial M}{\partial s}\mathbf{n}\cdot\frac{\partial \boldsymbol{\psi}}{\partial s}\,. \quad (5.6.13)$$

Integrating (5.6.10) over (s_1, s_2) and applying the fundamental theorem of calculus,

$$-\int_{s_1}^{s_2}\frac{\partial^2 M}{\partial s^2}\mathbf{n}\cdot\boldsymbol{\psi}\,\mathrm{d}s = -\left[\frac{\partial M}{\partial s}\mathbf{n}\cdot\boldsymbol{\psi}\right]_{s=s_1}^{s_2} - \int_{s_1}^{s_2}\frac{\partial \alpha}{\partial s}\frac{\partial M}{\partial s}\mathbf{t}\cdot\boldsymbol{\psi}\,\mathrm{d}s + \int_{s_1}^{s_2}\frac{\partial M}{\partial s}\mathbf{n}\cdot\frac{\partial \boldsymbol{\psi}}{\partial s}\,\mathrm{d}s\,. \quad (5.6.14)$$

Because the original term contains $\partial^2 M/\partial s^2$, we are still left with terms involving $\partial M/\partial s$. To eliminate these derivatives, we integrate by parts again. We identify the middle term on the right-hand side of (5.6.14) as (5.6.11). Thus, we have

$$-\int_{s_1}^{s_2}\frac{\partial^2 M}{\partial s^2}\mathbf{n}\cdot\boldsymbol{\psi}\,\mathrm{d}s = -\left[\frac{\partial M}{\partial s}\mathbf{n}\cdot\boldsymbol{\psi}\right]_{s=s_1}^{s_2}$$
$$-\left[\frac{\partial \alpha}{\partial s}M\mathbf{t}\cdot\boldsymbol{\psi}\right]_{s=s_1}^{s_2} + \int_{s_1}^{s_2}\frac{\partial^2\alpha}{\partial s^2}M\mathbf{t}\cdot\boldsymbol{\psi}\,\mathrm{d}s$$
$$+\int_{s_1}^{s_2}\left(\frac{\partial \alpha}{\partial s}\right)^2 M\mathbf{n}\cdot\boldsymbol{\psi}\,\mathrm{d}s + \int_{s_1}^{s_2}\frac{\partial \alpha}{\partial s}M\mathbf{t}\cdot\frac{\partial \boldsymbol{\psi}}{\partial s}\,\mathrm{d}s$$
$$+\int_{s_1}^{s_2}\frac{\partial M}{\partial s}\mathbf{n}\cdot\frac{\partial \boldsymbol{\psi}}{\partial s}\,\mathrm{d}s\,. \quad (5.6.15)$$

One of the remaining terms still contains $\partial M/\partial s$, namely the last term of (5.6.14). Similarly to before, we see that this is produced by

$$\frac{\partial}{\partial s}\left(M\mathbf{n}\cdot\frac{\partial\psi}{\partial s}\right) = \frac{\partial M}{\partial s}\mathbf{n}\cdot\frac{\partial\psi}{\partial s} + M\frac{\partial\mathbf{n}}{\partial s}\cdot\frac{\partial\psi}{\partial s} + M\mathbf{n}\cdot\frac{\partial^2\psi}{\partial s^2}$$
$$= \frac{\partial M}{\partial s}\mathbf{n}\cdot\frac{\partial\psi}{\partial s} - M\frac{\partial\alpha}{\partial s}\mathbf{t}\cdot\frac{\partial\psi}{\partial s} + M\mathbf{n}\cdot\frac{\partial^2\psi}{\partial s^2}\,,$$

whence by rearranging,

$$\frac{\partial M}{\partial s}\mathbf{n}\cdot\frac{\partial\psi}{\partial s} = \frac{\partial}{\partial s}\left(M\mathbf{n}\cdot\frac{\partial\psi}{\partial s}\right) + M\frac{\partial\alpha}{\partial s}\mathbf{t}\cdot\frac{\partial\psi}{\partial s} - M\mathbf{n}\cdot\frac{\partial^2\psi}{\partial s^2}\,.$$

Integrating and applying the fundamental theorem of calculus, we have

$$\int_{s_1}^{s_2}\frac{\partial M}{\partial s}\mathbf{n}\cdot\frac{\partial\psi}{\partial s}\,\mathrm{d}s = \left[M\mathbf{n}\cdot\frac{\partial\psi}{\partial s}\right]_{s=s_1}^{s_2} + \int_{s_1}^{s_2} M\frac{\partial\alpha}{\partial s}\mathbf{t}\cdot\frac{\partial\psi}{\partial s}\,\mathrm{d}s - \int_{s_1}^{s_2} M\mathbf{n}\cdot\frac{\partial^2\psi}{\partial s^2}\,\mathrm{d}s\,. \tag{5.6.16}$$

Finally, inserting (5.6.16) into (5.6.15), we obtain the result

$$\begin{aligned}-\int_{s_1}^{s_2}\frac{\partial^2 M}{\partial s^2}\mathbf{n}\cdot\psi\,\mathrm{d}s = &-\left[\frac{\partial M}{\partial s}\mathbf{n}\cdot\psi\right]_{s=s_1}^{s_2}\\ &-\left[\frac{\partial\alpha}{\partial s}M\mathbf{t}\cdot\psi\right]_{s=s_1}^{s_2} + \int_{s_1}^{s_2}\frac{\partial^2\alpha}{\partial s^2}M\mathbf{t}\cdot\psi\,\mathrm{d}s\\ &+\int_{s_1}^{s_2}\left(\frac{\partial\alpha}{\partial s}\right)^2 M\mathbf{n}\cdot\psi\,\mathrm{d}s + \int_{s_1}^{s_2}\frac{\partial\alpha}{\partial s}M\mathbf{t}\cdot\frac{\partial\psi}{\partial s}\,\mathrm{d}s\\ &+\left[M\mathbf{n}\cdot\frac{\partial\psi}{\partial s}\right]_{s=s_1}^{s_2} + \int_{s_1}^{s_2}M\frac{\partial\alpha}{\partial s}\mathbf{t}\cdot\frac{\partial\psi}{\partial s}\,\mathrm{d}s - \int_{s_1}^{s_2}M\mathbf{n}\cdot\frac{\partial^2\psi}{\partial s^2}\,\mathrm{d}s\,.\end{aligned} \tag{5.6.17}$$

Inserting the final results (5.6.7), (5.6.11) and (5.6.17) into (5.6.2), collecting terms with the same combination of unit vector (**t** or **n**) and test (ψ, $\partial\psi/\partial s$ or $\partial^2\psi/\partial s^2$), and performing cancellations, we are left with

$$\begin{aligned}&\int_{s_1}^{s_2}\frac{\partial\alpha}{\partial s}\mu\mathbf{t}\cdot\psi\,\mathrm{d}s - \int_{s_1}^{s_2}\frac{\partial\mu}{\partial s}\mathbf{n}\cdot\psi\,\mathrm{d}s\\ &+\int_{s_1}^{s_2}\left[\frac{\partial\alpha}{\partial s}M - N\right]\mathbf{t}\cdot\frac{\partial\psi}{\partial s}\,\mathrm{d}s - \int_{s_1}^{s_2}M\mathbf{n}\cdot\frac{\partial^2\psi}{\partial s^2}\,\mathrm{d}s\\ &+[N\mathbf{t}\cdot\psi]_{s=s_1}^{s_2} + \left[-\frac{\partial M}{\partial s}\mathbf{n}\cdot\psi\right]_{s=s_1}^{s_2} + \left[M\mathbf{n}\cdot\frac{\partial\psi}{\partial s}\right]_{s=s_1}^{s_2}\\ &+\int_{s_1}^{s_2}\mathbf{q}\cdot\psi\,\mathrm{d}s - \int_{s_1}^{s_2}m\frac{\partial^2\mathbf{x}}{\partial t^2}\cdot\psi\,\mathrm{d}s = 0\,.\end{aligned} \tag{5.6.18}$$

Finally, to allow for concentrated moment loads (i.e., Dirac deltas in $\mu = \mu(s)$), we integrate by parts in the second integral on the first line. We have

$$\frac{\partial}{\partial s}(\mu \mathbf{n} \cdot \boldsymbol{\psi}) = \frac{\partial \mu}{\partial s}\mathbf{n} \cdot \boldsymbol{\psi} + \mu \frac{\partial \mathbf{n}}{\partial s} \cdot \boldsymbol{\psi} + \mu \mathbf{n} \cdot \frac{\partial \boldsymbol{\psi}}{\partial s}$$
$$= \frac{\partial \mu}{\partial s}\mathbf{n} \cdot \boldsymbol{\psi} - \mu \frac{\partial \alpha}{\partial s}\mathbf{t} \cdot \boldsymbol{\psi} + \mu \mathbf{n} \cdot \frac{\partial \boldsymbol{\psi}}{\partial s} ,$$

whence by integration

$$[\mu \mathbf{n} \cdot \boldsymbol{\psi}]_{s=s_1}^{s_2} = \int_{s_1}^{s_2} \frac{\partial \mu}{\partial s}\mathbf{n} \cdot \boldsymbol{\psi}\, \mathrm{d}s - \int_{s_1}^{s_2} \mu \frac{\partial \alpha}{\partial s}\mathbf{t} \cdot \boldsymbol{\psi}\, \mathrm{d}s + \int_{s_1}^{s_2} \mu \mathbf{n} \cdot \frac{\partial \boldsymbol{\psi}}{\partial s}\, \mathrm{d}s ,$$

and rearranging,

$$\int_{s_1}^{s_2} \frac{\partial \mu}{\partial s}\mathbf{n} \cdot \boldsymbol{\psi}\, \mathrm{d}s = [\mu \mathbf{n} \cdot \boldsymbol{\psi}]_{s=s_1}^{s_2} + \int_{s_1}^{s_2} \mu \frac{\partial \alpha}{\partial s}\mathbf{t} \cdot \boldsymbol{\psi}\, \mathrm{d}s - \int_{s_1}^{s_2} \mu \mathbf{n} \cdot \frac{\partial \boldsymbol{\psi}}{\partial s}\, \mathrm{d}s . \quad (5.6.19)$$

Using (5.6.19) in (5.6.18), we obtain

$$\int_{s_1}^{s_2} \mu \mathbf{n} \cdot \frac{\partial \boldsymbol{\psi}}{\partial s}\, \mathrm{d}s + \int_{s_1}^{s_2} \left[\frac{\partial \alpha}{\partial s} M - N\right] \mathbf{t} \cdot \frac{\partial \boldsymbol{\psi}}{\partial s}\, \mathrm{d}s - \int_{s_1}^{s_2} M \mathbf{n} \cdot \frac{\partial^2 \boldsymbol{\psi}}{\partial s^2}\, \mathrm{d}s$$
$$+ [\, N \mathbf{t} \cdot \boldsymbol{\psi}\,]_{s=s_1}^{s_2} + \left[-\left(\mu + \frac{\partial M}{\partial s}\right) \mathbf{n} \cdot \boldsymbol{\psi} \right]_{s=s_1}^{s_2} + \left[M \mathbf{n} \cdot \frac{\partial \boldsymbol{\psi}}{\partial s} \right]_{s=s_1}^{s_2}$$
$$+ \int_{s_1}^{s_2} \mathbf{q} \cdot \boldsymbol{\psi}\, \mathrm{d}s - \int_{s_1}^{s_2} m \frac{\partial^2 \mathbf{x}}{\partial t^2} \cdot \boldsymbol{\psi}\, \mathrm{d}s = 0 . \quad (5.6.20)$$

Finally, by the moment balance Eq. (5.1.63),

$$\frac{\partial M}{\partial s} + \mu = Q ,$$

so we may write

$$\int_{s_1}^{s_2} \mu \mathbf{n} \cdot \frac{\partial \boldsymbol{\psi}}{\partial s}\, \mathrm{d}s + \int_{s_1}^{s_2} \left[\frac{\partial \alpha}{\partial s} M - N\right] \mathbf{t} \cdot \frac{\partial \boldsymbol{\psi}}{\partial s}\, \mathrm{d}s - \int_{s_1}^{s_2} M \mathbf{n} \cdot \frac{\partial^2 \boldsymbol{\psi}}{\partial s^2}\, \mathrm{d}s$$
$$+ [\, N \mathbf{t} \cdot \boldsymbol{\psi}\,]_{s=s_1}^{s_2} + [\, -Q \mathbf{n} \cdot \boldsymbol{\psi}\,]_{s=s_1}^{s_2} + \left[M \mathbf{n} \cdot \frac{\partial \boldsymbol{\psi}}{\partial s} \right]_{s=s_1}^{s_2}$$
$$+ \int_{s_1}^{s_2} \mathbf{q} \cdot \boldsymbol{\psi}\, \mathrm{d}s - \int_{s_1}^{s_2} m \frac{\partial^2 \mathbf{x}}{\partial t^2} \cdot \boldsymbol{\psi}\, \mathrm{d}s = 0 . \quad (5.6.21)$$

Equation (5.6.21) is the weak form of the force balance Eq. (5.1.40); it is required to hold for any admissible virtual displacement $\boldsymbol{\psi}$.

Let us now specialize (5.6.21) to the small-displacement regime. Recall the representation of the tangent and normal vectors in (x, z) coordinates in the small-displacement regime, Eq. (5.2.29) in Sect. 5.2:

$$\mathbf{t} \approx (1,\ \partial w/\partial x)\ , \quad \mathbf{n} \approx (-\partial w/\partial x,\ 1)\ .$$

The small-displacement version of the weak form (5.6.21) is then, in the axial (x) direction,

$$\begin{aligned}
&- \int_{x_1}^{x_2} \mu \frac{\partial w}{\partial x} \frac{\partial \phi}{\partial x}\, \mathrm{d}x \\
&+ \int_{x_1}^{x_2} \left[\frac{\partial^2 w}{\partial x^2} M - N \right] \frac{\partial \phi}{\partial x}\, \mathrm{d}x + \int_{x_1}^{x_2} M \frac{\partial w}{\partial x} \frac{\partial^2 \phi}{\partial x^2}\, \mathrm{d}x \\
&+ \left[(N + Q \frac{\partial w}{\partial x}) \phi \right]_{x=x_1}^{x_2} - \left[M \frac{\partial w}{\partial x} \frac{\partial \phi}{\partial x} \right]_{x=x_1}^{x_2} \\
&+ \int_{x_1}^{x_2} q_x \phi\, \mathrm{d}x - \int_{x_1}^{x_2} m \frac{\partial^2 u}{\partial t^2} \phi\, \mathrm{d}x = 0\ ,
\end{aligned} \tag{5.6.22}$$

and in the transverse (z) direction,

$$\begin{aligned}
&\int_{x_1}^{x_2} \mu \frac{\partial \psi}{\partial x}\, \mathrm{d}x \\
&- \int_{x_1}^{x_2} N \frac{\partial w}{\partial x} \frac{\partial \psi}{\partial x}\, \mathrm{d}x - \int_{x_1}^{x_2} M \frac{\partial^2 \psi}{\partial x^2}\, \mathrm{d}x \\
&+ \left[(N \frac{\partial w}{\partial x} - Q) \psi \right]_{x=x_1}^{x_2} + \left[M \frac{\partial \psi}{\partial x} \right]_{x=x_1}^{x_2} \\
&+ \int_{x_1}^{x_2} q_z \psi\, \mathrm{d}x - \int_{x_1}^{x_2} m \frac{\partial^2 w}{\partial t^2} \psi\, \mathrm{d}x = 0\ .
\end{aligned} \tag{5.6.23}$$

Second-order small terms have been discarded, and we have used Eq. (5.2.4) for $\partial^2 \mathbf{x}/\partial t^2$. Also note that for any vector $\mathbf{v}$ (especially $\mathbf{t}$, $\mathbf{n}$, $\mathbf{q}$ and $\partial^2 \mathbf{x}/\partial t^2$, which appear in the above equations),

$$\mathbf{v} \cdot \boldsymbol{\psi} = (v_x,\ v_z) \cdot (\phi,\ \psi) = v_x \phi + v_z \psi\ . \tag{5.6.24}$$

We may split sums of this form into contributions to two separate equations, because ϕ and ψ (the tangential and normal components of the virtual displacement) are arbitrary and independent.

In the small-displacement Eqs. (5.6.22) and (5.6.23), observe that both N and Q appear in the boundary terms of both x and z components, and M appears also in a boundary term of the x component. The reason is that in the general case, the boundary terms actually refer to the tangential and normal directions, which are not exactly aligned with x and z. For example, the normal vector $\mathbf{n}$, as given by (5.2.29), contains a small component in the x direction, which is rigorously accounted for by the boundary terms.

To fix this, let us approximate the geometry, at the boundary points x_1 and x_2 only, as perfectly horizontal, that is,

$$\mathbf{t} \approx (1,\, 0)\,, \quad \mathbf{n} \approx (0,\, 1)\,. \tag{5.6.25}$$

In the interior of the domain we may not make the approximation (5.6.25), because that would discard the term involving N in (5.6.23), which describes the transverse contribution of axial loading (e.g., the straightening effect of axial tension; buckling caused by axial compression), and the terms involving μ and M in (5.6.22).

Rewriting the boundary terms in (5.6.21) using the approximate tangent and unit vectors (5.6.25) instead of (5.2.29), the small-displacement Eqs. (5.6.22) and (5.6.23) become

$$\begin{aligned}
&-\int_{x_1}^{x_2} \mu \frac{\partial w}{\partial x}\frac{\partial \phi}{\partial x}\,\mathrm{d}x \\
&+\int_{x_1}^{x_2} \left[\frac{\partial^2 w}{\partial x^2} M - N\right] \frac{\partial \phi}{\partial x}\,\mathrm{d}x + \int_{x_1}^{x_2} M \frac{\partial w}{\partial x}\frac{\partial^2 \phi}{\partial x^2}\,\mathrm{d}x \\
&+ [\,N\phi\,]_{x=x_1}^{x_2} \\
&+\int_{x_1}^{x_2} q_x \phi\,\mathrm{d}x - \int_{x_1}^{x_2} m \frac{\partial^2 u}{\partial t^2}\phi\,\mathrm{d}x = 0\,,
\end{aligned} \tag{5.6.26}$$

$$\begin{aligned}
&\int_{x_1}^{x_2} \mu \frac{\partial \psi}{\partial x}\,\mathrm{d}x \\
&-\int_{x_1}^{x_2} N \frac{\partial w}{\partial x}\frac{\partial \psi}{\partial x}\,\mathrm{d}x - \int_{x_1}^{x_2} M \frac{\partial^2 \psi}{\partial x^2}\,\mathrm{d}x \\
&+ [\,-Q\psi\,]_{x=x_1}^{x_2} + \left[M \frac{\partial \psi}{\partial x}\right]_{x=x_1}^{x_2} \\
&+\int_{x_1}^{x_2} q_z \psi\,\mathrm{d}x - \int_{x_1}^{x_2} m \frac{\partial^2 w}{\partial t^2}\psi\,\mathrm{d}x = 0\,.
\end{aligned} \tag{5.6.27}$$

Finally, it may be instructive to stop for a moment and recognize which terms in the small-displacement weak forms (5.6.22) and (5.6.23), before making the boundary geometry approximation, correspond to which terms in the small-displacement strong forms (5.2.30) and (5.2.31), that is to say,

$$\frac{\partial N}{\partial x} + \frac{\partial^2 w}{\partial x^2}\frac{\partial M}{\partial x} + \frac{\partial^2 w}{\partial x^2}\mu + \frac{\partial w}{\partial x}\frac{\partial^2 M}{\partial x^2} + \frac{\partial w}{\partial x}\frac{\partial \mu}{\partial x} + q_x - m\frac{\partial^2 u}{\partial t^2} = 0\,,$$

$$\frac{\partial w}{\partial x}\frac{\partial N}{\partial x} + \frac{\partial^2 w}{\partial x^2}N - \frac{\partial^2 M}{\partial x^2} - \frac{\partial \mu}{\partial x} + q_z - m\frac{\partial^2 w}{\partial t^2} = 0\,.$$

The strong and weak small-displacement equations were specialized independently to the small-displacement regime, starting in each case from the general strong or weak form. The correspondences are seen easily by applying integration by parts in reverse while ignoring boundary terms. For example, the integral on the first line of the axial component equation (5.6.22) produces $[\,(\partial\mu/\partial x)(\partial w/\partial x) + \mu\,\partial^2 w/\partial x^2\,]\phi$, and the term involving N produces $(\partial N/\partial x)\phi$. Note the sign flips.

Some care must be taken. Concerning the terms involving M on the second line of the axial component equation (5.6.22), the strategy that works is to apply integration by parts (to transfer derivatives away from ϕ) only to the last term, twice. As a result of the first integration by parts, the moment term in the first integral on the same line cancels. The transverse Eq. (5.6.23) is treated similarly.

Let us now consider the weak form for axially moving materials. We have two frames: the laboratory frame, and a co-moving frame undergoing constant-velocity translation in the laboratory. The co-moving frame plays the role of the classical material frame. In the co-moving frame, we have the general force balance of a beam in weak form, Eq. (5.6.21):

$$\begin{aligned}
&\int_{s_1}^{s_2} \widetilde{\mu}\mathbf{n}\cdot\frac{\partial\widetilde{\psi}}{\partial s}\,\mathrm{d}s \\
&+\int_{s_1}^{s_2}\left[\frac{\partial\alpha}{\partial s}\widetilde{M} - \widetilde{N}\right]\mathbf{t}\cdot\frac{\partial\widetilde{\psi}}{\partial s}\,\mathrm{d}s - \int_{s_1}^{s_2}\widetilde{M}\mathbf{n}\cdot\frac{\partial^2\widetilde{\psi}}{\partial s^2}\,\mathrm{d}s \\
&+\left[\widetilde{N}\mathbf{t}\cdot\widetilde{\psi}\right]_{s=s_1}^{s_2} + \left[-\widetilde{Q}\mathbf{n}\cdot\widetilde{\psi}\right]_{s=s_1}^{s_2} + \left[\widetilde{M}\mathbf{n}\cdot\frac{\partial\widetilde{\psi}}{\partial s}\right]_{s=s_1}^{s_2} \\
&+\int_{s_1}^{s_2}\widetilde{\mathbf{q}}\cdot\widetilde{\psi}\,\mathrm{d}s - \int_{s_1}^{s_2} m\frac{\partial^2\widetilde{\mathbf{u}}}{\partial t^2}\cdot\widetilde{\psi}\,\mathrm{d}s = 0\,,
\end{aligned} \tag{5.6.28}$$

where $\widetilde{\psi}$ is the virtual displacement in the co-moving frame, and any time derivatives are taken at constant $\boldsymbol{\xi}$. We have explicitly written out the second time derivative in the inertial term, and used (5.2.4) for $\ddot{\mathbf{x}} = \ddot{\mathbf{u}}$. This is valid also for axially moving materials, provided that the drive velocity is a constant.

The last integral in (5.6.28), that is, the inertial term, has the same general form for any weak formulation involving axially moving materials. Recall the representation (5.5.24) in Sect. 5.5 for the second co-moving derivative. For an arbitrary constant drive velocity vector $\mathbf{v}$, we may write

$$\frac{\partial^2\widetilde{\mathbf{u}}}{\partial t^2} = \left(\frac{\partial}{\partial t} + (\mathbf{v}\cdot\nabla)\right)\left(\frac{\partial}{\partial t} + (\mathbf{v}\cdot\nabla)\right)\mathbf{u}(\mathbf{x},t) = \frac{\partial^2\mathbf{u}}{\partial t^2} + 2(\mathbf{v}\cdot\nabla)\frac{\partial\mathbf{u}}{\partial t} + (\mathbf{v}\cdot\nabla)(\mathbf{v}\cdot\nabla)\mathbf{u}\,.$$

The inertial term thus transforms into the laboratory frame as

$$-\int_{s_1}^{s_2} m \frac{\partial^2 \widetilde{\mathbf{u}}}{\partial t^2} \cdot \widetilde{\psi}\, \mathrm{d}s = -\int_{s_1}^{s_2} m\psi \cdot \frac{\partial^2 \mathbf{u}}{\partial t^2}\, \mathrm{d}s - \int_{s_1}^{s_2} 2m\psi \cdot \left[(\mathbf{v} \cdot \nabla) \frac{\partial \mathbf{u}}{\partial t} \right] \mathrm{d}s$$
$$- \int_{s_1}^{s_2} m\psi \cdot [(\mathbf{v} \cdot \nabla)(\mathbf{v} \cdot \nabla)\mathbf{u}]\, \mathrm{d}s \ . \tag{5.6.29}$$

On the left-hand side, each $\partial/\partial t$ is taken in the co-moving frame (at fixed $\boldsymbol{\xi}$); on the right-hand side, each $\partial/\partial t$ is taken in the laboratory frame (at fixed $\mathbf{x}$).

On the right-hand side, we have placed the test function ψ on the left as a matter of convenience for the calculation that we will perform next; the ordering does not matter because the dot product of two vectors is commutative. Note that $\mathbf{u}$ and ψ are vectors, while $\mathbf{v} \cdot \nabla$ is a scalar operator.

To complete the weak form, we will need an integration by parts formula for the second directional derivative $(\mathbf{v} \cdot \nabla)(\mathbf{v} \cdot \nabla)$ in the last integral in (5.6.29). The second directional derivative very rarely if ever comes up in other applications of mechanics, but for problems of axially moving materials, it is extremely important.

Let us treat this term for arbitrary constant $\mathbf{v}$. In the following, we will use both nabla notation and index notation, but will restrict the consideration to Cartesian coordinate systems only, avoiding the need to keep track of contravariant and covariant components (see e.g. [40, 129]).

For simplicity, we consider the case where m is a constant. Recall that for the beam, m is the linear density ($[m] = \mathrm{kg/m}$), and for the panel, it is the area density ($[m] = \mathrm{kg/m^2}$). If we wish to let m vary in space, it may be passed through the following derivation by defining $\varphi \equiv m\psi$. We then repeat the exact same steps, now with φ playing the role of the test function instead of ψ, and in the final result, substitute back. If necessary, one may then, for $\nabla\varphi = \nabla(m\psi)$, apply the identity

$$\nabla(m\psi) = \partial_i(m\psi_j) = (\partial_i m)\psi_j + m(\partial_i \psi_j) = (\nabla m) \otimes \psi + m\nabla\psi \ . \tag{5.6.30}$$

To begin, observe that

$$\Big((\mathbf{v} \cdot \nabla)(\mathbf{v} \cdot \nabla)\mathbf{u}\Big)_k = (v_i \partial_i)(v_j \partial_j) u_k = (v_i \partial_i)(v_j \partial_j u_k) = \Big((\mathbf{v} \cdot \nabla)(\mathbf{v} \cdot \nabla\mathbf{u})\Big)_k \ .$$

For any differentiable vector fields $\mathbf{a}$, $\mathbf{b}$ and $\mathbf{c}$ in any number of space dimensions, it holds that (see Appendix A)

$$\nabla \cdot (\mathbf{a} \cdot (\mathbf{b} \otimes \mathbf{c})) = \mathbf{b} \cdot (\mathbf{c} \cdot \nabla\mathbf{a}) \ + \ \mathbf{a} \cdot (\nabla \cdot (\mathbf{c} \otimes \mathbf{b})) \ , \tag{5.6.31}$$

where the notational conventions are

$$(\nabla\mathbf{a})_{ij} \equiv \partial_i a_j \ , \quad (\nabla \cdot \mathbf{A})_j \equiv \partial_i A_{ij} \ , \quad (\mathbf{a} \otimes \mathbf{b})_{ij} = a_i b_j \ . \tag{5.6.32}$$

The claim (5.6.31) is easily verified using index notation.

In (5.6.31) and (5.6.32), **a** and **b** are vectors, and **A** is a rank-2 tensor. The summation convention for repeated indices applies. Note the ordering of indices in the gradient; we use the transpose Jacobian convention. We make this choice mainly because this eliminates the need for special rules for the symbol ∇. For example, the directional derivative can be simply transliterated from left to right: $\mathbf{c}\cdot\nabla = c_i\partial_i$, and $\mathbf{c}\cdot\nabla\mathbf{a} = c_i\partial_i a_j$.

We integrate (5.6.31) over the domain Ω, obtaining

$$\int_\Omega \nabla\cdot(\mathbf{a}\cdot(\mathbf{b}\otimes\mathbf{c}))\,\mathrm{d}V = \int_\Omega \mathbf{b}\cdot(\mathbf{c}\cdot\nabla\mathbf{a})\,\mathrm{d}V + \int_\Omega \mathbf{a}\cdot(\nabla\cdot(\mathbf{c}\otimes\mathbf{b}))\,\mathrm{d}V\ . \quad (5.6.33)$$

Applying the Gauss–Green–Ostrogradsky divergence theorem (see calculus textbooks such as Adams and Essex [2, pp. 907–908]) to the left-hand side, we have

$$\int_{\partial\Omega} \mathbf{n}\cdot(\mathbf{a}\cdot(\mathbf{b}\otimes\mathbf{c}))\,\mathrm{d}A = \int_\Omega \mathbf{b}\cdot(\mathbf{c}\cdot\nabla\mathbf{a})\,\mathrm{d}V + \int_\Omega \mathbf{a}\cdot(\nabla\cdot(\mathbf{c}\otimes\mathbf{b}))\,\mathrm{d}V\ , \quad (5.6.34)$$

where **n** is the outer unit normal of the boundary surface $\partial\Omega$. Rearranging terms, we have the general result

$$\int_\Omega \mathbf{b}\cdot(\mathbf{c}\cdot\nabla\mathbf{a})\,\mathrm{d}V = \int_{\partial\Omega} \mathbf{n}\cdot(\mathbf{a}\cdot(\mathbf{b}\otimes\mathbf{c}))\,\mathrm{d}A - \int_\Omega \mathbf{a}\cdot(\nabla\cdot(\mathbf{c}\otimes\mathbf{b}))\,\mathrm{d}V\ , \quad (5.6.35)$$

which allows us to move the derivative that originally operates on **a**. Comparing the integrand on the left-hand side and the last integrand in (5.6.29), let us choose $\mathbf{a} = (\mathbf{v}\cdot\nabla\mathbf{u})$, $\mathbf{b} = \boldsymbol{\psi}$, and $\mathbf{c} = \mathbf{v}$. We obtain the result

$$\int_\Omega \boldsymbol{\psi}\cdot(\mathbf{v}\cdot\nabla)(\mathbf{v}\cdot\nabla\mathbf{u})\,\mathrm{d}\Omega = \int_{\partial\Omega} \mathbf{n}\cdot[(\mathbf{v}\cdot\nabla\mathbf{u})\cdot(\boldsymbol{\psi}\otimes\mathbf{v})]\,\mathrm{d}\Gamma - \int_\Omega (\mathbf{v}\cdot\nabla\mathbf{u})\cdot[\nabla\cdot(\mathbf{v}\otimes\boldsymbol{\psi})]\,\mathrm{d}\Omega\ . \quad (5.6.36)$$

A more symmetric form of the result can be obtained by observing that in any Cartesian coordinate system,

$$\begin{aligned}\mathbf{n}\cdot[(\mathbf{v}\cdot\nabla\mathbf{u})\cdot(\boldsymbol{\psi}\otimes\mathbf{v})] &= n_j\big[(v_k\partial_k u_i)(\psi_i v_j)\big]\\ &= v_k(\partial_k u_i)\psi_i n_j v_j\\ &= (\mathbf{v}\cdot\nabla\mathbf{u}\cdot\boldsymbol{\psi})(\mathbf{n}\cdot\mathbf{v})\\ &= (\mathbf{n}\cdot\mathbf{v})(\mathbf{v}\cdot\nabla\mathbf{u})\cdot\boldsymbol{\psi}\ , \end{aligned} \quad (5.6.37)$$

and

$$\begin{aligned}(\mathbf{v}\cdot\nabla\mathbf{u})\cdot[\nabla\cdot(\mathbf{v}\otimes\boldsymbol{\psi})] &= (v_k\partial_k u_j)\big[\,\partial_i(v_i\psi_j)\,\big]\\ &= (v_k\partial_k u_j)\big[\,\psi_j(\partial_i v_i) + v_i(\partial_i\psi_j)\,\big]\\ &= (\mathbf{v}\cdot\nabla\mathbf{u})\cdot[\,\boldsymbol{\psi}(\nabla\cdot\mathbf{v}) + \mathbf{v}\cdot\nabla\boldsymbol{\psi}\,]\\ &= (\nabla\cdot\mathbf{v})(\mathbf{v}\cdot\nabla\mathbf{u})\cdot\boldsymbol{\psi} \ +\ (\mathbf{v}\cdot\nabla\mathbf{u})\cdot(\mathbf{v}\cdot\nabla\boldsymbol{\psi})\ . \end{aligned} \quad (5.6.38)$$

Inserting (5.6.37) and (5.6.38) into (5.6.36) leads to

$$\int_{\Omega} \boldsymbol{\psi} \cdot (\mathbf{v} \cdot \nabla)(\mathbf{v} \cdot \nabla \mathbf{u}) \, \mathrm{d}\Omega = \int_{\partial\Omega} (\mathbf{n} \cdot \mathbf{v})(\mathbf{v} \cdot \nabla \mathbf{u}) \cdot \boldsymbol{\psi} \, \mathrm{d}\Gamma - \int_{\Omega} (\nabla \cdot \mathbf{v})(\mathbf{v} \cdot \nabla \mathbf{u}) \cdot \boldsymbol{\psi} \, \mathrm{d}\Omega - \int_{\Omega} (\mathbf{v} \cdot \nabla \mathbf{u}) \cdot (\mathbf{v} \cdot \nabla \boldsymbol{\psi}) \, \mathrm{d}\Omega \,. \tag{5.6.39}$$

The last term in (5.6.39) is of the form $\mathcal{L}(\mathbf{u}) \cdot \mathcal{L}(\boldsymbol{\psi})$, where $\mathcal{L} \equiv \mathbf{v} \cdot \nabla$. We may write

$$\int_{\Omega} \boldsymbol{\psi} \cdot \mathcal{L}(\mathcal{L}(\mathbf{u})) \, \mathrm{d}\Omega = \int_{\partial\Omega} (\mathbf{n} \cdot \mathbf{v}) \, \mathcal{L}(\mathbf{u}) \cdot \boldsymbol{\psi} \, \mathrm{d}\Gamma - \int_{\Omega} \mathrm{div}(\mathbf{v}) \, \mathcal{L}(\mathbf{u}) \cdot \boldsymbol{\psi} \, \mathrm{d}\Omega - \int_{\Omega} \mathcal{L}(\mathbf{u}) \cdot \mathcal{L}(\boldsymbol{\psi}) \, \mathrm{d}\Omega \,, \quad \mathcal{L} \equiv \mathbf{v} \cdot \nabla \,. \tag{5.6.40}$$

If $\mathbf{u}$ and $\boldsymbol{\psi}$ are scalar fields (but the domain Ω may have two or more space dimensions), Eq. (5.6.40) becomes

$$\int_{\Omega} \psi \, \mathcal{L}(\mathcal{L}(u)) \, \mathrm{d}\Omega = \int_{\partial\Omega} (\mathbf{n} \cdot \mathbf{v}) \, \mathcal{L}(u) \, \psi \, \mathrm{d}\Gamma - \int_{\Omega} \mathrm{div}(\mathbf{v}) \, \mathcal{L}(u) \, \psi \, \mathrm{d}\Omega - \int_{\Omega} \mathcal{L}(u) \, \mathcal{L}(\psi) \, \mathrm{d}\Omega \,, \quad \mathcal{L} \equiv \mathbf{v} \cdot \nabla \,. \tag{5.6.41}$$

In the sense of similarity in form, (5.6.39) is the closest possible analogue, for our $\mathcal{L}$, of Green's first integral identity for the Laplacian. Recall that Green's first integral identity, also known as Green's formula, is (see e.g. [34, p. 310], [37, Appendix C])

$$\int_{\Omega} \psi \, \Delta u \, \mathrm{d}\Omega \equiv \int_{\Omega} \psi \nabla \cdot \nabla u \, \mathrm{d}\Omega = \int_{\partial\Omega} (\mathbf{n} \cdot \nabla u) \psi \, \mathrm{d}\Gamma - \int_{\Omega} \nabla u \cdot \nabla \psi \, \mathrm{d}\Omega \,. \tag{5.6.42}$$

Equation (5.6.42) is a consequence of the identity

$$\nabla \cdot (\psi \nabla u) = \partial_i (\psi \partial_i u) = \partial_i \psi \partial_i u + \psi \partial_i \partial_i u = \nabla \psi \cdot \nabla u + \psi \nabla \cdot \nabla u \,, \tag{5.6.43}$$

and the Gauss–Green–Ostrogradsky divergence theorem. If we denote $\mathcal{L} \equiv \nabla$, Green's first integral identity becomes

$$\int_{\Omega} \psi \mathcal{L} \cdot (\mathcal{L}(u)) \, \mathrm{d}\Omega = \int_{\partial\Omega} (\mathbf{n} \cdot \mathcal{L}(u)) \psi \, \mathrm{d}\Gamma - \int_{\Omega} \mathcal{L}(u) \cdot \mathcal{L}(\psi) \, \mathrm{d}\Omega \,, \quad \mathcal{L} \equiv \nabla \,. \tag{5.6.44}$$

Equation (5.6.44) is expressed in a form analogous to (5.6.40), aside from the obvious difference: ∇ raises the tensor rank by one, whereas $\mathbf{v} \cdot \nabla$ is a scalar operator, that is, it does not affect the rank.

A second alternative representation for our result is obtained by observing the identities

$$\begin{aligned}(\mathbf{n}\cdot\mathbf{v})(\mathbf{v}\cdot\nabla\mathbf{u})\cdot\boldsymbol{\psi} &= (n_i v_i)((v_j\partial_j u_k)\psi_k) = n_i(v_i v_j \partial_j u_k)\psi_k = \mathbf{n}\cdot(\mathbf{v}\otimes\mathbf{v}\cdot\nabla\mathbf{u})\cdot\boldsymbol{\psi} \\ &= \boldsymbol{\psi}\cdot(\mathbf{v}\otimes\mathbf{v}\cdot\nabla\mathbf{u})^{\mathrm{T}}\cdot\mathbf{n} \qquad (5.6.45)\end{aligned}$$

and

$$\begin{aligned}(\mathbf{v}\cdot\nabla\mathbf{u})\cdot(\mathbf{v}\cdot\nabla\boldsymbol{\psi}) &= (v_i(\partial_i u_j))(v_k(\partial_k\psi_j)) = (\partial_k\psi_j)(v_k v_i \partial_i u_j) \\ &= (\partial_k\psi_j)(v_j v_i \partial_i u_k)^{\mathrm{T}} = \nabla\boldsymbol{\psi} : (\mathbf{v}\otimes\mathbf{v}\cdot\nabla\mathbf{u})^{\mathrm{T}} , \qquad (5.6.46)\end{aligned}$$

where we use the notation

$$(\mathbf{A}^{\mathrm{T}})_{ij} \equiv A_{ji} , \quad \mathbf{A} : \mathbf{B} \equiv A_{ij}B_{ji} . \qquad (5.6.47)$$

Note the convention where the double-dot operator uses swapped indices in its second operand. In (5.6.45), we have also used the identity

$$\mathbf{a}\cdot\mathbf{B}\cdot\mathbf{c} = a_i B_{ij} c_j = c_j B_{ij} a_i = c_j (B^{\mathrm{T}})_{ji} a_i = \mathbf{c}\cdot\mathbf{B}^{\mathrm{T}}\cdot\mathbf{a} , \qquad (5.6.48)$$

where **a** and **c** are vectors, and **B** is a rank-2 tensor. Inserting (5.6.45) and (5.6.46) into (5.6.39), and rearranging in the middle term, yields

$$\begin{aligned}\int_\Omega \boldsymbol{\psi}\cdot(\mathbf{v}\cdot\nabla)(\mathbf{v}\cdot\nabla\mathbf{u})\,\mathrm{d}\Omega &= \int_{\partial\Omega} \boldsymbol{\psi}\cdot(\mathbf{v}\otimes\mathbf{v}\cdot\nabla\mathbf{u})^{\mathrm{T}}\cdot\mathbf{n}\,\mathrm{d}\Gamma - \int_\Omega \boldsymbol{\psi}\cdot(\mathbf{v}\cdot\nabla\mathbf{u})(\nabla\cdot\mathbf{v})\,\mathrm{d}\Omega \\ &\quad - \int_\Omega \nabla\boldsymbol{\psi} : (\mathbf{v}\otimes\mathbf{v}\cdot\nabla\mathbf{u})^{\mathrm{T}}\,\mathrm{d}\Omega . \qquad (5.6.49)\end{aligned}$$

This form is especially useful in problems of linear elasticity, where $\nabla\boldsymbol{\psi} : (\dots)$ appears naturally; for an example, see Kurki et al. [69]. Equation (5.6.49) is, for our operator, the closest possible analogue to Green's first integral identity in a sense different from (5.6.39); it allows the conversion of terms into a form natural for a particular problem class under consideration.

Equations (5.6.36), (5.6.39) and (5.6.49) are equivalent (for constant **v**), and represent the general result in any number of space dimensions.

Specializing to the case where the domain is one-dimensional, $\Omega = (s_1, s_2)$, we just replace each $\int_\Omega \dots \,\mathrm{d}V \to \int_{s_1}^{s_2} \dots \,\mathrm{d}s$ and $\int_{\partial\Omega} \dots \,\mathrm{d}A \to \sum_{s=s_1,s_2} [\dots]$. In the boundary term, we must use summation instead of one-dimensional definite integral substitution $[\dots]_{s=s_1}^{s_2}$, because the direction of the outer normal **n**, which in one space dimension is $+1$ at s_2 and -1 at s_1, already accounts for the minus sign. Alternatively, we may absorb the **n** into the substitution:

$$\sum_{s=s_1,s_2} \mathbf{n}\cdot(\dots) \;=\; +1\cdot(\dots)\,|_{s=s_2} - 1\cdot(\dots)\,|_{s=s_1} \;\equiv\; [\dots]_{s=s_1}^{s_2} , \qquad (5.6.50)$$

provided that the boundary term has this form.

From (5.6.36), we have

$$\int_{s_1}^{s_2} \psi \cdot (\mathbf{v} \cdot \nabla)(\mathbf{v} \cdot \nabla \mathbf{u}) \, \mathrm{d}s = \Big[(\mathbf{v} \cdot \nabla \mathbf{u}) \cdot (\psi \otimes \mathbf{v}) \Big]_{s=s_1}^{s_2} - \int_{s_1}^{s_2} (\mathbf{v} \cdot \nabla \mathbf{u}) \cdot [\nabla \cdot (\mathbf{v} \otimes \psi)] \, \mathrm{d}s \,, \tag{5.6.51}$$

and from (5.6.39),

$$\begin{aligned}\int_{s_1}^{s_2} \psi \cdot (\mathbf{v} \cdot \nabla)(\mathbf{v} \cdot \nabla \mathbf{u}) \, \mathrm{d}s = & \sum_{s=s_1,s_2} (\mathbf{n} \cdot \mathbf{v})(\mathbf{v} \cdot \nabla \mathbf{u}) \cdot \psi - \int_{s_1}^{s_2} (\nabla \cdot \mathbf{v})(\mathbf{v} \cdot \nabla \mathbf{u}) \cdot \psi \, \mathrm{d}s \\ & - \int_{s_1}^{s_2} (\mathbf{v} \cdot \nabla \mathbf{u}) \cdot (\mathbf{v} \cdot \nabla \psi) \, \mathrm{d}s \,. \end{aligned} \tag{5.6.52}$$

In (5.6.51) we have used (5.6.50) to replace the boundary summation by definite integral substitution, as the boundary term is of the appropriate form, $\mathbf{n} \cdot (\ldots)$. In (5.6.52) this is not possible.

Using (5.6.51), the inertial term (5.6.29) becomes

$$\begin{aligned} - \int_{s_1}^{s_2} m \frac{\partial^2 \widetilde{\mathbf{u}}}{\partial t^2} \cdot \widetilde{\psi} \, \mathrm{d}s = & - \int_{s_1}^{s_2} m\psi \cdot \frac{\partial^2 \mathbf{u}}{\partial t^2} \, \mathrm{d}s - \int_{s_1}^{s_2} 2m\psi \cdot \left[(\mathbf{v} \cdot \nabla) \frac{\partial \mathbf{u}}{\partial t} \right] \mathrm{d}s \\ & - \int_{s_1}^{s_2} m\psi \cdot [(\mathbf{v} \cdot \nabla)(\mathbf{v} \cdot \nabla)\mathbf{u}] \, \mathrm{d}s \,. \\ = & - \int_{s_1}^{s_2} m\psi \cdot \frac{\partial^2 \mathbf{u}}{\partial t^2} \, \mathrm{d}s - \int_{s_1}^{s_2} 2m\psi \cdot \left[(\mathbf{v} \cdot \nabla) \frac{\partial \mathbf{u}}{\partial t} \right] \mathrm{d}s \\ & - \Big[m(\mathbf{v} \cdot \nabla \mathbf{u}) \cdot (\psi \otimes \mathbf{v}) \Big]_{s=s_1}^{s_2} + \int_{s_1}^{s_2} m(\mathbf{v} \cdot \nabla \mathbf{u}) \cdot [\nabla \cdot (\mathbf{v} \otimes \psi)] \, \mathrm{d}s \,, \end{aligned} \tag{5.6.53}$$

or equivalently, using (5.6.52),

$$\begin{aligned} - \int_{s_1}^{s_2} m \frac{\partial^2 \widetilde{\mathbf{u}}}{\partial t^2} \cdot \widetilde{\psi} \, \mathrm{d}s = & - \int_{s_1}^{s_2} m\psi \cdot \frac{\partial^2 \mathbf{u}}{\partial t^2} \, \mathrm{d}s - \int_{s_1}^{s_2} 2m\psi \cdot \left[(\mathbf{v} \cdot \nabla) \frac{\partial \mathbf{u}}{\partial t} \right] \mathrm{d}s \\ & - \sum_{s=s_1,s_2} m(\mathbf{n} \cdot \mathbf{v})(\mathbf{v} \cdot \nabla \mathbf{u}) \cdot \psi + \int_{s_1}^{s_2} m(\nabla \cdot \mathbf{v})(\mathbf{v} \cdot \nabla \mathbf{u}) \cdot \psi \, \mathrm{d}s \\ & + \int_{s_1}^{s_2} m(\mathbf{v} \cdot \nabla \mathbf{u}) \cdot (\mathbf{v} \cdot \nabla \psi) \, \mathrm{d}s \,. \end{aligned} \tag{5.6.54}$$

Inserting (5.6.53) or (5.6.54) into (5.6.28) completes the general weak form for a material subjected to constant-velocity translation. In the case of the beam and panel problems considered in this chapter, the caveats concerning lower-dimensional manifolds embedded in higher-dimensional ambient space, mentioned in the discussion on moving materials, still apply.

In the case where $\mathbf{u} = u$, $\mathbf{v} = V_0$ and $\psi = \phi$ are scalars (e.g. axial vibrations of an axially moving one-dimensional rod), Eq. (5.6.53) reduces to

$$-\int_{s_1}^{s_2} m\frac{\partial^2 \widetilde{u}}{\partial t^2}\phi \, ds = -\int_{s_1}^{s_2} m\phi\frac{\partial^2 u}{\partial t^2} \, ds - \int_{s_1}^{s_2} 2mV_0\frac{\partial^2 u}{\partial x \partial t} \, ds \\ -\left[[m(V_0\frac{\partial u}{\partial x}) \cdot (\phi V_0)] \right]_{s=s_1}^{s_2} + \int_{s_1}^{s_2} mV_0\frac{\partial u}{\partial x}\left[\frac{\partial}{\partial x}(V_0\phi) \right] ds \, .$$

In the same case, the form (5.6.54) reduces to

$$-\int_{s_1}^{s_2} m\frac{\partial^2 \widetilde{u}}{\partial t^2}\phi \, ds = -\int_{s_1}^{s_2} m\phi\frac{\partial^2 u}{\partial t^2} \, ds - \int_{s_1}^{s_2} 2mV_0\frac{\partial^2 u}{\partial x \partial t} \, ds - \left[m(V_0)(V_0\frac{\partial u}{\partial x})\phi \right]_{s=s_1}^{s_2} \\ + \int_{s_1}^{s_2} m\left(\frac{\partial V_0}{\partial x}\right)\left(V_0\frac{\partial u}{\partial x}\right) ds + \int_{s_1}^{s_2} m\left(V_0\frac{\partial u}{\partial x}\right)\left(V_0\frac{\partial \phi}{\partial x}\right) ds \, .$$

From either equation (note $\partial V_0/\partial x \equiv 0$), we obtain the scalar result

$$-\int_{s_1}^{s_2} m\frac{\partial^2 \widetilde{u}}{\partial t^2}\phi \, ds = -\int_{s_1}^{s_2} m\phi\frac{\partial^2 u}{\partial t^2} \, ds - \int_{s_1}^{s_2} 2mV_0\frac{\partial^2 u}{\partial x \partial t} \, ds - \left[mV_0^2\frac{\partial u}{\partial x}\phi \right]_{s=s_1}^{s_2} \\ + \int_{s_1}^{s_2} mV_0^2\frac{\partial u}{\partial x}\frac{\partial \phi}{\partial x} \, ds \, . \tag{5.6.55}$$

This result is also useful for treating components of the small-displacement equations for an axially traveling beam, panel or string. In that context, for the axial component, (5.6.55) is valid as-is, while for the transverse component, we simply replace $u \to w$ and $\phi \to \psi$.

As an example, consider now an axially moving panel in the small-displacement regime. We start from the small-displacement weak form components (5.6.26) and (5.6.27), where we have used the approximated tangent and normal vectors at the boundaries s_1 and s_2. We replace the resultant force and moment by those per unit width (to treat a panel instead of a beam). In the co-moving frame, we have

$$-\int_{\xi_1}^{\xi_2} \widetilde{\mu}\frac{\partial \widetilde{w}}{\partial \xi}\frac{\partial \widetilde{\phi}}{\partial \xi} \, d\xi + \int_{\xi_1}^{\xi_2} \left[\frac{\partial^2 \widetilde{w}}{\partial \xi^2}\widetilde{\widetilde{M}} - \widetilde{\widetilde{N}} \right]\frac{\partial \widetilde{\phi}}{\partial \xi} \, d\xi + \int_{\xi_1}^{\xi_2} \widetilde{\widetilde{M}}\frac{\partial \widetilde{w}}{\partial \xi}\frac{\partial^2 \widetilde{\phi}}{\partial x^2} \, d\xi + \left[\widetilde{\widetilde{N}}\widetilde{\phi} \right]_{\xi=\xi_1}^{\xi_2} \\ + \int_{\xi_1}^{\xi_2} \widetilde{q}_\xi\widetilde{\phi} \, d\xi - \int_{\xi_1}^{\xi_2} m\frac{\partial^2 \widetilde{u}}{\partial t^2}\widetilde{\phi} \, d\xi = 0 \, , \tag{5.6.56}$$

$$\int_{\xi_1}^{\xi_2} \widetilde{\mu}\frac{\partial \widetilde{\psi}}{\partial \xi} \, d\xi - \int_{\xi_1}^{\xi_2} \widetilde{\widetilde{N}}\frac{\partial \widetilde{w}}{\partial \xi}\frac{\partial \widetilde{\psi}}{\partial \xi} \, d\xi - \int_{\xi_1}^{\xi_2} \widetilde{\widetilde{M}}\frac{\partial^2 \widetilde{\psi}}{\partial \xi^2} \, d\xi + \left[-\widetilde{\widetilde{Q}}\,\widetilde{\psi} \right]_{\xi=\xi_1}^{\xi_2} \\ + \left[\widetilde{\widetilde{M}}\frac{\partial \widetilde{\psi}}{\partial \xi} \right]_{\xi=\xi_1}^{\xi_2} + \int_{\xi_1}^{\xi_2} \widetilde{q}_z\widetilde{\psi} \, d\xi - \int_{\xi_1}^{\xi_2} m\frac{\partial^2 \widetilde{w}}{\partial t^2}\widetilde{\psi} \, d\xi = 0 \, , \tag{5.6.57}$$

where $\widetilde{\phi}$ and $\widetilde{\psi}$ are the components of the virtual displacement in the co-moving frame, and the $\partial/\partial t$ are taken at constant ξ. In the laboratory frame, we obtain

$$
\begin{aligned}
-\int_{x_1}^{x_2} \mu \frac{\partial w}{\partial x}\frac{\partial \phi}{\partial x}\,\mathrm{d}x + \int_{x_1}^{x_2} \left[\frac{\partial^2 w}{\partial x^2}\widehat{M} - \widehat{N} \right] \frac{\partial \phi}{\partial x}\,\mathrm{d}x + \int_{x_1}^{x_2} \widehat{M} \frac{\partial w}{\partial x}\frac{\partial^2 \phi}{\partial x^2}\,\mathrm{d}x + \left[\widehat{N}\phi \right]_{x=x_1}^{x_2} \\
+ \int_{x_1}^{x_2} q_x \phi\,\mathrm{d}x - \int_{x_1}^{x_2} \phi m \left[\frac{\partial^2}{\partial t^2} + 2V_0 \frac{\partial^2}{\partial x \partial t} + V_0^2 \frac{\partial^2}{\partial x^2} \right] u\,\mathrm{d}x = 0\,, \qquad (5.6.58)
\end{aligned}
$$

$$
\begin{aligned}
\int_{x_1}^{x_2} \mu \frac{\partial \psi}{\partial x}\,\mathrm{d}x - \int_{x_1}^{x_2} \widehat{N} \frac{\partial w}{\partial x}\frac{\partial \psi}{\partial x}\,\mathrm{d}x - \int_{x_1}^{x_2} \widehat{M} \frac{\partial^2 \psi}{\partial x^2}\,\mathrm{d}x + \left[-\widehat{Q}\psi \right]_{x=x_1}^{x_2} \\
+ \left[\widehat{M} \frac{\partial \psi}{\partial x} \right]_{x=x_1}^{x_2} + \int_{x_1}^{x_2} q_z \psi\,\mathrm{d}x - \int_{x_1}^{x_2} \psi m \left[\frac{\partial^2}{\partial t^2} + 2V_0 \frac{\partial^2}{\partial x \partial t} + V_0^2 \frac{\partial^2}{\partial x^2} \right] w\,\mathrm{d}x = 0\,, \\
(5.6.59)
\end{aligned}
$$

where the $\partial/\partial t$ are now taken at constant x. Rewriting u and w in terms of u_{lab} and w_{lab}, there are no changes to the equations, so we may omit the label "lab" as was done in the strong form (with the same caveat regarding rigor and constitutive models), and just use these equations directly.

We still need to integrate by parts in the terms involving $\partial^2/\partial x^2$ in the last integral in both (5.6.58) and (5.6.59). For simplicity, let us consider the case where m is constant. We use the result (5.6.55) for each component. Grouping the generated terms into the appropriate integrals,

$$
\begin{aligned}
-\int_{x_1}^{x_2} \mu \frac{\partial w}{\partial x}\frac{\partial \phi}{\partial x}\,\mathrm{d}x + \int_{x_1}^{x_2} \left[\frac{\partial^2 w}{\partial x^2}\widehat{M} - (\widehat{N} - m V_0^2 \frac{\partial u}{\partial x}) \right] \frac{\partial \phi}{\partial x}\,\mathrm{d}x \\
+ \int_{x_1}^{x_2} \widehat{M} \frac{\partial w}{\partial x}\frac{\partial^2 \phi}{\partial x^2}\,\mathrm{d}x + \left[(\widehat{N} - m V_0^2 \frac{\partial u}{\partial x})\phi \right]_{x=x_1}^{x_2} \\
+ \int_{x_1}^{x_2} q_x \phi\,\mathrm{d}x - \int_{x_1}^{x_2} \phi m \left[\frac{\partial^2}{\partial t^2} + 2V_0 \frac{\partial^2}{\partial x \partial t} \right] u\,\mathrm{d}x = 0\,, \qquad (5.6.60)
\end{aligned}
$$

$$
\begin{aligned}
\int_{x_1}^{x_2} \mu \frac{\partial \psi}{\partial x}\,\mathrm{d}x - \int_{x_1}^{x_2} (\widehat{N} - m V_0^2) \frac{\partial w}{\partial x}\frac{\partial \psi}{\partial x}\,\mathrm{d}x - \int_{x_1}^{x_2} \widehat{M} \frac{\partial^2 \psi}{\partial x^2}\,\mathrm{d}x + \left[-(\widehat{Q} + m V_0^2 \frac{\partial w}{\partial x})\psi \right]_{x=x_1}^{x_2} \\
+ \left[\widehat{M} \frac{\partial \psi}{\partial x} \right]_{x=x_1}^{x_2} + \int_{x_1}^{x_2} q_z \psi\,\mathrm{d}x - \int_{x_1}^{x_2} \psi m \left[\frac{\partial^2}{\partial t^2} + 2V_0 \frac{\partial^2}{\partial x \partial t} \right] w\,\mathrm{d}x = 0\,. \qquad (5.6.61)
\end{aligned}
$$

Equations (5.6.60) and (5.6.61) are the general weak form of the force balance for small-displacement problems of axially moving panels (and beams). As before, for any specific constitutive model, any time derivatives in the expressions of $\widehat{N}$ and $\widehat{M}$ must be replaced by the co-moving derivative (5.5.5).

We see that in the boundary term, the axial resultant force $\widehat{N}$ has become replaced by an effective force

$$
\widehat{N}_{\mathrm{eff}} \equiv \widehat{N} - m V_0^2 \frac{\partial u}{\partial x}\,, \qquad (5.6.62)
$$

and in the transverse direction, $\widehat{Q}$ has become replaced by an effective shear force

$$\widehat{Q}_{\text{eff}} \equiv \widehat{Q} + m V_0^2 \frac{\partial w}{\partial x} . \tag{5.6.63}$$

The boundary term for the resultant moment remains as it was.

In the transverse component equation (5.6.61), the transverse projection of the axial force has changed from $\widehat{N}\, \partial w/\partial x$, of the classical stationary case, to $(\widehat{N} - m V_0^2)\, \partial w/\partial x$ in the axially moving case. This suggests that in the laboratory frame, the centrifugal effect resulting from the axial driving motion plays the role of a compressive force. Indeed, as is pointed out in the literature (e.g., [153]), increasing the drive velocity eventually causes the traveling beam (or panel) to lose stability analogously to the compressed Euler column.

5.7 Boundary Conditions

In problems of partial differential equations, the domain is seldom all of $\mathbb{R}^n$. Hence in order to completely specify the problem, information about the behavior of the solution at the edges of the considered region must be provided in the form of boundary conditions. In Eq. (5.6.21), we arrived at the weak form of the force balance. As a side effect, the integrations by parts produced boundary terms, which we may use now to read off boundary conditions.

We will first discuss the generated boundary terms, after which we will very briefly review the classical boundary conditions for the transverse motion of beams. We will then move on to the main focus of this section, namely Eulerian boundary conditions for axially moving materials. We will define the Eulerian control interval. We will see that analogously to fluid mechanics, inflow and outflow boundaries must be treated differently. Finally, we will consider a mixed formulation with the resultants N and M as auxiliary variables, and will see that this further mitigates the regularity requirements that must be imposed on the solution. This will also provide us with another perspective for the boundary conditions of axially moving materials.

Let us begin from the boundary terms generated in the derivation of Eq. (5.6.21). As is usual for weak forms, in the boundary terms, each integrand is a product of a physical quantity and a test. It is straightforward to prescribe either one of the factors of the product at the boundary. In each case, prescribing the physical quantity corresponds to a natural (Neumann) boundary condition, while prescribing the quantity indicated by the test corresponds to an essential (Dirichlet) boundary condition.

First, let us consider the tangential direction. In the first term on the third line of (5.6.21), viz.

$$[\, N\mathbf{t} \cdot \psi \,]_{s=s_1}^{s_2} \tag{5.7.1}$$

if we prescribe the axial force N, then its contribution to the virtual work, for any given virtual displacement ψ, is determined by evaluating the expression $N\mathbf{t} \cdot \psi$ at the boundary points s_1 and s_2. In the axially moving case the corresponding force that can be prescribed is N_{eff} or $\widehat{N}_{\mathrm{eff}}$, Eq. (5.6.62).

On the other hand, if we prescribe the tangential component of the displacement $\mathbf{t} \cdot \mathbf{u}$, then the corresponding component of the virtual displacement, $\mathbf{t} \cdot \psi$, must be zero. Physically, this is because the true displacement at the considered boundary point is then fixed, allowing no virtual displacement there. In this case, the contribution of the relevant boundary term at that boundary point is deleted from the equation, and the prescribed value of the tangential displacement $\mathbf{t} \cdot \mathbf{u}$ at the appropriate boundary point is inserted.

Along the normal direction, we have two independent boundary conditions, because the term involving $\partial^2 M/\partial s^2$ was integrated by parts twice. By the middle term on the third line of (5.6.21), viz.

$$[-Q\mathbf{n} \cdot \psi]_{s=s_1}^{s_2} \tag{5.7.2}$$

we may either prescribe the shear force Q, or the quantity corresponding to $\mathbf{n} \cdot \psi$, which is $\mathbf{n} \cdot \mathbf{u}$, the normal component of the displacement. In the axially moving case, Q becomes replaced by Q_{eff} or $\widehat{Q}_{\mathrm{eff}}$, Eq. (5.6.63).

Finally, by the last term on the third line of (5.6.21), namely

$$\left[M\mathbf{n} \cdot \frac{\partial \psi}{\partial s} \right]_{s=s_1}^{s_2} , \tag{5.7.3}$$

the second independent boundary condition along the normal direction is that we may prescribe either M, or the quantity corresponding to $\mathbf{n} \cdot \partial\psi/\partial s$, which is $\mathbf{n} \cdot \partial\mathbf{u}/\partial s$.

In the small-displacement regime, using the approximate tangent and normal vectors (5.6.25) at the boundary points, the natural boundary conditions become analogous to the general case, and it becomes explicit which boundary conditions belong to which component.

From the small-displacement weak forms (5.6.26) and (5.6.27), we read off the small-displacement boundary conditions as follows. In the axial (x) direction, at the boundary points x_1 and x_2 (corresponding to s_1 and s_2 of the general case), we may prescribe either the axial force N, or alternatively,

$$\mathbf{t} \cdot \mathbf{u} = (1, \frac{\partial w}{\partial x}) \cdot (u, w) = u + \frac{\partial w}{\partial x} w \approx u , \tag{5.7.4}$$

where in the last form we have used the fact that the displacement and its derivatives are assumed to be small, discarding the second-order small quantity.

In the transverse (z) direction, we may prescribe either Q, or

$$\mathbf{n} \cdot \mathbf{u} = (-\frac{\partial w}{\partial x}, 1) \cdot (u, w) = -\frac{\partial w}{\partial x} u + w \approx w . \tag{5.7.5}$$

Additionally, we may prescribe either M, or

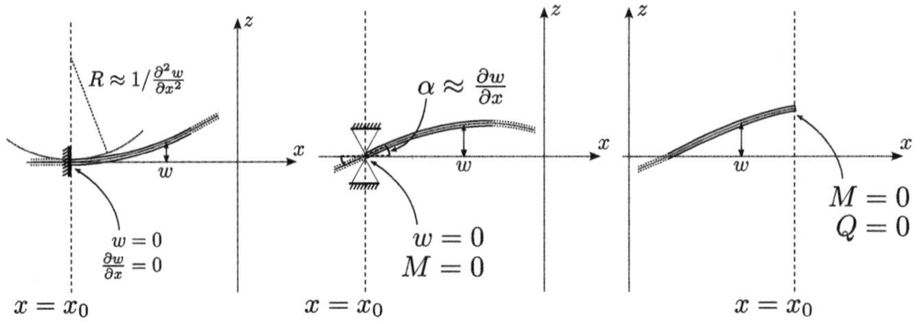

Fig. 5.21 Classical boundary conditions at an endpoint of the beam, for transverse displacement in the small-displacement range. *Left*: Clamped (built-in, C). The end is kinematically constrained. *Center*: Simply supported (hinged, pinned, S). The displacement is zero, but the end may rotate to satisfy the zero moment condition. Mechanically this corresponds to a pointlike hinge placed at the mid-surface. *Right*: Free (F). Both moment and shear force are zero

$$\mathbf{n} \cdot \frac{\partial \mathbf{u}}{\partial s} = (-\frac{\partial w}{\partial x}, 1) \cdot (\frac{\partial u}{\partial x}, \frac{\partial w}{\partial x}) = -\frac{\partial w}{\partial x}\frac{\partial u}{\partial x} + \frac{\partial w}{\partial x} \approx \frac{\partial w}{\partial x} . \tag{5.7.6}$$

Observe that the approximations (5.6.25) are equivalent with $\partial w/\partial x = 0$. At a clamped (C) end, see Fig. 5.21, we have $\partial w/\partial x = 0$, so for a clamped end (5.6.25) are exact. Also, if either $\partial w/\partial x = 0$ or $M = 0$ at an end of the beam, then the second boundary condition for the x component in Eq. (5.6.22) in Sect. 5.7 exactly vanishes at that end. This occurs for clamped (C), simply supported (S) and free (F) ends.

Three out of the four possible combinations of transverse boundary conditions at a given end of the beam correspond to classical small-displacement boundary conditions, shown schematically in Fig. 5.21. These classical cases are defined as follows:

- *Clamped* (C), also known as *built-in*:

$$w = 0 , \quad \partial w/\partial x = 0 ; \tag{5.7.7}$$

- *Simply supported* (S), also known as *hinged* or *pinned*:

$$w = 0 , \quad M = 0 ; \tag{5.7.8}$$

- *Free* (F):

$$M = 0 , \quad Q = 0 . \tag{5.7.9}$$

The fourth combination, $\partial w/\partial x = 0$ and $Q = 0$, is seldom seen in practice.

The simply supported (S) condition, for linear elastic Euler–Bernoulli beams and Kirchhoff–Love plates, is often written as

$$w = 0 , \quad \frac{\partial^2 w}{\partial x^2} = 0 , \tag{5.7.10}$$

but in the general case, this is not equivalent to the definition, Eq. (5.7.8). In the linear elastic case $\partial^2 w/\partial x^2$ corresponds to M, but in general this is not true. To obtain the correct simply supported condition at a point x_0, one must expand $M(x_0) = 0$ using the combination of constitutive and kinematic models for the specific case under consideration.

In the general case, the condition $\partial^2 w/\partial x^2 = 0$ is kinematic [119]. Recall that by Eqs. (5.1.47) and (5.1.61), in the small-displacement regime we have $\partial^2 w/\partial x^2 \approx 1/R$, so this condition actually requires the curvature $1/R$ to be zero.

Let us now move on to consider Eulerian boundary conditions for axially moving beams and panels. From the viewpoint of axially moving materials, the boundary conditions discussed above are actually written in the axially co-moving frame. Since our aim is to describe the motion of the material as it flows past the observer in the laboratory frame, we must first define an Eulerian domain, and then transform the boundary conditions into a form suitable for use with Eulerian quantities.

Because the axially moving material flows through the domain, the inflow boundary requires additional consideration. Similarly to the treatment of inflow boundaries in fluid mechanics, from the viewpoint of information flow, something must be known about the material particles arriving at the inflow end of the Eulerian domain in order to completely specify the problem.

As an illustrative example, we will consider the viscoelastic case, in the context of an axially traveling Kelvin–Voigt material, following and generalizing [41, 119]. As was previously discussed, when dealing with axially moving materials, we first work in the axially co-moving frame, and transform just the end result into the laboratory frame. Interpreting the above derivation for boundary conditions as being written in the co-moving frame, the co-moving boundary points ξ_1 and ξ_2 (corresponding to the earlier x_1 and x_2 in the case of the classical stationary beam) are in axial motion with respect to the laboratory.

Recall the coordinate transformation for axially moving materials, Eq. (5.5.1) in Sect. 5.7. The co-moving frame travels toward $+x$ at a constant velocity V_0. In order to describe boundary points x_1 and x_2, which are taken as stationary in the laboratory frame, we must choose our considered interval (ξ_1, ξ_2) of the beam, in the co-moving frame, dynamically by

$$\xi_j = x_j - V_0 t \;, \quad j = 1, 2 \;. \tag{5.7.11}$$

For fixed x_1 and x_2, this gives us a *control interval* $\Omega \equiv (x_1, x_2)$, which remains stationary in the laboratory. It is an Eulerian domain, analogous to the flow domain in fluid mechanics. In terms of the co-moving frame, (5.7.11) makes the considered interval of the beam depend on t. To emphasize: the set of material particles that comprise the considered section of the beam is different at each fixed value of t.

The dynamics are originally considered in the axially co-moving frame. It is admissible to define boundaries that are in motion with respect to this frame, because the original interval (ξ_1, ξ_2) in the derivation is arbitrary (there denoted by (x_1, x_2) as usual in the classical stationary case), and the weak form force balance equation (5.6.21) (and in the small-displacement regime, respectively Eqs. (5.6.26) and (5.6.27)), which produced the boundary terms, holds separately at each fixed time t.

It may appear that the rest is a simple matter of applying the coordinate transformation (5.5.1) and its consequences, mainly the co-moving derivative (5.5.5). Observe especially that if any $\partial/\partial t$ appear in the boundary conditions (e.g., M for viscoelastic materials), this derivation of the boundary conditions requires these time derivatives to be taken locally in the co-moving frame, that is, at constant ξ. In the transformation to laboratory coordinates, they will be transformed into co-moving derivatives (material derivatives).

We will first look at examples concerning the outflow boundary, and then discuss how to systematically treat the inflow boundary. Let us consider outflow boundary conditions corresponding to the classical C, S and F conditions for an axially traveling isotropic Kelvin–Voigt panel in the small-displacement regime, at the outflow point x_2 of the control interval (x_1, x_2).

The clamped (C) conditions appear as essential (Dirichlet) conditions in the beam problem. In the co-moving frame, we write

$$\widetilde{w}(\xi_2) = 0 \,, \quad \frac{\partial \widetilde{w}}{\partial \xi}(\xi_2) = 0 \,,$$

which holds separately at each fixed time t, at a different material point $\xi_2 = \xi_2(x_2, t)$ for each t. Explicitly, by the definition of the considered boundary, Eq. (5.7.11), we have $\xi_2 = x_2 - V_0 t$.

To proceed rigorously, we apply the coordinate transformation for the axial space derivative, Eq. (5.5.8), which in practice is the identity transformation. In the laboratory frame we thus have

$$w(x_2) = 0 \,, \quad \frac{\partial w}{\partial x}(x_2) = 0 \,.$$

Now consider the simply supported (S) conditions. The S conditions state that $w = 0$ and $M = 0$. In principle, the condition on the moment is specialized to a given combination of kinematic and constitutive models by substituting a specific expression for the resultant moment M.

To consider a panel as an example, we replace the resultant moment M by $\widehat{M}$, the resultant moment per unit width. For the isotropic Kelvin–Voigt panel, in the co-moving frame, $\widehat{M}$ is given by Eq. (5.4.24) as

$$\widetilde{\widehat{M}}(\xi) = \left[D + D_{\mathrm{VE}} \frac{\partial}{\partial t} \right] \frac{\partial^2 \widetilde{w}}{\partial \xi^2} \,,$$

where ξ is the co-moving axial coordinate, and the $\partial/\partial t$ is taken locally in the co-moving frame (i.e., keeping ξ fixed). The notation on the left-hand side makes it explicit that this quantity is $\widehat{M}$, as expressed in the co-moving frame (denoted by the tilde).

In the laboratory frame, by applying the transformations of the derivatives, Eqs. (5.5.5) and (5.5.8), we have

$$\begin{aligned}\widehat{M}(x) &= \left[D + D_{\mathrm{VE}}\left(\frac{\partial}{\partial t} + V_0\frac{\partial}{\partial x}\right)\right]\frac{\partial^2 w}{\partial x^2} ,\\ &= D\frac{\partial^2 w}{\partial x^2} + D_{\mathrm{VE}}\frac{\partial^3 w}{\partial x^2 \partial t} + V_0 D_{\mathrm{VE}}\frac{\partial^3 w}{\partial x^3}\\ &= \frac{\partial}{\partial x}\left[D\frac{\partial w}{\partial x} + D_{\mathrm{VE}}\frac{\partial^2 w}{\partial x \partial t} + V_0 D_{\mathrm{VE}}\frac{\partial^2 w}{\partial x^2}\right] ,\end{aligned} \tag{5.7.12}$$

provided that the material parameters D and D_{VE} do not depend on x (or ξ). Now the $\partial/\partial t$ is taken locally in the laboratory frame (i.e., keeping x fixed). Equation (5.7.12) is the actual physical moment experienced by the traveling material—i.e. the moment in the co-moving frame—expressed in terms of Eulerian (laboratory) quantities. We then express w in terms of w_{lab} using Eq. (5.5.17); in practice, this results in no changes to the equation, so we may drop the label lab and use (5.7.12) as-is.

When using finite differences, prescribing a value for the expression (5.7.12) requires some work to set up using, for example, the method of virtual points. However, for finite elements, recall that $M\,\partial\psi/\partial x$ appears in the weak form for the transverse component, Eq. (5.6.23). In other words, M appears in a natural (Neumann) boundary condition, independent of the constitutive model being used, and all we need to do is to substitute its known value.

With finite elements, we may thus prescribe a value for M directly, without having to specialize the moment boundary condition to a particular combination of kinematic and constitutive models. Furthermore, $M = 0$ is a zero Neumann boundary condition, which is also known as a do-nothing boundary condition, because it causes the boundary term to vanish, requiring no handling in the numerical implementation. Hence, when using finite elements, we enforce $M = 0$ without actually doing anything.

The same remark applies to the condition $Q = 0$. Hence, in terms of a finite element implementation, a free end (F) with $M = 0$ and $Q = 0$ is *do-nothing*.

Now, let us take a step back and consider the formulation of the complete problem. If we insert $\widehat{M}$ from (5.7.12) into the equation for the small transverse displacement of a traveling Kelvin–Voigt panel, (5.5.33) in Sect. 5.5, we obtain a fifth-order partial differential equation, with the term $\partial^2\widehat{M}/\partial x^2$ producing a term $V_0 D_{\mathrm{VE}}\,\partial^5 w/\partial x^5$. This is one order higher in $\partial/\partial x$ than for the corresponding stationary beam (or for the purely elastic beam, even with axial motion). This suggests that for the transverse displacement, an extra boundary condition is needed.

It must be emphasized that an axially traveling beam differs from a stationary one in one important qualitative aspect: new material particles are constantly arriving into the control interval at the inflow boundary x_1, while at the outflow boundary x_2, material particles that have travelled through the control interval are exiting.

This gives a hint as to where to obtain the additional boundary condition. Concerning the physical flow of information, how the traveling material behaves just before it reaches the control interval is important. This is analogous to the behavior of inflow boundaries in fluid mechanics. In order to completely specify the problem, some information is needed regarding the fluid particles arriving into the considered domain.

The following argument follows and expands on Saksa et al. [120], Saksa [119]. Starting from an idea of Flügge [41], we impose physically appropriate continuity conditions at the inflow point x_1. Let us denote by $w^* \equiv w^*(x)$ the transverse displacement outside the considered interval $[x_1, x_2]$ (hence especially for $x < x_1$).

We impose kinematic continuity at the inflow boundary at each fixed time t:

$$\lim_{\delta \to 0} \left[\frac{\partial w}{\partial x}(x_1 + \delta) - \frac{\partial w^*}{\partial x}(x_1 - \delta) \right] = 0 \,, \tag{5.7.13}$$

i.e. the slope of the beam (or panel) must remain continuous at x_1 at all times t; physically, this means there are no sharp creases. Obviously, the displacement itself is also required to remain continuous, but it turns out that the continuity of the slope is more useful for the purposes of deriving an inflow boundary condition for the transverse component.

Secondly, we impose the following continuity condition on the resultant moment M:

$$\lim_{\delta \to 0} \int_{x_1 - \delta}^{x_1 + \delta} M(x) \, \mathrm{d}x = 0 \,, \tag{5.7.14}$$

requiring that (5.7.14) holds at each fixed t. A similar condition was used in Flügge [41] for an infinite beam on a continuous support, with a concentrated load moving along the beam at a constant velocity.

To see more clearly what (5.7.14) represents, we rewrite it as

$$\lim_{\delta \to 0} [\ P(x_1 + \delta) - P(x_1 - \delta)\ ] = 0 \,, \tag{5.7.15}$$

where we have defined

$$P(x) \equiv \int M(x) \, \mathrm{d}x \,. \tag{5.7.16}$$

The condition (5.7.14) requires that the function $P(x)$ is continuous at x_1. Recall that for any function f, if $f(x)$ belongs to the class C^k (functions continuously differentiable k times), then $(\partial f / \partial x)(x)$ belongs to C^{k-1}. Choosing $f = P(x)$, Eq. (5.7.14) then says that at the point x_1, f is continuous, that is, at least C^0. This implies that $M(x) = \partial f / \partial x$ is allowed to be at worst C^{-1} i.e. finitely discontinuous at the point x_1.

A concentrated moment load located at a point x_0 (for a concentrated load of strength μ_0, we have $\mu(x) = \mu_0 \, \delta(x - x_0)$, where $\delta(\dots)$ is the Dirac delta) indeed causes a finite discontinuity in M at x_0, so Eq. (5.7.14) is the physically appropriate general continuity condition for the resultant moment M.

Considering the weak form (5.6.21), the condition (5.7.14) is also mathematically appropriate. Taking (5.6.21) as the definition of our problem, we observe that we do not need to take derivatives of N, Q and M. Hence those derivatives may as well not exist. If we only wish to avoid handling Dirac deltas in the integrals, it is enough if N, Q and M themselves are at most finitely discontinuous, which is exactly what (5.7.14) prescribes for M.

In the special case where the appropriate derivatives do exist, the weak form (5.6.21) is, of course, equivalent with the classical strong form (5.1.62) to (5.1.63), with the same boundary conditions. The point of, after the fact, taking a step back and declaring the weak form (5.6.21) as the definition of the problem, is that it is more general, allowing cases that are not admissible in the strong form (i.e. precisely those cases which lack the derivatives needed to define the strong form).

In the axial direction, we proceed similarly. Let $u^* \equiv u^*(x)$ denote the axial displacement outside the considered interval $[x_1, x_2]$. We impose kinematic continuity

$$\lim_{\delta \to 0} \left[\, u(x_1 + \delta) - u^*(x_1 - \delta) \, \right] = 0 \,, \tag{5.7.17}$$

in other words, physically, there are no gaps in the displacement. Second, we require that the internal axial resultant force N is at most finitely discontinuous:

$$\lim_{\delta \to 0} \int_{x_1 - \delta}^{x_1 + \delta} N(x) \, \mathrm{d}x = 0 \,. \tag{5.7.18}$$

Before explaining how to use the continuity conditions (5.7.13) and (5.7.14) to generate the additional boundary condition for a specific model, let us first make some general observations. In the case of an open system with multiple spans delimited by boundaries, the outflow boundary for span k is the inflow boundary for span $k + 1$. Hence the behavior of the material at the outflow point of span k will, by the above continuity conditions, affect its behavior at the inflow point of span $k + 1$. Because at the outflow point there is no additional boundary condition (when compared to the classical stationary case), this information comes from the solution itself at the end of span k.

This implies that there is an important difference between open systems and closed-loop systems of traveling materials. In a closed-loop system, the last span connects back to the first one. This suggests that a closed-loop system must be analyzed as a whole; the vibration behavior in any given span may differ from a corresponding open system due to this physical coupling. A vibration coupling effect between spans in a closed-loop system indeed appears, as has been reported by Mote and Wu [104].

As an example of how to use the continuity conditions (5.7.13) and (5.7.14), we slightly generalize the treatment of Saksa [119, Sect. 3]. Consider an axially traveling isotropic Kelvin–Voigt panel in the small-displacement regime, flowing across the control interval (x_1, x_2).

Recall that our task is to determine what must be known about the material points entering the domain in order to completely specify the problem, in the form of an additional boundary condition to be imposed at the inflow point x_1.

We already derived an expression for the resultant moment per unit width $\widehat{M}$, as expressed in the laboratory frame, namely (5.7.12). Inserting equation (5.7.12) into the moment continuity condition (5.7.14), and performing the integration to get rid of the common $\partial/\partial x$, we have

$$\lim_{\delta\to 0}\left[D\frac{\partial w}{\partial x}(x_1+\delta)+D_{\mathrm{VE}}\frac{\partial^2 w}{\partial x\partial t}(x_1+\delta)+V_0 D_{\mathrm{VE}}\frac{\partial^2 w}{\partial x^2}(x_1+\delta)\right.$$
$$\left.-D\frac{\partial w^*}{\partial x}(x_1-\delta)-D_{\mathrm{VE}}\frac{\partial^2 w^*}{\partial x\partial t}(x_1-\delta)-V_0 D_{\mathrm{VE}}\frac{\partial^2 w^*}{\partial x^2}(x_1-\delta)\right]=0\,. \quad (5.7.19)$$

In the limit $\delta \to 0$, Eq. (5.7.19) becomes

$$D\frac{\partial w}{\partial x}(x_1)+D_{\mathrm{VE}}\frac{\partial^2 w}{\partial x\partial t}(x_1)+V_0 D_{\mathrm{VE}}\frac{\partial^2 w}{\partial x^2}(x_1)$$
$$-D\frac{\partial w^*}{\partial x}(x_1)-D_{\mathrm{VE}}\frac{\partial^2 w^*}{\partial x\partial t}(x_1)-V_0 D_{\mathrm{VE}}\frac{\partial^2 w^*}{\partial x^2}(x_1)=0\,. \quad (5.7.20)$$

The kinematic continuity condition (5.7.13) says that

$$\frac{\partial w}{\partial x}(x_1)-\frac{\partial w^*}{\partial x}(x_1)=0$$

for all t. Assuming sufficient continuity so that we may write $\partial^2 w/\partial x\partial t = (\partial/\partial t)(\partial w/\partial x)$, we have also

$$\frac{\partial^2 w}{\partial x\partial t}(x_1)-\frac{\partial^2 w^*}{\partial x\partial t}(x_1)=\frac{\partial}{\partial t}\left[\frac{\partial w}{\partial x}(x_1)-\frac{\partial w^*}{\partial x}(x_1)\right]=0\,.$$

We are left with

$$V_0 D_{\mathrm{VE}}\left[\frac{\partial^2 w}{\partial x^2}(x_1)-\frac{\partial^2 w^*}{\partial x^2}(x_1)\right]=0\,, \quad (5.7.21)$$

which is the required additional boundary condition, representing the requirement (5.7.14) that the resultant moment per unit width $\widehat{M}(x)$ is, at most, finitely discontinuous.

Observe that if $V_0 D_{\mathrm{VE}} = 0$, that is, if the material is either fully elastic ($\eta = 0$) or not in axial motion ($V_0 = 0$), Eq. (5.7.21) vanishes identically, meaning in those cases no additional boundary condition is generated. We should indeed expect this.

The $V_0\,\partial^5/\partial x^5$ term only appears in the case $\eta \neq 0$, $V_0 \neq 0$, so only in that case is a fifth boundary condition needed.

If $V_0 D_{\mathrm{VE}} \neq 0$, we thus have

$$\frac{\partial^2 w}{\partial x^2}(x_1) = \frac{\partial^2 w^*}{\partial x^2}(x_1)\,. \tag{5.7.22}$$

We may interpret (5.7.22) as a kinematic condition, requiring that the curvature ($\partial^2 w/\partial x^2 \approx 1/R$) is continuous at the inflow point x_1. However, the motivation, and the reason behind the physical validity of (5.7.22) is that for this specific model, it is a consequence of the moment continuity condition (5.7.14).

In conclusion, we may write the additional transverse boundary condition for the axially moving Kelvin–Voigt panel (or beam) as

$$\frac{\partial^2 w}{\partial x^2}(x_1) = \kappa\,, \tag{5.7.23}$$

where $\kappa = 1/R$ is the local curvature of the panel at the inflow point x_1, considered given; its numeric value obviously depends on the geometry of the physical situation being modelled. For example, if the traveling panel represents a paper web traveling around a roller and then entering an open draw, then $\kappa = 1/r$, where r is the radius of the roller. If the traveling panel just passes the support in a straight line, then $\kappa = 0$.

To obtain (5.7.23), we have used only the continuity conditions (5.7.13) and (5.7.14), and the combination of kinematic and constitutive models. Hence it is valid for any choice of other boundary conditions at the inflow point x_1. If we have C conditions at the inflow point, then (5.7.23) is independent of them, and can be prescribed as an additional boundary condition. It is an essential (Dirichlet) boundary condition for the second derivative, which can be applied using Hermite C^2 elements.

This motivates the definition of the C^+ (C-plus, clamped-plus) inflow boundary conditions:

$$w(x_1) = 0\,, \quad \frac{\partial w}{\partial x}(x_1) = 0\,, \quad \frac{\partial^2 w}{\partial x^2}(x_1) = \kappa\,. \tag{5.7.24}$$

The name C^+ follows the nomenclature of Saksa [119], indicating that (5.7.24) is an extension of the classical clamped (C) conditions, but we allow for nonzero κ.

A trivial generalization are the kinematically constrained (K) inflow boundary conditions:

$$w(x_1) = 0\,, \quad \frac{\partial w}{\partial x}(x_1) = \theta\,, \quad \frac{\partial^2 w}{\partial x^2}(x_1) = \kappa\,, \tag{5.7.25}$$

where θ is the angle of $\mathbf{t}(x_1)$, the tangent of the panel surface at the inflow point x_1, measured counterclockwise from the $+x$ axis. We have used the small-displacement approximation, where $\tan\theta \approx \theta$.

The C^+ (and K) inflow boundary conditions (5.7.24) (respectively (5.7.25)) are valid at the inflow point x_1 for an axially traveling Kelvin–Voigt panel in the small-displacement regime.

The simply supported (S) conditions can be extended similarly. The general additional boundary condition for the Kelvin–Voigt model, Eq. (5.7.21), again requires continuity of $\partial^2 w/\partial x^2$. Note that for a Kelvin–Voigt material, this quantity is linearly independent of M.

Thus, the following three conditions, which we define as the S^+ (S-plus, simply supported plus) inflow boundary conditions, are independent of each other:

$$w(x_1) = 0 \,, \quad \frac{\partial^2 w}{\partial x^2}(x_1) = \kappa \,, \quad M(x_1) = 0 \,. \tag{5.7.26}$$

The conditions (5.7.26) are already specific to the Kelvin–Voigt constitutive model, because the Kelvin–Voigt model has been used in the derivation of (5.7.21). For other constitutive models, the above derivation of (5.7.21) must be repeated using the appropriate expression for M. Specifically for the linear elastic model, no additional condition is generated.

To specialize explicitly, we may use (5.7.12). The conditions (5.7.26) become

$$w(x_1) = 0 \,, \quad \frac{\partial^2 w}{\partial x^2}(x_1) = \kappa \,, \quad \left(D\frac{\partial^2 w}{\partial x^2} + D_{\mathrm{VE}}\frac{\partial^3 w}{\partial x^2 \partial t} + V_0 D_{\mathrm{VE}}\frac{\partial^3 w}{\partial x^3} \right)\bigg|_{x=x_1} = 0 \,. \tag{5.7.27}$$

In (5.7.26) and (5.7.27), the curvature at the inflow point, namely κ, has so far been treated as a prescribed quantity. The first two conditions are essential (Dirichlet) boundary conditions for w and $\partial^2 w/\partial x^2$, respectively, while the third condition is a natural (Neumann) condition on M.

Whereas we have used the kinematic continuity condition (5.7.13) for obtaining (5.7.26), now we do not impose a value for $\partial w/\partial x$ explicitly. The condition (5.7.13) just requires that the slope is continuous across the support at x_1—whatever its value (as a consequence of the dynamical equation and the other boundary conditions) happens to be.

In light of (5.7.26), imposing an explicit value for $\partial w/\partial x$ would be problematic in practice, because if we prescribe $\partial w/\partial x$, the boundary term that sets $M = 0$—using (5.7.26) as a Neumann condition without explicit specialization—will be lost from the transverse component of the small-displacement weak form, Eq. (5.6.27). Recall that the pairs that appear in the natural boundary conditions are $(w, \; Q)$ and $(\partial w/\partial x, \; M)$; we may prescribe only one component of each pair.

Thus, the conditions (5.7.26) and (5.7.27), while possibly useful for single-span setups, are inappropriate for setups with multiple spans, where the continuity conditions (5.7.13) and (5.7.21) at the start of each span (except the first) must be enforced by applying kinematic data from the end of the previous span.

This can be remedied as follows. Solving the last condition in (5.7.27) (that represents $M(x_1) = 0$) for $\partial^2 w/\partial x^2$ gives

$$\frac{\partial^2 w}{\partial x^2}\bigg|_{x=x_1} = -\frac{1}{D}\left(D_{\mathrm{VE}}\frac{\partial^3 w}{\partial x^2 \partial t} + V_0 D_{\mathrm{VE}}\frac{\partial^3 w}{\partial x^3} \right)\bigg|_{x=x_1} \,, \tag{5.7.28}$$

provided that $D \neq 0$ (i.e., $E \neq 0$, i.e. the material exhibits at least some elasticity and not only viscous behavior). On the other hand, by the second condition in (5.7.27), the left-hand side of (5.7.28) is exactly κ. Equation (5.7.28) thus eliminates κ, fixing it dynamically to a value that satisfies $M = 0$ at any given t.

In Hermite C^2 elements, $\partial^2 w/\partial x^2$ is available as a degree of freedom. Hence, it can be prescribed as an essential (Dirichlet) boundary condition, independent of the pairs $(w,\ Q)$ and $(\partial w/\partial x,\ M)$ that appear in the natural boundary conditions. For these elements, the third derivative that appears in (5.7.28) exists and is finitely discontinuous across element boundaries. We only need its value at the inflow point (i.e., the left endpoint of the first element).

Now that the condition $M = 0$ has been encoded into the equivalent kinematic condition (5.7.28), the third condition in (5.7.26) has become redundant. We no longer need to set the natural (Neumann) boundary condition on M; hence, we are free to introduce a condition on its pair $\partial w/\partial x$.

Applying the kinematic continuity condition (5.7.13) as before, we thus obtain an alternative representation for the S^+ inflow boundary conditions for Kelvin–Voigt material:

$$w(x_1) = 0\,, \quad \frac{\partial w}{\partial x}(x_1) = \theta\,, \quad \left.\frac{\partial^2 w}{\partial x^2}\right|_{x=x_1} = -\frac{1}{D}\left.\left(D_{\mathrm{VE}}\frac{\partial^3 w}{\partial x^2 \partial t} + V_0 D_{\mathrm{VE}}\frac{\partial^3 w}{\partial x^3}\right)\right|_{x=x_1} . \tag{5.7.29}$$

The conditions (5.7.29) enforce both kinematic continuity of $\partial w/\partial x$ at x_1 (Eq. (5.7.13)) and the zero moment condition $M(x_1) = 0$. The continuity of $\partial^2 w/\partial x^2$ at x_1 (Eq. (5.7.21)) is also enforced if $M = 0$ also at the endpoint of the previous span (that coincides with the start of this span), because then (5.7.28) must have a unique value at x_1 regardless of which span is being considered. At the outflow of the previous span, we do not need to evaluate $\partial^3 w/\partial x^3$, because there $M = 0$ can be enforced as a Neumann condition.

The right-hand side of (5.7.28) refers to the unknown field w; it depends linearly on (derivatives of) the solution. Thus, we cannot eliminate this degree of freedom from the discrete equation system by substituting a prescribed value, because the value of the right-hand side is not explicitly known. Instead, to enforce this boundary condition, we insert the Galerkin representation of w into the right-hand side, and then replace the appropriate row of the discrete equation system by the discrete representation of (5.7.28) thus obtained.

This procedure is analogous to the one that is used to treat Robin boundary conditions when solving the Poisson problem. The standard Robin boundary condition for the Poisson problem is $\mathbf{n} \cdot \nabla u = \beta u$, where β is a constant. The left-hand side appears in the natural boundary condition, while the right-hand side can be expressed using the Galerkin representation of u. Inserting the right-hand side, expressed in this way, as the known value into the natural boundary condition, then enforces the Robin boundary condition.

The only difference with our present case is that the left-hand side of (5.7.28) corresponds to a Dirichlet condition directly prescribing a degree of freedom, instead of being an expression that appears in a Neumann boundary condition.

In the axial direction, we may derive the additional boundary condition for the inflow point similarly. For the Kelvin–Voigt panel in the small-displacement regime, the resultant axial force per unit width $\widehat{N}$ is, in the co-moving frame, given by (5.4.23), namely

$$\widetilde{\widehat{N}}(\xi) = h\left[\frac{E}{1-\nu^2} + \frac{\eta}{1-\mu^2}\frac{\partial}{\partial t}\right]\frac{\partial u}{\partial \xi}\,,$$

where the $\partial/\partial t$ is taken at a fixed value of the co-moving coordinate ξ. Transforming into the laboratory frame, we have

$$\begin{aligned}\widehat{N}(x) &= h\left[\frac{E}{1-\nu^2} + \frac{\eta}{1-\mu^2}\left(\frac{\partial}{\partial t} + V_0\frac{\partial}{\partial x}\right)\right]\frac{\partial u}{\partial x}\\ &= h\left[\frac{E}{1-\nu^2}\frac{\partial u}{\partial x} + \frac{\eta}{1-\mu^2}\frac{\partial^2 u}{\partial x\partial t} + \frac{V_0\eta}{1-\mu^2}\frac{\partial^2 u}{\partial x^2}\right]\\ &= h\frac{\partial}{\partial x}\left[\frac{E}{1-\nu^2}u + \frac{\eta}{1-\mu^2}\frac{\partial u}{\partial t} + \frac{V_0\eta}{1-\mu^2}\frac{\partial u}{\partial x}\right]\,,\end{aligned} \tag{5.7.30}$$

where now the $\partial/\partial t$ is taken at fixed laboratory coordinate x, and we have assumed sufficient continuity to reorder derivatives. We then express u in terms of u_{lab} using Eq. (5.5.17); in practice, this results in no changes to the equation, so we may drop the label "lab" and use (5.7.30) as-is.

Inserting (5.7.30) into the axial force continuity condition (5.7.18), performing the integration to get rid of the common $\partial/\partial x$, and applying the kinematic continuity condition (5.7.17) (which holds at all t), we are left with

$$\frac{h}{1-\mu^2}V_0\eta\left[\frac{\partial u}{\partial x}(x_1) - \frac{\partial u^*}{\partial x}(x_1)\right] = 0\,. \tag{5.7.31}$$

Again, this generates an additional boundary condition if and only if $V_0\eta \neq 0$ In other words, the material must be both viscoelastic ($\eta \neq 0$) and in axial motion ($V_0 \neq 0$). This is as expected; the $\partial^3 u/\partial x^3$ term, requiring an additional boundary condition, only appears for the viscoelastic axially moving case, when we insert $\widehat{N}$ from (5.7.30) into the equation for the small axial displacement of a traveling Kelvin–Voigt panel, (5.5.32) in Sect. 5.5. It is produced by the term $\partial\widehat{N}/\partial x$, provided that $V_0\eta \neq 0$.

Thus, we see that the continuity condition for the axial force, Eq. (5.7.18), which allows the force to be at most finitely discontinuous, implies for the axially moving Kelvin–Voigt panel that the axial strain $\partial u/\partial x = \varepsilon_{xx}$ must be continuous.

Because in the axial direction, we may classically prescribe either u or N, we have two alternative sets of *axial inflow boundary conditions* for the traveling Kelvin–Voigt panel, namely

$$u = u_0 \,, \quad \frac{\partial u}{\partial x} = \varepsilon_0 \tag{5.7.32}$$

and

$$N = N_0 \,, \quad \frac{\partial u}{\partial x} = \varepsilon_0 \,. \tag{5.7.33}$$

The quantity ε_0 is a prescribed axial strain. For a multi-span system, it can be readily calculated from the axial drive velocities at the rollers of the previous span; recall Eq. (5.5.57) in Sect. 5.5. At the outflow boundary, one prescribes just u or N in the usual manner.

The natural boundary term in the axial component of the small-displacement weak form, Eq. (5.6.22), refers to only u and N. Hence, in principle, we may set $\partial u/\partial x$ at the boundary without disturbing the existing boundary condition. To do this in practice, it is possible to use beam elements for u, making $\partial u/\partial x$ appear as a degree of freedom, that can then be prescribed via a Dirichlet boundary condition at the inflow boundary.

The formulation discussed above has required C^2 continuity for w, and C^1 continuity for u. This is perhaps the appropriate point to take a step back and consider the resultant force and moment for axially moving linear viscoelastic materials. This will provide us with another perspective to the boundary conditions, along with reducing continuity requirements.

For the sake of generality, we will do this for an arbitrary linear viscoelastic material. Recall the general linear viscoelastic material law, Eq. (5.3.46) in Sect. 5.3. Constitutive models with $a_1 = 0$ (linear elastic and Kelvin–Voigt) can be treated in the classical manner, by explicitly representing N and M with the help of the resultant integrals (5.3.2), the constitutive law (5.3.46), and the small-displacement total axial strain (5.2.43).

For constitutive models with $a_1 \neq 0$ (Maxwell and SLS), there are two main options for practical numerical work. A classical approach, used in books such as those by Sobotka [128] and Marynowski [88] and in the study [123], is based on the following algebraic manipulation. Let us apply the operator $\Gamma(\ldots)$, appearing on the left-hand side of the constitutive law (5.3.46), to the resultant integrals (5.3.2), and then use the constitutive law to rewrite the integrand. We obtain

$$\Gamma N(x) = b \int_{-h/2}^{+h/2} \Gamma \sigma_{xx} \,\mathrm{d}z = b \int_{-h/2}^{+h/2} \Xi\, \varepsilon_{xx} \,\mathrm{d}z \,, \tag{5.7.34}$$

$$\Gamma M(x) = -b \int_{-h/2}^{+h/2} \Gamma \sigma_{xx}\, z \,\mathrm{d}z = -b \int_{-h/2}^{+h/2} \Xi\, \varepsilon_{xx}\, z \,\mathrm{d}z \,. \tag{5.7.35}$$

We have here used the fact that the operator $\Gamma(\ldots)$ does not depend on y or z, so it can be taken inside the integral. Sobotka [128] calls the quantity $\Gamma M(x)$ the equivalent bending moment. Now, by inserting the small-displacement total axial strain (5.2.43),

$$\varepsilon_{xx}(x, z) \approx \frac{\partial u}{\partial x}(x) - z\frac{\partial^2 w}{\partial x^2}(x) ,$$

into Eqs. (5.7.34) and (5.7.35), we have

$$\begin{aligned} \Gamma N(x) &= b \int_{-h/2}^{+h/2} \Xi \left(\frac{\partial u}{\partial x}(x) - z\frac{\partial^2 w}{\partial x^2}(x) \right) \mathrm{d}z \\ &= bh\,\Xi \left(\frac{\partial u}{\partial x} \right) \\ &= A\,\Xi \left(\frac{\partial u}{\partial x} \right) , \end{aligned} \tag{5.7.36}$$

$$\begin{aligned} \Gamma M(x) &= -b \int_{-h/2}^{+h/2} \Xi \left(\frac{\partial u}{\partial x}(x) - z\frac{\partial^2 w}{\partial x^2}(x) \right) z\,\mathrm{d}z \\ &= \frac{bh^3}{12}\,\Xi \left(\frac{\partial^2 w}{\partial x^2} \right) \\ &= I\,\Xi \left(\frac{\partial^2 w}{\partial x^2} \right) , \end{aligned} \tag{5.7.37}$$

where $A = bh$ is the cross-sectional area of the beam, and the second moment of area I (of the cross-section) is given by (5.3.11). Note the analogy with linear elasticity, where recalling Eqs. (5.3.9) and (5.3.10), we have

$$N(x) = AE\,\frac{\partial u}{\partial x} , \quad M(x) = IE\,\frac{\partial^2 w}{\partial x^2} .$$

Applying $\Gamma(\ldots)$ to the small-displacement force balance Eqs. (5.2.30) and (5.2.31) and reordering differentiations, provided sufficient continuity, will cause ΓN and ΓM to appear in the result, allowing us to insert (5.7.36) and (5.7.37), and then proceed as usual.

This approach has the drawback of increasing continuity requirements, because the operator $\Gamma(\ldots)$ introduces an additional $\partial/\partial t$. In the context of axially moving materials, this $\partial/\partial t$ is taken in the co-moving frame. Thus, in the laboratory coordinates it will become a co-moving derivative, introducing a term with another $\partial/\partial x$ to the already highest-order term in the force balance equation, namely $\partial^2 M/\partial x^2$. This was already observed above for the axially moving Kelvin–Voigt panel.

The second approach is to formulate the problem in a mixed form. We include a representation of the constitutive law (5.3.46) into the model along with the force balance equation. This leads to a system of differential equations, where the fields N and M play the role of auxiliary variables, analogous to the role of the flux variable in the solution of the Poisson equation in the mixed form. Then the system can be discretized simultaneously, removing the need to have explicit formulas for N and M.

In this second approach, Eqs. (5.7.36) and (5.7.37) are still useful, because ultimately the stress-related quantities appearing in the force balance equation are the resultants N and M. Instead of seeking to insert them into the force balance equations, we just include (5.7.36) and (5.7.37) in the system of differential equations to be solved numerically, and use the numerical representations of N and M in the force balance equation.

In either approach, for axially moving materials, one must keep in mind that the constitutive law (5.3.46) is written in the co-moving frame, i.e. each $\partial/\partial t$ is taken at a fixed value of the co-moving axial coordinate ξ. When the constitutive law is transformed into laboratory coordinates, the time derivatives become co-moving derivatives, and we have

$$\left[a_0 + a_1\left(\frac{\partial}{\partial t} + V_0\frac{\partial}{\partial x}\right)\right]\sigma = \left[b_0 + b_1\left(\frac{\partial}{\partial t} + V_0\frac{\partial}{\partial x}\right)\right]\varepsilon\,, \tag{5.7.38}$$

where V_0 is a constant axial drive velocity, and each $\partial/\partial t$ is now taken at a fixed value of the laboratory axial coordinate x. Equations (5.7.36) and (5.7.37) become, respectively,

$$\left[a_0 + a_1\left(\frac{\partial}{\partial t} + V_0\frac{\partial}{\partial x}\right)\right]N(x) = A\left[b_0 + b_1\left(\frac{\partial}{\partial t} + V_0\frac{\partial}{\partial x}\right)\right]\frac{\partial u}{\partial x}\,, \tag{5.7.39}$$

$$\left[a_0 + a_1\left(\frac{\partial}{\partial t} + V_0\frac{\partial}{\partial x}\right)\right]M(x) = I\left[b_0 + b_1\left(\frac{\partial}{\partial t} + V_0\frac{\partial}{\partial x}\right)\right]\frac{\partial^2 w}{\partial x^2}\,. \tag{5.7.40}$$

We are now ready to consider the weak form of the resultant force and moment for axially moving linear viscoelastic materials. By treating (5.7.39) and (5.7.40) in a weak form, we may allow for N and M that are not differentiable in space. We integrate by parts to transfer the $\partial/\partial x$ to the test function, thus interpreting the differentiation in the weak sense. Also, to avoid introducing higher derivatives of u and w than already occur in the force balance equations, we integrate by parts on the right-hand side.

Expanding the parentheses and brackets in (5.7.39) and (5.7.40), we have

$$a_0 N + a_1\frac{\partial N}{\partial t} + a_1 V_0\frac{\partial N}{\partial x} = Ab_0\frac{\partial u}{\partial x} + Ab_1\frac{\partial^2 u}{\partial x\partial t} + Ab_1 V_0\frac{\partial^2 u}{\partial x^2}\,, \tag{5.7.41}$$

$$a_0 M + a_1\frac{\partial M}{\partial t} + a_1 V_0\frac{\partial M}{\partial x} = Ib_0\frac{\partial^2 w}{\partial x^2} + Ib_1\frac{\partial^3 w}{\partial x^2\partial t} + Ib_1 V_0\frac{\partial^3 w}{\partial x^3}\,. \tag{5.7.42}$$

For the resultant axial force N, multiplying (5.7.41) by a test function $\varphi = \varphi(x)$ and integrating over the control interval (x_1, x_2) (in the laboratory frame), we have

$$a_0 \int_{x_1}^{x_2} N \varphi \,\mathrm{d}x + a_1 \int_{x_1}^{x_2} \frac{\partial N}{\partial t} \varphi \,\mathrm{d}x + a_1 V_0 \int_{x_1}^{x_2} \frac{\partial N}{\partial x} \varphi \,\mathrm{d}x$$
$$= A b_0 \int_{x_1}^{x_2} \frac{\partial u}{\partial x} \varphi \,\mathrm{d}x + A b_1 \int_{x_1}^{x_2} \frac{\partial^2 u}{\partial x \partial t} \varphi \,\mathrm{d}x + A b_1 V_0 \int_{x_1}^{x_2} \frac{\partial^2 u}{\partial x^2} \varphi \,\mathrm{d}x \,. \tag{5.7.43}$$

Equation (5.7.43) is a statement of the principle of virtual work; it is required to hold for any admissible φ. In physical terms, φ is an arbitrary axial virtual displacement. It is independent of the axial virtual displacement that appears in the weak form of the axial displacement Eq. (5.6.26), but it must conform to the same restrictions. Particularly, if u is prescribed at a boundary point, then at that point $\varphi = 0$.

Applying integration by parts on both sides of (5.7.43) yields

$$a_0 \int_{x_1}^{x_2} N \varphi \,\mathrm{d}x + a_1 \int_{x_1}^{x_2} \frac{\partial N}{\partial t} \varphi \,\mathrm{d}x - a_1 V_0 \int_{x_1}^{x_2} N \frac{\partial \varphi}{\partial x} \,\mathrm{d}x + a_1 V_0 \left[N \varphi \right]_{x=x_1}^{x_2}$$
$$= A b_0 \int_{x_1}^{x_2} \frac{\partial u}{\partial x} \varphi \,\mathrm{d}x + A b_1 \int_{x_1}^{x_2} \frac{\partial^2 u}{\partial x \partial t} \varphi \,\mathrm{d}x$$
$$- A b_1 V_0 \int_{x_1}^{x_2} \frac{\partial u}{\partial x} \frac{\partial \varphi}{\partial x} \,\mathrm{d}x + A b_1 V_0 \left[\frac{\partial u}{\partial x} \varphi \right]_{x=x_1}^{x_2} \,. \tag{5.7.44}$$

Equation (5.7.44) is the weak form which allows us to solve for $N(x)$ numerically using finite elements. Note that all nonclassical terms vanish if $V_0 = 0$.

In the case where $a_1 V_0 \neq 0$ (axially moving Maxwell or SLS), the boundary term that appears on the left-hand side makes it technically possible, at the endpoints of the considered control interval (x_1, x_2), to prescribe either N, or the quantity corresponding to the virtual axial displacement φ, which is the axial displacement u. These possibilities are exactly the same as in the small-displacement axial boundary condition (5.7.4), which was generated by (5.6.26), the weak form of the axial displacement equation. Thus, the same information must be fed into this boundary term, and it does not actually represent a new boundary condition.

On the right-hand side of the equation, if $b_1 V_0 \neq 0$ (axially moving Kelvin–Voigt or SLS), the boundary term makes it technically possible to prescribe either $\partial u/\partial x$ or u. Upon closer consideration, this boundary term appears only for the boundary point x_i, $i = 1, 2$, if in the axial displacement equation, N (instead of u) is prescribed at the point x_i. If u is prescribed at the point x_i in the axial displacement equation, this boundary term vanishes for that point, because then the virtual axial displacement φ is zero at x_i.

When N is prescribed at x_i, one must insert here an expression for $\partial u/\partial x = \varepsilon_{xx}$ at that point. Since this term physically describes the axial transport of $\partial u/\partial x$, one should prescribe $\partial u/\partial x$ only at the inflow point x_1, thus providing the necessary information on the behavior of the material when it arrives at the control interval. At the outflow point x_2, if this term appears, the Galerkin representation of u should be inserted. This retains the term in the equation without prescribing a value.

The resultant moment M is treated similarly. Multiplying (5.7.42) by a test function $\theta = \theta(x)$ and integrating over the domain (x_1, x_2), we have

$$
\begin{aligned}
a_0 \int_{x_1}^{x_2} M\theta \, \mathrm{d}x + a_1 \int_{x_1}^{x_2} \frac{\partial M}{\partial t}\theta \, \mathrm{d}x + a_1 V_0 \int_{x_1}^{x_2} \frac{\partial M}{\partial x}\theta \, \mathrm{d}x \\
= Ab_0 \int_{x_1}^{x_2} \frac{\partial^2 w}{\partial x^2}\theta \, \mathrm{d}x + Ab_1 \int_{x_1}^{x_2} \frac{\partial^3 w}{\partial^2 x \partial t}\theta \, \mathrm{d}x + Ab_1 V_0 \int_{x_1}^{x_2} \frac{\partial^3 w}{\partial x^3}\theta \, \mathrm{d}x \, . \quad (5.7.45)
\end{aligned}
$$

Equation (5.7.45) must hold for any admissible θ. In physical terms, the test function represents an arbitrary virtual rotation in the xz plane by an angle θ; also Eq. (5.7.45) is a statement of the principle of virtual work. Performing integration by parts, we have

$$
\begin{aligned}
a_0 \int_{x_1}^{x_2} M\theta \, \mathrm{d}x + a_1 \int_{x_1}^{x_2} \frac{\partial M}{\partial t}\theta \, \mathrm{d}x - a_1 V_0 \int_{x_1}^{x_2} M \frac{\partial \theta}{\partial x} \, \mathrm{d}x + a_1 V_0 \left[M\theta \right]_{x=x_1}^{x_2} \\
= Ab_0 \int_{x_1}^{x_2} \frac{\partial^2 w}{\partial x^2}\theta \, \mathrm{d}x + Ab_1 \int_{x_1}^{x_2} \frac{\partial^3 w}{\partial x^2 \partial t}\theta \, \mathrm{d}x \\
- Ab_1 V_0 \int_{x_1}^{x_2} \frac{\partial^2 w}{\partial x^2} \frac{\partial \theta}{\partial x} \, \mathrm{d}x + Ab_1 V_0 \left[\frac{\partial^2 w}{\partial x^2}\theta \right]_{x=x_1}^{x_2} \, . \quad (5.7.46)
\end{aligned}
$$

Equation (5.7.46) is the weak form which allows us to solve for $M(x)$ numerically. Here too the nonclassical terms vanish if $V_0 = 0$.

If $a_1 V_0 \neq 0$, on the left hand side it is technically possible to prescribe, at the endpoints, either the resultant moment M, or the quantity corresponding to θ, which in the small-displacement regime is $\partial w / \partial x$. The same information is already fed into the small-displacement transverse boundary condition (5.7.6), which was generated by (5.6.27), the weak form of the transverse displacement equation. Hence, this term does not introduce a new boundary condition.

If $b_1 V_0 \neq 0$, on the right-hand side, it is possible to prescribe either $\partial^2 w / \partial x^2$ or $\partial w / \partial x$. Note that whereas the test function θ represents a rotation (i.e., a difference of orientation), the quantity $\partial w / \partial x$ represents an orientation. Also keep in mind that when $b_1 \neq 0$, the term $\partial^2 w / \partial x^2$ is no longer equivalent to M; this is an independent kinematic condition. We conclude that for a clamped end (C), no new boundary condition is generated, but for a simply supported (S) or free (F) end, which does not prescribe $\partial w / \partial x$, an additional kinematic condition for $\partial^2 w / \partial x^2$ appears.

Similarly to the above, this boundary term represents the axial transport of $\partial^2 w / \partial x^2$, so the same notes apply. When this term appears ($\partial w / \partial x$ not prescribed at x_i), a prescribed value for $\partial^2 w / \partial x^2$ should only be supplied at x_1, and at x_2 the Galerkin series of w should be used to retain the term in the equation while not prescribing a value.

This mixed form, where we use N and M as auxiliary variables, has beneficial implications for solving the force balance equation. Particularly, it requires only a bare minimum of regularity, regardless of which of the discussed constitutive models is used.

In the weak forms (5.7.44) and (5.7.46), there is no need to take derivatives of the fields N and M. Hence, in a finite element solution, we may use any basis (including discontinuous bases) for both quantities.

In the mixed form, $\partial^2 w/\partial x^2$ is the highest space derivative that appears anywhere in the equations. This suggests that the Sobolev space H^2 is the natural space in which to look for the solution w. For the u component, the highest space derivative appearing anywhere in the equations is now $\partial u/\partial x$, suggesting H^1 as the natural space for u.

To convert this into practical requirements for finite elements, we will use one of the Sobolev embedding theorems. The standard reference covering this topic is Adams [1]. The mathematical reader may be interested also in Brezis [22], which discusses some variants. Brenner and Scott [21] discuss Sobolev inequalities, which express the embeddings in terms of relations between the norms in the different function spaces.

On a Lipschitz domain $\Omega \subset \mathbb{R}^n$, one has the embedding [1, p. 97]

$$W^{j+m,p}(\Omega) \to C^{j,\lambda}(\overline{\Omega}) , \quad 0 < \lambda \leq m - \frac{n}{p} . \tag{5.7.47}$$

Adams cautions that some care must be taken when interpreting the relation (5.7.47); in the reference, see the discussion following the definition of the target spaces. Strictly speaking, the elements $u \in W^{j,p}(\Omega)$ are not functions defined everywhere on Ω; rather, they are equivalence classes of such functions defined and equal up to sets of measure zero. The embedding should be interpreted as follows: by taking advantage of the freedom to choose a representative of the class, it is possible to pick a function u^* that belongs to the target space, and is bounded there by a constant times the $W^{j,p}(\Omega)$-norm of u; furthermore, the constant is independent of u.

Of course, this is precisely what we need to be able to represent functions in $W^{j,p}(\Omega)$ by finite elements, using basis functions that (also across inter-element boundaries) belong to a target space that is easier to represent in a practical implementation.

To understand the space $C^{j,\lambda}(\overline{\Omega})$, some definitions are needed. Following Adams [1], first let $C^j_B(\Omega)$ denote the space of functions having bounded, continuous derivatives up to order j on Ω, with the derivative of order 0 denoting the function itself. Then, let $C^j(\overline{\Omega})$ denote the closed subspace of $C^j_B(\Omega)$, whose functions have bounded, uniformly continuous derivatives up to order j on Ω. Finally, the space $C^{j,\lambda}(\overline{\Omega})$ is defined as the closed subspace of $C^j(\overline{\Omega})$, containing the functions whose derivatives up to order j are Hölder continuous with exponent λ.

Hölder continuity is defined as follows. For a function f that is Hölder continuous with exponent α on a domain Ω, there exist constants $\alpha \geq 0$, $C \geq 0$ such that

$$|f(x) - f(y)| \leq C \, \|x - y\|^{\alpha} \tag{5.7.48}$$

for all x, y in Ω. For $\alpha = 0$, Hölder continuity reduces to boundedness, and for $\alpha = 1$, to Lipschitz continuity.

Armed with (5.7.47), in our case we have $n = 1$, $\Omega = (x_1, x_2)$, and $p = 2$. As per standard notation for Sobolev spaces, $H^k \equiv W^{k,2}$. Thus, for any $j = 0, 1, 2, \ldots$, we see that we must choose $m = 1$. For the Hölder exponent, (5.7.47) gives the requirement $0 < \lambda \leq 1/2$, so we can choose $\lambda = \epsilon$, where $\epsilon > 0$ can be made arbitrarily small. In one space dimension, we thus have

$$H^{j+1}(\Omega) \to C^{j,\epsilon}(\overline{\Omega}) \;, \quad j = 0, 1 \;. \tag{5.7.49}$$

Beam elements provide $C^{1,\epsilon}$ Hölder continuity (also across element seams) for any sufficiently small ϵ. This indicates that beam elements are an appropriate choice to represent w. Similarly, any standard C^0 elements—such as linear elements, or hierarchical polynomial elements (Lobatto basis)—provide $C^{0,\epsilon}$ Hölder continuity for any sufficiently small ϵ, and are thus an appropriate choice to represent u.

Another important point is that beside allowing the embedding of the natural solution space H^{j+1}, having $C^{j,\epsilon}(\overline{\Omega})$ continuity implies that the $(j+1)$th space derivatives of the basis functions will be at worst finitely discontinuous (strictly, classical $C^j(\Omega)$ is sufficient for this remark). This makes the highest-order derivatives that appear in the equations to belong in $L^2(\Omega)$, that is, they will be square integrable. Since in classical Galerkin methods the sets of the basis functions and test functions are chosen to be the same, this observation holds for both sets of functions. This is important for the integrability of the weak form, when the integrals are split into their elementwise contributions (this is noted in e.g. [52]).

Finally, let us consider the continuity requirements for the classical formulation, where M and N are inserted into the small-displacement weak form force balance equations, e.g. (5.6.60) and (5.6.61) in Sect. 5.6 for an axially moving panel. For the sake of this example, let the traveling panel be made of Kelvin–Voigt material.

In the classical formulation, the final required degree of continuity depends on the representations of N and M introduced by the constitutive model, because they may contain higher derivatives than are already present in (5.6.60) and (5.6.61). As was discussed, an axially moving Kelvin–Voigt material introduces $\partial^3 u/\partial x^3$ and $\partial^5 w/\partial x^5$ in the strong form, which implies that after integration by parts, the highest space derivatives in the weak form will be $\partial^2 u/\partial x^2$ and $\partial^3 w/\partial x^3$. Thus u must belong to H^2 and w to H^3, which necessitates $C^{1,\epsilon}$ continuity for u, and $C^{2,\epsilon}$ continuity for w.

Thus, by taking M and N as auxiliary variables—which better corresponds to the underlying physics—the continuity requirements of the solutions u and w are reduced by one ($C^{j,\epsilon}$ instead of $C^{j+1,\epsilon}$).

5.8 Numerical Examples in Stability of Axially Moving Elastic and Viscoelastic Panels

To conclude this chapter, we present a summary and some numerical examples, demonstrating in practice how all the parts of the discussion fit together. Specifically, we will perform a linear stability analysis for axially moving, tensioned panels, using two different constitutive models. For simplicity, we choose the orthotropic linear elastic and Kelvin–Voigt models, which we consider in the context of paper materials. To make the problem suitable for a general parametric study, we will first convert it to a nondimensional form.

The stability of the axially moving, tensioned Kelvin–Voigt panel has been considered in the series of studies [119–121, 124], and expanded into the Poynting–Thomson variant of the SLS model in Saksa and Jeronen [123]. Here we will follow the same classical approach, without utilizing the mixed form.

The governing equations may be summarized as follows. The combined force balance Eq. (5.1.40), and its weak form (5.6.21), always holds for any beam of the Euler–Bernoulli type, regardless of the kinematic and constitutive models chosen. The small-displacement components (5.2.30) and (5.2.31), and their weak forms (5.6.22) and (5.6.23), hold in the small-displacement regime regardless of the constitutive model.

We will concentrate on the small-displacement regime. For an orthotropic linear elastic beam, we have the resultant axial force (5.3.9) and resultant moment (5.3.10). For the corresponding panel model, we respectively have (5.4.20) and (5.4.21). For an orthotropic Kelvin–Voigt viscoelastic beam, we have (5.3.19) and (5.3.20), and for the corresponding panel, (5.4.30) and (5.4.31).

For an axially moving material, all the aforementioned equations are written in the co-moving frame. Since the equations were originally derived in the classical stationary case, the notation changes slightly. Each x is interpreted as ξ, the co-moving axial coordinate. All fields, such as the displacement w, are interpreted as co-moving (tilded) quantities, such as $\widetilde{w}$. The transformation to the mixed Eulerian–Lagrangean description—written with respect to Eulerian x, but with the value of the displacement still described in Lagrangean terms—is then performed using the frame invariance principle and systematic application of the appropriate coordinate transformation.

In practice, each $\partial/\partial t$ in the co-moving frame (i.e., at fixed ξ) becomes a co-moving derivative $\mathrm{d}/\mathrm{d}t$ in the laboratory frame. The co-moving derivative is given by (5.5.4), and the second co-moving derivative by (5.5.6). Space derivatives are passed through as-is, also in the axial direction; $\partial/\partial\xi \to \partial/\partial x$.

As the sole exception for these conversions of derivatives, are cases that include an elastic or viscoelastic (stationary or axially moving) foundation. Terms that refer to the foundation must be treated in the correct co-moving frame. Generally, the co-moving frame of the foundation is different from the co-moving frame of the axially moving material. As an important special case, the co-moving frame of a stationary foundation is the laboratory frame.

Finally, the mixed Eulerian–Lagrangean description is transformed into a pure Eulerian description, by transforming the mixed Eulerian–Lagrangean displacement $\mathbf{u}(\mathbf{x}, t)$ into the laboratory displacement $\mathbf{u}_{\mathrm{lab}}(\mathbf{x}, t)$ using Eq. (5.5.17). For terms referring to a foundation, the corresponding relation is (5.5.37).

The boundary conditions are also transformed into the laboratory frame. The kinematic conditions for the displacement and its derivative transform trivially, as they do not involve time derivatives. In any boundary condition involving time derivatives,

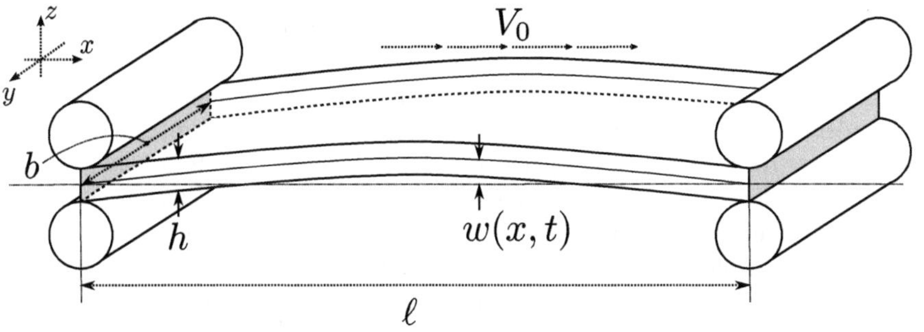

Fig. 5.22 Problem setup for small transverse deformations of a panel, driven axially at constant velocity V_0. The Eulerian control interval is $0 < x < \ell$. The function $w(x, t)$ describes the z-directional displacement of the mid-plane of the panel, measured with respect to the trivial equilibrium state at $z = 0$

such as those for M or Q in the Kelvin–Voigt case, the time derivatives become co-moving derivatives, because the constitutive law is written in the co-moving frame.

The axially moving Kelvin–Voigt beam or panel, beside the classical boundary conditions, obtains an additional inflow condition that results from considerations of physically appropriate continuity, and the physical transport of information. This additional boundary condition is generated only in the cases where it is actually needed, that is, when the material is both viscoelastic ($\eta \neq 0$) and in axial motion ($V_0 \neq 0$).

Let us begin with the strong form. We start from the small-displacement Eulerian equations for the axially moving panel, (5.5.32) and (5.5.33), and insert the constitutive models. Consider the problem setup in Fig. 5.22.

For orthotropic panels, we define the constants

$$C \equiv \frac{E_x h}{1 - \nu_{xy}\nu_{yx}} , \quad D \equiv \frac{E_x h^3}{12\,(1 - \nu_{xy}\nu_{yx})} = \frac{h^2}{12} C . \tag{5.8.1}$$

The quantity D is the orthotropic variant of the flexural rigidity (5.4.13). For the orthotropic linear elastic panel, in the co-moving frame, Eqs. (5.4.20) and (5.4.21) give

$$\widetilde{\widetilde{N}}\,(\xi) = C\frac{\partial \widetilde{u}}{\partial \xi} , \quad \widetilde{\widetilde{M}}\,(\xi) = D\frac{\partial^2 \widetilde{w}}{\partial \xi^2} . \tag{5.8.2}$$

In the laboratory frame, we have

$$\widehat{N}(x) = C\frac{\partial u}{\partial x} , \quad \widehat{M}(x) = D\frac{\partial^2 w}{\partial x^2} , \tag{5.8.3}$$

where u and w are, respectively, the axial and transverse mixed Eulerian–Lagrangean displacements.

Let us take a step back to consider the order of magnitude of the coefficients C and D. For paper materials, typically $E_x \approx 10^9$ Pa, and $h \approx 10^{-4}$ m. Ignoring the factor of $1/(1-\nu_{xy}\nu_{yx})$, which for physically reasonable Poisson's ratios does not affect the order of magnitude, we have $C \approx 10^5$ N/m and $D \approx 10^{-4}$ Nm, accounting for the $1/12 \approx 10^{-1}$. Hence, for paper materials, C is large enough to counteract the smallness of $\partial u/\partial x$, making the resultant axial force $\widehat{N}(x)$ at least the order of unity.

The resultant bending moment $\widehat{M}(x)$, however, is a small quantity, because D is small. At first glance, it may even be considered a negligible one, justifying the use of an ideal string model instead of the panel (or beam) model. However, since it is the term with the highest-order derivative (resulting in $\partial^4/\partial x^4$ in both (5.5.32) and (5.5.33)), it changes the qualitative behavior of the equation. Keep in mind that as was noted in the introduction to this chapter, a traveling ideal string remains stable at any axial drive velocity V_0, while a traveling beam (or panel) becomes unstable at its first critical velocity.

A differential equation with a small coefficient multiplying its highest-order term is called a singularly perturbed equation; for classical solution techniques, see the book by Carl et al. [12], and the review by Lagerstrom and Casten [73]. A famous example of singularly perturbed partial differential equations is encountered in Prandtl's boundary layer theory for fluids with small viscosity. Boundary layer effects similar to this typically occur in singularly perturbed equations. In the bulk of the domain, a singularly perturbed equation behaves almost as its lower-order counterpart, and the effect of the singular perturbation term localizes to near the boundaries (where the equation must satisfy more boundary conditions than its lower-order counterpart).

For problems of ideal strings, the resultant axial force $\widehat{N}(x)$ is often treated as a prescribed function. This can also be done in beam and panel problems. Inserting only $\widehat{M}(x)$ from Eq. (5.8.3) into (5.5.32) and (5.5.33), we have

$$\frac{\partial \widehat{N}}{\partial x} + D\frac{\partial^2 w}{\partial x^2}\frac{\partial^3 w}{\partial x^3} + \frac{\partial^2 w}{\partial x^2}\mu + D\frac{\partial w}{\partial x}\frac{\partial^4 w}{\partial x^4} + \frac{\partial w}{\partial x}\frac{\partial \mu}{\partial x} + q_x - m\left[\frac{\partial^2}{\partial t^2} + 2V_0\frac{\partial^2}{\partial x \partial t} + V_0^2\frac{\partial^2}{\partial x^2}\right]u = 0\,, \tag{5.8.4}$$

$$\frac{\partial w}{\partial x}\frac{\partial \widehat{N}}{\partial x} + \frac{\partial^2 w}{\partial x^2}\widehat{N} - D\frac{\partial^4 w}{\partial x^4} - \frac{\partial \mu}{\partial x} + q_z - m\left[\frac{\partial^2}{\partial t^2} + 2V_0\frac{\partial^2}{\partial x \partial t} + V_0^2\frac{\partial^2}{\partial x^2}\right]w = 0\,. \tag{5.8.5}$$

Note that Eq. (5.8.4) is nonlinear in w and linear in u, while (5.8.5) is linear in w and does not depend on u.

For prescribed $\widehat{N}(x)$, we may thus first solve the transverse displacement w from (5.8.7), and then insert this w into (5.8.6), solving for the axial displacement u. In many practical applications of small-displacement theory such as stability analysis, only the transverse displacement w is of interest; in such cases we will simply stop after obtaining w.

Because w is fully determined by Eq. (5.8.7), the (second-order small) singular perturbation terms involving w in Eq. (5.8.6) effectively behave as load terms for u. Hence, regarding u, these terms introduce no new behavior, and we may drop them.

Discarding the second-order small terms, we obtain the **governing equations** for the axially moving, tensioned, linear elastic panel:

$$\frac{\partial \widehat{N}}{\partial x} + \frac{\partial^2 w}{\partial x^2}\mu + \frac{\partial w}{\partial x}\frac{\partial \mu}{\partial x} + q_x - m\left[\frac{\partial^2}{\partial t^2} + 2V_0\frac{\partial^2}{\partial x \partial t} + V_0^2\frac{\partial^2}{\partial x^2}\right]u = 0\,, \tag{5.8.6}$$

$$\frac{\partial w}{\partial x}\frac{\partial \widehat{N}}{\partial x} + \frac{\partial^2 w}{\partial x^2}\widehat{N} - D\frac{\partial^4 w}{\partial x^4} - \frac{\partial \mu}{\partial x} + q_z - m\left[\frac{\partial^2}{\partial t^2} + 2V_0\frac{\partial^2}{\partial x \partial t} + V_0^2\frac{\partial^2}{\partial x^2}\right]w = 0\,. \tag{5.8.7}$$

Now both equations are linear. Equation (5.8.6) is linear in both u and w, and Eq. (5.8.7) is linear in w and does not depend on u. The solution strategy remains the same. Equations (5.8.6) and (5.8.7) are the formulation that is typically used for solving problems of axially moving, tensioned, linear elastic panels.

Regarding the appearance of w in Eq. (5.8.6), we see that the only terms that appear are the kinematic quantities $\partial^2 w/\partial x^2 \approx 1/R$ and $\partial w/\partial x \approx \alpha$. In other words, the one-way coupling from w to u, for the linear elastic panel, is purely geometric. In the absence of distributed moment loading (i.e., if $\mu = 0$), these terms vanish, and the small-displacement components u and w become fully decoupled.

As was discussed earlier, another option is to treat $\widehat{N}(x)$ consistently, as a quantity generated by the internal stresses of the beam, as implied by the chosen kinematic and constitutive models. Inserting both $\widehat{N}(x)$ and $\widehat{M}(x)$ from (5.8.3) into (5.5.32) and (5.5.33), we have

$$C\frac{\partial^2 u}{\partial x^2} + D\frac{\partial^2 w}{\partial x^2}\frac{\partial^3 w}{\partial x^3} + \frac{\partial^2 w}{\partial x^2}\mu + D\frac{\partial w}{\partial x}\frac{\partial^4 w}{\partial x^4} + \frac{\partial w}{\partial x}\frac{\partial \mu}{\partial x} + q_x - m\left[\frac{\partial^2}{\partial t^2} + 2V_0\frac{\partial^2}{\partial x \partial t} + V_0^2\frac{\partial^2}{\partial x^2}\right]u = 0\,, \tag{5.8.8}$$

$$C\frac{\partial w}{\partial x}\frac{\partial^2 u}{\partial x^2} + C\frac{\partial^2 w}{\partial x^2}\frac{\partial u}{\partial x} - D\frac{\partial^4 w}{\partial x^4} - \frac{\partial \mu}{\partial x} + q_z - m\left[\frac{\partial^2}{\partial t^2} + 2V_0\frac{\partial^2}{\partial x \partial t} + V_0^2\frac{\partial^2}{\partial x^2}\right]w = 0\,. \tag{5.8.9}$$

Now we may not drop the singular perturbation terms involving w, because both u and w appear in both equations, and thus the above argument does not apply. We have no reason to expect that the singular perturbation terms will not qualitatively affect the solution of w via both equations. Also, we may not drop any terms generated by replacing $\widehat{N}$, because $\widehat{N}$ is not a small quantity (the coefficient C cancelling at least one order of smallness).

Equations (5.8.8) and (5.8.9) are a system of nonlinear partial differential equations that determine both u and w simultaneously. The axial force is then obtained, if desired, by inserting the solution into (5.8.3).

Due to the nonlinearity, the sensible approach to solve (5.8.8) and (5.8.9) is to apply finite elements to the corresponding weak form (see Eqs. (5.8.52) and (5.8.53) below). With appropriate finite elements (i.e., $C^{0,\epsilon}$ continuous elements for u and $C^{1,\epsilon}$ continuous elements for w), and iterative linearization via the use of Picard iteration, a numerical solution can be obtained. If desired, convergence may be accelerated by switching to Newton–Raphson iteration after a few Picard iterates.

It should be pointed out that Picard iteration is also known, more descriptively, as fixed point iteration. It is based on the Banach fixed point theorem, for more on which see Rosenlicht [117, p. 170 ff.]. The related topic of the existence and uniqueness of solutions to initial value problems (nowadays commonly known as the Picard–Lindelöf theorem) is also discussed in the same book (p. 177 ff). The Picard–Lindelöf theorem constructively provides the iterative fixed point procedure that leads to the unique solution of one timestep in the often encountered task of direct time simulation (provided that Δt is taken small enough—in practice, the required timestep size is problem-dependent). The Picard–Lindelöf theorem can also be exploited further. If the load function is Lipschitz continuous with a known Lipschitz constant, a timestepping algorithm giving a guaranteed quantitative error bound for the solution has been developed by Matculevich et al. [96].

Before we move on, let us transform the governing equations (5.8.6) and (5.8.7) into a nondimensional form, which is especially convenient for parametric studies. It brings out the structure of the equations in terms of the problem parameters, introducing a new set of parameters that, from a mathematical viewpoint, are as independent as is possible.

For clarity of exposition, let us review the procedure. We start by introducing the nondimensional coordinates

$$x' \equiv \frac{x}{L} , \quad t' \equiv \frac{t}{T} , \tag{5.8.10}$$

where L and T are, respectively, a characteristic length and a characteristic time. Following the terminology of natural units used in some branches of physics, they may also be called a natural length and a natural time.

From a mathematical perspective, L and T are arbitrary constants, the values of which are chosen to make the scaling convenient for the particular problem under consideration. The quantities L and T are dimensional, $[L] = \mathrm{m}$ and $[T] = \mathrm{s}$, thus making x' and t' nondimensional.

For any dimensional function $F(x, t)$, we then define

$$F(x, t) = F(Lx', Tt') \equiv f(x', t') , \tag{5.8.11}$$

where in the first step we have used (5.8.10). This gives us a function f that accepts nondimensional coordinates as input, but returns the same output as the original F. We then proceed to replace F in the original PDE by f.

However, the differentiations in the PDE are still performed with respect to the dimensional x and t. Invoking the chain rule, for any differentiable function $f(x', t')$ we have

$$\frac{\partial f}{\partial x} = \frac{\partial f}{\partial x'}\frac{\partial x'}{\partial x} + \frac{\partial f}{\partial t'}\frac{\partial t'}{\partial x} = \frac{1}{L}\frac{\partial f}{\partial x'}, \tag{5.8.12}$$

$$\frac{\partial f}{\partial t} = \frac{\partial f}{\partial x'}\frac{\partial x'}{\partial t} + \frac{\partial f}{\partial t'}\frac{\partial t'}{\partial t} = \frac{1}{T}\frac{\partial f}{\partial t'}, \tag{5.8.13}$$

which lets us rewrite the differentiations. Thus, when we transform our governing equations to scaled (x', t') coordinates, each differentiation by x (respectively t) in the original equation introduces one multiplication by $1/L$ (resp. $1/T$).

Note also that

$$\int_0^L F(x)\,\mathrm{d}x = L\int_0^1 f(x')\,\mathrm{d}x'\,. \tag{5.8.14}$$

For deriving (5.8.14), physics-oriented texts and elementary mathematics textbooks often take the differential of both sides of (5.8.10) to formally convert $\mathrm{d}x = L\,\mathrm{d}x'$. Adams and Essex [2, p. 319] calls the one-dimensional procedure *substitution in a definite integral*, while the two- and three-dimensional versions are treated separately (p. 814 and 824, respectively) as *change of variables in a double (triple) integral.*

Mathematicians may prefer to invoke the *theorem of change of variable in an integral*, which introduces a factor of $\partial g/\partial x'$ to the integrand, where $x = g(x')$ is the coordinate transformation that defines the change of variable; the result is again (5.8.14). See Rosenlicht [117, p. 128], who notes, and shows, that the one-dimensional case follows from the fundamental theorem of calculus. The book also gives a proof for the n-dimensional version of the theorem (pp. 239–244), which is substantially more technical.

When developing the weak form, we may divide both sides of the equation by the constant L that appears in all terms due to (5.8.14). As long as there are no nested integrals in the weak form, this global scaling thus vanishes.

We then proceed to define

$$f'(x', t') \equiv \frac{1}{f_0} f(x', t')\,, \tag{5.8.15}$$

where f_0 is a nonzero constant (*normalization factor*) that has the same dimension as the output of the function f. This gives us a function f', for which both input and output are nondimensional. Inverting (5.8.15), we insert $f = f_0 f'$ into the equation being transformed.

The final steps are to perform appropriate algebraic manipulations on the resulting equation, and to extract definitions for the nondimensional problem parameters. The normalization factor for the equation itself, that nondimensionalizes the coefficients of the equation, is a design decision; obviously, it must be nonzero for the range of cases that the resulting nondimensional parametrization is desired to be able to access.

Let us now apply this procedure to the governing equations (5.8.6) and (5.8.7). We will use a common normalization q_0 for the q_x and q_z components of the distributed force loading. Leaving the choice of the normalization factors otherwise free for now, we have

$$\frac{\widehat{N}_0}{L}\frac{\partial\widehat{N'}}{\partial x'}+\frac{w_0\mu_0}{L^2}\left(\frac{\partial^2 w'}{\partial x'^2}\mu'+\frac{\partial w'}{\partial x'}\frac{\partial \mu'}{\partial x'}\right)+q_0q_x' \\ -mu_0\left[\frac{1}{T^2}\frac{\partial^2}{\partial t'^2}+\frac{2V_0}{LT}\frac{\partial^2}{\partial x'\partial t'}+\frac{V_0^2}{L^2}\frac{\partial^2}{\partial x'^2}\right]u'=0\,, \tag{5.8.16}$$

$$\frac{w_0\widehat{N}_0}{L^2}\left(\frac{\partial w'}{\partial x'}\frac{\partial\widehat{N}'}{\partial x'}+\frac{\partial^2 w'}{\partial x'^2}\widehat{N}'\right)-\frac{Dw_0}{L^4}\frac{\partial^4 w'}{\partial x'^4}-\frac{\mu_0}{L}\frac{\partial\mu'}{\partial x'}+q_0q_z' \\ -mw_0\left[\frac{1}{T^2}\frac{\partial^2}{\partial t'^2}+\frac{2V_0}{LT}\frac{\partial^2}{\partial x'\partial t'}+\frac{V_0^2}{L^2}\frac{\partial^2}{\partial x'^2}\right]w'=0\,. \tag{5.8.17}$$

The mass per unit area m is nonzero for all physically interesting cases. Also, the normalization constants u_0 and w_0 are always nonzero. In light of this, it will be convenient to normalize the equation by the coefficient of the $\partial^2/\partial t^2$ term. This is a good choice also because it scales the coefficient of the highest time derivative to unity, which is a common mathematical convention when discussing partial differential equations.

Multiplying (5.8.16) by $T^2/(mu_0)$ and (5.8.17) by $T^2/(mw_0)$ to cancel the coefficient of the $\partial^2/\partial t^2$ term, we obtain

$$\frac{T^2\widehat{N}_0}{Lmu_0}\frac{\partial\widehat{N'}}{\partial x'}+\frac{T^2w_0\mu_0}{L^2mu_0}\left(\frac{\partial^2 w'}{\partial x'^2}\mu'+\frac{\partial w'}{\partial x'}\frac{\partial \mu'}{\partial x'}\right)+\frac{T^2q_0}{mu_0}q_x' \\ -\left[\frac{\partial^2}{\partial t'^2}+\frac{2TV_0}{L}\frac{\partial^2}{\partial x'\partial t'}+\frac{T^2V_0^2}{L^2}\frac{\partial^2}{\partial x'^2}\right]u'=0\,, \tag{5.8.18}$$

$$\frac{T^2\widehat{N}_0}{L^2m}\left(\frac{\partial w'}{\partial x'}\frac{\partial\widehat{N}'}{\partial x'}+\frac{\partial^2 w'}{\partial x'^2}\widehat{N}'\right)-\frac{T^2D}{L^4m}\frac{\partial^4 w'}{\partial x'^4}-\frac{T^2\mu_0}{Lmw_0}\frac{\partial\mu'}{\partial x'}+\frac{T^2q_0}{mw_0}q_z' \\ -\left[\frac{\partial^2}{\partial t'^2}+\frac{2TV_0}{L}\frac{\partial^2}{\partial x'\partial t'}+\frac{T^2V_0^2}{L^2}\frac{\partial^2}{\partial x'^2}\right]w'=0\,. \tag{5.8.19}$$

The coefficient of each term (5.8.18) and (5.8.19) is nondimensional.

Now that we see the initial forms of the nondimensional coefficients, let us simplify them by choosing appropriate values for the normalization factors for this problem. First, observe the inertial terms, appearing in brackets in (5.8.18) and (5.8.19). We see that in order to nondimensionalize the V_0, we may take L/T equal to some reference velocity.

We know from the simpler problem of a traveling ideal string, that when subjected to a constant axial tension $\widehat{N}(x,t)\equiv\widehat{N}_0$, a critical value of the axial drive velocity exists at

$$V_0^* = \sqrt{\frac{\widehat{N}_0}{m}} \,. \tag{5.8.20}$$

This is a reasonable reference velocity also for the traveling panel problem, especially if we aim to consider the case of constant axial tension. Note the dimensions: for the string, $[\widehat{N}_0] = \mathrm{N}$, $[m] = \mathrm{kg/m}$, whereas for the panel, $[\widehat{N}_0] = \mathrm{N/m}$, $[m] = \mathrm{kg/m^2}$. Both cases lead to $[V_0^*] = \mathrm{m/s}$, as expected.

Motivated by this, we choose the ratio between L and T to be

$$\frac{L}{T} = V_0^* \,, \tag{5.8.21}$$

and proceed to define the nondimensional axial drive velocity

$$c \equiv \frac{V_0}{V_0^*} = \frac{T V_0}{L} \,. \tag{5.8.22}$$

Since we will consider the problem in an Eulerian domain $0 < x < \ell$, it is convenient to choose $L = \ell$, making $0 < x' < 1$. These considerations fix the characteristic time as

$$T = \frac{L}{V_0^*} = \frac{\ell}{\sqrt{\widehat{N}_0/m}} \,, \tag{5.8.23}$$

which is the time it takes to travel the span length ℓ at the critical velocity V_0^*.

In the following algebraic manipulation, we will only need the square of (5.8.23),

$$T^2 = \frac{\ell^2 m}{\widehat{N}_0} \,. \tag{5.8.24}$$

Observe that (5.8.23) requires $\widehat{N}_0 \neq 0$; the panel must be axially tensioned for our choice of the characteristic time T to be admissible. This is indeed the case that we will investigate, but it is important to keep in mind that with this normalization, the resulting nondimensional equations will be unable to represent the case where no tension is applied.

Considering that the displacements are small, let us choose their normalization as $u_0 = w_0 = h$, where h is the thickness of the panel. Applying the above definitions to Eqs. (5.8.18) and (5.8.19), we have

$$\frac{\ell}{h}\frac{\partial \widehat{N}'}{\partial x'} + \frac{\mu_0}{\widehat{N}_0}\left(\frac{\partial^2 w'}{\partial x'^2}\mu' + \frac{\partial w'}{\partial x'}\frac{\partial \mu'}{\partial x'}\right) + \frac{\ell}{h}\frac{q_0}{\widehat{N}_0/\ell}q_x' - \left[\frac{\partial^2}{\partial t'^2} + 2c\frac{\partial^2}{\partial x'\partial t'} + c^2\frac{\partial^2}{\partial x'^2}\right]u' = 0 \,, \tag{5.8.25}$$

$$\frac{\partial w'}{\partial x'}\frac{\partial \widehat{N}'}{\partial x'} + \frac{\partial^2 w'}{\partial x'^2}\widehat{N}' - \frac{D}{\ell^2 \widehat{N}_0}\frac{\partial^4 w'}{\partial x'^4} - \frac{\ell}{h}\frac{\mu_0}{\widehat{N}_0}\frac{\partial \mu'}{\partial x'} + \frac{\ell}{h}\frac{q_0}{\widehat{N}_0/\ell}q_z' - \left[\frac{\partial^2}{\partial t'^2} + 2c\frac{\partial^2}{\partial x'\partial t'} + c^2\frac{\partial^2}{\partial x'^2}\right]w' = 0 \,. \tag{5.8.26}$$

One must be careful with the dimensions and their interpretation. We have $[\mu_0] = (\mathrm{Nm})/(\mathrm{m}^2) = \mathrm{N}/\mathrm{m}$. One of the dividing m represents length (from the derivation of the combined force balance of the beam), the other width (from the conversion to the panel model). For the panel, μ represents *moment per unit area*. The resultant axial force happens to have the same dimension, $[\widehat{N}_0] = \mathrm{N}/\mathrm{m}$, but here the only dividing m is due to the conversion to the panel model, representing width; $\widehat{N}$ represents *force per unit width*.

For the distributed force, $[q_0] = (\mathrm{N}/\mathrm{m})/\mathrm{m} = \mathrm{N}/\mathrm{m}^2$, the term q represents *force per unit area* (i.e., a pressure). In the derivation of the force balance, q was given as force per unit length of the beam, and the other dividing m results from the conversion to the panel model.

From Eqs. (5.8.25) and (5.8.26) we see that the following normalizations for the distributed force and moment loads are particularly convenient:

$$q_0 \equiv \frac{\widehat{N}_0}{\ell} \,, \quad \mu_0 \equiv \widehat{N}_0 \,. \tag{5.8.27}$$

As the final step, we label the remaining nondimensional coefficients appearing in Eqs. (5.8.25) and (5.8.26), thereby defining the constants

$$\alpha \equiv \frac{\ell}{h} \,, \quad \beta' \equiv \frac{D}{\ell^2 \widehat{N}_0} \,, \tag{5.8.28}$$

We recognize α as the span aspect ratio, and β' as a nondimensional bending rigidity. However, these constants are not yet fully independent, due to the definition of D. Recall Eq. (5.8.1) defining the constants C and D, namely

$$C \equiv \frac{E_x h}{1 - \nu_{xy}\nu_{yx}} \,, \quad D \equiv \frac{h^2}{12} C \,.$$

We have

$$\beta' = \frac{D}{\ell^2 \widehat{N}_0} = \frac{h^2 C}{12\, \ell^2 \widehat{N}_0} = \frac{1}{\alpha^2} \frac{C}{12\, \widehat{N}_0} \,, \tag{5.8.29}$$

which explicitly depends on the span aspect ratio α. We see that bending resistance becomes more significant as $\alpha = \ell/h$ becomes smaller, that is, as the thickness h becomes larger with respect to the span length ℓ. Quantitatively, we see that this effect is quadratic in α.

To obtain a second independent parameter, we define

$$\beta \equiv \alpha^2 \beta' = \alpha^2 \frac{D}{\ell^2 \widehat{N}_0} = \frac{C}{12\, \widehat{N}_0} = \frac{E_x h}{12\, \widehat{N}_0 [1 - \nu_{xy}\nu_{yx}]} \,. \tag{5.8.30}$$

Note that β still includes a linear dependency on h, but since there is no corresponding ℓ in the denominator, β is independent of α.

Applying the above definitions, Eqs. (5.8.25) and (5.8.26) become

$$\alpha \frac{\partial \widehat{N}'}{\partial x'} + \frac{\partial^2 w'}{\partial x'^2} \mu' + \frac{\partial w'}{\partial x'} \frac{\partial \mu'}{\partial x'} + \alpha q'_x - \left[\frac{\partial^2}{\partial t'^2} + 2c \frac{\partial^2}{\partial x' \partial t'} + c^2 \frac{\partial^2}{\partial x'^2} \right] u' = 0 , \tag{5.8.31}$$

$$\frac{\partial w'}{\partial x'} \frac{\partial \widehat{N}'}{\partial x'} + \frac{\partial^2 w'}{\partial x'^2} \widehat{N}' - \alpha^{-2} \beta \frac{\partial^4 w'}{\partial x'^4} - \alpha \frac{\partial \mu'}{\partial x'} + \alpha q'_z - \left[\frac{\partial^2}{\partial t'^2} + 2c \frac{\partial^2}{\partial x' \partial t'} + c^2 \frac{\partial^2}{\partial x'^2} \right] w' = 0 . \tag{5.8.32}$$

Finally, it is customary to clean up the notation by omitting the prime; it is implicitly understood that $0 < x < 1$, and that all quantities are nondimensional. Omitting the primes, and rearranging terms to highlight the formal similarity between the obtained component equations, we obtain the final nondimensional governing equations as

$$\frac{\partial w}{\partial x} \frac{\partial \mu}{\partial x} + \frac{\partial^2 w}{\partial x^2} \mu + \alpha \left[q_x + \frac{\partial \widehat{N}}{\partial x} \right] - \left[\frac{\partial^2}{\partial t^2} + 2c \frac{\partial^2}{\partial x \partial t} + c^2 \frac{\partial^2}{\partial x^2} \right] u = 0 , \tag{5.8.33}$$

$$-\alpha^{-2} \beta \frac{\partial^4 w}{\partial x^4} + \frac{\partial w}{\partial x} \frac{\partial \widehat{N}}{\partial x} + \frac{\partial^2 w}{\partial x^2} \widehat{N} + \alpha \left[q_z - \frac{\partial \mu}{\partial x} \right] - \left[\frac{\partial^2}{\partial t^2} + 2c \frac{\partial^2}{\partial x \partial t} + c^2 \frac{\partial^2}{\partial x^2} \right] w = 0 . \tag{5.8.34}$$

Equations (5.8.33) and (5.8.34) describe the small vibrations of an axially moving, tensioned, orthotropic linear elastic panel. The nondimensional coefficients α, β and c are given by (5.8.28), (5.8.30) and (5.8.22), respectively.

When computing a solution, one must keep in mind the normalizations that were performed to obtain (5.8.33) and (5.8.34). For example, for a constant axial tension we must use $\widehat{N}'(x', t') \equiv 1$, because the magnitude was already absorbed into the normalization constant $\widehat{N}_0$ when applying (5.8.15) to represent $\widehat{N}(x, t)$ in nondimensional form. The distributed force loading components q_x and q_z must be given in units of q_0, and the distributed moment loading μ in units of μ_0 (both defined in Eq. (5.8.27)). The domain is $0 < x < 1$ regardless of the value of the span length ℓ, and the computed displacements will be obtained in units of h.

Next, let us consider the axially moving, tensioned, Kelvin–Voigt panel. Analogously to the constants C and D in (5.8.1), let us define

$$C_{\text{VE}} \equiv \frac{\eta_x h}{1 - \mu_{xy} \mu_{yx}} , \quad D_{\text{VE}} \equiv \frac{\eta_x h^3}{12 (1 - \mu_{xy} \mu_{yx})} = \frac{h^2}{12} C_{\text{VE}} , \tag{5.8.35}$$

related to the viscous behavior. For a given material, if it occurs that the viscous and elastic Poisson ratios coincide, $\nu_{ij} = \mu_{ij}$, then using the retardation time defined in Eq. (5.3.15) in Sect. 5.3, we may write

$$C_{\text{VE}} = \tau C , \quad D_{\text{VE}} = \tau D , \tag{5.8.36}$$

where $\tau \equiv \eta_x / E_x$.

From (5.4.30) and (5.4.31), in the co-moving frame we have the following resultants per unit width,

$$\widetilde{\widetilde{N}}(x) = \left[C + C_{\mathrm{VE}}\frac{\partial}{\partial t}\right]\frac{\partial \widetilde{u}}{\partial \xi}, \tag{5.8.37}$$

$$\widetilde{\widetilde{M}}(x) = \left[D + D_{\mathrm{VE}}\frac{\partial}{\partial t}\right]\frac{\partial^2 \widetilde{w}}{\partial \xi^2}. \tag{5.8.38}$$

Transforming to the laboratory frame, each $\partial/\partial t$ taken in the co-moving frame becomes a co-moving derivative, and we have

$$\widehat{N}(x) = \left[C + C_{\mathrm{VE}}\left(\frac{\partial}{\partial t} + V_0\frac{\partial}{\partial x}\right)\right]\frac{\partial u}{\partial x}, \tag{5.8.39}$$

$$\widehat{M}(x) = \left[D + D_{\mathrm{VE}}\left(\frac{\partial}{\partial t} + V_0\frac{\partial}{\partial x}\right)\right]\frac{\partial^2 w}{\partial x^2}, \tag{5.8.40}$$

where the $\partial/\partial t$ are now taken in the laboratory frame. Concerning the magnitude of the coefficients, η_x is typically much smaller than E_x; the axial drive velocity is approximately $V_0 \approx 20\,\mathrm{m/s} \approx 10^1\,\mathrm{m/s}$. Thus, because for paper materials D is small, D_{VE} (which is smaller) is also a small quantity. We again conclude that the resultant bending moment $\widehat{M}(x)$ is small, but it introduces a singular perturbation.

With (5.8.39) and (5.8.40), we may now specialize (5.5.32) and (5.5.33) to the Kelvin–Voigt case. Proceeding as before, treating $\widehat{N}(x)$ as prescribed, we insert only (5.8.40), obtaining

$$\begin{aligned}
&\frac{\partial \widehat{N}}{\partial x} + \frac{\partial^2 w}{\partial x^2}\left[D + D_{\mathrm{VE}}\left(\frac{\partial}{\partial t} + V_0\frac{\partial}{\partial x}\right)\right]\frac{\partial^3 w}{\partial x^3} + \frac{\partial^2 w}{\partial x^2}\mu \\
&+\frac{\partial w}{\partial x}\left[D + D_{\mathrm{VE}}\left(\frac{\partial}{\partial t} + V_0\frac{\partial}{\partial x}\right)\right]\frac{\partial^4 w}{\partial x^4} + \frac{\partial w}{\partial x}\frac{\partial \mu}{\partial x} + q_x \\
&\qquad -m\left[\frac{\partial^2}{\partial t^2} + 2V_0\frac{\partial^2}{\partial x \partial t} + V_0^2\frac{\partial^2}{\partial x^2}\right]u = 0,
\end{aligned} \tag{5.8.41}$$

$$\begin{aligned}
&\frac{\partial w}{\partial x}\frac{\partial \widehat{N}}{\partial x} + \frac{\partial^2 w}{\partial x^2}\widehat{N} - \left[D + D_{\mathrm{VE}}\left(\frac{\partial}{\partial t} + V_0\frac{\partial}{\partial x}\right)\right]\frac{\partial^4 w}{\partial x^4} - \frac{\partial \mu}{\partial x} + q_z \\
&\qquad -m\left[\frac{\partial^2}{\partial t^2} + 2V_0\frac{\partial^2}{\partial x \partial t} + V_0^2\frac{\partial^2}{\partial x^2}\right]w = 0,
\end{aligned} \tag{5.8.42}$$

where we have assumed sufficient continuity to reorder derivatives. Again, the second equation fully determines w, and the singular perturbation terms involving w in the first equation become load terms for u. Since they are second-order small, we may drop them.

Therefore, for an axially moving, tensioned, Kelvin–Voigt viscoelastic panel, with a prescribed resultant axial force per unit width $\widehat{N}(x)$, we have the governing equations

$$\frac{\partial \widehat{N}}{\partial x} + \frac{\partial^2 w}{\partial x^2}\mu + \frac{\partial w}{\partial x}\frac{\partial \mu}{\partial x} + q_x - m\left[\frac{\partial^2}{\partial t^2} + 2V_0\frac{\partial^2}{\partial x \partial t} + V_0^2\frac{\partial^2}{\partial x^2}\right]u = 0\,, \tag{5.8.43}$$

$$\frac{\partial w}{\partial x}\frac{\partial \widehat{N}}{\partial x} + \frac{\partial^2 w}{\partial x^2}\widehat{N} - \left[D + D_{\mathrm{VE}}\left(\frac{\partial}{\partial t} + V_0\frac{\partial}{\partial x}\right)\right]\frac{\partial^4 w}{\partial x^4} - \frac{\partial \mu}{\partial x} + q_z - m\left[\frac{\partial^2}{\partial t^2} + 2V_0\frac{\partial^2}{\partial x \partial t} + V_0^2\frac{\partial^2}{\partial x^2}\right]w = 0\,. \tag{5.8.44}$$

The second equation determines w, and using the solution in the first equation then determines u. Both equations are linear. The first equation is linear in both u and w, and the second equation is linear and w and does not depend on u. Also here, the coupling is one-way from w to u, purely geometric, and vanishes in the special case of no distributed moment loading. Equations (5.8.43) and (5.8.44) are the formulation that are typically used for solving problems of axially moving, tensioned, Kelvin–Voigt viscoelastic panels.

As an alternative, to treat $\widehat{N}(x)$ consistently, we insert both (5.8.39) and (5.8.40) into (5.5.32) and (5.5.33), obtaining

$$\left[C + C_{\mathrm{VE}}\left(\frac{\partial}{\partial t} + V_0\frac{\partial}{\partial x}\right)\right]\frac{\partial^2 u}{\partial x^2} + \frac{\partial^2 w}{\partial x^2}\left[D + D_{\mathrm{VE}}\left(\frac{\partial}{\partial t} + V_0\frac{\partial}{\partial x}\right)\right]\frac{\partial^3 w}{\partial x^3} + \frac{\partial^2 w}{\partial x^2}\mu + \frac{\partial w}{\partial x}\left[D + D_{\mathrm{VE}}\left(\frac{\partial}{\partial t} + V_0\frac{\partial}{\partial x}\right)\right]\frac{\partial^4 w}{\partial x^4} + \frac{\partial w}{\partial x}\frac{\partial \mu}{\partial x} + q_x - m\left[\frac{\partial^2}{\partial t^2} + 2V_0\frac{\partial^2}{\partial x \partial t} + V_0^2\frac{\partial^2}{\partial x^2}\right]u = 0\,, \tag{5.8.45}$$

$$\frac{\partial w}{\partial x}\left[C + C_{\mathrm{VE}}\left(\frac{\partial}{\partial t} + V_0\frac{\partial}{\partial x}\right)\right]\frac{\partial^2 u}{\partial x^2} + \frac{\partial^2 w}{\partial x^2}\left[C + C_{\mathrm{VE}}\left(\frac{\partial}{\partial t} + V_0\frac{\partial}{\partial x}\right)\right]\frac{\partial u}{\partial x} - \left[D + D_{\mathrm{VE}}\left(\frac{\partial}{\partial t} + V_0\frac{\partial}{\partial x}\right)\right]\frac{\partial^4 w}{\partial x^4} - \frac{\partial \mu}{\partial x} + q_z - m\left[\frac{\partial^2}{\partial t^2} + 2V_0\frac{\partial^2}{\partial x \partial t} + V_0^2\frac{\partial^2}{\partial x^2}\right]w = 0\,. \tag{5.8.46}$$

For the same reasons as before, no terms may be dropped. Equations (5.8.45) and (5.8.46) form a system of nonlinear partial differential equations simultaneously determining both u and w. The axial tension field can then be obtained from (5.8.39). Appropriate finite elements (now $C^{1,\epsilon}$ for u and $C^{2,\epsilon}$ for w) with iterative linearization can be used to obtain a numerical solution.

The nondimensional form is obtained as above. Starting from Eqs. (5.8.43) to (5.8.44), we again set $L/T = V_0^*$, $c = V_0/V_0^*$, $L = \ell$, $u_0 = w_0 = h$, and obtain T from (5.8.23). Due to the formal similarity of the linear elastic and Kelvin–Voigt models, it is likely that the rest of the definitions will carry over, but to be sure we perform the transformation explicitly. We have

$$\frac{\widehat{N}_0}{L}\frac{\partial \widehat{N}'}{\partial x'} + \frac{w_0\mu_0}{L^2}\left(\frac{\partial^2 w'}{\partial x'^2}\mu' + \frac{\partial w'}{\partial x'}\frac{\partial \mu'}{\partial x'}\right) + q_0 q_x' - m u_0\left[\frac{1}{T^2}\frac{\partial^2}{\partial t'^2} + \frac{2V_0}{LT}\frac{\partial^2}{\partial x'\partial t'} + \frac{V_0^2}{L^2}\frac{\partial^2}{\partial x'^2}\right]u' = 0\,,$$

$$\frac{w_0\widehat{N}_0}{L^2}\left(\frac{\partial w'}{\partial x'}\frac{\partial \widehat{N}'}{\partial x'} + \frac{\partial^2 w'}{\partial x'^2}\widehat{N}\right) - \frac{w_0}{L^4}\left[D + \frac{1}{T}D_{\mathrm{VE}}\left(\frac{\partial}{\partial t'} + \frac{TV_0}{L}\frac{\partial}{\partial x'}\right)\right]\frac{\partial^4 w'}{\partial x'^4} - \frac{\mu_0}{L}\frac{\partial \mu'}{\partial x'} + q_0 q_z'$$
$$-m w_0\left[\frac{1}{T^2}\frac{\partial^2}{\partial t'^2} + \frac{2V_0}{LT}\frac{\partial^2}{\partial x'\partial t'} + \frac{V_0^2}{L^2}\frac{\partial^2}{\partial x'^2}\right]w' = 0\,.$$

Multiplying by T^2/mu_0 and T^2/mw_0, respectively, we obtain

$$\frac{T^2\widehat{N}_0}{Lmu_0}\frac{\partial \widehat{N}'}{\partial x'} + \frac{T^2 w_0\mu_0}{L^2 mu_0}\left(\frac{\partial^2 w'}{\partial x'^2}\mu' + \frac{\partial w'}{\partial x'}\frac{\partial \mu'}{\partial x'}\right) + \frac{T^2 q_0}{mu_0}q_x' - \left[\frac{\partial^2}{\partial t'^2} + \frac{2TV_0}{L}\frac{\partial^2}{\partial x'\partial t'} + \frac{T^2V_0^2}{L^2}\frac{\partial^2}{\partial x'^2}\right]u' = 0\,,$$

$$\frac{T^2\widehat{N}_0}{L^2 m}\left(\frac{\partial w'}{\partial x'}\frac{\partial \widehat{N}'}{\partial x'} + \frac{\partial^2 w'}{\partial x'^2}\widehat{N}\right) - \frac{T^2}{L^4 m}\left[D + \frac{1}{T}D_{\mathrm{VE}}\left(\frac{\partial}{\partial t'} + \frac{TV_0}{L}\frac{\partial}{\partial x'}\right)\right]\frac{\partial^4 w'}{\partial x'^4} - \frac{T^2\mu_0}{Lmw_0}\frac{\partial \mu'}{\partial x'} + \frac{T^2 q_0}{mw_0}q_z'$$
$$-\left[\frac{\partial^2}{\partial t'^2} + \frac{2TV_0}{L}\frac{\partial^2}{\partial x'\partial t'} + \frac{T^2V_0^2}{L^2}\frac{\partial^2}{\partial x'^2}\right]w' = 0\,.$$

Using (5.8.23) to represent $T^2 = \ell^2 m/\widehat{N}_0$, and inserting the definitions of L and c, and $u_0 = w_0 = h$,

$$\frac{\ell}{h}\frac{\partial \widehat{N}'}{\partial x'} + \frac{\mu_0}{\widehat{N}_0}\left(\frac{\partial^2 w'}{\partial x'^2}\mu' + \frac{\partial w'}{\partial x'}\frac{\partial \mu'}{\partial x'}\right) + \frac{\ell}{h}\frac{q_0}{\widehat{N}_0/\ell}q_x'$$
$$-\left[\frac{\partial^2}{\partial t'^2} + 2c\frac{\partial^2}{\partial x'\partial t'} + c^2\frac{\partial^2}{\partial x'^2}\right]u' = 0\,,$$

$$\frac{\partial w'}{\partial x'}\frac{\partial \widehat{N}'}{\partial x'} + \frac{\partial^2 w'}{\partial x'^2}\widehat{N} - \frac{1}{\ell^2\widehat{N}_0}\left[D + \frac{1}{T}D_{\mathrm{VE}}\left(\frac{\partial}{\partial t'} + c\frac{\partial}{\partial x'}\right)\right]\frac{\partial^4 w'}{\partial x'^4} - \frac{\ell}{h}\frac{\mu_0}{\widehat{N}_0}\frac{\partial \mu'}{\partial x'} + \frac{\ell}{h}\frac{q_0}{\widehat{N}_0/\ell}q_z'$$
$$-\left[\frac{\partial^2}{\partial t'^2} + 2c\frac{\partial^2}{\partial x'\partial t'} + c^2\frac{\partial^2}{\partial x'^2}\right]w' = 0\,.$$

To simplify the viscous term, we will restrict to the case of materials with $\nu_{ij} = \mu_{ij}$. This allows us to use (5.8.36) to extract the D, leading to

$$\frac{\ell}{h}\frac{\partial \widehat{N}'}{\partial x'} + \frac{\mu_0}{\widehat{N}_0}\left(\frac{\partial^2 w'}{\partial x'^2}\mu' + \frac{\partial w'}{\partial x'}\frac{\partial \mu'}{\partial x'}\right) + \frac{\ell}{h}\frac{q_0}{\widehat{N}_0/\ell}q_x' - \left[\frac{\partial^2}{\partial t'^2} + 2c\frac{\partial^2}{\partial x'\partial t'} + c^2\frac{\partial^2}{\partial x'^2}\right]u' = 0\,,$$

$$\frac{\partial w'}{\partial x'}\frac{\partial \widehat{N}'}{\partial x'} + \frac{\partial^2 w'}{\partial x'^2}\widehat{N} - \frac{D}{\ell^2\widehat{N}_0}\left[1 + \frac{\tau}{T}\left(\frac{\partial}{\partial t'} + c\frac{\partial}{\partial x'}\right)\right]\frac{\partial^4 w'}{\partial x'^4} - \frac{\ell}{h}\frac{\mu_0}{\widehat{N}_0}\frac{\partial \mu'}{\partial x'} + \frac{\ell}{h}\frac{q_0}{\widehat{N}_0/\ell}q_z'$$
$$-\left[\frac{\partial^2}{\partial t'^2} + 2c\frac{\partial^2}{\partial x'\partial t'} + c^2\frac{\partial^2}{\partial x'^2}\right]w' = 0\,.$$

We have already seen almost all of the nondimensional coefficients here appear in the linear elastic case, with the only exception being τ/T. This motivates the definition

$$\gamma \equiv \frac{\tau}{T} = \frac{\tau}{\ell}\sqrt{\frac{\widehat{N}_0}{m}} = \frac{\tau}{\ell}V_0^* = \frac{V_0^*}{\ell/\tau}\,, \tag{5.8.47}$$

which we recognize as a relative timescale of viscous effects occurring, in a span of length ℓ, for a material with retardation time τ, when compared to the critical axial drive velocity V_0^*.

The result is thus

$$\alpha\frac{\partial \widehat{N}'}{\partial x'} + \frac{\partial^2 w'}{\partial x'^2}\mu' + \frac{\partial w'}{\partial x'}\frac{\partial \mu'}{\partial x'} + \alpha q'_x - \left[\frac{\partial^2}{\partial t'^2} + 2c\frac{\partial^2}{\partial x'\partial t'} + c^2\frac{\partial^2}{\partial x'^2}\right]u' = 0\,,$$

$$\frac{\partial w'}{\partial x'}\frac{\partial \widehat{N}'}{\partial x'} + \frac{\partial^2 w'}{\partial x'^2}\widehat{N} - \alpha^{-2}\beta\left[1 + \gamma\left(\frac{\partial}{\partial t'} + c\frac{\partial}{\partial x'}\right)\right]\frac{\partial^4 w'}{\partial x'^4} - \alpha\frac{\partial \mu'}{\partial x'} + \alpha q'_z$$
$$- \left[\frac{\partial^2}{\partial t'^2} + 2c\frac{\partial^2}{\partial x'\partial t'} + c^2\frac{\partial^2}{\partial x'^2}\right]w' = 0\,.$$

Again, omitting the primes and rearranging gives the final nondimensional governing equations:

$$\frac{\partial w}{\partial x}\frac{\partial \mu}{\partial x} + \frac{\partial^2 w}{\partial x^2}\mu + \alpha\left[q_x + \frac{\partial \widehat{N}}{\partial x}\right] - \left[\frac{\partial^2}{\partial t^2} + 2c\frac{\partial^2}{\partial x\partial t} + c^2\frac{\partial^2}{\partial x^2}\right]u = 0\,, \tag{5.8.48}$$

$$- \alpha^{-2}\beta\left[1 + \gamma\left(\frac{\partial}{\partial t} + c\frac{\partial}{\partial x}\right)\right]\frac{\partial^4 w}{\partial x^4} + \frac{\partial w}{\partial x}\frac{\partial \widehat{N}}{\partial x} + \frac{\partial^2 w}{\partial x^2}\widehat{N}$$
$$+ \alpha\left[q_z - \frac{\partial \mu}{\partial x}\right] - \left[\frac{\partial^2}{\partial t^2} + 2c\frac{\partial^2}{\partial x\partial t} + c^2\frac{\partial^2}{\partial x^2}\right]w = 0\,. \tag{5.8.49}$$

Equations (5.8.48) and (5.8.49) describe the small vibrations of an axially moving, tensioned, orthotropic Kelvin–Voigt panel. The coefficients α, β, γ and c are given by (5.8.28), (5.8.30), (5.8.47) and (5.8.22), respectively. We see that the viscoelasticity only appears in the w component equation, in the term with the nondimensional coefficient γ; this is because we treat $\widehat{N}$ as a prescribed function.

Equations (5.8.48) and (5.8.49) are valid for materials for which the viscous and elastic Poisson ratios coincide, $\nu_{ij} = \mu_{ij}$, as we used this property to define the nondimensional viscous coefficient γ.

The **weak forms** of the governing equations for the axially moving, tensioned, linear elastic and Kelvin–Voigt panels are developed similarly. We start from the small-displacement weak form for the axially moving panel, (5.6.60) and (5.6.61), and insert the corresponding constitutive models.

For the linear elastic orthotropic panel, we use (5.8.3), namely

$$\widehat{N}(x) = C\frac{\partial u}{\partial x}\,, \quad \widehat{M}(x) = D\frac{\partial^2 w}{\partial x^2}\,.$$

As before, we first insert $\widehat{M}$ only, obtaining

$$
\begin{aligned}
&- \int_{x_1}^{x_2} \mu \frac{\partial w}{\partial x} \frac{\partial \phi}{\partial x} \mathrm{d}x + \int_{x_1}^{x_2} \left[D \left(\frac{\partial^2 w}{\partial x^2} \right)^2 - (\widehat{N} - m V_0^2 \frac{\partial u}{\partial x}) \right] \frac{\partial \phi}{\partial x} \mathrm{d}x \\
&+ \int_{x_1}^{x_2} D \frac{\partial^2 w}{\partial x^2} \frac{\partial w}{\partial x} \frac{\partial^2 \phi}{\partial x^2} \mathrm{d}x + \left[(\widehat{N} - m V_0^2 \frac{\partial u}{\partial x}) \phi \right]_{x=x_1}^{x_2} \\
&+ \int_{x_1}^{x_2} q_x \phi \, \mathrm{d}x - \int_{x_1}^{x_2} \phi m \left[\frac{\partial^2 u}{\partial t^2} + 2V_0 \frac{\partial^2 u}{\partial x \partial t} \right] \mathrm{d}x = 0 \,, \qquad (5.8.50)
\end{aligned}
$$

$$
\begin{aligned}
&\int_{x_1}^{x_2} \mu \frac{\partial \psi}{\partial x} \mathrm{d}x - \int_{x_1}^{x_2} (\widehat{N} - m V_0^2) \frac{\partial w}{\partial x} \frac{\partial \psi}{\partial x} \mathrm{d}x - D \int_{x_1}^{x_2} \frac{\partial^2 w}{\partial x^2} \frac{\partial^2 \psi}{\partial x^2} \mathrm{d}x + \left[-(\widehat{Q} + m V_0^2 \frac{\partial w}{\partial x}) \psi \right]_{x=x_1}^{x_2} \\
&+ \left[\widehat{M} \frac{\partial \psi}{\partial x} \right]_{x=x_1}^{x_2} + \int_{x_1}^{x_2} q_z \psi \, \mathrm{d}x - \int_{x_1}^{x_2} \psi m \left[\frac{\partial^2 w}{\partial t^2} + 2V_0 \frac{\partial^2 w}{\partial x \partial t} \right] \mathrm{d}x = 0 \,. \qquad (5.8.51)
\end{aligned}
$$

In both equations, $\widehat{M}$ inside the domain has been replaced using (5.8.3), but the $\widehat{M}$ in the moment boundary condition in the transverse component has been left as-is. This is to highlight the fact that if $\widehat{M}$ is to be prescribed at the boundary points as part of the problem description, then its value there is known. Recall that the alternative is to prescribe $\partial w / \partial x$, in which case the moment boundary term is not needed at the boundary where $\partial w / \partial x$ is prescribed.

Similarly, $\widehat{Q}_{\text{eff}}$ and $\widehat{N}_{\text{eff}}$ (recall Eqs. (5.6.62) and (5.6.63) in Sect. 5.6) are also either known at the boundary points as part of the problem description, or not needed (if w is prescribed instead).

Similarly to the strong forms, Eq. (5.8.51) does not involve u. It fully determines w, and hence in (5.8.50), we need not consider the singular perturbation introduced by the moment terms (as before, they act as a second-order small load on u). Dropping the moment terms from (5.8.50), we obtain the weak form governing equations for the orthotropic linear elastic panel,

$$
\begin{aligned}
&- \int_{x_1}^{x_2} \mu \frac{\partial w}{\partial x} \frac{\partial \phi}{\partial x} \mathrm{d}x - \int_{x_1}^{x_2} \left[(\widehat{N} - m V_0^2 \frac{\partial u}{\partial x}) \right] \frac{\partial \phi}{\partial x} \mathrm{d}x + \left[(\widehat{N} - m V_0^2 \frac{\partial u}{\partial x}) \phi \right]_{x=x_1}^{x_2} \\
&+ \int_{x_1}^{x_2} q_x \phi \, \mathrm{d}x - \int_{x_1}^{x_2} \phi m \left[\frac{\partial^2 u}{\partial t^2} + 2V_0 \frac{\partial^2 u}{\partial x \partial t} \right] \mathrm{d}x = 0 \,, \qquad (5.8.52)
\end{aligned}
$$

$$
\begin{aligned}
&\int_{x_1}^{x_2} \mu \frac{\partial \psi}{\partial x} \mathrm{d}x - \int_{x_1}^{x_2} (\widehat{N} - m V_0^2) \frac{\partial w}{\partial x} \frac{\partial \psi}{\partial x} \mathrm{d}x - D \int_{x_1}^{x_2} \frac{\partial^2 w}{\partial x^2} \frac{\partial^2 \psi}{\partial x^2} \mathrm{d}x + \left[-(\widehat{Q} + m V_0^2 \frac{\partial w}{\partial x}) \psi \right]_{x=x_1}^{x_2} \\
&+ \left[\widehat{M} \frac{\partial \psi}{\partial x} \right]_{x=x_1}^{x_2} + \int_{x_1}^{x_2} q_z \psi \, \mathrm{d}x - \int_{x_1}^{x_2} \psi m \left[\frac{\partial^2 w}{\partial t^2} + 2V_0 \frac{\partial^2 w}{\partial x \partial t} \right] \mathrm{d}x = 0 \,. \qquad (5.8.53)
\end{aligned}
$$

Both equations are linear. Equation (5.8.52) is linear in both u and w, and Eq. (5.8.53) is linear in w and does not depend on u.

The only term in (5.8.52) in which w appears is related to the distributed moment load, and even then the coupling is purely geometric (depending on the kinematic quantity $\partial w / \partial x$ only). Unlike in the strong form (5.8.6), there is no dependence on

$\partial^2 w/\partial x^2$. This is because one differentiation has been transferred to the test function ϕ. Again, inthe absence of distributed moment loading, $\mu = 0$, the displacement components u and w become fully decoupled.

Equations (5.8.52) and (5.8.53) are the weak formulation that is typically used for solving problems of axially moving, tensioned, linear elastic panels. Required continuity across element seams is $C^{0,\epsilon}$ for u, and $C^{1,\epsilon}$ for w.

By comparing (5.8.52) and (5.8.53) to the strong form (5.8.33) to (5.8.34), we see that the **nondimensional weak form** is

$$
\begin{aligned}
&-\int_0^1 \mu \frac{\partial w}{\partial x}\frac{\partial \phi}{\partial x}\,\mathrm{d}x - \int_0^1 \left[(\alpha\widehat{N} - c^2\frac{\partial u}{\partial x})\right]\frac{\partial \phi}{\partial x}\,\mathrm{d}x + \left[(\alpha\widehat{N} - c^2\frac{\partial u}{\partial x})\phi\right]_{x=0}^{1} \\
&+ \alpha\int_0^1 q_x\phi\,\mathrm{d}x - \int_0^1 \phi\left[\frac{\partial^2 u}{\partial t^2} + 2c\frac{\partial^2 u}{\partial x\partial t}\right]\mathrm{d}x = 0\,,
\end{aligned}
\tag{5.8.54}
$$

$$
\begin{aligned}
&\alpha\int_0^1 \mu\frac{\partial \psi}{\partial x}\,\mathrm{d}x - \int_0^1 (\widehat{N} - c^2)\frac{\partial w}{\partial x}\frac{\partial \psi}{\partial x}\,\mathrm{d}x - \alpha^{-2}\beta\int_0^1 \frac{\partial^2 w}{\partial x^2}\frac{\partial^2 \psi}{\partial x^2}\,\mathrm{d}x + \left[-(\widehat{Q} + c^2\frac{\partial w}{\partial x})\psi\right]_{x=0}^{1} \\
&+ \alpha^{-2}\beta\left[\widehat{M}\frac{\partial \psi}{\partial x}\right]_{x=0}^{1} + \alpha\int_0^1 q_z\psi\,\mathrm{d}x - \int_0^1 \psi\left[\frac{\partial^2 w}{\partial t^2} + 2c\frac{\partial^2 w}{\partial x\partial t}\right]\mathrm{d}x = 0\,.
\end{aligned}
\tag{5.8.55}
$$

We have divided away the global multiplier of ℓ that is introduced by the change of variable in the integrals. In the transverse component equation (5.8.55), the boundary term involving the resultant shear force $\widehat{Q}$ requires some attention. In the small-displacement range, from the moment balance Eq. (5.1.63) we have

$$
\widehat{Q}(x) = \frac{\partial \widehat{M}}{\partial x}(x) + \mu(x)\,.
$$

By Eq. (5.8.3), a linear elastic panel experiences the resultant moment

$$
\widehat{M} = D\frac{\partial^2 w}{\partial x^2}\,,
$$

so we may write

$$
\widehat{Q} = D\frac{\partial^3 w}{\partial x^3} + \mu\,.
\tag{5.8.56}
$$

Observe that we may obtain the Eqs. (5.8.52) and (5.8.53) also by multiplying the strong form (5.8.33) to (5.8.34) by the test functions ϕ and ψ, integrating over the domain, and applying integration by parts. Following this procedure, in (5.8.55) the nondimensionalization causes $D\partial w^3/\partial x^3$ appear with the coefficient $\alpha^{-2}\beta$, while μ appears with coefficient α. Thus, before we are able to apply (5.8.56), we must derive its nondimensional form.

The same procedure shows that the global multiplier of ℓ also appears in the boundary terms, by first performing the change of variable into nondimensional x, and then applying the integration by parts.

Nondimensionalizing (5.8.56) in the same normalization as before, we have

$$\widehat{Q}_0\widehat{Q}' = \frac{Dw_0}{L^3}\frac{\partial^3 w'}{\partial x'^3} + \mu_0\mu' = \frac{Dh}{\ell^3}\frac{\partial^3 w'}{\partial x'^3} + \widehat{N}_0\mu' = \alpha^{-1}\frac{D}{\ell^2}\frac{\partial^3 w'}{\partial x'^3} + \widehat{N}_0\mu' , \tag{5.8.57}$$

where we have used $w_0 = h$, $L = \ell$ and $\mu_0 = \widehat{N}_0$. Recall Eqs. (5.8.28) and (5.8.30) defining the nondimensional coefficients α and β, namely

$$\alpha \equiv \frac{\ell}{h} , \quad \beta \equiv \alpha^2 \frac{D}{\ell^2 \widehat{N}_0} .$$

Multiplying (5.8.57) by $\alpha/\widehat{N}_0$, we have

$$\frac{\alpha\widehat{Q}_0}{\widehat{N}_0}\widehat{Q}' = \alpha^{-2}\beta\frac{\partial^3 w'}{\partial x'^3} + \alpha\mu' .$$

Finally, choosing the normalization of the shear force as

$$\widehat{Q}_0 \equiv \frac{\widehat{N}_0}{\alpha} = \frac{h}{\ell}\widehat{N}_0 , \tag{5.8.58}$$

we have

$$\widehat{Q}' = \alpha^{-2}\beta\frac{\partial^3 w'}{\partial x'^3} + \alpha\mu' ,$$

which is exactly the combination originally appearing in the boundary term in (5.8.55), thus justifying its replacement by the nondimensional resultant shear force $\widehat{Q}'$.

Similarly to before, a formulation with consistent treatment of $\widehat{N}$ is obtained by inserting both $\widehat{N}$ and $\widehat{M}$ from (5.8.3) into (5.6.60) and (5.6.61). We have

$$\begin{aligned} -\int_{x_1}^{x_2} \mu\frac{\partial w}{\partial x}\frac{\partial \phi}{\partial x}\,\mathrm{d}x + \int_{x_1}^{x_2}\left[D\left(\frac{\partial^2 w}{\partial x^2}\right)^2 - (C - mV_0^2)\frac{\partial u}{\partial x}\right]\frac{\partial \phi}{\partial x}\,\mathrm{d}x + \int_{x_1}^{x_2} D\frac{\partial^2 w}{\partial x^2}\frac{\partial w}{\partial x}\frac{\partial^2 \phi}{\partial x^2}\,\mathrm{d}x \\ + \left[(\widehat{N} - mV_0^2\frac{\partial u}{\partial x})\phi\right]_{x=x_1}^{x_2} + \int_{x_1}^{x_2} q_x\phi\,\mathrm{d}x - \int_{x_1}^{x_2}\phi m\left[\frac{\partial^2 u}{\partial t^2} + 2V_0\frac{\partial^2 u}{\partial x\partial t}\right]\mathrm{d}x = 0 , \end{aligned} \tag{5.8.59}$$

$$\begin{aligned} \int_{x_1}^{x_2}\mu\frac{\partial \psi}{\partial x}\,\mathrm{d}x - \int_{x_1}^{x_2}(C\frac{\partial u}{\partial x} - mV_0^2)\frac{\partial w}{\partial x}\frac{\partial \psi}{\partial x}\,\mathrm{d}x - D\int_{x_1}^{x_2}\frac{\partial^2 w}{\partial x^2}\frac{\partial^2 \psi}{\partial x^2}\,\mathrm{d}x + \left[-(\widehat{Q} + mV_0^2\frac{\partial w}{\partial x})\psi\right]_{x=x_1}^{x_2} \\ + \left[\widehat{M}\frac{\partial \psi}{\partial x}\right]_{x=x_1}^{x_2} + \int_{x_1}^{x_2} q_z\psi\,\mathrm{d}x - \int_{x_1}^{x_2}\psi m\left[\frac{\partial^2 w}{\partial t^2} + 2V_0\frac{\partial^2 w}{\partial x\partial t}\right]\mathrm{d}x = 0 . \end{aligned} \tag{5.8.60}$$

As before, Eqs. (5.8.52) and (5.8.53) form a system of nonlinear partial differential equations determining both u and w simultaneously, and no terms may be dropped. Equation (5.8.52) is nonlinear in w, while (5.8.53) involves a term of the form $(\partial u/\partial x)(\partial w/\partial x)$.

In the case of the Kelvin–Voigt panel, we use (5.8.39) and (5.8.40) for the resultants expressed in the laboratory frame. Inserting just $\widehat{M}$ from (5.8.40) into (5.6.60) and (5.6.61) yields

$$\begin{aligned}
&-\int_{x_1}^{x_2} \mu \frac{\partial w}{\partial x}\frac{\partial \phi}{\partial x}\,\mathrm{d}x + \int_{x_1}^{x_2} \left[\frac{\partial^2 w}{\partial x^2}\left[D + D_{\mathrm{VE}}\left(\frac{\partial}{\partial t} + V_0\frac{\partial}{\partial x}\right)\right]\frac{\partial^2 w}{\partial x^2} - (\widehat{N} - mV_0^2\frac{\partial u}{\partial x})\right]\frac{\partial \phi}{\partial x}\,\mathrm{d}x \\
&+ \int_{x_1}^{x_2} \frac{\partial w}{\partial x}\frac{\partial^2 \phi}{\partial x^2}\left[D + D_{\mathrm{VE}}\left(\frac{\partial}{\partial t} + V_0\frac{\partial}{\partial x}\right)\right]\frac{\partial^2 w}{\partial x^2}\,\mathrm{d}x + \left[(\widehat{N} - mV_0^2\frac{\partial u}{\partial x})\phi\right]_{x=x_1}^{x_2} \\
&+ \int_{x_1}^{x_2} q_x\phi\,\mathrm{d}x - \int_{x_1}^{x_2}\phi m\left[\frac{\partial^2 u}{\partial t^2} + 2V_0\frac{\partial^2 u}{\partial x\partial t}\right]\mathrm{d}x = 0\,,
\end{aligned} \tag{5.8.61}$$

$$\begin{aligned}
&\int_{x_1}^{x_2} \mu\frac{\partial \psi}{\partial x}\,\mathrm{d}x - \int_{x_1}^{x_2}(\widehat{N} - mV_0^2)\frac{\partial w}{\partial x}\frac{\partial \psi}{\partial x}\,\mathrm{d}x - \int_{x_1}^{x_2}\frac{\partial^2 \psi}{\partial x^2}\left[D + D_{\mathrm{VE}}\left(\frac{\partial}{\partial t} + V_0\frac{\partial}{\partial x}\right)\right]\frac{\partial^2 w}{\partial x^2}\,\mathrm{d}x \\
&+ \left[-(\widehat{Q} + mV_0^2\frac{\partial w}{\partial x})\psi\right]_{x=x_1}^{x_2} + \left[\widehat{M}\frac{\partial \psi}{\partial x}\right]_{x=x_1}^{x_2} + \int_{x_1}^{x_2} q_z\psi\,\mathrm{d}x \\
&- \int_{x_1}^{x_2}\psi m\left[\frac{\partial^2 w}{\partial t^2} + 2V_0\frac{\partial^2 w}{\partial x\partial t}\right]\mathrm{d}x = 0\,.
\end{aligned} \tag{5.8.62}$$

Making the same observations as before, we drop the second-order small moment terms from (5.8.61), obtaining the weak form governing equations for the orthotropic Kelvin–Voigt viscoelastic panel,

$$\begin{aligned}
&-\int_{x_1}^{x_2} \mu \frac{\partial w}{\partial x}\frac{\partial \phi}{\partial x}\,\mathrm{d}x - \int_{x_1}^{x_2}\left[(\widehat{N} - mV_0^2\frac{\partial u}{\partial x})\right]\frac{\partial \phi}{\partial x}\,\mathrm{d}x \\
&+ \left[(\widehat{N} - mV_0^2\frac{\partial u}{\partial x})\phi\right]_{x=x_1}^{x_2} + \int_{x_1}^{x_2} q_x\phi\,\mathrm{d}x - \int_{x_1}^{x_2}\phi m\left[\frac{\partial^2 u}{\partial t^2} + 2V_0\frac{\partial^2 u}{\partial x\partial t}\right]\mathrm{d}x = 0\,,
\end{aligned} \tag{5.8.63}$$

$$\begin{aligned}
&\int_{x_1}^{x_2} \mu\frac{\partial \psi}{\partial x}\,\mathrm{d}x - \int_{x_1}^{x_2}(\widehat{N} - mV_0^2)\frac{\partial w}{\partial x}\frac{\partial \psi}{\partial x}\,\mathrm{d}x - \int_{x_1}^{x_2}\frac{\partial^2 \psi}{\partial x^2}\left[D + D_{\mathrm{VE}}\left(\frac{\partial}{\partial t} + V_0\frac{\partial}{\partial x}\right)\right]\frac{\partial^2 w}{\partial x^2}\,\mathrm{d}x \\
&+ \left[-(\widehat{Q} + mV_0^2\frac{\partial w}{\partial x})\psi\right]_{x=x_1}^{x_2} + \left[\widehat{M}\frac{\partial \psi}{\partial x}\right]_{x=x_1}^{x_2} + \int_{x_1}^{x_2} q_z\psi\,\mathrm{d}x \\
&- \int_{x_1}^{x_2}\psi m\left[\frac{\partial^2 w}{\partial t^2} + 2V_0\frac{\partial^2 w}{\partial x\partial t}\right]\mathrm{d}x = 0\,.
\end{aligned} \tag{5.8.64}$$

Equations (5.8.63) and (5.8.64) are the weak formulation that is typically used for solving problems of axially moving, tensioned, Kelvin–Voigt panels. Required continuity across element seams is $C^{0,\epsilon}$ for u, and $C^{2,\epsilon}$ for w. The continuity requirement for u is particularly low, because $\widehat{N}$ is treated as a prescribed function; in the consistent treatment, $\widehat{N}$ is the source of the highest-order space derivative for u.

For w, we may use Hermite C^2 elements. These can be constructed using the same procedure as for the construction of the standard C^1-continuous beam elements (for more on this issue, we refer the interested reader to Appendix C). Likewise, their use requires similar care when dealing with the derivative (and second derivative) degrees of freedom. The global derivative degrees of freedom for w refer to $\partial w/\partial x$ in the global domain, whereas the basis functions N_j on the reference element give $\partial N_j/\partial x$ on the reference (master) element.

By comparing (5.8.63) and (5.8.64) to the strong form (5.8.48) to (5.8.49), the nondimensional weak form is

$$
\begin{aligned}
&-\int_0^1 \mu \frac{\partial w}{\partial x}\frac{\partial \phi}{\partial x}\,\mathrm{d}x - \int_0^1 \left[(\alpha\widehat{N} - c^2\frac{\partial u}{\partial x})\right]\frac{\partial \phi}{\partial x}\,\mathrm{d}x \\
&\quad + \left[(\alpha\widehat{N} - c^2\frac{\partial u}{\partial x})\phi\right]_{x=0}^{1} + \alpha\int_0^1 q_x\phi\,\mathrm{d}x - \int_0^1 \phi\left[\frac{\partial^2 u}{\partial t^2} + 2c\frac{\partial^2 u}{\partial x\partial t}\right]\mathrm{d}x = 0\,, \qquad (5.8.65)
\end{aligned}
$$

$$
\begin{aligned}
&\alpha\int_0^1 \mu\frac{\partial \psi}{\partial x}\,\mathrm{d}x - \int_0^1 (\widehat{N} - c^2)\frac{\partial w}{\partial x}\frac{\partial \psi}{\partial x}\,\mathrm{d}x - \alpha^{-2}\beta\int_0^1 \frac{\partial^2\psi}{\partial x^2}\left[1 + \gamma\left(\frac{\partial}{\partial t} + c\frac{\partial}{\partial x}\right)\right]\frac{\partial^2 w}{\partial x^2}\,\mathrm{d}x \\
&\quad + \left[-(\widehat{Q} + c^2\frac{\partial w}{\partial x})\psi\right]_{x=0}^{1} + \alpha^{-2}\beta\left[\widehat{M}\frac{\partial \psi}{\partial x}\right]_{x=0}^{1} \\
&\quad + \alpha\int_0^1 q_z\psi\,\mathrm{d}x - \int_0^1 \psi\left[\frac{\partial^2 w}{\partial t^2} + 2c\frac{\partial^2 w}{\partial x\partial t}\right]\mathrm{d}x = 0\,. \qquad (5.8.66)
\end{aligned}
$$

The same treatment as above applies to the boundary term with $\widehat{Q}$, but using the resultant moment expression $\widehat{M}$ for a Kelvin–Voigt material, as given in Eq. (5.8.40). Here, too, we have divided away the common multiplier of ℓ introduced by the change of variable in the integrals.

To finish this chapter, we will perform a linear stability analysis of the transverse component, for orthotropic linear elastic and Kelvin–Voigt viscoelastic materials, described by Eqs. (5.8.55) and (5.8.66), respectively. In fact, due to the formal similarity between these two models, we need only Eq. (5.8.66) because we can obtain (5.8.55) from it by going to the elastic limit $\gamma \to 0$. In our numerical examples, we consider the case with no distributed loading, that is, $\mu = q_x = q_z \equiv 0$.

From Eq. (5.8.66), we thus have our final nondimensional weak form governing equation:

$$
\begin{aligned}
&-\int_0^1 (\widehat{N} - c^2)\frac{\partial w}{\partial x}\frac{\partial \psi}{\partial x}\,\mathrm{d}x - \alpha^{-2}\beta\int_0^1 \frac{\partial^2\psi}{\partial x^2}\left[1 + \gamma\left(\frac{\partial}{\partial t} + c\frac{\partial}{\partial x}\right)\right]\frac{\partial^2 w}{\partial x^2}\,\mathrm{d}x \\
&\quad + \left[-(\widehat{Q} + c^2\frac{\partial w}{\partial x})\psi\right]_{x=0}^{1} + \alpha^{-2}\beta\left[\widehat{M}\frac{\partial \psi}{\partial x}\right]_{x=0}^{1} - \int_0^1 \psi\left[\frac{\partial^2 w}{\partial t^2} + 2c\frac{\partial^2 w}{\partial x\partial t}\right]\mathrm{d}x = 0\,. \\
&\qquad (5.8.67)
\end{aligned}
$$

For the sake of simplicity, we use the C^+ boundary conditions, introduced in Eq. (5.7.24) in Sect. 5.7:

$$w(0) = w(1) = 0 , \quad \frac{\partial w}{\partial x}(0) = \frac{\partial w}{\partial x}(1) = 0 , \quad \frac{\partial^2 w}{\partial x^2}(0) = \kappa . \tag{5.8.68}$$

The C^+ boundary conditions cause both boundary terms in (5.8.67) to vanish. The last condition in (5.8.68) is a kinematic inflow condition as discussed in Sect. 5.7, and is only present for the Kelvin–Voigt model. We will set the inflow curvature as $\kappa = 0$, which corresponds to the panel passing the inflow support point in a straight line.

Because we will solve for free vibrations, we will not set any initial conditions. It is also possible to consider the corresponding initial boundary value problem, describing a particular time evolution of the system, for which (5.8.67) and (5.8.68) require also an initial state for both w and $\partial w/\partial t$. Keep in mind that these actually are the nondimensional quantities w' and $\partial w'/\partial t'$. Nondimensionalizing the original w and $\partial w/\partial t$, we have

$$w = w_0 w' = h w' , \quad \frac{\partial w}{\partial t} = \frac{w_0}{T} \frac{\partial w'}{\partial t'} = \frac{h}{\ell} \sqrt{\widehat{N}_0/m} \frac{\partial w'}{\partial t'} . \tag{5.8.69}$$

We invert (5.8.69) to obtain the relations

$$w' = \frac{w}{h} , \quad \frac{\partial w'}{\partial t'} = \frac{\ell}{h\sqrt{\widehat{N}_0/m}} \frac{\partial w}{\partial t} , \tag{5.8.70}$$

which can be used to set the initial conditions for a particular physical setup.

To analyze a given physical setup using (5.8.67) and (5.8.68), one sets the physical parameters (such as ℓ and h), and then determines the nondimensional parameters (such as α) from them. On the other hand, a parametric study of the final weak form governing equation (5.8.67) can be performed without fixing the values of the physical parameters. The behavior of the equation is fully determined by the nondimensional parameters α, β, γ and c; the values of the physical parameters will only affect the interpretation of the results.

In the Eq. (5.8.67), the span aspect ratio α appears only in the terms related to bending. The nondimensional coefficients α, β, γ and c are given by (5.8.28), (5.8.30), (5.8.47) and (5.8.22), respectively. Keep in mind that as was cautioned before, the magnitude of the axial resultant force $\widehat{N}$ has already been absorbed into the constant $\widehat{N}_0$ (which appears in the definition of the nondimensional coefficient β); thus, for constant axial tension, we must set $\widehat{N}(x) \equiv 1$.

If $\beta \to 0$, Eq. (5.8.67) reduces to the weak form governing equation of the axially moving ideal string,

$$\int_0^1 (\widehat{N} - c^2) \frac{\partial w}{\partial x} \frac{\partial \psi}{\partial x} \, \mathrm{d}x + \int_0^1 \psi \left[\frac{\partial^2 w}{\partial t^2} + 2c \frac{\partial^2 w}{\partial x \partial t} \right] \mathrm{d}x = 0 , \tag{5.8.71}$$

Equation (5.8.71) has only one nondimensional parameter, namely the axial drive velocity c. The classical vibrating string (without axial motion) is obtained at $c \to 0$. The free vibrations of the classical traveling string can be determined analytically, already from the strong form; see Chap. 7.

Strictly speaking, when obtaining (5.8.71) from (5.8.67), one must be careful with the boundary term involving $\widehat{Q}$, but with the typical pinhole boundary conditions $w(0) = w(1) = 0$, this term vanishes. At first glance, it appears that the ideal string equation is also recovered from (5.8.67) at $\alpha \to \infty$, which corresponds to $h \to 0$. However, this other choice is not consistent with the considered formulation, because it makes the resultants $\widehat{N}$, $\widehat{Q}$ and $\widehat{M}$ all vanish. Recall that to obtain the resultants, integration was applied across the thickness direction.

With an eye toward allowing different choices for the boundary conditions, let us determine the normalization factor of the nondimensional resultant moment $\widehat{M}$, which appears as one of the boundary terms in (5.8.67). For a Kelvin–Voigt panel, the resultant moment is given by Eq. (5.8.40), namely

$$\widehat{M} = \left[D + D_{\mathrm{VE}} \left(\frac{\partial}{\partial t} + V_0 \frac{\partial}{\partial x} \right) \right] \frac{\partial^2 w}{\partial x^2} \,.$$

Nondimensionalizing, we have

$$\begin{aligned}
\widehat{M}_0 \widehat{M}' &= \frac{w_0}{L^2} \left[D + D_{\mathrm{VE}} \left(\frac{1}{T} \frac{\partial}{\partial t'} + V_0 \frac{1}{L} \frac{\partial}{\partial x'} \right) \right] \frac{\partial^2 w'}{\partial x'^2} \\
&= \frac{w_0}{L^2} \left[D + D_{\mathrm{VE}} \frac{1}{T} \left(\frac{\partial}{\partial t'} + \frac{T V_0}{L} \frac{\partial}{\partial x'} \right) \right] \frac{\partial^2 w'}{\partial x'^2} \\
&= \frac{w_0}{L^2} \left[D + D_{\mathrm{VE}} \frac{1}{T} \left(\frac{\partial}{\partial t'} + c \frac{\partial}{\partial x'} \right) \right] \frac{\partial^2 w'}{\partial x'^2} \\
&= \frac{w_0}{L^2} \left[D + D \frac{\tau}{T} \left(\frac{\partial}{\partial t'} + c \frac{\partial}{\partial x'} \right) \right] \frac{\partial^2 w'}{\partial x'^2} \\
&= \frac{D w_0}{L^2} \left[1 + \gamma \left(\frac{\partial}{\partial t'} + c \frac{\partial}{\partial x'} \right) \right] \frac{\partial^2 w'}{\partial x'^2} \,.
\end{aligned}$$

Dividing by $\widehat{M}_0$, we obtain

$$\begin{aligned}
\widehat{M}' &= \frac{D w_0}{L^2 \widehat{M}_0} \left[1 + \gamma \left(\frac{\partial}{\partial t'} + c \frac{\partial}{\partial x'} \right) \right] \frac{\partial^2 w'}{\partial x'^2} \\
&= \frac{D h}{\ell^2 \widehat{M}_0} \left[1 + \gamma \left(\frac{\partial}{\partial t'} + c \frac{\partial}{\partial x'} \right) \right] \frac{\partial^2 w'}{\partial x'^2} \,.
\end{aligned}$$

Thus, by choosing

$$\widehat{M}_0 \equiv h \widehat{N}_0 \,, \tag{5.8.72}$$

the boundary term obtains the coefficient $D/(\ell^2 \widehat{N}_0) = \alpha^{-2}\beta$, as required. The corresponding linear elastic case is recovered at $\gamma \to 0$.

Note the dimensions and interpretation; the resultant $[\widehat{N}_0] = \mathrm{N/m}$ is force per unit width of the panel, and $[\widehat{M}_0] = \mathrm{N} = (\mathrm{Nm})/\mathrm{m}$ is moment per unit width. In both, the m in the denominator arises in the conversion to the panel model. Looking at (5.8.72), we may also interpret $[\widehat{M}_0] = \mathrm{N} = \mathrm{m}\ (\mathrm{N/m})$; here the m in the numerator corresponds to the thickness direction, while the one in the denominator corresponds to the width direction. Thus, we interpret the lever arm of the moment as being aligned with the thickness direction.

Consider now the scaling properties of the governing equation (5.8.67). The question is which physical setups attain the same values for the nondimensional coefficients. Classically, such scaling properties were crucial for the physical construction of scale models (in the sense of a miniature experimental setup) that behave identically to an intended full-scale system. A classical example of this is the experimental investigation of flow patterns in fluid dynamics, where many nondimensional parameters (such as the Reynolds number) indeed appear. In the age of scientific computing, generally speaking the practical relevance of the construction of physical scale models has obviously decreased. However, scaling properties remain important for a mathematical understanding of the governing equation, which is our goal here.

In the scaling analysis, we must make explicit any dependencies on parameters that will take part in the scaling. Especially, recall that $[m] = \mathrm{kg/m^2}$, where one m in the denominator comes from the derivation of the force balance equation for the beam, representing length, while the other m comes from the conversion to the panel model, representing width, and m is the mass per unit area of the panel. On the other hand, we may write the mass per unit area as

$$m = \rho h \,, \tag{5.8.73}$$

where ρ is the density of the panel material, $[\rho] = \mathrm{kg/m^3}$.

From (5.8.28), (5.8.30), (5.8.47) and (5.8.22), we have

$$\alpha = \frac{\ell}{h} \,, \tag{5.8.74}$$

$$\beta = \frac{C}{12\,\widehat{N}_0} = \frac{E_x\,h}{12\,\widehat{N}_0[1 - \nu_{xy}\nu_{yx}]} \,, \tag{5.8.75}$$

$$\gamma = \frac{\tau}{T} = \frac{\tau V_0^*}{\ell} = \frac{\tau}{\ell}\sqrt{\frac{\widehat{N}_0}{m}} = \frac{\tau}{\ell}\sqrt{\frac{\widehat{N}_0}{\rho h}} \,, \tag{5.8.76}$$

$$c = \frac{V_0}{V_0^*} = \frac{T\,V_0}{\ell} = \frac{V_0}{\sqrt{\widehat{N}_0/m}} = V_0\sqrt{\frac{\rho h}{\widehat{N}_0}} \,. \tag{5.8.77}$$

For purely elastic materials, the retardation time $\tau = 0$, and in that case we will have $\gamma \equiv 0$ regardless of scaling. Only viscoelastic materials ($\tau \neq 0$) require scaling of γ. Similarly, if the material is not traveling axially, then $V_0 = 0$, which leads to $c \equiv 0$ regardless of scaling.

If the values of α, β, γ and c stay fixed, so does the behavior of the nondimensional governing equation; in the linear stability analysis, we will obtain the same nondimensional eigenfrequencies. However, when we later define the nondimensional frequency, we will make use of the characteristic time, so it may be desirable to keep also T fixed, in order to make the dimensional (actual physical) eigenfrequencies to stay fixed as well. From Eq. (5.8.23), we have

$$T = \frac{\ell}{V_0^*} = \frac{\ell}{\sqrt{\widehat{N}_0/m}} = \ell\sqrt{\frac{\rho h}{\widehat{N}_0}} \,. \tag{5.8.78}$$

The coefficient α stays fixed if we scale ℓ and h by the same factor, changing only the overall physical size of the system. Because α only depends on ℓ and h, this *fixed aspect ratio* is a requirement for *any* scaling of the panel model that is to keep all nondimensional coefficients fixed. The scaling of ℓ and h causes β, γ, c and T to change. This is as expected; the behavior of a physical system very rarely, if ever, remains invariant under a pure geometric scaling. Considering that we are still free to scale any of the other physical parameters, we may choose to balance this in a number of ways.

If we allow T to change, one option is to scale the material constants such that $E_x/[1 - \nu_{xy}\nu_{yx}] \propto 1/h$ and $\tau \propto \ell\sqrt{h}$ (the latter is only needed if originally $\tau \neq 0$), while keeping $\widehat{N}_0$ and ρ fixed, and scale $V_0 \propto 1/\sqrt{h}$ (if originally $V_0 \neq 0$). This gives a scaling that keeps the (dimensional) axial tension fixed. The dimensional eigenfrequencies s scale as $s \propto \ell^{-3/2}$. This is because $s \propto 1/T$ (see Eq. (5.8.81) and accompanying discussion below), $T \propto \ell\sqrt{h}$ (with $\widehat{N}_0$ and ρ fixed), and we require α to remain constant, i.e. we scale both ℓ and h by the same factor.

A second option is to scale the axial tension by $\widehat{N}_0 \propto h$, fixing β, provided that the ratio $E_x/[1 - \nu_{xy}\nu_{yx}]$ is kept fixed. For fixed ρ and V_0, this also fixes c. Finally, to fix γ, we scale the material constant τ such that $\tau \propto \ell$ (again, only needed if $\tau \neq 0$); the h in the expression of γ cancels due to the scaling chosen for $\widehat{N}_0$. This scaling keeps the (dimensional) axial drive velocity fixed, and the dimensional eigenfrequencies scale as $s \propto 1/\ell$.

As a third option, to obtain a scaling that keeps T fixed, after fixing α, we scale $\widehat{N}_0 \propto \rho\ell^2 h$. In order to fix γ, we must then fix also the (dimensional) retardation time τ, because $\gamma = \tau/T$. To fix c, we scale $V_0 \propto (\widehat{N}_0/\rho h)^{1/2} \propto \ell$ (if originally $V_0 \neq 0$). The final step is to fix β. Because now $\widehat{N}_0/h \propto \rho\ell^2$, this requires the remaining material properties to be chosen such that the ratio $E_x/(\rho[1 - \nu_{xy}\nu_{yx}]) \propto \ell^2$. This is a scaling that keeps the characteristic time fixed the dimensional eigenfrequencies s remain constant. It can also be considered a fixed proportional velocity scaling, as V_0/ℓ is now a constant.

Actual physical materials are only available for some discrete values of the material parameters, and for any given material, the parameters come as a set (i.e., they cannot be chosen individually). Furthermore, it is very unlikely for the set of actual physical materials to follow the pattern of parameter values required by the scalings considered here (when we start from the set of parameter values for any given material).

In the elastic special case, it is possible to construct physical scale models using the second scaling (fixed dimensional axial drive velocity), because then the material parameters are allowed to stay fixed. For viscoelastic materials, this is only possible if the material running through the span can be engineered such that the retardation time τ is customized without changing the ratio $E_x/[1-\nu_{xy}\nu_{yx}]$.

Thus, for this particular system, this may make infeasible in the general case the physical construction of scale models equivalent to an intended full-scale physical realization—i.e. miniature paper machines with vibration behavior quantitatively identical to a given full-scale setup do not exist. However, as mentioned, physical constructibility is not necessary for a mathematical understanding of the behavior of the governing equation.

To summarize, Eqs. (5.8.74)–(5.8.78) tell us—in a purely mathematical fashion, from the viewpoint of the governing equation for small transverse vibrations — which physical realizations of the considered system are equivalent, whether or not these realizations can be physically constructed.

In the special case of the ideal string, only c appears in the governing equation (5.8.71). In addition, we have the characteristic time T. From Eqs. (5.8.77) and (5.8.78), we see that in this case, scaling first $\widehat{N}_0/m \propto \ell^2$ and then $V_0 \propto \ell$ fixes T and c. Note that for the ideal string, $m = \rho A$, where A is the area of the cross-section. This scaling is similar to the third option above. We conclude that ideal string systems following this scaling behave identically.

In linear stability analysis, when we represent w using the standard time-harmonic trial function

$$w(x,t) = \exp(st)\, W(x) \,, \tag{5.8.79}$$

each differentiation with respect to t becomes a multiplication by the (complex-valued) Lyapunov exponent s, and we obtain a quadratic eigenvalue problem for the pair (s, W). For a given vibration mode, the real part of s governs stability, while the imaginary part gives the natural frequency. We will call s simply the eigenfrequency (as it is a mathematical eigenvalue that represents the physical natural frequency) or by its technical name, the Lyapunov exponent.

Nondimensionalizing, (5.8.79) becomes

$$w_0 w'(x',t') = W_0 \exp(Ss'\,Tt')\, W'(x') \,, \tag{5.8.80}$$

where we have defined the nondimensional frequency

$$s' \equiv \frac{s}{S} \,, \tag{5.8.81}$$

with the normalization factor S having dimension $[S] = \text{Hz}$. By choosing $W_0 = w_0$, and $S = 1/T$, we have

$$w'(x',t') = \exp(s't')\, W'(x') \,, \tag{5.8.82}$$

which has exactly the same form as the original dimensional trial function (5.8.79). The function (5.8.82) is inserted into the nondimensional governing equation (5.8.67) and boundary conditions (5.8.68). Omitting the primes and discarding the common exp factor, we have the eigenvalue problem for (s, W):

$$
\begin{aligned}
&-\int_0^1 (\widehat{N} - c^2)\frac{\partial W}{\partial x}\frac{\partial \psi}{\partial x}\,\mathrm{d}x - \alpha^{-2}\beta \int_0^1 \frac{\partial^2 \psi}{\partial x^2}\left[1 + \gamma\left(s + c\frac{\partial}{\partial x}\right)\right]\frac{\partial^2 W}{\partial x^2}\,\mathrm{d}x \\
&+ \left[-(\widehat{Q} + c^2\frac{\partial W}{\partial x})\psi\right]_{x=0}^1 + \alpha^{-2}\beta\left[\widehat{M}\frac{\partial \psi}{\partial x}\right]_{x=0}^1 - \int_0^1 \psi\left[s^2 W + 2cs\frac{\partial W}{\partial x}\right]\mathrm{d}x = 0\,,
\end{aligned}
\tag{5.8.83}
$$

$$
W(0) = W(1) = 0\,, \quad \frac{\partial W}{\partial x}(0) = \frac{\partial W}{\partial x}(1) = 0\,, \quad \frac{\partial^2 W}{\partial x^2}(0) = \kappa\,. \tag{5.8.84}
$$

Note that with these particular boundary conditions, both the boundary terms in (5.8.83) vanish.

When working with nondimensional variables, we will obtain the nondimensional eigenfrequencies s'. When interpreting results from the parametric study, keep in mind that the computed s' is obtained in units of $1/T$, where T depends on the values of the physical parameters, as was discussed above. For a particular physical setup, the corresponding dimensional frequencies s are obtained by

$$
s = Ss' = \frac{s'}{T} = \frac{1}{\ell}\sqrt{\widehat{N}_0/m}\cdot s'\,. \tag{5.8.85}
$$

Physical intuition suggests that, as far as the critical points are concerned, the sign of the nondimensional axial drive velocity c does not matter, because the sign only controls the direction of travel. To rigorously show that the problem (5.8.83) and (5.8.84) does indeed have this property, we introduce a new space coordinate X, defined by the coordinate transformation

$$
X \equiv 1 - x\,, \tag{5.8.86}
$$

mirroring the domain in-place, that is, with respect to the span midpoint $x = 1/2$. For any function f, we define its counterpart in the mirrored X coordinates by

$$
\overline{f}(X) = \overline{f}(X(x)) \equiv f(x(X)) = f(x)\,. \tag{5.8.87}
$$

Then, by the chain rule, for any differentiable function f it holds that

$$
\frac{\partial f(x)}{\partial x} = \frac{\partial \overline{f}(X(x))}{\partial X}\frac{\partial X}{\partial x} = -\frac{\partial \overline{f}(X)}{\partial X}\,. \tag{5.8.88}
$$

Generally, for any function differentiable at least n times,

$$\frac{\partial^k f}{\partial x^k} = (-1)^k \frac{\partial^k \overline{f}}{\partial X^k} , \quad k = 0, 1, 2, \ldots, n . \tag{5.8.89}$$

Due to our choice of boundary conditions, Eq. (5.8.84), the boundary terms on the second line of (5.8.83) vanish. Consider the remaining terms that involve derivatives of an odd order, which are the only ones that may flip their sign in the mirroring. Applying the coordinate transformation (5.8.86), we have

$$c^2 \frac{\partial W}{\partial x} \frac{\partial \psi}{\partial x} = c^2 \cdot (-1) \cdot \frac{\partial \overline{W}}{\partial X} \cdot (-1) \cdot \frac{\partial \overline{\psi}}{\partial X} = c^2 \frac{\partial \overline{W}}{\partial X} \frac{\partial \overline{\psi}}{\partial X} , \tag{5.8.90}$$

$$c \frac{\partial^2 \psi}{\partial x^2} \frac{\partial^3 W}{\partial x^3} = c \cdot (-1)^2 \cdot \frac{\partial^2 \overline{\psi}}{\partial X^2} \cdot (-1)^3 \cdot \frac{\partial^3 \overline{W}}{\partial X^3} = -c \frac{\partial^2 \overline{\psi}}{\partial X^2} \frac{\partial^3 \overline{W}}{\partial X^3} , \tag{5.8.91}$$

$$c\psi \frac{\partial W}{\partial x} = -c\overline{\psi} \frac{\partial \overline{W}}{\partial x} . \tag{5.8.92}$$

Note that c only appears in the terms considered in (5.8.90)–(5.8.92), and in a boundary term that is eliminated by our boundary conditions. Equations (5.8.90)–(5.8.92) thus suggest that if, in addition to mirroring x, we also define

$$\overline{c} \equiv -c , \tag{5.8.93}$$

then the form of the governing equation stays identical to the original Eq. (5.8.83). Indeed, the transformation (5.8.86) swaps the inflow boundary condition at $x = 0$ to $X = 1$, as required.

The conclusion is as follows. If we change the axial drive velocity c to $\overline{c} = -c$, then the solution $\overline{W}(X)$, as expressed in the mirrored X coordinates, is identical to the solution $W(x)$ of (5.8.83), as expressed in the original x coordinates. Interpreting the mirrored solution in terms of the original one, this means that flipping the sign of c causes the solution $W(x)$ to become mirrored with respect to the span midpoint; but otherwise the solution—including the Lyapunov exponent s—remains the same. In this exact technical sense, the sign of c does not matter. Thus, without loss of generality, we may restrict our study to positive c only.

Let us briefly review the steps to obtain the discrete equation system for the problem (5.8.83) and (5.8.84). Using the finite element method, we represent $W(x)$ as a Galerkin series,

$$W(x) \equiv \sum_{n=1}^{\infty} W_n \varphi_n(x) \equiv \lim_{N \to \infty} \sum_{n=1}^{N} W_n \varphi_n(x) , \tag{5.8.94}$$

where W_n are unknown coefficients, and $\varphi_n(x) \in H^3(\Omega)$ are the basis functions defined on the global domain $\Omega = (0, 1)$. The middle form in (5.8.94) is a notational shorthand for the last form, which explicitly displays the exact technical meaning. (Here N is just an index, unrelated to the axial tension.)

As was noted at the end of the last section, the Sobolev embedding theorem says we can use functions belonging to the space $C^{2,\epsilon}(\overline{\Omega})$. In practice, after we discretize the function space later, we may use Hermite C^2 elements.

We insert the series (5.8.94) into the final governing equation (5.8.83), obtaining

$$
\begin{aligned}
&-\int_0^1 (1-c^2)\frac{\partial}{\partial x}\left(\sum_{n=1}^{\infty} W_n\varphi_n(x)\right)\frac{\partial\psi}{\partial x}\,\mathrm{d}x - \alpha^{-2}\beta\int_0^1 \frac{\partial^2\psi}{\partial x^2}\left[1+\gamma\left(s+c\frac{\partial}{\partial x}\right)\right]\frac{\partial^2}{\partial x^2}\left(\sum_{n=1}^{\infty} W_n\varphi_n(x)\right)\mathrm{d}x \\
&-\int_0^1 \psi\left[s^2\left(\sum_{n=1}^{\infty} W_n\varphi_n(x)\right) + 2cs\frac{\partial}{\partial x}\left(\sum_{n=1}^{\infty} W_n\varphi_n(x)\right)\right]\mathrm{d}x = 0\,. \qquad (5.8.95)
\end{aligned}
$$

We have applied the boundary conditions (5.8.84) to eliminate the boundary terms, and restricted our consideration to the case with constant tension, $\widehat{N}(x) \equiv 1$. Recall that the numerical value of $\widehat{N}_0$ is already taken into account in the values of the nondimensional coefficients.

At this point, it is customary to take the summations in front, but strictly speaking, because the Galerkin series (5.8.94) involves a sum with an infinite number of terms, some care must be taken. For the real-valued case of the following two theorems, see Rosenlicht [117, Chap. VII, pp. 137–140]; the proofs are short and straightforward, and generalize to functions $f : \Omega \to \mathbb{C}$, where Ω is an open interval in $\mathbb{R}$, without major modification.

First, for taking each $\partial/\partial x$ inside the sums, we must justify the interchange of differentiation and infinite summation. We may use the theorem on differentiation of the limit of a sequence of differentiable functions. Let $f_1, f_2, f_3, \ldots$ be a sequence of functions defined on an open interval $U \in \mathbb{R}$, each f_n having a continuous derivative. If the sequence $\partial f_1/\partial x,\ \partial f_2/\partial x,\ \partial f_3/\partial x,\ \ldots$ converges uniformly on U and for some $a \in U$ the sequence $f_1(a),\ f_2(a),\ f_3(a),\ \ldots$ converges, then $\lim_{n\to\infty} f_n$ exists for all $x \in U$, is differentiable, and

$$
\frac{\partial}{\partial x}\left(\lim_{n\to\infty} f_n(x)\right) = \lim_{n\to\infty}\frac{\partial f_n}{\partial x}(x)\,. \qquad (5.8.96)
$$

The second step, for taking the sum in front of each integral, is to justify the interchange of definite integration and infinite summation. We call upon the theorem on integration of a limit of a sequence of integrable functions. Let $a, b \in \mathbb{R}$, $a < b$, and let $f_1, f_2, f_3, \ldots$ be a uniformly convergent sequence of (Riemann-)integrable functions on $[a, b]$. Then

$$
\int_a^b \left(\lim_{n\to\infty} f_n(x)\right)\mathrm{d}x = \lim_{n\to\infty}\int_a^b f_n(x)\,\mathrm{d}x\,. \qquad (5.8.97)
$$

In both theorems, uniform convergence is indeed required; pointwise convergence is not sufficient. Rosenlicht [117, p. 138] provides a simple counterexample demonstrating this.

To apply (5.8.96) in our case, as the sequence f_n, we take the partial sums of the Galerkin series (5.8.94),

$$f_n(x) \equiv \sum_{k=1}^{n} W_k \varphi_k(x) ,$$

and in the term with $\partial^2/\partial x^2$ in Eq. (5.8.95), we simply use (5.8.96) twice. This is the technical justification that allows us to move the differentiations inside the infinite sums.

Considering that we aim to apply finite elements, we will eventually partition our domain. This introduces a minor technical complication: typically, the highest-order derivative of $\varphi_k(x)$ (that appears in the governing equation) is discontinuous across the inter-element boundaries, which in the highest-order term disallows the use of Eq. (5.8.96) across all of $\Omega = (0, 1)$. However, upon closer inspection this turns out not to be an issue. We look for the solution $W(x)$ in the space $H^3(\Omega)$, so all the integrands in (5.8.95) are integrable without requiring any special handling at the inter-element boundaries. Thus, we may first perform the division into elements in Eq. (5.8.95) (without performing any interchanges), and only then apply (5.8.96) separately in each element Ω_i, relegating the interchange of $\partial/\partial x$ and $\lim_{n\to\infty}$ into an implementation detail.

In the second step, to use Eq. (5.8.97), we include into f_n all parts of the appropriate integrand that depend on the summation index or x. For example, for the first term in (5.8.95), we define

$$f_n(x) \equiv \sum_{k=1}^{n} W_k \frac{\partial \varphi_k}{\partial x}(x) \frac{\partial \psi}{\partial x}(x) ,$$

and for the last (viscous axially moving) part of the second term, which involves $c\partial/\partial x$,

$$f_n(x) \equiv \sum_{k=1}^{n} W_k \frac{\partial^3 \varphi_k}{\partial x^3}(x) \frac{\partial^2 \psi}{\partial x^2}(x) .$$

Although ψ does not depend on the summation index, it must be taken as part of $f_n(x)$, because it depends on x. We then apply (5.8.97), moving the infinite sums outside the integrals.

We conclude that, provided that the appropriate partial sums satisfy the technical requirements cited above, the interchanges, first of $\partial/\partial x$ and $\lim_{n\to\infty}$, and then of $\int_0^1 \ldots \,\mathrm{d}x$ and $\lim_{n\to\infty}$, are admissible; with the above caveat on the ordering of these operations and the partitioning of Ω into finite elements.

In practice, these interchanges are almost always possible, but in some corner cases it is extremely important to be aware of the exact requirements. For example, in problems involving Green's function solutions of some auxiliary problem with a singular kernel, it may turn out that the kernel, or one of its derivatives, does not satisfy the requirements (this will occur in Chap. 8, where we discuss fluid—structure interaction). In such cases, the requirements have important implications, especially concerning the correct definition of the strong form of the problem (keep in mind here that the weak form is the primary definition).

We then extract constants—including the unknown Galerkin coefficients W_n—out of the integrals. Performing the above steps, we have

$$(1-c^2)\sum_{n=1}^{\infty} W_n \int_0^1 \frac{\partial \varphi_n}{\partial x}\frac{\partial \psi}{\partial x}\,\mathrm{d}x + \alpha^{-2}\beta \sum_{n=1}^{\infty} W_n \int_0^1 \frac{\partial^2 \psi}{\partial x^2}\left[1+\gamma\left(s+c\frac{\partial}{\partial x}\right)\right]\frac{\partial^2 \varphi_n}{\partial x^2}\,\mathrm{d}x$$
$$+\sum_{n=1}^{\infty} W_n \int_0^1 \psi\left[s^2\varphi_n(x) + 2cs\frac{\partial \varphi_n}{\partial x}\right]\mathrm{d}x = 0\,. \tag{5.8.98}$$

We have multiplied the equation by -1, considering that all terms in (5.8.95) have a minus sign. Expanding parentheses, the result is

$$(1-c^2)\sum_{n=1}^{\infty} W_n \int_0^1 \frac{\partial \varphi_n}{\partial x}\frac{\partial \psi}{\partial x}\,\mathrm{d}x + \alpha^{-2}\beta \sum_{n=1}^{\infty} W_n \int_0^1 \frac{\partial^2 \psi}{\partial x^2}\frac{\partial^2 \varphi_n}{\partial x^2}\,\mathrm{d}x$$
$$+\alpha^{-2}\beta\gamma\left[s\sum_{n=1}^{\infty} W_n \int_0^1 \frac{\partial^2 \psi}{\partial x^2}\frac{\partial^2 \varphi_n}{\partial x^2}\,\mathrm{d}x + c\sum_{n=1}^{\infty} W_n \int_0^1 \frac{\partial^2 \psi}{\partial x^2}\frac{\partial^3 \varphi_n}{\partial x^3}\,\mathrm{d}x\right]$$
$$+s^2\sum_{n=1}^{\infty} W_n \int_0^1 \psi\varphi_n\,\mathrm{d}x + 2cs\sum_{n=1}^{\infty} W_n \int_0^1 \psi\frac{\partial \varphi_n}{\partial x}\,\mathrm{d}x = 0\,. \tag{5.8.99}$$

Now the problem is ready for discretization. Choosing a finite-dimensional subspace of $H^3(\Omega)$, we truncate the series at some finite value of the summation index n:

$$W(x) \approx W_h(x) \equiv \sum_{n=1}^{N} W_n\varphi_n(x)\,, \tag{5.8.100}$$

where N is the number of basis functions on the global domain $\Omega = (0, 1)$. The subscript h, denoting the mesh spacing, is just a reminder that we are now working with a discrete representation.

For the discretization, we must also choose a discrete set of test functions. In the classical Galerkin method, we use the same set as for the basis. Recall that the Eq. (5.8.99) must hold independently for each admissible ψ. We set $\psi = \varphi_j$, where $j = 1, 2, \dots, N$, creating a system of N discrete equations in N unknowns.

Replacing $W(x)$ with $W_h(x)$ in (5.8.99), that is, truncating the sums, and setting $\psi = \varphi_j$, we have the discrete equation system

$$
\begin{aligned}
&(1-c^2)\sum_{n=1}^{N} W_n \int_0^1 \frac{\partial \varphi_j}{\partial x}\frac{\partial \varphi_n}{\partial x}\,\mathrm{d}x + \alpha^{-2}\beta \sum_{n=1}^{N} W_n \int_0^1 \frac{\partial^2 \varphi_j}{\partial x^2}\frac{\partial^2 \varphi_n}{\partial x^2}\,\mathrm{d}x \\
&+ \alpha^{-2}\beta\gamma \left[s\sum_{n=1}^{N} W_n \int_0^1 \frac{\partial^2 \varphi_j}{\partial x^2}\frac{\partial^2 \varphi_n}{\partial x^2}\,\mathrm{d}x + c\sum_{n=1}^{N} W_n \int_0^1 \frac{\partial^2 \varphi_j}{\partial x^2}\frac{\partial^3 \varphi_n}{\partial x^3}\,\mathrm{d}x \right] \\
&+ s^2\sum_{n=1}^{N} W_n \int_0^1 \varphi_j \varphi_n\,\mathrm{d}x + 2cs\sum_{n=1}^{N} W_n \int_0^1 \varphi_j \frac{\partial \varphi_n}{\partial x}\,\mathrm{d}x = 0\,, \qquad (5.8.101)
\end{aligned}
$$

which must hold independently for $j = 1, 2, \ldots, N$. To compact the notation, we introduce the matrices $\mathbf{M}^{(a,b)}$, and the vector of discrete unknowns $\mathbf{W}$:

$$
M_{jn}^{(a,b)} \equiv \int_0^1 \frac{\partial^a \varphi_j}{\partial x^a}\frac{\partial^b \varphi_n}{\partial x^b}\,\mathrm{d}x\,, \quad j, n = 1, 2, \ldots, N\,, \qquad (5.8.102)
$$

$$
\mathbf{W} \equiv (W_1, W_2, \ldots, W_N)\,. \qquad (5.8.103)
$$

In the context of second-order partial differential equations, some of the matrices $\mathbf{M}^{(a,b)}$ have standard names; $\mathbf{M}^{(0,0)} = \mathbf{M}$ often appears as the *mass matrix*, and $\mathbf{M}^{(1,1)} = \mathbf{K}$ as the *stiffness matrix*. For problems that are of second order in time, $\mathbf{M}^{(0,1)}$ is the *gyroscopic matrix* $\mathbf{G}$. In the (fourth-order) beam equation, the matrix $\mathbf{M}^{(2,2)}$ appears in the bending term. The matrix $\mathbf{M}^{(2,3)}$ is new to the case of an axially moving Kelvin–Voigt beam.

We will here omit the details of partitioning the domain $\Omega = (0, 1)$ into finite elements, evaluating the matrix entries (5.8.102) with the help of a reference (master) element and the local-to-global coordinate transform, and the assembly of the global matrix (5.8.102) from the local contributions. Compressed into one sentence, this requires the theorem of change of variable in an integral, and the chain rule. For a brief introduction to specifically Hermite C^2 elements, which are relatively seldom seen in applications, see Appendix C in this book; for the standard finite element machinery (mostly concerning C^0 FEM), see e.g. one of the books by Zienkiewicz et al. [164], Fish and Belytschko [39], Hughes [52], Eriksson et al. [34], Allen et al. [4] or Johnson [55].

Inserting the definition (5.8.102), we rewrite (5.8.101) as

$$
\sum_{n=1}\Big((1-c^2)M_{jn}^{(1,1)} + \alpha^{-2}\beta M_{jn}^{(2,2)} + \alpha^{-2}\beta\gamma\left[sM_{jn}^{(2,2)} + cM_{jn}^{(2,3)}\right] + s^2 M_{jn}^{(0,0)} + 2csM_{jn}^{(0,1)} \Big) W_n = 0\,. \qquad (5.8.104)
$$

Further applying (5.8.103), we write

$$
\Big((1-c^2)\mathbf{M}^{(1,1)} + \alpha^{-2}\beta \mathbf{M}^{(2,2)} + \alpha^{-2}\beta\gamma\left[s\mathbf{M}^{(2,2)} + c\mathbf{M}^{(2,3)}\right] + s^2 \mathbf{M}^{(0,0)} + 2cs\mathbf{M}^{(0,1)} \Big) \mathbf{W} = 0\,. \qquad (5.8.105)
$$

Collecting by powers of s, we have the discrete quadratic eigenvalue problem

$$\left(\left[\mathbf{M}^{(0,0)}\right]s^2+\left[\alpha^{-2}\beta\gamma\mathbf{M}^{(2,2)}+2c\mathbf{M}^{(0,1)}\right]s+\left[(1-c^2)\mathbf{M}^{(1,1)}+\alpha^{-2}\beta\mathbf{M}^{(2,2)}+\alpha^{-2}\beta\gamma c\mathbf{M}^{(2,3)}\right]\right)\mathbf{W}=0\,, \tag{5.8.106}$$

where brackets have been used for clarity. To reduce (5.8.107) into a first-order form, we first reorganize the terms:

$$\left(-\left[\alpha^{-2}\beta\gamma\mathbf{M}^{(2,2)}+2c\mathbf{M}^{(0,1)}\right]s-\left[(1-c^2)\mathbf{M}^{(1,1)}+\alpha^{-2}\beta\mathbf{M}^{(2,2)}+\alpha^{-2}\beta\gamma c\mathbf{M}^{(2,3)}\right]\right)\mathbf{W}=\left[\mathbf{M}^{(0,0)}\right]s^2\mathbf{W}\,. \tag{5.8.107}$$

We then apply the companion form technique, where we start by defining [137]

$$\mathbf{v}\equiv\begin{bmatrix} s\mathbf{W} \\ \mathbf{W}\end{bmatrix}\,. \tag{5.8.108}$$

Note that although here we are only interested in s, if we wished to obtain $\mathbf{W}$ from the results, we may simply take the last N components of $\mathbf{v}$.

With the help of the definition (5.8.108), Eq. (5.8.107) transforms into the *companion form*, which is a first-order system of $2N$ equations in $2N$ unknowns:

$$\begin{bmatrix} -\mathbf{M}_1 & -\mathbf{M}_0 \\ \mathbf{I} & \end{bmatrix}\begin{bmatrix} s\mathbf{W} \\ \mathbf{W}\end{bmatrix}=s\begin{bmatrix} \mathbf{M}_2 & \\ & \mathbf{I}\end{bmatrix}\begin{bmatrix} s\mathbf{W} \\ \mathbf{W}\end{bmatrix}\,, \tag{5.8.109}$$

where empty entries in the block matrices denote blocks of zeroes, and $\mathbf{I}$ is the identity matrix of size $N\times N$. The upper block in (5.8.109) represents Eq. (5.8.107), while the lower block simply states $s\mathbf{W}=s\mathbf{W}$. We have defined the $N\times N$ matrices

$$\mathbf{M}_2\equiv\mathbf{M}^{(0,0)}\,, \tag{5.8.110}$$

$$\mathbf{M}_1\equiv\alpha^{-2}\beta\gamma\mathbf{M}^{(2,2)}+2c\mathbf{M}^{(0,1)}\,, \tag{5.8.111}$$

$$\mathbf{M}_0\equiv(1-c^2)\mathbf{M}^{(1,1)}+\alpha^{-2}\beta\mathbf{M}^{(2,2)}+\alpha^{-2}\beta\gamma c\mathbf{M}^{(2,3)}\,, \tag{5.8.112}$$

which, respectively, play the roles of the mass, combined damping and gyroscopic, and stiffness matrices for our problem. The matrix $\mathbf{M}_2$ is symmetric. The damping part of $\mathbf{M}_1$ involving $\mathbf{M}^{(2,2)}$ is symmetric; the gyroscopic part involving $\mathbf{M}^{(0,1)}$ is skew-symmetric (antisymmetric). The first two parts of $\mathbf{M}_0$ are symmetric, while the last part involving $\mathbf{M}^{(2,3)}$ is skew-symmetric.

At this point, we apply the boundary conditions (5.8.84). For Hermite C^2 elements, all of our boundary conditions are of the Dirichlet type, which can be applied using the method of elimination (see e.g. [164]). We eliminate those discrete unknowns (degrees of freedom) that correspond to known values. Since the right-hand sides of all our boundary conditions in (5.8.84) are zero, we only need to delete the appropriate rows and columns directly from the final matrices (5.8.110)–(5.8.112). Deleting a column sets the corresponding degree of freedom to zero on all rows, while deleting a row removes the discrete equation corresponding to that degree of freedom.

Note that the elimination causes a renumbering of the global degrees of freedom; this must be later accounted for if one wishes to use the resulting eigenvector $\mathbf{W}$ to construct a numerical representation of the eigenfunction.

Equation (5.8.109) is a standard generalized linear eigenvalue problem for the eigenvalue–eigenvector pair $(s, \mathbf{v})$:

$$\mathbf{A}\mathbf{v} = s\mathbf{B}\mathbf{v} , \qquad (5.8.113)$$

for which numerical solvers are readily available. If N is small enough to numerically invert $\mathbf{M}_2$ (once, at the start of the solution process), it is also possible to instead use a simpler eigenvalue solver, by transforming (5.8.109) to

$$\begin{bmatrix} -\mathbf{M}_2^{-1}\mathbf{M}_1 & -\mathbf{M}_2^{-1}\mathbf{M}_0 \\ \mathbf{I} & \end{bmatrix} \begin{bmatrix} s\mathbf{W} \\ \mathbf{W} \end{bmatrix} = s \begin{bmatrix} s\mathbf{W} \\ \mathbf{W} \end{bmatrix} , \qquad (5.8.114)$$

which is a standard linear eigenvalue problem for the eigenvalue–eigenvector pair $(s, \mathbf{v})$:

$$\mathbf{A}\mathbf{v} = s\mathbf{v} . \qquad (5.8.115)$$

This may or may not be advantageous, depending on practical details.

The final remaining practical issues are the sorting of the numerically obtained eigenvalues to obtain the lowest modes, and at adjacent values of the problem parameter c, connecting the correct solution points to form continuous curves for visualization, and to aid automatic detection of points of interest.

The sorting is important because the discretization cannot represent all of the countably infinite spectrum of eigenvalues. Typically only the first few numerical modes will be close to the true solutions of the continuum problem. How many modes are accurately obtained, depends on the ability of the discrete basis of $W(x)$ to represent the corresponding eigenfunctions. Practical experience suggests that, in a quadratic eigenvalue problem, even if the basis is chosen optimally, at best only one half of the returned numerical solutions are solutions of the continuum problem. (This can be seen in some simplified cases by using analytically obtained exact eigenfunctions as the basis, leading to a *spectral discretization*, and then comparing the numerically obtained eigenvalues to the analytical solution for s.)

The ordering issue, on the other hand, arises due to algorithmic reasons. Eigenvalue solvers usually return the solutions in a random order, which may be (and usually is) different at each value of the problem parameter c. To remedy this, see Jeronen [54] for a discussion and algorithms for postprocessing the data.

In the following numerical results, we display the three lowest pairs of nondimensional eigenfrequencies s' against the nondimensional axial drive velocity c, for fixed values of α, β and γ. Considering the focus of this chapter, the values of the problem parameters have been chosen as typical to some paper materials; see Table 5.2 for their values. For the in-plane Poisson ratios, for simplicity, we have taken $\nu_{xy} = \nu_{yx} = \nu$. In addition, we take $\mu_{ij} = \nu_{ij}$, which allows us to introduce the retardation time τ.

Table 5.2 Problem parameters used in the numerical examples

ℓ [m]	h [m]	E_x [Pa]	ν	τ [s]	$\widehat{N}_0$ [N/m]	m [kg/m^2]
1	10^{-4}	10^9	0.3	$0, 10^{-5}, 10^{-4}, 10^{-3}, 10^{-2}$	500	0.08
	D [Nm]	T [s]	α	β	γ	
	9.1575×10^{-5}	1.2649	10^4	18.315	$7.90569 \times \{0, 10^{-4}, 10^{-3}, 10^{-2}, 10^{-1}\}$	

We have chosen the range of the nondimensional axial drive velocity c to investigate the behavior near the first few bifurcation points. The lower end of the range has been set to 1, the critical point of the axially traveling ideal string. It is known that the introduction of bending rigidity ($\beta > 0$) stabilizes the system, increasing the critical value of c; hence at $c = 1$ the considered systems will be in the initial stable region that begins at $c = 0$. The upper end we have cut slightly before any interactions with the fourth pair of eigenfrequencies occur in the elastic case.

In Fig. 5.23, we display an overview of the behavior of the nondimensional Lyapunov exponents s (prime omitted) as a function of c, starting from the elastic case, and then increasing τ as indicated in Table 5.2. As an alternative overview, corresponding views in the $(\mathrm{Re}(s), \mathrm{Im}(s))$ plane are shown in Fig. 5.24. The solution curves belonging to different vibration modes in the elastic and viscoelastic cases are identified in Fig. 5.25.

The results from the elastic model ($\tau = 0$) are shown in detail in Fig. 5.26. Corresponding detail views of results from the viscoelastic Kelvin–Voigt model at the values of τ indicated in Table 5.2 are shown in Figs. 5.27, 5.28, 5.29 and 5.30. Some noteworthy details of the elastic and the first two viscoelastic cases are shown in Figs. 5.31, 5.32, 5.33, 5.34, 5.35 and 5.36. Numerical values at points of interest are collected into Tables 5.3, 5.4, 5.5, 5.6 and 5.7.

In the numerical solutions presented below, Hermite C^2 elements are used, with a uniform grid of 20 elements (21 mesh nodes). Each mesh node is associated with three unique global unknowns, namely the values of W, $\partial W/\partial x$ and $\partial^2 W/\partial x^2$ at the node. The Dirichlet boundary conditions eliminate five unknowns (and five equations) from the discrete equation system. The total number of discrete unknowns is thus $21 \cdot 3 - 5 = 58$ (in the elastic special case, $21 \cdot 3 - 4 = 59$). The companion form then doubles the number of unknowns to $2 \cdot 58 = 116$ (118 in the elastic case, respectively). This is the final size of each problem instance.

We have taken advantage of the small size of the problem instances by discretizing the c axis at a high resolution in order to obtain visually smooth curves for the detail close-ups as well and, especially, a more accurate representation near bifurcation points in the views in the $(\mathrm{Re}(s), \mathrm{Im}(s))$ plane. In the cases $\tau = 0$ and $\tau = 10^{-5}$ s, the studied range of c is discretized into a uniform grid of $10^5 + 1$ points, and in the other three cases, $2 \times 10^4 + 1$ points are used; the purpose of the one extra point being to make the c step a round number in base-10.

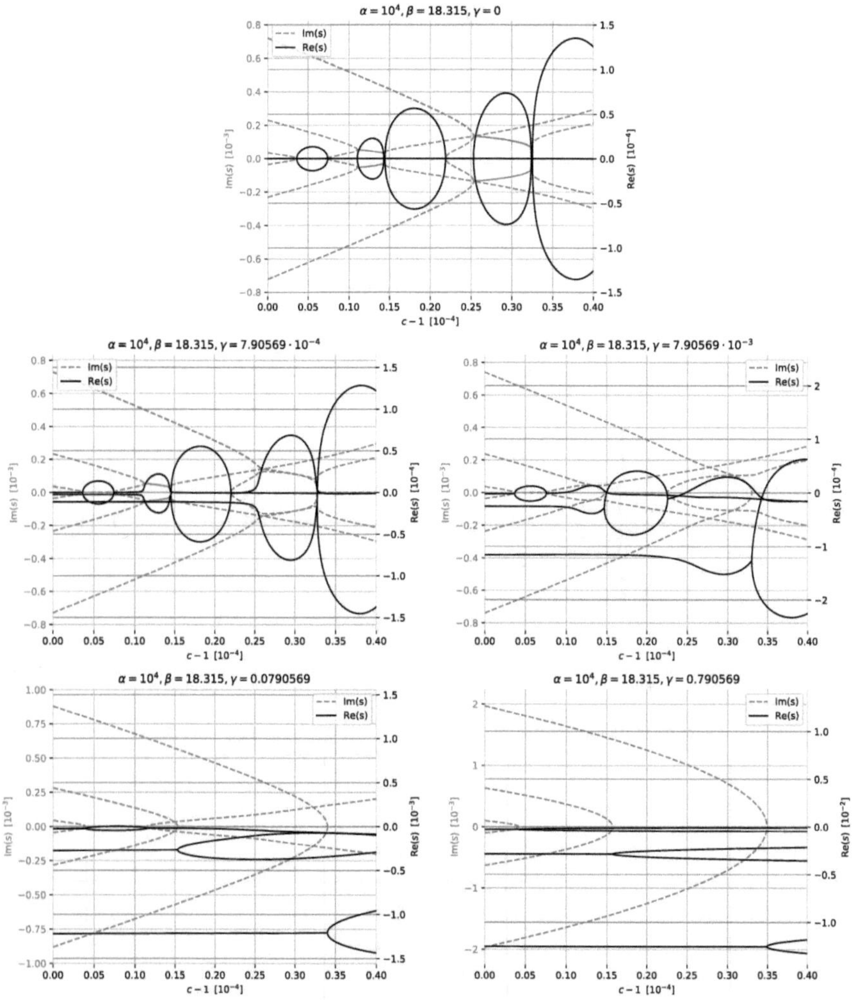

Fig. 5.23 Lyapunov exponents of a traveling panel near $c = 1$, overview of the effect of viscosity. Note the units and the shift on the horizontal axis. *Top*: elastic, $\tau = 0$. *Middle left*: $\tau = 10^{-5}$ s. *Middle right*: $\tau = 10^{-4}$ s. *Bottom left*: $\tau = 10^{-3}$ s. *Bottom right*: $\tau = 10^{-2}$ s

From the numerical results, we make the following observations. Both divergence and flutter modes can be seen. As we increase the nondimensional axial drive velocity c, starting from 1, divergence occurs first; this is typical for gyroscopic systems with symmetric boundary conditions, as in our elastic case here. We see that the same feature carries over to the viscoelastic case.

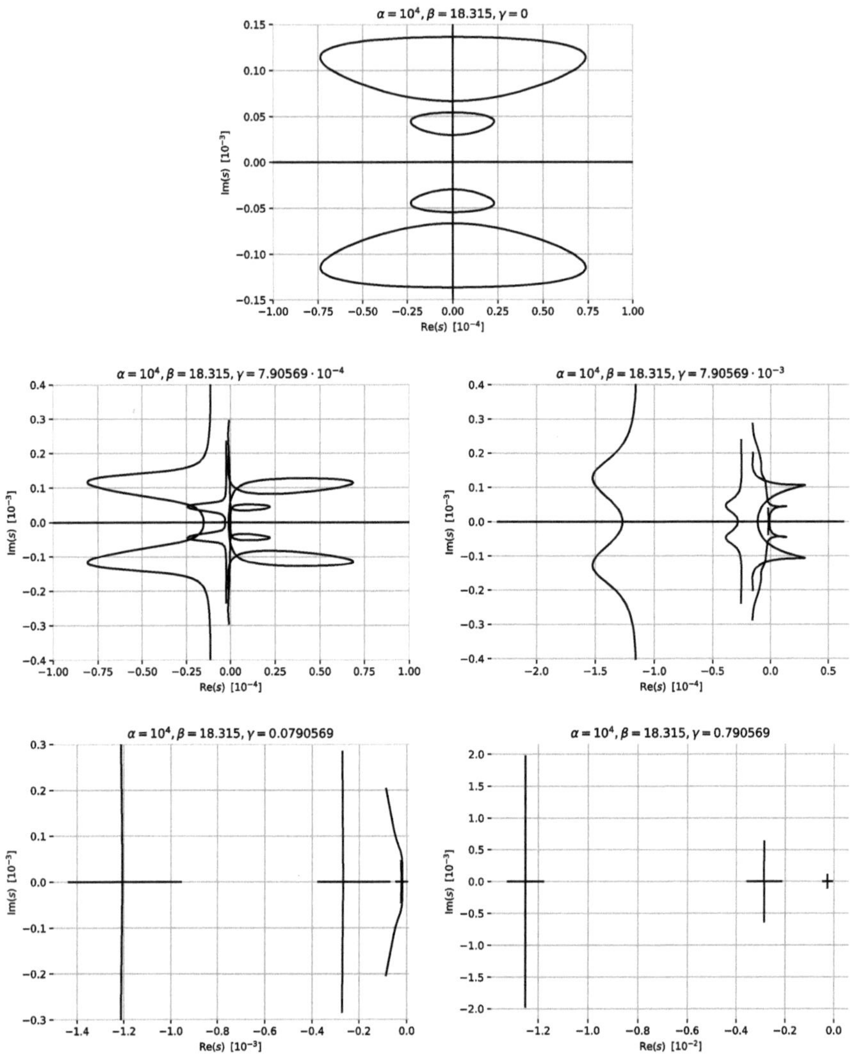

Fig. 5.24 Lyapunov exponents of a traveling panel near $c = 1$, view in the $(\mathrm{Re}(s),\ \mathrm{Im}(s))$ plane. The range of c from which the data are taken is the same as in the other figures, $c \in [1, 1 + 4 \cdot 10^{-5}]$. *Top*: elastic, $\tau = 0$. Note bifurcations of both divergence and flutter types. *Middle left*: $\tau = 10^{-5}$ s. *Middle right*: $\tau = 10^{-4}$ s. *Bottom left*: $\tau = 10^{-3}$ s. *Bottom right*: $\tau = 10^{-2}$ s. At the bottom row, the modes are identified from left to right as 3, 2, 1

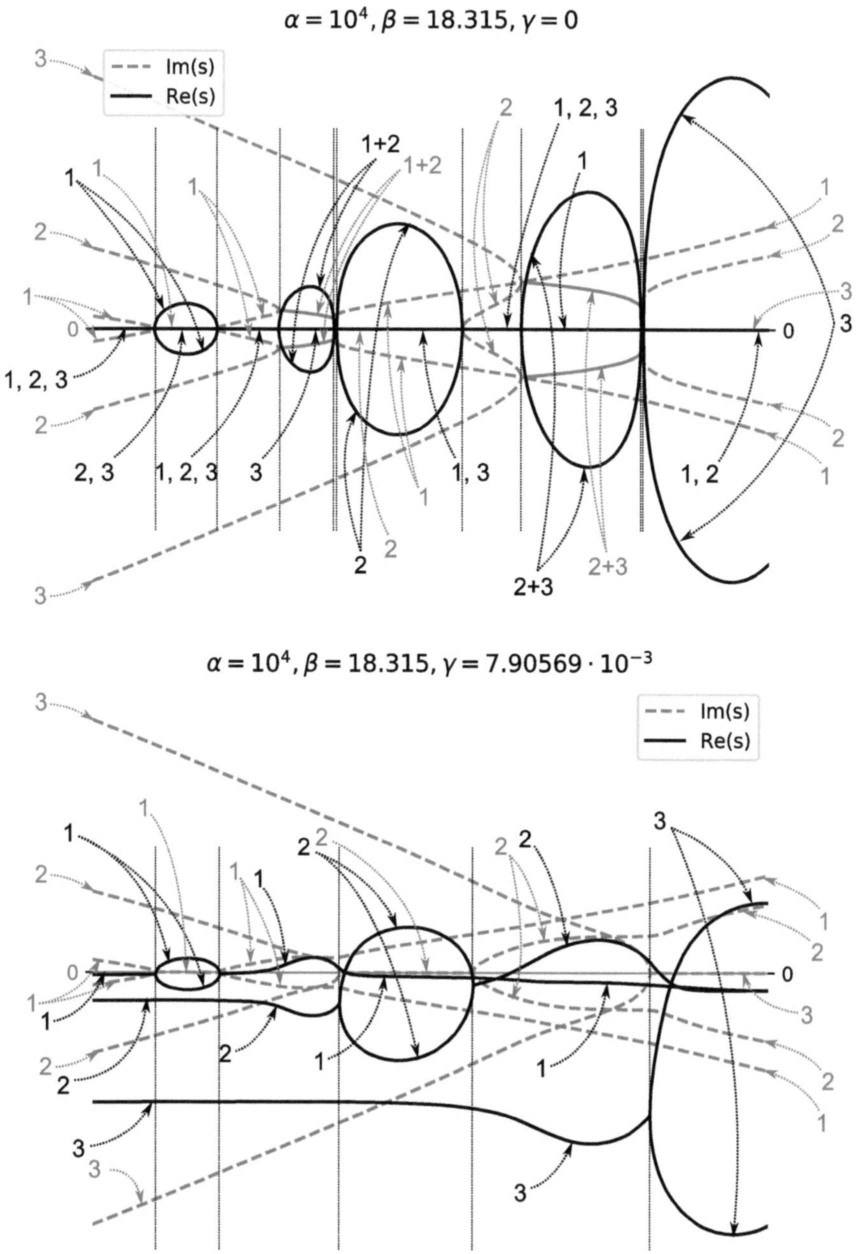

Fig. 5.25 Identification of the modes. Real parts indicated by black arrows and text, and imaginary parts in gray. The zeros indicate the coordinate axis. The vertical lines indicate bifurcation points. For numerical values, refer to Figs. 5.23, 5.26 and 5.28. *Top*: elastic panel. *Bottom*: viscoelastic panel with $\tau = 10^{-4}$ s, taken as representative of the viscoelastic case. Note the absence of mode interaction

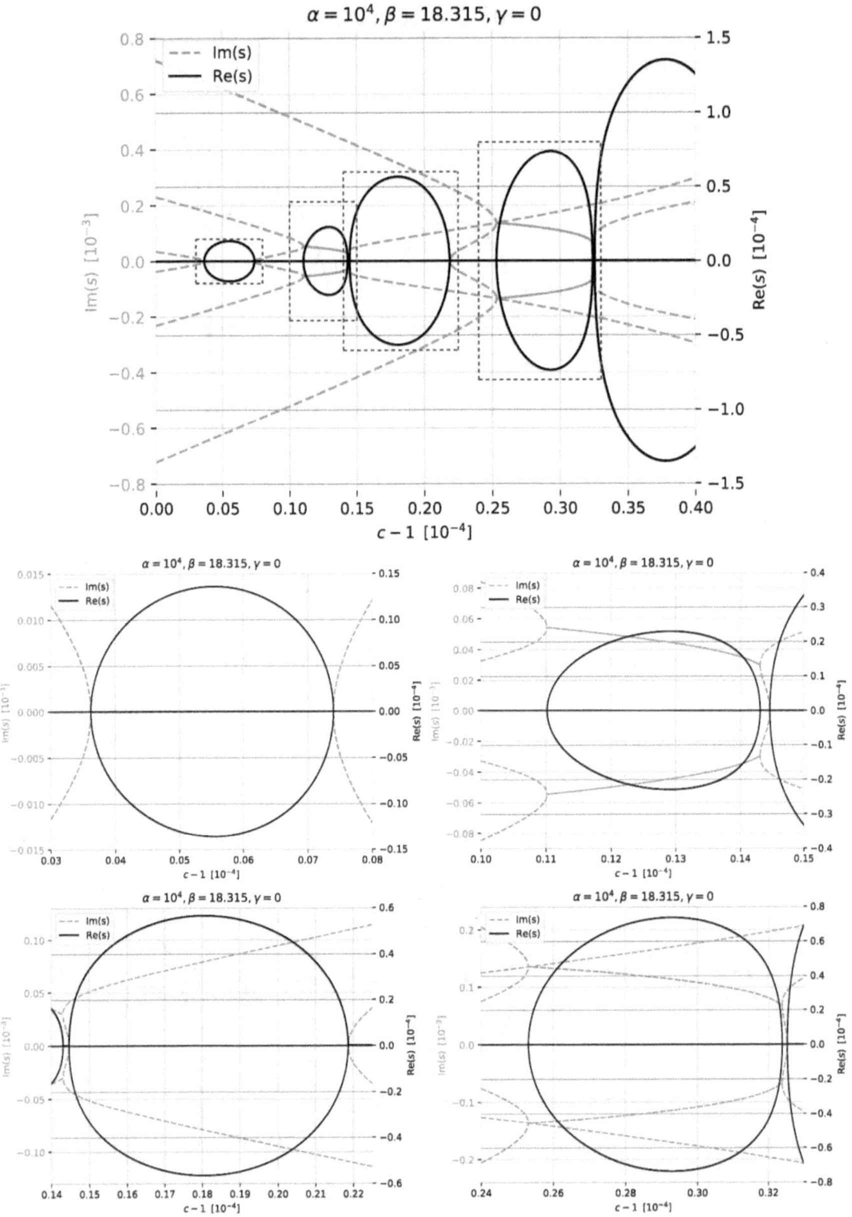

Fig. 5.26 Lyapunov exponents of an elastic traveling panel near $c = 1$. *Top*: overview. *Others*: detail of the first few bifurcation points. In each subfigure, the two vertical axes have been zoomed independently to bring out the detail. The zoom ranges of $\mathrm{Re}(s)$ are indicated in the overview

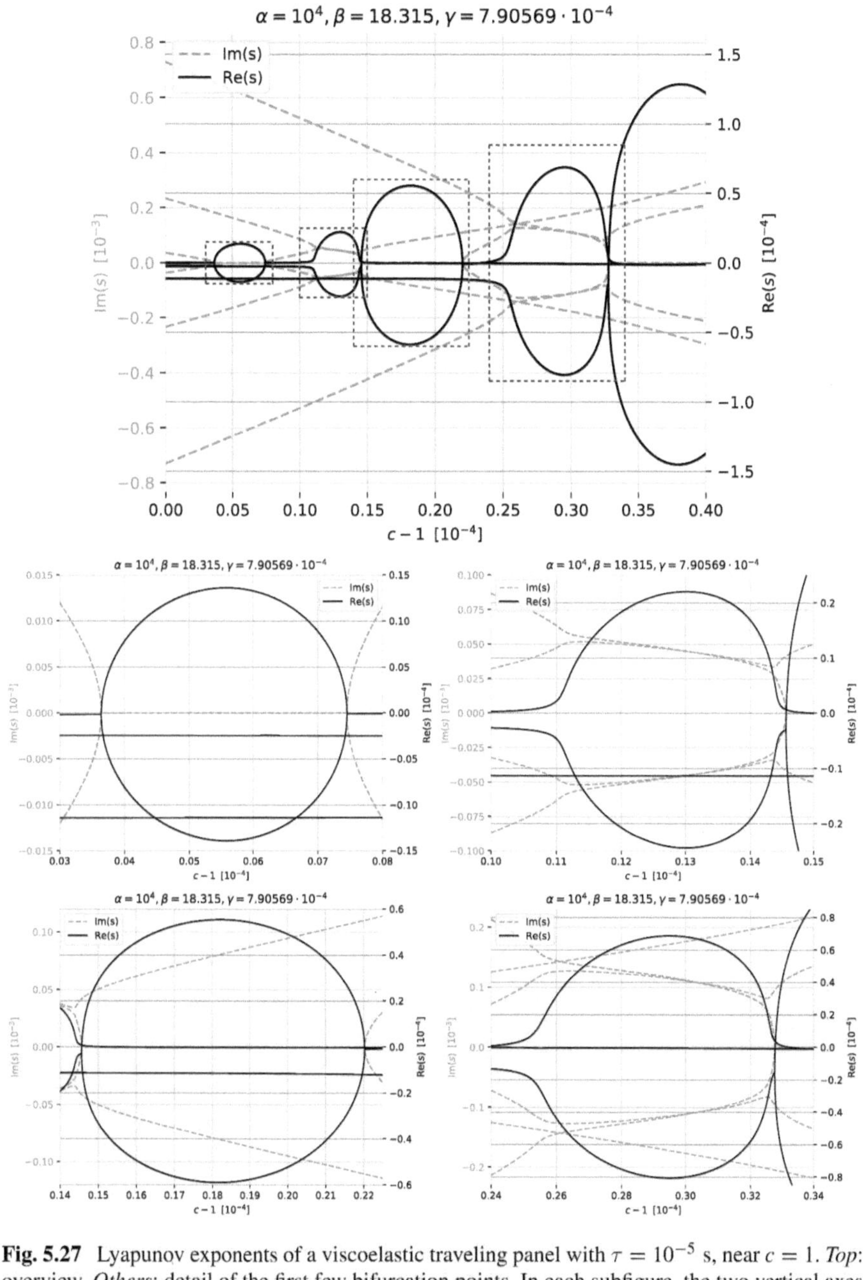

Fig. 5.27 Lyapunov exponents of a viscoelastic traveling panel with $\tau = 10^{-5}$ s, near $c = 1$. *Top*: overview. *Others*: detail of the first few bifurcation points. In each subfigure, the two vertical axes have been zoomed independently to bring out the detail. The zoom ranges of Re(s) are indicated in the overview

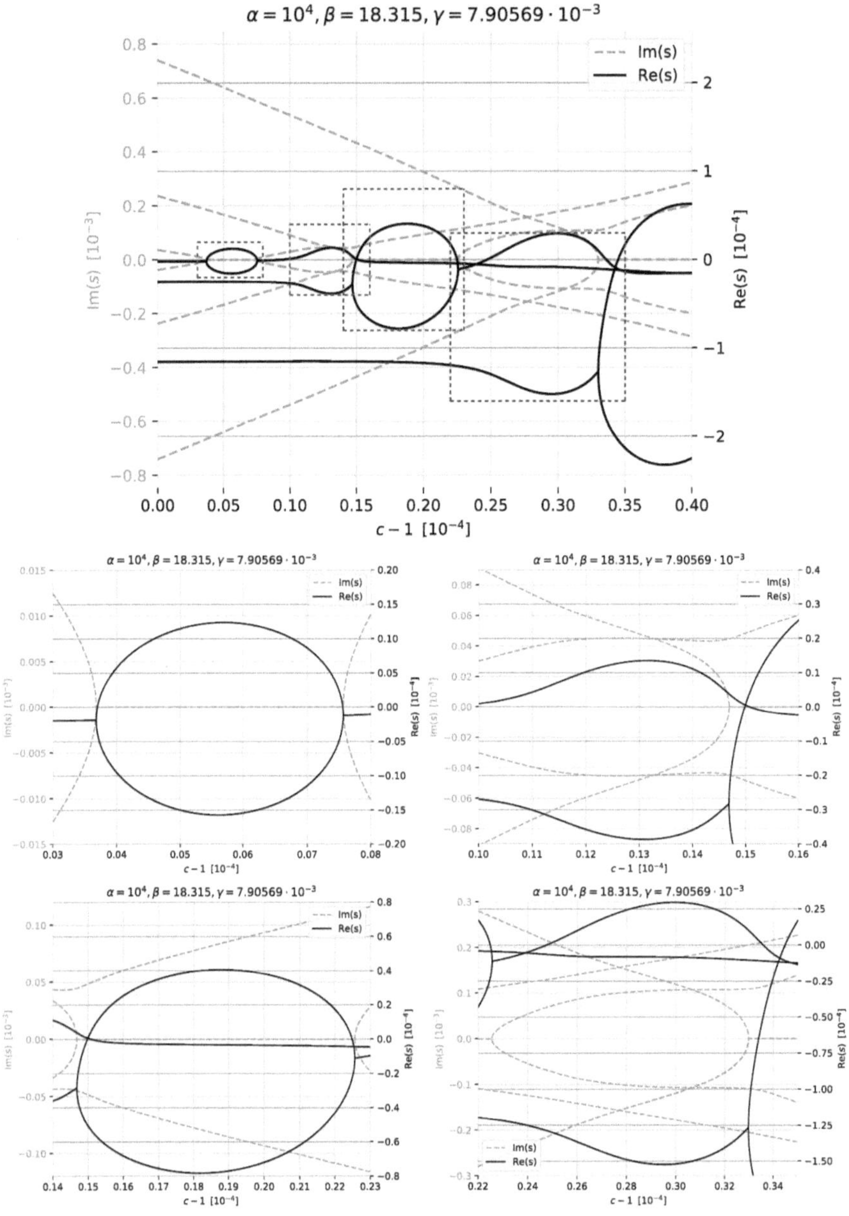

Fig. 5.28 Lyapunov exponents of a viscoelastic traveling panel with $\tau = 10^{-4}$ s, near $c = 1$. *Top*: overview. *Others*: detail of the first few bifurcation points. In each subfigure, the two vertical axes have been zoomed independently to bring out the detail. The zoom ranges of $\mathrm{Re}(s)$ are indicated in the overview

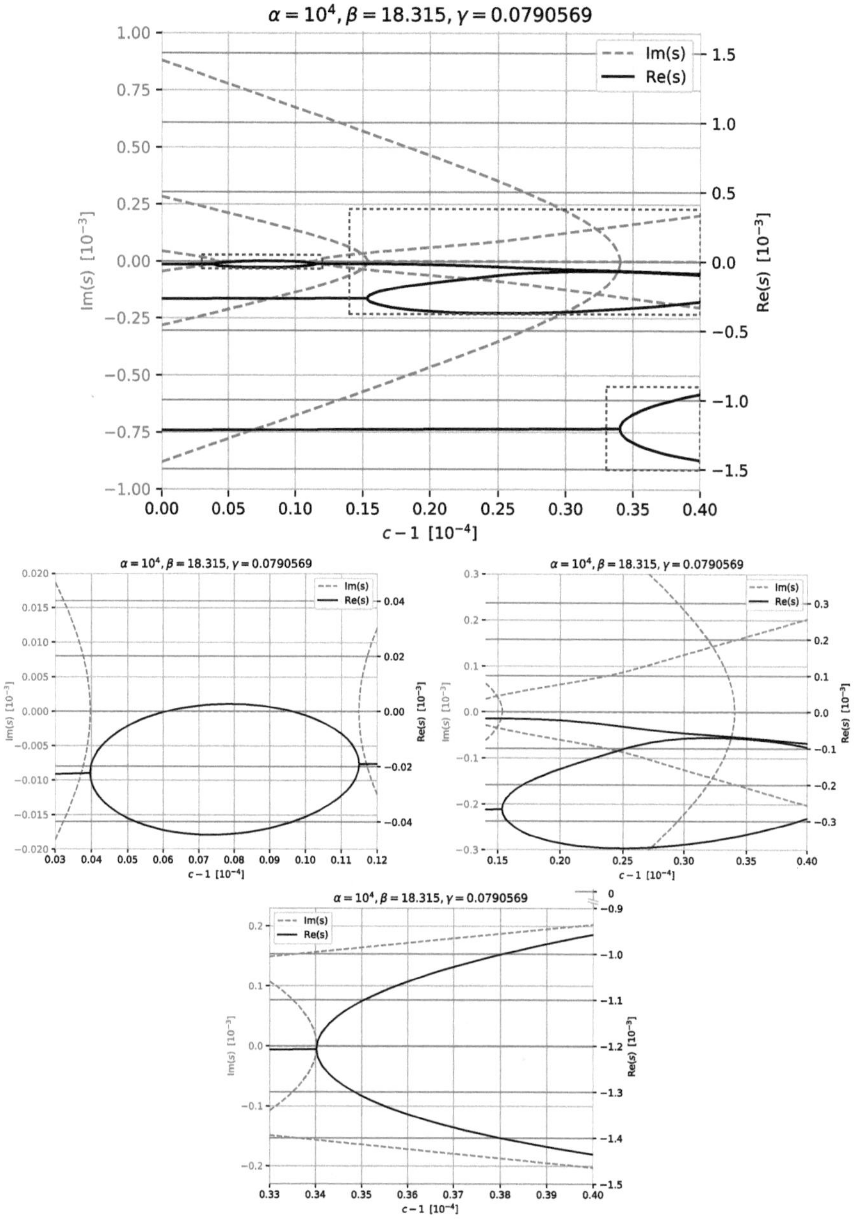

Fig. 5.29 Lyapunov exponents of a viscoelastic traveling panel with $\tau = 10^{-3}$ s, near $c = 1$. *Top*: overview. *Others*: detail of the first few bifurcation points. In each subfigure, the two vertical axes have been zoomed independently to bring out the detail. The zoom ranges of $\mathrm{Re}(s)$ are indicated in the overview

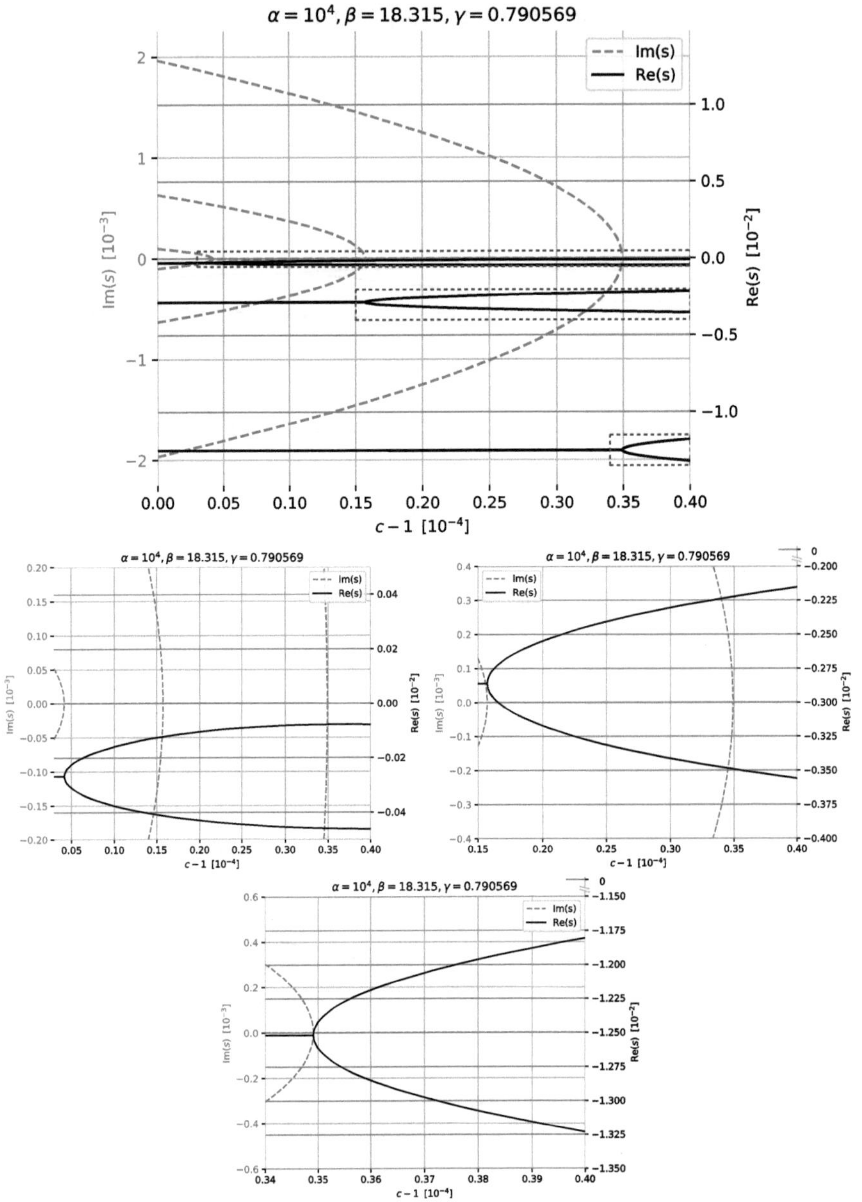

Fig. 5.30 Lyapunov exponents of a viscoelastic traveling panel with $\tau = 10^{-2}$ s, near $c = 1$. *Top*: overview. *Others*: detail of the first three bifurcation points. In each subfigure, the two vertical axes have been zoomed independently to bring out the detail. The zoom ranges of Re(s) are indicated in the overview

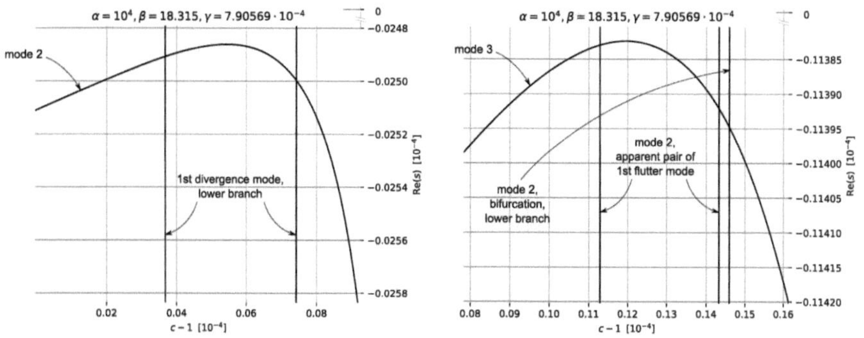

Fig. 5.31 Typical local maxima of Re(s) appearing in the viscoelastic case, with $\tau = 10^{-5}$ s. Only the real parts are shown. The vertical axis has been significantly zoomed to make this detail visible. The near-vertical lines belong to other modes as indicated on the figure. The corresponding overview is shown in Fig. 5.27. *Left*: mode 2, local maximum Re(s) $= -0.02486 \cdot 10^{-4}$ at $c - 1 = 0.0545 \cdot 10^{-4}$. *Right*: mode 3, local maximum Re(s) $= -0.1138 \cdot 10^{-4}$ at $c - 1 = 0.120 \cdot 10^{-4}$

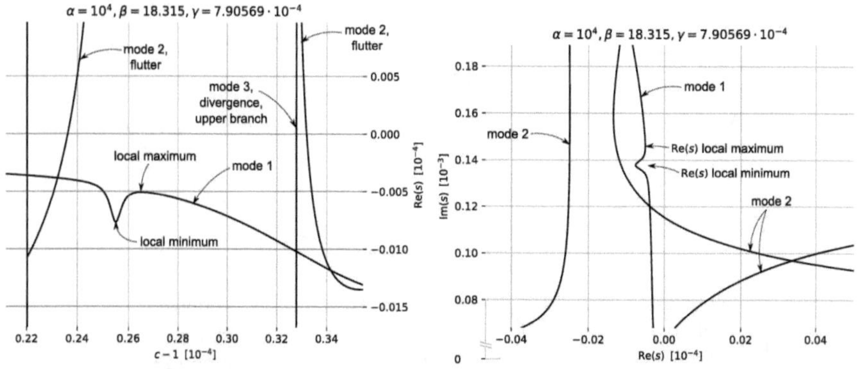

Fig. 5.32 A curious local extremum of Re(s), mode 1, viscoelastic case with $\tau = 10^{-5}$ s. The local minimum Re(s) $= -0.007679 \cdot 10^{-4}$ occurs at $c - 1 = 0.255 \cdot 10^{-4}$. The local maximum Re($s$) $= -0.005072 \cdot 10^{-4}$ occurs at $c - 1 = 0.265 \cdot 10^{-4}$. *Left*: Re($s$) as a function of c. *Right*: view in the (Re(s), Im(s)) plane

Higher-order stable regions exist beyond the first critical point, after the system regains stability. If the physical system can be somehow guided into these regions, it may be possible to have stable operation at post-critical axial drive velocities. From the viewpoint of the small-displacement model, it does not matter if the small-displacement assumption is broken while the guiding is performed, as long as the target state, in which the system is left (before we analyze it using the small-displacement model), again fulfills the small-displacement assumption.

In the elastic case, the following symmetry property holds. If $s_1 = a + ib$ is an eigenvalue, so are $s_2 = \text{conj}(s_1) = a - ib$, $s_3 = -a + ib$ and $s_4 = \text{conj}(s_3) = -a - ib$. (Some of these may be equal if either $a = 0$ or $b = 0$.) The introduction

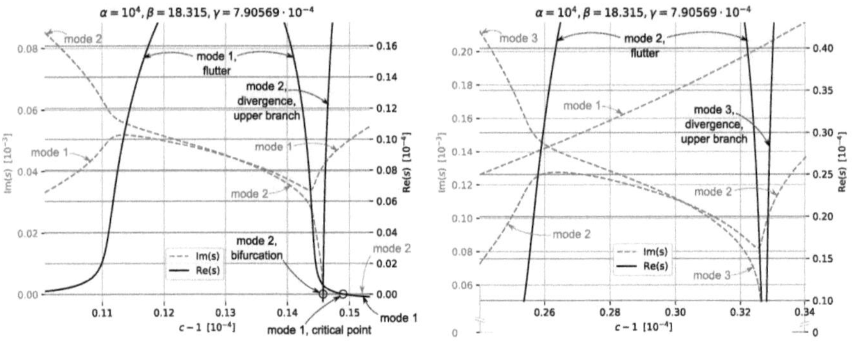

Fig. 5.33 Detail of flutter modes in the viscoelastic case, with $\tau = 10^{-5}$ s. The flutter is single-mode only; no mode interaction occurs, and no flutter-related bifurcations appear. *Left*: 1st flutter mode. *Right*: 2nd flutter mode

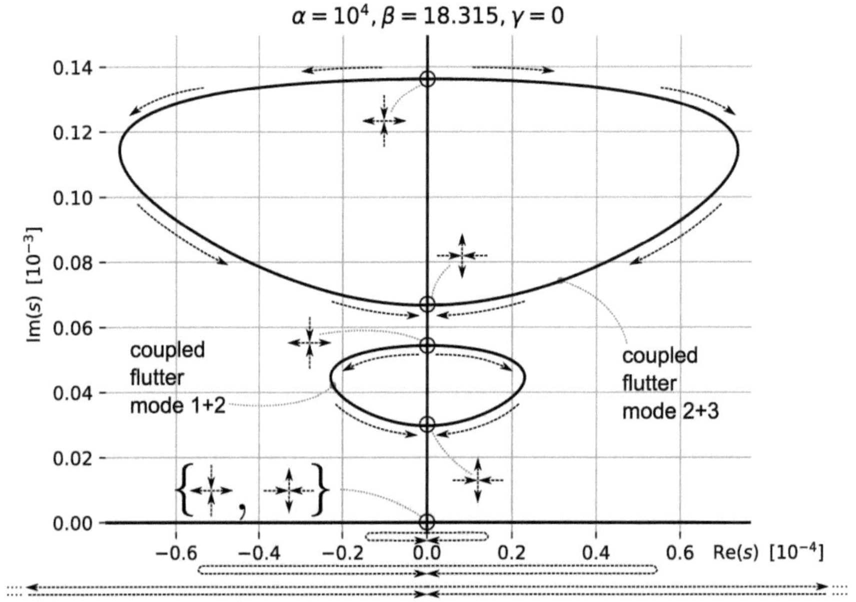

Fig. 5.34 The coupled flutter modes in the elastic case, $\tau = 0$. Due to symmetry, only the half with $\mathrm{Im}(s) \geq 0$ is shown. See the overview in Fig. 5.24, the data in Table 5.3, and the projection in Fig. 5.26

of viscosity causes some of the symmetry to be lost. In the viscoelastic case, if s_1 is a solution, so is $s_2 = \mathrm{conj}(s_1)$, but s_3 and s_4 generally are not.

In the viscoelastic case, all modes experience damping, with higher damping for the higher modes. This implies that in a real system, the lowest modes are more likely to be observed, as they decay more slowly.

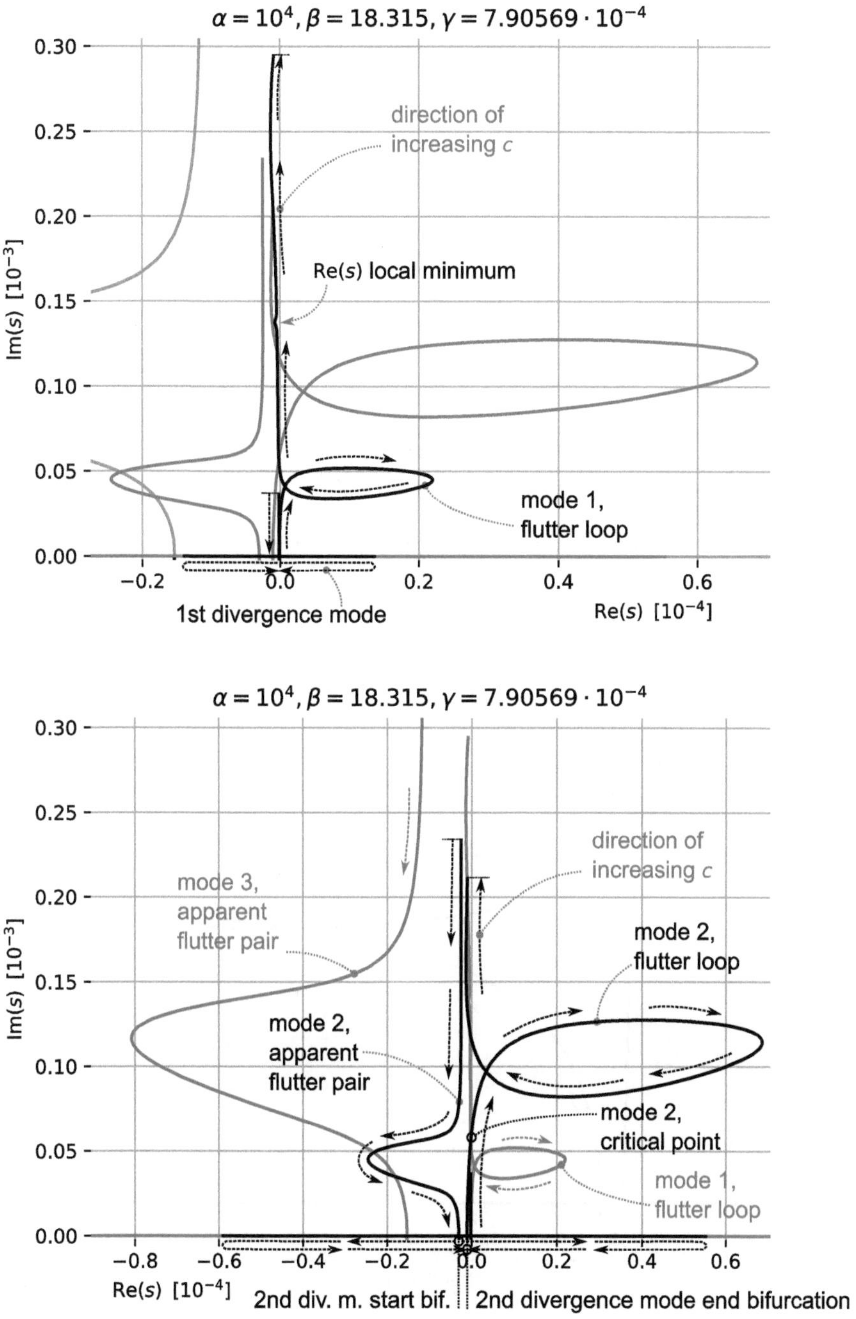

Fig. 5.35 Detail showing the behavior of the first two modes in the (Re(s), Im(s)) plane, viscoelastic case with $\tau = 10^{-5}$ s. Due to symmetry, only the half with Im(s) ≥ 0 is shown. See the overview in Fig. 5.24, the data in Table 5.4, and the projection in Fig. 5.27. *Top*: mode 1. *Bottom*: mode 2

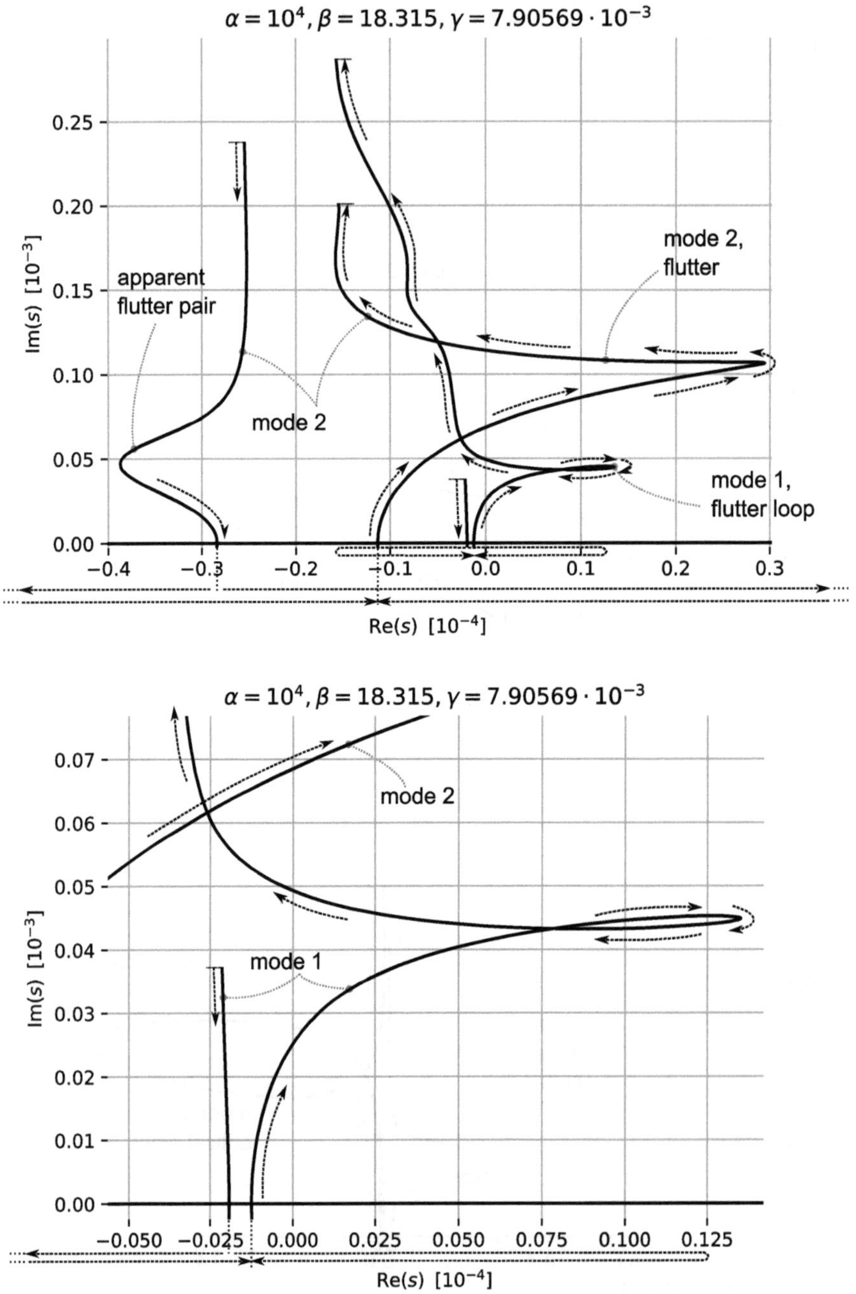

Fig. 5.36 Detail showing the behavior of the first two modes in the (Re(s), Im(s)) plane, viscoelastic case with $\tau = 10^{-4}$ s. Due to symmetry, only the half with Im(s) ≥ 0 is shown. See the overview in Fig. 5.24, the data in Table 5.5, and the projection in Fig. 5.28. *Top*: modes 1 and 2. *Bottom*: detail of mode 1

From a mathematical perspective, the most interesting point is that even at very small viscosities, features typical of the elastic case vanish. The qualitative behavior of the Lyapunov exponents for the viscoelastic case is radically different from the elastic case. Coupled-mode flutter does not occur for the Kelvin–Voigt panel, even at vanishingly small values of the viscosity. The modes do not interact; bifurcations occur for each mode separately. For the Kelvin–Voigt panel, the visually apparent pair of a flutter mode is actually a different, independent mode (refer to Figs. 5.25, 5.35 and 5.36).

For divergence modes, in the viscoelastic case the bifurcation point is no longer the critical point, at which s crosses $\mathrm{Re}(s) = 0$, although at small values of the viscosity, it is still a good approximation. See Fig. 5.27 and Table 5.4 for $\tau = 10^{-5}$ s; compare Fig. 5.28 and Table 5.5 for $\tau = 10^{-4}$ s. Flutter modes, on the other hand, no longer have bifurcation points associated with them; to find them, one must look at crossings of $\mathrm{Re}(s) = 0$. These features are typical of systems with damping (cf. [18, 19]).

Thus, in a practical sense, in the case of viscoelastic systems critical points are more important than bifurcation points. However, for the mathematical understanding of the governing equation, bifurcation points remain of interest, as they are obviously radically different from regular points on the solution curves.

On viscous systems, we make two final observations. First, high viscosity stabilizes the system across the range of c studied, as is intuitively expected. Finally, extra local extrema appear on the solution curves. Each solution wiggles on its own, without interacting with the other solutions. However, the present analysis does not tell how much of this is a physical effect and how much an artifact of numerical discretization.

It is time to conclude this chapter. Starting from fundamental principles, we systematically constructed a model for a one-dimensional moving material, developed its weak form, and performed a linear stability analysis to study its behavior with regard to self-amplification of small-amplitude free vibrations. Both the systematic construction and the obtained results shed light on the fundamental properties of mathematical models employed in process industry applications.

The present chapter, while covering the fundamentals, has barely scratched the surface of the modeling of axially moving materials and the study of bifurcation phenomena in the models. For example, the focus was on one-dimensional models, and the more complex cases of membranes and plates, where the transverse displacement may vary also in the width direction, were neglected. Furthermore, in order to keep the development simple, the influence of the surrounding air was not modelled. In practice, as paper is a lightweight material, the presence of the surrounding air is known to drastically reduce the natural frequencies of free vibrations [8, 42, 67, 109, 110]. To cover some of the related topics, we will look at some two-dimensional models of axially moving materials in Chap. 6, some analytical free-vibration solutions of simple cases in one-dimensional models in Chap. 7, and a simple fluid—structure interaction model in Chap. 8.

In most of the construction, mathematical rigor was emphasized in order to ensure strict reliability while also making explicit the exact assumptions upon which the model and its analysis depend. Elementary mathematics was preferred whenever possible, keeping the discussion straightforward and accessible. Emphasis was given also to providing general-purpose tools and methods, which are applicable not only to this particular system but also to a wide variety of different systems in mechanics.

A scaling argument was developed, showing the limitations of constructing physical scale models that behave identically to an intended full-scale setup. It was seen that for an elastic traveling material this is always possible, but the viscoelastic (Kelvin–Voigt) case requires very specific engineering of individual material constants of the material running through the span, and thus is likely not to be feasible. In the age of scientific computing, generally speaking the practical relevance of the construction of physical scale models has obviously decreased. The practical value of the analysis of the scaling properties now resides in the mathematical understanding it affords with regard to the behavior of the governing equation.

Table 5.3 Points of interest, for the first three vibration modes k, in the range of c shown in the figures, for an elastic panel ($\tau = 0$)

$c - 1$ [10^{-4}]	k	Re(s) [10^{-4}]	Im(s) [10^{-3}]	Type	Note
0.0361	1	0	0	Bifurcation; critical point	Loss of stability (divergence)
0.0554	1	±0.1360	0	−: min Re(s); +: max Re(s)	Extremum, 1st divergence mode
0.0740	1	0	0	Bifurcation; critical point	Regain of stability (divergence)
0.110	1 + 2	0	±0.05441	Bifurcation; critical point	Loss of stability (flutter)
0.129	1 + 2	±0.2295	±0.04461	−Re: min Re(s); +Re: max Re(s)	Extremum, coupled flutter mode; all combinations of ±
0.143	1 + 2	0	±0.02970	Bifurcation; critical point	Regain of stability (flutter)
0.145	2	0	0	Bifurcation; critical point	Loss of stability (divergence)
0.181	2	±0.5654	0	−: min Re(s); +: max Re(s)	Extremum, 2nd divergence mode
0.219	2	0	0	Bifurcation; critical point	Regain of stability (divergence)
0.253	2 + 3	0	±0.1363	Bifurcation; critical point	Loss of stability (flutter)
0.293	2 + 3	±0.7368	±0.1143	−Re: min Re(s); +Re: max Re(s)	Extremum, coupled flutter mode; all combinations of ±
0.324	2 + 3	0	±0.06674	Bifurcation; critical point	Regain of stability (flutter)
0.325	3	0	0	Bifurcation; critical point	Loss of stability (divergence)
0.378	3	±1.352	0	−: min Re(s); +: max Re(s)	Extremum, 3rd div. mode

Table 5.4 Points of interest, for the first three vibration modes k, in the range of c shown in the figures, for a viscoelastic panel with $\tau = 10^{-5}$ s. The pairs of divergence modes are actual eigenvalue pairs; viscosity causes the loss of symmetry of the positive and negative extremal values of $\mathrm{Re}(s)$. For the flutter modes, considering the real parts, the pairing is only apparent; the components of the pair belong to different vibration modes k. Furthermore, strictly speaking, the bifurcation points with nonzero $\mathrm{Re}(s)$ are not the exact critical points (at which $\mathrm{Re}(s) = 0$), but for this example these are separated by less than one step of c at the displayed precision. In these cases, we have preferred to show the value of s from the bifurcation point

$c - 1$ [10^{-4}]	k	Re(s) [10^{-4}]	Im(s) [10^{-3}]	Type	Note
0.0364	1	−0.001900	0	Bifurcation; critical point	Loss of stability (divergence)
0.0545	2	−0.02486	±0.15769	max Re(s)	Local maximum
0.0557	1	−0.1392	0	min Re(s)	Extremum, pair of 1st divergence mode
0.0558	1	+0.1359	0	max Re(s)	Extremum, 1st divergence mode
0.0745	1	−0.001229	0	Bifurcation; critical point	Regain of stability (divergence)
0.0955	1	0	±0.02754	Critical point	Loss of stability (flutter)
0.113	1	+0.1021	±0.05191	−: min Im(s); +: max Im(s)	No bifurcation
0.120	3	−0.1138	±0.4872	max Re(s)	Local maximum
0.130	2	−0.2448	±0.04521	min Re(s)	Extremum, apparent pair of 1st flutter mode
0.130	1	+0.2199	±0.04503	max Re(s)	Extremum, 1st flutter mode
0.144	1	+0.07212	±0.03388	−: max Im(s); +: min Im(s)	No bifurcation
0.146	2	−0.03041	0	Bifurcation; critical point	Loss of stability (divergence)
0.149	1	0	±0.04874	Critical point	Regain of stability (flutter)
0.182	2	−0.5907	0	min Re(s)	Extremum, pair of 2nd divergence mode
0.182	2	+0.5535	0	max Re(s)	Extremum, 2nd divergence mode
0.220	2	−0.01073	0	Bifurcation; critical point	Regain of stability (divergence)
0.236	2	0	±0.06150	Critical point	Loss of stability (flutter)
0.255	1	−0.007679	±0.1379	min Re(s)	Local minimum
0.264	2	+0.4131	±0.1276	−: min Im(s); +: max Im(s)	No bifurcation
0.265	1	−0.005072	±0.1462	max Re(s)	Local maximum
0.295	3	−0.8081	±0.1166	min Re(s)	Extremum, apparent pair of 2nd flutter mode
0.295	2	+0.6861	±0.1146	max Re(s)	Extremum, 2nd flutter mode
0.325	2	+0.2180	±0.08214	−: max Im(s); +: min Im(s)	No bifurcation
0.328	3	−0.1529	0	Bifurcation; critical point	Loss of stability (divergence)
0.332	2	0	±0.1156	Critical point	Regain of stability (flutter)
0.354	2	−0.01350	±0.1589	min Re(s)	Local minimum
0.370	1	−0.01374	±0.2537	min Re(s)	Local minimum
0.380	3	−1.453	0	min Re(s)	Extremum, pair of 3rd divergence mode
0.381	3	+1.285	0	max Re(s)	Extremum, 3rd divergence mode

Table 5.5 Points of interest, for the first three vibration modes k, in the range of c shown in the figures, for a viscoelastic panel with $\tau = 10^{-4}$ s. Viscosity causes the offset in the c values of extremal values of modes belonging to the same pair

$c - 1$ [10^{-4}]	k	Re(s) [10^{-4}]	Im(s) [10^{-3}]	Type	Note
0.0368	1	−0.01928	0	Bifurcation	
0.0370	1	0	0	Critical point	Loss of stability (divergence)
0.0544	2	−0.2516	±0.1611	max Re(s)	Local maximum
0.0561	1	−0.1575	0	min Re(s)	Extremum, pair of 1st divergence mode
0.0569	1	+0.1234	0	max Re(s)	Extremum, 1st divergence mode
0.0756	1	0	0	Critical point	Regain of stability (divergence)
0.0757	1	−0.01260	0	Bifurcation	
0.0946	1	0	±0.02524	Critical point	Loss of stability (flutter)
0.118	3	−1.151	±0.5003	max Re(s)	Local maximum
0.126	1	+0.1266	±0.04526	−: min Im(s); +: max Im(s)	No bifurcation
0.131	2	−0.3873	±0.04694	min Re(s)	Extremum, apparent pair of 1st flutter mode
0.132	1	+0.1351	±0.04486	max Re(s)	Extremum, 1st flutter mode
0.143	1	+0.08849	±0.04322	−: max Im(s); +: min Im(s)	No bifurcation
0.147	2	−0.2835	0	Bifurcation	
0.150	2	0	0	Critical point	Loss of stability (divergence)
0.151	1	0	±0.04928	Critical point	Regain of stability (flutter)
0.182	2	−0.7834	0	min Re(s)	Extremum, pair of 2nd divergence mode
0.188	2	+0.4049	0	max Re(s)	Extremum, 2nd divergence mode
0.225	2	0	0	Critical point	Regain of stability (divergence)
0.226	2	−0.1129	0	Bifurcation	
0.248	2	0	±0.06840	Critical point	Loss of stability (flutter)
0.272	1	−0.08229	±0.1518	min Re(s)	Local minimum
0.283	1	−0.08179	±0.1622	max Re(s)	Local maximum
0.295	3	−1.524	±0.1272	min Re(s)	Extremum, apparent pair of 2nd flutter mode
0.299	2	+0.2944	±0.1065	max Re(s)	Extremum, 2nd flutter mode
0.330	3	−1.269	0	Bifurcation	
0.336	2	0	±0.1142	Critical point	Regain of stability (flutter)
0.345	3	0	0	Critical point	Loss of stability (divergence)
0.370	2	−0.1579	±0.1682	min Re(s)	Local minimum
0.380	3	−2.327	0	min Re(s)	Extremum, pair of 3rd divergence mode
0.397	2	−0.1543	±0.1973	min Re(s)	Local minimum
0.398	3	+0.6267	0	max Re(s)	Extremum, 3rd divergence mode

Table 5.6 Points of interest, for the first three vibration modes k, in the range of c shown in the figures, for a viscoelastic panel with $\tau = 10^{-3}$ s. Note scaling of Re(s). The second divergence mode remains stable for the range of c studied

$c - 1$ [10^{-4}]	k	Re(s) [10^{-3}]	Im(s) [10^{-3}]	Type	Note
0.0396	1	−0.02246	0	Bifurcation	
0.0610	1	0	0	Critical point	Loss of stability (divergence)
0.0732	1	−0.04467	0	min Re(s)	Extremum, pair of 1st divergence mode
0.0787	1	0.002754	0	max Re(s)	Extremum, 1st divergence mode
0.0962	1	0	0	Critical point	Regain of stability (divergence)
0.115	1	−0.01912	0	Bifurcation	
0.144	1	−0.01840	±0.03177	max Re(s)	Local maximum
0.153	2	−0.2678	0	Bifurcation	
0.250	2	−0.3754	0	min Re(s)	Extremum, pair of 2nd divergence mode
0.321	2	−0.07033	0	max Re(s)	Extremum, 2nd divergence mode (stable)
0.340	3	−1.207	0	Bifurcation	

Table 5.7 Points of interest, for the first three vibration modes k, in the range of c shown in the figures, for a viscoelastic panel with $\tau = 10^{-2}$ s. Note scaling of Re(s). All three modes remain stable for the range of c studied

$c - 1$ [10^{-4}]	k	Re(s) [10^{-2}]	Im(s) [10^{-3}]	Type	Note
0.0419	1	−0.02673	0	Bifurcation	
0.157	2	−0.2861	0	Bifurcation	
0.349	3	−1.252	0	Bifurcation	
0.371	1	−0.007758	0	max Re(s)	Extremum, 1st divergence mode (stable)

In the numerical results, in terms of the eigenfrequencies, the small-viscosity case was seen to behave radically differently from the elastic case. Because no material is perfectly elastic, especially in papermaking, the behavior of the viscoelastic system at small values of the viscosity is important in practice for the correct qualitative understanding of real physical systems.

References

1. Adams RA (1975) Sobolev spaces. Academic Press
2. Adams RA, Essex C (2010) Calculus: a complete course, 7th edn. Pearson
3. Alava M, Niskanen K (2006) The physics of paper. Rep Prog Phys 69(3):669–723

4. Allen MB III, Herrera I, Pinder GF (1988) Numerical modeling in science and engineering. Wiley Interscience
5. Antman SS (2005) Nonlinear problems of elasticity. Applied mathematical sciences, vol 107. Springer. https://doi.org/10.1007/0-387-27649-1. ISBN 978-0-387-20880-0
6. Archibald FR, Emslie AG (1958) The vibration of a string having a uniform motion along its length. ASME J Appl Mech 25:347–348
7. Banichuk N, Jeronen J, Kurki M, Neittaanmäki P, Saksa T, Tuovinen T (2011) On the limit velocity and buckling phenomena of axially moving orthotropic membranes and plates. Int J Solids Struct 48(13):2015–2025. https://doi.org/10.1016/j.ijsolstr.2011.03.010. ISSN 0020-7683
8. Banichuk N, Jeronen J, Neittaanmäki P, Saksa T, Tuovinen T (2014) Mechanics of moving materials. Solid mechanics and its applications, vol 207. Springer. ISBN 978-3-319-01744-0 (print), 978-3-319-01745-7 (electronic)
9. Bathe K-J (1996) Finite element procedures. Prentice Hall. ISBN 0-13-301458-4
10. Baum GA, Brennan DC, Habeger CC (1981) Orthotropic elastic constants of paper. TAPPI J 64(8):97–101
11. Belytschko T, Liu WK, Moran B (2000) Nonlinear finite elements for continua and structures. Wiley. ISBN 978-0-471-98774-1
12. Bender CM, Orszag SA (1978) Advanced mathematical methods for scientists and engineers I: asymptotic methods and perturbation theory. Springer. 1999 reprint: ISBN 978-0-387-98931-0
13. Berdichevsky V (2009) Variational principles of continuum mechanics. Springer
14. Bernstein DS, Bhat SP (1995) Lyapunov stability, semistability, and asymptotic stability of matrix second-order systems. Trans ASME 117:145–153
15. Berryman JG, Pride SR (1998) Volume averaging, effective stress rules, and inversion for microstructural response of multicomponent porous media. Int J Solids Struct 35:4811–4843
16. Bisplinghoff RL (1962) Ashley H (1975) Principles of aeroelasticity, 2nd edn. Dover Publications Inc, New York
17. Bleecker D, Csordas G (1996) Basic partial differential equations. International Press, Cambridge, MA
18. Bolotin VV (1964) The dynamic stability of elastic systems. Holden-Day, Inc. Translated from the Russian (1956) and German (1961) editions
19. Bolotin VV (1963) Nonconservative problems of the theory of elastic stability. Pergamon Press, New York
20. Bower AF (2009) Applied mechanics of solids, 1st edn. CRC Press. ISBN 978-1439802472
21. Brenner SC, Scott LR (2010) The mathematical theory of finite element methods. Texts in applied mathematics, vol 15, 3rd edn. Springer
22. Brezis H (2011) Functional analysis. Springer, Sobolev spaces and partial differential equations. https://doi.org/10.1007/978-0-387-70914-7. ISBN 978-0-387-70913-0
23. Chen L-Q (2005) Analysis and control of transverse vibrations of axially moving strings. ASME Applied Mechanics Reviews 58:91–116. https://doi.org/10.1115/1.1849169
24. Chen L-Q (2005a) Principal parametric resonance of axially accelerating viscoelastic strings with an integral constitutive law. Proc R Soc A Math Phys Eng Sci 461(2061):2701–2720. https://doi.org/10.1098/rspa.2005.1471
25. Chen L-Q, Wang B (2009) Stability of axially accelerating viscoelastic beams: asymptotic perturbation analysis and differential quadrature validation. Eur J Mech - A/Solids 28(4):786–791. https://doi.org/10.1016/j.euromechsol.2008.12.002. ISSN 0997-7538
26. Chen L-Q, Yang X-D, Cheng C-J (2004) Dynamic stability of an axially accelerating viscoelastic beam. Eur J Mech - A/Solids 23(4):659–666. https://doi.org/10.1016/j.euromechsol.2004.01.002. ISSN 0997-7538
27. Chen L-Q, Zu JW, Wu J, Yang X-D (2004) Transverse vibrations of an axially accelerating viscoelastic string with geometric nonlinearity. J Eng Math 48(2):171–182. https://doi.org/10.1023/B:ENGI.0000011929.17902.87. ISSN 0022-0833

28. Chen L-Q, Chen H, Lim CW (2008) Asymptotic analysis of axially accelerating viscoelastic strings. Int J Eng Sci 46(10):976–985. https://doi.org/10.1016/j.ijengsci.2008.03.009. ISSN 0020-7225
29. Clelland JN, Vassiliou PJ (2013) Strings attached: new light on an old problem. arXiv:1302.6672 [math.AP]
30. Courant R, Hilbert D (1966) Methods of mathematical physics, vol I. Interscience Publishers Inc., New York
31. DeCrosta EF, Vennos SLN (1968) What happens in a dryer pocket. TAPPI J 51(7):289–298
32. Ding H, Chen L-Q (2008) Stability of axially accelerating viscoelastic beams: multi-scale analysis with numerical confirmations. Eur J Mech - A/Solids 27(6):1108–1120. https://doi.org/10.1016/j.euromechsol.2007.11.014. ISSN 0997-7538
33. Donea J, Huerta A (2003) Finite element methods for flow problems. Wiley. ISBN 0-471-49666-9
34. Eriksson K, Estep D, Hansbo P, Johnson C (1996) Computational Differential Equations. Studentlitteratur, Lund. ISBN 91-44-49311-8
35. Erkkilä A-L, Leppänen T, Hämäläinen J, Tuovinen T (2015) Hygro-elasto-plastic model for planar orthotropic material. Int J Solids Struct 62:66–80. https://doi.org/10.1016/j.ijsolstr.2015.02.001
36. Erkkilä A, Leppänen T, Hämäläinen J (2013) Empirical plasticity models applied for paper sheets having different anisotropy and dry solids content levels. Int J Solids Struct 50:2151–2179
37. Evans LC (1998) Partial differential equations. American Mathematical Society. ISBN 0-8218-0772-2
38. Feng K, Shi Z-C (1996) Mathematical theory of elastic structures. Springer. ISBN 3-540-51326-4, 0-387-51326-4
39. Fish J, Belytschko T (2007) A first course in finite elements. Wiley. ISBN 978-0-470-03580-1
40. Flügge W (1972) Tensor analysis and continuum mechanics. Springer
41. Flügge W (1975) Viscoelasticity, 2nd edn. Springer, New York
42. Frondelius T, Koivurova H, Pramila A (2006) Interaction of an axially moving band and surrounding fluid by boundary layer theory. J Fluids Struct 22(8):1047–1056
43. Fung YC (1965) Foundations of solid mechanics. Prentice-Hall
44. Ghayesh MH, Amabili M, Païdoussis MP (2013) Nonlinear dynamics of axially moving plates. J Sound Vib 332(2):391–406. https://doi.org/10.1016/j.jsv.2012.08.013. ISSN 0022-460X
45. Gibson P (1994) Governing equations for multiphase heat and mass transfer in hygroscopic porous media with applications to clothing materials. Document ADA286418 of Army Natick Research Development And Engineering Center, MA, available online
46. Gibson P, Charmchi M (1997) The use of volume-averaging techniques to predict temperature transients due to water vapor sorption in hygroscopic porous polymer materials. J Appl Polym Sci 64(3):493–505. https://doi.org/10.1002/(SICI)1097-4628(19970418)64:3<493::AID-APP6>3.0.CO;2-F
47. Holzapfel GA (2000) Nonlinear solid mechanics–a continuum approach for engineering. Wiley
48. Hosaka H, Crandall SH (1992) Self-excited vibrations of a flexible disk rotating on an air film above a flat surface. Acta Mech 3(Supplement):115–127
49. Huber MT (1914) Die Grundlagen einer rationellen Berechnung der kreuzweise bewehrten Eisenbetonplatten. Zeitschrift der Österreichische Ingeniur- und Architekten-Vereines 30:557–564
50. Huber MT (1923) Die Theorie des kreuzweise bewehrten Eisenbetonplatten. Der Bauingenieur 4:354–392
51. Huber MT (1926) Einige Andwendungen fer Biegungstheorie orthotroper Platten. Zeitschrift für angewandte Mathematik und Mechanik 6(3):228–232
52. Hughes TJR (2000) The finite element method. linear static and dynamic finite element analysis. Dover Publications Inc., Mineola, N.Y., USA. ISBN 0-486-41181-8

53. Itskov M, Aksel N (2002) Elastic constants and their admissible values for incompressible and slightly compressible anisotropic materials. Acta Mech 157:81–96
54. Jeronen J (2011) On the mechanical stability and out-of-plane dynamics of a travelling panel submerged in axially flowing ideal fluid: a study into paper production in mathematical terms. PhD thesis, Department of Mathematical Information Technology, University of Jyväskylä. http://urn.fi/URN:ISBN:978-951-39-4596-1. Jyväskylä studies in computing 148. ISBN 978-951-39-4595-4 (book), ISBN 978-951-39-4596-1 (PDF)
55. Johnson C (2009) Numerical solution of partial differential equations by the finite element method. Cambridge University Press, 1987. Reprint by Dover
56. Karlsson M (1984) On-line study and mathematical modeling of a paper dryer section. PhD thesis. Åbo Akademi, Faculty of Chemical Engineering, Plant Design Laboratory
57. Karlsson M (ed) (2009) Papermaking, Part 2 drying. Papermaking science and technology, vol 9. Paperi ja Puu Oy. ISBN 978-952-5216-37-0 (book 9)
58. Kasner E (1914) The ratio of the arc to the chord of an analytic curve need not approach unity. Bull Am Math Soc 20(10):524–531. http://projecteuclid.org/euclid.bams/1183422829
59. Kaviany M, Mittal M (1987) Funicular state in drying of a porous slab. Int J Heat Mass Transf 30(7):1407–1418. https://doi.org/10.1016/0017-9310(87)90172-4
60. Kim J, Cho J, Lee U, Park S (2003) Modal spectral element formulation for axially moving plates subjected to in-plane axial tension. Comput Struct 81:2011–2020. https://doi.org/10.1016/S0045-7949(03)00229-3
61. Kirillov ON (2013) Nonconservative stability problems of modern physics. de Gruyter. ISBN 978-3-11-027043-3
62. Kirillov ON, Verhulst F (2010) Paradoxes of dissipation-induced destabilization or who opened Whitney's umbrella? Zeitschrift für Angewandte Mathematik und Mechanik 90(6):462–488. https://doi.org/10.1002/zamm.200900315
63. Koivurova H, Salonen E-M (1999) Comments on non-linear formulations for travelling string and beam problems. J Sound Vib 225(5):845–856
64. Komkov V (1988) Variational principles of continuum mechanics with engineering applications. Vol. 1: critical points theory. Reidel Publishing Co., Dordrecht
65. Kong L, Parker RG (2004) Approximate eigensolutions of axially moving beams with small flexural stiffness. J Sound Vib 276:459–469. https://doi.org/10.1016/j.jsv.2003.11.027
66. Krizek M, Neittaanmäki P (1990) Finite element Approximation of variational problems and applications. Longman Scientific & Technical, Harlow, Copubl. Wiley, New York
67. Kulachenko A, Gradin P, Koivurova H (2007) Modelling the dynamical behaviour of a paper web. Part I. Comput Struct 85:131–147. https://doi.org/10.1016/j.compstruc.2006.09.006
68. Kurki M (2005) Modeling of kinematical and rheological web line behavior in a papermaking environment. Licentiate thesis, Lappeenranta University of Technology, Department of Mechanical Engineering. Lappeenranta, Finland
69. Kurki M, Jeronen J, Saksa T, Tuovinen T (2016) The origin of in-plane stresses in axially moving orthotropic continua. Int J Solids Struct. https://doi.org/10.1016/j.ijsolstr.2015.10.027
70. Kurki M, Jeronen J, Saksa T, Tuovinen T (2012) Strain field theory for viscoelastic continuous high-speed webs with plane stress behavior. In: Eberhardsteiner J, Böhm HJ, Rammerstorfer FG (eds) CD-ROM proceedings of the 6th European congress on computational methods in applied sciences and engineering (ECCOMAS (2012) Vienna University of Technology. Austria, Vienna. ISBN 978-3-9502481-9-7
71. Kurki M, Juppi K, Ryymin R, Taskinen P, Pakarinen P (1995) On the web tension dynamics on an open draw. In: Proceedings of the third international conference on web handling. Web Handling Research Center, Oklahoma State University, Stillwater, Oklahoma, pp 230–244
72. Kurki M, Lehtinen A (2009) In-plane strain field theory for 2-d moving viscoelastic webs. In: Papermaking research symposium 2009 (Kuopio. Finland). PRS
73. Lagerstrom PA, Casten RG (1972) Basic concepts underlying singular perturbation techniques. SIAM Rev 14(1):63–120

74. Lampinen MJ, Toivonen K (1984) Application of a thermodynamic theory to determine capillary pressure and other fundamental material properties affecting the drying process. Drying '84. Hemisphere, New York, pp 228–244
75. Landau LD, Lifshitz EM (1970) Teoriya uprugosti (Theory of elasticity, 2nd edn.). English 2nd edn. published by Pergamon Press, Oxford, 1965
76. Lee S-Y, Mote CD Jr (1996) Vibration control of an axially moving string by boundary control. ASME J Dyn Syst Meas Control 118:66–74. https://doi.org/10.1115/1.2801153
77. Lee U, Oh H (2005) Dynamics of an axially moving viscoelastic beam subject to axial tension. Int J Solids Struct 42(8):2381–2398. https://doi.org/10.1016/j.ijsolstr.2004.09.026. ISSN 0020-7683
78. Lepomäki H (1993) Rainan sisäisten kuivatusparametrien määritys parametrisovituksella (The determination of drying parameters inside the paper web by using parameter fitting). Master's thesis, Lappeenrannan teknillinen korkeakoulu (Lappeenranta University of Technology)
79. Lif JO, Östlund S, Fellers C (1999) Applicability of anisotropic viscoelasticity of paper at small deformations. Mech Time-Depend Mater 2(3):245–267. https://doi.org/10.1023/A:1009818022865. ISSN 1385-2000
80. Lin CC (1997) Stability and vibration characteristics of axially moving plates. Int J Solids Struct 34(24):3179–3190. https://doi.org/10.1016/S0020-7683(96)00181-3
81. Lin CC, Mote CD (1995) Equilibrium displacement and stress distribution in a two-dimensional, axially moving web under transverse loading. ASME J Appl Mech 62:772–779. https://doi.org/10.1115/1.2897013
82. Lin CC, Mote CD (1996) Eigenvalue solutions predicting the wrinkling of rectangular webs under non-linearly distributed edge loading. J Sound Vib 197(2):179–189. https://doi.org/10.1006/jsvi.1996.0524
83. Lu T, Shen SQ (2007) Numerical and experimental investigation of paper drying: Heat and mass transfer with phase change in porous media. Appl Therm Eng 27:1248–1258
84. Luo ACJ, Mote CD Jr (2000) An exact, closed-form solution for equilibrium of traveling, sagged, elastic cables under uniformly distributed loading. Commun Nonlinear Sci Numer Simul 5(1):6–11
85. Malvern LE (1969) Introduction to the mechanics of a continuous medium. Prentice-Hall
86. Mann RW, Baum GA, Habeger CC (1980) Determination of all nine orthotropic elastic constants for machine-made paper. TAPPI J 63(2):163–166
87. Marsden JE, Hughes TJR (1983) Mathematical foundations of elasticity. Dover
88. Marynowski K (2008) Dynamics of the axially moving orthotropic web. Lecture notes in applied and computational mechanics, vol 38. Springer, Germany. https://doi.org/10.1007/978-3-540-78989-5. ISBN 978-3-540-78988-8 (print), 978-3-540-78989-5 (online)
89. Marynowski K (2004) Non-linear vibrations of an axially moving viscoelastic web with time-dependent tension. Chaos Solitons Fractals 21(2):481–490. https://doi.org/10.1016/j.chaos.2003.12.020. ISSN 0960-0779
90. Marynowski K (2006) Two-dimensional rheological element in modelling of axially moving viscoelastic web. Eur J Mech - A/Solids 25(5):729–744. https://doi.org/10.1016/j.euromechsol.2005.10.005. ISSN 0997-7538
91. Marynowski K (2008a) Non-linear vibrations of the axially moving paper web. J Theor Appl Mech 46(3):565–580
92. Marynowski K (2010) Free vibration analysis of the axially moving Levy-type viscoelastic plate. Eur J Mech - A/Solids 29(5):879–886. https://doi.org/10.1016/j.euromechsol.2010.03.010. ISSN 0997-7538
93. Marynowski K, Kapitaniak T (2002) Kelvin-Voigt versus Bürgers internal damping in modeling of axially moving viscoelastic web. Int J Non-Linear Mech 37(7):1147–1161. https://doi.org/10.1016/S0020-7462(01)00142-1. ISSN 0020-7462
94. Marynowski K, Kapitaniak T (2007) Zener internal damping in modelling of axially moving viscoelastic beam with time-dependent tension. Int J Non-Linear Mech 42(1):118–131. https://doi.org/10.1016/j.ijnonlinmec.2006.09.006. ISSN 0020-7462

95. Marynowski K, Kapitaniak T (2014) Dynamics of axially moving continua. Int J Mech Sci 81:26–41. https://doi.org/10.1016/j.ijmecsci.2014.01.017. ISSN 0020-7403
96. Matculevich S, Neittaanmäki P, Repin S (2013) Guaranteed error bounds for a class of Picard–Lindelöf iteration methods. In Repin S, Tiihonen T, Tuovinen T (eds) Numerical methods for differential equations, optimization, and technological problems. Dedicated to Professor P. Neittaanmäki on his 60th Birthday. Computational methods in applied sciences, vol 27. Springer Netherlands, pp 175–189. ISBN 978-94-007-5287-0 (Print) 978-94-007-5288-7 (Online)
97. Mikhlin SG (1964) Variational methods in mathematical physics. Pergamon Press, New York
98. Miranker WL (1960) The wave equation in a medium in motion. IBM J Res Dev 4:36–42. https://doi.org/10.1147/rd.41.0036
99. Mote CD (1968) Divergence buckling of an edge-loaded axially moving band. Int J Mech Sci 10:281–295. https://doi.org/10.1016/0020-7403(68)90013-1
100. Mote CD (1968) Dynamic stability of an axially moving band. J Frankl Inst 285(5):329–346. https://doi.org/10.1016/0016-0032(68)90482-1
101. Mote CD (1972) Dynamic stability of axially moving materials. Shock Vib Dig 4(4):2–11
102. Mote CD (1975) Stability of systems transporting accelerating axially moving materials. ASME J Dyn Syst Meas Control 97:96–98. https://doi.org/10.1115/1.3426880
103. Mote CD Jr, Wickert JA (1991) Response and discretization methods for axially moving materials. Appl Mech Rev 44(11):S279–S284. https://doi.org/10.1115/1.3121365
104. Mote CD Jr, Wu WZ (1985) Vibration coupling in continuous belt and band systems. J Sound Vib 102(1):1–9. https://doi.org/10.1016/S0022-460X(85)80099-7
105. Ottosen NS, Ristinmaa M (2005) The mechanics of constitutive modeling. Elsevier. ISBN 0-008-44606-X
106. Paavola J, Salonen E-M (1999) Fundamental theory of curved structures from a non-tensorial point of view. Struct Eng Mech Int J 7(2):147–157 https://doi.org/10.12989/sem.1999.7.2.147
107. Parker RG (1998) On the eigenvalues and critical speed stability of gyroscopic continua. ASME J Appl Mech 65:134–140. https://doi.org/10.1115/1.2789016
108. Parker RG (1999) Supercritical speed stability of the trivial equilibrium of an axially-moving string on an elastic foundation. J Sound Vib 221(2):205–219. https://doi.org/10.1006/jsvi.1998.1936
109. Pramila A (1986) Sheet flutter and the interaction between sheet and air. TAPPI J 69(7):70–74
110. Pramila A (1987) Natural frequencies of a submerged axially moving band. J Sound Vib 113(1):198–203
111. Quintard M, Whitaker S (1994) Transport in ordered and disordered porous media I: the cellular average and the use of weighting functions. Transp Porous Media 14:163–177
112. Quintard M, Whitaker S (1994) Transport in ordered and disordered porous media II: generalized volume averaging. Transp Porous Media 14:179–206
113. Quintard M, Whitaker S (1994) Transport in ordered and disordered porous media III: closure and comparison between theory and experiment. Transp Porous Media 15:31–49
114. Quintard M, Whitaker S (1994) Transport in ordered and disordered porous media IV: computer generated porous media for three-dimensional systems. Transp Porous Media 15:51–70
115. Quintard M, Whitaker S (1994) Transport in ordered and disordered porous media V: geometrical results for two-dimensional systems. Transp Porous Media 15:183–196
116. Renshaw AA, Mote CD Jr (1996) Local stability of gyroscopic systems near vanishing eigenvalues. ASME J Appl Mech 63:116–120. https://doi.org/10.1115/1.2787185
117. Rosenlicht M (1985) Introduction to analysis. Dover. ISBN 978-0-486-65038-3. Republication of the edition published by Scott Foresman & Co., 1968
118. Sack RA (1954) Transverse oscillations in traveling strings. Br J Appl Phys 5:224–226. http://stacks.iop.org/0508-3443/5/i=6/a=307
119. Saksa T (2013) On modelling and stability of axially moving viscoelastic materials. PhD thesis, Department of Mathematical Information Technology, University of Jyväskylä. http://urn.fi/URN:ISBN:978-951-39-5543-4. Jyväskylä studies in computing 186. ISBN 978-951-39-5542-7 (book). ISBN 978-951-39-5543-4 (PDF)

120. Saksa T, Banichuk N, Jeronen J, Kurki M, Tuovinen T (2012) Dynamic analysis for axially moving viscoelastic panels. Int J Solids Struct 49(23–24):3355–3366. https://doi.org/10.1016/j.ijsolstr.2012.07.017
121. Saksa T, Jeronen J, Tuovinen T (2012) Stability of moving viscoelastic panels interacting with surrounding fluid. Rakenteiden mekaniikka (Finn J Struct Mech) 45(3):88–103
122. Saksa T, Banichuk N, Jeronen J, Kurki M, Tuovinen T (2013) Dynamic behaviour of a travelling viscoelastic band in contact with rollers. In Repin S, Tiihonen T, Tuovinen T (eds) Numerical methods for differential equations, optimization, and technological problems. Dedicated to Professor P. Neittaanmäki on his 60th Birthday. Computational methods in applied sciences, vol 27. Springer Netherlands, pp 393–408. ISBN 978-94-007-5287-0 (Print) 978-94-007-5288-7 (Online)
123. Saksa T, Jeronen J (2016) Dynamic analysis for axially moving viscoelastic Poynting–Thomson beams. In: Neittaanmäki P, Repin S, Tuovinen T (eds) Mathematical modeling and optimization of complex structures, dedicated to Prof. Nikolay Banichuk for his 70th Anniversary. Springer, pp 131–151. ISBN 978-3-319-23563-9
124. Saksa T, Jeronen J, Banichuk N, Kurki M (2013) On travelling web stability including material viscoelasticity and surrounding air. In Anson SJI' (ed) Advances in pulp and paper research, Cambridge 2013, vol 1. The Pulp & Paper Fundamental Research Society, pp 449–468. ISBN 978-0-9926163-0-4
125. Shin C, Chung J, Kim W (2005) Dynamic characteristics of the out-of-plane vibration for an axially moving membrane. J Sound Vib 286(4–5):1019–1031. https://doi.org/10.1016/j.jsv.2005.01.013
126. Simpson A (1973) Transverse modes and frequencies of beams translating between fixed end supports. J Mech Eng Sci 15:159–164. https://doi.org/10.1243/JMES_JOUR_1973_015_031_02
127. Skutch R (1897) Uber die Bewegung eines gespannten Fadens, weicher gezwungen ist durch zwei feste Punkte, mit einer constanten Geschwindigkeit zu gehen, und zwischen denselben in Transversal-schwingungen von gerlinger Amplitude versetzt wird. Annalen der Physik und Chemie 61:190–195
128. Sobotka Z (1984) Rheology of materials and engineering structures. Elsevier Science Ltd., Amsterdam
129. Sokolnikoff IS (1951) Tensor analysis: theory and applications. Wiley
130. Stenberg N, Fellers C (2002) Out-of-plane Poisson's ratios of paper and paperboard. Nord Pulp Pap Res J 17(4):387–394
131. Swope RD, Ames WF (1963) Vibrations of a moving threadline. J Frankl Inst 275:36–55. https://doi.org/10.1016/0016-0032(63)90619-7
132. Tang Y-Q, Chen L-Q (2013) Stability analysis and numerical confirmation in parametric resonance of axially moving viscoelastic plates with time-dependent speed. Eur J Mech - A/Solids 37:106–121. https://doi.org/10.1016/j.euromechsol.2012.05.010
133. Thorpe JL (1981) Paper as an orthotropic thin plate. TAPPI J 64(3):119–121
134. Thurman AL, Mote CD Jr (1969) Free, periodic, nonlinear oscillation of an axially moving strip. J Appl Mech 36(1):83–91. https://doi.org/10.1115/1.3564591
135. Timoshenko SP, Woinowsky-Krieger S (1959) Theory of plates and shells, 2nd edn. McGraw-Hill, New York, Tokyo. ISBN 0-07-085820-9
136. Timoshenko S, Goodier J (1987) Theory of elasticity, 3rd edn. McGraw-Hill
137. Tisseur F, Meerbergen K (2001) The quadratic eigenvalue problem. SIAM Rev 43:235–286
138. Uesaka T, Murakami K, Imamura R (1980) Two-dimensional linear viscoelasticity of paper. Wood Sci Technol 14:131–142
139. Ulson de Souza SMAG, Whitaker S (2003) Mass transfer in porous media with heterogeneous chemical reaction. Braz J Chem Eng 20(2):191–199. https://doi.org/10.1590/S0104-66322003000200013
140. Ulsoy AG, Mote CD (1980) Analysis of bandsaw vibration. Wood Sci 13:1–10
141. Ulsoy AG, Mote CD (1982) Vibration of wide band saw blades. ASME J Eng Ind 104:71–78. https://doi.org/10.1115/1.3185801

142. Ulsoy AG, Mote CD, Szymni R (1978) Principal developments in band saw vibration and stability research. Holz als Roh- und Werkstoff 36(7):273–280. https://doi.org/10.1007/BF02610748
143. Wahlström T, Fellers C (1999) Biaxial straining of paper during drying, relations between stresses, strains and properties. In: TAPPI engineering conference, Anaheim, CA, 12–16 Sept 1999
144. Wang B, Chen L-Q (2009) Asymptotic stability analysis with numerical confirmation of an axially accelerating beam constituted by the standard linear solid model. J Sound Vib 328(4–5):456–466. https://doi.org/10.1016/j.jsv.2009.08.016
145. Wang Y, Huang L, Liu X (2005a) Eigenvalue and stability analysis for transverse vibrations of axially moving strings based on Hamiltonian dynamics. Acta Mech Sin 21:485–494. https://doi.org/10.1007/s10409-005-0066-2
146. Washizu K (1982) Variational methods in elasticity and plasticity. Pergamon, Oxford
147. Weinstock R (2008) Calculus of variations—with applications to physics and engineering. Weinstock Press. ISBN 978-1443728812. Reprint of 1974 edition
148. Whitaker S (1969) Advances in theory of fluid motion in porous media. Ind Eng Chem 61(12):14–28
149. Whitaker S (1986) Flow in porous media I: a theoretical derivation of Darcy's law. Transp Porous Media 1:3–25
150. Whitaker S (1986) Flow in porous media II: the governing equations for immiscible, two-phase flow. Transp Porous Media 1:105–125
151. Wickert JA (1992) Non-linear vibration of a traveling tensioned beam. Int J Non-Linear Mech 27(3):503–517. https://doi.org/10.1016/0020-7462(92)90016-Z
152. Wickert JA, Mote CD (1988) Current research on the vibration and stability of axially-moving materials. Shock Vib Dig 20:3–13
153. Wickert JA, Mote CD (1990) Classical vibration analysis of axially moving continua. ASME J Appl Mech 57:738–744. https://doi.org/10.1115/1.2897085
154. Winkler E (1867) Die Lehre Von Elasticität Und Festigkeit, 1st edn. H. Dominicus, Prague
155. Yang X-D, Chen L-Q (2005b) Bifurcation and chaos of an axially accelerating viscoelastic beam. Chaos Solitons Fractals 23(1):249–258. https://doi.org/10.1016/j.chaos.2004.04.008
156. Yang SM, Mote CD Jr (1991) Stability of non-conservative linear discrete gyroscopic systems. J Sound Vib 147(3):453–464. https://doi.org/10.1016/0022-460X(91)90493-4
157. Yokoyama T, Nakai K (2007) Evaluation of in-plane orthotropic elastic constants of paper and paperboard. In: 2007 SEM annual conference & exposition on experimental and applied mechanics
158. Zhang N-H, Wang J-J, Cheng C-J (2007) Complex-mode galerkin approach in transverse vibration of an axially accelerating viscoelastic string. Appl Math Mech 28(1):1–9
159. Zhou Y-F, Wang Z-M (2007) Transverse vibration characteristics of axially moving viscoelastic plate. Appl Math Mech (English Edition) 28(2):209–218. https://doi.org/10.1007/s10483-007-0209-1
160. Zhou Y-F (2009) Wang Z-M (2009) Dynamic stability of axially accelerating viscoelastic plate. Sch Res Exch 2:1–6. https://doi.org/10.3814/2009/856320
161. Ziegler H (1952) Die stabilitätskriterien der elastomechanik. Ing Arch 20:49–56
162. Zienkiewicz OC, Taylor RL, Fox DD (2013) The finite element method for solid and structural mechanics, vol 2, 7th edn. Butterworth-Heinemann
163. Zienkiewicz OC, Taylor RL, Nithiarasu P (2013) The finite element method for fluid dynamics, vol 3, 7th edn. Butterworth-Heinemann
164. Zienkiewicz OC, Taylor RL, Zhu JZ (2013) The finite element method: its basis and fundamentals, vol 1, 7th edn. Butterworth-Heinemann

Chapter 6
Stability of Axially Moving Plates

This chapter focuses on the stability analysis of axially moving materials, concentrating on two-dimensional models. There are many similarities with the classical stability analysis of structures, such as the buckling of plates. However, as we have seen in the previous chapter, the axial motion introduces inertial effects. At first, we consider the stability of an axially moving elastic isotropic plate travelling at a constant velocity between two supports and experiencing small transverse vibrations. The model of a thin elastic plate subjected to bending and tension is used to describe the bending moment and the distribution of membrane forces. The stability of the plate is investigated with the help of an analytical approach. Then we will look at elastic orthotropic plates, and finally an isotropic plate subjected to an axial tension distribution that varies in the width direction.

6.1 Isotropic Plates

Travelling flexible strings, membranes, beams and plates are the most common models of axially moving materials. These models are frequently used for describing the mechanical behavior of, for example, moving paper webs, magnetic tape, film, transmission cables, and swimming fish [1, 2]. In many applications the models of axially moving materials can be used for evaluation of a critical transport velocity that leads to a loss of stability. This is important, because the occurrence of instability can cause, in particular, damage in a paper web and breakage of transmission cables. Many studies of these models focus on free vibrations, including the nature of wave propagation, in moving media and the effects of axial motion on the frequency spectrum and eigenfunctions (see, e.g., [3–9]).

Stability considerations were first reviewed in Mote [10]. The effects of axial motion on the frequency spectrum and eigenfunctions were investigated in Archibald and Emslie [11] and Simpson [12]. It was shown that the natural frequency of each

N. Banichuk et al., *Stability of Axially Moving Materials*, Solid Mechanics and Its Applications 259, https://doi.org/10.1007/978-3-030-23803-2_6

mode decreases when transport speed increases, and that the travelling string and beam both experience divergence instability at a sufficiently high speed. It was later shown in Wang et al. [13], using Hamiltonian mechanics, that an axially moving ideal string (with exactly zero bending rigidity) remains stable at any axial velocity. Although no physically existing material has exactly zero bending rigidity, this result is important for a correct fundamental understanding. Bending rigidity, no matter how small, is a crucial factor that contributes, not only quantitatively but also qualitatively, to the stability behavior of any real material. Indeed, the term describing the bending response is of the highest order in the governing equation, so if its coefficient is small but finite, it will introduce a singular perturbation (see e.g., [14–16]).

Response prediction has been made for particular cases when excitation assumes special forms such as a constant transverse point force [17] or a harmonic motion of the supports [18]. Arbitrary excitation and initial conditions have been analyzed with the help of modal analysis and a Green function method in Wickert and Mote [19]. As a result, the associated critical speeds have been determined explicitly.

In the article [20], the authors predicted the wrinkling instability and the corresponding wrinkled shape of a web with small flexural stiffness. The stability and the vibration characteristics of an axially moving plate were investigated by Lin [21]. The loss of stability was studied by applying dynamic as well as static approaches. Lin used Wickert's approach (see [22]) to derive the equation of motion for the plate in matrix form and then apply the Galerkin method. It was shown by means of numerical analysis that, for all cases investigated, dynamic instability is realized when the frequency of free vibrations becomes zero and the critical velocity coincides with the corresponding velocity obtained from a static analysis. In Shin et al. [23], the out-of-plane vibration of an axially moving membrane was studied. Shin also used numerical analysis to find that for a membrane with a no-friction boundary condition in the lateral direction along the rollers, the membrane remains dynamically stable until the critical speed, at which static instability occurs.

The dynamical properties of moving plates have been studied by Shen et al. [24] and by Shin et al. [23], and the properties of a moving paper web have been studied in the two-part article by Kulachenko et al. [25, 26]. Critical regimes and other problems of stability analysis have been studied by Wang [27] and Sygulski [28].

Often the mechanical behavior of a paper web in an open draw, under a nonfailure condition, is adequately described by the model of an elastic orthotropic plate. An orthotropic material is orthogonally anisotropic. The model was suggested by Huber [29–31] for steel-reinforced concrete, but it works well also for paper materials. The rigidity coefficients of the orthotropic plate model that describe the tension and bending of the paper sheet have been estimated for various types of paper in many publications. See, for example, the articles by Göttsching and Baumgarten [32], Mann et al. [33], Baum et al. [34], Thorpe [35], Skowronski and Robertson [36], Seo [37] and Erkkilä et al. [38]. As is noted by Stenberg and Fellers [39], paper is auxetic: stretching in the machine direction will cause the paper web to thicken in the out-of-plane direction. The Poisson ratio ν_{13} is negative, and $|\nu_{13}|$ can be as large as 3.0. Auxetic behavior arises because paper essentially consists of a network of fibers, see [40]. For isotropic materials, the result that a Poisson ratio of one-half describes an incompressible material is classical and very well known, but less attention has been devoted to the orthotropic case. For some considerations on incompressibility conditions for orthotropic materials, see [41, 42].

The free vibrations of classical stationary (not axially moving) orthotropic rectangular plates have been studied extensively. The classical reference work in this area is the book by Gorman [43]. More recently, Biancolini et al. [44] included in their study all combinations of simply supported and clamped boundary conditions on the edges. Xing and Liu [45] obtained exact solutions for the free vibrations of stationary rectangular orthotropic plates. They considered three combinations of simply supported (S) and clamped (C) boundary conditions: SSCC, SCCC, and CCCC. Kshirsagar and Bhaskar [46] studied vibrations and the buckling of loaded stationary orthotropic plates. They found critical loads of buckling for all combinations of boundary conditions S, C, and F.

In addition, in Hatami et al. [47], the free vibration of an axially moving orthotropic rectangular plate was studied at sub- and supercritical speeds, and its flutter and divergence instabilities were studied at supercritical speeds. The study was limited to simply supported boundary conditions at all edges. For the solution of equations of orthotropic moving materials, many necessary fundamentals can be found in the book by Marynowski [48].

In this chapter, we apply analytical methods to the stability analysis of an axially moving rectangular plate, and investigate the dependence of the solution on the problem parameters. In the frame of the general dynamic approach, a functional expression for the characteristic index of stability is found and can be effectively used for evaluation of the frequencies of free vibration. A static stability analysis is performed, and the possible buckled forms of the plate (symmetric and antisymmetric) are studied as functions of geometric and mechanical problem parameters. In particular, we observe that the buckling shape of the plate is symmetric and that the elastic deflections are localized in the vicinity of the free edges of the plate. This localization phenomenon is familiar from the eigenfunctions of stationary plates under an in-plane compressive load (see, e.g., [43]), and it can also be seen in the numerical results of Shin et al. [23] for an axially travelling membrane.

A detailed analysis of some of the topics presented in this chapter is presented in the book by Banichuk et al. [49].

To obtain a realistic model of a system where the axially moving material is thin and wide, we may use the differential equation for small transverse vibrations of an axially moving plate. Consider a rectangular part Ω of a moving plate in the Cartesian coordinate system. Let us define the part Ω as occupying the region

$$\Omega = \{0 < x < \ell, \ -b < y < b\} , \tag{6.1.1}$$

where b and ℓ are describe the geometry of the span, see Fig. 6.1.

One can represent the equation for small transverse vibrations of the moving plate in the following form:

$$m \frac{\mathrm{d}^2 w}{\mathrm{d}t^2} = \mathcal{L}^{\mathrm{M}}(w) - \mathcal{L}^{\mathrm{B}}(w) \ . \tag{6.1.2}$$

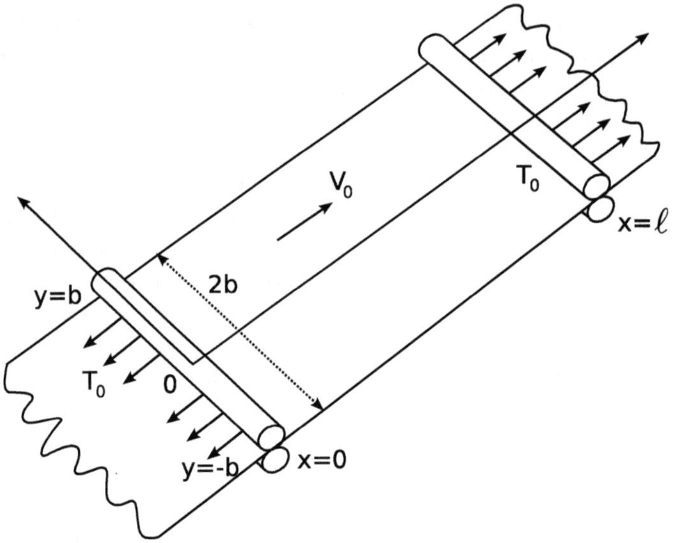

Fig. 6.1 An axially moving plate. The roller symbols represent simple supports, with presence of axial motion. The other two edges of the plate are assumed to be free of traction

Here m is the mass per unit area of the plate. For a constant axial velocity V_0, the total acceleration on the left-hand side of (6.1.2) is expressed as

$$\frac{\mathrm{d}^2 w}{\mathrm{d}t^2} = \frac{\mathrm{d}}{\mathrm{d}t}\left(\frac{\partial w}{\partial t} + V_0 \frac{\partial w}{\partial x}\right) = \frac{\partial^2 w}{\partial t^2} + 2V_0 \frac{\partial^2 w}{\partial x \partial t} + V_0^2 \frac{\partial^2 w}{\partial x^2} \,. \tag{6.1.3}$$

The right-hand side in (6.1.3) contains three terms, representing respectively the local inertia, the Coriolis effect, and the centrifugal effect. To justify the names, consider the small-displacement regime. We have $\partial w/\partial x = \tan\theta \approx \theta$, where θ is the angle between the plate surface and the horizontal axis, so $\partial^2 w/\partial x \partial t \approx \partial\theta/\partial t$, and hence in approximation, the middle term is proportional to the local rate of rotation just like the (in-plane) Coriolis effect. Similarly, curvature becomes $1/R = (\partial^2 w/\partial x^2)/(1 + (\partial w/\partial x)^2)^{3/2} \approx \partial^2 w/\partial x^2$, so in approximation the last term is proportional to the curvature just like the centrifugal effect.

However, the names should be interpreted loosely. The terms arise from the coordinate transformation for axially moving materials that was discussed in Chap. 5, and in this transformation, no approximations were made. Also, we did not use a rotating reference frame, so the analogy with the rotating case cannot be perfect. Instead, there is different, more general commonality. These terms describe how to construct the local transverse acceleration of a material point (the left-hand side of (6.1.3)), in the co-moving linearly translating frame, when we are given the apparent behavior of the material in the global Eulerian frame (the right-hand side of (6.1.3)).

This is analogous to the fictitious forces that arise in a rotating coordinate system, but care should be taken not to over-stretch the particular analogy.

As the presence of V_0 suggests, the additional terms describe the (fictitious) interaction between the axial motion and the local transverse acceleration, when viewed in a coordinate system whose axial velocity differs from that of the moving material. Finally, it should be stressed that regardless of the actual axial component of the motion of the material, which may include vibrations around the steady linear translation, the reference coordinate system is taken to undergo linear translation at a constant velocity; see [50].

The membrane operator $\mathcal{L}^{\mathrm{M}}$ on the right-hand side of (6.1.2) is

$$\mathcal{L}^{\mathrm{M}}(w) = T_{xx}\frac{\partial^2 w}{\partial x^2} + 2T_{xy}\frac{\partial^2 w}{\partial x \partial y} + T_{yy}\frac{\partial^2 w}{\partial y^2} \; . \tag{6.1.4}$$

The coefficients T_{xx}, T_{xy}, and T_{yy} of the linear operator $\mathcal{L}^{\mathrm{M}}$ are related to the corresponding in-plane stresses σ_{xx}, σ_{xy} and σ_{yy} by the expression

$$T_{ij} = h\sigma_{ij} \; , \tag{6.1.5}$$

where h is the thickness of the plate, which is taken as constant. The linear bending operator $\mathcal{L}^{\mathrm{B}}$ is given by the expression

$$\mathcal{L}^{\mathrm{B}}(w) = D\Delta^2 w = D\left(\frac{\partial^4 w}{\partial x^4} + 2\frac{\partial^4 w}{\partial x^2 \partial y^2} + \frac{\partial^4 w}{\partial y^4}\right) \tag{6.1.6}$$

in the case of an isotropic elastic plate. Here, Δ^2 is the biharmonic operator and D is the bending rigidity of an isotropic plate. If, instead of a plate, we consider a membrane, the bending rigidity is neglected and the entire operator $\mathcal{L}^{\mathrm{B}}$ is omitted, $\mathcal{L}^{\mathrm{B}} \equiv 0$.

When we later consider an orthotropic plate, the operator $\mathcal{L}^B$ will depend on three constants, and is written as

$$\mathcal{L}^{\mathrm{B}}(w) = D_1\frac{\partial^4 w}{\partial x^4} + 2D_3\frac{\partial^4 w}{\partial x^2 \partial y^2} + D_2\frac{\partial^4 w}{\partial y^4} \; . \tag{6.1.7}$$

We have the following expressions (see, e.g., [51] for the bending rigidities in 6.1.7):

$$D_1 = \frac{h^3}{12}C_{11} \; , \quad D_2 = \frac{h^3}{12}C_{22} \; , \quad D_3 = \frac{h^3}{12}\left(C_{12} + 2\,C_{66}\right) \; , \tag{6.1.8}$$

where C_{ij} are the elastic moduli. These can be expressed in terms of the Young moduli E_1, E_2 and Poisson ratios ν_{12}, ν_{21} as (see, e.g., [52])

$$C_{11} = \frac{E_1}{1 - \nu_{12}\nu_{21}} \,, \qquad C_{22} = \frac{E_2}{1 - \nu_{12}\nu_{21}} \,,$$

$$C_{12} = \frac{\nu_{21} E_1}{1 - \nu_{12}\nu_{21}} = \frac{\nu_{12} E_2}{1 - \nu_{12}\nu_{21}} = C_{21} \,, \tag{6.1.9}$$

$$C_{66} = G_{12} \,.$$

In (6.1.9), E_1 is the Young modulus in the x direction, and ν_{12} is the Poisson ratio in the xy plane when the stretching is applied in the x direction. Respectively, E_2 is the Young modulus in the y direction, and ν_{21} is the Poisson ratio in the xy plane when the stretching is applied in the y direction. G_{12} is the shear modulus in the xy plane. For the isotropic plate, $E_1 = E_2 = E$, $\nu_{12} = \nu_{21} = \nu$ and $G_{12} = G$.

Let the deflection function w and its partial derivatives be small, and such that they satisfy the boundary conditions. For an orthotropic plate, the boundary conditions read

$$(w)_{x=0,\ell} = 0 \,, \qquad \left(\frac{\partial^2 w}{\partial x^2}\right)_{x=0,\ell} = 0 \,, \qquad -b \le y \le b \,, \tag{6.1.10}$$

$$\left(\frac{\partial^2 w}{\partial y^2} + \beta_1 \frac{\partial^2 w}{\partial x^2}\right)_{y=\pm b} = 0 \,, \qquad 0 \le x \le \ell \,, \tag{6.1.11}$$

$$\left(\frac{\partial^3 w}{\partial y^3} + \beta_2 \frac{\partial^3 w}{\partial x^2 \partial y}\right)_{y=\pm b} = 0 \,, \qquad 0 \le x \le \ell \,, \tag{6.1.12}$$

representing two opposite edges simply supported and the other two edges traction-free.

The mechanical parameters β_1 and β_2 in the free-edge boundary conditions are defined as

$$\beta_1 = \nu_{12} \,,$$

$$\beta_2 = \nu_{12} + \frac{4 G_{12}}{E_2} (1 - \nu_{12}\nu_{21}) \,. \tag{6.1.13}$$

For the isotropic plate considered in this section, β_1 and β_2 simplify to

$$\beta_1 = \nu \quad \text{and} \quad \beta_2 = 2 - \nu \,. \tag{6.1.14}$$

This can be seen by setting $\nu_{12} = \nu_{21} = \nu$, $E_1 = E_2 = E$, $G_{12} = G$, and using the isotropic shear modulus relation $G = E/(2\,(1+\nu))$. Then, factoring $1 - \nu_{12}\nu_{21} = 1 - \nu^2 = (1+\nu)(1-\nu)$ and simplifying reduces (6.1.13) into (6.1.14).

Using the geometric average approximation G_{H} for the shear modulus G_{12},

$$G_{\mathrm{H}} \equiv \frac{\sqrt{E_1 E_2}}{2\left(1 + \sqrt{\nu_{12}\nu_{21}}\right)} \,, \tag{6.1.15}$$

also known as the Huber approximation, the equations for the orthotropic plate reduce to those of an isotropic plate. The geometric average shear modulus in (6.1.15) was introduced by Huber [30], generalizing the shear modulus for orthotropic materials.

It is easy to show that this reduction property of G_{H} remains valid for the time-dependent travelling plate problem as well. Because in the transformations producing this reduction, coordinate scaling is only required in the y direction (see [51], Chap. 11), the "Coriolis" term $\partial^2 w/\partial x \partial t$ generated by the axial motion of the plate does not alter the approach. However, note that especially for paper materials, if measured, the actual shear modulus generally differs from the geometric average (6.1.15).

In the case of a membrane, the boundary condition at the rollers reads

$$(w)_{x=0,\,\ell} = 0\,, \quad -b \leq y \leq b\,. \tag{6.1.16}$$

On the free edges, classical membrane theory asserts (see, e.g., [53, 54])

$$\left(\frac{\partial w}{\partial y}\right)_{y=\pm b} = 0\,, \quad 0 \leq x \leq \ell\,, \tag{6.1.17}$$

but this is not the only possible boundary condition. For zero traction, Shin et al. [23] have used a different condition, which does not contain the transverse displacement w. We will see below that in our case for a membrane, both choices of the boundary condition on the free edges are possible.

Let us continue to the analysis of in-plane tensions. The stability of moving materials is sensitive to in-plane tension. The higher the tension, the larger the region of transport velocities at which the system remains stable. We will restrict our consideration to the case of stationary in-plane forces, that is, we assume that the in-plane tensions do not depend on time t. The in-plane tensions T_{xx}, T_{xy} and T_{yy} are assumed to satisfy the equilibrium equations (see, e.g., [55])

$$\frac{\partial T_{xx}}{\partial x} + \frac{\partial T_{xy}}{\partial y} = 0\,, \quad \frac{\partial T_{xy}}{\partial x} + \frac{\partial T_{yy}}{\partial y} = 0\,. \tag{6.1.18}$$

The in-plane tensions T_{xx}, T_{xy} and T_{yy} are related to the corresponding stress tensor components σ_{xx}, σ_{xy} and σ_{yy} by the relation (6.1.5), that is,

$$T_{ij} = h\sigma_{ij}\,.$$

Note the system (6.1.18) is not yet closed; there are three unknowns T_{xx}, T_{xy} and T_{yy}, but only two equations. Kinematic and constitutive relations, that is, strain–displacement and stress–strain relations, are needed to make the system uniquely solvable.

Indeed, by themselves the Eqs. (6.1.18) do not even distinguish whether our material of interest is a fluid or a solid. Essentially they describe, in two space dimensions, the (pointwise) balance of linear momentum in continuum form, usually written in mechanics as $\rho\ddot{\boldsymbol{u}} - \nabla \cdot \boldsymbol{\sigma}^{\mathrm{T}} = \boldsymbol{F}$ (see, for example, the book by Myron [56]). Here the double-dot denotes the second time derivative, and the superscript T is the transpose of a rank-2 tensor. Equations (6.1.18) are specialized to the case where there are no body forces (such as gravity acting in the plane of the sheet), $\boldsymbol{F} \equiv 0$, and no in-plane accelerations, $\ddot{\boldsymbol{u}} = 0$.

Both of these conditions are approximations that simplify the analysis. In reality, as the sheet deforms in the out-of-plane direction, its surface will no longer be perfectly horizontal, and hence there will be a gravitational contribution (even if small) also in the plane of the sheet. Secondly, we assume that, as the sheet travels axially, the axial velocity stays constant and uniform: there is no rigid-body acceleration axially, no in-plane vibrations, and no nonuniform stretching.

Why no nonuniform stretching? Recall Chap. 5: when treating axially moving materials, we essentially first write the governing equations in the axially co-moving frame, and then transform them into the Eulerian frame. The second time derivative in the $\rho\ddot{\boldsymbol{u}}$ term must thus be taken in the co-moving frame, so due to the coordinate transformation, in the Eulerian frame it becomes $\mathrm{d}^2\boldsymbol{u}/\mathrm{d}t^2 = \partial^2\boldsymbol{u}/\partial t^2 + 2V_0\partial^2\boldsymbol{u}/\partial x\partial t + V_0^2\partial^2\boldsymbol{u}/\partial x^2$. Therefore, even in what appears as a steady state in the Eulerian frame—with the material flowing through the considered domain without any time-dependent changes in the motion—the material is still moving (much like in a steady state in fluid mechanics), $V_0 \neq 0$, and the $V_0^2\partial^2\boldsymbol{u}/\partial x^2$ term will be present, unless $\partial^2\boldsymbol{u}/\partial x^2 \equiv 0$. Because for small displacements $\partial u/\partial x = \varepsilon_{xx}$, it follows that in order for this term to vanish, we must have $\partial\varepsilon_{xx}/\partial x \equiv 0$. For elastic materials (such as those analyzed in this chapter), this will be satisfied, but if the axially moving material is viscoelastic, then with axial tension applied at the ends of the domain, its axial strain field will be nonuniform. See Kurki et al. [42] for an analysis of the in-plane behavior of an axially moving viscoelastic material of the Kelvin–Voigt type.

For a rectangular band of material stretched at opposite ends with no shear, and free of traction at the two other edges, the boundary conditions for the Eqs. (6.1.18) are

$$T_{xx} = k(y)\,, \quad T_{xy} = 0 \quad \text{at } x = 0,\ \ell\,, \quad -b \le y \le b\,, \tag{6.1.19}$$

$$T_{yy} = 0\,, \quad T_{xy} = 0 \quad \text{at } y = \pm b\,, \quad 0 \le x \le \ell\,. \tag{6.1.20}$$

Let us begin with a uniform tension profile at the supports, where $k(y)$ degenerates to a constant T_0. Taking into account the behavioral equation of the plane theory of elasticity and the boundary conditions (6.1.19) and (6.1.20), we have the following tension field for the band:

$$T_{xx} = T_0\,, \qquad T_{yy} = T_{xy} = 0\,, \qquad (x, y) \in \Omega\,, \tag{6.1.21}$$

where

$$\Omega : \{0 < x < \ell,\ -b < y < b\}\,.$$

For a linearly elastic orthotropic band, the in-plane displacements u, and v, oriented respectively along the axes x and y, are related to the stresses by means of the generalized Hooke's law:

$$\sigma_{xx} = C_{11}\frac{\partial u}{\partial x} + C_{12}\frac{\partial v}{\partial y}\,, \tag{6.1.22}$$

$$\sigma_{xy} = C_{66}\left(\frac{\partial u}{\partial y} + \frac{\partial v}{\partial x}\right), \tag{6.1.23}$$

$$\sigma_{yy} = C_{21}\frac{\partial u}{\partial x} + C_{22}\frac{\partial v}{\partial y}\,, \tag{6.1.24}$$

which plays the role of the constitutive model. Here C_{ij} are the elastic moduli, see Eq. (6.1.9), and we have used the linear strain-displacement relations (Cauchy strains, infinitesimal strains), valid for small displacements:

$$\begin{aligned} \varepsilon_{xx} &= \frac{\partial u}{\partial x}\,, \\ \gamma_{xy} &= \frac{\partial u}{\partial y} + \frac{\partial v}{\partial x}\,, \\ \varepsilon_{yy} &= \frac{\partial v}{\partial y}\,. \end{aligned} \tag{6.1.25}$$

In the following, we will also use the material parameter compatibility relation

$$E_1\nu_{21} = E_2\nu_{12}\,, \tag{6.1.26}$$

which is implied by the symmetricity of the elastic moduli, in other words, $C_{12} = C_{21}$.

It is possible to show that if instead of a prescribed axial tension T_0, we have a prescribed axial displacement u_0 at $x = \ell$, the generated tension field has the same form as the form given by (6.1.21). Using (6.1.5), (6.1.18) and the constitutive model (6.1.22)–(6.1.24), we obtain the stress equilibrium equations in terms of the in-plane displacement fields u and v:

$$\begin{aligned} C_{11}\frac{\partial^2 u}{\partial x^2} + (C_{12}+C_{66})\frac{\partial^2 v}{\partial x\partial y} + C_{66}\frac{\partial^2 u}{\partial y^2} &= 0\,, \\ C_{22}\frac{\partial^2 v}{\partial y^2} + (C_{12}+C_{66})\frac{\partial^2 u}{\partial x\partial y} + C_{66}\frac{\partial^2 v}{\partial x^2} &= 0\,. \end{aligned} \tag{6.1.27}$$

The boundary conditions in terms of u and v are

$$
\begin{aligned}
&\frac{\partial u}{\partial y} + \frac{\partial v}{\partial x} = 0 \qquad \text{at } x = 0,\, \ell\,, \quad -b \le y \le b\,, \\
&u = 0 \qquad \text{at } x = 0\,, \quad -b \le y \le b\,, \\
&u = u_0 \qquad \text{at } x = \ell\,, \quad -b \le y \le b\,,
\end{aligned}
\tag{6.1.28}
$$

and

$$
\begin{aligned}
&C_{12}\frac{\partial u}{\partial x} + C_{22}\frac{\partial v}{\partial y} = 0 \qquad \text{at } y = \pm b\,, \quad 0 \le x \le \ell\,, \\
&\frac{\partial u}{\partial y} + \frac{\partial v}{\partial x} = 0 \qquad \text{at } y = \pm b\,, \quad 0 \le x \le \ell\,.
\end{aligned}
\tag{6.1.29}
$$

From the equilibrium equations (6.1.27) and the boundary conditions (6.1.28) and (6.1.29), the in-plane displacement field in the domain Ω is described as

$$
u(x, y) = \frac{u_0}{\ell} x\,, \qquad v(x, y) = -\frac{C_{12}}{C_{22}}\frac{u_0}{\ell} y\,, \qquad (x, y) \in \Omega\,.
$$

The displacement field clearly exhibits the well-known Poisson effect: a stretch applied in the x direction causes compression in the y direction.

Inserting these into the constitutive model (6.1.22)–(6.1.24) and converting the stress to tension using (6.1.5), we see that the tension field has the form given by Eq. (6.1.21) with the constant value $T_{xx} = T_0$, where

$$
T_0 = h\frac{u_0}{\ell}\left(C_{11} - \frac{C_{12}^2}{C_{22}}\right) = h\frac{u_0}{\ell}E_1\,. \tag{6.1.30}
$$

From (6.1.9) and (6.1.26), the last form in (6.1.30) easily follows. We see that the only material parameter that affects the uniform tension field generated by the prescribed displacement is the Young modulus in the longitudinal direction. For an isotropic material, (6.1.30) becomes

$$
T_0 = h\frac{u_0}{\ell}E\,. \tag{6.1.31}
$$

In the case of a nonuniform tension profile $k(y)$ at the supports, one classical way to find compatible tension fields T_{xx}, T_{yy} and T_{xy} inside the domain is via the help of an Airy stress function Υ, which acts as a kind of potential. For the benefit of the rest of this chapter, we will first do this for the more general, orthotropic sheet, and then specialize the result to the isotropic sheet currently under consideration. We represent the tensions in terms of the second derivatives of the Airy stress function as

$$
\begin{aligned}
T_{xx} = h\sigma_{xx} &= \frac{\partial^2 \Upsilon}{\partial y^2} , \\
T_{yy} = h\sigma_{yy} &= \frac{\partial^2 \Upsilon}{\partial x^2} , \\
T_{xy} = h\sigma_{xy} &= -\frac{\partial^2 \Upsilon}{\partial x \partial y} .
\end{aligned}
\tag{6.1.32}
$$

Consider now the inverse relation of the chosen constitutive model, that is, the generalized Hooke's law (6.1.22)–(6.1.24). We can express it as

$$
\begin{aligned}
\varepsilon_{xx} &= \frac{1}{E_1}\sigma_{xx} - \frac{\nu_{21}}{E_2}\sigma_{yy} , \\
\gamma_{xy} &= \frac{1}{G_{12}}\sigma_{xy} , \\
\varepsilon_{yy} &= \frac{1}{E_2}\sigma_{yy} - \frac{\nu_{12}}{E_1}\sigma_{xx} .
\end{aligned}
\tag{6.1.33}
$$

Directly from the linear strain-displacement relations (6.1.25), it follows that the Cauchy strains satisfy the strain compatibility equation (see, e.g., [55])

$$
\frac{\partial^2 \varepsilon_{xx}}{\partial y^2} + \frac{\partial^2 \varepsilon_{yy}}{\partial x^2} - \frac{\partial^2 \gamma_{xy}}{\partial x \partial y} = 0 . \tag{6.1.34}
$$

Equation (6.1.34) is a purely kinematic (essentially, geometric) relation, with no constitutive assumptions, that holds for any situation with small displacements.

By inserting (6.1.32) and (6.1.33) into (6.1.34), noting that $T_{ij} = h\sigma_{ij}$, and using (6.1.26), we see that for an orthotropic linear elastic material, the Airy stress function Υ must satisfy the equation

$$
\frac{\partial^4 \Upsilon}{\partial x^4} + \left(\frac{E_2}{G_{12}} - 2\nu_{21} \right) \frac{\partial^4 \Upsilon}{\partial x^2 \partial y^2} + \frac{E_2}{E_1} \frac{\partial^4 \Upsilon}{\partial y^4} = 0 . \tag{6.1.35}
$$

In the isotropic case, by setting $E_1 = E_2 = E$, $G_{12} = G$, $\nu_{21} = \nu$, and using the isotropic shear modulus relation $G = E/(2(1+\nu))$, we observe that (6.1.35) reduces to the biharmonic equation

$$
\Delta^2 \Upsilon \equiv \frac{\partial^4 \Upsilon}{\partial x^4} + 2\frac{\partial^4 \Upsilon}{\partial x^2 \partial y^2} + \frac{\partial^4 \Upsilon}{\partial y^4} = 0 \quad \text{(for isotropic material)} . \tag{6.1.36}
$$

For an axially tensioned rectangular band of material, the boundary conditions satisfied by Υ, corresponding to (6.1.19) and (6.1.20), are

$$\left(\frac{\partial^2 \Upsilon}{\partial y^2}\right)_{x=0,\ell} = k(y)\,, \quad \left(\frac{\partial^2 \Upsilon}{\partial x \partial y}\right)_{x=0,\ell} = 0\,, \quad -b \le y \le b\,, \tag{6.1.37}$$

$$\left(\frac{\partial^2 \Upsilon}{\partial x^2}\right)_{y=\pm b} = 0\,, \quad \left(\frac{\partial^2 \Upsilon}{\partial x \partial y}\right)_{y=\pm b} = 0\,, \quad 0 \le x \le \ell\,. \tag{6.1.38}$$

The tensions expressed via the stress function Υ in (6.1.32) will satisfy the stress equilibrium equations (6.1.18) for any function Υ that is sufficiently smooth. The problem (6.1.35), (6.1.37) and (6.1.38), which must be solved, expresses the condition of compatibility for the tensions.

Even though in the isotropic case, (6.1.36) factors into two Poisson equations that can be solved in sequence, the problem is still very difficult to solve analytically for a general $k(y)$. A Fourier series based general approach, decomposing $k(y)$ into modes with different spatial frequencies, is found in the paper by Gorman and Singhal [57]. A numerical solution using C^1 continuous finite elements is another option, but this misses what is perhaps the greatest advantage of the Airy approach: if the setup is sufficiently simple, the approach may allow finding an analytical solution. If the tension fields are to be found numerically, it is mathematically more reasonable not to introduce extra continuity requirements, and rather solve the system (6.1.18)–(6.1.20) directly. (A compatibility relation between the tension components must be introduced also in this approach; as was mentioned, the system (6.1.18) has only two equations for three unknowns.)

Later, we will concentrate on a linear tension profile $k(y)$, and use the rigorous solution of the boundary value problem (6.1.35), (6.1.37) and (6.1.38) corresponding to the case where the tension profile function in (6.1.19), (6.1.37) is taken as

$$k(y) = T_0 + \alpha y\,. \tag{6.1.39}$$

Here $\alpha > 0$ is a given constant, characterizing the skew of the linear tension profile. We have

$$\Upsilon(x, y) = T_0 \frac{y^2}{2} + \alpha \frac{y^3}{6} + c_1 x + c_2 y + c_0\,, \quad (x, y) \in \Omega\,. \tag{6.1.40}$$

The corresponding tensions will be

$$T_{xx}(x, y) = T_0 + \alpha y\,, \qquad T_{xy}(x, y) = 0\,, \qquad T_{yy}(x, y) = 0\,, \tag{6.1.41}$$

where $(x, y) \in \Omega$, and c_0, c_1 and c_2 are arbitrary constants.

Let us next perform a dynamic analysis. It is known that the mechanical instability of a travelling paper web can arise at some critical velocities, and that the instability may occur in either dynamic or static forms. These critical velocities are of both theoretical and practical interest, as they set an upper limit for the running speed of a moving plate. Some previous investigations show that for an axially moving elastic plate under a uniform tension profile along the rollers and certain other conditions, the divergence velocity V_0^{div} is smaller than the flutter velocity V_0^{fl}, and hence the critical instability will be of the divergence type, that is, a static loss of stability.

To analyze dynamic stability, we will follow the method described by Bolotin [58]. Recall the equation of small transverse vibrations of the travelling plate subjected to a uniform tension profile, leading to a constant tension field, (6.1.2). We represent it as follows:

$$\frac{\partial^2 w}{\partial t^2} + 2V_0 \frac{\partial^2 w}{\partial x \partial t} + \left(V_0^2 - C^2\right) \frac{\partial^2 w}{\partial x^2} + \frac{D_0}{m} \mathcal{L}_0(w) = 0 \,, \qquad C = \sqrt{\frac{T_0}{m}} \,, \tag{6.1.42}$$

where $w = w(x, t)$ is the transverse displacement, and the orthotropic bending operator is

$$\mathcal{L}_0(w) = \frac{D_1}{D_0} \frac{\partial^4 w}{\partial x^4} + \frac{2D_3}{D_0} \frac{\partial^4 w}{\partial x^2 \partial y^2} + \frac{D_2}{D_0} \frac{\partial^4 w}{\partial y^4} \,. \tag{6.1.43}$$

Here D_j for $j = 1, 2, 3$ are the orthotropic bending rigidities

$$D_1 = \frac{h^3}{12} C_{11} \,, \quad D_2 = \frac{h^3}{12} C_{22} \,, \quad D_3 = \frac{h^3}{12} \left(C_{12} + 2\,C_{66}\right) \,.$$

The C_{ij} are the elastic moduli, (6.1.9). The quantity D_0 is a normalization constant, for which we have chosen the value $D_0 = D_1$.

The boundary value problem consisting of (6.1.42) and (6.1.43) with the boundary conditions (6.1.10)–(6.1.12) is homogeneous and invariant with respect to the symmetry operation $y \to -y$ and, consequently, all solutions of the problem are either symmetric or antisymmetric functions of y, that is,

$$w(x, y, t) = w(x, -y, t) \qquad \text{or} \qquad w(x, y, t) = -w(x, -y, t) \,. \tag{6.1.44}$$

In the following analysis, however, this symmetry property is not necessary.

We will use the time-harmonic trial function

$$w(x, t) = \exp(st)\, W(x) \,, \tag{6.1.45}$$

where s is the stability exponent (which is a complex number), $W(x)$ is an unknown eigenmode to be determined, and x is a scalar or a vector depending on the dimensionality of the problem. This trial function removes the time dependence from the partial differential equation, making it sufficient to solve a (pseudo-)steady-state problem including the unknown scalar s, the allowed values of which are determined implicitly by the boundary conditions and problem parameters. The resulting equation will be a partial differential equation in space, but polynomial with respect to s (6.1.45).

We can represent the solution of our dynamic boundary-value problem (6.1.42), (6.1.43), (6.1.10)–(6.1.12) as

$$w(x, y, t) = W(x, y)e^{i\omega t} = W(x, y)e^{st} , \tag{6.1.46}$$

where ω is the angular frequency of small transverse vibrations and $s = i\omega$. If s is purely imaginary and consequently ω is real, the membrane or plate performs harmonic vibrations of a small amplitude and its motion can be considered stable. If, for some values of the problem parameters, the real part of the stability exponent s becomes positive, the transverse vibrations grow exponentially and consequently the behavior is unstable.

To investigate the dynamic behavior of the plate, we insert the representation (6.1.46) into (6.1.42). Therefore, for small time-harmonic vibrations of the travelling plate subjected to uniform tension, we have the equation

$$s^2 W + 2s V_0 \frac{\partial W}{\partial x} + (V_0^2 - C^2)\frac{\partial^2 W}{\partial x^2} + \frac{D}{m}\Delta^2 W = 0 . \tag{6.1.47}$$

The boundary conditions are

$$(W)_{x=0,\ell} = 0 , \quad \left(\frac{\partial^2 W}{\partial x^2}\right)_{x=0,\ell} = 0 , \quad -b \le y \le b , \tag{6.1.48}$$

$$\left(\frac{\partial^2 W}{\partial y^2} + \beta_1 \frac{\partial^2 W}{\partial x^2}\right)_{y=\pm b} = 0 , \quad 0 \le x \le \ell , \tag{6.1.49}$$

$$\left(\frac{\partial^3 W}{\partial y^3} + \beta_2 \frac{\partial^3 W}{\partial x^2 \partial y}\right)_{y=\pm b} = 0 , \quad 0 \le x \le \ell . \tag{6.1.50}$$

For an orthotropic plate, we have

$$\begin{aligned} \beta_1 &= \nu_{12} , \\ \beta_2 &= \nu_{12} + \frac{4 G_{12}}{E_2}(1 - \nu_{12}\nu_{21}) . \end{aligned} \tag{6.1.51}$$

As was noted above, in the case of an isotropic plate, the parameters above become

$$\beta_1 = \nu \quad \text{and} \quad \beta_2 = 2 - \nu . \tag{6.1.52}$$

We multiply (6.1.47) by W and perform integration over the domain Ω to obtain

$$s^2 \int_\Omega W^2 \mathrm{d}\Omega + 2s V_0 \int_\Omega W \frac{\partial W}{\partial x} \mathrm{d}\Omega + (V_0^2 - C^2) \int_\Omega W \frac{\partial^2 W}{\partial x^2} \mathrm{d}\Omega + \frac{D}{m} \int_\Omega W \Delta^2 W \mathrm{d}\Omega = 0 . \tag{6.1.53}$$

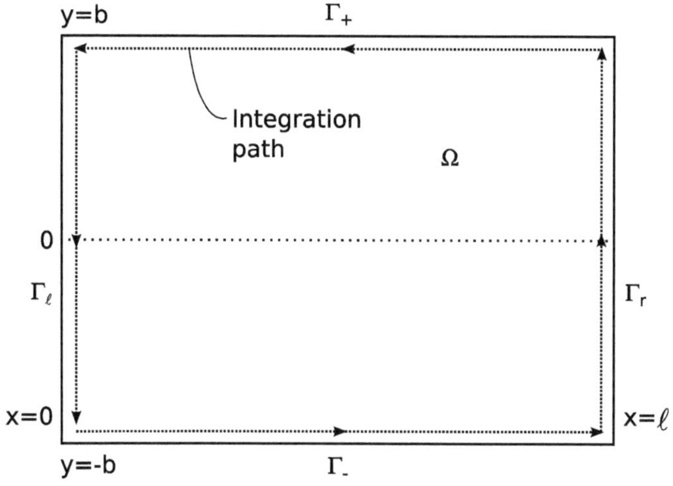

Fig. 6.2 Division of the boundary Γ for the investigated contour integral

Equation (6.1.53) can be seen as an eigenvalue problem for the pair (s, W) with the parameter V_0, producing a spectrum of complex eigenfrequencies s and eigenmodes W for the chosen value of V_0. Alternatively, (6.1.53) can be viewed as an eigenvalue problem for the pair (V_0, W) with the parameter s, when s is fixed to any such value that at least one complex eigenfrequency exists for at least one choice of V_0. For other choices of s, this second eigenvalue problem has no solution.

By Green's second identity, the last integral in (6.1.53) transforms into

$$\int_\Omega W\Delta^2 W \, \mathrm{d}\Omega = \int_\Omega (\Delta W)^2 \, \mathrm{d}\Omega + \int_\Gamma \left(W \frac{\partial}{\partial n} \Delta W - \Delta W \frac{\partial W}{\partial n} \right) \mathrm{d}\Gamma \,, \tag{6.1.54}$$

where n is the exterior unit normal to the boundary Γ of the domain Ω. We divide the boundary Γ into four parts, see Fig. 6.2:

$$\Gamma_- = \{0 \le x \le \ell,\ y = -b\} \,, \qquad \Gamma_r = \{x = \ell,\ -b \le y \le b\} \,,$$

$$\Gamma_+ = \{0 \le x \le \ell,\ y = b\} \,, \qquad \Gamma_\ell = \{x = 0,\ -b \le y \le b\} \,.$$

Admitting counterclockwise integration along Γ, we have

$$I = \int_\Gamma \left(W \frac{\partial}{\partial n} \Delta W - \Delta W \frac{\partial W}{\partial n} \right) \mathrm{d}\Gamma = I_- + I_r + I_+ + I_\ell \,. \tag{6.1.55}$$

Here

$$I_r = I_\ell = 0 \,, \tag{6.1.56}$$

$$
\begin{aligned}
I_- &= \int_{\Gamma_-} \left(W \frac{\partial}{\partial n} \Delta W - \Delta W \frac{\partial W}{\partial n} \right) \mathrm{d}\Gamma \\
&= - \int_0^{\ell} \left(W \frac{\partial}{\partial y} \Delta W - \Delta W \frac{\partial W}{\partial y} \right)_{y=-b} \mathrm{d}x \, , \qquad (6.1.57) \\
I_+ &= \int_{\Gamma_+} \left(W \frac{\partial}{\partial n} \Delta W - \Delta W \frac{\partial W}{\partial n} \right) \mathrm{d}\Gamma \\
&= - \int_{\ell}^{0} \left(W \frac{\partial}{\partial y} \Delta W - \Delta W \frac{\partial W}{\partial y} \right)_{y=b} \mathrm{d}x \\
&= \int_0^{\ell} \left(W \frac{\partial}{\partial y} \Delta W - \Delta W \frac{\partial W}{\partial y} \right)_{y=b} \mathrm{d}x \, . \qquad (6.1.58)
\end{aligned}
$$

where we have used the relations

$$
\mathrm{d}\Gamma = \mathrm{d}x \, , \quad \frac{\partial}{\partial n} = -\frac{\partial}{\partial y} \quad \text{for } (x, y) \in \Gamma_- \, , \tag{6.1.59}
$$

$$
\mathrm{d}\Gamma = -\mathrm{d}x \, , \quad \frac{\partial}{\partial n} = \frac{\partial}{\partial y} \quad \text{for } (x, y) \in \Gamma_+ \, , \tag{6.1.60}
$$

and

$$
W = \Delta W = 0 \quad \text{for } (x, y) \in \Gamma_\ell + \Gamma_r \, . \tag{6.1.61}
$$

We obtain

$$
I = I_- + I_+ = \int_0^{\ell} \left(Q(W, W)_{y=b} - Q(W, W)_{y=-b} \right) \mathrm{d}x \, , \tag{6.1.62}
$$

where

$$
Q(w, v) \equiv v \frac{\partial}{\partial y} \Delta w - \Delta w \frac{\partial v}{\partial y} \, , \tag{6.1.63}
$$

with arbitrary functions v and w. Using the boundary conditions for an isotropic plate, (6.1.49) and (6.1.50), we find that

$$
Q(W, W) = \frac{W \dfrac{\partial^3 W}{\partial y^3}}{\left(\dfrac{2 - \nu}{1 - \nu} \right)} + \frac{\dfrac{\partial W}{\partial y} \dfrac{\partial^2 W}{\partial y^2}}{\left(\dfrac{\nu}{1 - \nu} \right)} \, , \quad \text{at } y = \pm b \, . \tag{6.1.64}
$$

We can see from (6.1.64) that the function Q is antisymmetric with respect to the transformation $y \to -y$ for symmetric and antisymmetric functions W, and consequently,

$$
Q(W, W)_{y=b} = -Q(W, W)_{y=-b} \, . \tag{6.1.65}
$$

It follows that

$$I = 2 \int_0^{\ell} Q(W, W)_{y=b} \, \mathrm{d}x \; . \tag{6.1.66}$$

From (6.1.54)–(6.1.66), we obtain

$$\int_\Omega W \Delta^2 W \, \mathrm{d}\Omega = \int_\Omega (\Delta W)^2 \, \mathrm{d}\Omega + 2 \int_0^{\ell} Q(W, W)_{y=b} \, \mathrm{d}x \; , \tag{6.1.67}$$

and, furthermore

$$\omega^2 = -s^2 = \frac{(C^2 - V_0^2) \displaystyle\int_\Omega \left(\frac{\partial W}{\partial x} \right)^2 \mathrm{d}\Omega + \frac{D}{m} \left(\int_\Omega (\Delta W)^2 \, \mathrm{d}\Omega + 2 \int_0^{\ell} Q_{y=b} \, \mathrm{d}x \right)}{\displaystyle\int_\Omega W^2 \, \mathrm{d}\Omega} \; . \tag{6.1.68}$$

We now observe from representation (6.1.68) the following equation for the divergence mode, with $s = 0$:

$$\left(V_0^{\mathrm{div}}\right)^2 = C^2 + \frac{D}{m} \frac{\displaystyle\int_\Omega (\Delta W)^2 \mathrm{d}\Omega + 2 \int_0^{\ell} Q_{y=b} \, \mathrm{d}x}{\displaystyle\int_\Omega \left(\frac{\partial W}{\partial x} \right)^2 \mathrm{d}\Omega} \; . \tag{6.1.69}$$

In particular, it follows from (6.1.69) that when the bending rigidity D is negligible, the critical velocity is the same as for the axially travelling string (see, e.g., [59]).

Next we will consider the analytical solution of the buckling (divergence) problem of an axially moving isotropic plate. The problem is formulated as an eigenvalue problem of the partial differential equation

$$\left(m V_0^2 - T_0\right) \frac{\partial^2 W}{\partial x^2} + D \left(\frac{\partial^4 W}{\partial x^4} + 2 \frac{\partial^4 W}{\partial x^2 \partial y^2} + \frac{\partial^4 W}{\partial y^4} \right) = 0 \; , \tag{6.1.70}$$

with the boundary conditions (6.1.48)–(6.1.50) and (6.1.52). We focus on studying a static loss of stability, and therefore time-dependent terms are excluded from (6.1.42). In order to determine the minimal eigenvalue

$$\lambda = \gamma^2 = \frac{\ell^2}{\pi^2 D} \left(m V_0^2 - T_0\right) \tag{6.1.71}$$

of the problem (6.1.48)–(6.1.50), (6.1.70), and the corresponding eigenfunction $W = W(x, y)$, we apply the following representation:

$$W = W(x, y) = f\left(\frac{y}{b}\right) \sin\left(\frac{\pi x}{\ell}\right) , \qquad (6.1.72)$$

where $f(y/b)$ is an unknown function. It follows from (6.1.72) that the desired buckling mode (steady-state solution) W satisfies the boundary condition (6.1.48). The half-sine shape of the solution in the longitudinal direction is well-known (see, e.g., [21]). Using the nondimensional quantities

$$\eta = \frac{y}{b} , \qquad \mu = \frac{\ell}{\pi b} , \qquad (6.1.73)$$

and the relations (6.1.49), (6.1.50) and (6.1.70)–(6.1.73), we obtain the following eigenvalue problem for the unknown function $f(\eta)$:

$$\mu^4 \frac{\mathrm{d}^4 f}{\mathrm{d}\eta^4} - 2\mu^2 \frac{\mathrm{d}^2 f}{\mathrm{d}\eta^2} + (1 - \lambda) f = 0 , \qquad -1 < \eta < 1 , \qquad (6.1.74)$$

$$\mu^2 \frac{\mathrm{d}^2 f}{\mathrm{d}\eta^2} - \nu f = 0 , \qquad \eta = \pm 1 , \qquad (6.1.75)$$

$$\mu^2 \frac{\mathrm{d}^3 f}{\mathrm{d}\eta^3} - (2 - \nu) \frac{\mathrm{d} f}{\mathrm{d}\eta} = 0 , \qquad \eta = \pm 1 , \qquad (6.1.76)$$

where (6.1.75) and (6.1.76) represent the free-of-traction boundary conditions.

We will now present the solution process of the eigenvalue problem (6.1.74) and (6.1.76). We consider the problem as a spectral boundary value problem. The problem is invariant with respect to the symmetry operation $\eta \to -\eta$, and consequently, all its eigenfunctions can be classified as

$$f^{\mathrm{s}}(\eta) = f^{\mathrm{s}}(-\eta), \quad f^{\mathrm{a}}(\eta) = -f^{\mathrm{a}}(-\eta), \quad 0 \le \eta \le 1 . \qquad (6.1.77)$$

Here f^{s} and f^{a} are symmetric and antisymmetric (skew-symmetric) with respect to the x axis ($\eta = 0$). When $\gamma \le 1$, a divergence mode symmetric with respect to the x axis can be presented in the form

$$W = f^{\mathrm{s}}(\eta) \sin\left(\frac{\pi x}{\ell}\right) \qquad (6.1.78)$$

where

$$f^{\mathrm{s}}(\eta) = A^{\mathrm{s}} \cosh\left(\frac{\kappa_+ \eta}{\mu}\right) + B^{\mathrm{s}} \cosh\left(\frac{\kappa_- \eta}{\mu}\right) \qquad (6.1.79)$$

and

$$\kappa_+ = \sqrt{1 + \gamma} , \quad \kappa_- = \sqrt{1 - \gamma} . \qquad (6.1.80)$$

The function $f^s(\eta)$ is a symmetric solution of (6.1.74), and A^s and B^s are arbitrary constants. At first, we concentrate on the symmetric case and return to the antisymmetric case later.

By inserting (6.1.79) into the boundary conditions (6.1.75) and (6.1.76) at $\eta = +1$, we derive the linear algebraic equations for determining the constants A^s and B^s :

$$A^s\left(\kappa_+^2 - \nu\right)\cosh\left(\frac{\kappa_+}{\mu}\right) + B^s\left(\kappa_-^2 - \nu\right)\cosh\left(\frac{\kappa_-}{\mu}\right) = 0\,, \tag{6.1.81}$$

$$-A^s\kappa_+\left(\kappa_-^2 - \nu\right)\sinh\left(\frac{\kappa_+}{\mu}\right) - B^s\kappa_-\left(\kappa_+^2 - \nu\right)\sinh\left(\frac{\kappa_-}{\mu}\right) = 0\,. \tag{6.1.82}$$

The condition for a nontrivial solution to exist in the form (6.1.78)–(6.1.80) is that the determinant of the system (6.1.81) and (6.1.82) must vanish. This is seen by observing that (6.1.81) and (6.1.82) is a homogeneous system of linear equations in A^s, B^s:

$$\begin{bmatrix} K_{11} & K_{12} \\ K_{21} & K_{22} \end{bmatrix}\begin{bmatrix} A^s \\ B^s \end{bmatrix} = \begin{bmatrix} 0 \\ 0 \end{bmatrix}, \tag{6.1.83}$$

where the coefficients K_{ij} are given by the obvious identifications. From linear algebra, it is known that a nontrivial solution satisfying (6.1.83) can only exist if the matrix $\mathbf{K}$ is singular. Hence its determinant must be zero.

This zero determinant condition leads to the transcendental equation

$$\kappa_-\left(\kappa_+^2 - \nu\right)^2\cosh\left(\frac{\kappa_+}{\mu}\right)\sinh\left(\frac{\kappa_-}{\mu}\right) - \kappa_+\left(\kappa_-^2 - \nu\right)^2\sinh\left(\frac{\kappa_+}{\mu}\right)\cosh\left(\frac{\kappa_-}{\mu}\right) = 0, \tag{6.1.84}$$

which determines the eigenvalues

$$\lambda = \gamma^2 \tag{6.1.85}$$

implicitly. Equation (6.1.84) can be transformed into a more convenient form,

$$\Phi(\gamma, \mu) - \Psi(\gamma, \nu) = 0\,, \tag{6.1.86}$$

where we have defined

$$\Phi(\gamma, \mu) = \tanh\left(\frac{\sqrt{1-\gamma}}{\mu}\right)\coth\left(\frac{\sqrt{1+\gamma}}{\mu}\right) \tag{6.1.87}$$

and

$$\Psi(\gamma, \nu) = \frac{\sqrt{1+\gamma}\,(\gamma + \nu - 1)^2}{\sqrt{1-\gamma}\,(\gamma - \nu + 1)^2}\,. \tag{6.1.88}$$

Finally, let us consider the modes of buckling which are antisymmetric about the x axis:

$$W = f^{\mathrm{a}}(\eta)\, \sin\left(\frac{\pi x}{\ell}\right) , \tag{6.1.89}$$

where

$$f^{\mathrm{a}}(\eta) = A^{\mathrm{a}} \sinh\left(\frac{\kappa_{+}\eta}{\mu}\right) + B^{\mathrm{a}} \sinh\left(\frac{\kappa_{-}\eta}{\mu}\right) \tag{6.1.90}$$

for $\gamma \leq 1$. The values κ_{+} and κ_{-} are again defined by the expressions (6.1.80). Using the expression (6.1.90) for f^{a} and the boundary conditions on the free edges of the plate (6.1.75) and (6.1.76), we obtain the following transcendental equation for determining the quantity γ:

$$\Phi\,(\gamma, \mu) - \frac{1}{\Psi\,(\gamma, \nu)} = 0 . \tag{6.1.91}$$

In (6.1.91), $\Phi\,(\gamma, \mu)$ and $\Psi\,(\gamma, \nu)$ are again defined by the formulas (6.1.87) and (6.1.88). In the segment $0 < \gamma \leq 1$ being considered, the equation has two roots,

$$\gamma = \gamma_1 \quad \rightarrow \quad \gamma_0 < \gamma_1 < 1 \tag{6.1.92}$$

and

$$\gamma = \gamma_2 \quad \rightarrow \quad \gamma_2 = 1 , \tag{6.1.93}$$

for arbitrary values of the Poisson ratio ν and the geometric parameter μ. By using (6.1.92) and (6.1.93) and some properties described in the next section, it is possible to determine that

$$\gamma_* < \gamma_1 < \gamma_2 , \tag{6.1.94}$$

where γ_* is the minimal eigenvalue for the symmetric case. Thus, the critical buckling mode is symmetric with respect to the x axis, and corresponds to $\gamma = \gamma_*$, that is, to the solution of (6.1.86).

We have obtained an equation determining the minimal eigenvalue γ_*, namely (6.1.86). By relation (6.1.71), the corresponding critical velocity of the axially travelling plate is then represented as

$$(V_0^{\mathrm{div}})^2 = \frac{T_0}{m} + \frac{\gamma_*^2}{m}\left(\frac{\pi^2 D}{\ell^2}\right) . \tag{6.1.95}$$

In order to obtain the corresponding eigenmode, either A^{s} or B^{s} can be solved from either of the Eqs. (6.1.81) and (6.1.82), and the other one (either B^{s} or A^{s}, respectively) can be chosen arbitrarily; it is the free coefficient of the eigenvalue problem. Finally, inserting the obtained γ_*, A^{s} and B^{s} into (6.1.78) and (6.1.79) gives the eigenmode corresponding to the eigenvalue γ_*.

One of A^s or B^s is left free, because the zero determinant condition holds at the value of $\gamma = \gamma_*$, which is a solution of (6.1.86). Hence, at $\gamma = \gamma_*$, the Eqs. (6.1.81) and (6.1.82) become linearly dependent, providing only one condition.

6.2 Orthotropic Plates

The ratio of Young's moduli, that is, the degree of orthotropicity, defines the properties of the actual paper product, affecting its behavior. Different degrees of orthotropicity are desired for different applications. Using an orthotropic material model, we can bring the analysis closer to the real life situation that is being modelled.

The problem is formulated similarly to the isotropic eigenvalue problem, but now describing the divergence of the travelling orthotropic plate. The axial tension field is still uniform, described by a constant T_0. We have the partial differential equation

$$\left(m V_0^2 - T_0\right) \frac{\partial^2 W}{\partial x^2} + D_0 \mathcal{L}_0(W) = 0 \,, \tag{6.2.1}$$

with the boundary conditions (6.1.48)–(6.1.50). Here the bending operator $\mathcal{L}_0(W)$ is

$$\mathcal{L}_0(W) = \frac{D_1}{D_0} \frac{\partial^4 W}{\partial x^4} + \frac{2D_3}{D_0} \frac{\partial^4 W}{\partial x^2 \partial y^2} + \frac{D_2}{D_0} \frac{\partial^4 W}{\partial y^4} \,, \tag{6.2.2}$$

where the coefficients D_j for $j = 1, 2, 3$ are the orthotropic bending rigidities

$$D_1 = \frac{h^3}{12} C_{11} \,, \quad D_2 = \frac{h^3}{12} C_{22} \,, \quad D_3 = \frac{h^3}{12} \left(C_{12} + 2\, C_{66}\right) \,,$$

which were already introduced as (6.1.8), Sect. 6.1 (or see [51], Chap. 11). The C_{ij} are the elastic moduli, (6.1.9). In (6.2.1) and (6.2.2), the coefficient D_0 is an arbitrary normalization constant, which is convenient to take as $D_0 = D_1$.

We wish to determine the minimal eigenvalue,

$$\lambda = \gamma^2 = \frac{\ell^2}{\pi^2 D_0} \left(m V_0^2 - T_0\right) \,, \tag{6.2.3}$$

of the problem (6.1.48)–(6.1.50), (6.2.1) and (6.2.2). For the corresponding eigenfunction $W = W(x, y)$, we apply the same representation as before,

$$W = W\left(x, y\right) = f\left(\frac{y}{b}\right) \sin\left(\frac{\pi x}{\ell}\right) \,. \tag{6.2.4}$$

As was noted, the fact that the solution is a half-sine in the longitudinal direction is well-known in the isotropic case. It can be shown that the same form is applicable for the orthotropic plate. Again, what remains to be determined is the unknown cross-section $f\,(y/b)$.

It follows from (6.2.4) that the desired buckling form W (steady-state solution) satisfies the boundary condition (6.1.48). Using the nondimensional quantities

$$\eta = \frac{y}{b}\,, \quad \mu = \frac{\ell}{\pi b}\,, \tag{6.2.5}$$

and the relations (6.1.49), (6.1.50) and (6.2.1)–(6.2.4), we obtain the following eigenvalue problem for the unknown function $f(\eta)$:

$$\mu^4 H_2 \frac{\mathrm{d}^4 f}{\mathrm{d}\eta^4} - 2\mu^2 H_3 \frac{\mathrm{d}^2 f}{\mathrm{d}\eta^2} + (H_1 - \lambda)\, f = 0\,, \qquad -1 < \eta < 1\,, \tag{6.2.6}$$

$$\mu^2 \frac{\mathrm{d}^2 f}{\mathrm{d}\eta^2} - \beta_1 f = 0\,, \qquad \eta = \pm 1\,, \tag{6.2.7}$$

$$\mu^2 \frac{\mathrm{d}^3 f}{\mathrm{d}\eta^3} - \beta_2 \frac{\mathrm{d}f}{\mathrm{d}\eta} = 0\,, \qquad \eta = \pm 1\,, \tag{6.2.8}$$

where H_1, H_2 and H_3 are nondimensional bending rigidities, defined by

$$H_1 = \frac{D_1}{D_0}\,, \qquad H_2 = \frac{D_2}{D_0}\,, \qquad H_3 = \frac{D_3}{D_0}\,, \tag{6.2.9}$$

with D_0 the arbitrary normalization constant that was introduced above. In this book, we will use the choice $D_0 = D_1$, which will be convenient in the calculations to follow. The parameters β_1 and β_2 are given by (6.1.51). Equations (6.2.7) and (6.2.8) represent the free-of-traction boundary conditions.

The general solutions of the homogeneous ordinary differential equation (6.2.6) have the form

$$f = A e^{p\eta}\,, \qquad p = \frac{\kappa}{\mu}\,, \tag{6.2.10}$$

where A is an arbitrary constant, and κ is a solution of the following biquadratic algebraic characteristic equation:

$$H_2 \kappa^4 - 2 H_3 \kappa^2 + (H_1 - \lambda) = 0\,. \tag{6.2.11}$$

The solution can be written as

$$\kappa_\pm^2 = \frac{H_3}{H_2}\left(1 \pm \sqrt{1 - \frac{H_2 (H_1 - \lambda)}{H_3^2}}\right) = \frac{H_3}{H_2}\left(1 \pm \sqrt{1 - \frac{H_2 (1 - \lambda)}{H_3^2}}\right)\,, \tag{6.2.12}$$

where the upper, and respectively the lower, signs correspond to each other. In the last form on the right, we have used the choice $D_0 = D_1$, which leads to $H_1 = 1$.

Let us consider the range of λ where the solution is real-valued. The numbers $\kappa_\pm^2$ are real-valued if the expression under the square root in (6.2.12) is nonnegative. This implies the following lower limit for λ:

$$\lambda_\mathrm{m} \equiv 1 - \frac{H_3^2}{H_2} < \lambda \,, \tag{6.2.13}$$

corresponding to a real-valued eigenfunction f. Note that in the case $\lambda = \lambda_\mathrm{m}$, the solution of (6.2.11) is

$$\kappa = \pm\sqrt{\frac{H_3}{H_2}} \,, \tag{6.2.14}$$

where both solutions are double roots.

Furthermore, if we require not only $\kappa_\pm^2$, but also $\kappa_\pm$ to be real-valued, the whole parenthetical expression in (6.2.12) must then be nonnegative. This gives us an upper limit for λ:

$$\lambda \leq 1 \equiv \lambda_\mathrm{max} \,. \tag{6.2.15}$$

Equation (6.2.15) holds regardless of the values of the problem parameters.

For the lower limit given by (6.2.13), it holds that

$$\lambda_\mathrm{m} \leq 0 \qquad \text{when} \qquad G_{12} \geq G_\mathrm{H} \,, \tag{6.2.16}$$

where G_{12} is the in-plane shear modulus of the orthotropic material, and G_H is the geometric average shear modulus. The quantity G_H is given by (6.1.15) (in Sect. 6.1), repeated here for convenience:

$$G_\mathrm{H} \equiv \frac{\sqrt{E_1 E_2}}{2\left(1 + \sqrt{\nu_{12}\nu_{21}}\right)} \,. \tag{6.2.17}$$

Knowing that $\lambda \geq 0$ and (6.2.13)–(6.2.15), in the case that (6.2.16) holds, we may seek the lowest eigenvalue in the range $0 \leq \lambda \leq 1$, as was done in the isotropic case.

On the other hand, one can find examples of measurements of G_{12} for paper materials, which indicate $G_{12} < G_\mathrm{H}$. See, e.g., the articles of Mann et al. [33], Seo [37], Yokoyama and Nakai [60], and Bonnin et al. [61]. For such a material,

$$\lambda_\mathrm{m} > 0 \qquad \text{when} \qquad G_{12} < G_\mathrm{H} \,. \tag{6.2.18}$$

This will produce complex solutions $\kappa_\pm$ and complex eigenfunctions if λ is between zero and λ_m. In practice however, it has been numerically observed (from (6.2.25), presented further below) that this interval contains no solutions. Thus, in this case the search for the lowest eigenvalue can be performed in the range $\lambda_\mathrm{m} \leq \lambda \leq 1$.

These considerations motivate the definition

$$\lambda_{\text{min}} \equiv \max(\lambda_{\text{m}}, 0) \ , \tag{6.2.19}$$

enabling us to define the relevant range for solutions as

$$\lambda_{\text{min}} \leq \lambda \leq \lambda_{\text{max}} \tag{6.2.20}$$

regardless of the value of the shear modulus G_{12}. The quantities λ_{m} and λ_{max} are defined by (6.2.13) and (6.2.15), respectively.

From (6.2.10) and (6.2.12) in the case that $\lambda \neq \lambda_{\text{m}}$, we obtain that the general solution can be represented in the form

$$f(\eta) = A_1 e^{+\frac{\kappa_+ \eta}{\mu}} + A_2 e^{-\frac{\kappa_+ \eta}{\mu}} + A_3 e^{+\frac{\kappa_- \eta}{\mu}} + A_4 e^{-\frac{\kappa_- \eta}{\mu}} \tag{6.2.21}$$

with unknown constants A_1, A_2, A_3 and A_4.

The eigenvalue boundary value problem (6.2.6)–(6.2.8) is invariant under the symmetry operation $\eta \to -\eta$, and consequently the eigenforms can be classified into functions that are symmetric (f^{s}) or antisymmetric (f^{a}) with respect to the origin. Using the relations (6.2.6)–(6.2.8) and (6.2.21), we obtain a general representation for the function $f^{\text{s}}(\eta)$ and linear algebraic equations for determining the constants A^{s} and B^{s}:

$$f^{\text{s}}(\eta) = A^{\text{s}} \cosh \frac{\kappa_+ \eta}{\mu} + B^{\text{s}} \cosh \frac{\kappa_- \eta}{\mu} \ , \tag{6.2.22}$$

$$A^{\text{s}} \left(\kappa_+^2 - \beta_1\right) \cosh \frac{\kappa_+}{\mu} + B^{\text{s}} \left(\kappa_-^2 - \beta_1\right) \cosh \frac{\kappa_-}{\mu} = 0 \ , \tag{6.2.23}$$

$$A^{\text{s}} \kappa_+ \left(\kappa_+^2 - \beta_2\right) \sinh \frac{\kappa_+}{\mu} + B^{\text{s}} \kappa_- \left(\kappa_-^2 - \beta_2\right) \sinh \frac{\kappa_-}{\mu} = 0 \ , \tag{6.2.24}$$

where A^{s} and B^{s} are unknown constants. Due to the symmetry (or antisymmetry) of the solution f, we have only two independent unknown constants, instead of the four in the general representation (6.2.21), where the symmetry considerations had not yet been applied.

Proceeding in the same manner as in the isotropic case, the conditions for a nontrivial solution to exist in the form of (6.2.22)–(6.2.24) reduce to the requirement that the determinant of the homogeneous linear system (6.2.23) and (6.2.24) vanishes.

Again, at the solution point, the zero determinant condition leads to the linear dependence of the Eqs. (6.2.23) and (6.2.24), providing only one independent condition. Thus, we may solve either of (6.2.23) and (6.2.24) for either A^{s} or B^{s}, and choose the other (free) coefficient arbitrarily.

After rearrangement, the zero determinant condition can be expressed in the convenient form

$$\Phi(\gamma, \mu, \nu_{12}, E_1, E_2, G_{12}) - \Psi(\gamma, \nu_{12}, E_1, E_2, G_{12}) = 0 \,, \tag{6.2.25}$$

where

$$\Phi(\gamma, \mu, \nu_{12}, E_1, E_2, G_{12}) = \tanh\frac{\kappa_-}{\mu}\coth\frac{\kappa_+}{\mu} \,, \tag{6.2.26}$$

$$\Psi(\gamma, \nu_{12}, E_1, E_2, G_{12}) = \frac{\kappa_+(\kappa_+^2 - \beta_2)(\kappa_-^2 - \beta_1)}{\kappa_-(\kappa_+^2 - \beta_1)(\kappa_-^2 - \beta_2)} \,, \tag{6.2.27}$$

and

$$\kappa_+ = \kappa_+(\gamma, \nu_{12}, E_1, E_2, G_{12}) \,, \qquad \kappa_- = \kappa_-(\gamma, \nu_{12}, E_1, E_2, G_{12}) \,. \tag{6.2.28}$$

The obtained transcendental equation (6.2.25) can be used to determine the eigenvalues

$$\lambda = \gamma^2 \tag{6.2.29}$$

corresponding to symmetric eigenfunctions with different values of the parameters μ, ν_{12}, E_1, E_2 and G_{12}.

In the definitions of Φ and Ψ, (6.2.26) and (6.2.27), there is no dependence on the parameter ν_{21}, because it is fully determined by ν_{12}, E_1 and E_2 via the compatibility relation (6.1.26) in Sect. 6.1. The independent parameters in Φ and Ψ can be chosen also in a different way, by choosing any combination of exactly three parameters out of E_1, E_2, ν_{12} and ν_{21}. Relation (6.1.26) can then be used to eliminate the remaining parameter.

Similarly, using the relations (6.2.7) and (6.2.8), we obtain a representation for antisymmetric eigenfunctions $f^{\mathrm{a}}(\eta)$, the equation for determining the corresponding constants A^{a} and B^{a}, and the transcendental equation

$$\Phi - \frac{1}{\Psi} = 0 \,, \tag{6.2.30}$$

where Φ and Ψ are the functions defined in (6.2.26) and (6.2.27). These equations can be used for determining the eigenvalues corresponding to antisymmetric eigenforms. The representations differ from (6.2.22)–(6.2.24) through the replacements

$$\cosh \to \sinh \quad \text{and} \quad \sinh \to \cosh \,. \tag{6.2.31}$$

Again, it turns out that the minimal antisymmetric eigenvalue is higher than the minimal symmetric one, so we will only consider the symmetric case.

Table 6.1 Physical parameters used in the numerical examples

T_0	m	h
500 N/m	0.08 kg/m^2	10^{-4} m

In the special case that $\lambda = \lambda_{\mathrm{m}}$, the characteristic Eq. (6.2.11) has two double roots (6.2.14), and then, the general solution has the form

$$f(\eta) = A_1 e^{+\frac{\kappa\eta}{\mu}} + A_2 e^{-\frac{\kappa\eta}{\mu}} + A_3 \eta e^{+\frac{\kappa\eta}{\mu}} + A_4 \eta e^{-\frac{\kappa\eta}{\mu}} \,.$$

In this case, the symmetric solution has the form

$$f^{\mathrm{s}}(\eta) = A^{\mathrm{s}} \cosh \frac{\kappa\eta}{\mu} + B^{\mathrm{s}} \eta \sinh \frac{\kappa\eta}{\mu} \,. \tag{6.2.32}$$

For this solution, we will also have a zero determinant condition (different from (6.2.25) and (6.2.30)) but for a fixed κ. It can be calculated that the determinant condition does not hold for (6.2.32) with the boundary conditions (6.2.7) and (6.2.8), and thus, there is no symmetric solution of the form (6.2.32), and we will have no solution when $\lambda = \lambda_{\mathrm{m}}$. The antisymmetric case can be explored in a similar manner.

Similar remarks about finalizing the solution apply as before. Once (6.2.25) has been solved, obtaining the minimal symmetric eigenvalue γ_*, the corresponding critical velocity of the travelling orthotropic plate can be found from (6.2.3). The critical velocity is

$$(V_0^{\mathrm{div}})^2 = \frac{T_0}{m} + \frac{\gamma_*^2}{m} \left(\frac{\pi^2 D_0}{\ell^2} \right) \,. \tag{6.2.33}$$

Then, in order to obtain the corresponding eigenmode, we can solve for either A^{s} or B^{s}, picking either of the Eqs. (6.2.23), (6.2.24). Recall that the equations are linearly dependent at the solution point $\gamma = \gamma_*$, so it does not matter which one is used. The other coefficient (either B^{s} or A^{s}, respectively) is then the free coefficient of the eigenvalue problem, and can be assigned an arbitrary value. Finally, inserting the obtained γ_*, A^{s} and B^{s} into (6.2.4), (6.2.12) and (6.2.22) gives the eigenmode corresponding to the eigenvalue γ_*.

We will illustrate the critical divergence velocities and the corresponding buckling modes of axially moving orthotropic plates by giving some numerical examples. The physical parameters used are varied with the examples. The mass per unit area m, the strength of the uniform tension field T_0 and the plate thickness h are kept constant. Their values are given in Table 6.1.

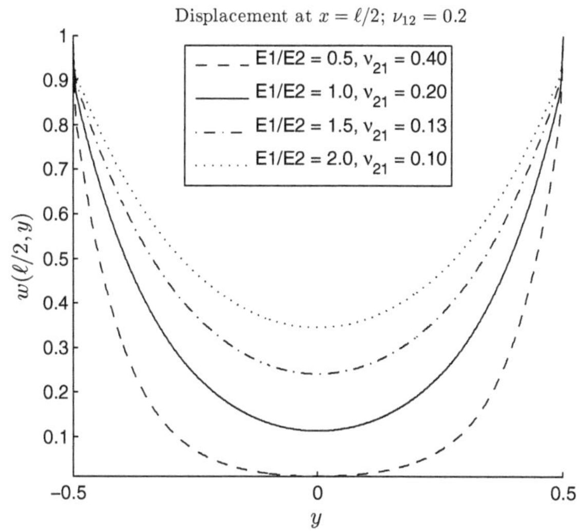

Fig. 6.3 Slices of buckling modes for different Young modulus ratios. Slices of the buckling modes at $x = \ell/2$ are shown. The ratio between the plate length and the plate width is $\ell/(2b) = 0.01$. The Young modulus in the x direction is $E_1 = 5$ GPa and the Poisson ratio ν_{12} is 0.2. The Poisson ratio ν_{21} is calculated from relation (6.1.26) on Sect. 6.1. For the shear modulus, the geometric average G_H from (6.1.15) Sect. 6.1 is used. (Reproduced from Banichuk et al. [62])

In Fig. 6.3, slices of buckling modes at $x = \ell/2$ are presented for four different Young modulus ratios E_1/E_2. We observe that the Young modulus ratio affects the localization of the buckling mode: the smaller the ratio is, the more the shape is localized near the edges.

The degree of localisation represents the variation of the displacement in the width (y) direction. Relative localisation is high, when most of the displacement occurs near the free edges. The problem parameters affecting the degree of localisation are the aspect ratio $\ell/(2b)$, the Young modulus ratio E_1/E_2, the Poisson ratio ν_{12}, and the in-plane shear modulus G_{12}.

In Fig. 6.4, we see three examples of complete buckling shapes for different values of the shear modulus G_{12}. The buckling shapes depend significantly on the in-plane shear modulus G_{12}. The figure also shows that if the ratio G_{12}/G_H is increased, then the degree of localisation decreases.

In Table 6.2, the values of critical velocities, defined in (6.2.33), are given for some selected values of the in-plane shear modulus G_{12} and the aspect ratio $\ell/(2b)$. The row $\ell/(2b) = 0.01$ corresponds to the buckling modes in Fig. 6.4. The effect of the increased in-plane shear modulus is that the value of the critical velocity slightly increases.

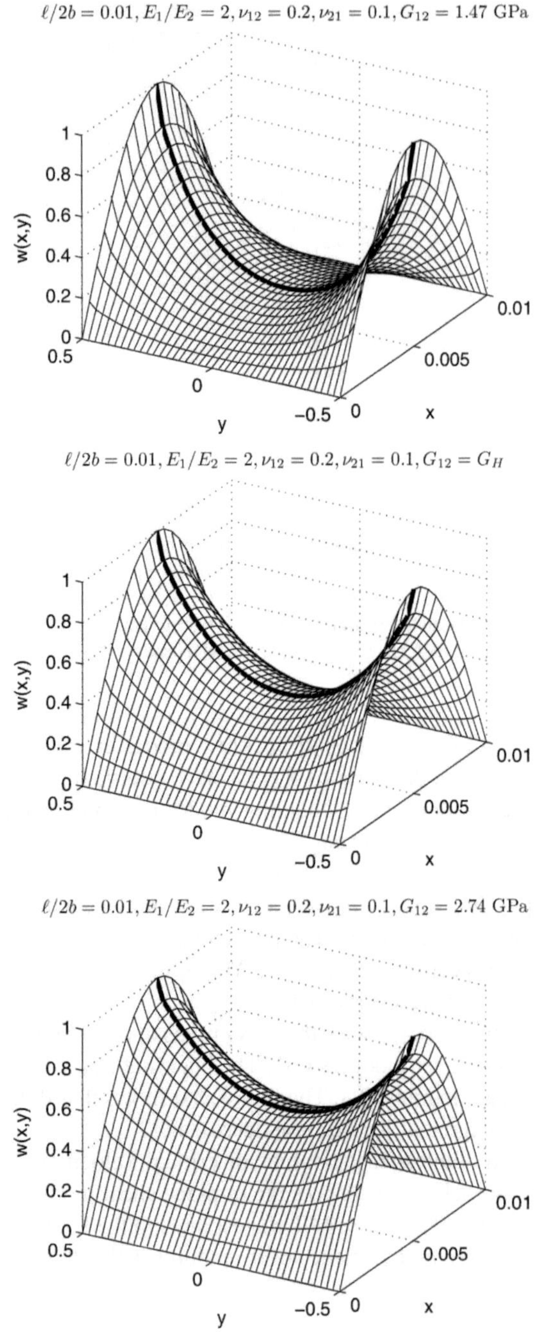

Fig. 6.4 Buckling modes for three different in-plane shear moduli. The ratio between the plate length and the plate width is $\ell/(2b) = 0.01$. *Top*: $G_{12} = 0.7G_{\mathrm{H}}$; *Middle*: $G_{12} = G_{\mathrm{H}}$; *Bottom*: $G_{12} = 1.3G_{\mathrm{H}}$, where G_{12} is the in-plane shear modulus and G_{H} is the geometric average shear modulus, (6.2.17). The Young moduli are $E_1 = 6.8$ GPa and $E_2 = 3.4$ GPa, and the Poisson ratio ν_{12} is 0.2. The Poisson ratio ν_{21} is calculated from relation (6.1.26) in Sect. 6.1, leading to $\nu_{21} = 0.1$. (Reproduced from Banichuk et al. [62])

Table 6.2 Critical velocities V_0^{div} (m/s) of an axially moving orthotropic plate for different values of in-plane shear modulus G_{12} and the ratio between the plate length and the plate width $\ell/(2b)$. G_{H} is the geometric average shear modulus, (6.2.17). The Young moduli are $E_1 = 6.8$ GPa and $E_2 = 3.4$ GPa, and the Poisson ratios ν_{12} is 0.2 and $\nu_{21} = 0.1$

$\ell/(2b)$	G_{12}		
	$0.7G_{\text{H}} \approx 1.47$ GPa (m/s)	$G_{\text{H}} \approx 2.11$ GPa (m/s)	$1.3G_{\text{H}} \approx 2.74$ GPa (m/s)
0.01	83.4456	83.4461	83.4463
0.1	79.1020	79.1020	79.1020
1	79.0574	79.0574	79.0574

6.3 Plates with a Nonuniform Axial Tension Distribution

In this section, we will look at the influence of a skewed tension profile on the divergence instability of an isotropic, axially travelling, thin elastic plate. The travelling plate is subjected to axial tension at the supports, but now the tension distribution along the supports is nonuniform. In order to look at the fundamental effects that a skewed tension profile introduces, while keeping the analysis simple, we will investigate the case with a linearly skewed tension profile.

First, we will perform a dynamic analysis of small time-harmonic vibrations, after which we will concentrate on the static stability (divergence) problem. We will see that even a small nonuniformity in the applied tension has a large effect on the divergence modes, and that a larger nonuniformity in the tension profile may significantly decrease the critical velocity of the plate.

Let a rectangular part of the plate

$$\Omega : \{0 < x < \ell, \ -b < y < b\}$$

be travelling at a constant velocity V_0 in the x direction between two rollers located at $x = 0$ and $x = \ell$, where ℓ and b are prescribed parameters. See Fig. 6.5. Let the considered part of the band be represented as an isotropic elastic plate, having constant thickness h, Poisson ratio ν, Young modulus E and bending rigidity D. We will make some notes on the orthotropic case later.

The plate is subjected to in-plane distributed forces

$$g = g(y) = T_0 + T(y) \tag{6.3.1}$$

applied at the plate boundaries $x = 0$ and $x = \ell$, acting in the x direction. The constant $T_0 > 0$ and the function $T(y)$, characterizing the nonuniformity of the in-plane tension of the axially moving plate, are considered given. The sides of the plate $\{x = 0, \ -b \le y \le b\}$ and $\{x = \ell, \ -b \le y \le b\}$ are simply supported, and the sides $\{y = -b, \ 0 \le x \le \ell\}$ and $\{y = b, \ 0 \le x \le \ell\}$ are free of tractions.

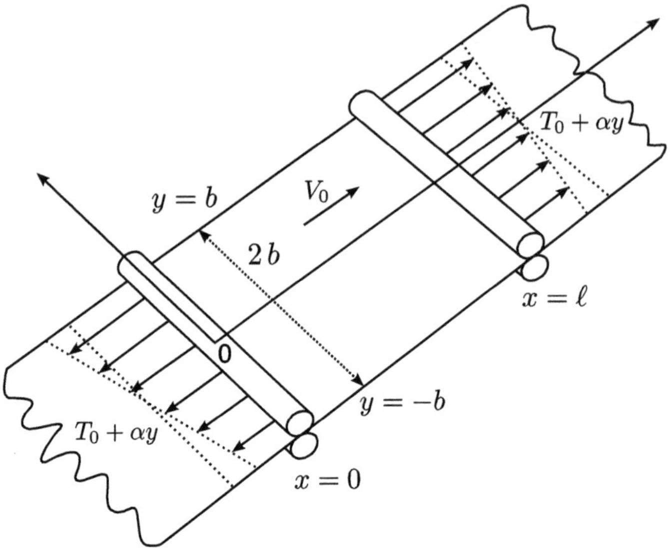

Fig. 6.5 Problem setup. A plate travelling at a constant velocity V_0 between two rollers placed at $x = 0$ and $x = \ell$. The tension profile at the rollers is nonuniform and the tension is positive everywhere. (Reproduced from Banichuk et al. [63])

The transverse displacement (out-of-plane deflection) of the travelling plate is described by the deflection function w, which depends on the space coordinates x and y, and time t. The differential equation for small transverse vibrations has the form

$$m\left(\frac{\partial^2 w}{\partial t^2} + 2V_0\frac{\partial^2 w}{\partial x \partial t} + V_0^2\frac{\partial^2 w}{\partial x^2}\right) = \mathcal{L}^{\mathrm{M}}(w) - \mathcal{L}^{\mathrm{B}}(w) , \qquad \text{in } \Omega . \tag{6.3.2}$$

The left-hand side in (6.3.2) contains the inertial terms (recall the discussion in Sect. 6.1). The membrane operator $\mathcal{L}^{\mathrm{M}}$ on the right-hand side of equation (6.3.2) is

$$\mathcal{L}^{\mathrm{M}}(w) = T_{xx}\frac{\partial^2 w}{\partial x^2} + 2\,T_{xy}\frac{\partial^2 w}{\partial x \partial y} + T_{yy}\frac{\partial^2 w}{\partial y^2} . \tag{6.3.3}$$

Again, the coefficients T_{xx}, T_{xy}, T_{yy} of the linear operator $\mathcal{L}^{\mathrm{M}}$ are related to the corresponding in-plane stresses σ_{xx}, σ_{xy} and σ_{yy} by the expressions

$$T_{ij} = h\sigma_{ij} .$$

For the isotropic elastic plate, the linear bending operator $\mathcal{L}^{\mathrm{B}}$ is given by the expression

$$\mathcal{L}^{\mathrm{B}}(w) = D\Delta^2 w = D\left(\frac{\partial^4 w}{\partial x^4} + 2\frac{\partial^4 w}{\partial x^2 \partial y^2} + \frac{\partial^4 w}{\partial y^4}\right) . \tag{6.3.4}$$

Boundary conditions for the deflection function w, corresponding to the simply supported boundaries and the free boundaries, can be written in the following form (see, e.g., [51])

$$(w)_{x=0,\,\ell} = 0\,, \qquad \left(\frac{\partial^2 w}{\partial x^2}\right)_{x=0,\,\ell} = 0\,, \qquad -b \le y \le b\,, \tag{6.3.5}$$

$$\left(\frac{\partial^2 w}{\partial y^2} + \nu \frac{\partial^2 w}{\partial x^2}\right)_{y=\pm b} = 0\,, \qquad 0 \le x \le \ell\,, \tag{6.3.6}$$

$$\left(\frac{\partial^3 w}{\partial y^3} + (2-\nu)\,\frac{\partial^3 w}{\partial x^2 \partial y}\right)_{y=\pm b} = 0\,, \qquad 0 \le x \le \ell\,. \tag{6.3.7}$$

Let us represent the in-plane tensions T_{xx}, T_{xy} and T_{yy} with the help of the Airy stress function Υ:

$$T_{xx} = \frac{\partial^2 \Upsilon}{\partial y^2}\,, \qquad T_{yy} = \frac{\partial^2 \Upsilon}{\partial x^2}\,, \qquad T_{xy} = -\frac{\partial^2 \Upsilon}{\partial x \partial y}\,. \tag{6.3.8}$$

As was seen in Sect. 6.1, in the case of an isotropic plate, the Airy stress function Υ satisfies the biharmonic equation

$$\Delta^2 \Upsilon \equiv \frac{\partial^4 \Upsilon}{\partial x^4} + 2\frac{\partial^4 \Upsilon}{\partial x^2 \partial y^2} + \frac{\partial^4 \Upsilon}{\partial y^4} = 0\,. \tag{6.3.9}$$

The boundary conditions for the tension are (6.1.19) and (6.1.20), repeated here for convenience:

$$T_{xx} = g(y)\,, \qquad T_{xy} = 0 \quad \text{at } x = 0,\ \ell, \quad -b \le y \le b\,,$$

$$T_{yy} = 0\,, \qquad T_{xy} = 0 \quad \text{at } y = \pm b, \quad 0 \le x \le \ell\,.$$

The boundary conditions satisfied by Υ, corresponding to (6.1.19) and (6.1.20) are

$$\left(\frac{\partial^2 \Upsilon}{\partial y^2}\right)_{x=0,\ell} = g(y)\,, \quad \left(\frac{\partial^2 \Upsilon}{\partial x \partial y}\right)_{x=0,\ell} = 0\,, \quad -b \le y \le b\,, \tag{6.3.10}$$

$$\left(\frac{\partial^2 \Upsilon}{\partial x^2}\right)_{y=\pm b} = 0\,, \quad \left(\frac{\partial^2 \Upsilon}{\partial x \partial y}\right)_{y=\pm b} = 0\,, \quad 0 \le x \le \ell\,. \tag{6.3.11}$$

Recall that the tensions expressed via the stress function Υ in (6.3.8) will satisfy the equilibrium of in-plane tensions for any function Υ that is smooth enough. The equilibrium equations are (6.1.18), repeated here for convenience:

$$\frac{\partial T_{xx}}{\partial x} + \frac{\partial T_{xy}}{\partial y} = 0\,, \quad \frac{\partial T_{xy}}{\partial x} + \frac{\partial T_{yy}}{\partial y} = 0\,.$$

Equation (6.3.9), which must be solved, with boundary conditions (6.3.11), expresses the condition of compatibility for the tensions.

Let us apply a linear profile for the axial tension at the supports:

$$g(y) = T_0 + \alpha y \equiv T_0 + T(y)\,. \tag{6.3.12}$$

Here $\alpha > 0$ is a given constant that will be called the *tension profile skew parameter*. It is easy to see that

$$\Upsilon(x, y) = T_0 \frac{y^2}{2} + \alpha \frac{y^3}{6} + c_1 x + c_2 y + c_0\,, \quad (x, y) \in \Omega \tag{6.3.13}$$

is then a solution of (6.3.9)–(6.3.11). Here c_0, c_1 and c_2 are arbitrary constants. The corresponding tensions will be

$$T_{xx}(x, y) = T_0 + \alpha y, \quad T_{xy}(x, y) = 0, \quad T_{yy}(x, y) = 0, \quad (x, y) \in \Omega\,. \tag{6.3.14}$$

In this case, the dynamic equation describing the out-of-plane displacement w takes the form

$$\frac{\partial^2 w}{\partial t^2} + 2\, V_0 \frac{\partial^2 w}{\partial x \partial t} + (V_0^2 - C^2) \frac{\partial^2 w}{\partial x^2} - \frac{T(y)}{m} \frac{\partial^2 w}{\partial x^2}$$

$$+ \frac{D}{m} \left(\frac{\partial^4 w}{\partial x^4} + 2 \frac{\partial^4 w}{\partial x^2 \partial y^2} + \frac{\partial^4 w}{\partial y^4} \right) = 0\,, \quad (x, y) \in \Omega\,, \tag{6.3.15}$$

where

$$C = \sqrt{\frac{T_0}{m}}\,, \quad T(y) = \alpha y\,.$$

Again following the dynamic stability analysis approach of Bolotin [58], let us represent the solution of the nonstationary boundary value problem for the partial differential equation (6.3.15) with the boundary conditions (6.3.5)–(6.3.7) using the time-harmonic trial function

$$w(x, y, t) = W(x, y)\, e^{st}\,, \qquad s = i\omega\,. \tag{6.3.16}$$

Here, ω is the frequency of the small transverse vibrations, and s is the stability exponent, which is a complex number. Using this (complex-valued) representation we have the pseudo steady state problem

$$s^2 W + 2sV_0 \frac{\partial W}{\partial x} + \left(V_0^2 - C^2\right) \frac{\partial^2 W}{\partial x^2} - \frac{T(y)}{m} \frac{\partial^2 W}{\partial x^2} + \frac{D}{m} \Delta^2 W = 0\,. \tag{6.3.17}$$

The boundary conditions for W follow from (6.3.5)–(6.3.7), by inserting (6.3.16). We obtain

$$(W)_{x=0,\ell} = 0\,, \qquad \left(\frac{\partial^2 W}{\partial x^2}\right)_{x=0,\ell} = 0\,, \qquad -b \le y \le b\,, \tag{6.3.18}$$

$$\left(\frac{\partial^2 W}{\partial y^2} + \nu \frac{\partial^2 W}{\partial x^2}\right)_{y=\pm b} = 0\,, \qquad 0 \le x \le \ell\,, \tag{6.3.19}$$

$$\left(\frac{\partial^3 W}{\partial y^3} + (2-\nu) \frac{\partial^3 W}{\partial x^2 \partial y}\right)_{y=\pm b} = 0\,, \qquad 0 \le x \le \ell\,. \tag{6.3.20}$$

Compare (6.1.48)–(6.1.50) and (6.1.52), in Sect. 6.1.

We multiply (6.3.17) by W and integrate over the domain Ω to obtain

$$s^2 \int_\Omega W^2 \, \mathrm{d}\Omega + 2sV_0 \int_\Omega W \frac{\partial W}{\partial x} \, \mathrm{d}\Omega + (V_0^2 - C^2) \int_\Omega W \frac{\partial^2 W}{\partial x^2} \, \mathrm{d}\Omega$$
$$- \frac{T(y)}{m} \int_\Omega W \frac{\partial^2 W}{\partial x^2} \, \mathrm{d}\Omega + \frac{D}{m} \int_\Omega W \Delta^2 W \, \mathrm{d}\Omega = 0\,. \tag{6.3.21}$$

Using the boundary conditions (6.3.18)–(6.3.20) and performing integration by parts, we find

$$\begin{aligned}
\int_\Omega W \frac{\partial W}{\partial x} \, \mathrm{d}\Omega &= \int_{-b}^{b} \int_0^\ell W \frac{\partial W}{\partial x} \, \mathrm{d}x \, \mathrm{d}y \\
&= \int_{-b}^{b} \left[\frac{W^2(\ell, y)}{2} - \frac{W^2(0, y)}{2} \right] \mathrm{d}y \\
&= 0\,,
\end{aligned}$$

$$\int_\Omega W \frac{\partial^2 W}{\partial x^2} \, \mathrm{d}\Omega = - \int_\Omega \left(\frac{\partial W}{\partial x}\right)^2 \mathrm{d}\Omega\,.$$

The term related to the nonuniform tension admits the following representation:

$$\int_\Omega y W \frac{\partial^2 W}{\partial x^2} \, \mathrm{d}\Omega = - \int_\Omega y \left(\frac{\partial W}{\partial x}\right)^2 \mathrm{d}\Omega\,. \tag{6.3.22}$$

We have

$$s^2 \int_\Omega W^2 \, \mathrm{d}\Omega + (C^2 - V_0^2) \int_\Omega \left(\frac{\partial W}{\partial x} \right)^2 \mathrm{d}\Omega + \\ + \frac{\alpha}{m} \int_\Omega y \left(\frac{\partial W}{\partial x} \right)^2 \mathrm{d}\Omega + \frac{D}{m} \int_\Omega W \Delta^2 W \, \mathrm{d}\Omega = 0 \,. \tag{6.3.23}$$

Two special cases, from which it is possible to draw further conclusions, will be considered. First, let $\alpha = 0$ and $T_{xx}(x, y) = T_0$, in other words, let us for the moment return to the case of uniform tension. In this case,

$$\int_\Omega W \, \Delta^2 W \, \mathrm{d}\Omega = \int_\Omega (\Delta W)^2 \, \mathrm{d}\Omega + 2 \int_0^\ell Q_{y=b} \, \mathrm{d}x \,, \tag{6.3.24}$$

in which we have abbreviated

$$Q = W \frac{\partial}{\partial y} (\Delta W) - \Delta W \frac{\partial W}{\partial y} \,. \tag{6.3.25}$$

Recall from Sect. 6.1 that in (6.3.24), the symmetry properties of the original partial differential equation were used to obtain this form of the Q integral. Consequently, one has

$$\omega^2 = -s^2 = \\ \frac{(C^2 - V_0^2) \int_\Omega \left(\frac{\partial W}{\partial x} \right)^2 \mathrm{d}\Omega + \frac{D}{m} \left[\int_\Omega (\Delta W)^2 \, \mathrm{d}\Omega + 2 \int_0^\ell Q_{y=b} \, \mathrm{d}x \right]}{\int_\Omega W^2 \, \mathrm{d}\Omega} \,. \tag{6.3.26}$$

At the critical velocity, it is again seen from (6.3.26) that the following relation between the critical velocity and the divergence mode holds:

$$\left(V_0^{\mathrm{div}}\right)^2 = C^2 + \frac{D}{m} \frac{\int_\Omega (\Delta W)^2 \, \mathrm{d}\Omega + 2 \int_0^\ell Q_{y=b} \, \mathrm{d}x}{\int_\Omega \left(\frac{\partial W}{\partial x} \right)^2 \mathrm{d}\Omega} \,. \tag{6.3.27}$$

In order to determine that $Q_{y=b} > 0$ at this point, one needs to use the solution from the corresponding static problem, described below for the general case. With that observation, we see that all integrals on the right side of (6.3.27) are positive, and it holds that

$$\left(V_0^{\mathrm{div}}\right)^2 > C^2 \,. \tag{6.3.28}$$

It follows from (6.3.27) that if the bending rigidity D is negligibly small, then

$$\left(V_{0\,\mathrm{mem}}^{\mathrm{div}}\right)^2 = C^2 = \frac{T_0}{m} \,. \tag{6.3.29}$$

In the one-dimensional case of axially travelling strings, this is a well-known result. See, for example, Chang and Moretti [59]. From (6.3.29), we see that the same value of the critical velocity also applies to an ideal membrane, with zero bending rigidity. The expression for $V_{0\,\mathrm{mem}}^{\mathrm{div}}$, (6.3.29), does not depend on W. Thus, for the special case of an ideal membrane under homogeneous tension, any combination of modes may occur at the critical velocity.

Consider now a second special case, where the bending rigidity of the axially moving plate is negligibly small, and the in-plane tension in the x direction is positive, avoiding the need to consider compression and possible wrinkling. Illustration can be seen in Fig. 6.5 in Sect. 6.3. In other words, let

$$D = 0 \,, \qquad T_0 > \alpha\, b \,, \tag{6.3.30}$$

where the latter condition comes from the constraints

$$T_{xx}(x, y) = T_0 + \alpha\, y > 0 \,, \quad y \geq -b \,. \tag{6.3.31}$$

In this case, the stability exponent s is evaluated as

$$\omega^2 = -s^2 = \frac{(C^2 - V_0^2) \displaystyle\int_\Omega \left(\frac{\partial W}{\partial x}\right)^2 \mathrm{d}\Omega + \frac{\alpha}{m} \int_\Omega y \left(\frac{\partial W}{\partial x}\right)^2 \mathrm{d}\Omega}{\displaystyle\int_\Omega W^2 \, \mathrm{d}\Omega} \,. \tag{6.3.32}$$

If a steady-state solution (divergence) exists, it will occur at velocity

$$\left(V_0^{\mathrm{div}}\right)^2 = C^2 + \frac{\alpha}{m} \frac{\displaystyle\int_\Omega y \left(\frac{\partial W}{\partial x}\right)^2 \mathrm{d}\Omega}{\displaystyle\int_\Omega \left(\frac{\partial W}{\partial x}\right)^2 \mathrm{d}\Omega} \,. \tag{6.3.33}$$

Let us assume that the divergence mode W is a real-valued function. Taking into account the expression in (6.3.33), and the fact that $y \geq -b$, we can estimate the divergence velocity (from below) as

$$\left(V_0^{\mathrm{div}}\right)^2 \geq C^2 - \frac{\alpha b}{m} = \frac{T_0 - \alpha b}{m} \,. \tag{6.3.34}$$

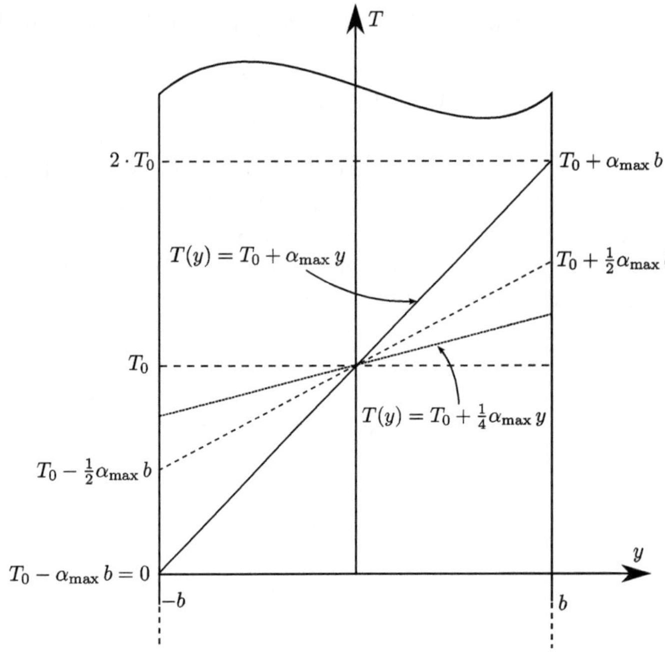

Fig. 6.6 The definition of α_{max}. It is the maximum skew that retains $T(y) \geq 0$ across the whole domain, avoiding compression and possible wrinkling

We see from (6.3.34) that as long as the condition for T_0 in (6.3.30) is fulfilled, we have $(V_0^{\text{div}})^2 \geq 0$, that is, the value of V_0^{div} is physically meaningful (Fig. 6.6).

We will next consider the static stability problem of the travelling thin plate subjected to a linearly skewed tension profile. The treatment of the problem follows the same approach as previously.

We begin with the transformation to an ordinary differential equation. The stationary eigenvalue problem of elastic stability consists of finding a nontrivial solution (mode) and the corresponding minimal eigenvalue of the following boundary-value problem. Consider the steady-state equation, corresponding to $s = 0$ in the nonstationary problem in (6.3.17),

$$\begin{aligned}(V_0^2 - C^2)\frac{\partial^2 W}{\partial x^2} - \frac{T(y)}{m}\frac{\partial^2 W}{\partial x^2} + \\ + \frac{D}{m}\left(\frac{\partial^4 W}{\partial x^4} + 2\frac{\partial^4 W}{\partial x^2 \partial y^2} + \frac{\partial^4 W}{\partial y^4}\right) = 0\,, \quad (x, y) \in \Omega\,. \qquad (6.3.35)\end{aligned}$$

with the boundary conditions for W in (6.3.18)–(6.3.20). From the latter condition in (6.3.30), requiring that the tension remains positive in all of the domain, we obtain a constraint for α:

$$T(y) = \alpha y \qquad \text{and} \qquad \alpha < T_0/b\,. \qquad (6.3.36)$$

To determine the minimal eigenvalue λ (see 6.1.71 in Sect. 6.1) of the problem (6.3.35) with boundary conditions (6.3.18)–(6.3.20), and the corresponding eigenfunction, we apply the same representation as before:

$$W = W(x, y) = f\left(\frac{y}{b}\right)\sin\left(\frac{\pi x}{\ell}\right), \tag{6.3.37}$$

where $f\,(y/b)$ is an unknown function. It follows from (6.3.37) that the divergence form W satisfies the boundary conditions (6.3.18).

As before, let us define the nondimensional quantities η and μ, given by (6.1.73) in Sect. 6.1,

$$\eta = \frac{y}{b}, \qquad \mu = \frac{\ell}{\pi b}, \tag{6.3.38}$$

and the eigenvalue λ as per (6.1.71),

$$\lambda = \gamma^2 = \frac{\ell^2}{\pi^2 D}\left(mV_0^2 - T_0\right). \tag{6.3.39}$$

By using the free-of-traction boundary conditions (6.3.19) and (6.3.20), the static stability equation (6.3.35) and the definition of W, (6.3.37), we obtain the following eigenvalue problem for the unknown function $f(\eta)$:

$$\mu^4\frac{\mathrm{d}^4 f}{\mathrm{d}\eta^4} - 2\mu^2\frac{\mathrm{d}^2 f}{\mathrm{d}\eta^2} + (1 - \lambda + \tilde{\alpha}\eta)f = 0, \quad -1 < \eta < 1, \tag{6.3.40}$$

where

$$\tilde{\alpha} = \frac{b\ell^2}{\pi^2 D}\alpha = \frac{b^3\mu^2}{D}\alpha. \tag{6.3.41}$$

Equation (6.3.40) is considered with the boundary conditions

$$\mu^2\frac{\mathrm{d}^2 f}{\mathrm{d}\eta^2} - \nu f = 0, \quad \eta = \pm 1 \quad \text{and} \tag{6.3.42}$$

$$\mu^2\frac{\mathrm{d}^3 f}{\mathrm{d}\eta^3} - (2 - \nu)\frac{\mathrm{d}f}{\mathrm{d}\eta} = 0, \quad \eta = \pm 1, \tag{6.3.43}$$

which correspond to the free-of-traction boundary conditions of the original problem.

Equation (6.3.40) with the boundary conditions (6.3.42) and (6.3.43) constitutes a linear eigenvalue problem for f with polynomial coefficients.

For an orthotropic material, it is possible to use problem (6.3.40), (6.3.42) and (6.3.43) in a straightforward way by setting the orthotropic in-plane shear modulus G_{12} as the geometric average shear modulus (6.1.15) and reducing the orthotropic problem into the isotropic one (see [51], Chap. 11).

Alternatively, if one wishes to keep G_{12} as an independent material parameter, which is more accurate for some materials, it is possible to derive the corresponding eigenvalue problem for the orthotropic plate following the same procedure that was used above for the isotropic plate. Again, let the axial in-plane tension (6.3.1) take the form (6.3.12), with the value of α in (6.3.12) constrained by (6.3.36). We have the following partial differential equation:

$$\left(m V_0^2 - T_0\right) \frac{\partial^2 W}{\partial x^2} - \frac{T(y)}{m} \frac{\partial^2 W}{\partial x^2} + D_0 \mathcal{L}_0(w) = 0 , \tag{6.3.44}$$

where the differential operator $\mathcal{L}_0(w)$ is given by (6.1.43) in Sect. 6.1,

$$\mathcal{L}_0(w) = \frac{D_1}{D_0} \frac{\partial^4 w}{\partial x^4} + \frac{2 D_3}{D_0} \frac{\partial^4 w}{\partial x^2 \partial y^2} + \frac{D_2}{D_0} \frac{\partial^4 w}{\partial y^4} ,$$

and D_0 is an arbitrary normalization constant, which is convenient to take as $D_0 = D_1$. The coefficients D_j for $j = 1, 2, 3$ are the orthotropic bending rigidities

$$D_1 = \frac{h^3}{12} C_{11} , \quad D_2 = \frac{h^3}{12} C_{22} , \quad D_3 = \frac{h^3}{12} \left(C_{12} + 2\, C_{66}\right) ,$$

which were already given as (6.1.8). The C_{ij} are the elastic moduli, (6.1.9).

The boundary conditions for W are given in (6.3.18)–(6.3.20). However, in the free edge boundary conditions (6.3.19) and (6.3.20), instead of the isotropic free boundary coefficients ν and $2 - \nu$, we must now use the orthotropic coefficients β_1 and β_2 (respectively) defined in (6.1.13) in the same way as before.

As previously, the in-plane tension field for an orthotropic plate in the case of a linear tension profile at the supports can be solved with the help of the Airy stress function. Refer to Eqs. (6.1.40) and (6.1.41) for the result.

To determine the minimal eigenvalue λ (see Eq. (6.3.39) for its meaning in terms of physical quantities), and the corresponding eigenfunction, of the problem (6.3.44) with boundary conditions (6.3.18)–(6.3.20), we apply the representation (6.3.37). By using the nondimensional quantities η and μ in (6.3.38), the free-of-traction boundary conditions (6.3.19) and (6.3.20), with the relations (6.3.39), (6.3.44) and (6.3.37), we obtain the eigenvalue problem for the orthotropic case:

$$\mu^4 H_2 \frac{\mathrm{d}^4 f}{\mathrm{d}\eta^4} - 2\mu^2 H_3 \frac{\mathrm{d}^2 f}{\mathrm{d}\eta^2} + (H_1 - \lambda - \bar{\alpha}\eta)\, f = 0 , \quad -1 < \eta < 1 , \tag{6.3.45}$$

$$\mu^2 \frac{\mathrm{d}^2 f}{\mathrm{d}\eta^2} - \beta_1 f = 0 , \quad \eta = \pm 1 , \tag{6.3.46}$$

$$\mu^2 \frac{\mathrm{d}^3 f}{\mathrm{d}\eta^3} - \beta_2 \frac{\mathrm{d} f}{\mathrm{d}\eta} = 0 , \quad \eta = \pm 1 . \tag{6.3.47}$$

In (6.3.45), the nondimensional tension profile skew parameter is defined as

$$\bar{\alpha} = \frac{b\ell^2}{\pi^2 D_0}\alpha = \frac{b^3\mu^2}{D_0}\alpha \tag{6.3.48}$$

and the H_j are the nondimensional bending rigidities defined by (6.2.9). As before, D_0 is the normalization constant for the bending rigidities and can be chosen arbitrarily. A convenient choice is $D_0 = D_1$. In (6.3.46) and (6.3.47), the β_j are defined in Eq. (6.1.13), Sect. 6.1. Again, we have a linear eigenvalue problem with polynomial coefficients. For the rest of the analysis, we will concentrate on the isotropic case.

Let us proceed with a numerical solution of the eigenvalue problem for the isotropic elastic plate. Finite differences will be used, with virtual points added to the ends of the domain to enforce the boundary conditions. As the considered problem is linear in f, the discretization will lead to a standard discrete linear eigenvalue problem representing (6.3.40):

$$\mathbf{Af} = \lambda\mathbf{f}\ . \tag{6.3.49}$$

Equation (6.3.49) does not yet include the boundary conditions (6.3.42) and (6.3.43). Because the boundary conditions are homogeneous, it is possible to add them to the discrete system by rewriting the original discrete problem (6.3.49) as a generalized linear eigenvalue problem,

$$\mathbf{Af} = \lambda\mathbf{Bf}\ , \tag{6.3.50}$$

where $\mathbf{B}$ is an identity matrix with the first two and last two rows zeroed out. In (6.3.50), the first two and the last two rows of A contain discrete representations of the boundary conditions (6.3.42) and (6.3.43).

Let us now perform this in detail. Equations (6.3.40)–(6.3.43) are to be discretized. The standard central difference formulas for the first four derivatives on a uniform grid are

$$\frac{\partial f_j}{\partial\eta} \approx \frac{f_{j+1} - f_{j-1}}{2\,(\Delta\eta)}\ , \tag{6.3.51}$$

$$\frac{\partial^2 f_j}{\partial\eta^2} \approx \frac{f_{j+1} - 2\,f_j + f_{j-1}}{(\Delta\eta)^2}\ , \tag{6.3.52}$$

$$\frac{\partial^3 f_j}{\partial\eta^3} \approx \frac{f_{j+2} - 2\,f_{j+1} + 2\,f_{j-1} - f_{j-2}}{2\,(\Delta\eta)^3}\ , \tag{6.3.53}$$

$$\frac{\partial^4 f_j}{\partial\eta^4} \approx \frac{f_{j+2} - 4\,f_{j+1} + 6\,f_j - 4\,f_{j-1} + f_{j-2}}{(\Delta\eta)^4}\ , \tag{6.3.54}$$

where $f \equiv f(x)$ is the function to be differentiated, $f_j \equiv f(\eta_j)$, and $\Delta\eta$ is the grid spacing. Concerning the truncation error, the asymptotic accuracy of (6.3.51)–(6.3.54) is $\mathrm{O}((\Delta\eta)^2)$. In any practical numerical implementation using finite-precision floating point numbers, there is another error source arising from catastrophic cancellation (see e.g. [64]), when nearly equal numbers are subtracted.

Thus there is an optimal $\Delta\eta$, below which cancellation error dominates, and above which truncation error dominates.

In practice, one usually just picks a reasonably sized grid spacing (such as 10^{-3} or 10^{-4} for the domain $-1 < \eta < 1$). We have to keep in mind that we are solving an eigenvalue problem, for which known algorithms have time complexity $O(n^3)$ (or $O(n^2)$ for just the eigenvalues, no eigenvectors), so the number of degrees of freedom should be kept fairly small.

When the derivatives in (6.3.40) are replaced by the discrete approximations (6.3.52) and (6.3.54) for each grid point η_j, we obtain the discrete equation system for the interior of the domain. The $\bar{\alpha}\eta$ term is handled by substituting in the coordinate of the jth grid point, $\eta_j = j\,(\Delta\eta)$. Then the discrete equations are collected into matrix form, and the $\lambda\mathbf{f}$ term is moved to the right-hand side.

The boundary conditions (6.3.42) and (6.3.43) are handled by adding two virtual points at each end of the domain. Applying (6.3.51)–(6.3.54) to the boundary conditions produces discrete equations connecting the function values at the virtual points to those inside the domain. In other words, these discrete equations implicitly define the values of f at the virtual points. It is of course possible to manipulate the relations into an explicit form, and then substitute them into the discrete equations describing the interior; this is the elimination method for treating boundary conditions. However, here we will simply introduce additional equations into the discrete equation system.

If we number the points starting at 1 at the first (outermost) virtual point at the left end of the domain, the final left-hand side matrix becomes

$$\mathbf{A} \equiv \mathbf{A}_4 + \mathbf{A}_2 + \mathbf{A}_0 + \mathbf{L}_1 + \mathbf{L}_2 + \mathbf{L}_3 + \mathbf{L}_4 \;,$$

where the terms $\mathbf{A}_m$ correspond to Eq. (6.3.40), and are given by

$$\mathbf{A}_4 \equiv \frac{\mu^4}{(\Delta\eta)^4} \begin{bmatrix} 0 & \dots & & & & \dots & 0 \\ 0 & \dots & & & & \dots & 0 \\ 1 & -4 & 6 & -4 & 1 & & \\ & \ddots & \ddots & \ddots & \ddots & \ddots & \\ & & 1 & -4 & 6 & -4 & 1 \\ 0 & \dots & & & & \dots & 0 \\ 0 & \dots & & & & \dots & 0 \end{bmatrix},$$

$$\mathbf{A}_2 \equiv -\frac{2\mu^2}{(\Delta\eta)^2} \begin{bmatrix} 0 \dots & & & & & \dots 0 \\ 0 \dots & & & & & \dots 0 \\ 0 & 1 & -2 & 1 & & \\ & & \ddots & \ddots & \ddots & \\ & & & 1 & -2 & 1 \; 0 \\ 0 \dots & & & & & \dots 0 \\ 0 \dots & & & & & \dots 0 \end{bmatrix},$$

$$\mathbf{A}_0 \equiv \begin{bmatrix} 0 \ldots & & & \ldots 0 \\ 0 \; 0 \; \ldots & & & \ldots 0 \\ 0 \; 0 \; a_k & & & \ldots 0 \\ & & \ddots & \\ 0 \ldots & & a_k & 0 \; 0 \\ 0 \ldots & & \ldots & 0 \; 0 \\ 0 \ldots & & & \ldots 0 \end{bmatrix},$$

where

$$a_k \equiv 1 + \bar{\alpha} \left[-1 + (k-3)\,(\Delta\eta)\right]$$

and k denotes the row number of the matrix $\mathbf{A}_0$. The first contribution in $\mathbf{A}_0$ is $1 - \bar{\alpha}$ (on row 3, corresponding to $\eta = -1$), and the last is $1 + \bar{\alpha}$ (third last row, corresponding to $\eta = +1$). Empty entries in the matrices denote zeroes; some zeroes are displayed explicitly to show more clearly where the nonzero entries belong.

The terms $\mathbf{L}_m$ correspond to the boundary conditions (6.3.42) and (6.3.43), and are given by

$$\mathbf{L}_1 \equiv \begin{bmatrix} 0 & \ldots & & & \ldots 0 \\ 0 & \mu^2/(\Delta\eta)^2 & -2\mu^2/(\Delta\eta)^2 - \nu & \mu^2/(\Delta\eta)^2 & \ldots 0 \\ 0 & \ldots & \ldots & \ldots & \ldots 0 \\ 0 & \ldots & \ldots & \ldots & \ldots 0 \\ 0 & \ldots & \ldots & \ldots & \ldots 0 \\ 0 & \ldots & & & \ldots 0 \\ 0 & \ldots & & & \ldots 0 \end{bmatrix},$$

$$\mathbf{L}_2 \equiv \begin{bmatrix} -\mu^2/[2(\Delta\eta)^3] & \chi & 0 & -\chi & \mu^2/[2(\Delta\eta)^3] & 0 \ldots 0 \\ 0 & \ldots & & & & \ldots 0 \\ 0 & \ldots & & & & \ldots 0 \\ 0 & \ldots & & & & \ldots 0 \\ 0 & \ldots & & & & \ldots 0 \\ 0 & \ldots & & & & \ldots 0 \\ 0 & \ldots & & & & \ldots 0 \end{bmatrix},$$

$$\mathbf{L}_3 \equiv \begin{bmatrix} 0 \ldots & & & \ldots & 0 \\ 0 \ldots & & & \ldots & 0 \\ 0 \ldots & \ldots & \ldots & \ldots & 0 \\ 0 \ldots & \ldots & \ldots & \ldots & 0 \\ 0 \ldots & \ldots & \ldots & \ldots & 0 \\ 0 \ldots & \mu^2/(\Delta\eta)^2 & -2\mu^2/(\Delta\eta)^2 - \nu & \mu^2/(\Delta\eta)^2 & 0 \\ 0 \ldots & & & \ldots & 0 \end{bmatrix},$$

$$\mathbf{L}_4 \equiv \begin{bmatrix} 0\dots & & & & & \dots & 0 \\ 0\dots & & & & & \dots & 0 \\ 0\dots & & & & & \dots & 0 \\ 0\dots & & & & & \dots & 0 \\ 0\dots & & & & & \dots & 0 \\ 0\dots & & & & & \dots & 0 \\ 0\dots & -\mu^2/[2(\Delta\eta)^3] & \chi & 0 & -\chi & & \mu^2/[2(\Delta\eta)^3] \end{bmatrix},$$

where in $\mathbf{L}_2$ and $\mathbf{L}_4$ we have defined

$$\chi \equiv \mu^2/(\Delta\eta)^3 + (2-\nu)/[2(\Delta\eta)] .$$

The matrices $\mathbf{L}_1$ and $\mathbf{L}_3$ correspond to the boundary condition (6.3.42) at the left and right endpoints of the domain, respectively, while $\mathbf{L}_2$ and $\mathbf{L}_4$ correspond to (6.3.43).

Finally, the discrete problem (6.3.50) is completed by defining

$$\mathbf{B} \equiv \begin{bmatrix} 0\dots & & & & \dots 0 \\ 0 & 0 & \dots & & \dots 0 \\ 0 & 0 & 1 & & \dots 0 \\ & & & \ddots & \\ 0\dots & & & 1 & 0 \; 0 \\ 0\dots & & & \dots & 0 \; 0 \\ 0\dots & & & & \dots 0 \end{bmatrix},$$

which enforces the homogeneous boundary conditions (6.3.42) and (6.3.43), represented by the first two and last two rows of the discrete equation system.

In order to solve the original problem, we compute the solution of (6.3.50), discard eigenvalues of infinite magnitude, which result from our way of handling the boundary conditions, and then extract the smallest eigenvalue λ and its corresponding eigenvector $\mathbf{f}$. The first two and last two components of the eigenvector are discarded, because they represent the function values at virtual points that were generated from the boundary conditions. Finally, the buckling mode $W(x, y)$ is constructed using the equation

$$W(x, y) = f\left(\frac{\pi y}{\ell}\right) \sin\left(\frac{\pi x}{\ell}\right) .$$

Below, numerical results are shown for some practically interesting choices of problem parameters. The physical parameters used in the examples are presented in Table 6.3. These parameter values approximately correspond to some paper materials, within the limitations of the isotropic model.

Various values of the Poisson ratio ν and the tension profile skew parameter $\tilde{\alpha}$ are used in the examples. For the Poisson ratio, the values 0, 0.1, 0.3 and 0.5 are used. The values of $\alpha/\alpha_{\max}$ (or $\tilde{\alpha}/\tilde{\alpha}_{\max}$) are 0, 10^{-6}, 10^{-4} and 10^{-2}, where $\alpha_{\max}$ corresponds to the upper limit imposed by the constraint (6.3.36), $\alpha < T_0/b$. Note that $\tilde{\alpha}_{\max}$ depends on ν, via D. In Table 6.4, critical divergence velocities are presented for

Table 6.3 Physical parameters used in the numerical examples

T_0 (tension at $y = 0$)	m	ℓ	$2b$	h	E
500 N/m	0.08 kg/m^2	0.1 m	1 m	10^{-4} m	10^9 N/m^2

Table 6.4 Critical divergence velocities V_0^{div} for example cases. Note that $\tilde{\alpha}_{\max}$ is different for each value of ν [63]

ν	$\tilde{\alpha}$			
	0	$10^{-6}\tilde{\alpha}_{\max}$	$10^{-4}\tilde{\alpha}_{\max}$	$10^{-2}\tilde{\alpha}_{\max}$
0	79.0634	79.0634	79.0605	78.6892
0.1	79.0635	79.0635	79.0605	78.6886
0.3	79.0640	79.0640	79.0609	78.6876
0.5	79.0652	79.0652	79.0618	78.6870

these cases. The analytical solution for $\tilde{\alpha} = 0$ for the same geometric and material parameters (see (6.1.71), (6.1.86)–(6.1.88) in Sect. 6.1) matches the values in the first column of the table.

The results for the transverse displacement are shown in Figs. 6.8, 6.9 and 6.10. In each figure, ν is fixed. Figure 6.8 is divided into two parts. Both parts of the figure are further divided into four subfigures. Each of these four subfigures shows the results for a different value of the skew parameter $\tilde{\alpha}$. In the upper four subfigures, $f(\eta)$ is plotted, showing a slice of the out-of-plane displacement from one free edge to the other at $x = \ell/2$. Tension increases toward positive η. The total out-of-plane displacement in the whole domain $\Omega = [0, \ell] \times [-b, b]$, from equation $W = f(\pi y/\ell)\sin(\pi x/\ell)$, is shown in the lower four subfigures. Note the orientation of the axes. In Figs. 6.9 and 6.10, the four subfigures show the slices of the out-of-plane displacement at $x = \ell/2$ for the limit cases $\nu = 0$ and $\nu = 0.5$, in analogous order.

From Figs. 6.7, 6.8, 6.9 and 6.10 and Table 6.4, three conclusions are apparent. First, it is seen that a significant skew in the tension profile may significantly decrease the critical velocities. Up to a 20% difference in axial tension between the midpoint and edges of the supports causes a decrease of 10% in the critical velocity. It is also seen that a wider plate is more sensitive to a nonuniform tension profile. Secondly, by comparing Figs. 6.8, 6.9 and 6.10, it is observed that materials with a larger Poisson ratio tend to exhibit a higher degree of sensitivity to a nonuniform tension profile.

Finally, we see that even for the smallest skew tested in the examples (one part in 10^6), for the problem parameters considered the divergence mode changes completely. Thus, from a practical point of view, although studies using a uniform tension profile can predict the critical velocity relatively accurately, if one wishes to predict the divergence shape, even a small skew in the tension profile at the supports must be accounted for.

Furthermore, the divergence shapes in Fig. 6.8 tell us that in a paper production environment, if the paper web happens to break due to a loss of elastic stability,

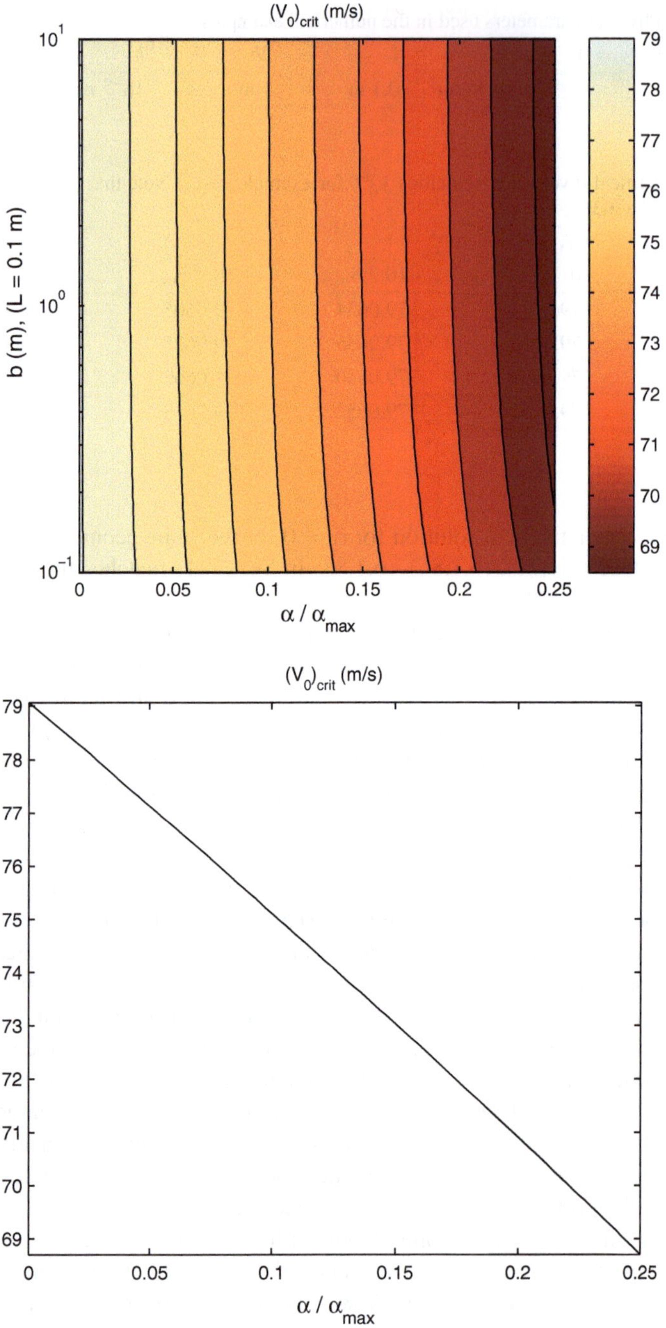

Fig. 6.7 *Top*: Critical plate velocity $(V_0)_{\text{crit}}$ with respect to the tension profile skew parameter α and plate half-width b. Note the logarithmic scale for b. The plate length is constant ($\ell = 0.1$ m). *Bottom*: The critical velocity plotted with respect to the tension profile skew parameter ($\ell = 0.1$ m, $2\,b = 1$ m). (Reproduced from Banichuk et al. [63])

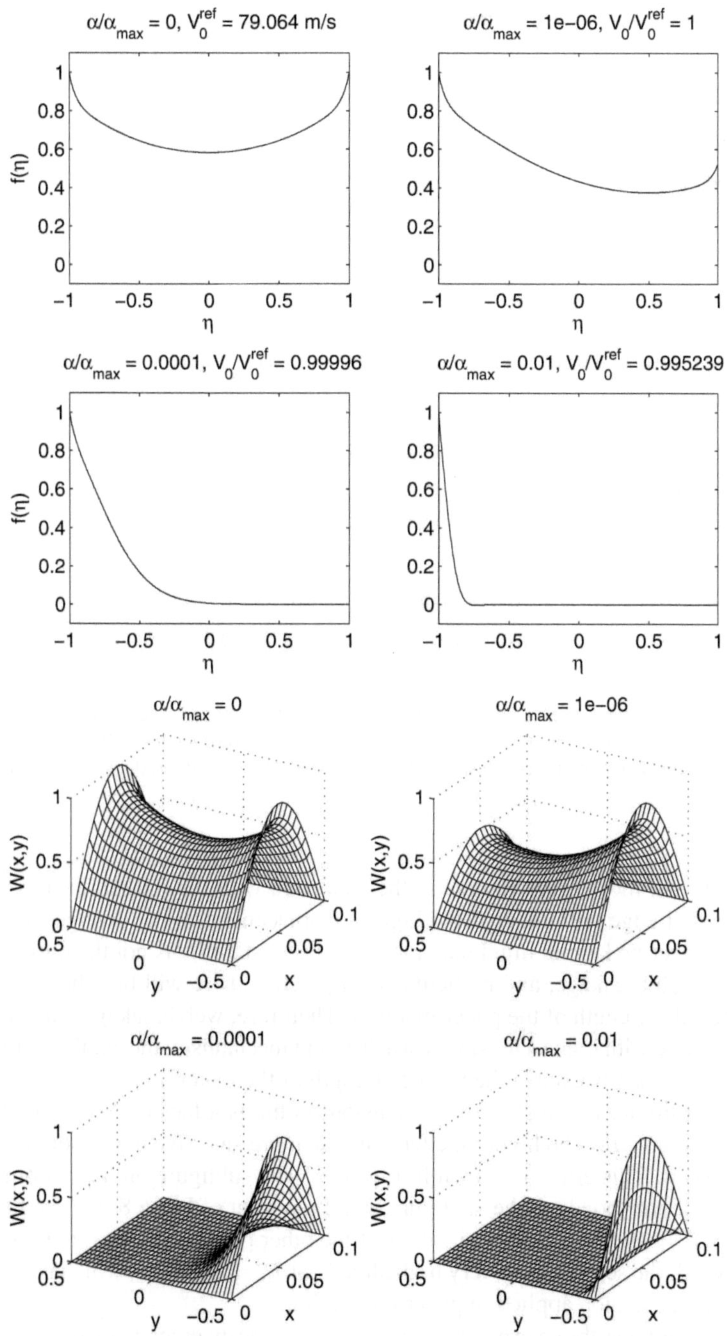

Fig. 6.8 Out-of-plane displacement of an axially travelling pinned-free plate with dimensions $\ell = 0.1$ m (length), $2b = 1$ m (width), $h = 10^{-4}$ m (thickness). Poisson ratio $\nu = 0.3$. Tension profile skew parameter $\alpha/\alpha_{\max} = 0,\ 10^{-6},\ 10^{-4},\ 10^{-2}$. Tension increases toward positive η (y in the lower subfigure). (Reproduced from Banichuk et al. [63])

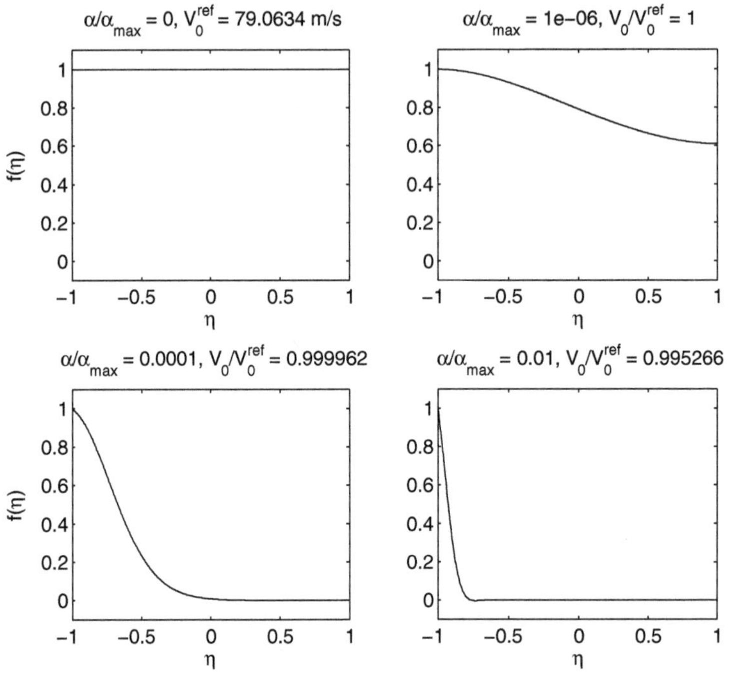

Fig. 6.9 Out-of-plane displacement of an axially travelling pinned-free plate at $x = \ell/2$ with dimensions $\ell = 0.1$ m (length), $2b = 1$ m (width), $h = 10^{-4}$ m (thickness). Poisson ratio $\nu = 0$. Tension profile skew parameter $\alpha/\alpha_{\max} = 0,\ 10^{-6},\ 10^{-4},\ 10^{-2}$. (Reproduced from Banichuk et al. [63])

the crack will most likely initiate at the free edge with the *lowest* tension. This is because that edge will undergo the highest displacement in the divergence state. It must however be kept in mind that a loss of elastic stability is not the only possible cause of web breakage, and particularly, in practice there will be inhomogeneities in the tensile strength of the paper material. Therefore, web breakage may also start from the edge with highest tension, via a different mechanism: the axial tension being applied may locally exceed the tensile strength of the material.

The sensitivity to the skew in the tension profile is affected also by the tension at the midpoint, T_0. The higher the tension, the more sensitive the system is to even a small skew. This effect is shown in Fig. 6.11. The subfigure on the bottom left of this figure corresponds to the subfigure at the top right of Fig. 6.8. We see that with $\nu = 0.3$, $\tilde{\alpha} = 10^{-6}\tilde{\alpha}_{\max}$, and the values of the other parameters fixed to those given in Table 6.3, the sensitivity is very high already at $T_0 = 500$ N/m, which is a realistic level of axial tension applied in paper production.

Finally, as far as the span geometry is concerned, the divergence shape is a function of not only the aspect ratio $\ell/2b$, but also of the overall scale. Even for the same aspect ratio, scaling ℓ (and also b to keep the same aspect ratio) changes the divergence

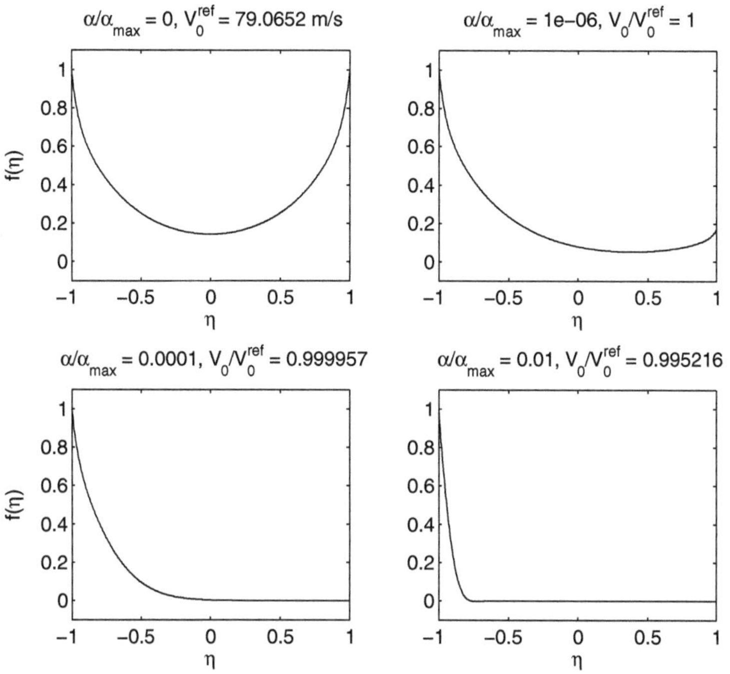

Fig. 6.10 Out-of-plane displacement of an axially travelling pinned-free plate at $x = \ell/2$ with dimensions $\ell = 0.1$ m (length), $2b = 1$ m (width), $h = 10^{-4}$ m (thickness). Poisson ratio $\nu = 0.5$. Tension profile skew parameter $\alpha/\alpha_{\max} = 0,\ 10^{-6},\ 10^{-4},\ 10^{-2}$. Tension increases toward positive η. (Reproduced from Banichuk et al. [63])

shape. This effect occurs even if h is scaled by the same amount as ℓ and b. Thus, it should be emphasized that, fully quantitatively speaking, the results in Figs. 6.8, 6.9, 6.10 and 6.11 only represent the specific case of plates with the dimensions $\ell \times 2b \times h = 0.1 \text{ m} \times 1 \text{ m} \times 10^{-4}$ m. Qualitative features of the solution of course remain rather similar even if the plate dimensions are changed.

In this chapter, we looked at the elastic stability of axially moving elastic plates, both for isotropic and orthotropic materials. Special attention was devoted to the case where the axial tension distribution varies in the width direction. In the next chapter, we will simplify the setting slightly, and concentrate on one-dimensional models of axially moving materials. We will analytically solve the free vibrations of an axially moving string with and without damping, and comment on analytical solutions for an axially moving elastic beam. We will also look at two possible extensions to the basic one-dimensional model. The first is a long (mathematically, infinite) beam traveling along a system of elastic rollers, and the second considers how a gravitational field affects the behavior of the classical traveling material.

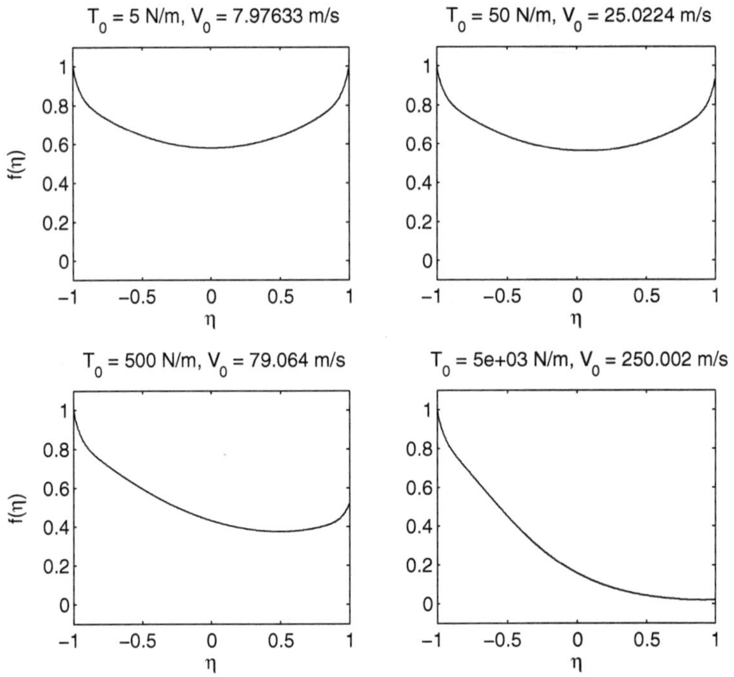

Fig. 6.11 Out-of-plane displacement of an axially travelling pinned-free plate at $x = \ell/2$ with dimensions $\ell = 0.1$ m (length), $2b = 1$ m (width), $h = 10^{-4}$ m (thickness). Poisson ratio $\nu = 0.3$, tension profile skew parameter $\alpha/\alpha_{\max} = 10^{-6}$. Tension increases toward positive η. Midpoint tension $T_0 = 5,\ 50,\ 500$ and 5000 N/m. (Reproduced from Banichuk et al. [63])

References

1. Ulsoy AG, Mote CD, Szymni R (1978) Principal developments in band saw vibration and stability research. Holz als Roh-und Werkstoff 36(7):273–280. https://doi.org/10.1007/BF02610748
2. Lighthill MJ (1960) Note on the swimming of slender fish. J Fluid Mech 9:305–317
3. Swope RD, Ames WF (1963) Vibrations of a moving threadline. J Frankl Inst 275:36–55. https://doi.org/10.1016/0016-0032(63)90619-7
4. Mote CD (1968a) Divergence buckling of an edge-loaded axially moving band. Int J Mech Sci 10:281–295. https://doi.org/10.1016/0020-7403(68)90013-1
5. Mote CD (1975) Stability of systems transporting accelerating axially moving materials. ASME J Dyn Syst Meas Control 97:96–98. https://doi.org/10.1115/1.3426880
6. Ulsoy AG, Mote CD (1980) Analysis of bandsaw vibration. Wood. Science 13:1–10
7. Ulsoy AG, Mote CD (1982) Vibration of wide band saw blades. ASME J Eng Ind 104:71–78. https://doi.org/10.1115/1.3185801
8. Lin CC, Mote CD (1995) Equilibrium displacement and stress distribution in a two-dimensional, axially moving web under transverse loading. ASME J Appl Mech 62:772–779. https://doi.org/10.1115/1.2897013
9. Kong L, Parker RG (2004) Approximate eigensolutions of axially moving beams with small flexural stiffness. J Sound Vib 276:459–469. https://doi.org/10.1016/j.jsv.2003.11.027
10. Mote CD (1972) Dynamic stability of axially moving materials. Shock Vib Dig 4(4):2–11

11. Archibald FR, Emslie AG (1958) The vibration of a string having a uniform motion along its length. ASME J Appl Mech 25:347–348
12. Simpson A (1973) Transverse modes and frequencies of beams translating between fixed end supports. J Mech Eng Sci 15:159–164. https://doi.org/10.1243/JMES_JOUR_1973_015_031_02
13. Wang Y, Huang L, Liu X (2005a) Eigenvalue and stability analysis for transverse vibrations of axially moving strings based on Hamiltonian dynamics. Acta Mech Sin 21:485–494. https://doi.org/10.1007/s10409-005-0066-2
14. Lagerstrom PA, Casten RG (1972) Basic concepts underlying singular perturbation techniques. SIAM Rev 14(1):63–120
15. Bender CM, Orszag SA (1978) Advanced mathematical methods for scientists and engineers I: asymptotic methods and perturbation theory. Springer. 1999 reprint: ISBN 978-0-387-98931-0
16. Chen L-Y, Goldenfeld N, Oono Y (1996) Renormalization group and singular perturbations: multiple scales, boundary layers, and reductive perturbation theory. Phys Rev E 54(1):376–394
17. Chonan S (1986) Steady state response of an axially moving strip subjected to a stationary lateral load. J Sound Vib 107:155–165. https://doi.org/10.1016/0022-460X(86)90290-7
18. Miranker WL (1960) The wave equation in a medium in motion. IBM J Res Dev 4:36–42. https://doi.org/10.1147/rd.41.0036
19. Wickert JA, Mote CD (1990) Classical vibration analysis of axially moving continua. ASME J Appl Mech 57:738–744. https://doi.org/10.1115/1.2897085
20. Lin CC, Mote CD (1996) Eigenvalue solutions predicting the wrinkling of rectangular webs under non-linearly distributed edge loading. J Sound Vib 197(2):179–189. https://doi.org/10.1006/jsvi.1996.0524
21. Lin CC (1997) Stability and vibration characteristics of axially moving plates. Int J Solids Struct 34(24):3179–3190. https://doi.org/10.1016/S0020-7683(96)00181-3
22. Wickert JA (1992) Non-linear vibration of a traveling tensioned beam. International Journal of Non-Linear Mechanics 27(3):503–517. https://doi.org/10.1016/0020-7462(92)90016-Z
23. Shin C, Chung J, Kim W (2005a) Dynamic characteristics of the out-of-plane vibration for an axially moving membrane. J Sound Vib 286(4–5):1019–1031. https://doi.org/10.1016/j.jsv.2005.01.013
24. Shen JY, Sharpe L, McGinley WM (1995) Identification of dynamic properties of plate-like structures by using a continuum model. Mech Res Commun 22(1):67–78. https://doi.org/10.1016/0093-6413(94)00043-D
25. Kulachenko A, Gradin P, Koivurova H (2007a) Modelling the dynamical behaviour of a paper web. Part I. Comput Struct 85:131–147. https://doi.org/10.1016/j.compstruc.2006.09.006
26. Kulachenko A, Gradin P, Koivurova H (2007b) Modelling the dynamical behaviour of a paper web. Part II. Comput Struct 85:148–157. https://doi.org/10.1016/j.compstruc.2006.09.007
27. Wang X (2003b) Instability analysis of some fluid-structure interaction problems. Comput Fluids 32(1):121–138. https://doi.org/10.1016/S0045-7930(01)00103-7
28. Sygulski R (2007) Stability of membrane in low subsonic flow. Int J Non-Linear Mech 42(1):196–202. https://doi.org/10.1016/j.ijnonlinmec.2006.11.012
29. Huber MT (1914) Die Grundlagen einer rationellen Berechnung der kreuzweise bewehrten Eisenbetonplatten. Zeitschrift der Österreichische Ingeniur-und Architekten-Vereines 30:557–564
30. Huber MT (1923) Die Theorie des kreuzweise bewehrten Eisenbetonplatten. Der Bauingenieur 4:354–392
31. Huber MT (1926) Einige Andwendungen fer Biegungstheorie orthotroper Platten. Zeitschrift für angewandte Mathematik und Mechanik 6(3):228–232
32. Göttsching L, Baumgarten HL (1976) Triaxial deformation of paper under tensile load. In: The fundamental properties of paper related to its uses, vol. 1, pp 227–252. Technical division of the British Paper and board industry federation
33. Mann RW, Baum GA, Habeger CC (1980) Determination of all nine orthotropic elastic constants for machine-made paper. TAPPI J 63(2):163–166

34. Baum GA, Brennan DC, Habeger CC (1981) Orthotropic elastic constants of paper. TAPPI J 64(8):97–101
35. Thorpe JL (1981) Paper as an orthotropic thin plate. TAPPI J 64(3):119–121
36. Skowronski J, Robertson AA (1985) A phenomenological study of the tensile deformation properties of paper. J Pulp Pap Sci 11(1):J21–J28
37. Seo YB (1999) Determination of in-plane shear properties by an off-axis tension method and laser speckle photography. J Pulp Pap Sci 25(9):321–325
38. Erkkilä A-L, Leppänen T, Hämäläinen J (2013b) Empirical plasticity models applied for paper sheets having different anisotropy and dry solids content levels. Int J Solids Struct 50(14–15):2151–2179. https://doi.org/10.1016/j.ijsolstr.2013.03.004
39. Stenberg N, Fellers C (2002) Out-of-plane Poisson's ratios of paper and paperboard. Nord Pulp Pap Res J 17(4):387–394
40. Alava M, Niskanen K (2006) The physics of paper. Rep Prog Phys 69(3):669–723
41. Itskov M, Aksel N (2002) Elastic constants and their admissible values for incompressible and slightly compressible anisotropic materials. Acta Mech 157:81–96
42. Kurki M, Jeronen J, Saksa T, Tuovinen T (2016) The origin of in-plane stresses in axially moving orthotropic continua. Int J Solids Struct. https://doi.org/10.1016/j.ijsolstr.2015.10.027
43. Gorman DJ (1982) Free vibration analysis of rectangular plates. Elsevier North Holland, Inc. ISBN 0-444-00601-X
44. Biancolini ME, Brutti C, Reccia L (2005) Approximate solution for free vibrations of thin orthotropic rectangular plates. J Sound Vib 288(1–2):321–344. ISSN 0022-460X. https://doi.org/10.1016/j.jsv.2005.01.005
45. Xing Y, Liu B (2009a) New exact solutions for free vibrations of rectangular thin plates by symplectic dual method. Acta Mech Sin 25:265–270
46. Kshirsagar S, Bhaskar K (2008) Accurate and elegant free vibration and buckling studies of orthotropic rectangular plates using untruncated infinite series. J Sound Vib 314(3–5):837–850. https://doi.org/10.1016/j.jsv.2008.01.013. ISSN 0022-460X
47. Hatami S, Azhari M, Saadatpour MM, Memarzadeh P (2009) Exact free vibration of webs moving axially at high speed. In: AMATH'09: Proceedings of the 15th American Conference on Applied Mathematics, pp. 134–139, Stevens Point, Wisconsin, USA. World Scientific and Engineering Academy and Society (WSEAS). ISBN 978-960-474-071-0. Houston, USA
48. Marynowski K (2008) Dynamics of the Axially Moving Orthotropic Web, vol 38. Lecture Notes in Applied and Computational Mechanics. Springer, Germany. https://doi.org/10.1007/978-3-540-78989-5. ISBN 978-3-540-78988-8 (print), 978-3-540-78989-5 (online)
49. Banichuk N, Jeronen J, Neittaanmäki P, Saksa T, Tuovinen T (2014) Mechanics of moving materials, vol 207. Solid mechanics and its applications. Springer. ISBN: 978-3-319-01744-0 (print), 978-3-319-01745-7 (electronic)
50. Koivurova H, Salonen E-M (1999) Comments on non-linear formulations for travelling string and beam problems. J Sound Vib 225(5):845–856
51. Timoshenko SP, Woinowsky-Krieger S (1959) Theory of plates and shells, 2nd edn. New York, Tokyo, McGraw-Hill. ISBN 0-07-085820-9
52. Kikuchi N (1986) Finite element methods in mechanics. Cambridge University Press, Cambridge, UK. ISBN 0 521 33972 3
53. Sagan H (1989) Boundary and eigenvalue problems in mathematical physics. Wiley, Inc. Slightly corrected reprint by Dover Publications, Inc
54. Weinstock R (2008) *Calculus of variations: with applications to physics and engineering*. Weinstock Press. ISBN 978-1443728812. Reprint of 1974 edition
55. Timoshenko S, Goodier J (1951) Theory of elasticity, 2nd edn. McGraw–Hill
56. Allen MB, Herrera I, Pinder GF (1988) Numerical modeling in science and engineering. Wiley Interscience
57. Gorman DJ, Singhal RK (1993) A superposition-rayeigh-ritz method for free vibration analysis of non-uniformly tensioned membranes. J Sound Vib 162(3):489–501
58. Bolotin VV (1963) Nonconservative problems of the theory of elastic stability. Pergamon Press, New York

59. Chang YB, Moretti PM (1991) Interaction of fluttering webs with surrounding air. TAPPI J 74(3):231–236
60. Yokoyama T, Nakai K (2007) Evaluation of in-plane orthotropic elastic constants of paper and paperboard. In 2007 SEM Annual Conference & Exposition on Experimental and Applied Mechanics
61. Bonnin A, Huchon R, Deschamps M (2000) Ultrasonic waves propagation in absorbing thin plates: application to paper characterization. Ultrasonics 37(8):555–563. https://doi.org/10.1016/S0041-624X(99)00106-7. ISSN 0041-624X
62. Banichuk N, Jeronen J, Kurki M, Neittaanmäki P, Saksa T, Tuovinen T (2011) On the limit velocity and buckling phenomena of axially moving orthotropic membranes and plates. Int J Solids Struct 48(13):2015–2025. https://doi.org/10.1016/j.ijsolstr.2011.03.010. ISSN 0020-7683
63. Banichuk N, Jeronen J, Neittaanmäki P, Saksa T, Tuovinen T (2013d) Theoretical study on travelling web dynamics and instability under non-homogeneous tension. Int J Mech Sci 66C:132–140. https://doi.org/10.1016/j.ijmecsci.2012.10.014
64. Press WH, Teukolsky SA, Vetterling WT, Flannery BP (2007) Numerical recipes: the art of scientific computing, 3rd edn. Cambridge University Press, New York

Chapter 7
Stability of Axially Moving Strings, Beams and Panels

In this chapter, problems of dynamics and stability of a moving web, modelled as an elastic rod or string, and axially traveling between rollers (supports) at a constant velocity, are studied using analytical approaches. In the first two sections, we consider the transverse, longitudinal and torsional vibrations of the moving web using a hyperbolic second-order partial differential equation, corresponding to the string and rod models. It is shown that in the framework of a quasistatic eigenvalue analysis, for these models, the critical point is always stable. The critical velocities of one-dimensional webs, and the arising nontrivial solution of free vibrations, are studied analytically. The analysis is then extended into the case with damping. The critical points of both static and dynamic types are found analytically. It is observed that if external friction is present, then for mode numbers sufficiently high, dynamic critical points may exist. Graphical examples of eigenvalue spectra are given for both the ideal (undamped) and damped systems. In the examples, it is seen that external friction leads to stabilization, whereas internal friction in the traveling material will destabilize the system in a dynamic mode at the static critical point. The last three sections of this chapter concentrate on a fourth-order model, where we include the bending rigidity. We consider exact eigensolutions for axially moving beams and panels, the effect of elastic supports at the boundaries to the vibration behavior on the beam (based on Banichuk et al. [1]), and the stability of the traveling beam in a gravitational field. The theory and results summarize and extend theoretical knowledge of the class of models studied, and can be used in various applications of moving materials, such as paper making processes.

N. Banichuk et al., *Stability of Axially Moving Materials*, Solid Mechanics and Its Applications 259, https://doi.org/10.1007/978-3-030-23803-2_7

7.1 Unified Model and Exact Eigensolutions for Torsional, Longitudinal and Transverse Vibration Types

Consider a narrow thin web supported by rollers at $x = 0$ and $x = \ell$ (see Fig. 7.1), modelled by a continuous one-dimensional elastic element (rod or) string, having rectangular cross-section of given width and thickness.

At first, consider a small amplitude torsional vibration of the web, described by the angle function $\vartheta(x, t)$, which represents the angle of torsion per unit length of the rod in the segment $0 \le x \le \ell$. The dynamics of free torsional small-amplitude vibrations of a classical, stationary (nontraveling) ideal elastic rod is governed by the partial differential equation

$$\rho I_0 \frac{\partial^2 \vartheta}{\partial t^2} - G I_k \frac{\partial^2 \vartheta}{\partial x^2} = 0 \,, \quad 0 < x < \ell \tag{7.1.1}$$

with boundary conditions

$$(\vartheta)_{x=0} = 0 \,, \quad (\vartheta)_{x=\ell} = 0 \,, \tag{7.1.2}$$

where ρ is the material density, I_0 the moment of inertia of the web cross-section around the x-axis, GI_k the torsional rigidity, G the shear modulus, I_k the polar moment of inertia, and $\vartheta = \vartheta(x, t)$ the angle of torsion.

Consider now an ideal straight rod, which moves in the x direction and performs vibrations torsional. The rod travels at a constant velocity V_0 between supports, which are fixed at $x = 0$ and $x = \ell$. We will set up the problem for the moving rod using the Euler laboratory coordinate system (x, t), see Fig. 7.1, and the axially co-moving

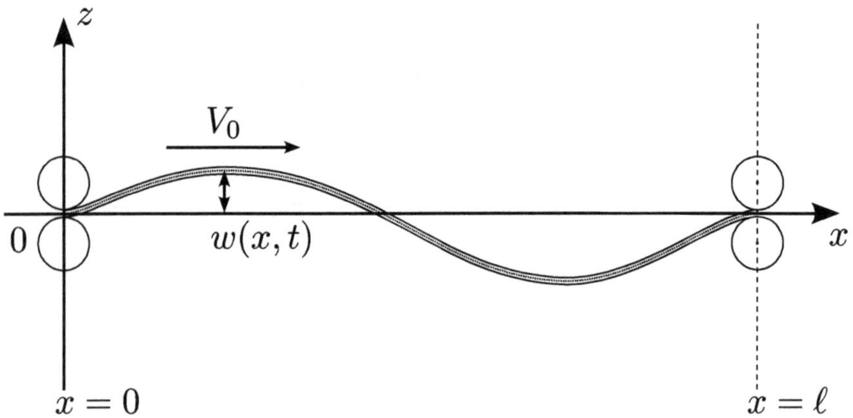

Fig. 7.1 Thin axially moving narrow web modelled as one-dimensional elastic element (rod or string) and supported at $x = 0$ and $x = \ell$. Here the transverse displacement w is shown

coordinate system $(\tilde{x}, t)$, and the notion of the material derivative (also known as the Lagrange derivative or the total derivative). We have

$$x = \tilde{x} + V_0 t \,, \quad \frac{\partial \vartheta\,(x(\tilde{x}, t)\,,\, t\,)}{\partial \tilde{x}} = \frac{\partial \vartheta}{\partial x}\frac{\partial x}{\partial \tilde{x}} = \frac{\partial \vartheta}{\partial x} \,, \tag{7.1.3}$$

$$\frac{\mathrm{d}\vartheta}{\mathrm{d}t} \equiv \left(\frac{\partial \vartheta}{\partial t}\right)_{\tilde{x}=\mathrm{const.}} = \frac{\partial \vartheta}{\partial t} + V_0 \frac{\partial \vartheta}{\partial x} \,, \tag{7.1.4}$$

$$\frac{\mathrm{d}^2\vartheta}{\mathrm{d}t^2} \equiv \left(\frac{\partial}{\partial t} + V_0 \frac{\partial}{\partial x}\right)\left(\frac{\partial \vartheta}{\partial t} + V_0 \frac{\partial \vartheta}{\partial x}\right) = \frac{\partial^2 \vartheta}{\partial t^2} + 2V_0 \frac{\partial^2 \vartheta}{\partial x \partial t} + V_0^2 \frac{\partial^2 \vartheta}{\partial x^2} \,. \tag{7.1.5}$$

In the Eqs. (7.1.4) and (7.1.5), and also in the rest of this chapter, $\partial/\partial t$ without subscript denotes partial differentiation with respect to time in the Euler coordinate system, that is, $\partial/\partial t \equiv (\partial/\partial t)_{x=\mathrm{const.}}$.

Applying transformations (7.1.3)–(7.1.5), we obtain the equation of small torsional vibrations of an axially traveling rod

$$\rho I_0 \frac{\partial^2 \vartheta}{\partial t^2} + 2\rho I_0 V_0 \frac{\partial^2 \vartheta}{\partial x \partial t} + \rho I_0 V_0^2 \frac{\partial^2 \vartheta}{\partial x^2} - G I_k \frac{\partial^2 \vartheta}{\partial x^2} = 0 \,. \tag{7.1.6}$$

The boundary conditions are those stated in the Eq. (7.1.2), that is,

$$(\vartheta)_{x=0} = 0 \,, \quad (\vartheta)_{x=\ell} = 0 \,. \tag{7.1.7}$$

Vibrations can also arise in the longitudinal direction. If the traveling web performs small-amplitude longitudinal vibrations (independent of the torsional vibrations and superposed onto the axial traveling motion), then the dynamics of the rod can be described by the analogous equation and boundary conditions

$$\rho S \frac{\partial^2 u}{\partial t^2} + 2\rho S V_0 \frac{\partial^2 u}{\partial x \partial t} + \rho S V_0^2 \frac{\partial^2 u}{\partial x^2} - E S \frac{\partial^2 u}{\partial x^2} = 0 \,, \tag{7.1.8}$$

$$(u)_{x=0} = 0 \,, \quad (u)_{x=\ell} = 0 \,, \tag{7.1.9}$$

where S is the cross-sectional area of the web, E the Young modulus of the material, and $u = u(x, t)$ the longitudinal displacement. Defined in the Euler coordinate system, the value $u(x, t)$ describes the longitudinal displacement, at time instant t, that the web experiences at laboratory coordinate x, compared to the reference state where the only axial motion is the uniform translation at velocity V_0.

The third and perhaps the most often considered type of small-amplitude vibrations of an axially moving elastic web are transverse vibrations, described by the transverse displacement function $w = w(x, t)$, which is governed by the following second-order partial differential equation and boundary conditions

Table 7.1 Definitions of state variable U, coefficients a, b and c, and critical velocities C, in the three considered cases. The coefficients and critical velocities in each case satisfy the relation $b^2 - ac = a^2C^2$

Vibration type	Equations	U	a	b	c	C
Torsional	(7.1.6)	ϑ	ρI_0	$\rho I_0 V_0$	$\rho I_0 V_0^2 - GI_k$	$\sqrt{GI_k/\rho I_0}$
Longitudinal	(7.1.8)	u	ρS	$\rho S V_0$	$\rho S V_0^2 - ES$	$\sqrt{E/\rho}$
Transverse	(7.1.10)	w	m	$m V_0$	$m V_0^2 - T_0$	$\sqrt{T_0/m}$

$$m\frac{\partial^2 w}{\partial t^2} + 2m V_0 \frac{\partial^2 w}{\partial x \partial t} + m V_0^2 \frac{\partial^2 w}{\partial x^2} - T_0 \frac{\partial^2 w}{\partial x^2} = 0 \,, \tag{7.1.10}$$

$$(w)_{x=0} = 0 \,, \quad (w)_{x=\ell} = 0 \,. \tag{7.1.11}$$

Here $m \equiv \rho S$ is the mass per unit length of the web. This is the classical one-dimensional string model of the elastic web, axially moving in the x direction, subjected to a constant tensile load $T_0 > 0$, and having zero bending rigidity. This model was first investigated by Skutch [2]. Note the string model ignores bending rigidity; the corresponding system with finite bending rigidity (a beam or a panel) will be discussed in Sect. 7.3.

The first three terms in (7.1.10) come from the second material derivative (7.1.5), and the term $T_0 \partial^2 w/\partial x^2$ represents the restoring force of the axial tension. As was noted in Chap. 6, the first three terms can be loosely interpreted to physically represent, respectively, the accelerations of local inertia, the Coriolis effect, and the centrifugal effect.

All three cases (7.1.6), (7.1.8) and (7.1.10) are of the same mathematical form. It is convenient to write them as a general second-order constant-coefficient partial differential equation

$$a\frac{\partial^2 U}{\partial t^2} + 2b\frac{\partial^2 U}{\partial x \partial t} + c\frac{\partial^2 U}{\partial x^2} = 0 \,, \quad 0 < x < \ell \,. \tag{7.1.12}$$

See Table 7.1 for the definitions of U, a, b and c in each particular case.

Because the coefficients are constants, the type of the equation (elliptic, parabolic or hyperbolic) holds globally regardless of x and t. The discriminants

$$D \equiv b^2 - ac \tag{7.1.13}$$

are, respectively,

$$D_\vartheta = \rho I_0 G I_k > 0 \,, \quad D_u = \rho S^2 E > 0 \,, \quad D_w = m T_0 > 0 \,, \tag{7.1.14}$$

and hence the Eqs. (7.1.6), (7.1.8) and (7.1.10) are always hyperbolic regardless of the value of axial velocity V_0. This observation reflects the physical nature of the problem: the introduction of axial velocity should not change the basic vibrational nature of the mechanical response. Indeed, in the co-moving (Lagrange) coordinate system, the mechanics is identical to that of the stationary string or rod subjected to moving boundaries.

Equation (7.1.12) requires two initial and two boundary conditions. In the following free vibration analysis, we set only the boundary conditions, and the solution will have two free parameters. Free vibration analysis is only concerned with determining possible motions of the unloaded system, in other words,. the nontrivial solutions of the homogeneous partial differential equation (7.1.12).

The considered boundary conditions (7.1.2), (7.1.9) and (7.1.11) are, in each case, zero Dirichlet:

$$(U)_{x=0} = 0\,, \quad (U)_{x=\ell} = 0\,. \tag{7.1.15}$$

Now we are ready to consider the mechanical response at the critical velocity. From Table 7.1, we observe that for the problems studied here, the coefficients a and b in (7.1.12) are always positive, but the sign of c is indeterminate. The special case where the coefficient c vanishes is of special interest. If the other problem parameters are considered fixed, this can always be realized by choosing a particular value for V_0 such that the two different contributions to the coefficient c cancel out.

In the following, we will call this special value of V_0 the critical velocity, denoted with the symbol C. See the corresponding column in Table 7.1 for the critical velocities in the different cases.

At the critical velocity, $V_0 = C$, Eq. (7.1.12) simplifies into (after dividing by a and noting $b/a = V_0$)

$$\frac{\partial^2 U}{\partial t^2} + 2C\frac{\partial^2 U}{\partial x \partial t} = 0\,. \tag{7.1.16}$$

The Eq. (7.1.16) is still globally hyperbolic, since $C^2 > 0$. Because C is a constant, Eq. (7.1.16) factors as

$$\frac{\partial}{\partial t}\left(\frac{\partial U}{\partial t} + 2C\frac{\partial U}{\partial x}\right) = 0\,. \tag{7.1.17}$$

Introducing the notation

$$v \equiv \frac{\partial U}{\partial t}\,, \tag{7.1.18}$$

Equation (7.1.17) can be rewritten as

$$\frac{\partial v}{\partial t} + 2C\frac{\partial v}{\partial x} = 0\,. \tag{7.1.19}$$

Thus, the velocity-like quantity v obeys the (homogeneous) first-order transport equation along the rod or string, with the transport of this quantity occurring at constant velocity $2C$. Hence, we have

$$v(x,t) = g(x - 2Ct) \tag{7.1.20}$$

for some differentiable function g. The function $U(x, t)$ is then

$$U(x,t) = f(x) + h(x - 2Ct) . \tag{7.1.21}$$

with some function $f(x)$ that is constant in time, and a differentiable function $h(x - 2Ct)$ related to $g(x - 2Ct)$ by the equation

$$v = \frac{\partial U}{\partial t} = (-2C)h'(x - 2Ct) \equiv g(x - 2Ct) . \tag{7.1.22}$$

The boundary conditions (7.1.15) imply, by (7.1.18), that

$$(v)_{x=0} = 0 , \quad (v)_{x=\ell} = 0 \tag{7.1.23}$$

for all t. Hence, in order to avoid violation of the boundary conditions, we must have $v(x, t) = g(x - 2Ct) \equiv 0$, and the considered rod or string, if traveling at the critical velocity, must stay in a steady-state configuration. This very compact way to solve the critical velocity special case was discussed by Wang et al. [3].

Furthermore, it is possible to find the function $f(x)$ explicitly via a direct approach. First, let us integrate Eq. (7.1.17) with respect to t, obtaining

$$\frac{\partial U}{\partial t} + 2C\frac{\partial U}{\partial x} = h(x) . \tag{7.1.24}$$

Equation (7.1.24) is the standard nonhomogeneous transport equation. Its solution is Polyanin et al. [4]

$$U(x,t) = \frac{1}{2C}\int h(x)\,\mathrm{d}x + g(x - 2Ct) . \tag{7.1.25}$$

As above, this is a linear superposition of two components: a steady-state one, and one being transported toward the $+x$ direction at velocity $2C$. Following the same argument regarding boundary conditions as above, we find $g(x - 2Ct) \equiv 0$. By differentiating (7.1.25) with respect to x, and then setting $t = 0$, we determine

$$h(x) = 2C\frac{\partial U}{\partial x}(x, 0) \tag{7.1.26}$$

and finally, by substituting (7.1.26) back into (7.1.25), obtain

$$U(x,t) = U(x,0) . \tag{7.1.27}$$

Hence, when the axial motion occurs at the critical velocity C given in Table 7.1, the initial condition for the position, namely

$$(U)_{t=0} = f_1(x) \tag{7.1.28}$$

(where $f_1(x)$ is a given function), completely determines the solution for all t. Because the solution (7.1.27) does not allow for time-dependent changes, the other initial condition, namely

$$(v)_{t=0} \equiv \left(\frac{\partial U}{\partial t}\right)_{t=0} = f_2(x)\ , \tag{7.1.29}$$

must have $f_2(x) \equiv 0$ in order to be compatible with the solution (7.1.27).

The conclusion is that if the compatibility condition $(v)_{t=0} \equiv 0,\ 0 < x < \ell$ holds for our initial conditions, then, upon a quasistatic transition to the limit state $V_0 = C$, the state variable profile of a freely vibrating axially traveling elastic element (rod or string) will "freeze" into the shape it had when the limit state was reached. By *state variable profile* we mean the function $U(x, t)$, describing the state U as seen in the laboratory coordinates (that is, by a stationary observer). The material still undergoes axial translation, but this profile will remain static in time. The observer sees a steady-state motion, akin to a steady-state fluid flow.

In practice, the compatibility condition $(v)_{t=0} \equiv 0,\ 0 < x < \ell$ is reasonable for many physically admissible situations for the considered model near $V_0 = C$, because as we will see below, all eigenfrequencies of the traveling elastic element tend to zero in the limit $V_0 \to C$.

Nevertheless, the quasistatic analysis has its limitations. For example, consider the case where we would like to initially set $V_0 = C - \varepsilon$ (with $\varepsilon > 0$ small), $(U)_{t=0} \equiv 0$, $(v)_{t=0} = f_2(x) \not\equiv 0$, and then perform a transition to $V_0 \to C$. The quasistatic analysis is not applicable, because the given initial condition for v violates the compatibility condition $f_2(x) \equiv 0$. If this category of cases is to be analyzed, a more general treatment of the dynamics including the effects of accelerating motion is required.

In next we will continue for the transformation of equations into canonical form. For the rest of the discussion in this section, we will concentrate on the general case $V_0 \neq C$. Let us briefly go through the derivation of an analytical solution for the free vibrations of the considered traveling elastic element. We will proceed in a manner similar to Swope and Ames [5].

Consider Eq. (7.1.12). A systematic way to derive the solution is to diagonalize the principal part of the operator (see, e.g., [6, Chap. 2], [7, Chap. 5], or [8] and references therein). Because Eq. (7.1.12) contains only second-order derivatives, we have only the principal part to consider.

It is known that the second-order partial differential equation in two variables can always be transformed into one of the canonical forms, depending on its type. We start with the characteristic equation of (7.1.12), namely ([6, Chap. 2])

$$a\,(\mathrm{d}x)^2 - 2b\mathrm{d}x\mathrm{d}t + c\,(\mathrm{d}t)^2 = 0\ . \tag{7.1.30}$$

The coefficients a, b and c are constants. Because the discriminant (7.1.13) is positive for the problems under consideration, and $c \neq 0$ (because $V_0 \neq C$), Eq. (7.1.30) can be understood as a second-order algebraic equation in the variable $\mathrm{d}t/\mathrm{d}x$, with the real-valued solutions

$$\frac{\mathrm{d}t}{\mathrm{d}x} = \frac{1}{c}\left(b \pm \sqrt{b^2 - ac}\right) . \tag{7.1.31}$$

From (7.1.31), we formally obtain

$$c\mathrm{d}t - \left(b \pm \sqrt{b^2 - ac}\right)\mathrm{d}x = 0 , \tag{7.1.32}$$

and by integrating both sides,

$$\varphi_\pm(x, t) \equiv ct - \left(b \pm \sqrt{b^2 - ac}\right)x = \kappa_\pm \tag{7.1.33}$$

for the $+$ and $-$ terms, respectively. Assigning a value to the constants $\kappa_\pm$ picks one specific curve from the family of characteristics; Eq. (7.1.33) represents the whole family. It is seen that the characteristics $\varphi_\pm$ of Eq. (7.1.12) are straight lines, as expected.

Performing a change of variables (leaving $\kappa_\pm$ free, taking only the left-hand side of (7.1.33))

$$X \equiv \varphi_+(x, t) , \quad Y \equiv \varphi_-(x, t) , \tag{7.1.34}$$

the original Eq. (7.1.12) transforms into the first canonical form of the wave equation (i.e., hyperbolic second-order partial differential equation),

$$\frac{\partial^2 U}{\partial X \partial Y} = 0 . \tag{7.1.35}$$

Changing variables again, now to ξ and η such that

$$X = \xi - \eta , \quad Y = \xi + \eta , \tag{7.1.36}$$

that is,

$$\xi = \frac{1}{2}(X - Y) , \quad \eta = \frac{1}{2}(X + Y) , \tag{7.1.37}$$

we arrive at the second canonical form of the wave equation,

$$\frac{\partial^2 U}{\partial \eta^2} - \frac{\partial^2 U}{\partial \xi^2} = 0 . \tag{7.1.38}$$

The mixed derivative has been eliminated. The form (7.1.38) is particularly convenient, because it admits a separable solution in the form $U(\xi, \eta) \equiv g_1(\xi)g_2(\eta)$. The standard separation technique can be used to carry out the rest of the solution process; however, there is a more compact approach that we will use below.

We will next derive the analytical solution of free vibrations of a traveling elastic element. Let us use the nondimensional variables

$$\hat{x} = x/\lambda \,, \quad \hat{t} = t/\tau \,, \quad \hat{U} = U/\delta \,, \tag{7.1.39}$$

where the hat indicates a nondimensional quantity. Here λ is a characteristic length, τ is a characteristic time, and δ is a characteristic value of the state variable U. The characteristic values are arbitrary, and are usually chosen in some convenient manner for each problem under discussion. A typical choice for the coordinate scalings is $\lambda = \ell$, $\tau = \ell/C$, but for the moment, we will leave these scalings free to be chosen later.

As for the state variable $\hat{U}$, we may insert $U = \hat{U}\delta$ into (7.1.12), and cancel δ, since the equation is linear in U and the right-hand side is zero. Hence δ can be dropped from further consideration.

Let us insert the nondimensional variables (7.1.39) into (7.1.12), and multiply the equation by τ^2/a. We will omit the hat from the notation. We obtain another equation of the form (7.1.12),

$$\alpha \frac{\partial^2 U}{\partial t^2} + 2\beta \frac{\partial^2 U}{\partial x \partial t} + \gamma \frac{\partial^2 U}{\partial x^2} = 0 \,, \quad 0 < x < 1 \,, \tag{7.1.40}$$

where U, x, and t are now the nondimensional variables (7.1.39), and the constant coefficients are

$$\alpha = 1 \,, \quad \beta = \frac{\tau b}{\lambda a} = \frac{\tau}{\lambda} V_0 \,, \quad \gamma = \frac{\tau^2 c}{\lambda^2 a} = \frac{\tau^2}{\lambda^2} \left(V_0^2 - C^2 \right) \,. \tag{7.1.41}$$

The last forms follow from Table 7.1. For the nondimensional coordinates x and t, we have the coordinate transforms

$$\xi = -\left(\sqrt{\beta^2 - \gamma}\right) x \,, \quad \eta = \gamma t - \beta x \,, \tag{7.1.42}$$

where (7.1.34)–(7.1.37) and $\alpha = 1$ have been used. The transforms (7.1.42) bring (7.1.40) directly into the form (7.1.38). Lines of constant x transform into lines of constant ξ, so a suitable separable solution in (ξ, η) coordinates will satisfy boundary conditions at constant x.

It should be mentioned that this solution approach is only useful for free vibration analysis, because initial conditions cannot be easily enforced. This is due to the fact that η is a linear combination of x and t, and hence lines of constant η are not parallel to either of the axes in the original (x, t) coordinate system. See Fig. 7.2.

A compact way to find the free vibration solution of (7.1.38) in the transformed (ξ, η) coordinates is to use the standard complex-valued time-harmonic trial function

$$U(\xi, \eta) = \exp(s\eta) W(\xi) \,. \tag{7.1.43}$$

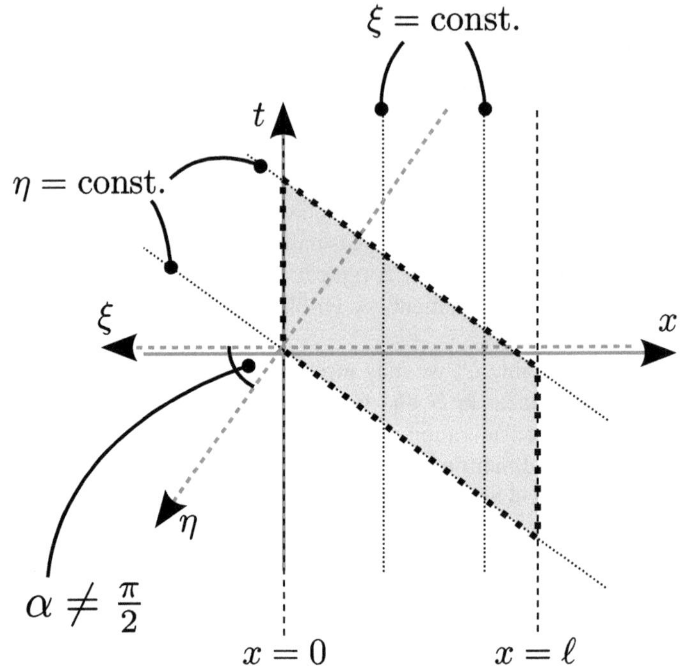

Fig. 7.2 The (x, t) and (ξ, η) coordinate systems in the plane. The shaded parallelogram indicates an area delimited by constant values of the (ξ, η) coordinates. On the edges of such shapes, boundary conditions (but not initial conditions) can be easily enforced by a solution separable in (ξ, η)

Substituting (7.1.43) into (7.1.38) will result in

$$s^2 W - W'' = 0\ , \quad \text{i.e.,,} \quad W'' = s^2 W\ . \tag{7.1.44}$$

This is a linear eigenvalue problem for (s^2, W), where s^2 is the eigenvalue and $W \equiv W(\xi)$ is the eigenfunction. Considering (7.1.44) and the corresponding boundary conditions, we see that the eigenfunction must be

$$W(\xi) = \sin(\omega \xi)\ . \tag{7.1.45}$$

To obtain ω, recall the coordinate transform (7.1.42) between x and ξ. Observe that

$$\beta^2 - \gamma = \frac{\tau^2}{\lambda^2 a^2}\left(b^2 - ac\right)\ , \quad b^2 - ac = a^2 C^2\ , \tag{7.1.46}$$

which leads to

$$\sqrt{\beta^2 - \gamma} = \frac{C\tau}{\lambda}\ , \quad \xi = -\frac{C\tau}{\lambda} x\ . \tag{7.1.47}$$

In order for (7.1.45) to satisfy the boundary conditions, for dimensional $x = \ell$, that is, nondimensional $x = \ell/\lambda$, we must have $\omega\xi = k\pi$, where k is a free nonzero integer. In other words, for the kth vibration mode it must hold that

$$\omega = -\frac{k\pi}{\left(\sqrt{\beta^2 - \gamma}\right) \cdot (\ell/\lambda)} = -\frac{k\pi\lambda}{C\tau} \cdot \frac{\lambda}{\ell} = -\frac{k\pi\ell}{C\tau} \,. \tag{7.1.48}$$

The last form is obtained by choosing the characteristic length as $\lambda = \ell$. Furthermore, in order to satisfy the partial differential equation (7.1.38), we must have

$$s^2 = -\omega^2 \,, \quad \text{i.e.,,} \quad s = \pm \mathrm{i}\omega \,. \tag{7.1.49}$$

Thus, the complex-valued solution for the kth mode, expressed in (ξ, η) coordinates is

$$U(\xi, \eta) = \exp(s_k \eta) \sin(\omega_k \xi) \,, \tag{7.1.50}$$

where the subscript k denotes the dependence on the mode number k of both the exponent $s = s_k$ and the mode shape parameter $\omega = \omega_k$.

By transforming back using (7.1.49), (7.1.48) and (7.1.42), we obtain the solution in the nondimensional $(x,\ t)$ coordinates. To make the solution easier to follow, note that by using (7.1.41) and $\lambda = \ell$, Eq. (7.1.40) can be rewritten as

$$\frac{\partial^2 U}{\partial t^2} + 2V_0 \frac{\tau}{\ell} \frac{\partial^2 U}{\partial x \partial t} + (V_0^2 - C^2) \frac{\tau^2}{\ell^2} \frac{\partial^2 U}{\partial x^2} = 0 \,, \quad 0 < x < 1 \,. \tag{7.1.51}$$

The complex-valued solution for the kth vibration mode, with the considered zero Dirichlet boundary conditions at $x = 0$ and $x = 1$, is

$$U(x, t) = E \exp\left(\pm \mathrm{i} \frac{k\pi\ell}{C\tau} \left[\left(V_0^2 - C^2\right) \frac{\tau^2}{\ell^2} t - V_0 \frac{\tau}{\ell} x \right] \right) \sin(k\pi x) \,, \tag{7.1.52}$$

where the $\pm$ corresponds to the $\pm$ in (7.1.49), $\mathrm{i} \equiv \sqrt{-1}$, and the constant coefficient E is free. As will be observed below, the real and imaginary parts of (7.1.52) are real-valued solutions.

It is seen that there is an exp factor in the solution (7.1.52), and it admits separation in x and t. That is, $U(x, t) = \chi_1(x)\chi_2(t)$ for some functions χ_1 and χ_2; specifically, the form of both functions is exp.

Thus, Eq. (7.1.52) represents time-harmonic behavior, and we can apply Bolotin's approach for determining the dynamic stability (see [9]). The stability exponent of the kth mode is the coefficient of t in the exp. For the kth vibration mode, we have

$$s_k^* = \pm \mathrm{i} \frac{k\pi\tau}{C\ell} \left(V_0^2 - C^2\right) \,. \tag{7.1.53}$$

We see that if $V_0 \to C$ (or $V_0 \to -C$), $s_k^* \to 0$ for all k. Hence the free vibration solution agrees with the steady state solution obtained for the special case.

For all $V_0 \neq C$, the stability exponent s_k^* is always pure imaginary. Thus, the axially traveling elastic element will undergo undamped harmonic vibrations, as expected. The fact that this is the case also for any $|V_0| > C$ suggests that, at least when judged by the initial postcritical behavior, the critical points $V_0 = \pm C$ for the considered models are neutrally stable.

The star in the symbol s_k^* is a reminder of the fact that (7.1.53) is expressed in the original (nondimensional) (x, t) coordinate system, instead of the transformed (ξ, η) coordinates used in (7.1.49). The exponent s_k^* is the one that determines the elastic stability behavior. Because $\eta = \gamma t - \beta x$, and no additional transformations were needed to produce the solution $U(x, t)$, the relation between the stability exponents in the two coordinate systems is simply $s_k^* = \gamma s_k$.

Now consider the problem (7.1.12), (7.1.15), augmented with some initial conditions matching a free vibration solution, as an initial boundary value problem. Because a purely time-harmonic vibration ($\operatorname{Re} s_k^* \equiv 0$) never exceeds its original maximum amplitude (defined as $M \equiv \max |U|$ taken over one period), we conclude that if the original maximum amplitude at $V_0 = 0$ was "small" ($M = \varepsilon$ for $\varepsilon > 0$ small), so is the maximum amplitude at every V_0, as V_0 is increased quasistatically. This holds especially in the limit of the dynamic free vibration solution as $V_0 \to C$ (and also as $V_0 \to -C$). On the other hand, we have seen that the limit solution is a steady state. Because the limit solution does not depend on time, it cannot become unstable in the Bolotin sense.

Gathering these facts together, we conclude that by the performed quasistatic analysis, the critical point of the presently considered model at $V_0 = \pm C$ cannot be unstable. This contrasts classical wisdom (e.g., [10]), but Wang et al. [3] have obtained a similar result for the ideal string via a completely different route; their argument was based on Hamiltonian mechanics.

This result is of fundamental theoretical importance. However, since no practical physical system has exactly zero bending rigidity, its main practical relevance is as an illuminating example underlining the significance of including in the model all relevant aspects of system behavior; in this case, bending rigidity, however small. If we introduce nonzero bending rigidity, however small, into the model, the critical point becomes unstable, and the results then agree with those of Wickert and Mote [10]. This is not surprising; the bending term has a fourth-order space derivative, so it is the leading-order term in the modified model. With a small coefficient, it introduces a singular perturbation (see, e.g., [11–13]), which are known to change the qualitative behavior of models based on partial differential equations.

If one is interested only in an approximate value for the critical velocity for materials with small bending rigidity, the string model considered here is already sufficient. The introduction of small bending rigidity has no major effect on the value of the critical velocity, even though it completely changes the qualitative behavior of the model around the critical point. See, for example, [14] for an analytical approach in the case with small bending rigidity.

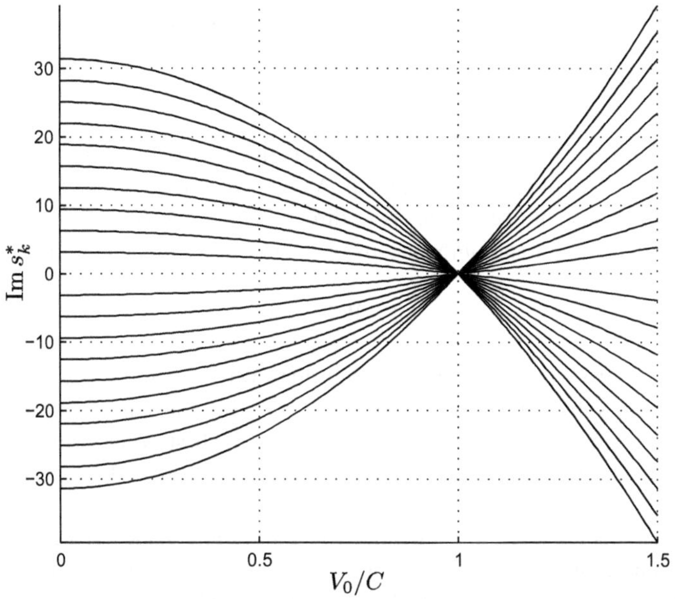

Fig. 7.3 Eigenvalues of the axially moving ideal one-dimensional web, up to vibration mode $k = 10$. Analytical result, Eq. (7.1.53). $\operatorname{Re} s_k^* \equiv 0$ for all k and V_0. At the critical point $V_0 = C$, $\operatorname{Im} s_k^* = 0$ for all k

A plot of the eigenfrequency spectrum described by Eq. (7.1.53), as a function of the axial velocity V_0, is provided in Fig. 7.3 up to mode number $k = 10$. The parameter values correspond to the model of a narrow panel in the membrane limit with $T_0 = 500\,\mathrm{N/m}$, $m = 0.08\,\mathrm{kg/m^2}$, and $\ell = 1\,\mathrm{m}$. The characteristic time τ was chosen as $\tau = \ell/C$, which leads to $\sqrt{\beta^2 - \gamma} = 1$, see Eq. (7.1.47).

We observe that the eigenvalues (7.1.53) come in pairs. This is a common feature of undamped gyroscopic systems, and is due to the structure of the characteristic equation. If s is an eigenvalue of such a system, then $-s$ is also an eigenvalue. For the relation of axially moving materials to classical gyroscopic systems, see [10].

From (7.1.52), the real and imaginary parts of the solution follow easily, by using the fact $\exp(a + b\mathrm{i}) = \exp(a)\exp(b\mathrm{i})$ and Euler's formula $\exp(\mathrm{i}\phi) = \cos\phi + \mathrm{i}\sin\phi$. We have

$$\operatorname{Re} U(x,t) = E\cos\left(\pm\frac{k\pi\ell}{C\tau}\left[\left(V_0^2 - C^2\right)\frac{\tau^2}{\ell^2}t - V_0\frac{\tau}{\ell}x\right]\right)\sin(k\pi x)\,, \qquad (7.1.54)$$

$$\operatorname{Im} U(x,t) = E\sin\left(\pm\frac{k\pi\ell}{C\tau}\left[\left(V_0^2 - C^2\right)\frac{\tau^2}{\ell^2}t - V_0\frac{\tau}{\ell}x\right]\right)\sin(k\pi x)\,. \qquad (7.1.55)$$

That these are indeed real-valued solutions of (7.1.51) can be easily verified by substitution. By comparing (7.1.54) and (7.1.55), it is obvious that their only difference is a phase shift with regard to the variable $\eta = \gamma t - \beta x$, because $\cos\phi = \sin(\phi + \pi/2)$ for all $\phi \in \mathbb{R}$. Observe that while the original complex-valued solution (7.1.52) is separable in x and t, that is, $U(x,t) = \chi_1(x)\chi_2(t)$, its real and imaginary parts are not.

To finish this section, let us list some alternative representations for the solution, and observe certain properties from them. A classical way to solve the free vibration problem with constant coefficients is to directly substitute $U(x,t) = \exp(st)W(x)$ into (7.1.40), leading to a homogeneous ordinary differential equation in $W(x)$,

$$s^2 W + 2\beta s W' + \gamma W'' = 0 \,. \tag{7.1.56}$$

Equation (7.1.56) with the boundary conditions $W(0) = W(1) = 0$ is then solved. One way is to substitute $W(x) = \exp(qx)Z(x)$ with constant q, and then choose the value of q such that the substitution eliminates the first-order term. The resulting equation for the new unknown $Z(x)$ then becomes similar to (7.1.44), which is easily solved.

Once the general solution is obtained, the admissible values of s are determined from the condition that a nontrivial solution $W(x) \not\equiv 0$ must be possible. If the two linearly independent solutions of (7.1.56) are denoted $W_1(x)$ and $W_2(x)$, the zero Dirichlet boundary conditions require that

$$\begin{aligned} W(0) &= a_1 W_1(0) + a_2 W_2(0) = 0 \,, \\ W(1) &= a_1 W_1(1) + a_2 W_2(1) = 0 \end{aligned} \quad \text{i.e.,} \quad \begin{bmatrix} W_1(0) & W_2(0) \\ W_1(1) & W_2(1) \end{bmatrix} \begin{bmatrix} a_1 \\ a_2 \end{bmatrix} = 0 \,, \tag{7.1.57}$$

which, as is well known, can have nontrivial solutions $(a_1, a_2) \neq (0,0)$ if and only if the determinant vanishes,

$$W_1(0)W_2(1) - W_2(0)W_1(1) = 0 \,. \tag{7.1.58}$$

The end result is

$$W(x) = E\left[\exp(-\mathrm{i}k\pi\,[1+\nu]\,x) - \exp(+\mathrm{i}k\pi\,[1-\nu]\,x)\right] \,, \quad 0 < x < 1 \,, \tag{7.1.59}$$

where

$$\nu \equiv \frac{V_0}{C} \,, \tag{7.1.60}$$

the complex-valued amplitude coefficient E is free, and x is the nondimensional space coordinate. From (7.1.59), it is readily apparent that if $V_0 = 0$, the space component reduces to $\sin(k\pi x)$, by choosing $E = -1/2\mathrm{i}$ and again applying Euler's formula. Nonzero V_0 introduces a shift into the exponent in both terms, and in this general case, the space component $W(x)$ can no longer be reduced into a real-valued form.

After applying some trigonometric identities, (7.1.59) can be rewritten as

$$W(x) = F\sin(k\pi x)\exp(-\mathrm{i}k\pi\nu x)\;, \quad 0 < x < 1\;, \tag{7.1.61}$$

where the coefficient $F = -2\mathrm{i}E$. From this representation, a velocity-dependent phase shift along the string (as reported by, e.g., [10]) is apparent.

When the time component is included into (7.1.59), the full solution is

$$\begin{aligned} U(x,t) &= E\exp(\mathrm{i}C\frac{\mathcal{T}}{\ell}\left[\nu^2-1\right]k\pi t)\left[\exp(-\mathrm{i}k\pi\left[1+\nu\right]x)-\exp(+\mathrm{i}k\pi\left[1-\nu\right]x)\right] \\ &\equiv E\exp(\mathrm{i}\tilde{C}t)\left[\exp(\mathrm{i}\tilde{A}x)-\exp(\mathrm{i}\tilde{B}x)\right] \\ &= E\left[\exp(\mathrm{i}[\tilde{A}x+\tilde{C}t])-\exp(\mathrm{i}[\tilde{B}x+\tilde{C}t])\right] \\ &= E\left[\exp\left(\mathrm{i}\left\langle (x,t),(\tilde{A},\tilde{C})\right\rangle\right)-\exp\left(\mathrm{i}\left\langle (x,t),(\tilde{B},\tilde{C})\right\rangle\right)\right]\;, \end{aligned} \tag{7.1.62}$$

where we have defined the constants

$$\tilde{A} = -k\pi\left[1+\nu\right]\;, \quad \tilde{B} = +k\pi[1-\nu]\;, \quad \tilde{C} = C[\mathcal{T}/\ell][\nu^2-1]k\pi\;, \tag{7.1.63}$$

and $\langle\cdot,\cdot\rangle$ denotes the usual scalar product. From this form, it is easily seen that when considered in the space-time plane, the complex-valued solution is a superposition of two plane waves with different directions of propagation.

From (7.1.62) it is also directly obvious why the real and imaginary parts of the solution satisfy Eq. (7.1.40) separately. The differentiation in the Coriolis term is of the second order when considered in the (x,t) plane, which conveniently eliminates the imaginary unit that would result from first-order differentiation of (7.1.62) with respect to only either x or t.

An alternative, much more general way to make this final observation is as follows. Let us write the complex-valued function as $U = \mathrm{Re}\,(U) + \mathrm{i}\,\mathrm{Im}\,(U)$. For any linear differential operator $\mathcal{L}$ with real coefficients, we observe that

$$\mathrm{Re}\;(\mathcal{L}(U)) = \mathrm{Re}\;[\mathcal{L}\,(\mathrm{Re}\,(U)+\mathrm{i}\,\mathrm{Im}\,(U))] \tag{7.1.64}$$

$$= \mathrm{Re}\;[\mathcal{L}\,(\mathrm{Re}\,(U))+\mathrm{i}\mathcal{L}\,(\mathrm{Im}\,(U))] \tag{7.1.65}$$

$$= \mathcal{L}\,(\mathrm{Re}\,U) \tag{7.1.66}$$

and similarly for the imaginary part. For example, for (7.1.64), $\mathcal{L}$ is given by

$$\mathcal{L}(U) \equiv \alpha\frac{\partial^2 U}{\partial t^2} + 2\beta\frac{\partial^2 U}{\partial x\partial t} + \gamma\frac{\partial^2 U}{\partial x^2}\;, \tag{7.1.67}$$

and we observe that in this case the requirement of real-valued coefficients holds. Note that the requirement does **not** hold for (7.1.56), where $\mathcal{L}$ is given by

$$\mathcal{L}(W) \equiv s^2 W + 2\beta s \frac{\partial W}{\partial x} + \gamma \frac{\partial^2 W}{\partial x^2} , \tag{7.1.68}$$

because s_k^* is imaginary by Eq. (7.1.53).

Let us make one final remark on the pseudo steady state Eq. (7.1.56), which can be written as

$$s^2 W + 2\beta s \frac{\partial W}{\partial x} + \gamma \frac{\partial^2 W}{\partial x^2} = 0 .$$

In a steady state, the linear reaction–advection–diffusion equation for a scalar field u is

$$\sigma u + a \cdot \nabla u - \nabla \cdot (\nu \cdot \nabla u) = f , \tag{7.1.69}$$

where σ is the reaction coefficient, a is a velocity field (vector, rank-1), and ν is the diffusivity tensor (rank-2). In one space dimension, this specializes to

$$\sigma u + a \frac{\partial u}{\partial x} - \frac{\partial}{\partial x}\left(\nu \frac{\partial u}{\partial x}\right) = f . \tag{7.1.70}$$

If there is no source term, $f = 0$, then (7.1.56) is a special case of (7.1.70) with

$$\sigma = s^2 , \quad a = 2\beta s , \quad \nu = -\gamma . \tag{7.1.71}$$

The reaction coefficient σ and the advection velocity a are complex-valued because in general, Lyapunov exponents s are. But we know that by (7.1.53), for an ideal string s is purely imaginary, so $\sigma < 0$ and a is imaginary. Using (7.1.41) in (7.1.71),

$$\nu = \frac{\tau^2}{\lambda^2}\left(C^2 - V_0^2\right) ,$$

so with regard to the quantity W, axial tension corresponds to diffusion, and the centrifugal effect corresponds to negative diffusion. At the critical velocity, the diffusive effect vanishes, and at supercritical velocities, negative diffusion takes over.

The diffusive effect is balanced by two mode-dependent (via s) terms, a reaction and an advection. The reaction term behaves like a linear spring with spring constant σ, although here the spring constant is negative.

As for the advection, from another viewpoint the pseudo steady state displacement function W is an Eulerian field (recall Chap. 5). The advection term pushes on this field at a uniform constant velocity, which (is imaginary and) depends on the mode number (via s) and on the axial drive velocity β. This has the same form as a fluid flow field pushing on a steady-state temperature field. (The presence of the advection term will affect the shape of the steady-state solution. The Eulerian temperature field is not advected at the fluid velocity, only the heat-carrying fluid itself is).

7.2 Exact Eigensolutions of the Traveling String with Damping

Internal or external friction is often modelled by adding first-order damping terms to Eq. (7.1.40). See e.g. [15] for the case of a traveling plate, where the external friction caused by the viscosity of the surrounding medium plays a role.

It is fairly simple to generalize the above analysis to cover the case with damping. We will also add a linear reaction term (elastic foundation of the classical Winkler type) to the equation, because as we will see, the solution process would already generate one. Consider the partial differential equation

$$a\frac{\partial^2 U}{\partial t^2} + 2b\frac{\partial^2 U}{\partial x \partial t} + c\frac{\partial^2 U}{\partial x^2} + A_1\frac{\partial U}{\partial t} + A_2\frac{\partial U}{\partial x} + BU = 0\,, \quad 0 < x < \ell\,. \tag{7.2.1}$$

Equation (7.2.1) follows the form of (7.1.40), but with damping and linear reaction terms added. It is the general constant-coefficient second-order autonomous homogeneous partial differential equation in two variables.

External friction in its most basic form appears as a nonzero value for A_1. This occurs, for example, when the traveling elastic element vibrates in a stationary surrounding medium. If the surrounding medium is in axial motion, independent of the axial velocity V_0 of the traveling material, we have the case where both A_1 and A_2 are nonzero.

If the traveling material experiences internal friction, we will have $A_2 = V_0 A_1$, because physical phenomena internal to the traveling material operate in the co-moving coordinate system. When we look at the traveling material in the Euler coordinate system, the $\mathrm{d}/\mathrm{d}t \equiv (\partial/\partial t)_{\bar{x}=\text{const.}}$, related to internal, will friction be transformed into the material derivative, as was discussed at the beginning of the chapter (Eq. (7.1.4)).

As above, we will use the zero Dirichlet boundary conditions for U, (7.1.15). Again, let us introduce nondimensional variables and multiply (7.2.1) by τ^2/a to obtain

$$\alpha\frac{\partial^2 U}{\partial t^2} + 2\beta\frac{\partial^2 U}{\partial x \partial t} + \gamma\frac{\partial^2 U}{\partial x^2} + \psi_1\frac{\partial U}{\partial t} + \psi_2\frac{\partial U}{\partial x} + \psi_3 U = 0\,, \quad 0 < x < 1\,, \tag{7.2.2}$$

where α, β and γ are given by (7.1.41) and

$$\psi_1 = A_1\frac{\tau}{a}\,, \quad \psi_2 = A_2\frac{\tau^2}{\lambda a}\,, \quad \psi_3 = B\frac{\tau^2}{a}\,. \tag{7.2.3}$$

Equation (7.2.2) can be solved using the same techniques as above. We begin by eliminating the mixed second derivative, as was done for (7.1.40) earlier. It is easy to confirm that our previous coordinate transformation (7.1.42) works also here. The first-order terms become, by the chain rule,

$$\frac{\partial U}{\partial t} = \frac{\partial U}{\partial \eta}\frac{\partial \eta}{\partial t} + \frac{\partial U}{\partial \xi}\frac{\partial \xi}{\partial t} = \gamma \frac{\partial U}{\partial \eta} , \tag{7.2.4}$$

$$\frac{\partial U}{\partial x} = \frac{\partial U}{\partial \eta}\frac{\partial \eta}{\partial x} + \frac{\partial U}{\partial \xi}\frac{\partial \xi}{\partial x} = -\beta \frac{\partial U}{\partial \eta} - \sqrt{\beta^2 - \gamma} \frac{\partial U}{\partial \xi} . \tag{7.2.5}$$

Note that $\partial U / \partial x$ will contribute to both $\partial U / \partial \eta$ and $\partial U / \partial \xi$. By defining

$$\varkappa_1 \equiv \psi_1 \gamma - \psi_2 \beta , \quad \varkappa_2 \equiv \psi_2 \sqrt{\beta^2 - \gamma} , \tag{7.2.6}$$

we have

$$\frac{\partial^2 U}{\partial \eta^2} - \frac{\partial^2 U}{\partial \xi^2} + \varkappa_1 \frac{\partial U}{\partial \eta} - \varkappa_2 \frac{\partial U}{\partial \xi} + \psi_3 U = 0 . \tag{7.2.7}$$

The signs of $\varkappa_1$ and $\varkappa_2$ have been chosen so that the signs of the first-order terms match the second-order ones. Next, to remove the first-order terms, we apply the same technique that reduces the telegraph equation into the Klein–Gordon equation [16]. Let

$$U(\xi, \eta) \equiv \exp(p\eta) Y(\xi, \eta) \tag{7.2.8}$$

with $p = -\varkappa_1/2$, and $Y(\xi, \eta)$ an unknown function to be determined. This eliminates $\partial U / \partial \eta$. A similar substitution can be used to eliminate $\partial U / \partial \xi$. The result is

$$U(\xi, \eta) \equiv \exp\left(-\frac{1}{2} \left[\varkappa_1 \eta + \varkappa_2 \xi \right] \right) Z(\xi, \eta) , \tag{7.2.9}$$

and (7.2.7) becomes

$$\frac{\partial^2 Z}{\partial \eta^2} - \frac{\partial^2 Z}{\partial \xi^2} = rZ , \tag{7.2.10}$$

where $Z = Z(\xi, \eta)$ is the unknown function to be determined, and

$$r \equiv \frac{1}{4} \left(\varkappa_1^2 - \varkappa_2^2 - 4\psi_3 \right) . \tag{7.2.11}$$

If $r = 0$, the rest of the solution process reduces to the undamped case analyzed further above. Therefore, let us now restrict consideration to the case $r \neq 0$. We aim to find a free vibration solution for (7.2.10). We will again use the classical time-harmonic trial function

$$Z(\xi, \eta) = \exp(s\eta) G(\xi) , \tag{7.2.12}$$

obtaining

$$s^2 G - G'' = rG \;, \quad \text{i.e.,} \quad G'' = (s^2 - r)G \;. \tag{7.2.13}$$

Equation (7.2.13) differs from the previous eigenvalue problem only in that the eigenvalue has shifted from s^2 to $s^2 - r$. Thus, by (7.2.13) and the boundary conditions (7.1.15) (which are inherited by $G(\xi)$) we again have

$$G = \sin(\omega\xi) \;, \tag{7.2.14}$$

where ω is to be determined using (7.1.15). By the original argument, and again choosing $\lambda = \ell$, we have

$$\omega = -\frac{k\pi\ell}{C_T} \;, \tag{7.2.15}$$

where k is any nonzero integer. Substituting (7.2.14) into (7.2.13) produces

$$-\omega^2 G = (s^2 - r)G \;, \tag{7.2.16}$$

and hence

$$s^2 = r - \omega^2 \;, \quad \text{i.e.,} \quad s = \pm\sqrt{r - \omega^2} \;. \tag{7.2.17}$$

Similarly as before,

$$s^2 = r - \frac{k^2\pi^2\ell^2}{C^2 T^2} \;, \tag{7.2.18}$$

and

$$Z(\xi, \eta) = \exp(s\eta)\sin(\omega\xi) \;. \tag{7.2.19}$$

Before we can make any elastic stability conclusions, we must transform back to (x, t) coordinates by using (7.2.18), (7.2.19), (7.2.11), (7.2.9), (7.2.6), (7.1.48) and the coordinate transformations (7.1.42).

Let us transform the stability exponent only. We start by combining (7.2.19) and (7.2.9),

$$U(\xi, \eta) = \exp\left(-\frac{1}{2}\left[\varkappa_1\eta + \varkappa_2\xi\right]\right)\exp(s\eta)\sin(\omega\xi) \;, \tag{7.2.20}$$

which separates as

$$U(\xi, \eta) = \exp\left(\left[s - \frac{1}{2}\varkappa_1\right]\eta\right) \cdot \exp(-\frac{1}{2}\varkappa_2\xi)\sin(\omega\xi) \;. \tag{7.2.21}$$

Due to the fact that we use the coordinate transformations (7.1.42), only the part involving η will contribute to the component that will eventually involve t. Because again we have $\eta = \gamma t - \beta x$, we see that for this problem the coefficient of t in the exp in the final complex-valued solution in (x, t) coordinates is

$$s_* = \gamma \left[s - \frac{1}{2}\varkappa_1 \right] = \gamma \left[\pm\sqrt{r - \omega^2} - \frac{1}{2}\varkappa_1 \right] , \tag{7.2.22}$$

where the last form follows from (7.2.17), r is given by (7.2.11), $\varkappa_1$ by (7.2.6), ω by (7.1.48), and γ by (7.1.41).

Again, the star indicates that (7.2.22) refers to the stability behavior in (x, t) coordinates instead of the transformed (ξ, η) coordinates. Due to the $-\varkappa_1/2$ term, we see a global (but V_0-dependent) shift along the real axis for all eigenvalues. This is typical for damped systems (see [9]). As before, the eigenvalues come in pairs, corresponding to the choice of sign in the $\pm$, but due to the shift term $-\varkappa_1/2$, the pairs are no longer of the type $\pm s$. Instead, each pair is centered on the shift value, with a pure real or pure imaginary offset depending on the sign of $r - \omega^2$. The contribution of the damping term is always purely real.

Let us investigate the external friction case, where the damping coefficients A_1 and A_2 do not depend on V_0. In the subcritical regime, $\gamma < 0$, and if $A_1 > 0$ and $A_2 \geq 0$, then $\varkappa_1 < 0$ for positive V_0 (see Eqs. (7.2.6) and (7.1.41)). If $A_2 = 0$, this holds also for negative V_0. If $A_2 > 0$, negative V_0 will reduce the magnitude of $\varkappa_1$, but for V_0 sufficiently near the origin, $\varkappa_1$ will remain negative. Hence $\gamma\varkappa_1 > 0$, and $-\gamma\varkappa_1 < 0$, at least near $V_0 = 0$. Thus for many practically interesting cases, $-\gamma\varkappa_1 < 0$. If, in addition, $r < \omega^2$ (we will investigate this condition below), then the square root in (7.2.22) is imaginary, and (7.2.21) represents exponentially decaying time-harmonic vibrations.

In the limit $V_0 \to C$ (or $V_0 \to -C$), we have $s_* \to 0$ just as for the undamped system, which suggests the existence of a steady state there. However, the present analysis is only applicable if $\gamma \neq 0$, that is, $|V_0| \neq C$, because at that point, the $\partial^2 U/\partial x^2$ term vanishes from (7.2.2) and the coordinate transformation (7.1.42) becomes invalid. To determine the situation at $|V_0| = C$, a separate analysis for the case $\gamma = 0$ would be needed, as was done above for the undamped case. We will however omit it for brevity.

Note the dependence of the quantities in (7.2.22) on the axial velocity: $\gamma = \gamma(V_0)$, $\varkappa_1 = \varkappa_1(\beta, \gamma) = \varkappa_1(V_0)$, and $r = r(V_0)$. Using (7.2.11), (7.2.6) and (7.1.41), the function $r(V_0)$ explicitly expands as

$$\begin{aligned} r(V_0) &= \frac{1}{4}\left(\varkappa_1^2 - \varkappa_2^2 - 4\psi_3\right) \\ &= \frac{\tau^2}{4\ell^2}\left[V_0^2 - C^2\right]\left[\psi_1^2\left(V_0^2 - C^2\right)\frac{\tau^2}{\ell^2} - 2\psi_1\psi_2 V_0 \frac{\tau}{\ell} + \psi_2^2\right] - \psi_3 \, , \end{aligned} \tag{7.2.23}$$

which is a fourth-order polynomial in V_0. Expression (7.2.23) is written for the case of external friction, where A_1 and A_2 (hence also ψ_1 and ψ_2) do not depend on V_0. In the case of internal friction, substituting $\psi_2 = V_0 [\tau/\ell] \psi_1$ obtains the final form.

The quantity $\omega = \omega(k)$ does not depend on the velocity, but it depends on the vibration mode number k. Thus each mode k has a different base frequency at $V_0 = 0$.

These frequencies also differ from the ones of the undamped system; if $r > 0$, the frequencies of the damped system are lower than those of the undamped one; and if $r < 0$, higher.

If $r(V_0) < \omega(k)^2$, the quantity under the square root in (7.2.22) is negative, and hence the contribution of the square root is purely imaginary (representing time-harmonic vibrations). From the expression of ω, Eq. (7.1.48), we see that $|\omega(k)|$ increases monotonically as the mode number k increases. Thus at $V_0 = 0$, the criterion $r(V_0 = 0) < \omega(k)^2$ is most easily violated for the first few modes. If such an integer $k_0 \geq 1$ exists that for all $k = 1, \ldots, k_0$ it occurs that $\omega(k)^2 < r(V_0 = 0)$, these modes will not vibrate, but will either just decay or grow exponentially, depending on the sign of the resulting s_*. The sign of s_* is also affected by the sign and relative magnitude of $\varkappa_1$.

The other possibility for the criterion $r(V_0) < \omega(k)^2$ to be violated is when $|V_0|$ is large, due to the $\varkappa_1^2$ term in $r(V_0)$, which (in the case of external friction) leads to a fourth-degree polynomial with a positive leading coefficient. This suggests that asymptotically, we will have $\operatorname{Im} s_* = 0$ for large $|V_0|$.

We are now ready to continue stability analysis of the system with damping. Let us first show that the damped system may have steady-state solutions only at $V_0 = \pm C$. Upon expanding (7.2.22) by inserting (7.2.11) for r, (7.2.6) for $\varkappa_1$ and $\varkappa_2$, (7.1.48) for ω and (7.1.47) for $\sqrt{\beta^2 - \gamma}$, we have

$$s_* = \gamma \, f^{\pm}(V_0;\, k) \,, \tag{7.2.24}$$

where

$$f^{\pm}(V_0;\, k) \equiv \pm\sqrt{\frac{1}{4}\left[(\psi_1\gamma - \psi_2\beta)^2 - (\psi_2 C\frac{\tau}{\ell})^2 - 4\psi_3\right] - \frac{k^2\pi^2\ell^2}{\tau^2 C^2}} - \frac{1}{2}\,(\psi_1\gamma - \psi_2\beta) \,. \tag{7.2.25}$$

Expression (7.2.24) may be zero if $\gamma = 0$ $(V_0 = \pm C)$, or if $f^{\pm} = 0$. Equation (7.2.25) defines a set of two functions, each belonging to a different eigenvalue s_*.

Because the term outside the square root is always real-valued for real V_0, we see that $f^{\pm} = 0$ is possible only if the square root term is real. Thus, a steady-state solution can only exist if the square root term is nonnegative, $r(V_0) - \omega(k)^2 \geq 0$. The sign of the square root term depends only on the sign chosen in the $\pm$.

We observe that (7.2.25) is of the form

$$f^{\pm}(a_1, a_2) = \pm\sqrt{a_1^2 + a_2} - a_1 \,, \tag{7.2.26}$$

where $a_1 \equiv \varkappa_1/2 = (\psi_1\gamma - \psi_2\beta)/2$, and we have collected the rest of the terms under the square root into a_2. If $a_1 > 0$, then $f^{\pm} = 0$ is possible only for f^+, and additionally then it must hold that $a_2 = 0$, which (in general) is not the case here. Similarly, if $a_1 < 0$, then $f^{\pm} = 0$ is possible only for f^-, and again we must have $a_2 = 0$, which (in general) does not occur. In either case, $f^{\pm} = 0$ has no solution.

Finally, if $r(V_0) - \omega(k)^2 < 0$, the square root becomes imaginary and (7.2.25) has no (real) solution. Therefore, in either case, we conclude that (in general) Eq. (7.2.25) has no (real) solutions at all in terms of V_0. We have thus shown that for the damped system, s_* may pass through the origin only at $V_0 = \pm C$. (The only possible exception are specially tuned combinations of problem parameters that lead to $a_2 = 0$ in (7.2.26), then requiring further analysis).

This also completes the stability analysis of the case $r(V_0) - \omega(k)^2 \geq 0$, where the whole expression (7.2.24) is real-valued. At parameter values where $r(V_0) - \omega(k)^2 < 0$, dynamic critical points may still exist. The question of finding such critical points becomes that of determining the critical values of V_0 where the real part $\operatorname{Re} s_*$ becomes zero.

When $r(V_0) - \omega(k)^2 < 0$, the square root term in (7.2.25) is purely imaginary, and it will not affect the real part. We have

$$\begin{aligned} \operatorname{Re} s_* &= -\frac{\gamma}{2}\left(\psi_1\gamma - \psi_2\beta\right) \\ &= -\frac{\tau^2}{2\ell^2}\left[V_0^2 - C^2\right]\left[\left(\psi_1\frac{\tau^2}{\ell^2}\right)V_0^2 + \left(-\psi_2\frac{\tau}{\ell}\right)V_0 + \left(-\psi_1 C^2\frac{\tau^2}{\ell^2}\right)\right], \end{aligned} \tag{7.2.27}$$

where we have used (7.1.41) and $\lambda = \ell$. We have obtained a fourth-order polynomial in V_0. Two of its zeroes are, obviously, $V_0 = \pm C$. This is as expected; if $s_* = 0$, then also $\operatorname{Re} s_* = 0$.

For the case with external friction (ψ_2 independent of V_0), solving for the zeroes of the second expression in brackets produces

$$V_0 = V_0^{\mathrm{dyn}_\pm} \equiv M \pm \sqrt{M^2 + C^2}\,, \tag{7.2.28}$$

where

$$M \equiv \frac{\psi_2\ell}{2\psi_1\tau} = \frac{A_2}{2A_1}\,. \tag{7.2.29}$$

The values of V_0 given by (7.2.28) are the dynamic critical points. These are the only other points, in addition to the static critical points $V_0 = \pm C$, where $\operatorname{Re} s_* = 0$ may occur for the damped system with friction external.

If $A_2 = 0$, as is the case for friction caused by a stationary external medium, we observe from (7.2.28) that then the dynamic critical points coincide with the static ones at $V_0 = \pm C$. If $A_2 \neq 0$, they will be distinct.

The existence of any distinct dynamic critical points is conditional on the validity of (7.2.27) as a representation for $\operatorname{Re} s_*$ at the particular values of V_0 given by (7.2.28). The only exception is if the dynamic critical points coincide with the static ones, because the static critical points always exist by the analysis above.

If the dynamic critical points exist, their location on the V_0 axis depends on the ratio of the damping coefficients, and on those problem parameters which affect the value of C. It does not depend on the mode number k, because (7.2.27) does not involve $\omega(k)$.

However, it may occur that the dynamic critical points do not exist for all modes, but only for mode numbers higher than some limiting number. Because $|\omega(k)|$ increases monotonically with increasing $|k|$, it follows that for any given value of V_0, we can always choose $|k_{\min}|$ such that the condition $r(V_0) - \omega(k)^2 < 0$ (with the restriction $|k| \geq |k_{\min}|$) is satisfied at the given V_0.

Summarizing, the system with external friction has two kinds of critical points: the static ones at $V_0 = \pm C$, which are independent of mode number, and up to two dynamic ones. The dynamic critical points may only exist for sufficiently high mode numbers, but their location on the V_0 axis is the same for all modes for which they exist.

Now let us consider friction internal. We insert $\psi_2 = V_0\,[\tau/\ell]\,\psi_1$ into (7.2.27), obtaining (after cancellation)

$$\operatorname{Re} s_* = \frac{\psi_1 C^2 \tau^4}{2\ell^4}\left[V_0^2 - C^2\right] , \tag{7.2.30}$$

which is valid whenever $r(V_0) - \omega(k)^2 < 0$. When this condition holds, we see that the real parts of all modes coincide, because (7.2.30) does not depend on k. It is seen that for the system with friction internal, there are no dynamic critical points; the only zeroes of (7.2.30) are $V_0 = \pm C$.

We observe from (7.2.30) that for $V_0^2 < C^2$, the real parts are all negative, while for $V_0^2 > C^2$ (provided that the validity condition $r(V_0) - \omega(k)^2 < 0$ still holds), the real parts are positive. This strongly suggests that for the system with internal friction, the static critical points $V_0 = \pm C$ are unstable.

Finally, even if $r(V_0) - \omega(k)^2 > 0$, we see that the real parts $\operatorname{Re} s_*$ in the internal friction case will be centered on the value (7.2.30), separated from it by plus and minus the contribution of the square root term that has become real-valued.

One final remark concerning both external and internal friction cases is in order. We have determined all zeroes of the real part $\operatorname{Re} s_*$ for the damped system. It is obvious that s_* is continuous with respect to V_0. Hence, $\operatorname{Re} s_*$ can change sign only at its zeroes.

Therefore, if we determine s_* at $V_0 = 0$ (obtaining the sign of $\operatorname{Re} s_*$ in the initial state), and examine all the critical points, we will know the sign of $\operatorname{Re} s_*$ for any real V_0. This implies that the graphical examples below are exhaustive; for the modes visualized, there cannot be any stability surprises outside the plotting range.

To finish this section, we will show some examples of eigenvalues (7.2.24) for various types of damping. The problem parameter values again correspond to the model of a narrow panel in the membrane limit with $T_0 = 500\,\mathrm{N/m}$, $m = 0.08\,\mathrm{kg/m^2}$, and $\ell = 1\,\mathrm{m}$. The characteristic time τ is chosen as $\tau = \ell/C$, which leads to $\sqrt{\beta^2 - \gamma} = 1$, see Eq. (7.1.47).

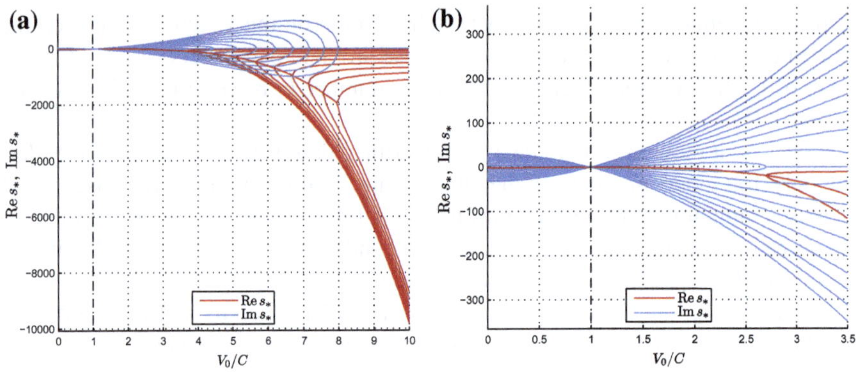

Fig. 7.4 Eigenvalues of the axially moving one-dimensional web with damping, up to vibration mode $k = 10$. Analytical result, Eq. (7.2.24). Typical basic case with only a time-dependent damping term present; damping parameters $\psi_1 = 1$, $\psi_2 = 0$. Physical interpretation is external friction in a stationary surrounding medium. **a** and **b**: different zoom levels

In the examples presented here, the reaction coefficient $B = \psi_3 = 0$. It can be observed numerically that the elastic foundation (reaction term) has only a minor effect on the eigenvalues. If $B > 0$, the imaginary parts pack much closer together, and there is more space between the smallest imaginary part and the real axis. Otherwise the results with $B > 0$ are very similar to the ones shown with $B = 0$, and are omitted for brevity.

Figure 7.4 represents the typical basic case of a moderate amount of purely time-dependent damping, such as that generated by external friction when the elastic element vibrates in a stationary surrounding medium. We see that $\operatorname{Re} s_* \leq 0$ for all V_0. When $\operatorname{Im} s_*$ vanishes, each eigenvalue pair settles onto a pair of purely real values diverging from each other.

In Fig. 7.5, we have a heavy amount of purely time-dependent damping. The physical interpretation is the same as in the first case, only the amount of external friction is higher. Also here $\operatorname{Re} s_* \leq 0$ for all V_0. This figure illustrates $\omega(k)^2 < r(V_0)$ for the first few modes near $V_0 = 0$. Because the condition $r(V_0) - \omega(k)^2 < 0$ is not fulfilled, the imaginary part is zero, and these first few modes simply decay exponentially without vibrating.

Constant time- and space-dependent damping is illustrated in Fig. 7.6. This type may arise due to external friction in an axially moving surrounding medium, where the axial velocity of the medium is independent of the axial velocity V_0 of the traveling elastic element.

In the final example, Figs. 7.7 and 7.8, we have $\psi_2 = V_0\psi_1$, which is of the form that is generated by the material derivative. As was discussed at the beginning of the analysis of the damped system, this case arises if the traveling material experiences friction internal.

In the case of internal friction, we observe an interesting phenomenon. Figures 7.7 and 7.8 indicate that the real parts $\operatorname{Re} s_*$ of the eigenvalues cross the real axis at

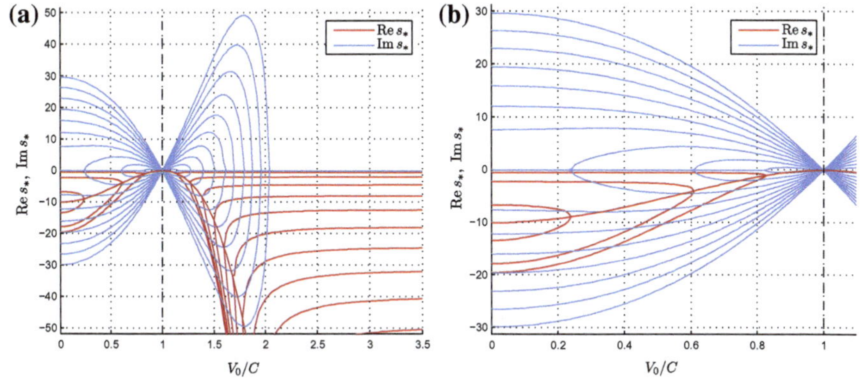

Fig. 7.5 Eigenvalues of the axially moving one-dimensional web with damping, up to vibration mode $k = 10$. Analytical result, Eq. (7.2.24). Typical case with heavy but only time-dependent damping; $\psi_1 = 20$, $\psi_2 = 0$. Note purely real eigenvalues near $V_0 = 0$. **a** and **b**: different zoom levels

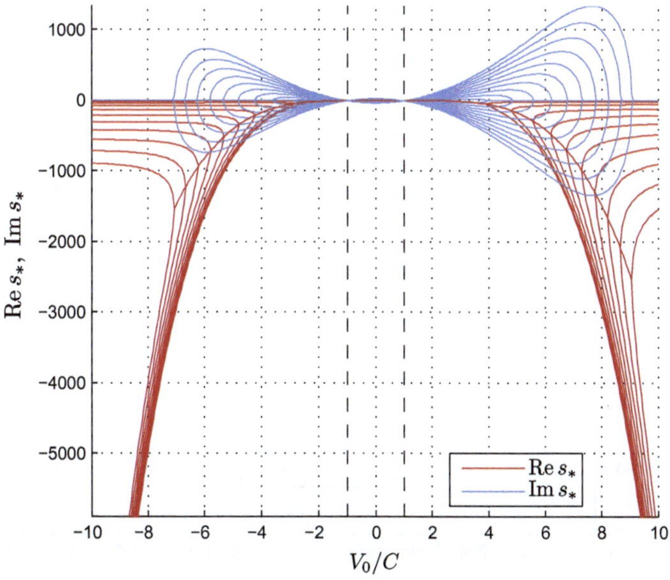

Fig. 7.6 Eigenvalues of the axially moving one-dimensional web with damping, up to vibration mode $k = 10$. Analytical result, Eq. (7.2.24). Typical case with time- and space-dependent damping; $\psi_1 = 1$, $\psi_2 = 2$. Nonzero ψ_2 makes the spectrum asymmetric with respect to positive and negative values for V_0. Physical interpretation is external friction in a surrounding medium which is in axial motion, independent of the axial velocity V_0 of the traveling elastic material

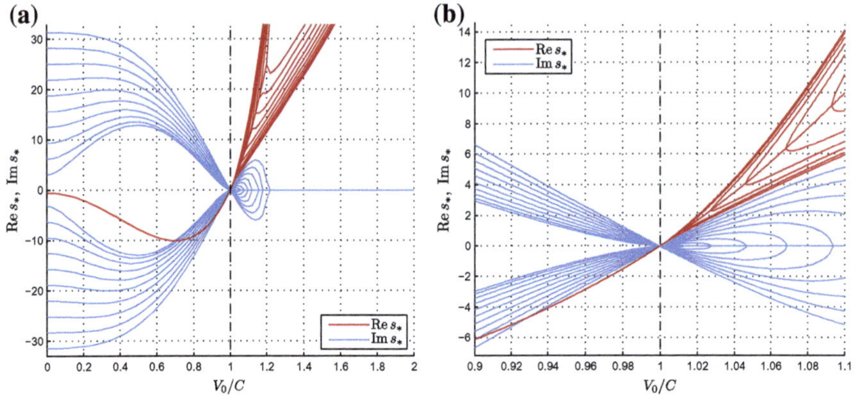

Fig. 7.7 Eigenvalues of the axially moving one-dimensional web with damping, up to vibration mode $k = 10$. Analytical result, Eq. (7.2.24). Case with $\psi_1 = 1$, $\psi_2 = V_0\psi_1$. Physical interpretation is internal friction in the traveling material itself. **a** and **b**: different zoom levels

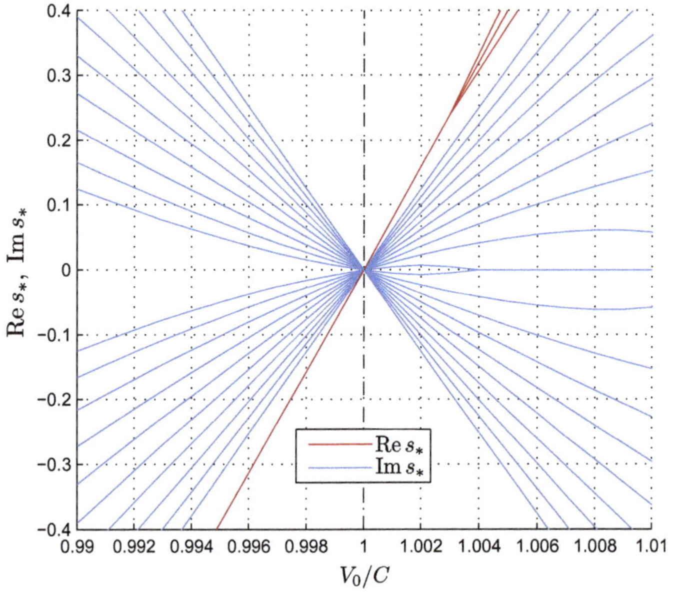

Fig. 7.8 Close-up view centered on the static critical point in Fig. 7.7. Because the mode $k = 1$ is nearest the imaginary axis, it is seen that immediately after the critical point, the imaginary part $\text{Im}\, s_* \neq 0$ for all k

$V_0 = C$, moving into the unstable region $\text{Re}\, s_* > 0$. The presence of internal friction has introduced an instability, which did not exist in the undamped system!

The importance of including damping effects (however small) in the model being analyzed is discussed in several monographs by Bolotin; for example, see [9, pp.

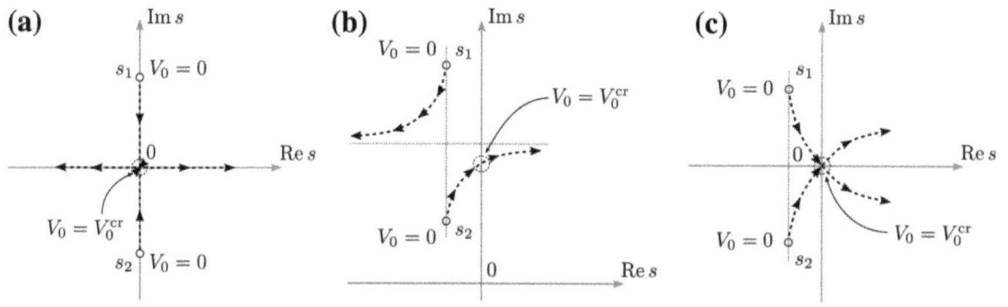

Fig. 7.9 **a** and **b**: Behavior of the stability exponent s for the two types of instability, after Bolotin [9]. **a**: static (or divergence) instability. **b**: dynamic (or flutter) instability. **c**: schematic of the present case with internal friction. Dynamic instability at a static critical point; eigenvalues pass through each other without interacting. Both obtain a positive real part as they pass the critical point. Compare Figs. 7.7 and 7.8

99–100], [17, pp. 290–291]. The general effect of dissipation-induced destabilization (in the sense of a drastically reduced first critical load) was first observed by Ziegler [18], and is well-known. For the history of this effect, see [19]. It is known that this particular kind of dissipation-induced destabilization occurs exclusively in nonconservative systems. For conservative systems, the *Kelvin–Tait–Chataev theorem* guarantees that damping always has a stabilizing effect on the first critical load; see [20]. Bolotin [9, p. 99] refers to this as *one of Kelvin's theorems*.

However, what we observe here is a different phenomenon. The critical load has not changed, but the initial postcritical behavior has become unstable. We will see this effect again in Chap. 8, where we will observe that a surrounding air mass that moves along with the axially traveling string has a qualitatively similar effect on the postcritical behavior of the string.

The type of the particular instability observed here is curious in that it does not follow either of the classical typical behaviors as described by Bolotin [9]. See Fig. 7.9. As was explained in Chap. 4, often the kind of instability which is produced by s_* passing through the origin leads to purely real values of s_* slightly above the critical point. This is called the divergence instability or static instability. Slightly above the critical point, the state variable of the system will grow exponentially without oscillating.

The other classical type is the flutter instability, also known as dynamic instability, where s_* crosses the imaginary axis at some nonzero value of $\operatorname{Im} s_*$, and $\operatorname{Re} s_*$ becomes positive. Slightly above the critical point, the state variable of the system will oscillate at an exponentially increasing amplitude.

In the system analyzed here, the critical point is located at the origin, but the initial postcritical behavior is oscillatory. Unlike in the usual static or dynamic instability types, the eigenvalues simply pass through each other at this point without interacting. This behavior is similar to that of the undamped system, except that now the real parts of the eigenvalues are nonzero. Compare the behavior near the static critical point in Figs. 7.3 and 7.8.

In a sense, the loss of stability is of the dynamic type, because of the initial postcritical behavior. However, it is possible to find the critical value for V_0 by a static stability analysis (in which we insert the time-harmonic trial function, set $s_* = 0$, and solve the resulting steady-state problem), because at the critical point $s_* = 0$.

In conclusion, for the system considered, the presence of internal friction destabilizes the static critical point, leading to a dynamic instability there; whereas external friction causes the vibrations to decay as expected.

Let us summarize. In this chapter so far, we have considered classical axially moving ideal one-dimensional elastic elements with zero Dirichlet (pinhole) boundary conditions. Both undamped and damped cases were studied.

For both cases, at the critical velocity $|V_0| = C$, one of the terms in the partial differential equation is eliminated, requiring a separate analysis. This analysis was performed for the undamped case, and it was shown that at the critical velocity, a steady state occurs, for any state function satisfying the boundary conditions is a solution.

The free vibrations for $|V_0| \neq C$ were solved following a classical approach. The eigenfrequencies were found explicitly for both systems, and for the undamped system also the eigenfunctions were determined. It was seen that for both systems, the eigenvalues come in pairs.

Based on quasistatic transition and the maximum-amplitude preserving property of time-harmonic vibration, it was argued that the ideal elastic web experiences no instability at the critical velocity. The initial postcritical behavior of the stability exponent was seen to point to the same conclusion. This contrasts classical wisdom, but agrees with the result of Wang et al. [3], obtained via a completely different approach (Hamiltonian mechanics).

This result is of fundamental theoretical importance. However, since no practical physical system has exactly zero bending rigidity, its main practical relevance is as an illuminating example underlining the significance of including in the model all relevant aspects of system behavior; in this case, bending rigidity, however small.

But if one is interested only in an approximate value for the critical velocity for materials with small bending rigidity, the string model considered here is already sufficient. The introduction of small bending rigidity has no major effect on the value of the critical velocity (see, e.g., [14] for an analytical approach), even though it completely changes the qualitative behavior of the model around the critical point.

The present results lend themselves to the following interpretation. After a quasistatic transition from an initial state ($V_0 = 0$) to the present state (V_0 nonzero), if the assumption of small displacement has not been violated anywhere up to that point, the model considered determines the behavior of free vibrations in the present state.

Especially, the system is (usually) seen to be stable at $V_0 = 0$. Hence, the governing partial differential equation is valid at least for all $|V_0| < C$. Whether the equation is valid at supercritical velocities, $|V_0| > C$, is therefore a question of whether there exists an instability at the critical velocity, where the stability exponents s_k^* vanish.

The argument given in this text leaves open two possibilities for instability at the critical point. First, it is easy to construct such dynamic free vibration solutions that are not compatible with the quasistatic transition into the steady state at $V_0 = \pm C$. An analysis accounting for axial acceleration—in order to perform a fully dynamic transition—may be needed to remove this limitation. The second possibility is the so far undetermined behavior of the global coefficient multiplying the solution as V_0 is changed; determining this requires techniques beyond eigenvalue analysis.

Damping can be used to model both internal and external friction. It was found that also for the damped system, static critical points always exist at $V_0 = \pm C$. In the case of the system with external friction, it was additionally shown that for mode numbers k sufficiently high, dynamic critical points may exist. It was also observed, and demonstrated in the examples, that if the traveling material is subjected to heavy external friction, the first few modes k, near $V_0 = 0$, only decay exponentially without vibrating.

It was seen that the presence of internal friction in the traveling material itself destabilizes the static critical point, leading to a dynamic instability there, while external friction (such as caused by the viscosity of an external medium in which the traveling elastic element vibrates) exhibits no such effect.

The theory and results presented in this section can be used as simplified but powerful tools of analysis in various applications of moving materials, for example paper making.

7.3 Exact Eigensolutions of Axially Moving Beams and Panels

In this section we derive an exact analytical solution for the natural frequencies of an axially moving elastic beam or panel having any value of bending rigidity. The solution will be obtained in an implicit parametric form. The analysis proceeds by examining time-harmonic vibrations, mapping the dependence between the axial transport velocity and the system's natural frequencies. Analytical implicit expressions for the solution curves are derived for velocity ranges where the natural frequencies are real-valued, corresponding to stable vibrations. Both axially tensioned and nontensioned traveling beams are considered. The classical special cases of the traveling beam with no axial tension (first analytically solved by Simpson [21]), and the tensioned stationary (nontraveling) beam are also discussed, and special-case solutions given. Numerical evaluation of the obtained general analytical results is discussed. Numerical examples are given for traveling beams subjected to two different tension levels, and for the non-tensioned traveling beam.

Let us first consider the analysis of eigensolutions. Linear partial differential equations that are second order in time, and have constant coefficients, are well known to exhibit time-harmonic solutions, which characterize the free vibration behavior of the modelled physical system. Time-harmonic solutions are important also in the analysis of stability of elastic systems, appearing in the dynamic method of Bolotin [9]. The appearance of complex-valued frequencies is interpreted as a loss

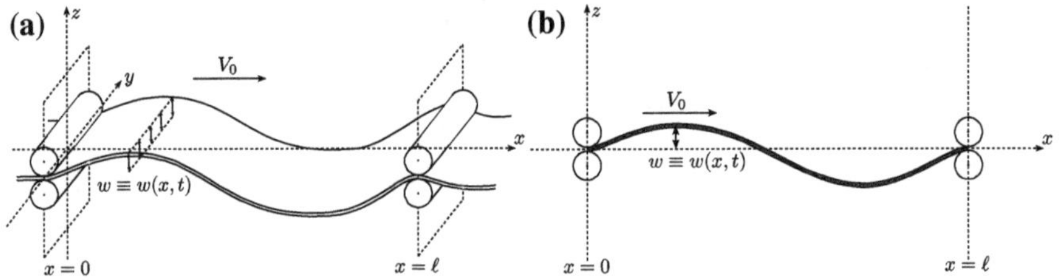

Fig. 7.10 Axially traveling panel, i.e., a plate undergoing cylindrical deformation. The pairs of rollers denote simple supports, and the finite thickness depicts the presence of bending rigidity. **a**: Problem setup. **b**: Projection of the setup to the xz plane

of stability in a dynamic form, corresponding to the loss of Lyapunov stability. A convergence of the natural frequencies to zero corresponds to a loss of stability in a static form, meeting the criteria of Euler buckling.

We will derive analytical implicit expressions for the natural frequency curves with respect to the problem parameters, for ranges of the parameter space where the natural frequencies are real-valued, corresponding to stable vibrations. Previously, Kong and Parker [14] have derived explicit expressions for the natural frequencies of axially moving beams in the case of small bending rigidity. We will allow the bending rigidity to be arbitrarily large.

Simple models such as beams or panels, where applicable, are extremely useful in a modern setting, because they easily lend themselves to highly efficient and extremely fast solvers. The solution procedure presented here is computationally light, which allows the development of very efficient and very fast solvers for determining the natural frequencies of traveling beams and panels. Realtime solvers, in turn, have applications such as online optimal control, parametric studies of complex physical situations, modelling-based measurements, statistical quantification of uncertainty, agile modelling, and industrial optimization, at a very low computational cost.

Let us begin with the consideration of an axially traveling rectangular plate undergoing cylindrical deformation, as shown in Fig. 7.10. The equation of small transverse vibrations is

$$m\frac{\partial^2 w}{\partial t^2} + 2mV\frac{\partial^2 w}{\partial x \partial t} + (mV^2 - T)\frac{\partial^2 w}{\partial x^2} + D\frac{\partial^4 w}{\partial x^4} = 0\,, \quad 0 < x < \ell\,, \quad (7.3.1)$$

where $m = \rho S$ is the mass of the panel per unit area, ρ is the density of the material, S the area of the cross section, and V is a constant axial transport velocity. The constant axial tension T has the dimension of force per unit length; for a panel, it can be expressed as $T = h\sigma_{xx}$, where h is the thickness of the panel and σ_{xx} is the axial stress, which is constant along the length for an elastic material. (The elastic constitutive model is important here; for a viscoelastic traveling material, it turns out that this classical constant-stress property fails. For an analysis of the axial stress field of a traveling Kelvin–Voigt material, see [22]). The quantity D is the bending rigidity, which for an isotropic elastic material it follows the relation [23]

$$D = \frac{Eh^3}{12(1-\nu^2)} , \tag{7.3.2}$$

where E is the Young's modulus of the material and ν its Poisson ratio. The symbol ℓ denotes the length of the free span between the mechanical supports. The transverse displacement of the panel, as it appears in laboratory coordinates (in other words, in the Eulerian frame), is described by the function $w \equiv w(x, t)$.

For transverse vibrations of beams, D is replaced by EI, the area mass m by the linear mass (SI unit: kg/m), and the unit of the tension T is adjusted accordingly (becoming just a force); otherwise the equations are the same.

Equation (7.3.1) is of the fourth order in x, so four boundary conditions are needed in total. In what follows, the simply supported boundary conditions of the Kirchhoff plate are used, that is,

$$w(0, t) = w(\ell, t) = 0 , \tag{7.3.3}$$

$$D\frac{\partial^2 w}{\partial x^2}(0, t) = D\frac{\partial^2 w}{\partial x^2}(\ell, t) = 0 . \tag{7.3.4}$$

The boundary conditions (7.3.3) and (7.3.4) arise by requiring that the transverse displacements and the bending moments for the elastic material at the boundary points $x = 0$ and $x = \ell$ are zero. Again, the elastic constitutive model influences the form of (7.3.4); recall the discussion in Chap. 5.

Harmonic vibrations of the system are represented as

$$w(x, t) = e^{i\omega t} u(x) , \tag{7.3.5}$$

where $u(x)$ is the amplitude function and ω is the frequency of vibration. It will be convenient to work in nondimensional variables. Let

$$\begin{aligned} x = \ell\tilde{x} , \quad \frac{\rho S \omega^2 \ell^4}{D} = \tilde{\omega}^2 , \quad \frac{\rho S \ell^2}{D} V^2 = \tilde{V}^2 , \\ \frac{\rho S \ell^2}{D} C^2 = \tilde{C}^2 , \quad C = \sqrt{\frac{T}{\rho S}} . \end{aligned} \tag{7.3.6}$$

From the chain rule, $\partial(\cdot)/\partial x \to (1/\ell)\partial(\cdot)/\partial\tilde{x}$. In the following, the tilde will be omitted.

We formulate the boundary-value problem for the amplitude function $u(x)$ as

$$\frac{\mathrm{d}^4 u}{\mathrm{d}x^4} + (V^2 - C^2)\frac{\mathrm{d}^2 u}{\mathrm{d}x^2} + 2i\omega V \frac{\mathrm{d}u}{\mathrm{d}x} - \omega^2 u = 0 , \quad 0 < x < 1 , \tag{7.3.7}$$

$$u(0) = u(1) = 0 , \quad \left(\frac{\mathrm{d}^2 u}{\mathrm{d}x^2}\right)_{x=0} = \left(\frac{\mathrm{d}^2 u}{\mathrm{d}x^2}\right)_{x=1} = 0 . \tag{7.3.8}$$

We continue our analysis by choosing the solution strategy. The amplitude function $u(x)$ is found via the help of the fundamental solution of the linear homogeneous ordinary differential equation (7.3.7):

$$u_g(x) = e^{i\gamma x} , \quad 0 < x < 1 , \tag{7.3.9}$$

where γ is the wave number. Substituting (7.3.9) into (7.3.7), we obtain the characteristic equation

$$\varphi \equiv \gamma^4 - (V^2 - C^2)\gamma^2 - 2\omega V \gamma - \omega^2 = 0 , \tag{7.3.10}$$

where we have defined the polynomial $\varphi \equiv \varphi(\gamma)$.

Let γ_1, γ_2, γ_3 and γ_4 be the roots of the characteristic Eq. (7.3.10). Then we can represent the solution of Eq. (7.3.7) as

$$u(x) = \sum_{k=1}^{4} A_k \exp(i\gamma_k(x - \frac{1}{2})) , \tag{7.3.11}$$

where A_k ($k = 1, 2, 3, 4$) are constants, which can be determined with the help of the boundary conditions (7.3.8). The shift by $-1/2$ is just a constant convenience factor that is accounted for in the definition of the constants A_k. (Strictly speaking, the solution (7.3.11) only works without modification if all the roots of the polynomial (7.3.10) are distinct. As is known from the theory of ordinary differential equations, for example in the case of double roots, the solution will have terms $e^{i\gamma_k x}$ and $xe^{i\gamma_k x}$, where γ_k is the double root).

In terms of the roots γ_1, γ_2, γ_3 and γ_4, we rewrite the characteristic Eq. (7.3.10) in the following form, by the fundamental theorem of algebra:

$$\begin{aligned}
\varphi &= (\gamma - \gamma_1)(\gamma - \gamma_2)(\gamma - \gamma_3)(\gamma - \gamma_4) \\
&= \gamma^4 - (\gamma_1 + \gamma_2 + \gamma_3 + \gamma_4)\gamma^3 + [\gamma_1\gamma_2 + \gamma_3\gamma_4 + (\gamma_1 + \gamma_2)(\gamma_3 + \gamma_4)]\gamma^2 \\
&\quad - [(\gamma_1 + \gamma_2)\gamma_3\gamma_4 + (\gamma_3 + \gamma_4)\gamma_1\gamma_2]\gamma + \gamma_1\gamma_2\gamma_3\gamma_4 = 0 .
\end{aligned} \tag{7.3.12}$$

If we compare the expressions (7.3.10) and (7.3.12) for φ, and equate the coefficients for like powers of γ, we obtain a system of four algebraic equations

$$\gamma_1 + \gamma_2 + \gamma_3 + \gamma_4 = 0 , \tag{7.3.13}$$

$$\gamma_1\gamma_2 + \gamma_3\gamma_4 + (\gamma_1 + \gamma_2)(\gamma_3 + \gamma_4) = -(V^2 - C^2) , \tag{7.3.14}$$

$$(\gamma_1 + \gamma_2)\gamma_3\gamma_4 + (\gamma_3 + \gamma_4)\gamma_1\gamma_2 = 2\omega V , \tag{7.3.15}$$

$$\gamma_1\gamma_2\gamma_3\gamma_4 = -\omega^2 . \tag{7.3.16}$$

Let us now concentrate on the case where ω is real-valued. It is a general property of the polynomial equation (7.3.10), which in this case has real-valued coefficients, that if there exists a complex root of the Eq. (7.3.10), then there exists also a complex conjugate root. In accordance with this observation, it is convenient to introduce new variables s_1, σ_1, s_2 and σ_2 using the relations

$$s_1 = \gamma_1 + \gamma_2 \,, \quad \sigma_1 = \gamma_1\gamma_2 \,, \quad s_2 = \gamma_3 + \gamma_4 \,, \quad \sigma_2 = \gamma_3\gamma_4 \,, \tag{7.3.17}$$

and then choose γ_2 and γ_4 to be the complex conjugate values with respect to γ_1 and γ_3, respectively, that is, $\gamma_2 = \gamma_1^*$ and $\gamma_4 = \gamma_3^*$. Then it follows that the new variables s_1, σ_1, s_2 and σ_2 are always real.

Such a choice is always possible, because the left-hand side of (7.3.14) contains all two-element products from the set $\{\gamma_1, \gamma_2, \gamma_3, \gamma_4\}$, and the left-hand side of (7.3.15) contains all three-element products. Hence, the particular arrangement of factors used in (7.3.14) and (7.3.15) is arbitrary. Choosing which of the γ_k represents which root of the characteristic equation is equivalent with first taking some fixed ordering of the roots as given, then rewriting the factorizations in the manner appropriate for that ordering, and finally renumbering the γ_k (and possibly reordering terms) so that the particular form (7.3.13)–(7.3.16) is obtained.

The roots γ_1, γ_2, γ_3 and γ_4 are expressed in terms of the new variables as

$$\gamma_{1,2} = \frac{1}{2}(s_1 \pm a_1) \,, \quad a_1 = \sqrt{s_1^2 - 4\sigma_1} \,, \tag{7.3.18}$$

$$\gamma_{3,4} = \frac{1}{2}(s_2 \pm a_2) \,, \quad a_2 = \sqrt{s_2^2 - 4\sigma_2} \,. \tag{7.3.19}$$

Equations (7.3.18), (7.3.19) give us a condition for the roots to be distinct: it must hold that $a_1 \neq 0$ and $a_2 \neq 0$.

Using the new variables, Eq. (7.3.12) becomes

$$\varphi = \gamma^4 - (s_1 + s_2)\gamma^3 + (\sigma_1 + \sigma_2 + s_1 s_2)\gamma^2 - (\sigma_1 s_2 + \sigma_2 s_1)\gamma + \sigma_1\sigma_2 = 0 \,. \tag{7.3.20}$$

From (7.3.13) and (7.3.17) it follows that $s_1 + s_2 = 0$, and consequently we may eliminate one variable by defining

$$s \equiv s_1 = -s_2 \,. \tag{7.3.21}$$

The relations (7.3.13)–(7.3.16) are thus transformed into

$$\sigma_1 + \sigma_2 - s^2 = -(V^2 - C^2) \,, \tag{7.3.22}$$

$$(\sigma_2 - \sigma_1)s = 2\omega V \,, \tag{7.3.23}$$

$$\sigma_1\sigma_2 = -\omega^2 \,. \tag{7.3.24}$$

Equations (7.3.22), (7.3.24) are always valid, regardless of whether the roots of the characteristic equation are distinct, because they follow directly from the characteristic equation.

Let us first deal with a special case. Let V and C be free. If $s = 0$, at first glance it seems we have the trivial solution $\omega = 0$, $\sigma_1 = 0$, $\sigma_2 = 0$, $V^2 = C^2$, but this leads to a quadruple root $\gamma = 0$, so we must be careful. In this special case, the solution $u(x)$ is a cubic polynomial (because now $\gamma = 0$, the exp factor is just a constant). Refer to the original partial differential equation (7.3.7); we are left with $\mathrm{d}^4 u/\mathrm{d}x^4 = 0$ and the original boundary conditions. All of the coefficients of $u(x)$ must vanish to satisfy the boundary conditions, whence the only solution is the trivial one. From other investigations it is known that for traveling panels and beams with finite bending rigidity, static ($\omega = 0$) critical points do exist, but at $|V| > C$. At such points, the fourth- and the second-order terms of (7.3.7) balance out, and (with simply supported boundary conditions) the buckling mode is a sine.

We will now concentrate on the general case, with $s \neq 0$. As was mentioned above, the solution (7.3.11) for the amplitude function $u(x)$ contains the arbitrary constants A_k ($k = 1, 2, 3, 4$), which can be determined with the help of the boundary conditions (7.3.8). Using (7.3.11) and (7.3.8), we obtain the following system of linear algebraic equations written in matrix form:

$$\mathbf{R}A = 0 \,, \tag{7.3.25}$$

where

$$\mathbf{R} = \begin{bmatrix} \psi_1^- & \psi_2^- & \psi_3^- & \psi_4^- \\ -\gamma_1^2\psi_1^- & -\gamma_2^2\psi_2^- & -\gamma_3^2\psi_3^- & -\gamma_4^2\psi_4^- \\ \psi_1^+ & \psi_2^+ & \psi_3^+ & \psi_4^+ \\ -\gamma_1^2\psi_1^+ & -\gamma_2^2\psi_2^+ & -\gamma_3^2\psi_3^+ & -\gamma_4^2\psi_4^+ \end{bmatrix} \,, \quad A = \begin{bmatrix} A_1 \\ A_2 \\ A_3 \\ A_4 \end{bmatrix} \,, \tag{7.3.26}$$

and

$$\psi_k^+ = \exp(i\frac{\gamma_k}{2}) \,, \quad \psi_k^- = \exp(-i\frac{\gamma_k}{2}) \,, \quad k = 1, 2, 3, 4 \,. \tag{7.3.27}$$

The terms involving ψ_k^- follow from boundary conditions at $x = 0$, while the ψ_k^+ terms come from boundary conditions at $x = 1$.

As is known from linear algebra, a nontrivial solution $A \not\equiv 0$ of the homogeneous linear equation system (7.3.25) exists if and only if the determinant is equal to zero, that is,

$$\Delta = \det \mathbf{R} = 0 \,. \tag{7.3.28}$$

The minus signs on the second and fourth rows of $\mathbf{R}$ can be eliminated by multiplying those rows by -1, leaving the value of the determinant unchanged.

Using (7.3.18), (7.3.19) and (7.3.25)–(7.3.27), and performing the necessary transformations, it follows that the solvability condition (7.3.28) for the spectral problem (7.3.7), (7.3.8) can be represented as

$$\Delta(\omega, V) = a_1 a_2 s^2 \left(\cos s - \cos\frac{a_1}{2}\cos\frac{a_2}{2}\right) + \\ \left(s^4 - 2(\sigma_1+\sigma_2)s^2 - 2(\sigma_1-\sigma_2)^2\right)\sin\frac{a_1}{2}\sin\frac{a_2}{2} = 0\,, \tag{7.3.29}$$

where

$$a_{1,2} = \sqrt{s^2 - 4\sigma_{1,2}}\,. \tag{7.3.30}$$

The quantities σ_1 and σ_2, and $s \equiv s_1$ are given by (7.3.17). The dependence of Δ on the frequency ω is implicit, via $s = s(\omega, V)$, $\sigma_1 = \sigma_1(\omega, V)$ and $\sigma_2 = \sigma_2(\omega, V)$.

With the same procedure, the solvability condition can be derived for cases with other boundary conditions. As an example, if clamped boundary conditions $u(0) = u(1) = (\mathrm{d}u/\mathrm{d}x)(0) = (\mathrm{d}u/\mathrm{d}x)(1) = 0$ are used, then the matrix corresponding to $\mathbf{R}$ in (7.3.26) is

$$\mathbf{R}_{\mathrm{CC}} = \begin{bmatrix} \psi_1^- & \psi_2^- & \psi_3^- & \psi_4^- \\ i\gamma_1\psi_1^- & i\gamma_2\psi_2^- & i\gamma_3\psi_3^- & i\gamma_4\psi_4^- \\ \psi_1^+ & \psi_2^+ & \psi_3^+ & \psi_4^+ \\ i\gamma_1\psi_1^+ & i\gamma_2\psi_2^+ & i\gamma_3\psi_3^+ & i\gamma_4\psi_4^+ \end{bmatrix}, \tag{7.3.31}$$

and the solvability condition becomes

$$\Delta_{\mathrm{CC}}(\omega, V) \equiv \cos s + \frac{2(\sigma_1+\sigma_2)+s^2}{a_1 a_2}\sin\frac{a_1}{2}\sin\frac{a_2}{2} - \cos\frac{a_1}{2}\cos\frac{a_2}{2} = 0\,. \tag{7.3.32}$$

The subscript CC indicates that the quantities (7.3.31) and (7.3.32) correspond to the case with the clamped boundary condition at each end of the domain. The symbols ψ_k^+, ψ_k^-, s, σ_1, σ_2, a_1 and a_2 have the exact same definitions as in the simply supported case. The imaginary units on the second and fourth rows of (7.3.31) can be cancelled out by multiplying those rows by $-i$, which (in total) only scales the value of the determinant by the constant $(-i)^2 = -1$. In this study, we will restrict our consideration to the simply supported case, represented by (7.3.26) and (7.3.29).

One must be aware that when using the derivation presented here, Eq. (7.3.29) depends on the particular form of the solution (7.3.11), and thus requires that the roots of the characteristic equation are distinct. It is possible to use Eq. (7.3.29) also at points (ω, V) where multiple roots occur (points where $a_1 = 0$ or $a_2 = 0$), but then we must require also $\partial\Delta/\partial\omega = 0$ to exclude spurious solutions. In the eigenvalue analysis of a vibration problem, solutions typically stay distinct, meeting only at bifurcation points. Here we have formulated the condition for a bifurcation point to exist for the curve $\omega(V)$, which is the unknown being solved here. We require the conditions of the implicit function theorem to be violated, taking $\Delta(\omega, V)$ as the function for which the equation $\Delta(\omega, V) = 0$ implicitly describes the relation $\omega(V)$. (In general there are several such curves, one for each vibration mode). If there exists a point (ω, V) at which $\Delta = 0$ and $\partial\Delta/\partial\omega = 0$, then the uniqueness of the representation $\omega(V)$ fails in a small neighborhood of that point

Whether or not (7.3.11) gives us the correct vibration mode under such exceptional circumstances is a separate issue; what we regardless gain from (7.3.29) is the correct bifurcation value for $\omega(V)$. We are essentially playing with the continuity of $\Delta(\omega, V)$ to extract the correct ω even at a point where the assumptions underlying (7.3.29) fail. Arbitrarily close to such a point (7.3.29) remains valid, and the solution curve is continuous, so the solution must have a unique limit there.

Equation (7.3.29) represents a constraint for triples (σ_1, σ_2, s) that give rise to nontrivial solutions of (7.3.25). It implicitly eliminates one of σ_1, σ_2 or s. Considered together with the equation system (7.3.22)–(7.3.24), the remaining unknowns are, in principle, ω and any two of σ_1, σ_2 and s. The axial velocity V is considered a prescribed problem parameter. To obtain the frequency spectrum as a function of V, the velocity can be varied quasistatically in the standard manner. Thus, considering the task of determining the wave number parameters γ_k $(k = 1, 2, 3, 4)$ and the corresponding free vibration frequency ω, we have three equations remaining, with three remaining unknowns. The consideration of the boundary conditions, in the form of (7.3.25)–(7.3.27), has closed the algebraic system, as indeed expected.

The amplitude function $u(x)$ can be assembled using (7.3.11), once a frequency ω and its corresponding σ_1, σ_2 and s (and hence all four γ_k, via Eqs. (7.3.18), (7.3.19) and (7.3.21)) are known. When the solvability condition (7.3.29) is fulfilled, the matrix $\mathbf{R}$ is singular. Thus, in order to actually determine the values of the A_k from (7.3.25), one must eliminate some of the A_k algebraically, until the remaining matrix has full rank (and hence the linear equation system yields a unique solution). In this study, we will concentrate only on the frequencies ω.

In what follows we will consider (7.3.22)–(7.3.24) as a system of nonlinear equations for the real variables σ_1, σ_2 and s. Suppose that $\sigma_1(\omega, V)$, $\sigma_2(\omega, V)$ and $s(\omega, V)$ are two-parameter solutions of the considered system, corresponding to a given value of C. If we substitute the corresponding expressions into (7.3.29), we will obtain the implicit equation

$$\Delta(\omega, V) = 0 \tag{7.3.33}$$

for determining the frequencies ω of the moving panel, as a function of the panel axial velocity V. Note that as pointed out above, Eq. (7.3.33) determines a set of solutions $\omega_j(V)$. Also, keep in mind that our solution is valid for the range of velocities at which the frequency ω_j is real-valued.

Recall that we restrict consideration to the general case, where $s \neq 0$. From Eqs. (7.3.22), (7.3.23), we have

$$\begin{aligned} \sigma_1 &= \frac{1}{2}\left[s^2 - (V^2 - C^2)\right] - \omega\frac{V}{s} \, . \\ \sigma_2 &= \frac{1}{2}\left[s^2 - (V^2 - C^2)\right] + \omega\frac{V}{s} \, . \end{aligned} \tag{7.3.34}$$

These equations follow directly from the characteristic equation, under only the assumption that ω is real-valued. Thus, they are valid whenever $s \neq 0$ and $\omega \in \mathbb{R}$; the roots of the characteristic equation need not be distinct.

After substituting (7.3.34) into (7.3.24) and multiplying the equation by $4s^2$, we find the relation connecting (ω, V, s):

$$s^2\left[s^2 - (V^2 - C^2)\right]^2 = 4\omega^2(V^2 - s^2)\,. \tag{7.3.35}$$

The same comment as for (7.3.34) applies.

Because in our solution range, ω is real-valued, from (7.3.35) we also find the following constraint for s:

$$0 < s^2 \leq V^2\,. \tag{7.3.36}$$

Equality at the lower limit is not possible in our present solution, because (7.3.35) was derived from Eq. (7.3.34), which are valid if $s \neq 0$. Equality at the upper limit holds in the special case $C = 0$; then we have $s = V$. This allows us to simplify (7.3.29) somewhat; this special case will be handled later.

Recalling that C is a known problem parameter, and keeping ω free for now, relation (7.3.35) allows us to eliminate one of V or s. Eliminating $s = s(\omega, V)$ allows us to write the expression $\Delta(\omega, V)$, originally given in terms of s as Eq. (7.3.29), in terms of ω and V. However, it should be noted that (7.3.35) is a cubic polynomial with respect to the variable s^2; hence its explicit solution is unwieldy to write out, and it is much more convenient in practice to employ a numerical root finder.

The other possibility is to eliminate $V = V(\omega, s)$. The polynomial is only quadratic in V^2, allowing a short explicit solution (valid where $s \neq 0$ and ω real-valued):

$$V^2 = s^2 + C^2 + 2\left(\frac{\omega}{s}\right)^2 \pm 2\sqrt{\left(\frac{\omega}{s}\right)^2\left(C^2 + \left(\frac{\omega}{s}\right)^2\right)}\,. \tag{7.3.37}$$

In practice, the solution with the minus sign for the square root term is the physically relevant one. This allows us, if we wish, to explicitly find the value of (7.3.29) at any point in the (ω, s) plane (with the help of (7.3.37), (7.3.34) and (7.3.29), in that order). Keep in mind, however, that the physical interpretation of solution curves in the (ω, s) plane is more difficult than for solution curves in the (ω, V) plane. We will show both kinds of curves in our numerical results below.

Let us return to the task of eliminating $s = s(\omega, V)$. We introduce a new positive variable

$$\tau = s^2\,. \tag{7.3.38}$$

Using (7.3.38), Eq. (7.3.35) becomes a cubic polynomial equation in τ:

$$\tau^3 - 2(V^2 - C^2)\tau^2 + \left[(V^2 - C^2)^2 + 4\omega^2\right]\tau - 4\omega^2 V^2 = 0\,. \tag{7.3.39}$$

This equation is valid whenever $\tau \neq 0$ (i.e., $s \neq 0$) and ω is real.

Observe that a positive solution $\tau > 0$ of (7.3.39) exists for any nonnegative values of ω^2 and V^2. The constant term of the polynomial depends only on the squares ω^2, V^2, and its sign is negative. Hence at $\tau = 0$, the left-hand side of (7.3.39) will be ≤ 0, with equality only if $\omega = V = 0$. The sign of the cubic term (which dominates for large $|\tau|$), on the other hand, is positive. The polynomial is continuous as a function of τ. Thus, as τ increases, the polynomial on the left-hand side will inevitably eventually cross zero at least once. Therefore, because $\tau = s^2$, at least one positive solution of (7.3.35) always exists in terms of s^2. Thus (7.3.39) determines the dependence $s = s(\omega, V)$ (we pick the smallest positive solution for s^2).

For any point in the (ω, V) plane, we first determine s from (7.3.39). (If $C = 0$, and we nevertheless wish to use the general solution procedure, it is possible to use the fact about this special case that $s = V$, skipping this step). Then we use (7.3.34) to obtain $\sigma_1(\omega, V)$ and $\sigma_2(\omega, V)$, and finally, determine the value of $\Delta(\omega, V)$ via (7.3.29). This allows us to look for solutions of (7.3.33) in the (ω, V) plane.

Above, we have treated the general case having $s \neq 0$, for any value for V, and with C nonzero. The solution given above is also applicable if $C = 0$ (no axial tension or compression), but in this special case it is possible to simplify the formulas, which we will do next.

We observed that if $C = 0$, then in (7.3.36) we have equality at the upper limit, in other words, it holds that $s = V$. This allows us to provide the following special-case result. Inserting $C = 0$ and $s = V$ into (7.3.34), we immediately obtain

$$\left.\begin{aligned} \sigma_1 &= -\omega \\ \sigma_2 &= +\omega \end{aligned}\right\} , \quad \text{if } C = 0 . \tag{7.3.40}$$

Equation (7.3.34) were derived under the assumption $s \neq 0$, that is, in this special case, it would seem we must have $V \neq 0$ in order for (7.3.40) to be applicable. However, we may alternatively insert $C = 0$ and $s = V$ directly into (7.3.22)–(7.3.24), which are always valid. We obtain

$$\left.\begin{aligned} \sigma_1 + \sigma_2 &= 0 \\ (\sigma_2 - \sigma_1)V &= 2\omega V \\ \sigma_1\sigma_2 &= -\omega^2 \end{aligned}\right\} , \quad \text{if } C = 0 . \tag{7.3.41}$$

If $V \neq 0$ (i.e., $s \neq 0$), we may proceed to derive the relations (7.3.34) as before, and obtain (7.3.40). If $V = 0$, the second equation of (7.3.41) vanishes identically, and it is easily seen that (7.3.40) is a solution satisfying the remaining two equations. Thus we may omit the requirement on V.

Substituting (7.3.40) into (7.3.29) produces first

$$\begin{aligned} a_1 &= \sqrt{V^2 + 4\omega} , \\ a_2 &= \sqrt{V^2 - 4\omega} , \end{aligned} \tag{7.3.42}$$

and then, for $\Delta(\omega, V)$ (after multiplication of both sides by $2/a_1 a_2$),

$$\Delta(\omega, V) = 2V^2 \left(\cos V - \cos \frac{\sqrt{V^2 + 4\omega}}{2} \cos \frac{\sqrt{V^2 - 4\omega}}{2} \right) + \left(V^4 - 8\omega^2\right) \frac{\sin \frac{\sqrt{V^2 + 4\omega}}{2}}{\frac{\sqrt{V^2 + 4\omega}}{2}} \frac{\sin \frac{\sqrt{V^2 - 4\omega}}{2}}{\frac{\sqrt{V^2 - 4\omega}}{2}} = 0 \,. \tag{7.3.43}$$

Equation (7.3.43) explicitly shows how Δ depends on ω and V in the special case $C = 0$. As for its range of applicability, (7.3.43) requires $a_1 \neq 0$ and $a_2 \neq 0$ as before; refer to (7.3.42). This is because Eq. (7.3.43) follows from (7.3.29), which already has this requirement. Especially, looking at the expression of a_2, we expect Eq. (7.3.29) not to be valid on the curve $\omega = (1/4)V^2$; this will be observed in the numerical results below.

Consider now another special case, where $V = 0$, $\omega \neq 0$ and C free, that corresponds to harmonic vibrations of a stationary (as opposed to axially moving) panel, subjected to extension or compression with load C (which may be zero or nonzero).

In this case, the nonlinear algebraic equation system (7.3.22)–(7.3.24) takes the form

$$\begin{aligned} \sigma_1 + \sigma_2 &= s^2 + C^2 \,, \\ (\sigma_2 - \sigma_1)s &= 0 \,, \\ \sigma_1 \sigma_2 &= -\omega^2 \,. \end{aligned} \tag{7.3.44}$$

Starting from the second equation of the system (7.3.44) and then proceeding to the other two equations, we obtain two distinct possibilities, namely either

$$s = 0 \,, \quad \sigma_1 + \sigma_2 = C^2 \,, \quad \sigma_1 \sigma_2 = -\omega^2 \tag{7.3.45}$$

or

$$\sigma_1 = \sigma_2 \equiv \sigma \,, \quad \sigma^2 = -\omega^2 \,, \quad s^2 = 2\sigma - C^2 \,. \tag{7.3.46}$$

If $C \neq 0$, the second possibility (7.3.46) is not applicable, because ω, σ_1, σ_2 and s are real-valued, and hence $\sigma^2 = -\omega^2$ has no solution except $\sigma = \omega = 0$. This, in turn, leads to $s^2 = -C^2$, which has no real-valued solution for $C \neq 0$.

If $C = 0$, then $\sigma = \omega = s = 0$ is a solution of (7.3.46). But this leads to $a_1 = a_2 = 0$, in which case Eq. (7.3.29) is not applicable, and for a full analysis, a new solvability condition must be derived. However, this case is not very interesting, since it implies $\gamma_k = 0$ for all $k = 1, 2, 3, 4$; recall (7.3.18), (7.3.19) and (7.3.21). This case will be omitted for brevity. (Recall that at the beginning, we already analyzed a similar case where V and C were both free).

Thus we see that in general, we must pick the first possibility (7.3.45). The stationary problem, $V = 0$, leads to $s = 0$. Observe that Eq. (7.3.45) are valid regardless of whether $C \neq 0$ or $C = 0$.

From the last two equations of the system (7.3.45), for $s = 0$ we have

$$\begin{aligned} \sigma_1 &= \frac{1}{2}\left[C^2 - \sqrt{C^4 + 4\omega^2}\right] < 0\,, \\ \sigma_2 &= \frac{1}{2}\left[C^2 + \sqrt{C^4 + 4\omega^2}\right] > 0\,. \end{aligned} \tag{7.3.47}$$

The inequalities hold, because we are considering the case $\omega \neq 0$. Observe that

$$a_{1,2} = \sqrt{s^2 - 4\sigma_{1,2}} = \sqrt{-4\sigma_{1,2}} \neq 0\,,$$

and hence the solvability condition (7.3.29) is valid also in this case. From (7.3.29), we obtain after inserting $s = 0$ that

$$\begin{aligned} \Delta(\omega,\, V{=}0) &= -2\left((\sigma_1 - \sigma_2)^2 \sin\frac{a_1}{2} \sin\frac{a_2}{2}\right) = 0\,, \\ a_{1,2} &= \sqrt{-4\sigma_{1,2}}\,. \end{aligned} \tag{7.3.48}$$

This can be simplified as

$$\begin{aligned} a_1 &= 2\sqrt{-\sigma_1}\,, \quad a_2 = 2i\sqrt{\sigma_2}\,, \\ \sin\frac{a_1}{2} &= \sin\sqrt{-\sigma_1}\,, \quad \sin\frac{a_2}{2} = i\sinh\sqrt{\sigma_2}\,. \end{aligned} \tag{7.3.49}$$

Summarizing, Eqs. (7.3.47)–(7.3.49) give the solution for the special case $V = 0$, $\omega \neq 0$ and C free.

In the following we will consider numerical solution of the auxiliary polynomial problem. Noting that Eq. (7.3.29) is implicit, it will be necessary to be able to quickly evaluate it at a large set of points in the (ω, V) plane in order to numerically find the zeroes of $\Delta(\omega, V)$. This, in turn, requires finding the roots of the cubic polynomial (7.3.39) at each point where $\Delta(\omega, V)$ is being evaluated.

The same consideration applies also to the case of evaluating (7.3.29) in the (ω, s) plane, which requires finding the roots of a quadratic polynomial, but of course there the solution algorithm is shorter. Despite the simplicity, we caution the interested reader to look at Press et al. [24, p. 227], because the standard analytical quadratic formula (which gave us the analytical result (7.3.37)) is numerically inaccurate for one of the roots (here the physically relevant one!) due to catastrophic cancellation in floating-point (i.e., finite precision) arithmetic. A small modification produces an algorithm that gives accurate results for both roots.

We conclude that in either case, a numerical solver for (7.3.29) requires using a polynomial equation solver. Since the details can be implemented as a subroutine, from a practical perspective it does not matter whether one computes solutions in the (ω, V) or the (ω, s) plane. It also does not matter (from the viewpoint of the solution algorithm of our primary problem of interest) how the polynomial solver works internally. This algorithmic viewpoint abstracts away also the process of choosing the physically relevant root. To the main solver algorithm, the polynomial solver appears as a mathematical function that takes in problem parameters and returns a unique value.

An approach providing low computational cost is to use an explicit analytical solution algorithm. (The difference between an explicit algorithm and a classical closed-form solution is mainly the length of the procedure, and the use of named intermediate results instead of inserting everything into one expression). One such algorithm, using trigonometric functions (originally due to work by François Viète, dating from the 16th century), is documented in Press et al. [24, pp. 227–229] (given also in the older edition [25, p. 179]) and is recommended for its numerical stability. The companion matrix method given in Press et al. [26, pp. 146–147], which is a general numerical method for solving arbitrary-degree polynomial equations, is another option offering good numerical stability.

Let us now consider the full panel problem in the general case. The equations to be solved to obtain a numerical solution are (7.3.29), (7.3.34) (two equations) and (7.3.35). In the numerical examples to follow, we will present the behavior of the first four frequencies ω, as a function of the panel axial velocity V, at some fixed values of the tension parameter C.

At any given point (ω, V), we evaluate the sequence (7.3.39), (7.3.34) and (7.3.29), in that order. When instead working in the (ω, s) plane, we replace (7.3.39) by (7.3.37) (in principle; in practice the quadratic polynomial should be solved by an algorithm avoiding cancellation error) to determine V from s; the rest of the sequence is identical. This evaluation produces a numerical value for the parametric expression (7.3.29), which we know to be zero at points that correspond to a solution.

One must be aware that in general, (7.3.29) is complex-valued. Although each quantity under the square roots in (7.3.29) is real-valued, there is no guarantee that these quantities are nonnegative. Indeed, complex values appeared already in the special case $s = 0$, in Eq. (7.3.49).

At closer observation of (7.3.29), it is seen that at any point (ω, V), either $\operatorname{Re} \Delta(\omega, V) = 0$ or $\operatorname{Im} \Delta(\omega, V) = 0$. Because the quantities under the square roots are always real-valued, the square roots are always either purely real or purely imaginary. In practice, in the numerical examples a_1 was always real, and a_2 obtained both real and imaginary values. Looking at each term of (7.3.29), the outcome is that $\Delta(\omega, V)$ itself is always either purely real or purely imaginary. Along the curve where $a_2 = 0$, spurious solutions will appear, where both the real and imaginary parts are zero. Solutions along this curve are not valid, because along that curve two of the roots of the characteristic equation coincide, and thus (7.3.29) is not applicable there (by itself without the additional condition on the derivative).

In order for a point (ω, V) to be a solution of (7.3.29), both the real and imaginary parts of $\Delta(\omega, V)$ must be zero at that point. Thus, it is convenient to shift our attention to the squared complex norm

$$\|\Delta\|^2 = \Delta\overline{\Delta} = (\operatorname{Re} \Delta)^2 + (\operatorname{Im} \Delta)^2 \ , \tag{7.3.50}$$

which is zero at only such points.

As actually computing some values of (7.3.50) quickly shows, the values of $\|\Delta\|^2$ are often very large. For example, in the case $C = 10$ with the plotting range set as $10^{-2} \leq \omega \leq 160$ and $10^{-2} \leq V \leq 15$, this expression obtains values up to 10^{23} (approximately). This range typically increases when the load parameter C is increased.

Thus, for the purposes of visualization of the values of $\|\Delta\|^2$ and numerical root-finding, it is more convenient to look at,for example,

$$g(\omega, V) \equiv \log(1 + \|\Delta\|^2) , \tag{7.3.51}$$

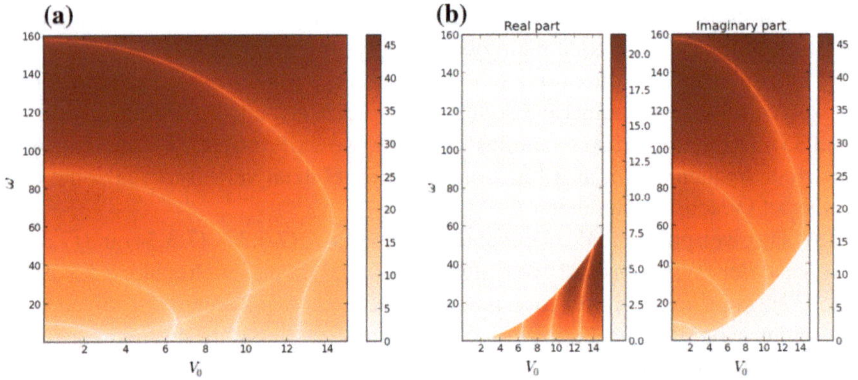

Fig. 7.11 Example with $C = 0$. The natural frequency curves are the zero level sets of the plotted expressions, with the exception of the seam between the real and imaginary regions (refer to subfigure (**b**)), where Eq. (7.3.29) (alone) is not valid. With $C = 0$, this seam follows the curve $\omega = (1/4)V^2$; see discussion following Eq. (7.3.43). **a**: Expression $g(\omega, V) = \log(1 + \|\Delta\|^2)$. **b**: Expressions $\log(1 + [\mathrm{Re}\,\Delta]^2)$ and $\log(1 + [\mathrm{Im}\,\Delta]^2)$

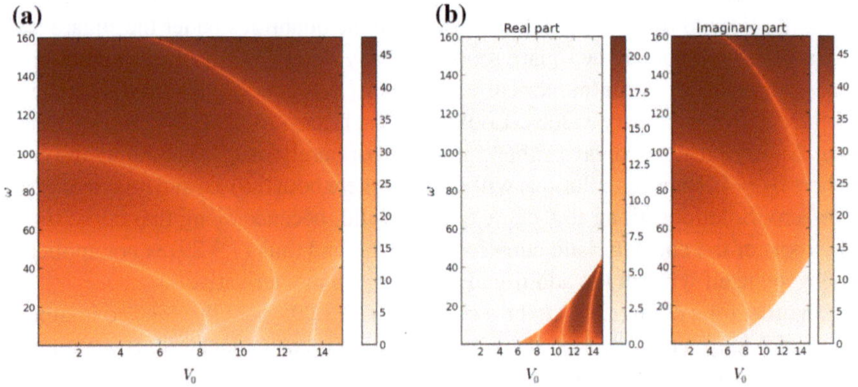

Fig. 7.12 Example with $C = 5$. The natural frequency curves are the zero level sets of the plotted expressions, with the exception of the seam between the real and imaginary regions (refer to subfigure (**b**)), where Eq. (7.3.29) (alone) is not valid. **a**: Expression $g(\omega, V) = \log(1 + \|\Delta\|^2)$. **b**: Expressions $\log(1 + [\mathrm{Re}\,\Delta]^2)$ and $\log(1 + [\mathrm{Im}\,\Delta]^2)$

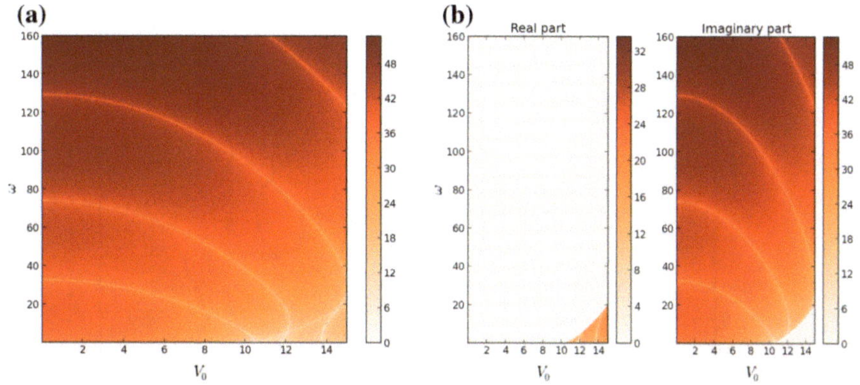

Fig. 7.13 Example with $C = 10$. The natural frequency curves are the zero level sets of the plotted expressions, with the exception of the seam between the real and imaginary regions (refer to subfigure (**b**)), where Eq. (7.3.29) (alone) is not valid. **a**: Expression $g(\omega, V) = \log(1 + \|\Delta\|^2)$. **b**: Expressions $\log(1 + [\mathrm{Re}\,\Delta]^2)$ and $\log(1 + [\mathrm{Im}\,\Delta]^2)$

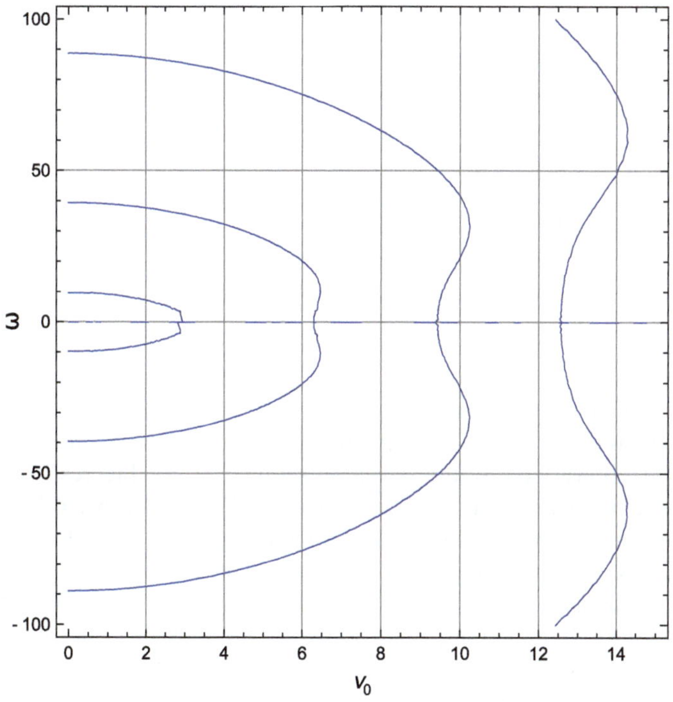

Fig. 7.14 Zero-level set of (7.3.29) with $C = 0$, showing the four lowest natural frequencies ω as a function of panel velocity V. Produced by directly searching for the zero level set of the complex-valued expression (7.3.43), using a contour plotter. Note that for $C = 0$, we have $V = s$ (see discussion following relation (7.3.36))

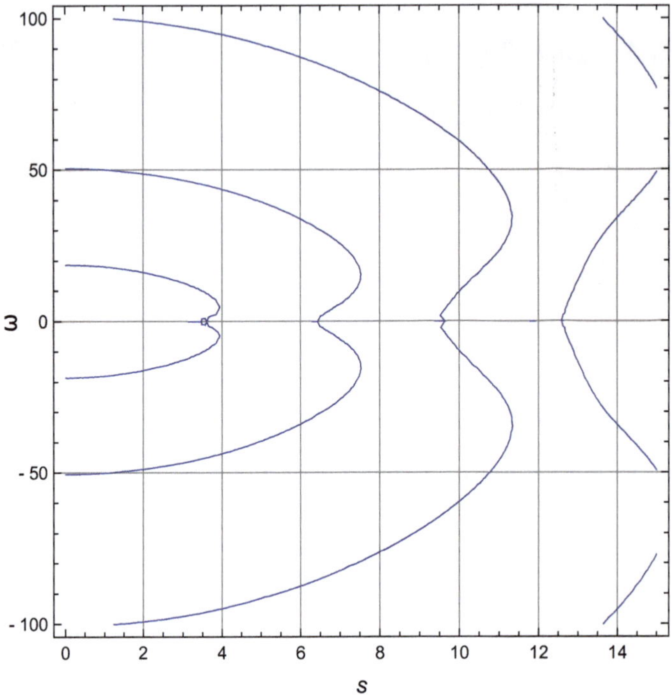

Fig. 7.15 Zero-level set of (7.3.29) with $C = 5$, showing the four lowest natural frequencies ω as a function of s. Produced by directly searching for the zero level set of the complex-valued expression (7.3.29), using a contour plotter

which reduces the output range to $0 \leq g < 52$ in the same example case. This also makes the gradients of the expression steeper near the solutions. Plots of (7.3.51) for $C = 0$, $C = 5$ and $C = 10$ are shown in Figs. 7.11, 7.12 and 7.13.

With a high-quality contour plotter, one strategy to find the solution curves $\omega(V)$ of (7.3.29) is to just look for the zero level set of $\Delta(\omega, V)$. Figures 7.14, 7.15 and 7.16, visualizing the solution in the (ω, s) plane for $C = 0$, $C = 5$ and $C = 10$, were produced via this approach. Note good quality in most part of the plotting area, and some accuracy issues near the V axis especially in the lowest mode.

Another approach, requiring only a simple numerical optimizer, is to search for the minima of (7.3.51). For any norm, $\|\cdot\| \geq 0$ for any value of the argument, and thus some of the minima can be expected to lie at the zeroes. After the minima are found, their values are easy to check against a prescribed tolerance. Points at which the value is smaller than the tolerance are then declared to be the solution points, and any other (spurious) minima are discarded. The minimization can be performed with regard to just one real variable. We track each solution curve, estimating its local tangent. To find the next point along the curve, we take a step in the direction of the

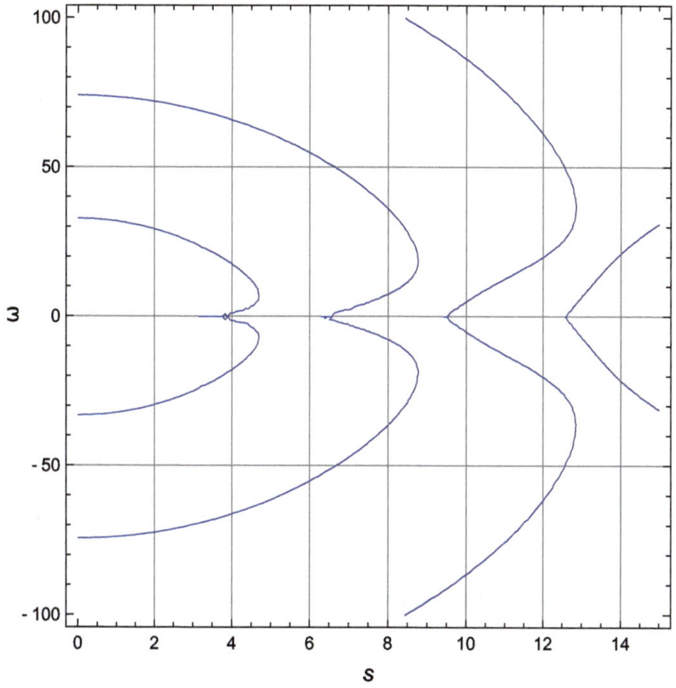

Fig. 7.16 Zero-level set of (7.3.29) with $C = 10$, showing the four lowest natural frequencies ω as a function of s. Produced by directly searching for the zero level set of the complex-valued expression (7.3.29), using a contour plotter

tangent, and minimize parametrically in the direction orthogonal to the tangent. See Fig. 7.20 for an illustration of the tracking process.

The final results in the (ω, V) plane for $C = 0$, $C = 5$ and $C = 10$ are shown in Figs. 7.17, 7.18 and 7.19, respectively. To produce these figures, we first plotted (7.3.51) on a Cartesian grid of 801×801 points (as shown in Figs. 7.11a, 7.12a and 7.13a), and visually determined the points where the solution curves cross the V and ω axes. Then, using these points as starting values, we numerically tracked the solutions, using the optimization approach. For the purposes of tracking, the plot area shown in the figures was scaled to have square aspect ratio, and the tracking step size was set to 0.25% of plot area width.

Each solution point was obtained by local minimization of (7.3.51) along a bounded line segment orthogonal to the local tangent of the curve. The bounds for the optimization were chosen as $\pm 5\ \Delta x$, where Δx is the tracking step size. Because the starting values are well-chosen and the curves are continuous, it is known a priori that the local minimum found at each step will lie at a zero of (7.3.51), and no check for the value of $\Delta(\omega, V)$ is needed.

The tangent of the curve to be tracked was initially taken to be orthogonal to the axis. This is a known property of the eigenfrequency curves for the panel/beam

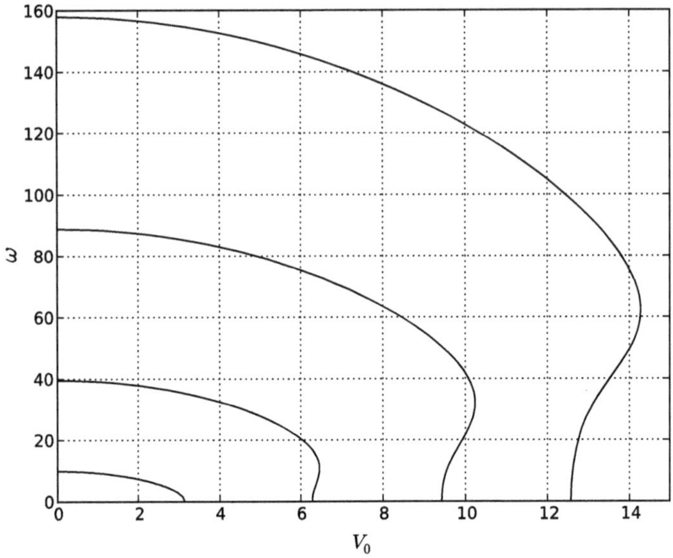

Fig. 7.17 Four lowest natural frequencies ω as a function of panel velocity V_0. Tension load parameter $C = 0$

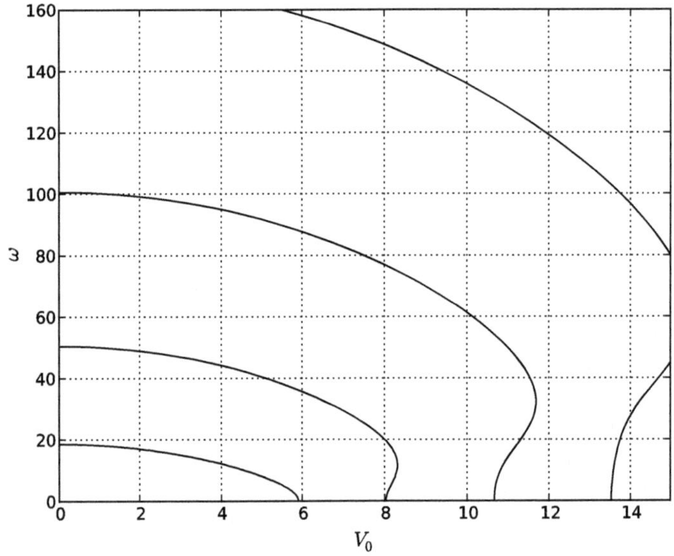

Fig. 7.18 Four lowest natural frequencies ω as a function of panel velocity V_0. Tension load parameter $C = 5$

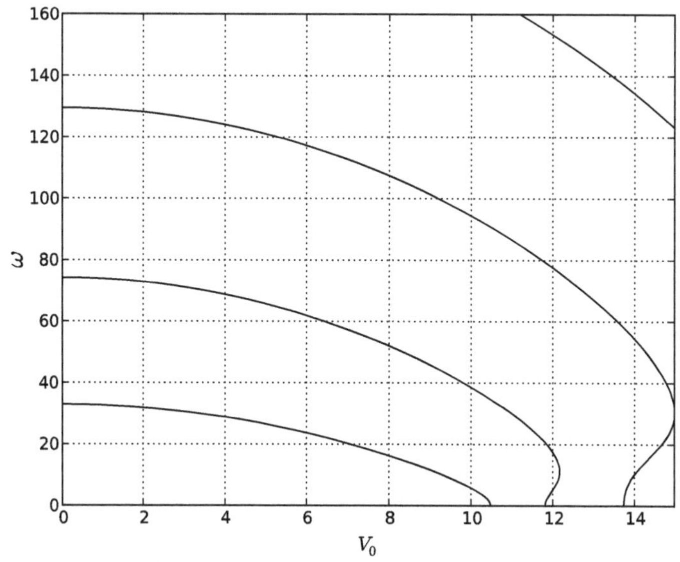

Fig. 7.19 Four lowest natural frequencies ω as a function of panel velocity V_0. Tension load parameter $C = 10$

model, and can be derived via the implicit function theorem; see [27]. At all further steps, the tangent direction was approximated as the difference vector between the latest two known solution points. Technically, this gives the first-order *backward* difference of the tangent at the last *known* solution point, but in practice this was found to be accurate enough for setting the search range for the optimizer.

For each solution curve, the tracking proceeded independently from each axis, and stopped at the seam where $\Delta(\omega, V)$ switches from real to imaginary or vice versa. This produced two independent solution curve segments for each solution curve.

When approaching the seam, the tracking process often caught a point or two on the spurious solution curve. These points were filtered out in a separate postprocessing step, utilizing the fact that the solution curves are at least C^1 continuous. In practice, this was implemented as a difference-based local tangent check. Each solution curve segment produced by the tracking phase was checked such that if the angle between the direction vectors determined by two successive point pairs (p_m, p_{m-1}) and (p_{m+1}, p_m) was larger than 5°, the points from $m + 1$ onward (inclusive) were discarded, terminating the postprocessing for that curve segment. After both segments belonging to the same solution curve were postprocessed, the segments were joined and plotted. The solutions behave smoothly enough near the seam that linear interpolation (which the plotter performs) between the last retained points is accurate enough to produce a smooth-looking visualization. Refer to Figs. 7.17, 7.18 and 7.19 for examples.

Let us conclude this section. The problem of free vibrations of a traveling beam (or panel) subjected to axial tension was investigated via the study of the dependences of

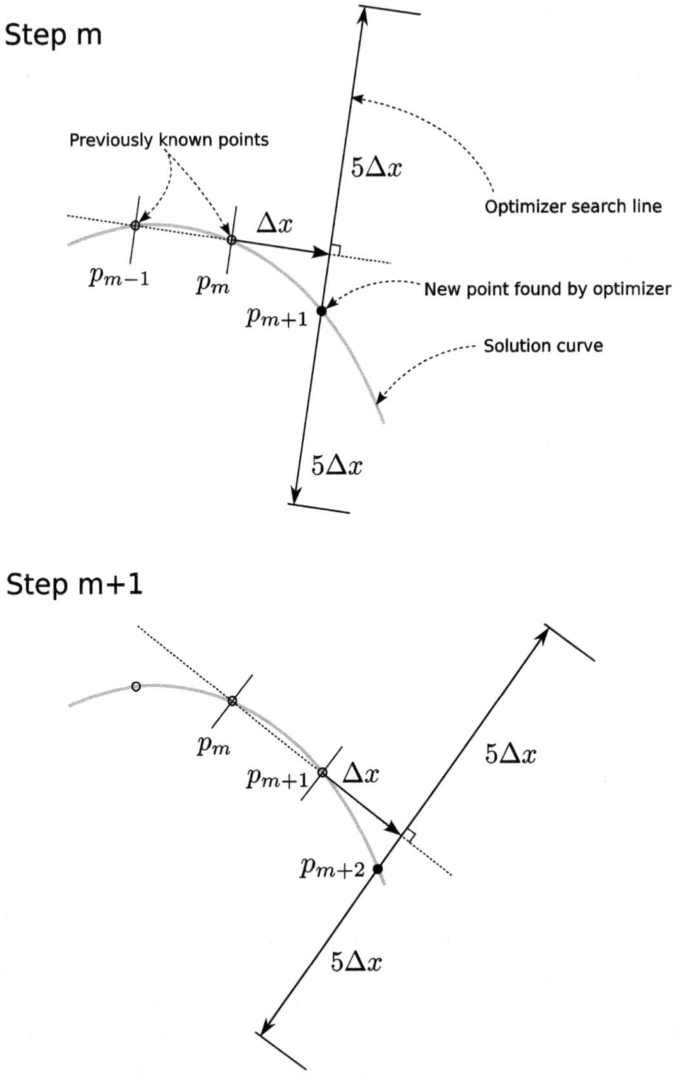

Fig. 7.20 Solution curve tracking strategy. The latest two known solution points are used to determine a direction to step in. A step of prescribed constant length Δx is taken, starting from the latest known solution point, and one-dimensional bounded minimization of (7.3.51) is performed along the orthogonal direction to locate the next solution point. The process then repeats

the system's natural frequencies on the problem parameters. The bending rigidity was allowed to be arbitrarily large. Analytical implicit expressions for the solution curves, with respect to problem parameters, were derived for ranges of the parameter space where the natural frequencies are real-valued, corresponding to stable vibrations (Fig. 7.20).

The classical special cases of the traveling panel without axial tension, and the tensioned stationary (nontraveling) panel were discussed, and special-case solutions given. Numerical evaluation of the obtained general analytical results was discussed, and numerical examples were given for panels subjected to two different tension levels, and for the nontensioned panel.

Simple models such as beams or panels, where applicable, are extremely useful in a modern setting, because they easily lend themselves to highly efficient and extremely fast solvers. The solution procedure presented here is computationally light, which allows the development of very efficient and very fast solvers for determining the natural frequencies of traveling beams and panels. Realtime solvers, in turn, have applications such as online optimal control, parametric studies of complex physical situations, modelling-based measurements, statistical quantification of uncertainty, agile modelling, and industrial optimization, at a very low computational cost.

7.4 Long Axially Moving Beam with Periodic Elastic Supports

In this chapter so far, we have looked at the free vibrations of traveling strings and rods with and without damping, and of traveling beams and panels. These are the most classical models of axially moving materials, but the picture is by no means complete. Different boundary conditions are perhaps the most obvious omission; we have only considered the most classical variant, namely pinhole conditions for the string and simply supported conditions for beams and panels. But this is not the only way in which to extend the models. It is possible to develop different variants, accounting for different physical problem setups, or including effects so far neglected. In the final two sections of this chapter, as a very small sampling of possible extensions, we will look at two very different ideas. In this section, we will consider a long (mathematically, infinite) beam traveling along a system of elastic rollers. In the final section, we will look at how a gravitational field affects the behavior of the classical traveling material.

We will now consider, using analytical approaches, an elastic beam of unlimited length and axially traveling between an infinite system of rollers (elastic supports) at a constant velocity. This is a highly simplified, theoretically fundamental model of the interior part of any system where an axially moving material travels through a system of many rollers. A typical example is a cylinder-based dryer section in a paper machine, although this model ignores the geometry.

Transverse elastic displacements of the beam are described by a fourth-order differential equation that includes the centrifugal effect, in-plane tension, bending resistance and the reaction of the elastic supports. The stability of the beam is investigated with the help of a small periodic transverse displacement. The multipoint spectral stability problem of the infinite traveling beam with elastic supports is formulated for the periodic interval. In the frame of spectral analysis, it is shown that the loss of stability takes place in the form of divergence. Perhaps the most interesting result from the analysis to follow is that the stability behavior of the infinite traveling beam with elastic supports coincides with that of the same beam with absolutely rigid supports, when the stiffness of the supports exceeds a critical value.

An infinite multi-span setup has received markedly less attention than the classical single-span setup. In Yurddas et al. [28], the nonlinear vibrations of an axially moving multisupported string have been investigated. They have studied a very similar case as here, with nonideal supports allowing minimal deflections between ideal supports at both ends of the string, but they used a Hamiltonian approach and concentrated on nonlinear dynamics. Moreover, in Chen [29] there is an extensive literature review of studies related to moving strings. Somewhat related literature includes also, for example, the studies concerning traveling strings and beams on an elastic foundation, by Bhat [30], Perkins [31], Wickert [32], Parker [33].

Let us start with the governing equations. The equation of unforced small transverse vibrations of a web traveling at a constant velocity V_0 along the axis x, and interacting with elastic supports at $x_n = n\ell$ $(n = 0, \pm 1, \pm 2, \ldots)$, has the form

$$m\frac{\partial^2 w}{\partial t^2} + 2mV_0\frac{\partial^2 w}{\partial x \partial t} + mV_0^2\frac{\partial^2 w}{\partial x^2} - T_0\frac{\partial^2 w}{\partial x^2} + D\frac{\partial^4 w}{\partial x^4} = 0\,, \tag{7.4.1}$$

where m is the mass per unit length of the beam, $D = EI$ is the bending rigidity of the beam (E is Young's modulus, I is the moment of inertia), T_0 is tension along the $x-$axis, and w is the small displacement in the $z-$direction, see Fig. 7.21. The Eq. (7.4.1) is written with respect to the fixed (Eulerian) reference frame xz.

We will concentrate on a static stability analysis of this system. In this case, the time derivatives in the Eq. (7.4.1) vanish. We will have

$$\left(mV_0^2 - T_0\right)\frac{\mathrm{d}^2 w}{\mathrm{d}x^2} + D\frac{\mathrm{d}^4 w}{\mathrm{d}x^4} = 0\,. \tag{7.4.2}$$

Introducing the notations

$$\lambda = \gamma^2 = \frac{1}{D}\left(mV_0^2 - T_0\right), \qquad \varkappa = \frac{k}{EI}\,, \tag{7.4.3}$$

we formulate the following multipoint spectral problem of elastic stability for the ordinary differential equation

$$\lambda\frac{\mathrm{d}^2 w}{\mathrm{d}x^2} + \frac{\mathrm{d}^4 w}{\mathrm{d}x^4} = 0\,, \qquad x_{j-1} < x < x_j\,, \tag{7.4.4}$$

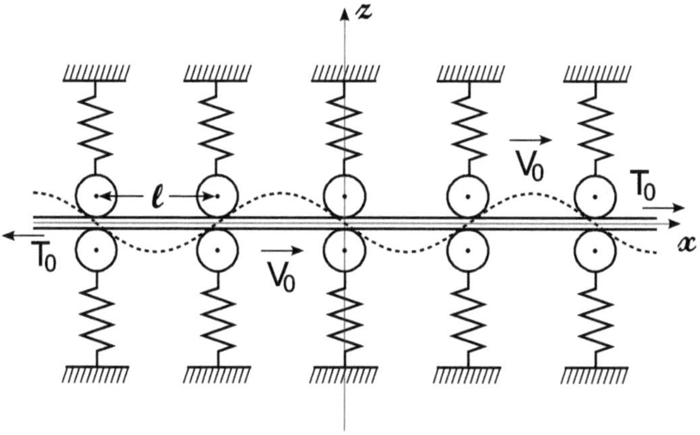

Fig. 7.21 Beam of unlimited length, axially traveling at velocity V_0, supported by an infinite system of elastic supports with spacing ℓ, and subjected to constant axial tension T_0. The straight lines represent the beam in the trivial equilibrium configuration, while the dashed line shows one possible deformed shape

where $x_j = j\ell,\ j = 0, \pm1, \pm2, ...$, with the conjunction conditions

$$(w)^+_{x_j} = (w)^-_{x_j}\,, \qquad \left(\frac{\mathrm{d}w}{\mathrm{d}x}\right)^+_{x_j} = \left(\frac{\mathrm{d}w}{\mathrm{d}x}\right)^-_{x_j}\,,$$

$$\left(\frac{\mathrm{d}^2w}{\mathrm{d}x^2}\right)^+_{x_j} = \left(\frac{\mathrm{d}^2w}{\mathrm{d}x^2}\right)^-_{x_j}\,, \qquad j = 0, \pm1, \pm2... \tag{7.4.5}$$

$$\left(\frac{\mathrm{d}^3w}{\mathrm{d}x^3}\right)^+_{x_j} - \left(\frac{\mathrm{d}^3w}{\mathrm{d}x^3}\right)^-_{x_j} = -\varkappa\,(w)_{x_j}\,,$$

where the upper symbols + and − denote right and left limiting values, respectively, and k (and its nondimensional counterpart $\varkappa$) is a constant usually called the modulus of the support (i.e., an elastic foundation). This constant denotes the reaction of the support per unit length when the deflection w is equal to unity. The parameter λ in the Eq. (7.4.4) plays the role of the eigenvalue of the considered spectral problem.

This problem is described by the ordinary differential equation (7.4.4) with constant coefficients and periodic boundary conditions (7.4.5). Consequently, its solution can be represented with the help of Floquet's theorem, as Floquet [34], Jkubovich and Starjinsky [35]

$$w\,(x, \alpha) = w_0\,(x)\, e^{i\alpha x}\,, \qquad -\infty < x < \infty\,, \qquad \alpha \in [\,0,\ 2\pi/\ell\,]\,. \tag{7.4.6}$$

Here $i \equiv \sqrt{-1}$ is the imaginary unit, α is a real parameter, and w_0 is a periodic function with period ℓ, that is,

$$w_0\,(x + s\ell) \equiv w_0\,(x)\,, \qquad s = \pm 1, \pm 2, \ldots\,. \tag{7.4.7}$$

Using the representation (7.4.6) and (7.4.7), it is required to find $w_0\,(x)$ in the interval $[0,\,\ell]$. Considering $w(x,\alpha)$ at $x = 0$ and $x = \ell$, and using the equality $w\,(0,\alpha) = w_0\,(0)$, we have

$$w\,(\ell,\alpha) = w_0\,(\ell)\,e^{i\alpha\ell} = w_0\,(0)\,e^{i\alpha\ell} = w\,(0,\alpha)\,e^{i\alpha\ell}\,. \tag{7.4.8}$$

Similarly, we obtain

$$\left(\frac{\mathrm{d}^s w}{\mathrm{d}x^s}\,(\ell,\alpha)\right)^{\pm} = \left(\frac{\mathrm{d}^s w}{\mathrm{d}x^s}\,(0,\alpha)\right)^{\pm} e^{i\alpha\ell}\,, \qquad s = 0, 1, 2, \ldots, \qquad \alpha \in [\,0,\;2\pi/\ell\,]\,. \tag{7.4.9}$$

Thus the multipoint periodic spectral problem (7.4.4), (7.4.5) is reduced to the eigenvalue problem formulated on the interval $[0,\,\ell]$:

$$\frac{\mathrm{d}^4 w}{\mathrm{d}x^4} + \lambda \frac{\mathrm{d}^2 w}{\mathrm{d}x^2} = 0, \qquad 0 < x < \ell\,, \tag{7.4.10}$$

$$w\,(\ell,\alpha) = w\,(0,\alpha)\,e^{i\alpha\ell}\,, \qquad \frac{\mathrm{d}w}{\mathrm{d}x}\,(\ell,\alpha) = \frac{\mathrm{d}w}{\mathrm{d}x}\,(0,\alpha)\,e^{i\alpha\ell}\,, \tag{7.4.11}$$

$$\frac{\mathrm{d}^2 w}{\mathrm{d}x^2}\,(\ell,\alpha) = \frac{\mathrm{d}^2 w}{\mathrm{d}x^2}\,(0,\alpha)\,e^{i\alpha\ell}\,, \qquad 0 \le \alpha\ell \le 2\pi\,, \tag{7.4.12}$$

$$\frac{\mathrm{d}^3 w}{\mathrm{d}x^3}\,(\ell,\alpha) = e^{i\alpha\ell}\left[\frac{\mathrm{d}^3 w}{\mathrm{d}x^3}\,(0,\alpha) - \varkappa w\,(0,\alpha)\right]. \tag{7.4.13}$$

In the case of absolutely rigid supports we have $\varkappa = \infty$, and the condition (7.4.13) is excluded from consideration. It is seen from the presented formulation (7.4.10)–(7.4.13) that the values of λ and w depend continuously on α in the interval $\alpha \in [\,0,\;2\pi/\ell\,]$, that is,

$$\lambda = \lambda\,(\alpha)\,, \qquad w = w\,(x,\alpha)\,. \tag{7.4.14}$$

One very important feature of the problem is that although α looks like a parameter, it is not part of the problem setup (unlike, for example, T_0 and D). Although for the purposes of analysis, we may temporarily fix the value of α, once we obtain a solution we must account for the fact that the problem admits solutions for any value of $\alpha \in [0, 2\pi/\ell]$. Each fixed value of α will lead to a discrete spectrum of eigenvalues, but when we allow α to vary continuously, the spectrum as a whole will be continuous. Thus, in order to determine the mechanical stability of the system, we must look for the minimum of $\lambda(\alpha)$ over the admissible values for α, because it is the minimal $\lambda(\alpha)$ that corresponds to the critical velocity, and its buckling mode will be the critical mode.

We may also write a variational formulation corresponding to the spectral problem (7.4.10)–(7.4.13). We omit the details here; for how to do this, see the treatment of a similar variational formulation in Sect. 7.5. It is required to find a minimal value

$$\lambda(\alpha) = \min_{w,w^*} \frac{\int_0^\ell \frac{\mathrm{d}^2 w}{\mathrm{d}x^2}(x,\alpha)\frac{\mathrm{d}^2 w^*}{\mathrm{d}x^2}(x,\alpha)\,\mathrm{d}x + \varkappa w(0,\alpha)\, w^*(0,\alpha)}{\int_0^\ell \frac{\mathrm{d}w}{\mathrm{d}x}(x,\alpha)\frac{\mathrm{d}w^*}{\mathrm{d}x}(x,\alpha)\,\mathrm{d}x}, \tag{7.4.15}$$

under the constraints

$$w = w(x,\alpha) \in \Lambda\,, \qquad w^* = w^*(x,\alpha) \in \Lambda^*\,, \tag{7.4.16}$$

where Λ and Λ^* are sets of admissible functions satisfying, respectively, boundary conditions (7.4.11)–(7.4.13) for w and complex conjugate boundary conditions for w^*.

In the limiting case of absolutely rigid supports, when $\varkappa = \infty$, the corresponding boundary conditions take the form

$$w(0,\alpha) = w(\ell,\alpha) = 0\,, \qquad \frac{\mathrm{d}w}{\mathrm{d}x}(\ell,\alpha) = \frac{\mathrm{d}w}{\mathrm{d}x}(0,\alpha)\, e^{i\alpha\ell}\,, \tag{7.4.17}$$

$$\frac{\mathrm{d}^2 w}{\mathrm{d}x^2}(\ell,\alpha) = \frac{\mathrm{d}^2 w}{\mathrm{d}x^2}(0,\alpha)\, e^{i\alpha\ell}\,, \qquad 0 \le \alpha\ell \le 2\pi\,.$$

In this case the variational formulation (7.4.15), (7.4.16) is reduced to the minimization

$$\lambda(\alpha) = \min_{w,w^*} \frac{\int_0^\ell \frac{\mathrm{d}^2 w}{\mathrm{d}x^2}(x,\alpha)\frac{\mathrm{d}^2 w^*}{\mathrm{d}x^2}(x,\alpha)\,\mathrm{d}x}{\int_0^\ell \frac{\mathrm{d}w}{\mathrm{d}x}(x,\alpha)\frac{\mathrm{d}w^*}{\mathrm{d}x}(x,\alpha)\,\mathrm{d}x}, \tag{7.4.18}$$

under the constraints that $w(x,\alpha)$ and $w^*(x,\alpha)$ satisfy, respectively, boundary conditions (7.4.17) for w and complex conjugate boundary conditions for w^*.

It follows from the considered formulation (7.4.10)–(7.4.13) that for each real eigenvalue $\lambda = \lambda(\alpha)$, beside the eigenfunction $w(x,\alpha)$ also the complex conjugate function w^* satisfies the differential equation. The boundary conditions for w^* are derived by passing on to complex conjugate values in (7.4.11)–(7.4.13). Taking into account that

$$e^{-i\alpha\ell} = e^{i2\pi}e^{-i\alpha\ell} = e^{i(2\pi/\ell-\alpha)\ell}\,, \tag{7.4.19}$$

we obtain

$$w^*(x,\alpha) = w(x, 2\pi/\ell - \alpha)\,. \tag{7.4.20}$$

Thus, for the special case $\alpha = \pi/\ell$ we have

$$w^* (x, {}^{\pi}/_{\ell}) = w (x, {}^{\pi}/_{\ell}) \ , \tag{7.4.21}$$

and in this case $w (x, \pi/\ell)$ is a real solution.

Let us use the property (7.4.20) and the variational representation (7.4.15) for $\lambda ({}^{2\pi}/_{\ell} - \alpha)$, and pass from the functions $w (x, {}^{2\pi}/_{\ell} - \alpha)$ to the functions $w (x, \alpha)$ in the integrands. As a result we will have the equality

$$\lambda (\alpha) = \lambda ({}^{2\pi}/_{\ell} - \alpha) \ , \tag{7.4.22}$$

so the two functions

$$w (x, \alpha) \, , \, w^* (x, \alpha) = w (x, {}^{2\pi}/_{\ell} - \alpha) \ , \tag{7.4.23}$$

correspond to one and the same eigenvalue $\lambda = \lambda (\alpha)$.

Thus, in the general case where $\alpha \neq \pi/\ell$, the eigenvalues are double. Taking into account that the differential equation (7.4.10) is invariant with respect to the operation $x \to \ell - x$, and introducing the notation

$$\widetilde{w} (x, \alpha) = w (\ell - x, \alpha) \ , \tag{7.4.24}$$

we obtain that the functions $\widetilde{w} (x, \alpha)$ and $w (x, \alpha)$ satisfy the differential equation for one and the same eigenvalue $\lambda (\alpha)$. In addition, the boundary conditions for the functions $\widetilde{w} (x, \alpha)$ correspond to $\widetilde{\alpha} = {}^{2\pi}/_{\ell} - \alpha$, and according to (7.4.22), (7.4.23) we will have $\widetilde{w} (x, \alpha) \equiv w^* (x, \alpha)$. As a result we obtain

$$w (\ell - x, \alpha) \equiv w^* (x, \alpha) \, , \qquad 0 \leq x \leq \ell, \qquad 0 \leq \alpha\ell < 2\pi \, . \tag{7.4.25}$$

Consequently, the complete analysis can be restricted to the interval $0 \leq \alpha \leq \pi/\ell$.

Taking into account the condition (7.4.25) and integrating twice we represent the Eq. (7.4.4) in the form

$$\lambda w + \frac{\mathrm{d}^2 w}{\mathrm{d}x^2} = 2i D_1 \left(x - \frac{\ell}{2} \right) + D_2 \, , \qquad 0 < x < \ell \, . \tag{7.4.26}$$

Here $D_1, \ D_2$ are unknown real constants of integration.

Let us continue to the stability analysis. For that, we will use the nondimensional variables

$$\tilde{x} = \frac{x}{\ell}, \qquad \tilde{\lambda} = \ell^2 \lambda, \qquad \tilde{\varkappa} = \varkappa \ell^3 = \frac{k \ell^3}{EI} \, . \tag{7.4.27}$$

In these nondimensional variables we have

$$\lambda (2\pi - \alpha) = \lambda (\alpha) \, , \qquad w (x, \ 2\pi - \alpha) = w^* (x, \alpha) \ , \tag{7.4.28}$$

and the boundary value problem (7.4.10)–(7.4.13) is written as

$$\frac{d^4 w}{dx^4} + \lambda \frac{d^2 w}{dx^2} = 0 , \qquad 0 < x < 1 , \tag{7.4.29}$$

$$(w)_{x=1} = e^{i\alpha} (w)_{x=0} , \qquad \left(\frac{dw}{dx}\right)_{x=1} = e^{i\alpha} \left(\frac{dw}{dx}\right)_{x=0} ,$$

$$\left(\frac{d^2 w}{dx^2}\right)_{x=1} = e^{i\alpha} \left(\frac{d^2 w}{dx^2}\right)_{x=0} , \tag{7.4.30}$$

$$\left(\frac{d^3 w}{dx^3}\right)_{x=1} = e^{i\alpha} \left(\left(\frac{d^3 w}{dx^3}\right)_{x=0} + \varkappa (w)_{x=0} \right), \qquad 0 \le \alpha \le \pi . \tag{7.4.31}$$

Here and in what follows the tilde is omitted. General solutions of (7.4.29) have the form

$$w = C_1 \cos\left(\sqrt{\lambda} x\right) + C_2 \sin\left(\sqrt{\lambda} x\right) + C_3 x + C_4 , \tag{7.4.32}$$

where the eigenvalue λ and the arbitrary constants $C_1,\ C_2,\ C_3,\ C_4$ are determined with the help of the boundary conditions (7.4.30), (7.4.31).

Let us analyze separately the solutions of the considered problem in the cases of elastic supports and absolutely rigid supports. In the case of rigid supports, $\varkappa = \infty$, and the boundary conditions are written as

$$(w)_{x=0} = (w)_{x=1} = 0 , \qquad 0 \le \alpha \le \pi , \tag{7.4.33}$$

$$\left(\frac{dw}{dx}\right)_{x=1} = e^{i\alpha} \left(\frac{dw}{dx}\right)_{x=0} , \qquad \left(\frac{d^2 w}{dx^2}\right)_{x=1} = e^{i\alpha} \left(\frac{d^2 w}{dx^2}\right)_{x=0} .$$

Inserting the general solution (7.4.32) into the boundary conditions (7.4.33), we have

$$\boldsymbol{MC} \equiv \begin{bmatrix} 1 & 0 & 0 & 1 \\ \cos\sqrt{\lambda} & \sin\sqrt{\lambda} & 1 & 1 \\ -\sqrt{\lambda}\sin\sqrt{\lambda} & \sqrt{\lambda}\left[\cos(\sqrt{\lambda}) - e^{i\alpha}\right] & 1 - e^{i\alpha} & 0 \\ \lambda\left[e^{i\alpha} - \cos\sqrt{\lambda}\right] & -\lambda \sin\sqrt{\lambda} & 0 & 0 \end{bmatrix} \begin{bmatrix} C_1 \\ C_2 \\ C_3 \\ C_4 \end{bmatrix} = 0 .$$

This has a nontrivial solution $\boldsymbol{C} \not\equiv 0$ when $\boldsymbol{M}$ is singular. The determinant of $\boldsymbol{M}$ becomes, after collecting terms,

$$\det \boldsymbol{M} = -\lambda^{3/2}\left[e^{2i\alpha} - 2e^{i\alpha}\cos(\sqrt{\lambda}) + 1\right] + \lambda\left[e^{2i\alpha} - 2e^{i\alpha} + 1\right]\sin\sqrt{\lambda} .$$

Taking a common factor of $e^{i\alpha}$, we have

$$\det \boldsymbol{M} = -\lambda^{3/2} e^{i\alpha} \left[e^{i\alpha} - 2\cos(\sqrt{\lambda}) + e^{-i\alpha} \right] + \lambda e^{i\alpha} \left[e^{i\alpha} - 2 + e^{-i\alpha} \right] \sin\sqrt{\lambda} \,.$$

Because $\cos a = \left[e^{ia} + e^{-ia} \right]/2$ for any complex number a, we have

$$\det \boldsymbol{M} = -\lambda^{3/2} e^{i\alpha} \left[2\cos(\alpha) - 2\cos(\sqrt{\lambda}) \right] + \lambda e^{i\alpha} \left[2\cos(\alpha) - 2 \right] \sin\sqrt{\lambda} \,.$$

Writing all terms with respect to a common factor of $2\lambda e^{i\alpha}$ gives

$$\det \boldsymbol{M} = 2\lambda e^{i\alpha} \left[\sqrt{\lambda}\cos(\sqrt{\lambda}) - \sqrt{\lambda}\cos(\alpha) + (\cos(\alpha) - 1)\sin\sqrt{\lambda} \right] .$$

Finally, regrouping yields

$$\det \boldsymbol{M} = 2\lambda e^{i\alpha} \left[\left(\sin(\sqrt{\lambda}) - \sqrt{\lambda} \right) \cos(\alpha) + \left(\sqrt{\lambda}\cos(\sqrt{\lambda}) - \sin\sqrt{\lambda} \right) \right] .$$

Requiring that this is zero, we obtain the following transcendental equation,

$$\cos\alpha = \varphi(\lambda) \,, \qquad \varphi(\lambda) \equiv \frac{\sqrt{\lambda}\cos\sqrt{\lambda} - \sin\sqrt{\lambda}}{\sqrt{\lambda} - \sin\sqrt{\lambda}} \,, \qquad 0 \leq \alpha \leq \pi \,, \tag{7.4.34}$$

for finding the eigenvalues in the case of rigid supports. We see by some trigonometric simplification that $\varphi(\lambda)$ can also be represented as

$$\varphi(\lambda) = \cos(\sqrt{\lambda}) + \frac{[\cos(\sqrt{\lambda}) - 1]\sin\sqrt{\lambda}}{\sqrt{\lambda} - \sin\sqrt{\lambda}} \,, \tag{7.4.35}$$

which tells us that for large values of λ, the function $\varphi(\lambda)$ asymptotically tends to $\cos(\sqrt{\lambda})$. The function $\varphi(\lambda)$ is visualized in Fig. 7.22, which also summarizes the properties we will observe below.

As a special case that sheds some light into the structure of the function $\varphi(\lambda)$, consider $\cos\alpha = 0$, that is $\alpha = \pi/2$. Looking at Eq. (7.4.34), and observing that $a\cos(a) - \sin(a) = 0$ is equivalent to $\tan(a) = a$ for any real a in the domain of tan, we see the zeroes of $\varphi(\lambda)$ occur at λ such that $\tan(\sqrt{\lambda}) = \sqrt{\lambda}$, because at these points the numerator is zero. We must separately check the points $a = \pi/2 + k\pi$, $k = 0, 1, 2, \ldots$, because they do not belong to the domain of tan. We see that at these points, $\cos a = 0$, $\sin a \neq 0$, so there the numerator is nonzero. Thus in the special case $\alpha = \pi/2$, the eigenvalues are determined by the equation $\tan(\sqrt{\lambda}) = \sqrt{\lambda}$.

In general, the eigenvalues are determined taking into account the condition $-1 \leq \cos\alpha \leq 1$, and consequently, when considering all admissible values of α, the eigenvalues fill the continuous interval where $-1 \leq \varphi(\lambda) \leq +1$.

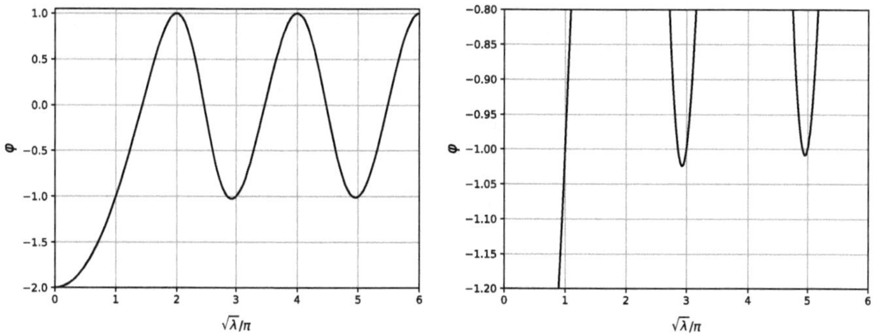

Fig. 7.22 The transcendental function $\varphi(\lambda)$ of Eq. (7.4.34), determining the eigenvalues λ in the case with rigid supports ($\varkappa = \infty$). *Left*: Overview. Geometrically speaking, for any fixed α, the eigenvalues are given by the intersection of this curve and a horizontal line at height $\cos\alpha$. Then, to obtain the whole spectrum, we allow α to vary over all admissible values. The local maxima of $\varphi(\lambda)$ are at $\sqrt{\lambda} = 2k\pi$, $k = 1, 2, 3, \ldots$; at each maximum, $\varphi(\lambda) = 1$. The zeroes of $\varphi(\lambda)$ are located at λ such that $\tan(\sqrt{\lambda}) = \sqrt{\lambda}$. *Right*: Zoom containing the first two regions of λ where $\varphi(\lambda) < -1$, implying that in these regions Eq. (7.4.34) has no solutions. Whenever $\varphi(\lambda)$ is increasing, it crosses the value -1 at $\sqrt{\lambda} = (1 + 2k)\pi$, $k = 0, 1, 2, \ldots$ When decreasing, it crosses the value -1 at λ such that $\tan(\sqrt{\lambda}/2) = \sqrt{\lambda}/2$

The function $\varphi(\lambda)$ tends to -2 (i.e., $\varphi(\lambda) \to -2$) when λ tends to zero ($\lambda \to 0$). Therefore, due to continuity of $\varphi(\lambda)$, there is a range $0 < \lambda < \lambda_0$ that admits no solutions of Eq. (7.4.34). Consequently, the lower bound of the continuous spectrum is determined with the help of the equation $\varphi(\lambda) = \cos\alpha = -1$, which corresponds to $\alpha = \pi$. Requiring $\varphi(\lambda) = -1$, we have

$$\sin(\sqrt{\lambda}) - \sqrt{\lambda} = \sqrt{\lambda}\cos(\sqrt{\lambda}) - \sin(\sqrt{\lambda}) .$$

Reorganizing terms,

$$2\sin\sqrt{\lambda} = \sqrt{\lambda} + \sqrt{\lambda}\cos(\sqrt{\lambda}) ,$$

which yields (for any λ for which $1 + \cos(\sqrt{\lambda}) \neq 0$)

$$\frac{\sin\sqrt{\lambda}}{1 + \cos(\sqrt{\lambda})} = \frac{\sqrt{\lambda}}{2} .$$

The left-hand side has become one of the half-angle formulas for the tangent, namely

$$\tan\left(\frac{a}{2}\right) = \frac{\sin a}{1 + \cos a} ,$$

which holds for any real a. Hence we have

$$\tan\left(\frac{\sqrt{\lambda}}{2}\right) = \frac{\sqrt{\lambda}}{2} .$$

We must treat separately the special case where $1 + \cos(\sqrt{\lambda}) = 0$. If $\cos\sqrt{\lambda} = -1$, then $\sin\sqrt{\lambda} = 0$. Referring to Eq. (7.4.34), we see this case also leads to $\varphi(\lambda) = -1$, so $1 + \cos(\sqrt{\lambda}) = 0$ is also a solution of (7.4.34) when $\alpha = \pi$. To bring this closer in form to the general-case solution, we may use the half-angle formula for cosine,

$$\cos\left(\frac{a}{2}\right) = -1^{\lfloor (a+\pi)/(2\pi) \rfloor} \sqrt{\frac{1+\cos a}{2}} \,,$$

which holds for any real a. We obtain

$$\cos\left(\frac{\sqrt{\lambda}}{2}\right) = 0 \,.$$

Summarizing, at $\alpha = \pi$, the Eq. (7.4.34) is reduced to the following equations:

$$\cos\left(\frac{\sqrt{\lambda}}{2}\right) = 0 \,, \qquad \tan\left(\frac{\sqrt{\lambda}}{2}\right) = \frac{\sqrt{\lambda}}{2} \,. \tag{7.4.36}$$

These are to be taken independently; solving each equation in (7.4.36) for λ gives a solution of (7.4.34) with $\alpha = \pi$, but the same solution does not need to satisfy both equations in (7.4.36). The smallest positive solution of the first equation in (7.4.36) is $\sqrt{\lambda}/2 = \pi/2$; for the second equation, $\sqrt{\lambda}/2 \approx 1.43\pi$. As λ grows, the solution types then alternate. This is because the positive solutions of the first equation are $\sqrt{\lambda}/2 = (1+2k)\pi/2$, $k = 0, 1, 2, \ldots$, spaced π units apart. The second equation contains one solution in each period of tan, the length of which is also π.

Hence the solution of the first equation in (7.4.36),

$$\lambda_k = \pi^2 \, (1+2k)^2 \,, \qquad k = 0, 1, 2, \ldots \tag{7.4.37}$$

determines the lower bound of the continuous spectrum, which is observed to occur when $\alpha = \pi$. The values $\lambda_k = \pi^2 \, (1+2k)^2$ are simple eigenvalues, and the corresponding eigenfunctions are

$$w_k \, (x, \pi) = C \sin\left[(1+2k)\,\pi x\right] \,, \qquad 0 \le x \le 1 \,. \tag{7.4.38}$$

In other words, as the parameter α varies, the value of $\lambda = \lambda_k$ given by (7.4.37) is the smallest possible value that, considered over all admissible values of α, the kth eigenvalue $\lambda(\alpha)$ may have. Refer to Fig. 7.22.

It can be shown that $\varphi(\lambda)$ never exceeds the value $+1$ for any $\lambda > 0$, so in fact only the points where the continuous function $\varphi(\lambda)$ crosses the value -1 may contribute to the existence of "forbidden bands", that is, ranges of λ where no solutions of (7.4.34) may appear. Hence the upper bound for the kth band of the continuous spectrum is given by the kth positive solution of the second equation in (7.4.36). For example, the first band of eigenvalues ends at $\sqrt{\lambda} \approx 2.86\pi$, while the second band starts at $\sqrt{\lambda} = 3\pi$. This leaves a forbidden band at $\sqrt{\lambda} \in (2.86\pi, 3\pi)$. In Fig. 7.22, the forbidden bands are the regions where $\varphi(\lambda)$ obtains values smaller than -1.

To justify the above claim, start by choosing $\cos\alpha = +1$, that is, $\alpha = 0$. Requiring $\varphi(\lambda) = +1$ in Eq. (7.4.34), we have

$$\sqrt{\lambda} - \sin(\sqrt{\lambda}) = \sqrt{\lambda}\cos(\sqrt{\lambda}) - \sin(\sqrt{\lambda})\,,$$

from which we obtain

$$\sqrt{\lambda}\left[1 - \cos(\sqrt{\lambda})\right] = 0\,.$$

Thus at $\alpha = 0$, the Eq. (7.4.34) reduces to $\cos\sqrt{\lambda} = 1$, with $\lambda \neq 0$. (The solution $\lambda = 0$ is not admissible, because it would change the form of our governing Eq. (7.4.29)). Let $a = \sqrt{\lambda}$. By the chain rule,

$$\varphi'(\lambda) = \frac{\partial\varphi}{\partial a}\frac{\partial a}{\partial\lambda} = \frac{1}{2\sqrt{\lambda}}\frac{\partial\varphi}{\partial a}\,, \quad \text{where } a = \sqrt{\lambda}\,. \tag{7.4.39}$$

Differentiation and some rearrangement yields

$$\frac{\partial\varphi}{\partial a} = \frac{[1 - \cos a]\,a + \sin(a)\left[1 - a^2 - \cos a\right]}{[a - \sin a]^2}\,. \tag{7.4.40}$$

Above, we exhaustively obtained all points where $\varphi(\lambda) = +1$; these are the solutions of the equation $\cos\sqrt{\lambda} = +1$. Equations (7.4.39), (7.4.40) tell us that these points are also among the solutions of $\varphi'(\lambda) = 0$, that is, all of them are among the extremum points of $\varphi(\lambda)$. Now, because the function $\varphi(\lambda)$ is continuous, if we can furthermore show that all these points are local maxima, there is no possibility for the function $\varphi(\lambda)$ to ever obtain a value larger than 1 (because due to continuity, it would have to pass through the value 1). A straightforward calculation shows that at any point where $\cos\sqrt{\lambda} = 1$, we have $\partial^2\varphi/\partial a^2 = -1$, which indicates a local maximum. By the chain rule,

$$\varphi''(\lambda) = \frac{\partial}{\partial\lambda}\left[\frac{\partial\varphi}{\partial a}\frac{\partial a}{\partial\lambda}\right] = \frac{\partial a}{\partial\lambda}\frac{\partial}{\partial\lambda}\left(\frac{\partial\varphi}{\partial a}\right) + \frac{\partial\varphi}{\partial a}\frac{\partial^2 a}{\partial\lambda^2} = \left(\frac{\partial a}{\partial\lambda}\right)^2\frac{\partial^2\varphi}{\partial a^2} + \frac{\partial\varphi}{\partial a}\frac{\partial^2 a}{\partial\lambda^2}\,. \tag{7.4.41}$$

At any point where $\cos\sqrt{\lambda} = 1$, we have seen that $\partial\varphi/\partial a = 0$, so at these points the second term in the final form of (7.4.41) vanishes. For $\lambda > 0$, the coefficient $\partial a/\partial\lambda$ is real, so its square is positive. Thus at $\cos\sqrt{\lambda} = 1$, the function $\varphi''(\lambda)$ has the same sign as $\partial^2\varphi/\partial a^2$. Hence these points are indeed local maxima of $\varphi(\lambda)$, and the conclusion is that $\varphi(\lambda) \leq 1$ for any $\lambda > 0$.

Let us now move on to the case of elastic supports, for which $\varkappa \neq \infty$. Armed with the same strategy as above,we will use the general solution (7.4.32) and boundary conditions (7.4.30) and (7.4.31) to find the unknown constants C_1, C_2, C_3, C_4. Substituting (7.4.32) into (7.4.30) and (7.4.31), we derive a homogeneous system of equations with respect to the constants C_1, C_2, C_3, C_4. To obtain a nontrivial solution of this system, we require that its determinant $\psi(\lambda, \varkappa, \alpha)$ become zero. Thus, we arrive at the following equation to determine $\lambda = \lambda(\varkappa, \alpha)$:

$$\psi(\lambda, \varkappa, \alpha) = 0\,, \tag{7.4.42}$$

where

$$\psi(\lambda, \varkappa, \alpha) \equiv \tag{7.4.43}$$

$$\frac{\varkappa}{\sqrt{\lambda^3}}\left(\cos\sqrt{\lambda} - \cos\alpha\right) + \frac{(1-\cos\alpha)}{\sqrt{\lambda}}\left[2\left(\cos\alpha - \cos\sqrt{\lambda}\right) - \frac{\varkappa \sin\sqrt{\lambda}}{\sqrt{\lambda^3}}\right].$$

Given a fixed value of the support stiffness parameter $\varkappa$, Eq. (7.4.43) describes the connection between λ and α.

Recall from the discussion above, when we used Floquet's theorem to set up our solution approach, that for any fixed value of $\varkappa$, the loss of elastic stability of the beam occurs when the eigenvalue $\lambda(\alpha)$, or strictly speaking, $\lambda(\varkappa, \alpha)$, achieves a minimum value:

$$\lambda_* = \min_{\alpha} \lambda(\varkappa, \alpha)\,, \qquad 0 \le \alpha \le \pi\,. \tag{7.4.44}$$

Thus, the critical values $\lambda = \lambda_*$ are found with the help of the equations

$$\psi(\lambda, \varkappa, \alpha) = 0\,, \qquad \frac{\partial \psi(\lambda, \varkappa, \alpha)}{\partial \alpha} = 0\,, \tag{7.4.45}$$

and the condition

$$\frac{\partial \psi(\lambda, \varkappa, \alpha)}{\partial \lambda} \neq 0\,. \tag{7.4.46}$$

The conditions (7.4.45), (7.4.46) arise from the implicit function theorem. Regarding the function ψ as describing an implicit dependence between λ, $\varkappa$ and α, Eq. (7.4.45) represents the bifurcation condition with respect to $\alpha = \alpha(\lambda, \varkappa)$. Recall that a bifurcation point exists when the derivative condition in the theorem is violated. Then, Eq. (7.4.46) says that we accept only points that, for λ, satisfy the derivative condition. The theorem tells us that at such points, a local representation exists for $\lambda = \lambda(\varkappa, \alpha)$. Together these conditions describe, for a fixed value of $\varkappa$, a local extremum of λ in the set $\psi(\lambda, \varkappa, \alpha) = 0$.

To see this, for illustration, think of the unit circle in the xy plane, described implicitly by $F(x, y) = 0$, where $F(x, y) = x^2 + y^2 - 1$. On the circle, at the extrema of x (namely $x = \pm 1$), we have $\partial F/\partial y = 0$, indicating a bifurcation of the local explicit representation $y = y(x)$, and $\partial F/\partial x \neq 0$, indicating that $x = x(y)$ is a locally admissible representation (and indeed it has an extremum there). In the case of the circle, there are two such local representations $x = x(y)$, one near the point $(x, y) = (-1, 0)$ and another near $(x, y) = (+1, 0)$. However, we do not actually need to discover and write out such explicit local representations; given just the implicit representation F, the implicit function theorem gives us a tool to locate all local extrema of x (or alternatively y) in the set that satisfies $F(x, y) = 0$.

Using the expression (7.4.43) for $\psi(\lambda, \varkappa, \alpha)$, we write the second equation in (7.4.45) as

$$\frac{\partial\psi(\lambda, \varkappa, \alpha)}{\partial\alpha} = \frac{\sin\alpha}{\sqrt{\lambda}}\left[4\cos\alpha - 2\left(1+\cos\sqrt{\lambda}\right) + \frac{\varkappa}{\lambda}\left(1 - \frac{\sin\sqrt{\lambda}}{\sqrt{\lambda}}\right)\right] = 0\,. \tag{7.4.47}$$

This equation is satisfied either when $\sin\alpha = 0$, that is, $\alpha = 0$ or $\alpha = \pi$, or when the expression in square brackets vanishes, that is,

$$\cos\alpha = \frac{1}{2}\left(1+\cos\sqrt{\lambda}\right) - \frac{\varkappa}{4\lambda}\left(1 - \frac{\sin\sqrt{\lambda}}{\sqrt{\lambda}}\right)\,. \tag{7.4.48}$$

Let us investigate all three cases. First, if $\alpha = 0$, then the system of equations is reduced to a single equation:

$$\psi(\lambda, \varkappa, 0) = \frac{\varkappa}{\sqrt{\lambda^3}}\left(\cos\sqrt{\lambda} - 1\right) = 0\,, \tag{7.4.49}$$

which directly follows from (7.4.43). This leads to the same solution $\cos\sqrt{\lambda} = 1$ as in the case with rigid supports (at $\alpha = 0$). This equation has double roots $\lambda_j = 4j^2\pi^2$, $j = 1, 2, \ldots$, and corresponding two-parameter eigenmodes

$$w_j = C_{1j}\cos(2j\pi x) + C_{2j}\sin(2j\pi x)\,.$$

That we should expect a double eigenvalue here is hinted at by the fact that in the case with rigid supports ($\varkappa = \infty$), in each band, the two different solution branches for the same $\cos\alpha$ merge at the point where $\varphi(\lambda) = +1$ (the local maximum). Refer to Fig. 7.22.

Next, consider the case $\alpha = \pi$. The system of Eq. (7.4.45) is reduced to the equation

$$\psi(\lambda, \varkappa, \pi) = \frac{\varkappa\left(1+\cos\sqrt{\lambda}\right)}{\sqrt{\lambda^3}} - \frac{2}{\sqrt{\lambda}}\left[2\left(1+\cos\sqrt{\lambda}\right) + \frac{\varkappa\sin\sqrt{\lambda}}{\sqrt{\lambda^3}}\right] = 0\,, \tag{7.4.50}$$

which can be written as

$$\left[\frac{\varkappa}{\lambda\sqrt{\lambda}}\left(\cos\frac{\sqrt{\lambda}}{2} - \frac{2}{\sqrt{\lambda}}\sin\frac{\sqrt{\lambda}}{2}\right) - \frac{4}{\sqrt{\lambda}}\cos\frac{\sqrt{\lambda}}{2}\right]\cos\frac{\sqrt{\lambda}}{2} = 0\,. \tag{7.4.51}$$

The first set of solutions of the Eq. (7.4.51) is given by $\cos(\sqrt{\lambda}/2) = 0$, and is written as $\lambda_j = (2j+1)^2\pi^2$, $j = 0, 1, 2, \ldots$. This set of solutions does not depend on the rigidities of the elastic supports; indeed, this set of solutions was already encountered in the case with rigid supports.

The second set of solutions of the Eq. (7.4.51) is obtained if we consider the case where $\cos(\sqrt{\lambda}/2) \neq 0$, which requires the expression in square brackets in (7.4.51) to vanish.

In the special case of rigid supports, i.e., asymptotically for large $\varkappa$, the bracketed expression in Eq. (7.4.51) reduces to the previous rigid-case solution $\tan(\sqrt{\lambda}/2) = \sqrt{\lambda}/2$. To see this, divide (7.4.51) by $\varkappa$, and then take the limit $\varkappa \to \infty$.

In the general case for any $\varkappa \geq 0$, for the second set of solutions we have the relation

$$\varkappa = \frac{4\lambda}{1 - \frac{2}{\sqrt{\lambda}} \tan\left(\frac{\sqrt{\lambda}}{2}\right)} . \tag{7.4.52}$$

Because $\lambda > 0$ and $\varkappa \geq 0$, in Eq. (7.4.52) the denominator must be positive. The denominator is essentially of the form

$$1 - \frac{\tan a}{a} > 0 , \quad \text{where } a = \frac{\sqrt{\lambda}}{2} .$$

For positive λ, this is equivalent with

$$\tan\left(\frac{\sqrt{\lambda}}{2}\right) < \frac{\sqrt{\lambda}}{2} . \tag{7.4.53}$$

The function $\tan a$ sweeps continuously from $-\infty$ to $+\infty$ in each interval $(-\pi/2, \pi/2) + k\pi$, where k is an integer. Hence, in each period of tan, the admissible range for eigenvalues λ is from $\sqrt{\lambda}/2 = k\pi - \pi/2$ to the solution point of (7.4.53) in that period. It will be in the interval $(0, \pi/2) + k\pi$, because tan is zero in the midpoint of each period.

Especially, the smallest positive solution of $\tan a = a$ occurs in the interval $(\pi, 3\pi/2)$. Thus we may state that in this set of solutions, for the minimal admissible λ, it must hold that $\pi < \sqrt{\lambda}/2 < 3\pi/2$. Consequently, we will have $\lambda > 4\pi^2$ for the minimal value of λ. Since this value exceeds the minimal value of λ for the other set of solutions ($\lambda_j = \pi^2$ for $j = 0$), this solution is not needed for determining the mechanical stability.

Finally, consider now the system (7.4.45) in the general case where $\alpha \neq 0$ and $\alpha \neq \pi$. In this case, as was noted before, the Eq. (7.4.47) will be fulfilled if (7.4.48) is satisfied. Substituting (7.4.48) into (7.4.42), thus eliminating $\cos\alpha$, we derive a nonlinear equation for finding λ as a function of $\varkappa$:

$$p_1^2 - \frac{p_1}{\lambda}(1 + p_3)\,\varkappa + \frac{1}{4\lambda^2}(1 - p_3)^2\,\varkappa^2 = 0 . \tag{7.4.54}$$

Here and in what follows we will use, for brevity, the notations

$$p_1 = 1 - \cos\sqrt{\lambda}, \quad p_2 = 1 + \cos\sqrt{\lambda}, \quad p_3 = \frac{\sin\sqrt{\lambda}}{\sqrt{\lambda}}, \quad z = \frac{\varkappa}{2\lambda}(1 - p_3) \geq 0\,. \tag{7.4.55}$$

This equation is solved under the constraint

$$\left|\frac{1}{2}p_2 - \frac{\varkappa}{4\lambda}(1 - p_3)\right| \leq 1\,, \tag{7.4.56}$$

which is a consequence of the inequality $-1 \leq \cos\alpha \leq 1$ and the representation (7.4.48) for $\cos\alpha$. Observe that by requiring (7.4.48), we have moved away from accepting any solution of $\psi(\lambda, \varkappa, \alpha) = 0$, into considering only solutions that, for a fixed $\varkappa$, lie on a point where $\lambda(\alpha)$ takes on a locally extremal value on the set described by $\psi(\lambda, \varkappa, \alpha) = 0$. Thus, we will directly find the minimal eigenvalue λ_* described by Eq. (7.4.44).

The Eq. (7.4.54) and the constraint (7.4.56) can be transformed into the following form:

$$z^2 - \frac{2p_1(1 + p_3)}{1 - p_3}z + p_1^2 = 0\,, \tag{7.4.57}$$

$$0 \leq z \leq \chi(\lambda) \equiv 2 + p_2\,. \tag{7.4.58}$$

The inequality (7.4.58) follows from transforming the constraint (7.4.56) into the form $|p_2 - z| \leq 2$. A priori, we only know that $0 \leq p_2 \leq 2$, and $z \geq 0$; we do not know whether z or p_2 is larger. If it occurs that $z < p_2$, then the condition becomes $p_2 - z \leq 2$, or equivalently, $z \geq p_2 - 2$. The most demanding case is $p_2 = 2$, which yields the condition $z \geq 0$, which is always true. Thus we see that whenever $z < p_2$, then the constraint (7.4.56) holds regardless of the value of p_2. On the other hand, if it happens that $z > p_2$, then we have the condition $z - p_2 \leq 2$, which leads to $z \leq 2 + p_2$, that is, the inequality (7.4.58). Thus, a sufficient (but not always necessary) condition for a solution λ to be admissible is that $z \leq 2 + p_2$.

Consider the two solutions z_1 and z_2 of the Eq. (7.4.57),

$$z_1 = \frac{p_1\left(1 + \sqrt{p_3}\right)}{1 - \sqrt{p_3}}\,, \qquad z_2 = \frac{p_1\left(1 - \sqrt{p_3}\right)}{1 + \sqrt{p_3}}\,, \tag{7.4.59}$$

on the interval $\lambda \in \left[0, \pi^2\right]$.

For the solution z_1, we observe $z_1\,(\lambda = 0) = 12$, $z_1\left(\lambda = \pi^2\right) = 2$, and that $z_1\,(\lambda)$ is a monotonically decreasing function in the interval $\lambda \in \left[0, \pi^2\right]$. Hence, the inequality (7.4.58) cannot be satisfied in this interval for the solution $z_1\,(\lambda)$. As for the solution z_2, we have $z_2(\lambda = 0) = 0$, $z_2(\lambda = \pi^2) = 2$, and $z_2(\lambda)$ monotonically increasing. The solution z_2 satisfies the inequality (7.4.58) in the whole interval.

Both solutions, the auxiliary function $\chi(\lambda) = 2 + p_2 = 3 + \cos\sqrt{\lambda}$ that appears in the constraint (7.4.58), and the function $p_2 = 1 + \cos\sqrt{\lambda}$ from (7.4.55), are

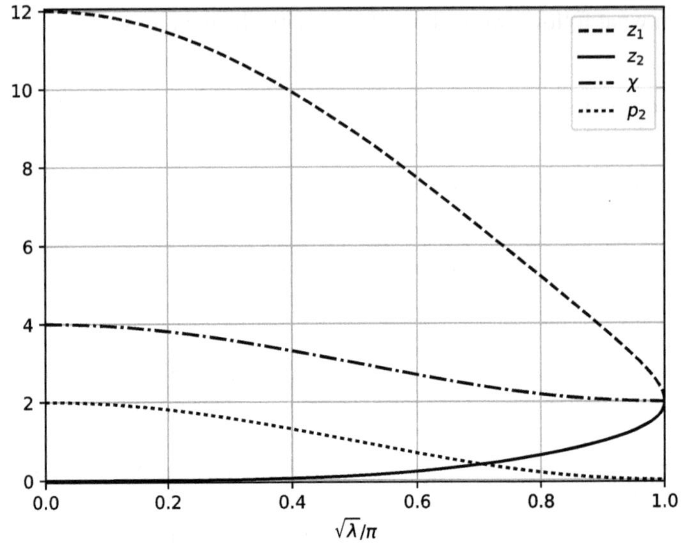

Fig. 7.23 The solutions $z_1(\lambda)$ and $z_2(\lambda)$, defined in Eq. (7.4.59), the auxiliary function $\chi(\lambda) = 3 + \cos\sqrt{\lambda}$ that appears in the constraint (7.4.58), and $p_2 = 1 + \cos\sqrt{\lambda}$

visualized in Fig. 7.23. Regarding the above discussion of the constraint $|p_2 - z| \leq 2$, we see that only the solution $z_2(\lambda)$ is ever smaller than p_2. Wherever $z \geq p_2$, we must require $z \leq 2 + p_2$ for the solution to be admissible. The solution z_2 is seen to be admissible.

Equating the expression (7.4.59) for the admissible solution z_2 with the definition (7.4.55) of z, we obtain

$$\frac{p_1\left(1 - \sqrt{p_3}\right)}{\left(1 + \sqrt{p_3}\right)} = \frac{\varkappa(1 - p_3)}{2\lambda}\,.$$

Cancelling the common factor and solving with respect to $\varkappa$, we have

$$\varkappa = \frac{2\lambda p_1}{\left(1 + \sqrt{p_3}\right)^2} = 2\lambda\left(1 - \cos\sqrt{\lambda}\right)\left(1 + \sqrt{\frac{\sin\sqrt{\lambda}}{\sqrt{\lambda}}}\right)^{-2}. \tag{7.4.60}$$

Equation (7.4.60) furnishes a connection between $\varkappa$ and λ, while also enforcing the requirement (7.4.48). Hence it is in fact a connection between $\varkappa$ and the critical λ_*. The corresponding value for the parameter α is then given by Eq. (7.4.48). This result is valid for the general case where $\alpha \neq 0$ and $\alpha \neq \pi$.

Recall that the support stiffness parameter $\varkappa$ is a real number. We see from (7.4.60) that this requires p_3 to be positive, which occurs when $\sin \sqrt{\lambda} \geq 0$. Consequently, the critical eigenvalues may appear only in the following intervals:

$$4j^2\pi^2 \leq \lambda_j \leq (2j+1)^2\,\pi^2\,, \qquad j = 0, 1, 2, \ldots\,. \tag{7.4.61}$$

The loss of elastic stability occurs at the minimal critical velocity, and thus the corresponding critical eigenvalue belongs to the first interval,

$$0 < \lambda \leq \pi^2\,. \tag{7.4.62}$$

From the variational formulation of the considered spectral problem, Eq. (7.4.15), it follows that $\lambda\,(\varkappa)$ is a monotonically increasing function, $\mathrm{d}\lambda(\varkappa)/\mathrm{d}\varkappa \geq 0$. From (7.4.60), we find $\varkappa = \varkappa_* = 4\pi^2$ for $\lambda = \lambda_* = \pi^2$. Now, the inequality (7.4.62) tells us that $\lambda_* = \pi^2$ is the maximal value that the smallest positive critical eigenvalue may take. Consequently, the critical eigenvalue λ must remain the same whenever $\varkappa \geq \varkappa_*$.

We conclude that if $\varkappa \geq \varkappa_*$, the infinite traveling beam with elastic supports loses stability in the same manner as the infinite traveling beam with rigid supports, for which $\varkappa = \infty$. This means, in particular, that the buckling mode of the traveling beam with elastic supports is characterized by zero transverse (vertical) displacement at the elastic supports when $\varkappa \geq \varkappa_* = 4\pi^2$. In other words, $\varkappa_*$ is a *critical support rigidity*, above which the stability behavior of the infinite traveling beam no longer changes.

The basic relation (7.4.60) between $\varkappa$ and λ can be simplified and written in an asymptotic manner. For small elastic support rigidities ($\varkappa \ll 1$) we have $\lambda \ll 1$, and the following asymptotic formula holds:

$$\lambda(\varkappa) = \sqrt{4\varkappa}\,, \qquad \varkappa \ll 1\,. \tag{7.4.63}$$

The dependence of the critical eigenvalue λ on the rigidity parameter $\varkappa$, obtained in accordance with the exact relation (7.4.60) and the asymptotic formula (7.4.63), are shown in Fig. 7.24. The figure illustrates the good precision of the asymptotic formula (7.4.63) in the interval $0 < \varkappa \leq 10$.

Finally, the dependence of the smallest positive eigenvalue λ on the parameter $\alpha \in [0, \pi]$ is shown in Fig. 7.25 for some fixed values of the support rigidity parameter $\varkappa$. In this final figure we return to considering the set of all solutions of $\psi(\lambda, \varkappa, \alpha) = 0$, for any α (that is, not only the value that minimizes $\lambda(\varkappa, \alpha)$ for given $\varkappa$). The figure is based on Eq. (7.4.42), by numerically assigning values to $\varkappa$ and α, and then applying the bisection method to find λ such that $\psi(\lambda, \varkappa, \alpha) = 0$. The critical values of the parameter $\alpha = \alpha_*$ correspond to the minimal values of $\lambda = \lambda_*$. Comparing the locations of the minima in Fig. 7.25, it is observed that α_* is a monotonically increasing function of $\varkappa$.

In this section, we examined the loss of stability of a moving web, modelled as an elastic beam (panel) of unlimited length and traveling between an infinite system

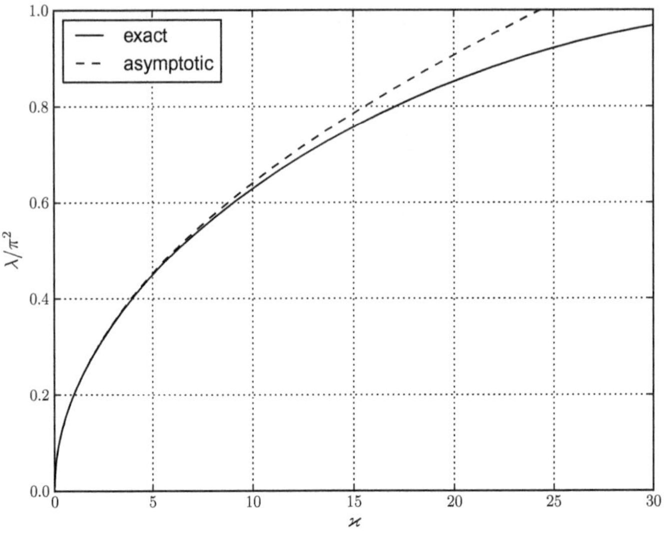

Fig. 7.24 The dependence of the critical eigenvalue λ on the elastic support rigidity parameter $\varkappa$, obtained in accordance with the exact relation (7.4.60) and the asymptotic formula (7.4.63)

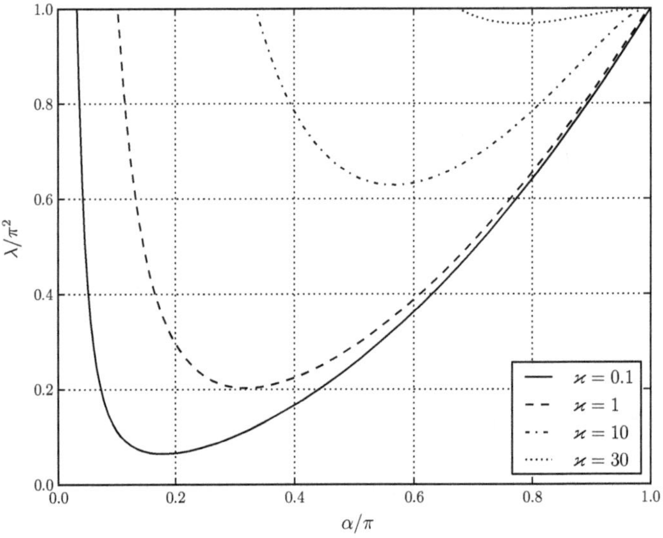

Fig. 7.25 The dependence of the minimal eigenvalue λ on the parameter $\alpha \in [0, \pi]$ for some fixed values of the elastic support rigidity parameter $\varkappa$

of rollers (elastic supports) at a constant velocity. Transverse elastic displacements of the web were described by a fourth-order differential equation that included the centrifugal effect, in-plane tension (axial tension), bending resistance and the reaction of the elastic supports.

The stability of the beam was investigated with the help of analysis of small periodic transverse displacements. A multipoint spectral stability problem was formulated for the periodic interval, and Floquet's theorem was used to represent the solution. As a result, the basic relations characterizing the behavior of the web at the onset of instability were found in an analytical form.

The critical velocity, that corresponds to the onset of instability in the form of divergence (buckling), was estimated in the frame of the spectral analysis. The obtained dependence of critical velocity on the support rigidity parameter $\varkappa$ was analyzed and, in particular, it was shown that the loss of stability of the infinite traveling beam with elastic supports coincides with the loss of stability of the same beam with absolutely rigid supports ($\varkappa = \infty$), when the (nondimensional) rigidity of the elastic supports exceeds a critical value, $\varkappa \geq \varkappa_* = 4\pi^2$.

This is as far as we will proceed in this particular direction; it is time to move on to generalize a completely different aspect of the fundamental single-span model.

7.5 Stability of a Traveling Beam in a Gravitational Field

All of the models discussed in this chapter so far have concentrated on the case with no body forces. However, in practice in engineering science applications a gravitational field is almost always present. In this section, we extend the fundamental single-span model to take into account the effect of gravity on the traveling material. We will also slightly vary our mathematical approach, with a brief excursion into the calculus of variations.

In previous studies concerning vibrations of (moving) materials, the effect of gravity has usually been neglected, because its magnitude is minor compared to the magnitude of the axial tension. Recently, [36] have studied the equilibrium of traveling elastic, sagging cables under uniformly distributed loading. For a three-dimensional model, they derived exact, closed-form solutions for the equilibrium configuration of and the tension distribution in the cables.

In this section, we concentrate on the stability of a thin, elastic panel (or beam), and specifically on estimating its critical velocity where a loss of mechanical stability occurs. We will perform a static stability analysis for an axially moving elastic band traveling at a constant velocity between two supports, and subjected to a uniform gravitational field. As before, the loss of elastic stability (buckling) of the traveling material is investigated in the framework of small transverse deflections. We investigate the case where the band is moving parallel to the uniform gravitational field (that is, vertically), but the same analysis is easily adapted to cases in which the direction of motion is not aligned with the gravitational field.

The critical transport velocity and the corresponding buckling mode are found numerically by solving directly the original differential equation via the Fourier–Galerkin method. To demonstrate another possible approach, they are also solved by minimizing a functional corresponding to the energy. The obtained nonlinear optimization problem is solved numerically using the Rayleigh–Ritz method.

Let us begin with basic relations. We consider an elastic band, traveling at a constant velocity between two supports, parallel to a uniform gravitational field (Earth's) in a cartesian xz coordinate system. We study the transverse displacement w assuming that the displacement is cylindrical, in other words, the displacement does not vary in the cross section (in the y direction). The distance between the two supports is constant, denoted by ℓ.

The transport velocity of the panel (band undergoing cylindrical deformation) is assumed to be a constant, V_0, and the panel travels along the x axis. It is tensioned at the endpoints of the span, and the tension T depends on the space coordinate x, that is, $T = T(x)$. The acceleration of the uniform gravitational field is denoted by g, and the thickness of the panel by h. The panel is assumed to have a constant mass per unit area, m.

The dynamic equation for the transverse displacement of the panel can be written as

$$m\frac{\partial^2 w}{\partial t^2} + 2mV_0\frac{\partial^2 w}{\partial x \partial t} + mV_0^2\frac{\partial^2 w}{\partial x^2} = T\frac{\partial^2 w}{\partial x^2} + \frac{\partial T}{\partial x}\frac{\partial w}{\partial x} - D\frac{\partial^4 w}{\partial x^4}\,, \quad 0 < x < \ell\,. \tag{7.5.1}$$

Here D describes the bending rigidity of the panel. As usual when treating axially moving materials, Eq. (7.5.1) is written with respect to a stationary (Eulerian) frame.

Because the panel is subjected to a uniform gravitational field that acts in the direction parallel to the axial motion, the tension T varies as

$$T(x) = mgx + T_0\,, \tag{7.5.2}$$

where T_0 is a constant tension at the lower edge of the panel (see Fig. 7.26). This is easily seen by a thought experiment. Before applying the axial motion, consider the static balance of forces if we were to cut the panel at some x_0. For static equilibrium to hold, the point at x_0 must support the weight of all the points below it. This does not change even after we impose a uniform traveling motion. Hence the tension increases linearly in the upward direction.

Substituting (7.5.2) into (7.5.1), we obtain

$$m\frac{\partial^2 w}{\partial t^2} + 2mV_0\frac{\partial^2 w}{\partial x \partial t} + mV_0^2\frac{\partial^2 w}{\partial x^2} = (T_0 + mgx)\frac{\partial^2 w}{\partial x^2} + mg\frac{\partial w}{\partial x} - D\frac{\partial^4 w}{\partial x^4}\,, \quad 0 < x < \ell\,. \tag{7.5.3}$$

In the more general case, where the direction of motion is inclined at some nonzero angle θ with respect to the uniform gravitational field, we can use the expression

$$T(x) = mgx\cos\theta + T_0 \tag{7.5.4}$$

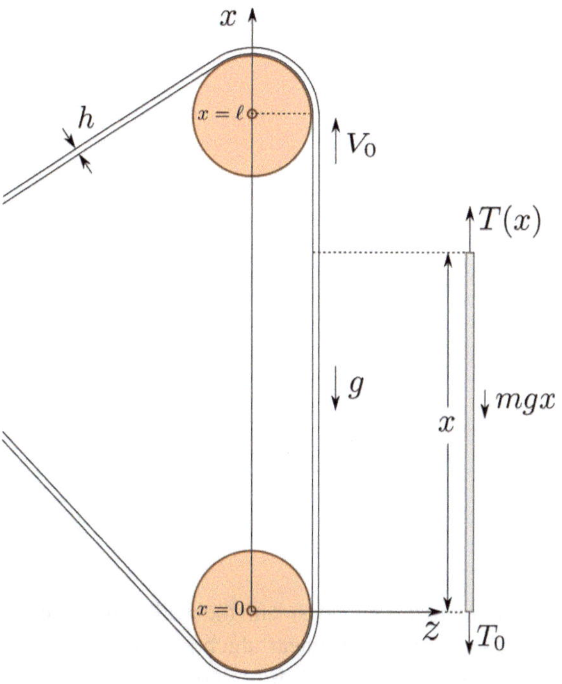

Fig. 7.26 Elastic band traveling in a uniform gravitational field

instead of (7.5.2), and replace each g by $g\cos\theta$ in the right-hand side of (7.5.3). See Fig. 7.27. In the following analysis, we will absorb the $\cos\theta$ into the value of g, so by analyzing just the fully vertical case ($\theta = 0$) we will have analyzed all inclinations. The only exception is a fully horizontal configuration ($\theta = \pi/2$), which has already been treated in the other sections of this chapter.

Strictly speaking, in any configuration except the fully vertical one ($\theta = 0$), there will be a vertical component to the gravitational field, $-mg\sin\theta$. In a direct time simulation of the dynamics, this would appear as a distributed load on the right-hand side of (7.5.3):

$$m\frac{\partial^2 w}{\partial t^2} + 2mV_0\frac{\partial^2 w}{\partial x \partial t} + mV_0^2\frac{\partial^2 w}{\partial x^2} = (T_0 + mgx)\frac{\partial^2 w}{\partial x^2} + mg\frac{\partial w}{\partial x} - D\frac{\partial^4 w}{\partial x^4} + q \ ,$$

where $q = -mg\sin\theta$. However, since we will perform only a stability analysis, we may ignore this term. The tension $T(x)$ is the only place in Eq. (7.5.3) where, for the purposes of stability analysis, we must account for gravity.

The reason is purely mathematical. The solution of a linear partial differential equation is the superposition of two components, the homogeneous solution (the solution for the same equation with $q = 0$), and the particular solution (which accounts for q). It happens that the stability behavior is fully governed by the homogeneous part; buckling essentially corresponds to the appearance of nontrivial solutions for the homogeneous equation.

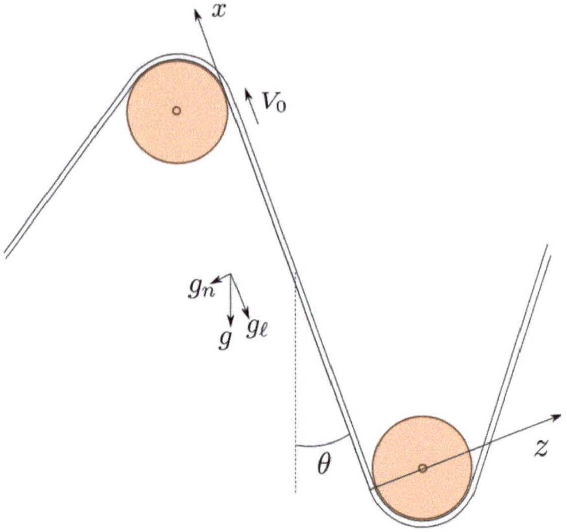

Fig. 7.27 Elastic band traveling in a direction inclined with respect to the uniform gravitational field

Compare this situation with the well-known theorems collectively known as *Fredholm's alternative* in linear algebra, in the theory of integral equations, and in the theory of elliptic partial differential equations. Roughly speaking, these theorems state that either the nonhomogeneous equation being considered has a unique solution for any given load (here q), or the corresponding homogeneous equation has at least one nontrivial solution. From the viewpoint of mathematics, the second situation is what gives rise to the loss of mechanical stability. However, keep in mind that our problem, being a description of vibrations, most closely resembles the hyperbolic type. As we have seen, it is in fact hyperbolic if $D = 0$, so we cannot directly apply those theorems here.

To study the mechanical stability of the panel, we perform a buckling analysis. This will give us configurations that are steady states in the Eulerian frame, although the material is in constant traveling motion. The appearance of a nontrivial steady-state eigensolution will indicate a buckling mode, in which the system loses its stability at the critical velocity corresponding to the eigenvalue.

We study the static form of the dynamic Eq. (7.5.3) as a spectral boundary value problem with simply supported boundary conditions. Equilibrium for the transverse displacement w is described by the differential equation

$$m V_0^2 \frac{\partial^2 w}{\partial x^2} - (T_0 + mgx)\frac{\partial^2 w}{\partial x^2} - mg\frac{\partial w}{\partial x} + D\frac{\partial^4 w}{\partial x^4} = 0\,, \quad 0 < x < \ell\,. \tag{7.5.5}$$

The boundary conditions are chosen as

$$w(0) = \frac{\partial^2 w}{\partial x^2}(0) = 0\,, \quad w(\ell) = \frac{\partial^2 w}{\partial x^2}(\ell) = 0\,. \tag{7.5.6}$$

We represent (7.5.5), (7.5.6) in a nondimensional form

$$\left[\frac{V_0^2}{T_0/m} - 1 - \frac{mg\ell}{T_0}x\right]\frac{\partial^2 w}{\partial x^2} - \frac{mg\ell}{T_0}\frac{\partial w}{\partial x} + \frac{D}{\ell^2 T_0}\frac{\partial^4 w}{\partial x^4} = 0\,, \quad 0 < x < 1\,,$$

$$w(0) = \frac{\partial^2 w}{\partial x^2}(0) = 0, \qquad w(1) = \frac{\partial^2 w}{\partial x^2}(1) = 0\,, \tag{7.5.7}$$

where $x \in [0, 1]$ is the nondimensional axial coordinate. By defining

$$c_0 := \frac{V_0}{\sqrt{T_0/m}}\,, \tag{7.5.8}$$

$$\alpha := \frac{D}{\ell^2 T_0}\,, \tag{7.5.9}$$

$$\beta := \frac{mg\ell}{T_0}\,, \tag{7.5.10}$$

and substituting (7.5.8)–(7.5.10) into (7.5.7), we obtain

$$\alpha\frac{\partial^4 w}{\partial x^4} + \left(c_0^2 - 1 - \beta x\right)\frac{\partial^2 w}{\partial x^2} - \beta\frac{\partial w}{\partial x} = 0\,, \tag{7.5.11}$$

$$w(0) = \frac{\partial^2 w}{\partial x^2}(0) = 0, \qquad w(1) = \frac{\partial^2 w}{\partial x^2}(1) = 0\,. \tag{7.5.12}$$

Consider the boundary value problem (7.5.11), (7.5.12). To obtain the weak form, we multiply (7.5.11) by a test function v that satisfies the boundary conditions (7.5.12), and integrate the obtained equation over the domain (0, 1):

$$\alpha\int_0^1 \frac{\partial^4 w}{\partial x^4} v\,\mathrm{d}x + \left(c_0^2 - 1\right)\int_0^1 \frac{\partial^2 w}{\partial x^2} v\,\mathrm{d}x - \beta\int_0^1 x\frac{\partial^2 w}{\partial x^2} v\,\mathrm{d}x - \beta\int_0^1 \frac{\partial w}{\partial x} v\,\mathrm{d}x = 0\,. \tag{7.5.13}$$

Integration by parts yields, after applying the boundary conditions,

$$\int_0^1 \frac{\partial^2 w}{\partial x^2} v\,\mathrm{d}x = -\int_0^1 \frac{\partial w}{\partial x}\frac{\partial v}{\partial x}\mathrm{d}x, \qquad \int_0^1 x\frac{\partial^2 w}{\partial x^2} v\,\mathrm{d}x = -\int_0^1 x\frac{\partial w}{\partial x}\frac{\partial v}{\partial x}\mathrm{d}x - \int_0^1 \frac{\partial w}{\partial x} v\,\mathrm{d}x,$$

$$\int_0^1 \frac{\partial^4 w}{\partial x^4} v\,\mathrm{d}x = \int_0^1 \frac{\partial^2 w}{\partial x^2}\frac{\partial^2 v}{\partial x^2}\mathrm{d}x\,. \tag{7.5.14}$$

Substituting (7.5.14) into (7.5.13), we obtain the weak form

$$\alpha \int_0^1 \frac{\partial^2 w}{\partial x^2}\frac{\partial^2 v}{\partial x^2}\mathrm{d}x + \left(1 - c_0^2\right) \int_0^1 \frac{\partial w}{\partial x}\frac{\partial v}{\partial x}\mathrm{d}x + \beta \int_0^1 x \frac{\partial w}{\partial x}\frac{\partial v}{\partial x}\mathrm{d}x = 0\,. \tag{7.5.15}$$

Choosing $v = w$ in (7.5.15), we have the variational formulation

$$\alpha \int_0^1 \left(\frac{\partial^2 w}{\partial x^2}\right)^2 \mathrm{d}x + \left(1 - c_0^2\right) \int_0^1 \left(\frac{\partial w}{\partial x}\right)^2 \mathrm{d}x + \beta \int_0^1 x \left(\frac{\partial w}{\partial x}\right)^2 \mathrm{d}x = 0\,. \tag{7.5.16}$$

The left-hand side of the Eq. (7.5.16) corresponds to the energy of the system. We want to find the minimal velocity, c_0, at which Eq. (7.5.16) has a nontrivial solution. Thus, we solve c_0^2 from (7.5.16), and minimize the other side of the obtained equation. This minimal velocity is the critical velocity for the buckling problem [37]:

$$\left(c_0^*\right)^2 = \min_{w \in K} \left(1 + \alpha \frac{\int_0^1 \left(\frac{\partial^2 w}{\partial x^2}\right)^2 \mathrm{d}x}{\int_0^1 \left(\frac{\partial w}{\partial x}\right)^2 \mathrm{d}x} + \beta \frac{\int_0^1 x \left(\frac{\partial w}{\partial x}\right)^2 \mathrm{d}x}{\int_0^1 \left(\frac{\partial w}{\partial x}\right)^2 \mathrm{d}x} \right), \tag{7.5.17}$$

where

$$K = \left\{ w \in C^{1+\epsilon}([0, 1]) : \; w(0) = 0, \; w(1) = 0 \right\}\,. \tag{7.5.18}$$

The function w minimizing (7.5.17), (7.5.18) is the buckling mode corresponding to the critical velocity c_0^*. Note that (7.5.17) has an infinite number of solutions; if w is a minimizer, then also cw is for any real $c \neq 0$. The continuity requirement imposed on w in (7.5.18) is chosen to make the most demanding term in (7.5.17), namely $(\partial^2 w/\partial x^2)^2$, integrable. Recall the brief discussion of Sobolev embedding theorems in Chap. 5; for the full mathematical details, see [38].

To justify the name *variational formulation*, let us for the moment forget that there even was another formulation to begin with, and show that the minimization problem *defined* by (7.5.17), (7.5.18) leads to the original problem (7.5.11), (7.5.12). To make it more practical to work with the expression to be minimized in (7.5.17), let us define

$$I(w) = 1 + \alpha \frac{I_1(w)}{I_2(w)} + \beta \frac{I_3(w)}{I_2(w)}\,,$$

where

$$I_1(w) = \int_0^1 \left(\frac{\partial^2 w}{\partial x^2}\right)^2 \mathrm{d}x\,, \quad I_2(w) = \int_0^1 \left(\frac{\partial w}{\partial x}\right)^2 \mathrm{d}x\,, \quad I_3(w) = \int_0^1 x \left(\frac{\partial w}{\partial x}\right)^2 \mathrm{d}x\,. \tag{7.5.19}$$

Obviously, if $w \not\equiv 0$, then

$$I_1(w) > 0\,, \qquad I_2(w) > 0\,, \qquad I_3(w) > 0\,.$$

Let w^* be a minimizer of (7.5.17), (7.5.18). To prove that w^* satisfies the necessary condition for a minimum, let us derive the Euler equation for w^*. Since w^* is a minimizer, the first variation of $I(w)$ at the point w^* is zero, that is,

$$\delta I(w^*) = 0\,.$$

Recall that the first variation is defined as

$$\delta J(u) = \lim_{\varepsilon \to 0} \frac{1}{\varepsilon}\left[J(u+\varepsilon v) - J(u)\right]\,, \tag{7.5.20}$$

for admissible u and v. The details of what exactly makes v admissible depend on the boundary conditions imposed on u, because the function $u + \varepsilon v$ must be admissible for $\varepsilon > 0$. Keeping in mind that our aim is to impose, in particular, the boundary conditions (7.5.12), we may simply take $u, v \in K$.

The first variation of I is

$$\delta I = \frac{1}{I_2}\left(\alpha \delta I_1 + \beta \delta I_3\right) - \frac{1}{I_2^2}\left(\alpha I_1 + \beta I_3\right)\delta I_2\,. \tag{7.5.21}$$

The division by I_2 is admissible, because as was noted, for any $w \not\equiv 0$, $I_2 > 0$. To evaluate (7.5.21), we compute the variations δI_k, $k = 1, 2, 3$. Observe that for any functional J of the form

$$J(w) = \int_0^1 f(x)\,(\mathcal{L}w)^2\,\mathrm{d}x\,, \tag{7.5.22}$$

where $\mathcal{L}$ is a linear differential operator and f does not depend on w, we have

$$\begin{aligned}
\delta J(w) &= \lim_{\varepsilon \to 0} \frac{1}{\varepsilon}\left[\int_0^1 f(x)\,(\mathcal{L}[w+\varepsilon\delta w])^2\,\mathrm{d}x - \int_0^1 f(x)\,(\mathcal{L}w)^2\,\mathrm{d}x\right] \\
&= \lim_{\varepsilon \to 0} \frac{1}{\varepsilon}\left[\int_0^1 f(x)\,(\mathcal{L}w + \varepsilon\mathcal{L}(\delta w))^2\,\mathrm{d}x - \int_0^1 f(x)\,(\mathcal{L}w)^2\,\mathrm{d}x\right] \\
&= \lim_{\varepsilon \to 0} \frac{1}{\varepsilon}\left[2\varepsilon\int_0^1 f(x)\mathcal{L}(w)\mathcal{L}(\delta w)\mathrm{d}x + \varepsilon^2\int_0^1 f(x)\,(\mathcal{L}(\delta w))^2\,\mathrm{d}x\right]
\end{aligned}$$

$$= \lim_{\varepsilon \to 0} \left[2 \int_0^1 f(x) \mathcal{L}(w) \mathcal{L}(\delta w) \mathrm{d}x + \varepsilon \int_0^1 f(x) \left(\mathcal{L}(\delta w) \right)^2 \mathrm{d}x \right]$$

$$= 2 \int_0^1 f(x) \mathcal{L}(w) \mathcal{L}(\delta w) \mathrm{d}x \; . \tag{7.5.23}$$

For I_1 and I_2, we choose $f(x) \equiv 1$ in (7.5.22); for I_3, we choose $f(x) = x$. The variation of I_2 becomes

$$\begin{aligned}
\delta I_2(w) &= 2 \int_0^1 \frac{\partial w}{\partial x} \frac{\partial (\delta w)}{\partial x} \mathrm{d}x \\
&= -2 \int_0^1 \frac{\partial^2 w}{\partial x^2} \delta w \, \mathrm{d}x + 2 \left[\frac{\partial w}{\partial x} \delta w \right]_{x=0}^1 , \\
&= -2 \int_0^1 \frac{\partial^2 w}{\partial x^2} \delta w \, \mathrm{d}x \; ,
\end{aligned} \tag{7.5.24}$$

where we have applied integration by parts, and foreshadowing the development to come later, our boundary conditions (7.5.12). For I_1, we obtain

$$\begin{aligned}
\delta I_1(w) &= 2 \int_0^1 \frac{\partial^2 w}{\partial x^2} \frac{\partial^2 (\delta w)}{\partial x^2} \mathrm{d}x \\
&= -2 \int_0^1 \frac{\partial^3 w}{\partial x^3} \frac{\partial (\delta w)}{\partial x} \mathrm{d}x + 2 \left[\frac{\partial^2 w}{\partial x^2} \frac{\partial (\delta w)}{\partial x} \right]_{x=0}^1 \\
&= 2 \int_0^1 \frac{\partial^4 w}{\partial x^4} \delta w \, \mathrm{d}x + 2 \left[\frac{\partial^2 w}{\partial x^2} \frac{\partial (\delta w)}{\partial x} \right]_{x=0}^1 - 2 \left[\frac{\partial^3 w}{\partial x^3} \delta w \right]_{x=0}^1 \\
&= 2 \int_0^1 \frac{\partial^4 w}{\partial x^4} \delta w \, \mathrm{d}x \; .
\end{aligned} \tag{7.5.25}$$

Finally, for I_3, first observing that

$$\frac{\partial}{\partial x} \left(x \frac{\partial w}{\partial x} \delta w \right) = \frac{\partial w}{\partial x} \delta w + x \frac{\partial^2 w}{\partial x^2} \delta w + x \frac{\partial w}{\partial x} \frac{\partial (\delta w)}{\partial x} \; ,$$

we determine that

$$\begin{aligned}\delta I_3(w) &= 2\int_0^1 x\frac{\partial w}{\partial x}\frac{\partial(\delta w)}{\partial x}\mathrm{d}x \\ &= -2\int_0^1 \frac{\partial w}{\partial x}\delta w\mathrm{d}x - 2\int_0^1 x\frac{\partial^2 w}{\partial x^2}\delta w\mathrm{d}x + \left[x\frac{\partial w}{\partial x}\delta w\right]_{x=0}^1 \\ &= -2\int_0^1 \frac{\partial w}{\partial x}\delta w\mathrm{d}x - 2\int_0^1 x\frac{\partial^2 w}{\partial x^2}\delta w\mathrm{d}x\,. \end{aligned} \tag{7.5.26}$$

Now, inserting (7.5.24)–(7.5.26) into (7.5.21), we obtain

$$\begin{aligned}\frac{1}{2}I_2(w^*)\,\delta I(w^*) &= \frac{1}{2}\int_0^1\left[\alpha\,\delta I_1 + \beta\,\delta I_3 - \left(\alpha\frac{I_1}{I_2} + \beta\frac{I_3}{I_2}\right)\delta I_2\right]\mathrm{d}x \\ &= \frac{1}{2}\int_0^1\left[\alpha\,\delta I_1 + \beta\,\delta I_3 - \left(\left(1 + \alpha\frac{I_1}{I_2} + \beta\frac{I_3}{I_2}\right) - 1\right)\delta I_2\right]\mathrm{d}x \\ &= \frac{1}{2}\int_0^1\left[\alpha\,\delta I_1 + \beta\,\delta I_3 - \left(I(w^*) - 1\right)\delta I_2\right]\mathrm{d}x \\ &= \int_0^1\left[\alpha\frac{\partial^4 w^*}{\partial x^4} - \beta\left(\frac{\partial w^*}{\partial x} + x\frac{\partial^2 w^*}{\partial x^2}\right) - \left(I(w^*) - 1\right)\frac{\partial^2 w^*}{\partial x^2}\right]\delta w\,\mathrm{d}x\,.\end{aligned}$$

Now, since w^* minimizes $I(w)$, we have

$$\frac{1}{2}I_2(w^*)\,\delta I(w^*) = \int_0^1\left[\alpha\frac{\partial^4 w^*}{\partial x^4} - \beta\left(\frac{\partial w^*}{\partial x} + x\frac{\partial^2 w^*}{\partial x^2}\right) - \left(\left(c_0^*\right)^2 - 1\right)\frac{\partial^2 w^*}{\partial x^2}\right]\delta w\,\mathrm{d}x$$

because this is how we defined $\left(c_0^*\right)^2$ in Eq. (7.5.17). Hence, as a result from the variation in (7.5.21), we obtain

$$\frac{2}{I_2\left(w^*\right)}\int_0^1\left(\alpha\frac{\partial^4 w^*}{\partial x^4} + \left(\left(c_0^*\right)^2 - 1\right)\frac{\partial^2 w^*}{\partial x^2} - \beta x\frac{\partial^2 w^*}{\partial x^2} - \beta\frac{\partial w^*}{\partial x}\right)\delta w\,\mathrm{d}x = 0\,, \tag{7.5.27}$$

for any function $\delta w \in K$. Because δw is arbitrary, the only way the Eq. (7.5.27) can be satisfied—provided w^* has enough continuity for the equation itself to be meaningful—is that the expression in the parentheses vanishes pointwise for all $x \in (0, 1)$. In addition, recall that in the derivation of (7.5.24)–(7.5.26) we imposed the boundary conditions (7.5.12), so we must now require w^* to satisfy them. (The general case can be handled similarly, by just keeping the boundary terms generated by the integrations by parts when deriving δI_k, $k = 1, 2, 3$). Thus, the resulting Euler equation is

$$\alpha\frac{\partial^4 w^*}{\partial x^4}x + \left(\left(c_0^*\right)^2 - 1\right)\frac{\partial^2 w^*}{\partial x^2} - \beta x\frac{\partial^2 w^*}{\partial x^2} - \beta\frac{\partial w^*}{\partial x} = 0\,,$$

which is exactly (7.5.11). In other words, a solution w^* of the minimization problem (7.5.17), (7.5.18) is a weak solution of the original problem (7.5.11), (7.5.12). We have derived our original problem from the presented variational formulation, justifying the name.

It should be emphasized that in this derivation, we started by taking Eqs. (7.5.17), (7.5.18) as the *definition* of our problem, and then investigated what follows *if* we have a w^* that minimizes $I(w)$. This change in perspective is what logically allowed us to postulate that $\delta I(w^*) = 0$ (a general property of minimizers), and to use Eq. (7.5.17) to substitute $\left(c_0^*\right)^2$ for $I(w^*)$. The result of the investigation was that if we have a minimizer w^*, it *must* be a solution of the problem (7.5.11), (7.5.12), provided that it has sufficient continuity to admit writing the problem in that form.

Now we are ready to estimate the critical velocity. First, we will derive lower and upper bounds analytically. We estimate the critical velocity c_0^* corresponding to (7.5.11), (7.5.12) given by the relation

$$\left(c_0^*\right)^2 = I(w^*) = 1 + \alpha\frac{I_1(w^*)}{I_2(w^*)} + \beta\frac{I_3(w^*)}{I_2(w^*)}\,. \tag{7.5.28}$$

For the lower bound of (7.5.28), we obtain

$$\left(c_0^*\right)^2 \geq 1 + \alpha\min_{\psi\in K}\frac{I_1(w)}{I_2(w)}\,, \tag{7.5.29}$$

because for any nonzero w, the functionals $I_k(w)$, $k = 1, 2, 3$ are positive and $\beta > 0$. For the upper bound of (7.5.28), we have

$$\left(c_0^*\right)^2 \leq 1 + \alpha\frac{I_1(w^a)}{I_2(w^a)} + \beta\frac{I_3(w^a)}{I_2(w^a)}\,, \tag{7.5.30}$$

where w^a is any function in K. (This holds because w^* is the minimizer of $I(w)$, so $I(w^a) \geq I(w^*)$ for any $w^a \in K$).

Combining (7.5.29) and (7.5.30), we can estimate c_0^* as follows:

$$1 + \alpha\min_{w\in K}\frac{I_1(w)}{I_2(w)} \leq \left(c_0^*\right)^2 \leq 1 + \alpha\frac{I_1(w^a)}{I_2(w^a)} + \beta\frac{I_3(w^a)}{I_2(w^a)}\,. \tag{7.5.31}$$

To find the lower bound accurately, we solve the minimization problem

$$\min_{w\in K}\frac{I_1(w)}{I_2(w)}\,. \tag{7.5.32}$$

Following the strategy explained above, the problem (7.5.32) can be transformed into the following boundary value problem (Euler equation with boundary conditions):

$$\frac{\partial^4 w}{\partial x^4} + \lambda \frac{\partial^2 w}{\partial x^2} = 0 \, ,$$
$$w(0) = \frac{\partial^2 w}{\partial x^2}(0) = 0 \, , \qquad w(1) = \frac{\partial^2 w}{\partial x^2}(1) = 0 \, . \tag{7.5.33}$$

Solutions to the eigenvalue problem (7.5.33) are known to be

$$w_k(x) = A \sin(k\pi\ell) \, , \quad k = 1, 2, 3, \dots \, ,$$
$$\lambda_k = (k\pi)^2 \; .$$

The normalized solution ($A = 1$) corresponding to the minimum in (7.5.32) can be shown to be

$$w_{\min}(x) = w_1(x) = \sin(\pi x) \, ,$$
$$\lambda_{\min} = \lambda_1 = \pi^2 \, .$$

Inserting w_1 into (7.5.19), we obtain

$$I_1(w_1) = \pi^2 \int_0^1 \cos^2(\pi x) \, \mathrm{d}x = \frac{\pi^2}{2} \, , \qquad I_2(w_1) = \int_0^1 \sin^2(\pi x) \, \mathrm{d}x = \frac{1}{2}$$
$$I_3(w_1) = \int_0^1 x \, \sin^2(\pi x) \, \mathrm{d}x = \frac{1}{4} \, . \tag{7.5.34}$$

Choosing $w^a = w_1 \in K$ in (7.5.30) and inserting (7.5.34) into (7.5.31), we obtain the estimate

$$1 + \alpha\pi^2 \le \left(c_0^*\right)^2 \le 1 + \alpha\pi^2 + \frac{\beta}{2} \, . \tag{7.5.35}$$

The estimate (7.5.35) gives guaranteed lower and upper bounds for $\left(c_0^*\right)^2$. Here, c_0, α, and β are defined in (7.5.8) and in (7.5.9), (7.5.10).

We will next show the numerical solution by the Rayleigh–Ritz and the Fourier–Galerkin methods. The minimization problem is discretized using the Rayleigh–Ritz method and solved using the interior point method. The differential equation is solved via the Fourier–Galerkin method. We start with the numerical solution of the minimization problem.

Consider the minimization problem (7.5.17). The constant term does not effect the location of the optimum and it can be omitted. Divide (7.5.17) by the constant α. The obtained minimization problem is equivalent to (7.5.17):

$$\frac{\left(c_0^*\right)^2 - 1}{\alpha} = \min_{w \in K} \left(\frac{\int_0^1 \left(\frac{\partial^2 w}{\partial x^2}\right)^2 \mathrm{d}x + a \int_0^1 x \left(\frac{\partial w}{\partial x}\right)^2 \mathrm{d}x}{\int_0^1 \left(\frac{\partial w}{\partial x}\right)^2 \mathrm{d}x} \right) , \tag{7.5.36}$$

where

$$a = \frac{\beta}{\alpha} = \frac{gm\ell^3}{D} .$$

The problem (7.5.36), just like (7.5.17), has an infinite number of solutions. If w is a solution, then also cw, where c is a constant, is a solution. The problem (7.5.36) can be solved as [37, p. 273]:

$$\min_{w \in K} \int_0^1 \left(\frac{\partial^2 w}{\partial x^2} \right)^2 \mathrm{d}x + a \int_0^1 x \left(\frac{\partial w}{\partial x} \right)^2 \mathrm{d}x \quad \text{subject to} \quad \int_0^1 \left(\frac{\partial w}{\partial x} \right)^2 \mathrm{d}x = 1 . \tag{7.5.37}$$

In addition, w must satisfy the boundary conditions $w(0) = w(1) = 0$. Now, the condition $\int_0^1 \left(\frac{\partial w}{\partial x} \right)^2 \mathrm{d}x = 1$ sets the absolute value of the constant c mentioned above. Note that the normalization constant, which is now chosen to be one, can be chosen freely.

Let us discretize (7.5.37) using the Rayleigh–Ritz method. We present the function w as a series

$$w(x) = \sum_{j=1}^{\infty} v_j \varphi_j(x) \tag{7.5.38}$$

in the basis

$$\varphi_j(x) \equiv \sin(j\pi x), \qquad x \in [0, 1] . \tag{7.5.39}$$

The basis (7.5.39) fulfills the boundary conditions $w(0) = w(1) = 0$ naturally. We fix a finite positive integer n_0 (number of modes), and approximate (7.5.38) with its finite analog

$$w(x) = \sum_{j=1}^{n_0} v_j \varphi_j(x) . \tag{7.5.40}$$

Inserting (7.5.39), (7.5.40) into (7.5.37), we obtain

$$\min_{\mathbf{v}} \mathbf{v}^{\mathrm{T}} \mathbf{A} \mathbf{v} + a \mathbf{v}^{\mathrm{T}} \mathbf{B} \mathbf{v} \quad \text{subject to} \quad \mathbf{v}^{\mathrm{T}} \mathbf{C} \mathbf{v} - 1 = 0 , \tag{7.5.41}$$

where $\mathbf{v} = (v_1, \ldots, v_{n_0})^{\mathrm{T}} \in \mathbb{R}^{n_0}$. (Note that the index variable in (7.5.40) is a dummy. To keep the two instances of w separate, we insert (7.5.40) once using the index j, and once using i. The result is a double summation, which in the matrix formalism can be written in the form (7.5.41)).

For our basis (7.5.39), the elements of the matrices in (7.5.41) can be found analytically. They are

$$A_{ij} := \int_0^1 \frac{\partial^2 \varphi_j}{\partial x^2}\frac{\partial^2 \varphi_i}{\partial x^2}\,\mathrm{d}x = \frac{j^4\pi^4}{2}\delta_{ij}$$

$$B_{ij} := \int_0^1 x\frac{\partial \varphi_j}{\partial x}\frac{\partial \varphi_i}{\partial x}\,\mathrm{d}x = -\int_0^1 x\frac{\partial^2 \varphi_j}{\partial x^2}\varphi_i \mathrm{d}x - \int_0^1 \frac{\partial \varphi_j}{\partial x}\varphi_i \mathrm{d}x = -E_{ij} - D_{ij},$$

$$C_{ij} := \int_0^1 \frac{\partial \varphi_j}{\partial x}\frac{\partial \varphi_i}{\partial x}\,\mathrm{d}x = \frac{j^2\pi^2}{2}\delta_{ij}\ ,$$

$$D_{ij} := \int_0^1 \frac{\partial \varphi_j}{\partial x}\varphi_i\,\mathrm{d}x = \begin{cases} 0 & , \quad i = j \\ \dfrac{ij\left[(-1)^{i+j} - 1\right]}{j^2 - i^2} & , \quad i \neq j \end{cases}\ ,$$

$$E_{ij} := \int_0^1 x\frac{\partial^2 \varphi_j}{\partial x^2}\varphi_i \mathrm{d}x = \begin{cases} -\dfrac{j^2\pi^2}{4} & , \quad i = j \\ -\dfrac{2ij^3\left[(-1)^{i+j} - 1\right]}{[i-j]^2[i+j]^2} & , \quad i \neq j \end{cases}\ , \tag{7.5.42}$$

where δ_{ij} is the Kronecker delta. The discretized problem (7.5.41) is a nonlinear optimization problem with a nonlinear objective function and a nonlinear constraint. The optimization problem variable is vector v. Note that the size of the problem depends on n_0. Note also that the boundary conditions are included in the matrices **A**, **B** and **C**.

The numerical solution of the original boundary value problem (7.5.5), (7.5.6) is performed by using the Fourier–Galerkin method. Let us define

$$\lambda = 1 - c_0^2\,. \tag{7.5.43}$$

Consider the nondimensional problem (7.5.11), (7.5.12). The eigenvalue problem for the eigenvalue–eigenfunction pair (λ, w) is

$$\alpha\frac{\partial^4 w}{\partial x^4} - \beta\left(x\frac{\partial^2 w}{\partial x^2} - \frac{\partial w}{\partial x}\right) = \lambda\frac{\partial^2 w}{\partial x^2}, \tag{7.5.44}$$

$$w(0) = \frac{\partial^2 w}{\partial x^2}(0) = 0, \qquad w(1) = \frac{\partial^2 w}{\partial x^2}(1) = 0\,. \tag{7.5.45}$$

By solving c_0 from (7.5.43), we see that the largest eigenvalue $\lambda_{\max}$ corresponds to the minimal critical velocity. It is easy to show that all eigenvalues $\lambda \leq 0$. To do this, we insert λ from (7.5.43) into the weak form (7.5.15), obtaining

$$\alpha\int_0^1 \frac{\partial^2 w}{\partial x^2}\frac{\partial^2 v}{\partial x^2}\mathrm{d}x - \beta\int_0^1 x\frac{\partial w}{\partial x}\frac{\partial v}{\partial x}\mathrm{d}x = \lambda\int_0^1 \frac{\partial w}{\partial x}\frac{\partial v}{\partial x}\mathrm{d}x\,. \tag{7.5.46}$$

Let us define the bilinear form

$$L(w, v) = -\alpha \int_0^1 \frac{\partial^2 w}{\partial x^2} \frac{\partial^2 v}{\partial x^2} \mathrm{d}x - \beta \int_0^1 x \frac{\partial w}{\partial x} \frac{\partial v}{\partial x} \mathrm{d}x \,. \tag{7.5.47}$$

Choosing $v = w$, we have $L(w, w) \leq 0$, but on the other hand

$$L(w, w) = \lambda \int_0^1 \left(\frac{\partial w}{\partial x} \right)^2 \mathrm{d}x \,. \tag{7.5.48}$$

Thus, $\lambda \leq 0$, and a physically meaningful solution ($c_0^2 > 0$) always exists. Furthermore, we see that $c_0^2 \geq 1$, in other words, the critical velocity cannot be smaller than that of a traveling string (for which it is $\sqrt{T_0/m}$).

To compute a numerical solution, we apply the Fourier–Galerkin method to the problem (7.5.44), (7.5.45). Once we have the eigenvalue λ, we obtain V_0^* from (7.5.43) and (7.5.8).

We present the function w as a Galerkin series (7.5.38) in the basis (7.5.39). As before, the basis (7.5.39) fulfills the boundary conditions (7.5.45) naturally. Inserting (7.5.39), (7.5.40) into the weak form (7.5.46), we obtain the matrix equation

$$(-\alpha \mathbf{A} - \beta \mathbf{B})\, \mathbf{v} = \lambda \mathbf{C} \mathbf{v} \,. \tag{7.5.49}$$

The elements of matrices in (7.5.49) are the same as above, and are given by (7.5.42). Equation (7.5.49) is a standard generalized linear eigenvalue problem for the pair $(\lambda, \mathbf{v})$, to which any standard solver may be applied.

To end this section, and indeed this chapter, we will now show some numerical results. The physical parameters used are given in Table 7.2. The number of basis functions used is $n_0 = 200$. By Eqs. (7.5.9) and (7.5.10), the given values lead to the nondimensional parameter values $\alpha = 1.8315 \cdot 10^{-7}$, $\beta = 1.5696 \cdot 10^{-3}$ and $a = \beta/\alpha = 8.5700 \cdot 10^3$, which was used as a reference case. The problem (7.5.41) was solved using the Matlab Optimization Toolbox. Optimization method was chosen to be the interior point method. The problem (7.5.49) was solved using Matlab's generalized linear eigenvalue solver.

Table 7.2 Physical parameters for the reference case

g	T_0	m	ℓ	h	E	ν
9.81 m/s^2	500 N/m	0.08 kg/m^2	1 m	10^{-4} m	10^9 N/m^2	0.3

$\Rightarrow$	$D = Eh^3/(12 \cdot (1 - \nu^2))$
	$9.1575 \cdot 10^{-5}$ Nm

Figure 7.28 presents the buckling modes and the corresponding critical velocities for different values of $a = \beta/\alpha$ in the case of the Rayleigh–Ritz and the Fourier–Galerkin methods. In each figure, a solid line corresponds to the solution given by the optimizer, and a dash-dot line corresponds to the critical mode given by the differential equation solver. The buckling modes and the critical velocities (in figure titles) were the same in the results given by the optimizer and the direct solver. (The solution plots for the calculated buckling modes overlap).
From the numerical results, we may make some qualitative observations. First, the effect of gravity on the eigenmode is very large, see Figs. 7.28 and 7.29. The extremum of the eigenmode occurs near the start of the span. This result is as expected, because a positive x axis was chosen to point up in the gravitational field. The effect is rather strong even at small angles with respect to the horizontal.

The effect on the critical velocity is very minor, typically less than 0.01%. See Fig. 7.30. In this respect, the results resemble those from Chap. 6, where we investigated the effect of a linear tension distribution at the rollers on the buckling behavior of an axially moving plate.

The strength of the effect that places the extremum near the start of the span depends on the ratio of the nondimensional parameters β and α, in other words, on the quantity $a = \beta/\alpha = mg\ell^3/D$. The larger this parameter is, the stronger is the effect. In the limit $a \to 0$ (ideally rigid material, or very long span) the effect vanishes. In the other limit $a \to \infty$ (the limit of zero bending stiffness), the eigenmode approaches a characteristic shape that resembles a sawtooth function (but starts smoothly from 0). See Fig. 7.29.

In the case where the direction of motion is inclined with respect to the gravity, we show solutions of the buckling problem for different values of the angle θ in Eq. (7.5.4). In Fig. 7.31 on the left, the graphs of the buckled shapes are shown for the values $\theta = 0,\ \pi/8,\ \pi/4,\ 3\pi/8,\ \pi/2$, and the displacement maxima are marked by $\star$.

In Fig. 7.31 on the right, the buckling modes are represented for the values $[0, \pi/2]$ of θ as a color sheet. It is seen that the buckling mode rapidly becomes nonsymmetric when the direction of motion of the panel becomes nonorthogonal to the gravity, and the nonsymmetric modes are quite similar for small values of θ. This behavior is illustrated more clearly in Fig. 7.32.

Note that the sign of the axial velocity V_0 does not affect the buckling problem (7.5.5), (7.5.6). Therefore, the range $\theta \in [0, \pi/2]$ covers all possible band orientations with respect to the gravity.

In this section we investigated the loss of elastic stability of an axially moving band, taking into account that the material travels in a gravitational field. The onset of instability in a divergence form, for some critical value of the transport velocity, was estimated using a variational principle to develop variational inequalities. Analytical lower and upper bounds for the critical velocity were derived. Additionally, it was shown that a critical velocity always exists.

Nonsymmetric buckling modes, characteristic to the behavior of the traveling band whenever the gravitational field has a nonzero axial component, were illustrated with the help of numerical examples. As a result, the large influence of the gravity force

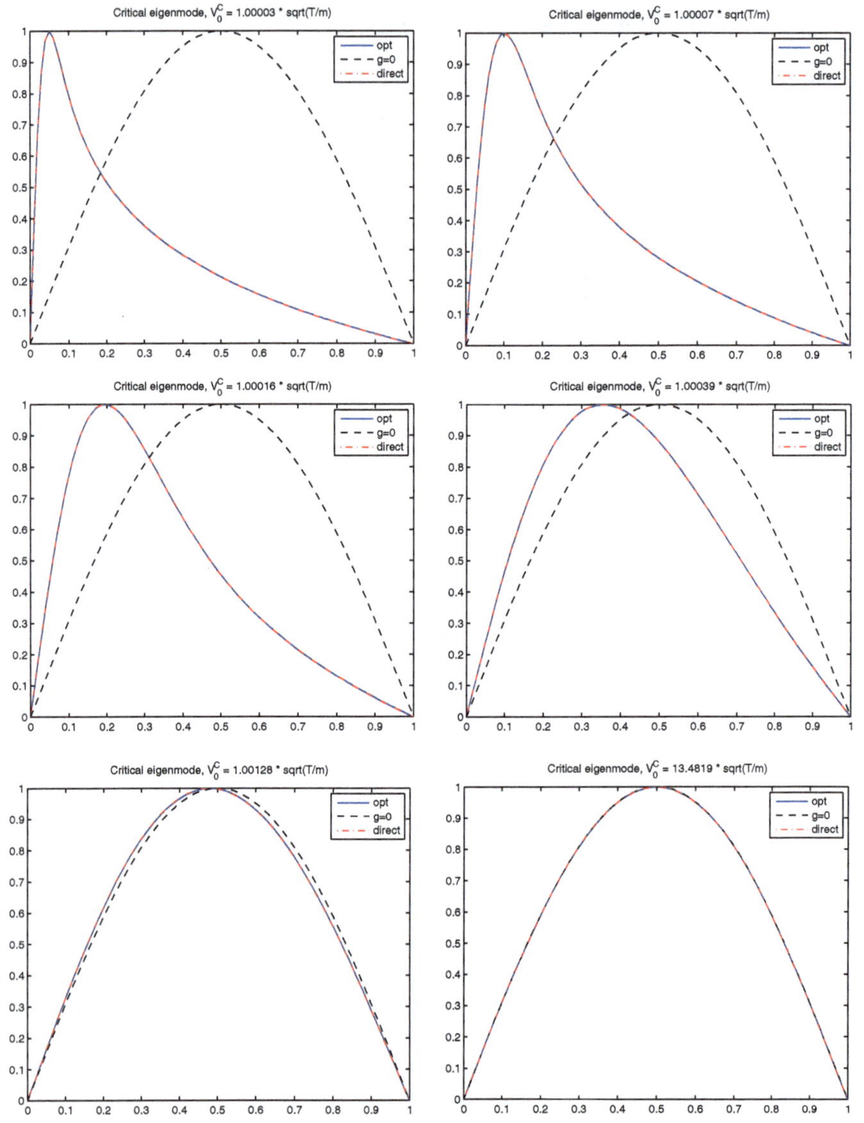

Fig. 7.28 Critical buckling mode for some values of the parameters. The band moves toward the right and gravity points toward the left. Comparison of the solutions given by the energy minimization (*opt*, solid line) and by the direct solving of a differential equation (*direct*, dash-dot line). The dashed line corresponds to a reference solution with no gravity in the axial direction ($g = 0$). The top right picture represents the reference case given in the text, for which $a = a_0 = 8.57 \cdot 10^3$. In the pictures, $a/a_0 = 10,\ 1,\ 0.1,\ 0.01,\ 0.001,\ 10^{-8}$ (from left to right, top to bottom, in that order)

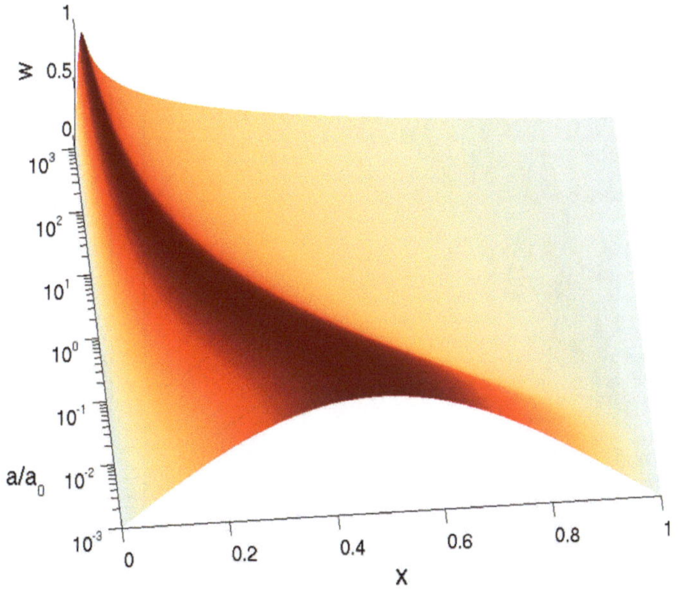

Fig. 7.29 Critical buckling mode for different parameter values. The band moves toward the right and gravity points toward the left. The reference value for β/α is $a = a_0 = 8.57 \cdot 10^3$. Note the logarithmic scale of a/a_0

on the buckling mode, and a small effect on the critical transport velocity, were established and discussed. Optimization methods were shown to give reliable results for solving this type of differential equations.

Compared to the case where the gravity is orthogonal to the axial motion and does not affect the buckling mode, it was seen that even a small angle is enough to produce a notably nonsymmetric shape. Similar behavior was seen earlier when we introduced a linear tension distribution at the rollers in the noncylindrical deformation case in Chap. 6. This suggests a general property of the classical single-span model of axially moving materials: the buckling form is highly sensitive to qualitative changes in the model, even if the modeled situation is close to an ideal one that omits certain features, whereas the critical velocity is relatively stable against such changes.

It is now time to close this chapter. In the first three sections, we looked at eigenvalue problems for various fundamental single-span models of axially moving materials, and analytically obtained exact eigensolutions for several of them. This explained their stability behavior in a very compact form. The last two sections explored two very different extensions of the single-span model, via various different approaches. First, we considered a long beam, and found the interesting result that there exists a critical support rigidity, above which the supports could as well be perfectly rigid (as far as the stability behavior of the traveling material is concerned). In this final section, we used both calculus of variations and a direct numerical approach to solve the weak form of the problem of a beam traveling in a gravitational field. Besides

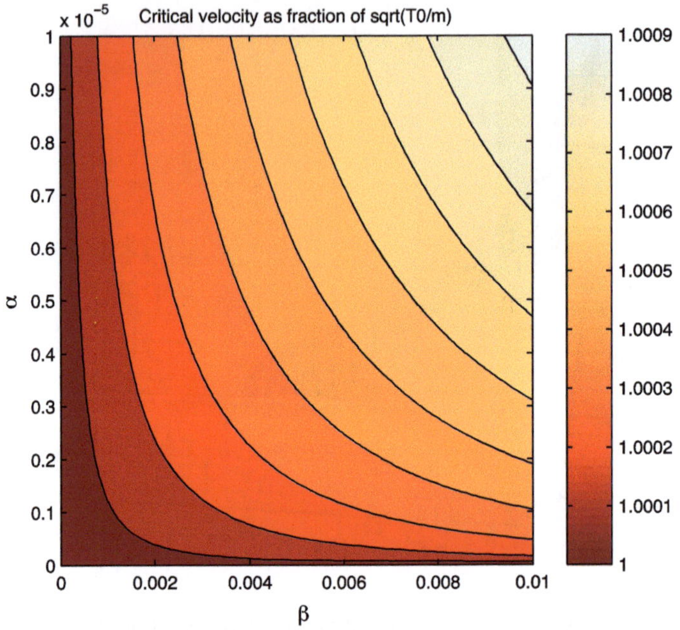

Fig. 7.30 Effect of the nondimensional parameters α and β (defined by (7.5.9) and (7.5.10)) on the critical velocity. Note that the reference case is at the lower edge of the figure at $\beta \approx 0.0016$. The scaling for the axes was chosen to show the structure of the data

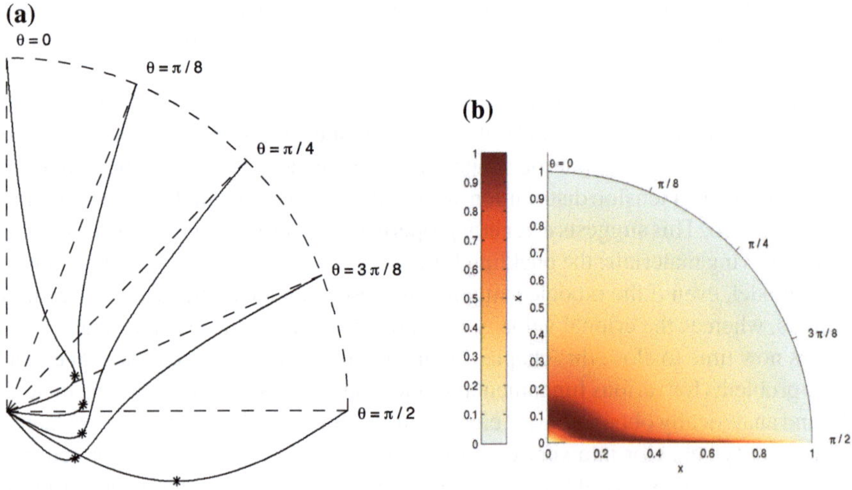

Fig. 7.31 Buckling modes of the original problem when the direction of motion of the band is at an angle to the gravity. **a** Graphs of buckling modes for some selected cases. Displacement maxima are marked by $\star$. **b** Color sheet of the buckling mode for values of θ between 0 and $\pi/2$

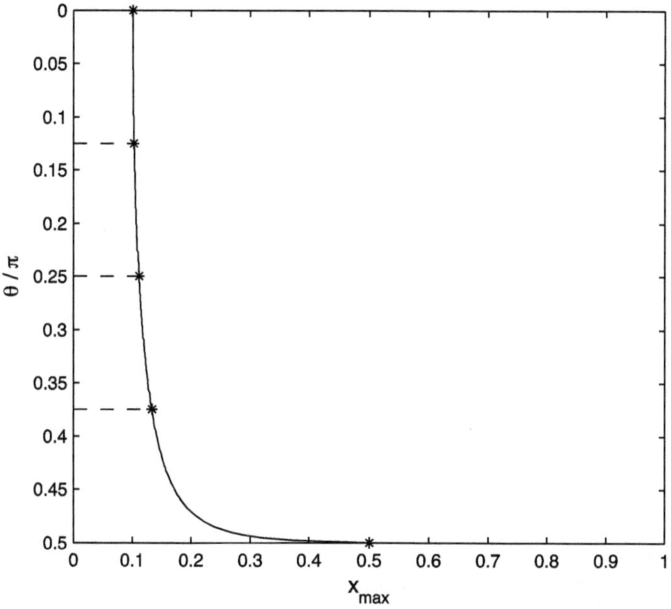

Fig. 7.32 The location of the maximum displacement for θ in $[0, \pi/2]$. The stars ($\star$) correspond to those shown in Fig. 7.31

axially moving materials, there is another overarching theme that connects all these topics: the implicit function theorem, and sometimes a careful change of perspective, provide valuable tools for analysis. In the next chapter, we will continue in the same spirit, considering some simple models of fluid–structure interaction for axially moving materials, with applications in stability analysis.

References

1. Banichuk Nikolay, Jeronen Juha, Ivanova Svetlana, Tuovinen Tero (2015b) Analytical approach for the problems of dynamics and stability of a moving web. Rakenteiden Mekaniikka (J Struct Mech) 48(3):136–163
2. Skutch Rudolf (1897) Uber die Bewegung eines gespannten Fadens, weicher gezwungen ist durch zwei feste Punkte, mit einer constanten Geschwindigkeit zu gehen, und zwischen denselben in Transversal-schwingungen von gerlinger Amplitude versetzt wird. Ann Phys Chem 61:190–195
3. Wang Y, Huang L, Liu X (2005a) Eigenvalue and stability analysis for transverse vibrations of axially moving strings based on Hamiltonian dynamics. Acta Mech Sin 21:485–494. https://doi.org/10.1007/s10409-005-0066-2
4. Polyanin AD, Zaitsev VF, Moussiaux A (2002) Handbook of first order partial differential equations. Taylor & Francis, London
5. Swope RD, Ames WF (1963) Vibrations of a moving threadline. J Frankl Inst 275:36–55. https://doi.org/10.1016/0016-0032(63)90619-7

6. John F (1982) Partial differential equations, 4th edn. Springer, New York
7. Courant R, Hilbert D (1962) Methods of mathematical physics, volume II: Partial differential equations. Wiley-VCH Verlag GmbH & Co. KGaA
8. Polyanin AD, Schiesser WE, Alexei IZ (2008) Partial differential equation/second-order partial differential equations. Scholarpedia. Revision, 121514
9. Bolotin VV (1963) Nonconservative problems of the theory of elastic stability. Pergamon Press, New York
10. Wickert JA, Mote CD (1990) Classical vibration analysis of axially moving continua. ASME J Appl Mech 57:738–744. https://doi.org/10.1115/1.2897085
11. Lagerstrom PA, Casten RG (1972) Basic concepts underlying singular perturbation techniques. SIAM Rev 14(1):63–120
12. Bender CM, Orszag SA (1978) Advanced mathematical methods for scientists and engineers I: asymptotic methods and perturbation theory. Springer, New York. 1999 reprint: ISBN 978-0-387-98931-0
13. Chen L-Y, Goldenfeld N, Oono Y (1996) Renormalization group and singular perturbations: multiple scales, boundary layers, and reductive perturbation theory. Phys Rev E 54(1):376–394
14. Kong L, Parker RG (2004) Approximate eigensolutions of axially moving beams with small flexural stiffness. J Sound Vib 276:459–469. https://doi.org/10.1016/j.jsv.2003.11.027
15. Vaughan M, Raman A (2010) Aeroelastic stability of axially moving webs coupled to incompressible flows. ASME J Appl Mech 77, 021001-1–021001-17. https://doi.org/10.1115/1.2910902
16. Polyanin A (2002) handbook of linear partial differential equations for engineers and scientists. Chapman & Hall/CRC
17. Bolotin VV (1964) The dynamic stability of elastic systems. Holden–Day, Inc. Translated from the Russian (1956) and German (1961) editions
18. Ziegler Hans (1952) Die stabilitätskriterien der elastomechanik. Ing Arch 20:49–56
19. Kirillov Oleg N, Verhulst Ferdinand (2010) Paradoxes of dissipation-induced destabilization or who opened Whitney's umbrella? Z Angew Math Mech 90(6):462–488. https://doi.org/10.1002/zamm.200900315
20. Kirillov ON (2013) Nonconservative stability problems of modern physics. de Gruyter. ISBN 978-3-11-027043-3
21. Simpson A (1973) Transverse modes and frequencies of beams translating between fixed end supports. J Mech Eng Sci 15:159–164. https://doi.org/10.1243/JMES_JOUR_1973_015_031_02
22. Kurki M, Jeronen J, Saksa T, Tuovinen T (2016) The origin of in-plane stresses in axially moving orthotropic continua. Int J Solids Struct. https://doi.org/10.1016/j.ijsolstr.2015.10.027
23. Timoshenko SP, Woinowsky-Krieger S (1959) Theory of plates and shells, 2nd edn. McGraw-Hill, New York, Tokyo. ISBN 0-07-085820-9
24. Press WH, Teukolsky SA, Vetterling WT, Flannery BP (2007) Numerical recipes: the art of scientific computing, 3rd edn. Cambridge University Press, New York. ISBN 978-0-521-88068-8
25. Press WH, Vetterling WT (1992) Numerical recipes in fortran 77: the art of scientific computing. Cambridge University Press. ISBN 0-521-43064-X
26. Horn RA, Johnson CR. (1999) Matrix analysis. Cambridge University Press. ISBN 978-0521386326
27. Banichuk N, Barsuk A, Tuovinen T, Jeronen J (2014c) Variational approach for analysis of harmonic vibration and stability of moving panels. Rakenteiden Mekaniikka (Finn J Struct Mech) 47(4):148–162
28. Yurddas A, Özkaya E, Boyaci H (2013) Nonlinear vibrations of axially moving multi-supported strings having non-ideal support conditions. Nonlinear Dyn 73(3):1223–1244.https://doi.org/10.1007/11071-012-0650-5. ISSN 0924-090X
29. Chen Li-Qun (2005b) Analysis and control of transverse vibrations of axially moving strings. ASME Appl Mech Rev 58:91–116. https://doi.org/10.1115/1.1849169

30. Bhat RB, Xistris GD, Sankar TS (1982) Dynamic behavior of a moving belt supported on elastic foundation. J Mech Design 104(1):143–147. https://doi.org/10.1115/1.3256304
31. Perkins NC (1990) Linear dynamics of a translating string on an elastic foundation. J Vib Acoust 112(1):2–7. https://doi.org/10.1115/1.2930094
32. Wickert JA (1994) Response solutions for the vibration of a traveling string on an elastic foundation. J Vib Acoust 116(1):137–139. https://doi.org/10.1115/1.2930389
33. Parker RG (1999) Supercritical speed stability of the trivial equilibrium of an axially-moving string on an elastic foundation. J Sound Vib 221(2):205–219. https://doi.org/10.1006/jsvi.1998.1936
34. Floquet G (1883) Sur les equations differentielles a coefficients periodiques. Ann l'Ecole Norm 12:47–88. http://www.numdam.org/item?id=ASENS_1883_2_12__47_0
35. Jkubovich VA, Starjinsky VM (1972) Linear differential equations with periodic coefficients and some applications (in Russian)
36. Luo ACJ, Mote CD Jr (2000) An exact, closed-form solution for equilibrium of traveling, sagged, elastic cables under uniformly distributed loading. Commun Nonlinear Sci Numer Simul 5(1):6–11
37. Courant R, Hilbert D (1966) Methods of mathematical physics, vol I. Interscience Publishers Inc, New York
38. Adams RA (1975) Sobolev spaces. Academic Press

Chapter 8
Stability in Fluid—Structure Interaction of Axially Moving Materials

In fluid—structure interaction problems, vibrations of the structure are still the main topic of interest, but now the motion is understood as being affected by the flow of the surrounding medium, such as air or water. The coupling is bidirectional and dynamic, because the motion of the structure also affects the flow. In this chapter, we first review some basic concepts of fluid mechanics, and then systematically derive a Green's function based analytical solution of the flow component of a simple fluid—structure interaction problem in two space dimensions. An added-mass approximation is also derived. As the structure component in the fluid—structure interaction problem we consider traveling ideal strings and panels. In the numerical results, we examine bifurcations in the natural frequencies of the coupled system. Whereas the natural frequencies of the traveling ideal string have no bifurcation points, bifurcations appear in the model with fluid—structure interaction whenever there is an axial free-stream flow.

8.1 Basic Concepts

Let us start this chapter by introducing some basic concepts related to fluid mechanics. At first, we will consider the balance of linear momentum for a general inviscid fluid, which leads to the Euler equation of motion. For the derivation of fundamental conservation laws for continua, see Allen et al. [1], Currie [2], Anderson [3]. Concerning the Gromeka—Lamb form of Euler's equation given below, see Sedov [4], pp. 166–167 or Kiselev [5], Chap. 4. As for tensor notations, see Appendix A.

In laboratory coordinates, the balance of linear momentum in continuum form reads

$$\rho\left(\frac{\partial \mathbf{v}}{\partial t} + \mathbf{v} \cdot \nabla \mathbf{v}\right) - \nabla \cdot \boldsymbol{\sigma}^{\mathrm{T}} = \mathbf{f} \, , \tag{8.1.1}$$

where ρ is the density of the continuum, $\mathbf{v}$ is the velocity field, $\boldsymbol{\sigma}$ is the stress tensor, $(\dots)^{\mathrm{T}}$ denotes the transpose of a rank-2 tensor, $(A^{\mathrm{T}})_{ij} = A_{ji}$, and $\mathbf{f}$ represents body

N. Banichuk et al., *Stability of Axially Moving Materials*, Solid Mechanics and Its Applications 259, https://doi.org/10.1007/978-3-030-23803-2_8

forces. Equation (8.1.1) is a fundamental law of mechanics valid for all fluids and solids. The feature that distinguishes these classes of continua is the form of the stress tensor $\boldsymbol{\sigma}$.

In this chapter, we choose $\boldsymbol{\sigma}$ as the stress tensor of an inviscid fluid,

$$\sigma_{ij} = -\delta_{ij} p \,, \tag{8.1.2}$$

where δ_{ij} is the Kronecker delta, and p is the pressure. An inviscid fluid only resists compression isotropically; there is no shear resistance. Note that choosing *inviscid fluid* is a constitutive assumption, which partially specifies the constitutive model for the continuum being modeled. Then

$$(\nabla \cdot \boldsymbol{\sigma}^{\mathrm{T}})_j = \partial_i \sigma_{ji} = -\partial_i (\delta_{ji} p) = -\partial_j p = -(\nabla p)_j \,, \tag{8.1.3}$$

and we obtain the equation for the flow of an inviscid fluid, known as *Euler's equation* (for fluid flow; not to be confused with the concept of *Euler equation* in the calculus of variations):

$$\frac{\partial \mathbf{v}}{\partial t} + \mathbf{v} \cdot \nabla \mathbf{v} = -\frac{1}{\rho} \nabla p + \mathbf{g} \,, \quad \text{where } \mathbf{g} \equiv \frac{1}{\rho} \mathbf{f} \,. \tag{8.1.4}$$

Inserting a tensor identity from Appendix A into Eq. (8.1.4), we have

$$\frac{\partial \mathbf{v}}{\partial t} + \frac{1}{2} \nabla (v^2) - \mathbf{v} \times (\nabla \times \mathbf{v}) = -\frac{1}{\rho} \nabla p + \mathbf{g} \,.$$

By defining the vorticity

$$\omega \equiv \frac{1}{2} (\nabla \times \mathbf{v}) \,, \tag{8.1.5}$$

and using the fact that $\mathbf{b} \times \mathbf{a} = -\mathbf{a} \times \mathbf{b}$ for any three-dimensional vectors $\mathbf{a}$ and $\mathbf{b}$, we obtain the *Gromeka—Lamb form* of Euler's equation for fluid flow:

$$\frac{\partial \mathbf{v}}{\partial t} + \frac{1}{2} \nabla \left(v^2\right) + 2\omega \times \mathbf{v} = -\frac{1}{\rho} \nabla p + \mathbf{g} \,. \tag{8.1.6}$$

The left-hand side represents the total acceleration $d\mathbf{v}/dt$ of a continuous medium, split into local time-dependent, potential and vorticity contributions. Contrast this with the split, in Eq. (8.1.4), into local time-dependent and inertial contributions. Note that we will use the Gromeka—Lamb form later below because it is particularly convenient for treating potential flows.

At this point, there is only one equation, (8.1.4) or alternatively (8.1.6), but there are three unknown fields: the velocity $\mathbf{v}$, the density ρ and the pressure p. In mechanics, there are four fundamental quantities that satisfy balance laws: linear momentum, angular momentum, mass, and internal energy. The balance of linear momentum produced the one equation we already have. The angular momentum balance only requires the stress tensor to be symmetric; this was already applied when choosing the stress tensor.

Therefore, let us turn our attention to the mass balance. In the Eulerian frame, local dynamic mass balance is described by the general conservation law of a scalar field,

$$\frac{\partial \rho}{\partial t} + \nabla \cdot (\rho \mathbf{v}) = 0 . \tag{8.1.7}$$

Here the conserved scalar field is the fluid density ρ. The divergence term in (8.1.7) can be expanded, yielding

$$\frac{\partial \rho}{\partial t} + \mathbf{v} \cdot \nabla \rho + \rho \nabla \cdot \mathbf{v} = 0 . \tag{8.1.8}$$

Moving one of the terms to the right-hand side gives a form that lends itself to an intuitive physical interpretation of (8.1.7):

$$\frac{\partial \rho}{\partial t} + \mathbf{v} \cdot \nabla \rho = -\rho \nabla \cdot \mathbf{v} . \tag{8.1.9}$$

Equation (8.1.9) is a scalar first-order transport equation with a source term. The operator on the left-hand side,

$$\left(\frac{\partial}{\partial t} + \mathbf{v} \cdot \nabla \right) (\ldots) \equiv \frac{\mathrm{d}}{\mathrm{d}t} (\ldots) , \tag{8.1.10}$$

is the material derivative for a field of material parcels, whose motion is described by the velocity field $\mathbf{v}$. The density field ρ thus undergoes advection by the velocity field $\mathbf{v}$. In addition, each material parcel experiences a source of strength $-\rho \nabla \cdot \mathbf{v}$. In other words, at points where $\nabla \cdot \mathbf{v} > 0$, that is, where fluid flows outward from the point, the density decreases proportionally to the existing density.

A simple way to complete the constitutive model, closing the equation system, is to take a *barotropic* equation of state: $\rho = \rho(p)$. A special case of this is $\rho = \text{const.}$, reducing the mass balance Eq. (8.1.7) into

$$\nabla \cdot \mathbf{v} = 0 , \tag{8.1.11}$$

which states that such a flow is incompressible. A typical case is any liquid.

If we had instead begun by requiring incompressibility, $\nabla \cdot \mathbf{v} = 0$, this transforms Eq. (8.1.7) into

$$\frac{\mathrm{d}\rho}{\mathrm{d}t} = \frac{\partial \rho}{\partial t} + \mathbf{v} \cdot \nabla \rho = 0 , \tag{8.1.12}$$

where $\mathrm{d}/\mathrm{d}t$ is the material derivative. Hence, in any incompressible flow, the density of each fluid parcel remains constant in time. The initial distribution of ρ is simply transported by the flow. Physically, in a single-phase flow of a liquid of a single type, it is reasonable to then take $\rho = \text{const.}$, but this is not strictly implied by just the property of incompressibility.

For an incompressible flow, no constitutive equation is needed to actually connect p and ρ; the pressure term in the balance of linear momentum, Eqs. (8.1.4) (or (8.1.6)), acts as a Lagrange multiplier for the incompressibility constraint. The pressure field instantaneously adjusts itself such that the velocity field remains divergence-free at each fixed time t. In computational fluid dynamics, predictor-corrector methods rely on this property. A physical interpretation is that because the pressure adjustment occurs instantaneously across the whole domain, the speed of sound in an incompressible fluid is infinite.

As a simple barotropic model specifically for gas flows, which are typically compressible, an option is to use the ideal gas equation of state:

$$pV = nRT \ , \tag{8.1.13}$$

where $p$is the pressure, V is the volume being considered, n (mol) is the amount of ideal gas in the volume V, the quantity $R = 8.3144598\,\text{J/mol K}$ is the universal gas constant, and T is the temperature (in Kelvins). For the molar amount, it holds that

$$n = \frac{m}{\mu} \ , \tag{8.1.14}$$

where m is the total mass of ideal gas in the volume V, and μ (kg/mol) is its mean molecular weight. For air, $\mu = 2.897 \cdot 10^{-2}$ kg/mol. This is accounting for a composition of approximately 78% N_2, 21% O_2, and 1% Ar, with negligible amounts of other constituents.

Inserting (8.1.14) into (8.1.13) and dividing both sides by $V \neq 0$, we have

$$p = \frac{m}{V}\frac{1}{\mu}RT \ . \tag{8.1.15}$$

Taking the limit as the control volume $V \to 0^+$, we have $m/V \to \rho$ and thus for the continuum, we obtain the pointwise relation

$$p = \rho\frac{1}{\mu}RT \ . \tag{8.1.16}$$

In order to make (8.1.16) barotropic, $p = p(\rho)$, we may regard the temperature as constant, for example, $T = 293.15$ K. Moreover, (8.1.16) is linear in ρ, the constant of proportionality being

$$\frac{RT}{\mu} = \frac{8.3144598\,\text{J/mol K} \cdot 293.15\,\text{K}}{2.897 \cdot 10^{-2}\text{kg/mol}} \approx 8.413 \cdot 10^4 \frac{\text{J}}{\text{kg}} \ . \tag{8.1.17}$$

In physical terms, the constant-temperature assumption is valid if any compression and expansion that occurs in the gas can be considered small enough to cause no significant temperature changes. In this variant of the model, we obtain a closed system of two equations in the two unknowns ρ and $\mathbf{v}$; Eqs. (8.1.4) (or (8.1.6)) and (8.1.7). The third unknown field, the pressure p, is given by (8.1.16).

For flows in which compression or expansion cause significant temperature changes, another way to complete the model is needed. Let us look at the fourth and final conserved quantity in mechanics: internal energy. Local dynamic balance of specific internal energy e (SI unit: J/kg), in continuum form, is given by

$$\frac{\partial e}{\partial t} + \mathbf{v} \cdot \nabla e + \frac{p}{\rho} \nabla \cdot \mathbf{v} = 0 . \tag{8.1.18}$$

Considering the internal energy adds one more equation, but it also adds a new unknown field, e. It is useful in cases where the equation of state is non-barotropic.

Similar to the mass balance, (8.1.18) is also a scalar transport equation with a source (or sink) term. As above, if we have an incompressible flow, the internal energy of each fluid parcel is conserved, and the initial internal energy field is just transported by the flow. Hence Eq. (8.1.18) is mainly of interest for compressible flows.

As an example of a non-barotropic, compressible flow, let us briefly consider the flow of an ideal gas allowing for temperature changes due to significant compression or expansion of the gas.

The total internal energy of an ideal gas is given by

$$U = c_V n T , \tag{8.1.19}$$

where U (SI unit: J) is the total internal energy for mass m, that is, $U/m = e$, and c_V (J/kg K) is the specific heat capacity of the ideal gas at constant volume. Dividing both sides of (8.1.19) by m and using (8.1.14), we obtain the relation between the specific internal energy and the temperature,

$$e = \frac{c_V}{\mu} T . \tag{8.1.20}$$

For air, the specific heat capacity itself depends on temperature; in the range 250 ... 300 K, engineering tables give the value $c_V \approx 0.717\,\mathrm{kJ/kg\,K} = 7.17 \cdot 10^2\,\mathrm{J/kg\,K}$. Solving Eq. (8.1.20) for T and inserting the result into (8.1.16) yields

$$p = \rho \frac{1}{\mu} R \left[\frac{e\mu}{c_V} \right] = \rho \frac{R}{c_V} e , \tag{8.1.21}$$

which explicitly gives p in terms of the other unknowns ρ and e. Also from (8.1.16), we have

$$\rho = \frac{\mu}{RT} p = \frac{c_V}{R} \frac{p}{e} . \tag{8.1.22}$$

Equations (8.1.20) and (8.1.22) are useful for initializing a simulation. It is sufficient to give the ambient pressure p_∞ and temperature T_∞; the corresponding specific internal energy e_∞ and density ρ_∞ are then determined from them.

Equations (8.1.7)–(8.1.18) and (8.1.21) form a closed system of four equations in the four unknowns ρ, $\mathbf{v}$, e and p. If we wish, we may explicitly eliminate p by inserting (8.1.21) into (8.1.4) and (8.1.18), obtaining

$$\begin{aligned}\frac{\partial \mathbf{v}}{\partial t} + \mathbf{v} \cdot \nabla \mathbf{v} + \frac{R}{c_V} \frac{\nabla(\rho e)}{\rho} &= \frac{\partial \mathbf{v}}{\partial t} + \mathbf{v} \cdot \nabla \mathbf{v} + \frac{R}{c_V} \frac{e \nabla \rho + \rho \nabla e}{\rho} \\ &= \frac{\partial \mathbf{v}}{\partial t} + \mathbf{v} \cdot \nabla \mathbf{v} + \frac{R}{c_V} \left[e \frac{\nabla \rho}{\rho} + \nabla e \right] = \mathbf{g} \end{aligned} \tag{8.1.23}$$

and

$$\frac{\partial e}{\partial t} + \mathbf{v} \cdot \nabla e + \frac{R}{c_V} \frac{\rho e}{\rho} \nabla \cdot \mathbf{v} = \frac{\partial e}{\partial t} + \mathbf{v} \cdot \nabla e + \frac{R}{c_V} e \nabla \cdot \mathbf{v} = 0\,. \tag{8.1.24}$$

The first equation describes the linear momentum balance, while the second gives the balance of internal energy. Equations (8.1.7) (for the mass balance), (8.1.23) and (8.1.24) now form a closed system of three equations in the three unknowns ρ, $\mathbf{v}$ and e. If desired, the pressure p can then be determined from (8.1.21).

Returning now to the simplest barotropic model $\rho = \text{const.}$, it is possible to derive a physically motivated equation for p in this setting. We start by taking the divergence of the balance of linear momentum, Eq. (8.1.4). We have

$$\nabla \cdot \left(\frac{\partial \mathbf{v}}{\partial t} + \mathbf{v} \cdot \nabla \mathbf{v} + \frac{\nabla p}{\rho} \right) = \nabla \cdot \mathbf{g}\,. \tag{8.1.25}$$

In index notation,

$$\partial_i \left(\partial_t v_i + v_j \partial_j v_i + \frac{1}{\rho} \partial_i p \right) = \partial_i g_i\,,$$

and expanding the parentheses, we obtain

$$\partial_i (\partial_t v_i) + \partial_i (v_j \partial_j v_i) + \partial_i \left(\frac{1}{\rho} \partial_i p \right) = \partial_i g_i\,.$$

Then, further expanding in the second term, we have

$$\partial_i (\partial_t v_i) + (\partial_i v_j)(\partial_j v_i) + v_j (\partial_i \partial_j v_i) + \partial_i \left(\frac{1}{\rho} \partial_i p \right) = \partial_i g_i\,.$$

Assuming sufficient continuity so that reordering derivatives is allowed, we may rewrite this as

$$\partial_t(\partial_i v_i) + (\partial_i v_j)(\partial_j v_i) + v_j \partial_j(\partial_i v_i) + \partial_i \left(\frac{1}{\rho} \partial_i p \right) = \partial_i g_i \ .$$

Using $\partial_i v_i = 0$ (incompressibility) and $\rho = \text{const.}$ yields

$$(\partial_i v_j)(\partial_j v_i) + \frac{1}{\rho} \partial_i \partial_i p = \partial_i g_i \ .$$

Converting back to nabla notation, the pressure field in a constant-density incompressible flow satisfies the equation

$$\nabla \mathbf{v} : \nabla \mathbf{v} + \frac{1}{\rho} \Delta p = \nabla \cdot \mathbf{g} \ , \tag{8.1.26}$$

where the double-dot product is defined as $\mathbf{A} : \mathbf{B} = A_{ij} B_{ji}$. Note the ordering of the indices in the operands. Equation (8.1.26) is a Poisson equation for the pressure field.

Some short remarks are in order here. Some authors write the first term, equivalently, as

$$(\partial_i v_j)(\partial_j v_i) = \text{tr}\left((\nabla \mathbf{v})^2 \right) \ , \tag{8.1.27}$$

where $\text{tr}\,(\mathbf{A}) = A_{ii}$ is the trace of a rank-2 tensor, and $\mathbf{A}^2 = \mathbf{A} \cdot \mathbf{A}$, i.e. $(\mathbf{A}^2)_{ij} = A_{ik} A_{kj}$. Note that if $\mathbf{g}$ is, for example, a uniform gravitational field, the right-hand side of (8.1.26) vanishes.

A physical interpretation for (8.1.26) is obtained by rearranging the original form from which we started, Eq. (8.1.25). We have

$$\frac{1}{\rho} \Delta p = \nabla \cdot \left(\mathbf{g} - \frac{\mathrm{d}\mathbf{v}}{\mathrm{d}t} \right) \ . \tag{8.1.28}$$

Local divergences in $\mathbf{g}$ and in the material acceleration $\mathrm{d}\mathbf{v}/\mathrm{d}t$ act as sources (and sinks) of pressure. The pressure satisfies the Poisson equation (i.e., a steady-state diffusion equation) with a coefficient of diffusivity of $1/\rho$. In other words, the lower the density of the flowing medium, the more diffuse the pressure variations caused by the sources and sinks. The reverse is also true: the higher the density, the sharper the variations in the pressure. Because there is no $\partial p/\partial t$ term, we observe that separately for each fixed t, the steady state of this diffusion process is reached instantaneously, which shows that the speed of sound in a constant-density flow is infinite.

We are now ready to move forward to the potential flow model and the Cauchy—Lagrange integral. For more on the Cauchy—Lagrange and Bernoulli integrals, see Sedov [6], pp. 155–156 or Kiselev et al. [5], Chap. 4. Potential flow in general is treated in, for example, Lighthill [7], Batchelor [8], Lamb [9], Acheson [10], Anderson [3]. A good exposition on elementary potential flows can be found in Currie [2]. For a focus on rotating fluids, see Vanyo [11].

Let us use the simplest, constant-density constitutive model. As stated above, this leads to incompressibility of the flow, that is, $\nabla \cdot \mathbf{v} = 0$. In addition, let us introduce the restriction that the flow is irrotational, $\nabla \times \mathbf{v} \equiv 0$, whence, by (8.1.5), the vorticity $\omega = 0$.

The fundamental theorem of vector calculus states that any smooth vector field in $\mathbb{R}^3$ can be decomposed into the sum of divergence-free (solenoidal) and curl-free (irrotational) parts:

$$\mathbf{v} = \mathbf{w} + \nabla\phi\,, \quad \nabla \cdot \mathbf{w} = 0\,. \tag{8.1.29}$$

Here ϕ is a scalar potential; recall that $\nabla \times \nabla(\ldots) \equiv 0$ for any operand in $C^2(\mathbb{R}^3)$. Equation (8.1.29) is known as the Helmholtz decomposition.

For irrotational flow, in order for the requirement $\nabla \times \mathbf{v} = 0$ to hold, the part which allows nonzero vorticity, that is, $\mathbf{w}$ in (8.1.29), must be zero. Thus, we are left with

$$\mathbf{v} = \nabla\Phi\,, \tag{8.1.30}$$

where $\Phi = \Phi(x, y, z, t)$ is the velocity potential. Thus we have potential flow, where the velocity field is the gradient of a scalar potential field.

One must be careful with the sign. Instead of (8.1.30), some authors define $\mathbf{v} = -\nabla\Phi$, so that the fluid moves "downhill" in the potential field, in analogy with the gravitational potential in classical physics, where the force vector points downhill. In general, a potential field is an abstract mathematical object, so it does not matter which definition one uses, as long as the usage is consistent.

Although the vorticity of a potential flow5 is zero, this does not completely exclude vortices. Inviscid vortices, which have zero vorticity everywhere except at a singularity at the center, are representable in potential flow theory. The singularity occurs in a set of spatial dimensionality lower than the flow domain itself. In a two-dimensional inviscid vortex, the singularity is a point, while in the three-dimensional case, it is a line (or a curve).

In an inviscid vortex, the differential fluid parcels move along the streamlines, circling around the center of the vortex, but they do not rotate; see Fig. 8.1. This is the physical meaning of *irrotational*; the term refers to the behavior of the differential fluid parcels, not to the macroscopic behavior of the flow itself. See, for example, Vanyo [11], pp. 132–134.

In classical treatments, the singularity is located outside the flow domain, inside an obstacle. However, allowing singularities to occur at isolated points inside the flow domain leads to an interesting expansion of flow configurations that can be described with the potential flow model. For example, two-dimensional potential flows exhibiting flow separation were investigated in Verhoff [12]. It was found that Prandtl's classical experimental results for flow past a cylinder (see e.g. Schlichting and Gersten [13], Schlichting [14]), before the instability (that starts the von Kármán vortex street) sets in, can be reproduced also in the potential flow model (with two empirically tuned parameters).

For constant density only, we can write

$$\frac{1}{\rho}\nabla p = \nabla\left(\frac{1}{\rho}p\right)\,. \tag{8.1.31}$$

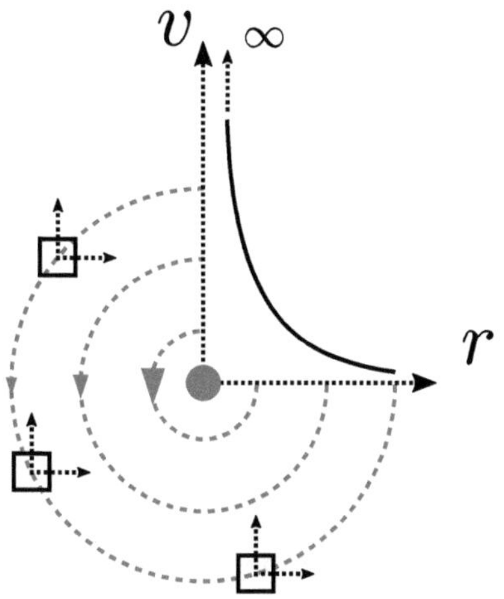

Fig. 8.1 Schematic illustration of an inviscid vortex. The differential fluid parcels (small squares) move along the circular streamlines surrounding the center of the vortex, but their orientation remains constant. At the center of the vortex, there is a singularity at which the fluid velocity is undefined. As a function of the distance from the center (r), the magnitude of the velocity (v) approaches infinity as $r \to 0$. Vorticity is zero everywhere except at the singularity. Inviscid vortices may occur in the potential flow model

By inserting (8.1.30) and (8.1.31) into the Gromeka—Lamb form of Euler's equation of fluid flow, (8.1.6), and rearranging the terms, we obtain

$$\nabla \left[\frac{\partial \Phi}{\partial t} + \frac{1}{2} (\nabla \Phi)^2 + \frac{1}{\rho} p \right] = \mathbf{g} . \tag{8.1.32}$$

Let us additionally restrict our consideration to the case where the external body forces $\mathbf{g}$ have a scalar potential g. This includes several cases of practical interest, such as a nonuniform gravitational field and the special case with no external forces. We have

$$\mathbf{g} = \nabla g , \tag{8.1.33}$$

and Eq. (8.1.32) becomes

$$\nabla \left[\frac{\partial \Phi}{\partial t} + \frac{1}{2} (\nabla \Phi)^2 + \frac{1}{\rho} p - g \right] = 0 . \tag{8.1.34}$$

Because (8.1.34) holds at all points of the domain, the operand in the brackets must be a function of time only; let us call it $h(t)$. We thus obtain the *Cauchy—Lagrange integral* that describes time-dependent flow of an irrotational, incompressible, inviscid fluid:

$$\frac{\partial \Phi}{\partial t} + \frac{1}{2} (\nabla \Phi)^2 + \frac{1}{\rho} p - g = h(t) . \tag{8.1.35}$$

In the case of a steady state, (8.1.35) becomes the corresponding *Bernoulli's integral*:

$$\frac{1}{2}(\nabla\Phi)^2 + \frac{1}{\rho}p - g = C\,, \tag{8.1.36}$$

where $h(t) \equiv C$ is a constant.

In conclusion, Eqs. (8.1.35) and (8.1.36) follow from the linear momentum balance. On the other hand, because $\mathbf{v} = \nabla\Phi$ and $\nabla\cdot\mathbf{v} = 0$, we must have also

$$0 = \nabla\cdot\mathbf{v} = \nabla\cdot(\nabla\Phi) = (\nabla\cdot\nabla)\Phi \equiv \Delta\Phi\,.$$

The velocity potential thus fulfills the Laplace equation

$$\Delta\Phi = 0\,, \tag{8.1.37}$$

which is commonly described as the equation that represents potential flow. We now recognize it as the incompressibility constraint in disguise (see e.g. Anderson [3], p. 131). Note that the irrotational part $\nabla\Phi$ in the fundamental theorem of vector calculus, Eq. (8.1.29), is not automatically divergence-free; it is necessary to require (8.1.37) to enforce this additional condition.

Equation (8.1.37) holds inside the fluid domain. Augmented with the appropriate boundary conditions, we may use it to solve the velocity field. Once the fluid velocity field has been obtained, equation (8.1.35) can then be used to determine the pressure if needed.

Because the Laplace Eq. (8.1.37) is linear in Φ, a linear combination of solutions is also a solution. Hence new velocity potentials, and thus also new velocity fields in the potential flow model, can be formed by superposition of elementary flows. Common elementary flows are the free-stream (axial) flow, the source/sink, the inviscid vortex, and the doublet. For more details, see Currie [2], Chap. 4 or Anderson [3], Sect. 3.9. The source/sink and the doublet are also mentioned by Ashley and Landahl [15], p. 28 and Vanyo [11], pp. 48–50; these flows are useful for obtaining the unique nonlifting potential flow past a cylinder. The inviscid vortex (line vortex) is also mentioned in Acheson [10], pp. 125–126.

In the special case of two-dimensional potential flow, conformal (i.e., locally angle-preserving) mappings can be used to treat other geometries when the velocity field in one geometry is known, because conformal mappings preserve the Laplacian. Conformal mappings are in fact the classical method to obtain the velocity field for potential flow past a cylinder, or that past some classical aerofoil shapes (see Ashley and Landahl [15], Sects. 2–10, Currie [2], Sects. 4.16–4.19, Acheson [10], Chap. 4, Batchelor [8], Sect. 6.7, Lighthill [7], Sects. 9.3, 9.4, and Vanyo [11], pp. 269–270).

A general approach for conformal mappings for polygonal domains is the Schwarz—Christoffel formula. For the fundamental theory and basic examples, see Nehari [16], pp. 193–196 and, for an advanced example, the study by Verhoff [12]. For practical considerations for more complicated geometries, see the numerical algorithm by Driscoll [17].

The classical choice for the boundary condition at surfaces of solid obstacles is that flow cannot cross the surface (no-flow condition; no-penetration condition). Provided that the obstacle remains stationary in the coordinate system where we consider the fluid motion, we impose the boundary condition that the normal velocity of the fluid is zero:

$$0 = \mathbf{n} \cdot \mathbf{v} = \mathbf{n} \cdot \nabla\Phi \quad \text{on } \Gamma\,, \tag{8.1.38}$$

where $\mathbf{n}$ is the unit outer normal of the obstacle surface Γ. Note that by the principle of Galilean relativity, objects in steady motion through a fluid, such as aeroplanes, can be considered as stationary structures subjected to an axial flow.

Equation (8.1.38) is also called the slip boundary condition, because it allows for arbitrary tangential fluid motion along the surface of the obstacle. In physical terms, an inviscid flow experiences no friction.

In the context of the Laplace Eq. (8.1.37), the condition (8.1.38) represents a Neumann boundary condition. This can be seen from the weak form of the problem. Multiplying (8.1.37) by a scalar test function ψ and integrating over the fluid domain Ω, we have

$$\int_\Omega (\Delta\Phi)\psi \,\mathrm{d}V = 0\,, \tag{8.1.39}$$

where $\mathrm{d}V$ represents a differential volume. Equation (8.1.39) transforms into its weak form by applying Green's first integral identity, but let us first take a small detour, and consider where the identity comes from. This will give us a general procedure for constructing formulas for multidimensional integration by parts, which, beside the specific case of Laplace and Poisson equations, works for many partial differential equations in mechanics.

The overall aim is to lower the continuity requirements on Φ by transferring differentiations to a test function. Like in previous chapters of this book, once we obtain a weak form, we may then declare it as the new *definition* of our problem. In situations where the extra continuity is available, the derivation of the weak form (applied in reverse) then shows it to be equivalent with the strong form.

Because $\Delta(\ldots) \equiv \nabla \cdot \nabla(\ldots)$, the integrand in (8.1.39) is of the form $(\nabla \cdot \mathbf{q})\psi$ for a vector field $\mathbf{q}$ and a scalar field ψ. The integrand thus involves a divergence as the outermost differential operator in the expression, suggesting that the Gauss–Green–Ostrogradsky divergence theorem is the appropriate tool for this particular problem.

Thus we would like to find an expression, whose divergence generates the integrand in (8.1.39) plus some extra terms. We see that the following choice works:

$$\nabla \cdot (\mathbf{q}\psi) = \partial_i(q_i\psi) = (\partial_i q_i)\psi + q_i(\partial_i\psi) = (\nabla \cdot \mathbf{q})\psi + \mathbf{q} \cdot \nabla\psi\,, \tag{8.1.40}$$

as our integrand appears on its right-hand side. Integrating (8.1.40) over the domain Ω, we obtain

$$\int_\Omega \nabla \cdot (\mathbf{q}\psi) \, \mathrm{d}V = \int_\Omega (\nabla \cdot \mathbf{q})\psi \, \mathrm{d}V + \int_\Omega \mathbf{q} \cdot \nabla\psi \, \mathrm{d}V \, . \tag{8.1.41}$$

Now, because the integrand on the left-hand side is a divergence, we may invoke the Gauss–Green–Ostrogradsky divergence theorem (also known as Gauss's theorem, or the divergence theorem), which states that

$$\int_\Omega \nabla \cdot \mathbf{f} \, \mathrm{d}V = \int_{\partial\Omega} \mathbf{n} \cdot \mathbf{f} \, \mathrm{d}A \, , \tag{8.1.42}$$

where $\mathbf{f}$ is any differentiable vector field, $\partial\Omega$ is the boundary surface of Ω, $\mathbf{n}$ is the unit outer normal of $\partial\Omega$, and $\mathrm{d}A$ represents a differential surface area. The proof of the theorem can be found in calculus textbooks, such as Adams and Essex [18], pp. 907–908.

Using (8.1.42) on the left-hand side of (8.1.41), we obtain

$$\int_{\partial\Omega} \mathbf{n} \cdot (\mathbf{q}\psi) \, \mathrm{d}A = \int_\Omega (\nabla \cdot \mathbf{q})\psi \, \mathrm{d}V + \int_\Omega \mathbf{q} \cdot \nabla\psi \, \mathrm{d}V \, .$$

Observing that ψ is a scalar field and rearranging terms gives the desired integration by parts formula

$$\int_\Omega (\nabla \cdot \mathbf{q})\psi \, \mathrm{d}V = -\int_\Omega \mathbf{q} \cdot \nabla\psi \, \mathrm{d}V + \int_{\partial\Omega} \psi \, \mathbf{n} \cdot \mathbf{q} \, \mathrm{d}A \, , \tag{8.1.43}$$

which is the end result of the general procedure for our case. Choosing $\mathbf{q} = \nabla\Phi$ in (8.1.43) yields Green's first integral identity:

$$\int_\Omega (\Delta\Phi)\psi \, \mathrm{d}V \equiv \int_\Omega (\nabla \cdot \nabla\Phi)\psi \, \mathrm{d}V = -\int_\Omega \nabla\Phi \cdot \nabla\psi \, \mathrm{d}V + \int_{\partial\Omega} \psi \, \mathbf{n} \cdot \nabla\Phi \, \mathrm{d}A \, . \tag{8.1.44}$$

Applying (8.1.44) in (8.1.39), we obtain the weak form of the Laplace problem:

$$\int_{\partial\Omega} \psi \, \mathbf{n} \cdot \nabla\Phi \, \mathrm{d}A - \int_\Omega \nabla\Phi \cdot \nabla\psi \, \mathrm{d}V = 0 \, . \tag{8.1.45}$$

We see that the natural (Neumann) boundary condition (i.e., the integrand in the boundary term produced by the integration by parts) involves the normal velocity of the fluid, $\mathbf{n} \cdot \nabla\Phi$.

In flows past an obstacle, it is customary to require that, at far away from the obstacle, the fluid returns to its undisturbed free-stream state. Without loss of generality, let us choose $+x$ as the direction of free-stream flow. We may split the fluid velocity potential into two parts,

$$\Phi\,(x, y, z, t) = x v_\infty + \varphi\,(x, y, z, t) \ , \tag{8.1.46}$$

where the first term represents the free-stream flow, and $\varphi(x, y, z, t)$ is a disturbance term representing the effect of the obstacle on the flow. The complete fluid velocity field is thus

$$\mathbf{v} = \nabla\Phi = v_\infty + \nabla\varphi \,. \tag{8.1.47}$$

The free-stream part of (8.1.46) is trivially harmonic in all of space, that is, it trivially satisfies Eq. (8.1.37). In order for the fluid to return to its undisturbed state at infinity, the effect of the disturbance field on the fluid velocity must vanish there, in all directions. Hence we impose the condition

$$\lim_{|\mathbf{r}|\to\infty} \frac{\mathbf{r}}{|\mathbf{r}|} \cdot \nabla\varphi = 0 \,, \tag{8.1.48}$$

where $\mathbf{r} = (x, y, z)$, and $\mathbf{r}/|\mathbf{r}| = \mathbf{n}$. Equation (8.1.48) is a Neumann boundary condition at the surface at infinity . Rigorously speaking, this means we take the domain Ω as a large sphere centered on the origin, define the Neumann boundary condition at its surface, and require that as the radius of the sphere grows without bound, the condition (8.1.48) holds at the limit.

In the condition (8.1.48) we use only the disturbance term φ, whereas in (8.1.38), we use the full velocity potential Φ. This is because the behavior of the free stream at infinity is fine as it is, but on the surface of the obstacle, if there is to be no flow through the surface,the normal component of the complete fluid velocity must be zero.

Because we now have a Laplace problem with pure Neumann boundary conditions, the value of the potential Φ is unique only up to an additive constant. This ambiguity can be resolved in an arbitrary manner because the value of the velocity potential is of no physical interest. To obtain the velocity field, we need only its gradient, on which a global additive constant clearly has no effect.

For those solving the flow problem numerically, strategies for resolving the ambiguity are a practical necessity. One standard approach is to use a reaction term $\varepsilon\Phi$, where ε is a small constant, to regularize the problem, replacing (8.1.37) by

$$\Delta\Phi + \varepsilon\Phi = 0 \,. \tag{8.1.49}$$

The new term requires no special treatment. Multiplying (8.1.49) by a scalar test function ψ, integrating over Ω and applying (8.1.44) in the first term, we have the weak form

$$\int_{\partial\Omega} \psi\, \mathbf{n} \cdot \nabla\Phi \, \mathrm{d}A - \int_\Omega \nabla\Phi \cdot \nabla\psi \, \mathrm{d}V + \varepsilon \int_\Omega \Phi\psi \, \mathrm{d}V = 0 \,. \tag{8.1.50}$$

The regularization introduces a small error, but makes the solution unique (even with pure Neumann boundary conditions).

Another approach is to solve the original problem, but in two steps, as was done for the Neumann—Poisson problem arising in the solution process for the Navier—

Stokes equations in the studies Min and Gibou [19, 20]. First, in addition to the original Neumann boundary conditions, we set a Dirichlet condition for any one degree of freedom (DOF) inside the domain in the discretization, fixing it to any arbitrarily chosen value. This makes the solution unique. Then we solve this modified problem, obtaining an intermediate solution, where the values in the region near the Dirichlet DOF are not meaningful, but the rest of the solution, especially near the original boundaries, is valid. Then, as a second step, we solve the problem again in all or part of the domain, using the data obtained from the first step as Dirichlet boundary conditions at the original boundaries. In this second step, we ignore the original Neumann boundary conditions, and drop the additional condition that was used to make the solution of the first step unique. Finally, we overwrite the solution in the region that was updated in the second step, obtaining an approximate full solution for the original problem.

8.2 Analytical Solution of Two-dimensional Potential Flow

For solving the flow problem of the fluid component, we will apply classical analytical techniques originally developed for two-dimensional flows over thin aerofoils, constructing a Green's function type solution via complex analysis. The idea of a Green's function solution is to find the response, of the problem under study, specifically in the geometry under study, for an ideal point load (a Dirac delta load). This produces a kernel function. Then, integrating the kernel over the actual shape of the load under study, one obtains the desired response. Typically the Green's function is very difficult if not impossible to find; hence, the approach is used only for very simple problems, such as the Poisson equation in the half-plane (e.g. Kornecki et al. [21]) or in a spherically symmetric region. For an easily approachable mathematical presentation of the theory of Green's function solutions, with fundamental examples, see the book on partial differential equations by Evans [22].

Complex analysis is useful for the Laplace equation in particular, because harmonic functions are solutions of the Laplace equation. For more on this topic, see the books by Silverman [23, 24]. Note that although we have the Laplace equation—with no load inside the domain—a nonzero load occurs at the boundary.

We will utilize the results for the Laplace problem in a plane with a slit, originally derived by Silverman [25] in the context of a problem in elasticity; see also similar developments in fluid mechanics in Flanigan [15]. The developments to adapt these solutions to the presently considered case are original, first published in Banichuk et al. [26, 27], Jeronen [28], Banichuk et al. [29], Jeronen et al. [30].

Because we treat flow past an obstacle in two space dimensions, topologically the flow domain becomes doubly connected. This introduces an additional complication. From potential flow theory, it is known that in any doubly connected domain, $\nabla\varphi$ is unique if and only if one prescribes the value of the circulation Γ around the hole in the domain. Note that this consideration is separate from the uniqueness of φ

itself, where we have the arbitrary global additive constant even for singly connected topologies.

To justify the statement on the uniqueness of $\nabla\varphi$, suppose that we have found a two-dimensional velocity potential of the noncirculatory flow, $\Gamma = 0$, satisfying the slip boundary condition on a given surface S. Now, it is possible to add to this flow an inviscid vortex of arbitrary strength Γ_0, for which one of the circular streamlines has been conformally transformed into the surface S, hence satisfying the slip boundary condition on S. The general mapping theorem (also known as Riemann's mapping theorem; see e.g. Nehari [16], p. 175 for details) guarantees that this can always be done. By superposition of the two flows, a new irrotational flow has been created, still satisfying the slip boundary condition on S, but now $\Gamma = \Gamma_0$, which was arbitrary. Hence, the circulation Γ must be specified if we wish to have a unique irrotational flow in the doubly connected two-dimensional domain ([15], pp. 42–43).

Lighthill [7] discusses this topic in Sect. 6.4, and the related topic of flows over rotating cylinders in Chap. 10. Anderson [3], p. 156 also mentions this while discussing spinning cylinders and the Magnus effect, considering a lifting flow over a cylinder as a superposition of a nonlifting flow and an inviscid vortex of strength Γ.

This ambiguity is commonly resolved by imposing the Kutta—Zhukowski condition at the trailing edge. This condition picks a unique solution, makes the pressure continuous at the trailing edge, and also moves the trailing stagnation point of the flow to the tip of the aerofoil, removing a singularity in the flow velocity field that would otherwise occur there for shapes with sharp trailing edges. It also makes the predicted lift, which is of interest for the analysis of aerofoils, match values obtained from physical experiments (see e.g. Ashley and Landahl [15] and Lighthill [7]). Below, we will use a different regularity condition, by Sherman [25], originating in a problem in elasticity sharing some of its mathematical form with our flow problem.

Let us now consider the pressure field of a two-dimensional potential flow past an obstacle. We will need the pressure field in our fluid—structure interaction problem, to determine the interaction between the flow and the vibrations of the structure. For simplicity, we restrict our consideration to the case where there is no variation in the flow in the y direction; $\mathbf{v}(x, y, z, t) = \mathbf{v}(x, z, t)$, and proceed with a two-dimensional model. Following the splitting approach of Eq. (8.1.46), the velocity potential of the fluid is

$$\Phi(x, z, t) = x v_\infty + \varphi(x, z, t) \ , \tag{8.2.1}$$

where the first term is the free-stream axial flow occurring in the $+x$ direction, and $\varphi(x, z, t)$ is a disturbance term representing the effect of the obstacle on the flow. The complete fluid velocity is thus

$$\mathbf{v} = \nabla\Phi = v_\infty + \nabla\varphi \ . \tag{8.2.2}$$

Let us further restrict to the case of no external body forces, $g = 0$. With the fluid velocity potential (8.2.1), the Cauchy—Lagrange integral (8.1.35) becomes

$$\frac{\partial}{\partial t}\left(x v_\infty + \varphi\right) + \frac{1}{2}\left(\nabla\left(x v_\infty + \varphi\right)\right)^2 + \frac{1}{\rho} p = h\left(t\right) ,$$

where in the second term

$$\begin{aligned}
(\nabla\Phi)^2 &= \left(\nabla\left(x v_\infty + \varphi\right)\right)^2 \\
&= \left((v_\infty, 0) + \nabla\varphi\right)^2 \\
&= \left((v_\infty, 0) + \left(\frac{\partial\varphi}{\partial x}, \frac{\partial\varphi}{\partial z}\right)\right)^2 \\
&= (v_\infty, 0)^2 + 2\left(v_\infty, 0\right) \cdot \left(\frac{\partial\varphi}{\partial x}, \frac{\partial\varphi}{\partial z}\right) + \left(\frac{\partial\varphi}{\partial x}, \frac{\partial\varphi}{\partial z}\right)^2 \\
&= v_\infty^2 + 2 v_\infty \frac{\partial\varphi}{\partial x} + \left(\frac{\partial\varphi}{\partial x}\right)^2 + \left(\frac{\partial\varphi}{\partial z}\right)^2 .
\end{aligned} \tag{8.2.3}$$

Provided that the components of the disturbance velocity $\nabla\varphi$ are first-order small, the last two terms in (8.2.3) are second-order small and can thus be ignored. Therefore, we obtain the following approximate equation of motion for the fluid:

$$\frac{\partial\varphi}{\partial t} + \frac{1}{2} v_\infty^2 + v_\infty \frac{\partial\varphi}{\partial x} + \frac{1}{\rho} p = h\left(t\right) , \tag{8.2.4}$$

which holds at all points in the flow domain, and is accurate up to first order in the small quantities. Solving (8.2.4) for the pressure p, we have:

$$p = p\left(x, z, t\right) = \rho\left[h\left(t\right) - \frac{\partial\varphi}{\partial t} - \frac{1}{2} v_\infty^2 - v_\infty \frac{\partial\varphi}{\partial x}\right] . \tag{8.2.5}$$

Because $h(t)$ does not vary with respect to x and z, it cancels out in any pressure difference across space. A physical interpretation is

$$h(t) = \frac{1}{\rho} p_\infty , \tag{8.2.6}$$

that is, the function $h(t)$ describes the hydrostatic ambient pressure, normalized by the constant density of the fluid. Note that the quantity $q = q(x, z, t)$,

$$q = \frac{1}{2} \rho v^2 , \tag{8.2.7}$$

represents a dynamic pressure. In general, the total air pressure is

$$p = p_\infty - q = p_\infty - \frac{1}{2} \rho v^2 , \tag{8.2.8}$$

and specifically at infinity, $v = v_\infty$. Hence the expression $\rho h(t) - (1/2)\rho v_\infty$ in (8.2.5) is the total (hydrostatic plus dynamic) ambient pressure at infinity.

Consider now a slip boundary condition for an axially traveling obstacle. We aim to solve the fluid flow in the (x, z) plane region exterior to our obstacle. On the surface of the obstacle, a boundary condition is needed for the velocity field of the fluid. We will impose the slip boundary condition that was discussed above. In other words, we require that the flow cannot cross the surface of our axially traveling panel.

Differing from the above general treatment, we now allow axial motion for both the obstacle and the free stream. The boundaries of the setup, e.g. a paper machine, stay fixed in the Eulerian frame—whereas with regard to the axially co-moving frame, the boundaries are in motion. Therefore, the setup for fluid—structure interaction of axially moving materials is qualitatively different from the classical one in fluid—structure interaction, where a stationary structure is subjected to a flow, or a moving object is considered in the Lagrangean frame thus reducing the problem to the previous case. To avoid dealing with moving boundaries, in problems of axially moving materials, the standard approach is to use an Eulerian description also for the solid.

The Eulerian description, as is well known from fluid mechanics, has the further advantage of allowing for a *steady state* that is not static. In the case of axially moving materials, this is a steady state where the solid material flows through the control volume in a steady manner, the Eulerian description of its motion remaining constant in time. Such steady states are of practical interest, for example, for analyzing the steady operation of a paper machine.

Let us work through the derivation of the slip boundary condition to illustrate how the axial motion of the obstacle affects it. We begin by stating the boundary condition in terms of the velocity fields of the fluid and the obstacle (travelling panel), both expressed in the laboratory (Eulerian) coordinates:

$$\mathbf{n} \cdot \mathbf{v} = \mathbf{n} \cdot \frac{\mathrm{d}\mathbf{U}}{\mathrm{d}t} \quad \text{on } \Gamma\,, \tag{8.2.9}$$

where $\mathbf{n}$ is the unit surface normal vector of the panel, $\mathbf{v}$ is the fluid velocity field, and $\mathbf{U}$ is the displacement field of the panel. The symbol $\mathrm{d}/\mathrm{d}t$ denotes the Lagrange derivative; $\mathrm{d}\mathbf{U}/\mathrm{d}t$ describes the rate of change of $\mathbf{U}$ at a fixed location in the Euler coordinates, accounting for the motion of the panel. We will return to the exact definition of $\mathbf{U}$ momentarily.

The boundary Γ is the surface of the panel. For describing the vibrations of the panel, we use a small-displacement approximation, which is sufficient for stability analysis of the trivial equilibrium position. When defining the domain of the flow problem, we will approximate the panel geometrically as the straight line segment $z = 0,\ -1 \le x \le 1$; this is approximately valid when the panel is near the trivial equilibrium position. However, in the term $\mathrm{d}\mathbf{U}/\mathrm{d}t$, we must account for the vibrations of the panel, in order to have nonzero load for the fluid velocity field at the panel surface.

We find that the local normal vector of the panel surface is (see Fig. 8.2)

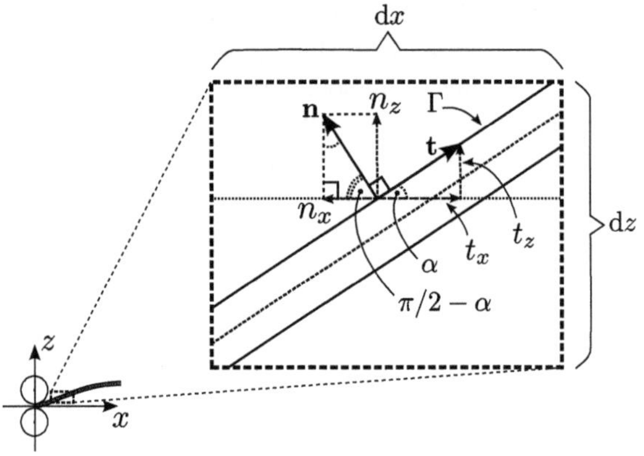

Fig. 8.2 Close-up of the panel surface Γ showing the local normal (**n**) and tangent (**t**) vectors

$$\mathbf{n} = \begin{pmatrix} -\sin\alpha \\ \cos\alpha \end{pmatrix} , \tag{8.2.10}$$

where α is the counterclockwise angle between the positive x axis and the local tangent vector of the panel. The direction of **n** is chosen such that it points toward positive z when $\alpha = 0$. It does not matter which choice we use, as long as we use the same choice on both sides of the Eq. (8.2.9).

To show (8.2.10), consider a small element of the panel, and in it, the upper 180° angle between the positive and negative x axis, which is divided by the panel surface and **n** into α, a straight angle and $(\pi/2) - \alpha$. Then consider the triangle formed by **n**, n_z and n_x, and the angles in this triangle. These must be the same three angles, but in a different order (see Fig. 8.2).

Alternatively, the normal vector can be found algebraically. Take the positively oriented tangent vector of the panel and the counterclockwise 2D rotation matrix:

$$\mathbf{t} = \begin{pmatrix} \cos\alpha \\ \sin\alpha \end{pmatrix} , \qquad \mathbf{R}(\theta) = \begin{pmatrix} \cos\theta & -\sin\theta \\ \sin\theta & \cos\theta \end{pmatrix} . \tag{8.2.11}$$

Then, evaluate the geometric relation connecting **n** and **t**:

$$\mathbf{n} = R(\pi/2)\, \mathbf{t} , \tag{8.2.12}$$

obtaining (8.2.10).

By the definitions of the tangent function, and on the other hand the derivative, we have

$$\tan\alpha = \frac{\partial w}{\partial x} \tag{8.2.13}$$

at the limit where dz and dx simultaneously tend to zero. The relation (8.2.13) is exact for arbitrarily large $\partial w/\partial x$; we have not yet applied the modeling assumption of small displacements.

Beside undergoing vibrations, the panel moves axially at a constant velocity V_0. It is thus convenient to introduce a coordinate system traveling toward positive x at a constant velocity of V_0; let us call this coordinate system the Lagrange (co-moving) coordinates. This is sometimes called the ; see Koivurova and Salonen [31]. Since the reference coordinate system moves at a constant velocity, it does not affect strains.

The displacement field of the panel, written in the Euler coordinates, is then

$$\mathbf{U}(x,t) = (V_0 t + \tilde{u}(\xi(x,t),t)\,,\ \tilde{w}(\xi(x,t),t))\,, \tag{8.2.14}$$

where $\tilde{u}$ and $\tilde{w}$ are, respectively, the in-plane and out-of-plane displacement functions defined in terms of the Lagrange coordinates, and the $V_0 t$ term accounts for the global axial motion (see Chap. 5). Here $\xi = \xi(x,t)$ is the horizontal coordinate in the co-moving coordinate system. The reference state for $\mathbf{U}$, with respect to which it measures the displacement, is the Eulerian position of the material particles at $t = 0$.

The form of (8.2.14) is critically important for deriving the slip boundary condition correctly. We have used the small-displacement modeling assumption in (8.2.14). It is obvious that in the general case,

$$\tilde{u} = \tilde{u}(s,t) \quad \text{and} \quad \tilde{w} = \tilde{w}(s,t),$$

where s is the longitudinal coordinate along the panel. The displacements are, generally speaking, functions of s and not of ξ (or x). However, in the small displacement regime, we can approximate

$$s \approx \xi\,.$$

In addition to being valid for small displacements only, this approximation has a further important property: describing shapes which are not single-valued functions of ξ is not possible in terms of the functions $\tilde{u}(\xi,t)$ and $\tilde{w}(\xi,t)$. This limits the class of shapes that can be described, but the limitation also induces an advantage: self-intersection of the panel surface is automatically prevented without the need for further constraints.

We have defined the displacement functions $\tilde{u}$ and $\tilde{w}$ as being concerned with the vibration behaviour only. Mathematically, they are the solutions of the partial differential equations governing the vibrations of the panel in the longitudinal and out-of-plane directions.

If we wrote the displacement field in the Lagrange coordinates, we would have

$$\tilde{\mathbf{U}}(\xi,t) = (\tilde{u}(\xi,t)\,,\ \tilde{w}(\xi,t))\,. \tag{8.2.15}$$

This is because at any given point of time t, each particle is only displaced from its original position *in the co-moving coordinate system* by the vibrations. The Lagrange coordinate system for this problem is defined precisely such that it accounts for the uniform axial motion. This is why a $V_0 t$ term appears in (8.2.14), but not in (8.2.15). In other words,

$$\mathbf{U}(\,x(\xi,t),\ t\,) = (\,V_0 t,\ 0\,) + \tilde{\mathbf{U}}(\xi,t)\,. \tag{8.2.16}$$

Before we proceed, it is worth pointing out that the axial tension in a paper machine is in practice generated by imposing a velocity difference between the rollers at the ends of each free span. This causes the web to stretch, which induces an x-dependent longitudinal strain Kurki et al. [32, 33]. Typically, these strains are small; we have neglected this effect in (8.2.14). If one wishes to take it into account, a new x-dependent term is needed in the axial component of $\mathbf{U}$. Although this term is easier to add into the Euler version $\mathbf{U}$ than the Lagrange version $\tilde{\mathbf{U}}$ (where it will depend on both ξ and t), from the physics of the situation it is evident that this term, if added, must appear in both coordinate systems.

From (8.2.14), we obtain in (x, t) coordinates the velocity field of the panel,

$$\frac{\mathrm{d}\mathbf{U}}{\mathrm{d}t} = (V_0 + \frac{\partial u}{\partial t} + V_0 \frac{\partial u}{\partial x} , \frac{\partial w}{\partial t} + V_0 \frac{\partial w}{\partial x}) \,. \tag{8.2.17}$$

For the normal component of the motion of the panel, from (8.2.10) and (8.2.17) we obtain

$$\mathbf{n} \cdot \frac{\mathrm{d}\mathbf{U}}{\mathrm{d}t} = -\sin(\alpha) \left[V_0 + \frac{\partial u}{\partial t} + V_0 \frac{\partial u}{\partial x} \right] + \cos(\alpha) \left[\frac{\partial w}{\partial t} + V_0 \frac{\partial w}{\partial x} \right] \,. \tag{8.2.18}$$

Similarly, from (8.2.10) and (8.2.2), for the normal component of the velocity of the fluid we have

$$\mathbf{n} \cdot \mathbf{v} = -\sin(\alpha) \left[v_\infty + \frac{\partial \varphi}{\partial x} \right] + \cos(\alpha) \frac{\partial \varphi}{\partial z} \,. \tag{8.2.19}$$

Subtracting (8.2.18) from (8.2.19) and using the slip boundary condition (8.2.9) to eliminate the left-hand side gives

$$-\sin(\alpha) \left[v_\infty - V_0 - \frac{\partial u}{\partial t} - V_0 \frac{\partial u}{\partial x} + \frac{\partial \varphi}{\partial x} \right] + \cos(\alpha) \left[\frac{\partial \varphi}{\partial z} - \frac{\partial w}{\partial t} - V_0 \frac{\partial w}{\partial x} \right] = 0 \,.$$

Dividing by $\cos \alpha$, substituting (8.2.13) for $\tan \alpha$, and multiplying the equation by -1 (for convenience) yields

$$\frac{\partial w}{\partial x} \left[v_\infty - V_0 - \frac{\partial u}{\partial t} - V_0 \frac{\partial u}{\partial x} + \frac{\partial \varphi}{\partial x} \right] - \left[\frac{\partial \varphi}{\partial z} - \frac{\partial w}{\partial t} - V_0 \frac{\partial w}{\partial x} \right] = 0 \,. \tag{8.2.20}$$

The terms with $V_0 \partial w / \partial x$ cancel exactly, and all other terms remain. In the small-displacement regime, the quantities

$$\frac{\partial w}{\partial x}, \ \frac{\partial u}{\partial t}, \ \frac{\partial u}{\partial x}, \ \text{and} \ \frac{\partial \varphi}{\partial x}$$

are considered small, and we discard second-order small terms. This will approximate the panel as nearly horizontal, which is a useful simplification in order to obtain an analytical solution for the flow problem when the structure undergoes only small

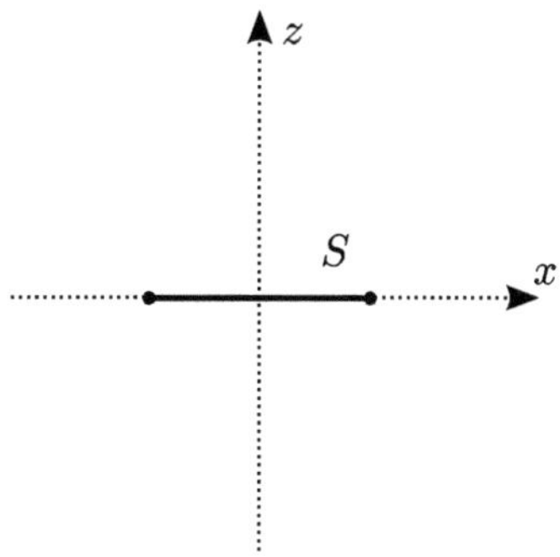

Fig. 8.3 Domain of the problem for the surrounding two-dimensional airflow. The panel is geometrically approximated as the infinitely thin linear cut $S \equiv \{z = 0,\ -1 \le x \le 1\}$

displacements. Rearranging terms, we obtain the linear approximation

$$\frac{\partial \varphi}{\partial z} = \frac{\partial w}{\partial t} + v_\infty \frac{\partial w}{\partial x} = \left[\frac{\partial}{\partial t} + v_\infty \frac{\partial}{\partial x} \right] w \equiv \gamma(x, t) \,, \tag{8.2.21}$$

which represents the small-displacement (approximate) slip boundary condition $\mathbf{n} \cdot \nabla \Phi = 0$ on the linearized panel surface S, where $\mathbf{n} = (0, \pm 1)$, see Fig. 8.3. It ensures that the flow will not cross the surface of the panel, accounting for up to first-order small terms. Note that the operator that appears is the material derivative for a coordinate system moving toward $+x$ at the free-stream velocity v_∞; it is the same operator that appeared in Eq. (8.2.25).

Equation (8.2.21) is more convenient to use than (8.2.20), because it does not require considering the longitudinal displacement u at all. This is another manifestation of the general phenomenon that in the small-displacement approximation, the in-plane and out-of-plane components of the motion of the structure become decoupled.

Note especially that the boundary condition (8.2.21) does not have a term that refers to V_0, although in Lagrange coordinates, what the system experiences is indeed the axial velocity difference $v_\infty - V_0$. The is easily confirmed by repeating the steps (8.2.14)–(8.2.21) in the Lagrange (co-moving) coordinates. When doing this, one must be careful to use

$$\frac{\partial \xi}{\partial t} = -V_0$$

instead of

$$\frac{\partial x}{\partial t} = V_0 \,,$$

which was used in the Euler coordinates. The $V_0 t$ term, which caused the cancellation, is not present in $\tilde{\mathbf{U}}$, and the velocity coefficient in the corresponding Lagrange boundary condition will indeed be $(v_\infty - V_0)$.

On the other hand, the lack of a V_0 term in (8.2.21) is not surprising, because both axial motions (panel and fluid) were accounted for independently in the Euler coordinate system. The free-stream fluid velocity was given in the Euler coordinates to begin with. By transforming the mechanics of the panel from the Lagrange into the Euler system, we have written both axial velocities in the same coordinate system. The approach of writing the boundary condition in the Lagrange coordinate system, on the other hand, transforms the fluid velocity into the coordinates axially moving with the panel, introducing a velocity shift by $-V_0$.

We will now evaluate the pressure difference across the panel. From this point on, for the purposes of flow geometry, we approximate the panel in the small-displacement range as the infinitely thin linear cut

$$S \equiv \{\, z = 0,\ -1 \leq x \leq 1 \,\}\,, \tag{8.2.22}$$

as shown in Fig. 8.3. The aerodynamic reaction pressure that the panel experiences is the jump of the total pressure across the cut S:

$$q_f \equiv p^- - p^+\,, \tag{8.2.23}$$

where p^+ (respectively p^-) indicates the total pressure on the $z > 0$ ($z < 0$) side of the cut, understood as a one-sided limit from the corresponding direction:

$$p^{\pm}(x, t) \equiv \lim_{z \to 0^{\pm}} p(x, z, t)\,. \tag{8.2.24}$$

Observe the signs in (8.2.23); an excess of pressure on the $-z$ side will push the panel upward (toward $+z$).

We obtain the aerodynamic reaction by inserting the total pressure, Eq. (8.2.5), into the expression for the aerodynamic reaction pressure, Eq. (8.2.23):

$$\begin{aligned} q_f &= p^- - p^+ \\ &= \rho \left[\cancel{h(t)} - \left(\frac{\partial \varphi}{\partial t}\right)^- - \cancel{\frac{1}{2} v_\infty^2} - v_\infty \left(\frac{\partial \varphi}{\partial x}\right)^- \right] \\ &\quad - \rho \left[\cancel{h(t)} - \left(\frac{\partial \varphi}{\partial t}\right)^+ - \cancel{\frac{1}{2} v_\infty^2} - v_\infty \left(\frac{\partial \varphi}{\partial x}\right)^+ \right] \\ &= \rho \left[\left(\frac{\partial \varphi}{\partial t}\right)^+ - \left(\frac{\partial \varphi}{\partial t}\right)^- + v_\infty \left\{ \left(\frac{\partial \varphi}{\partial x}\right)^+ - \left(\frac{\partial \varphi}{\partial x}\right)^- \right\} \right] \\ &= \rho \left[\frac{\partial}{\partial t} (\varphi^+ - \varphi^-) + v_\infty \frac{\partial}{\partial x} (\varphi^+ - \varphi^-) \right] \\ &\equiv \rho \left[\frac{\partial}{\partial t} + v_\infty \frac{\partial}{\partial x} \right] \psi\,, \end{aligned} \tag{8.2.25}$$

where in the last step an auxiliary function

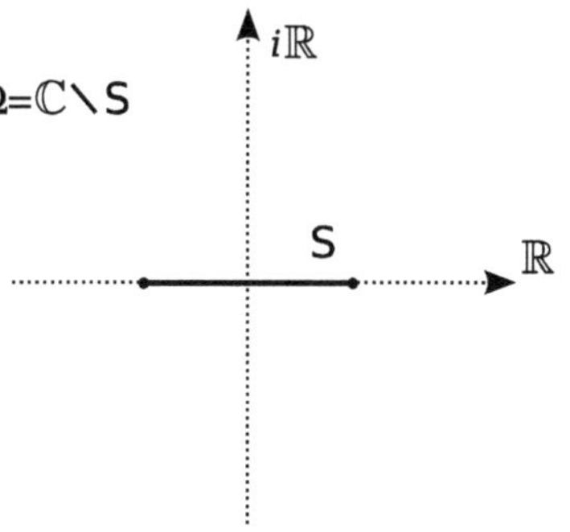

Fig. 8.4 Domain of the problem for the surrounding two-dimensional airflow. The Laplace problem will be solved using techniques of complex analysis. To this end, we identify the complex plane $\eta = x + iz$ with the two-dimensional space where the flow occurs. The panel is geometrically approximated as the infinitely thin linear cut $S \equiv \{z = 0, \ -1 \leq x \leq 1\}$. The symbol i denotes the imaginary unit, $i \equiv \sqrt{-1}$

$$\psi \equiv \varphi^+ - \varphi^- \tag{8.2.26}$$

has been defined. The superscript notation in (8.2.25)–(8.2.26) is defined analogously to (8.2.24).

As was presented in the Eq. (8.2.1), we consider the flow as a superposition of a constant-velocity free stream and a disturbance term representing the effect of the obstacle. The free stream moves toward positive x at the velocity v_∞. The free-stream potential is linear in x; hence it is trivially harmonic.

The approximate slip boundary condition for the geometrically linearized panel surface S (see Figs. 8.3, 8.4 and definition (8.2.22)) was derived above, the result being Eq. (8.2.21).

We only need to solve the flow problem for the disturbance potential φ. This problem can be stated as

$$\Delta\varphi \equiv \frac{\partial^2 \varphi}{\partial x^2} + \frac{\partial^2 \varphi}{\partial z^2} = 0 \qquad \text{in } \Omega \tag{8.2.27}$$

$$\left(\frac{\partial \varphi}{\partial z}\right)^{\pm} = \gamma(x, t)\,, \qquad \text{on } S \tag{8.2.28}$$

$$(\nabla\varphi)_\infty = 0\,, \tag{8.2.29}$$

where (8.2.27) is the Laplace equation for the disturbance potential φ, (8.2.28) is the linearized slip boundary condition where $\gamma(x, t)$ is given by (8.2.21), and (8.2.29) represents the boundary condition at infinity, Eq. (8.1.48), requiring that the disturbance in the fluid velocity field vanishes far away from the obstacle. The $\pm$ notation is defined as earlier in (8.2.24), and the subscript infinity denotes the limit

$$(\cdot)_\infty \equiv \lim_{|\eta|\to\infty} (\cdot) \,, \tag{8.2.30}$$

where $|\eta| \equiv \sqrt{x^2 + z^2}$ is the complex modulus. Note that in the boundary condition (8.2.28), we must use limits because the fluid velocity potential is not defined on the cut S. Furthermore, these limits must be one-sided, because the function φ may be discontinuous across the cut; and indeed must be, if it is to induce any aerodynamic reaction in the form (8.2.25)–(8.2.26).

Before we move on to solving (8.2.27)–(8.2.29), let us summarize our assumptions. The problem (8.2.27)–(8.2.29) involves several approximations. Some are obvious, such as the two-dimensional problem setup and the potential flow model. Thus, width-directional variation both in the flow and in the behavior of the traveling panel are neglected, as is the rotation of differential fluid elements and the viscosity of the fluid. A further approximation is that the boundary condition (8.2.28) is only valid up to first-order small terms, as was discussed during the derivation of (8.2.21).

The domain of the aerodynamic problem is infinite. It consists of the whole xz plane with the exception of the cut S, which is our linearized representation of the space (approximately) occupied by the panel (see Fig. 8.4) in the small-displacement regime. Thus, although we consider an axially moving panel, for the purposes of the aerodynamic problem the panel only exists on the interval $-1 \le x \le 1$. Effectively, in the panel domain, if we assume $V_0 > 0$ (without loss of generality), there is a material source at $x = -1$ and a material sink at $x = +1$. Also the rollers, which in a physical system would drive the panel, are ignored in the flow problem, being represented only in the boundary conditions for the panel displacement.

The way we have proceeded is of course just one possible choice to build a model for the situation considered, even within the context of the potential flow model. Instead of our setup of the plane with a slit, it could be assumed that the elastic panel is embedded in a rigid baffle of infinite extent, splitting the xz plane into two parts, as was done in the study by Kornecki et al. [21]. This choice leads to a singly connected flow topology, eliminating the need for a circulation condition.

Now we will present an analytical solution of (8.2.27)–(8.2.29). To start with, observe that because potential flow is memory-free, the flow field reconfigures itself instantly at each time t, independent of its history. Therefore, as far as the flow problem is concerned, the time t in the function $\gamma(x, t)$ in boundary condition (8.2.28) is just a parameter. This is the only time-dependent part in the fluid flow problem (8.2.27)–(8.2.29). Hence, in the following, we will consider an arbitrary fixed value for t, and treat only x and z as variables.

In accordance with the complex analysis approach to two-dimensional potential flows, we introduce an auxiliary analytic function

$$\Pi\,(\eta, t) = \Psi\,(x, z, t) + i\varphi\,(x, z, t) \tag{8.2.31}$$

of the complex variable

$$\eta = x + iz \,,$$

where $i^2 = -1$. The Cauchy—Riemann equations from complex analysis, and the boundary condition (8.2.28) for the flow on the panel surface, together imply that

$$\frac{\partial \Psi}{\partial x}|_{z=0} = \frac{\partial \varphi}{\partial z}|_{z=0} = \gamma(x,t) \ . \tag{8.2.32}$$

Let us denote $\Psi(x,t) \equiv \Psi(x,0,t)$. We have

$$\Psi(x,t) = \chi(x,t) + C(t) \ , \tag{8.2.33}$$

where

$$\chi(x,t) = \int_{-1}^{x} \gamma(\xi,t) \, d\xi \ , \tag{8.2.34}$$

and $C(t)$ is a real constant of integration for each fixed t. Here ξ is a dummy variable for integration (unrelated to the co-moving axial coordinate discussed earlier). Thus, finding the velocity potential φ reduces to the computation of the imaginary part of the analytic function (8.2.31), whose real part on $[-1, 1]$ is (8.2.33).

The idea of introducing the auxiliary function Π is that we may use the Neumann boundary data (8.2.28) for the original unknown potential φ to generate Dirichlet boundary data for the (also unknown) stream function Ψ, as per (8.2.33) and (8.2.34). Note that unlike the usual convention, the real part of Π is here the stream function, and the imaginary part is the real-valued potential for which the original problem was formulated.

We use the results given by Sherman [25] (compare also Ashley and Landahl [15], Chaps. 5–3) and represent the solution of this problem as

$$\Pi(\eta,t) = \frac{1}{2\pi i} \left(\frac{\eta - 1}{\eta + 1}\right)^{1/2} \int_{-1}^{1} \left(\frac{\xi + 1}{\xi - 1}\right)^{1/2} \frac{\chi(\xi,t) + C(t)}{\xi - \eta} \, d\xi \ . \tag{8.2.35}$$

Here also ξ is a dummy variable for integration. The real constant $C(t)$ (for any fixed t) is determined with the help of the following equation:

$$\frac{1}{2\pi i} \int_{-1}^{1} \frac{\chi(\xi,t) + C(t)}{\sqrt{\xi^2 - 1}} \, d\xi = 0 \ , \tag{8.2.36}$$

which represents a regularity condition for the function Π. The condition (8.2.36) is provided in the study by Sherman [25]. To motivate this condition, take the limit $\eta \to -1$ in (8.2.35), and use the identity

$$(\xi - 1)(\xi + 1) = \xi^2 - 1 \tag{8.2.37}$$

in the denominator in the integral. At the limit $\eta \to -1$, the denominator becomes $\xi + 1$, which cancels the $\sqrt{\xi + 1}$ in the numerator, leaving $\sqrt{(\xi - 1)(\xi + 1)}$ and allowing us to apply (8.2.37). This makes the integral in (8.2.35) have the same form as in (8.2.36).

Thus the condition (8.2.36) requires that the integral factor in (8.2.35) vanishes at $\eta = -1$, leading to (8.2.35) becoming an indefinite form (0/0) at the limit. Due to the coefficient $1/\sqrt{\eta + 1}$ in (8.2.35), which becomes zero at $\eta = -1$, requiring (8.2.36) is the only possibility that can lead to $|\Pi(-1, t)| < \infty$.

The next step is to solve (8.2.36) for the integration constant $C(t)$. Splitting the sum, using the fact that $C(t)$ is a constant (with respect to the integration variable ξ) and multiplying the equation by $2\pi i$, we have

$$\int_{-1}^{1} \frac{\chi(\xi, t)}{\sqrt{\xi^2 - 1}} \, d\xi + C(t) \int_{-1}^{1} \frac{1}{\sqrt{\xi^2 - 1}} \, d\xi = 0 \, .$$

We may explicitly evaluate the integral that multiplies $C(t)$:

$$\int_{-1}^{1} \frac{1}{\sqrt{\xi^2 - 1}} \, d\xi = -\pi i \, .$$

Transferring terms, we are left with

$$C(t) = \frac{1}{\pi i} \int_{-1}^{1} \frac{\chi(\xi, t)}{\sqrt{\xi^2 - 1}} \, d\xi \, . \tag{8.2.38}$$

The expression (8.2.38) is indeed real-valued; observing that $\sqrt{\xi^2 - 1} = i\sqrt{1 - \xi^2}$, we have the alternative representation

$$C(t) = -\frac{1}{\pi} \int_{-1}^{1} \frac{\chi(\xi, t)}{\sqrt{1 - \xi^2}} \, d\xi \, , \tag{8.2.39}$$

where all quantities are explicitly real-valued. By complex analysis, it can be shown that

$$\frac{1}{2\pi i} \int_{-1}^{1} \left(\frac{\xi + 1}{\xi - 1} \right)^{1/2} \frac{1}{\xi - \eta} \, d\xi = \frac{1}{2} \left[\left(\frac{\eta + 1}{\eta - 1} \right)^{1/2} - 1 \right] . \tag{8.2.40}$$

For a detailed treatment of an integral similar to (8.2.40), see Ashley and Landahl [15], pp. 94–95. Splitting the sum in (8.2.35) and using (8.2.40) in the second term, we have

$$\begin{aligned} \Pi(\eta, t) = \frac{1}{2\pi i} \left(\frac{\eta - 1}{\eta + 1} \right)^{1/2} \int_{-1}^{1} \left(\frac{\xi + 1}{\xi - 1} \right)^{1/2} & \frac{\chi(\xi, t)}{\xi - \eta} \, d\xi \\ + \frac{1}{2} C(t) \left[1 - \left(\frac{\eta - 1}{\eta + 1} \right)^{1/2} \right] & . \end{aligned}$$

Expanding the brackets on the second line and inserting (8.2.38) in the last term yields

$$\Pi(\eta, t) = \frac{1}{2\pi i}\left(\frac{\eta-1}{\eta+1}\right)^{1/2}\int_{-1}^{1}\left(\frac{\xi+1}{\xi-1}\right)^{1/2}\frac{\chi(\xi, t)\,\mathrm{d}\xi}{\xi-\eta} + \frac{1}{2}\,C(t)$$
$$-\frac{1}{2\pi i}\left(\frac{\eta-1}{\eta+1}\right)^{1/2}\int_{-1}^{1}\frac{\chi(\xi, t)}{\sqrt{\xi^2-1}}\,\mathrm{d}\xi\,.$$

Combining the integrals, we have

$$\begin{aligned}
\Pi(\eta, t) &= \frac{1}{2\pi i}\left(\frac{\eta-1}{\eta+1}\right)^{1/2}\int_{-1}^{1}\left[\left(\frac{\xi+1}{\xi-1}\right)^{1/2}\frac{1}{\xi-\eta} - \frac{1}{\sqrt{\xi^2-1}}\right]\chi(\xi, t)\,\mathrm{d}\xi + \frac{1}{2}C(t) \\
&= \frac{1}{2\pi i}\left(\frac{\eta-1}{\eta+1}\right)^{1/2}\int_{-1}^{1}\left[\frac{\sqrt{\xi+1}}{\sqrt{\xi-1}}\frac{\sqrt{\xi+1}}{\sqrt{\xi+1}(\xi-\eta)} - \frac{(\xi-\eta)}{(\xi-\eta)\sqrt{\xi+1}\sqrt{\xi-1}}\right]\chi(\xi, t)\,\mathrm{d}\xi + \frac{1}{2}C(t) \\
&= \frac{1}{2\pi i}\left(\frac{\eta-1}{\eta+1}\right)^{1/2}\int_{-1}^{1}\left[\frac{1+\eta}{(\xi-\eta)\sqrt{\xi+1}\sqrt{\xi-1}}\right]\chi(\xi, t)\,\mathrm{d}\xi + \frac{1}{2}C(t) \\
&= \frac{1}{2\pi i}\left(\frac{\eta-1}{\eta+1}\right)^{1/2}(1+\eta)\int_{-1}^{1}\frac{\chi(\xi, t)}{(\xi-\eta)\sqrt{\xi^2-1}}\,\mathrm{d}\xi + \frac{1}{2}C(t)\,.
\end{aligned}$$

The result is

$$\Pi(\eta, t) = \frac{\sqrt{\eta^2-1}}{2\pi i}\int_{-1}^{1}\frac{\chi(\xi, t)}{(\xi-\eta)\sqrt{\xi^2-1}}\,\mathrm{d}\xi + \frac{1}{2}\,C(t)\,. \tag{8.2.41}$$

Recalling (8.2.31), we can compute the quantity $\varphi^{+}(x, t)$ from the representation (8.2.41):

$$\begin{aligned}
\varphi^{+}(x, t) &= \lim_{z\to 0^{+}}\left[\operatorname{Im}\Pi(x+iz,\, t)\right] \\
&= \text{p.v.}\left(-\frac{\sqrt{1-x^2}}{2\pi}\int_{-1}^{1}\frac{\chi(\xi, t)}{(\xi-x)\sqrt{1-\xi^2}}\,\mathrm{d}\xi\right).
\end{aligned} \tag{8.2.42}$$

Because the constant $C(t)$ in (8.2.41) is real, it does not contribute to the limit of the imaginary part in (8.2.42). The divergent improper integral in (8.2.42) is understood in the sense of *Cauchy's principal value*, here denoted p.v.(·).

Observe that for any real x, it holds that $\sqrt{x^2-1} = i\sqrt{1-x^2}$. This has been applied in both the coefficient in front (canceling the existing i) and in the denominator in the integral (producing a new i). Finally, we have used $1/i = -i$; this is the source of the minus sign. Observe that all the quantities appearing in (8.2.42) are real-valued; at $z \to 0$ (leading to $\eta \to x$, because $\eta = x + iz$), the first term of (8.2.41) is purely imaginary.

Because the flow is antisymmetric with respect to the linearized panel surface S, it holds for the disturbance potential that

$$\varphi^+ - \varphi^- = 2\varphi^+ \, . \tag{8.2.43}$$

This is a general property of antisymmetric potential flow configurations; see Païdoussis [34, 35], and for a similar example, Eloy et al. [36]. Alternatively, we can take the corresponding limit of (8.2.41) on the side with negative z,

$$\eta = x - iz \;\rightarrow\; x - i \cdot 0 \;\left(z \rightarrow 0^-\right) \, ,$$

and obtain the same result.

By definition of Cauchy's principal value, we have

$$\begin{aligned} 2\varphi^+ = \quad & \text{p.v.} \left(-\frac{1}{\pi} \int_{-1}^{1} \left(\frac{1-x^2}{1-\xi^2} \right)^{1/2} \frac{\chi(\xi, t) \, \mathrm{d}\xi}{\xi - x} \right) \\ \equiv \quad & \lim_{\varepsilon \to 0} \; -\frac{1}{\pi} \left[\int_{-1}^{x-\varepsilon} \left(\frac{1-x^2}{1-\xi^2} \right)^{1/2} \frac{\chi(\xi, t) \, \mathrm{d}\xi}{\xi - x} \right. \\ & \left. + \int_{x+\varepsilon}^{1} \left(\frac{1-x^2}{1-\xi^2} \right)^{1/2} \frac{\chi(\xi, t) \, \mathrm{d}\xi}{\xi - x} \right] . \end{aligned} \tag{8.2.44}$$

Let us integrate by parts and substitute expression (8.2.34) for $\chi(x, t)$. We have the result

$$\begin{aligned} 2\varphi^+ = \quad & \lim_{\varepsilon \to 0} \left[N(x - \varepsilon, x) \int_{-1}^{x-\varepsilon} \gamma(\xi, t) \, \mathrm{d}\xi \right. \\ & - N(x + \varepsilon, x) \int_{-1}^{x+\varepsilon} \gamma(\xi, t) \, \mathrm{d}\xi \\ & - \int_{-1}^{x-\varepsilon} N(\xi, x) \gamma(\xi, t) \, \mathrm{d}\xi \\ & \left. - \int_{x+\varepsilon}^{1} N(\xi, x) \gamma(\xi, t) \, \mathrm{d}\xi \right] , \end{aligned} \tag{8.2.45}$$

where we have defined

$$N(\xi, x) \equiv \frac{1}{\pi} \ln \left| \frac{1 + \Lambda(\xi, x)}{1 - \Lambda(\xi, x)} \right| , \qquad \text{where}$$

$$\Lambda(\xi, x) \equiv \left[\frac{(1-x)(1+\xi)}{(1-\xi)(1+x)} \right]^{1/2} . \tag{8.2.46}$$

This is obtained by setting

$$\frac{\partial u}{\partial \xi} = \frac{\sqrt{\frac{1-x^2}{1-\xi^2}}}{\xi - x} , \qquad v = \chi(\xi, t)$$

in the integration by parts formula

$$\int_a^b \frac{\partial u}{\partial \xi} v \, \mathrm{d}\xi = [uv]_{\xi=a}^b - \int_a^b u \frac{\partial v}{\partial \xi} \, \mathrm{d}\xi .$$

The first term in (8.2.45) is the boundary term uv, evaluated at the upper limit $x - \varepsilon$ of the first integral in (8.2.44). The lower limit at -1 produces no term, because $\chi(x, t)$ is defined with the help of an integral from -1 to x; hence, $\chi(-1, t) = 0$. The second term in (8.2.45) is uv, evaluated at the *lower* limit $x + \varepsilon$ of the second integral in (8.2.44). The limits of integration from -1 to $x + \varepsilon$ are due to evaluating $\chi(x + \varepsilon, t)$. Now the upper limit at $+1$ produces no term, because

$$N(\xi, x) \to 0 \quad \text{as} \quad \xi \to 1 .$$

The last two terms in (8.2.45) are the straightforward integrated-by-parts terms of the form $u \, \partial v / \partial \xi$. See Figs. 8.5 and 8.6 for a qualitative illustration of the functions Λ and N, and Fig. 8.7 for contour plots.

We observe that all terms on the right-hand side of (8.2.45) are finite; therefore the integration by parts is legitimate. As $\varepsilon \to 0$, the sum of the first two terms in (8.2.45) approaches zero. As we will show below, the last two integrals converge as $\varepsilon \to 0$.

Therefore, the required functional dependence is of the form

$$2\varphi^+(x, t) = -\int_{-1}^{1} N(\xi, x)\, \gamma(\xi, t) \, \mathrm{d}\xi . \tag{8.2.47}$$

With the help of Eq. (8.2.47); the expression for the aerodynamic reaction, Eqs. (8.2.25)–(8.2.26); the boundary condition providing the function $\gamma(x, t)$, equation (8.2.21); and the definition of $N(\xi, x)$ in (8.2.46), we arrive at the analytical solution for the aerodynamic reaction. Writing out the coordinate scaling factors τ and ℓ explicitly, we have:

$$\begin{aligned} q_{\mathrm{f}}(x, t) &= p^-(x, t) - p^+(x, t) \\ &= \rho_{\mathrm{f}} \left(\frac{1}{\tau} \frac{\partial}{\partial t} + \frac{1}{\ell} v_\infty \frac{\partial}{\partial x} \right) \left(\varphi^+(x, t) - \varphi^-(x, t) \right) \\ &= \rho_{\mathrm{f}} \left(\frac{1}{\tau} \frac{\partial}{\partial t} + \frac{1}{\ell} v_\infty \frac{\partial}{\partial x} \right) \left(2\varphi^+(x, t) \right) \end{aligned}$$

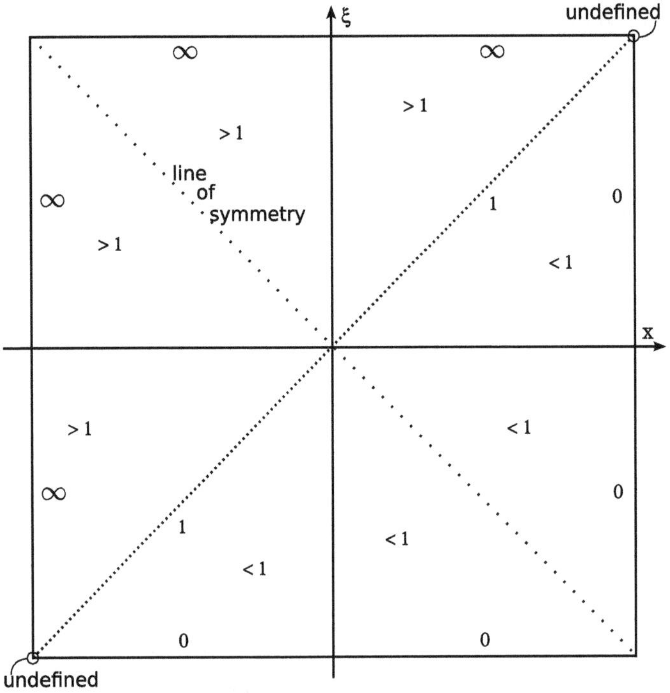

Fig. 8.5 Auxiliary function $\Lambda(\xi, x)$ in $[-1, 1] \times [-1, 1]$. Qualitative illustration. The infinities should be understood in the sense of limits

$$
\begin{aligned}
&= -\rho_{\mathrm{f}} \left(\frac{1}{\tau} \frac{\partial}{\partial t} + \frac{1}{\ell} v_\infty \frac{\partial}{\partial x} \right) \int_{-1}^{1} N(\xi, x)\, \gamma(\xi, t)\, \mathrm{d}\xi \qquad (8.2.48) \\
&= -\rho_{\mathrm{f}} \left(\frac{1}{\tau} \frac{\partial}{\partial t} + \frac{1}{\ell} v_\infty \frac{\partial}{\partial x} \right) \int_{-1}^{1} N(\xi, x) \left(\frac{\ell}{\tau} \frac{\partial}{\partial t} + v_\infty \frac{\partial}{\partial x} \right) w(\xi, t)\, \mathrm{d}\xi \\
&= -\rho_{\mathrm{f}} \frac{1}{\ell} \left(\frac{\ell}{\tau} \frac{\partial}{\partial t} + v_\infty \frac{\partial}{\partial x} \right) \int_{-1}^{1} N(\xi, x) \left(\frac{\ell}{\tau} \frac{\partial}{\partial t} + v_\infty \frac{\partial}{\partial x} \right) w(\xi, t)\, \mathrm{d}\xi \,.
\end{aligned}
$$

The scaling factor ℓ is the half-length of the span, and τ is an arbitrary scaling factor for nondimensionalization of the time coordinate. For a physically meaningful scaling, one can choose, for example, $\tau = \ell \,/\, V_{\mathrm{C}}$, where $V_{\mathrm{C}} = \sqrt{T/m}$, the critical velocity of an axially traveling tensioned ideal string. The unit of τ is $[\tau] = \mathrm{s}$.

In conclusion of this section, we have obtained the result we need to compute numerical results, but it remains to be shown that the expression (8.2.48) is mathematically meaningful. One way to do this is to note that $N(\xi, x)$ is a Green's function of Laplace's equation. This is sufficient to guarantee that the improper integral (8.2.47) converges (see the book by Evans [22]).

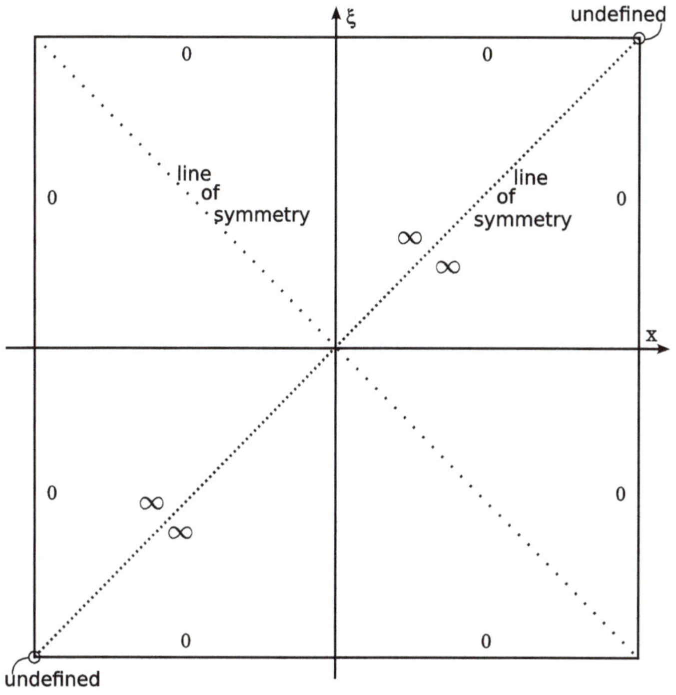

Fig. 8.6 Aerodynamic kernel $N(\xi, x)$ in $[-1, 1] \times [-1, 1]$. Qualitative illustration. The infinities, and the upper and left edges, which are outside the domain of the auxiliary function $\Lambda(\xi, x)$, needed by $N(\xi, x)$, should be understood in the sense of limits

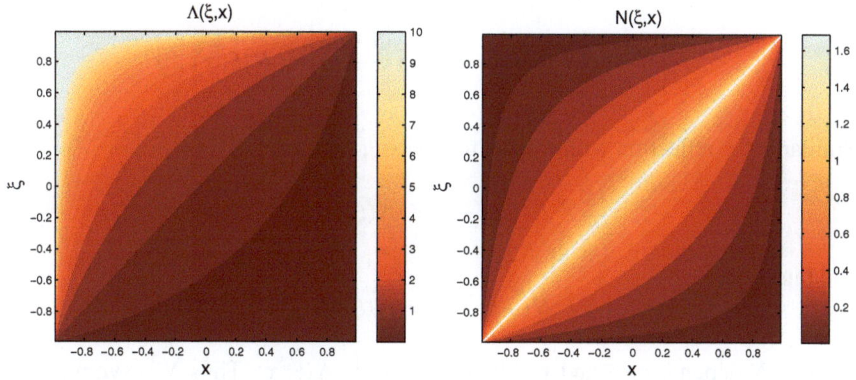

Fig. 8.7 Contour plots of the functions $\Lambda(\xi, x)$ and $N(\xi, x)$. *Left*: Λ. *Right*: N. Note that both functions grow without bound toward the singularities marked in Figs. 8.5 and 8.6. In both subfigures, the upper end of the color scale has been chosen arbitrarily to show the structure away from the singularity

But the convergence can also be established in a more direct manner. Let us show that the aerodynamic kernel function $N(\xi, x)$ is integrable. First, as an auxiliary result, we will show $N(\xi, x)$ is symmetric with respect to reflection by the lines $x = \pm\xi$. The symmetricity can be obtained by inspection of (8.2.46) and algebraic manipulation.

Note that the domain of $\Lambda(\xi, x)$ is $(-1, 1) \times (-1, 1)$, and that of $N(\xi, x)$ is

$$D_0 \equiv \{(-1, 1) \times (-1, 1)\} \setminus \{\xi = x\} .$$

Consider $\Lambda(\xi, x)$, defined in (8.2.46). Reflecting the point (ξ, x) with respect to $x = \xi$, we evaluate $\Lambda(x, \xi)$:

$$\begin{aligned} \Lambda(x, \xi) &= \left[\frac{(1-\xi)(1+x)}{(1-x)(1+\xi)}\right]^{1/2} \\ &= \left[\frac{(1-x)(1+\xi)}{(1-\xi)(1+x)}\right]^{-1/2} = \frac{1}{\Lambda(\xi, x)} . \end{aligned} \tag{8.2.49}$$

Then, using (8.2.49) we have

$$\begin{aligned} N(x, \xi) &= \frac{1}{\pi} \ln \left| \frac{1 + \frac{1}{\Lambda}}{1 - \frac{1}{\Lambda}} \right| = \frac{1}{\pi} \ln \left| \frac{\Lambda + 1}{\Lambda - 1} \right| \\ &= \frac{1}{\pi} \ln \left| -\frac{1+\Lambda}{1-\Lambda} \right| = \frac{1}{\pi} \ln \left| \frac{1+\Lambda}{1-\Lambda} \right| = N(\xi, x) . \end{aligned} \tag{8.2.50}$$

Similarly, for reflection with respect to $x = -\xi$, we have

$$\Lambda(-x, -\xi) = \Lambda(\xi, x) ,$$

and thus also

$$N(-x, -\xi) = N(\xi, x) ,$$

because N depends on ξ and x only implicitly via $\Lambda(\xi, x)$. Thus N is symmetric in reflection with respect to the lines $x = \pm\xi$.

Now we are in a position to consider the integrability. For the aerodynamic problem, we need to show that the integral (8.2.47) converges. Let $x \in (-1, 1)$ and $t \in [0, \infty)$ be fixed. Let

$$I_1(x) \equiv \int_{-1}^{1} N(\xi, x) f(\xi)\, \mathrm{d}\xi , \tag{8.2.51}$$

where

$$f(\xi) \equiv f(\xi;\ t)\ ,$$

that is, f is allowed to depend on t, but this dependence is omitted from the notation since we hold t fixed, so it can be treated as a parameter. We require that $f(\xi)$ is bounded for $-1 \leq \xi \leq 1$.

We estimate (8.2.51) from above by

$$\begin{aligned} I_1(x) \leq \left|I_1(x)\right| &\equiv \left|\int_{-1}^{1} N(\xi, x)\, f(\xi)\, \mathrm{d}\xi\right| \leq \int_{-1}^{1} \left|N(\xi, x)\, f(\xi)\right| \mathrm{d}\xi \\ &\leq \int_{-1}^{1} \left|N(\xi, x)\right| \mathrm{d}\xi \cdot \max_{\xi \in [-1,1]} |f(\xi)|\ , \end{aligned} \tag{8.2.52}$$

where the last form follows from Hölder's inequality (max being the ∞-norm). Because $N(\xi, x) \geq 0$ over the whole domain, we can omit the absolute value in the integral on the last line above. Thus, it is sufficient to show that the integral

$$I_2(x) \equiv \int_{-1}^{1} N(\xi, x)\ \mathrm{d}\xi \tag{8.2.53}$$

converges. Furthermore, since $N(\xi, x)$ is symmetric with respect to the lines $\xi = x$ and $\xi = -x$, it is sufficient to consider

$$I_3(x) \equiv \int_{-1}^{x} N(\xi, x)\ \mathrm{d}\xi\ , \tag{8.2.54}$$

because due to the symmetries, it holds that

$$I_2(x) = I_3(x) + I_3(-x) \quad \text{for all } x \in [-1, 1]\ .$$

See Fig. 8.8 for an illustration. The integral (8.2.54) is of the form of the third integral in (8.2.45), when taken to the limit, so the following argument will prove the convergence of that limit, too. Due to the symmetry, it is also sufficient for proving the convergence of the fourth integral in the same equation.

To show that (8.2.54) converges, we can use the sandwich theorem, which is also called the comparison theorem for improper integrals (e.g. by Adams and Essex [18]), from analysis. The outline of the argument is as follows (for details, see Jeronen [28]).

Due to the symmetry of $N(\xi, x)$ with respect to $\xi = x$, it is sufficient to consider just one half of the domain. Hence, let $\xi < x$. We leave out strict equality to avoid the singularity of $N(\xi, x)$ at $x = \xi$. Consider the set

$$D_1 \equiv \left\{ (x, \xi) \in \mathbb{R}^2 \ \mid\ -1 < x < 1,\ -1 < \xi < x \right\}\ ,$$

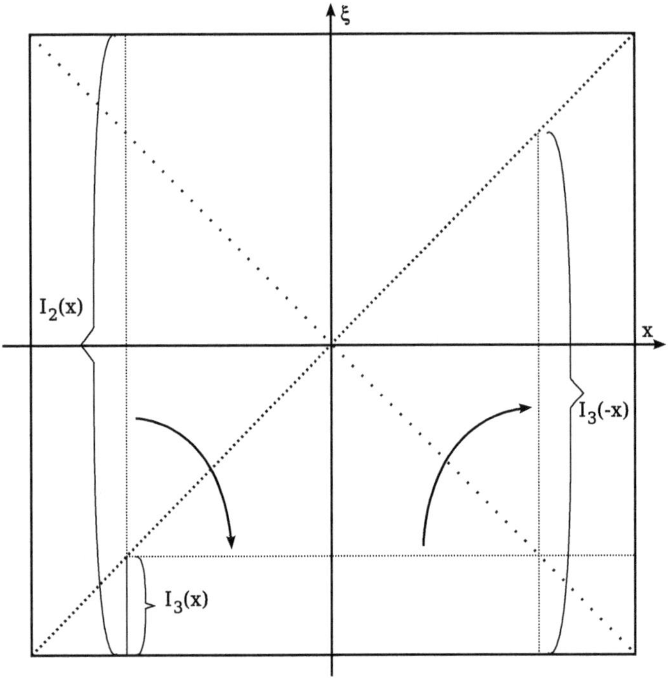

Fig. 8.8 Effect of the symmetries of $N(\xi, x)$ on the integral $I_2(x)$

which is a triangular half of the original domain D_0, see Fig. 8.8. We observe that

$$0 \leq \Lambda(\xi, x) < 1 \quad \text{for all } (x, \xi) \in D_1 \ .$$

Taking this into account, we can estimate N from above, simplifying the expression slightly:

$$\begin{aligned} N(\xi, x) &\equiv \frac{1}{\pi} \ln \left| \frac{1+\Lambda}{1-\Lambda} \right| \\ &\overset{\text{for } \xi < x}{=} \frac{1}{\pi} \ln \frac{1+\Lambda}{1-\Lambda} \\ &< \ln \frac{2}{1-\Lambda} \equiv s(\xi, x) \ , \end{aligned} \tag{8.2.55}$$

where $s(\xi, x)$ is defined as indicated. This simplifies the numerator and gets rid of the absolute value. Let us define transformed coordinates $\eta \equiv x - \xi$ and $\zeta \equiv x + \xi$. Let

$$r(\eta) \equiv \ln\left(\frac{4}{\eta}\right) ,$$

and

$$r(\xi, x) \equiv r(\eta) \equiv r(x - \xi) .$$

In the set D_1, it holds that $\eta \in (0, 2)$. In this range, the function r is positive and integrable:

$$\int r(x - \xi)\, \mathrm{d}\xi < \infty .$$

Once we show that on lines parallel with $x = -\xi$, for the function s it holds that

$$s(\eta, \zeta) \leq s(\eta, 0) ,$$

the argument reduces into one dimension. Then, it is sufficient to show that for $\eta \in (0, 2)$, we have

$$s(\eta, 0) < r(\eta) .$$

Finally, applying the sandwich theorem concludes the argument.

As a final remark, we see that $N(\xi, x)$ decreases quickly. Quantitatively speaking, it decreases faster than $\ln(4/\eta)$, as the distance $\eta \equiv |\xi - x|$ from the singularity $\xi = x$ increases.

8.3 Added-Mass Approximation

The integral in (8.2.48) must be handled carefully in a numerical implementation, due to the singularity of $N(\xi, x)$. Although the aerodynamic kernel $N(\xi, x)$ itself is integrable (L^1), its derivative $\partial N(\xi, x)/\partial x$ is not, so to correctly compute the integral in (8.2.48), one must either evaluate the integral first before applying the outer $\partial/\partial x$, or transform the problem into a weak form and transfer the outer $\partial/\partial x$ to the test function.

This section addresses a simplified way to include the effect of fluid in a simulation that avoids numerically dealing with singular integrals in the problem matrices. Added-mass models are sometimes used for taking into account the inertial effects of fluid—structure interaction (see, e.g., Pramila [34, 35, 37, 38]). We will construct an added-mass approximation for our model, and compare the result to some existing added-mass models. This section is additional material that will not be used in the rest of the book.

Above, it was observed that $N(\xi, x)$ has the following properties:

1. N has a singularity at $\xi = x$:

$$N(\xi, x) \to \infty \text{ as } \xi \to x ,$$

2. N is integrable:

$$\int_{-1}^{1} \left| N(\xi, x) \right| \mathrm{d}\xi < \infty \,, \quad \text{and}$$

3. the value of $N(\xi, x)$ decreases quickly as the distance from the singularity increases.

Motivated by these properties, let us make the following simplifying approximation:

$$\int_{-1}^{1} N(\xi, x)\, f(\xi)\, \mathrm{d}\xi \approx \mu \int_{-1}^{1} \delta(\xi, x)\, f(\xi)\, \mathrm{d}\xi \,, \tag{8.3.1}$$

where $\delta(\xi, x)$ is the Dirac delta distribution, $f(\xi) = f(\xi; t)$ is any admissible function, and μ is the constant

$$\mu \equiv \underset{x \in (-1,1)}{\text{mean}} \int_{-1}^{1} N(\xi, x)\, \mathrm{d}\xi = \frac{1}{2} \int_{-1}^{1} \left[\int_{-1}^{1} N(\xi, x)\, \mathrm{d}\xi \right] \mathrm{dx} \,. \tag{8.3.2}$$

The symbol mean$(\cdot)$ denotes the average of the indicated quantity over the indicated interval.

The approximated integral in (8.3.1) is easy to evaluate. Using the definition of the Dirac delta distribution, we have

$$\mu \int_{-1}^{1} \delta(\xi, x)\, f(\xi)\, \mathrm{d}\xi = \mu\, f(x) \quad \forall x \in [-1, 1] \,. \tag{8.3.3}$$

Numerically, we find that the required mean value is $\mu \approx \pi/4$. To do this, the formula (8.3.2) can be evaluated in several ways. One way is to use the Monte Carlo method: take the middle form of the expression (8.3.2), replace the mean by the sample mean, and sample the integral for, say, $n = 1000$ uniformly distributed random values of x. The sample mean then gives an approximation for μ.

Alternatively, it is possible to use the rightmost form of (8.3.2), and evaluate the double integral directly by applying a quadrature method to both the outer and inner integrals. This approach requires many evaluations of the inner integral, and is thus fairly slow, but it gives a more accurate result.

Whichever method is used, due to the singularity of $N(\xi, x)$ at $\xi = x$, in practice the (inner) integral must be numerically approximated as

$$\int_{-1}^{1} N(\xi, x)\, \mathrm{d}\xi \approx \int_{-1}^{x-\varepsilon} N(\xi, x)\, \mathrm{d}\xi + \int_{x+\varepsilon}^{1} N(\xi, x)\, \mathrm{d}\xi \,, \tag{8.3.4}$$

where ε is small.

It is important to notice that the added-mass model, by replacing the exact aerodynamic kernel by the Dirac delta approximation as per (8.3.1), the fluid—structure interaction is approximated as pointwise local. Thus, the added-mass model can be seen as performing mass lumping on the original model; the factor μ approximates the total strength of the aerodynamic reaction on one point of the panel surface.

A global mean value is not the only way to perform mass lumping. Alternatively, one can approximate

$$\int_{-1}^{1} N(\xi, x)\, f(\xi)\, \mathrm{d}\xi \approx S(x) \int_{-1}^{1} \delta(\xi, x)\, f(\xi)\, \mathrm{d}\xi\ ,$$

where

$$S(x) \equiv \int_{-1}^{1} N(\xi, x)\, \mathrm{d}\xi\ . \tag{8.3.5}$$

This choice is slightly more sophisticated in that it accounts for differences in the strength of the coupling at different values of x. It approximates the effect of the kernel with the local mean at each fixed x, instead of the global mean $\pi/4$. For a plot of $S(x)$, see Fig. 8.9. If fast computation is desired, a polynomial approximation can be used. A linear least squares fit of the form

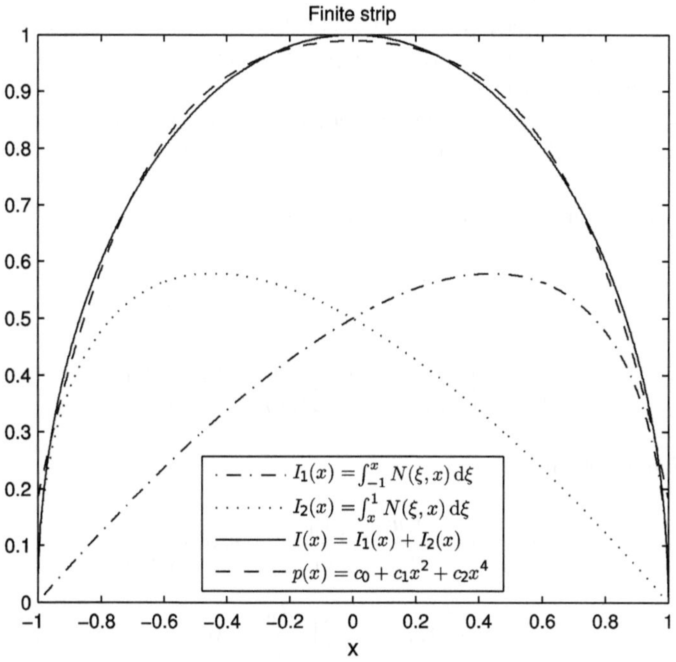

Fig. 8.9 The function $S(x)$ defined by (8.3.5), for local approximation of the aerodynamic reaction. Both the original function and its three-term polynomial approximation in the basis $\phi_k(x) = \{1,\ x^2,\ x^4\}$, $k = 0, 1, 2$, are shown. The polynomial fit coefficients are $c_0 = 0.9892$, $c_1 = -0.3195$, and $c_2 = -0.4872$. Note that the original function touches zero at both ends of the domain, whereas the polynomial approximation does not

$$S(x) \approx \sum_k c_k\, \phi_k(x) \tag{8.3.6}$$

can be performed, for example, by the method of normal equations. Note that *linear* refers to linearity with respect to the coefficients c_k; the basis functions $\phi_k(x)$ are allowed to be nonlinear. With the basis

$$\phi_0(x) \equiv 1\,, \qquad \phi_1(x) = x^2\,, \qquad \phi_2(x) = x^4\,,$$

we obtain the values of the coefficients c_k as $c_0 = 0.9892$, $c_1 = -0.3195$, and $c_2 = -0.4872$. This approach leads to a variable-coefficient partial differential equation, and is in spirit similar to the approach of Frondelius et al. [38], where an added-mass model was used, with the added masses taken as functions of x derived from boundary layer theory.

In any case, how accurate the added-mass approximation is depends on the physical situation being considered. In Païdoussis [39], it is pointed out that the problem of plates subjected to axial flow is more complex than the otherwise very similar, canonical problem of the fluid-carrying pipe, due to the nonlocal coupling effect of the aerodynamic reaction.

Let us work out the added masses predicted by our model. For simplicity, consider only the inertial terms from the travelling panel model, and the aerodynamic reaction q_{f}. Let us consider the case with $T = 0$, $D = 0$ and $g \equiv 0$ for simplicity, as we can easily add these terms back when finished. We approximate q_{f} by inserting (8.3.1) and $\mu = \pi/4$ into (8.2.48). Then, we evaluate the approximated aerodynamic integral by (8.3.3).

In nondimensional coordinates, we have the result

$$\begin{aligned} &\frac{m}{\tau^2}\frac{\partial^2 w}{\partial t^2} + 2\frac{m V_0}{\ell\tau}\frac{\partial^2 w}{\partial x \partial t} + \frac{m V_0^2}{\ell^2}\frac{\partial^2 w}{\partial x^2} = \\ &-\rho_{\mathrm{f}}\frac{\pi}{4}\left[\frac{\ell}{\tau^2}\frac{\partial^2 w}{\partial t^2} + 2\frac{v_\infty}{\tau}\frac{\partial^2 w}{\partial x \partial t} + \frac{v_\infty^2}{\ell}\frac{\partial^2 w}{\partial x^2}\right], \end{aligned} \tag{8.3.7}$$

from which we obtain

$$\begin{aligned} &\frac{1}{\tau^2}\left[m + m_{\mathrm{a}}\right]\frac{\partial^2 w}{\partial t^2} + 2\frac{V_0}{\ell\tau}\left[m + m_{\mathrm{a}} r_{\mathrm{v}}\right]\frac{\partial^2 w}{\partial x \partial t} + \\ &\frac{V_0^2}{\ell^2}\left[m + m_{\mathrm{a}} r_{\mathrm{v}}^2\right]\frac{\partial^2 w}{\partial x^2} = 0\,, \end{aligned} \tag{8.3.8}$$

where

$$m_{\mathrm{a}} \equiv \frac{\ell \rho_{\mathrm{f}} \pi}{4} \quad \text{and} \quad r_{\mathrm{v}} \equiv \frac{v_\infty}{V_0}\,. \tag{8.3.9}$$

The result reduces to a classical one- or three-term single-parameter added-mass model by choosing $r_\mathrm{v} = 0$, that is, $v_\infty = 0$, no free-stream flow in laboratory coordinates, or $r_\mathrm{v} = 1$, that is, $v_\infty = V_0$, where the whole air mass moves axially with the panel.

The prediction for the added mass m_a thus derived, in (8.3.9), agrees with Eq. (12A) of Pramila [37]:

$$m_\mathrm{a} = \alpha \pi \rho a / 4 \,,$$

if we take $\alpha = 0.5$ in Pramila's equation. Our ℓ denotes the span half-length and Pramila's a denotes the full length; hence $a = 2\ell$. The symbol ρ represents the fluid density (the same as our ρ_f). According to Pramila [37], the choice $\alpha = 0.5$ corresponds to an aspect ratio slightly larger than 1.0, that is, the span is slightly longer than wide.

Compare also Eq. (13) in Pramila [40], there attributed to T. Y.-T. Wu, which has been reported to hold for long and narrow spans:

$$q_1 = -\pi \rho (b^2/4) \left(\frac{\partial^2 w}{\partial t^2} + 2v \frac{\partial^2 w}{\partial x \partial t} + v^2 \frac{\partial^2 w}{\partial x^2} \right) .$$

Here q_1 denotes the lift force per unit length, and the term is situated on the right-hand side of the dynamical equation, in the same way as our q_f. Hence, upon transferring the lift force to the left-hand side of the equation, the sign becomes positive. The velocity v corresponds to our v_∞ and the fluid density ρ to our ρ_f. When we consider the force per unit surface area, by dividing the given expression by the span width b, the corresponding added mass becomes

$$m_\mathrm{a} = \frac{b \rho_\mathrm{f} \pi}{4} .$$

This formulation, instead of ℓ, uses the span width b as the length scale. When it is compared to our approximation in (8.3.9), the best agreement is obtained when $\ell / b \approx 1$ (i.e., when the span is twice as long as wide).

Finally, in the special case $V_0 = 0$, the effect of fluid cannot be included in the mass only, because for $V_0 = 0$ the quantity r_v in (8.3.9) becomes undefined. However, the added mass m_a is still well-defined. The added mass m_a and the fluid axial velocity v_∞ together fully characterize the approximated effect of the fluid.

In the general case, instead of writing an added-mass approximation in the form (8.3.8), we can make the following definitions:

$$m_\mathrm{total} = m + m_\mathrm{a} \,, \tag{8.3.10}$$

$$p_\mathrm{total} = m V_0 + m_\mathrm{a} v_\infty \,, \tag{8.3.11}$$

$$2K_\mathrm{total} = m V_0^2 + m_\mathrm{a} v_\infty^2 \,, \tag{8.3.12}$$

where m_{a} is the added mass, $m_{\mathrm{a}} v_{\infty}$ is the added linear momentum, and $\frac{1}{2} m_{\mathrm{a}} v_{\infty}^2$ is the added kinetic energy. We have

$$\frac{1}{\tau^2} m_{\mathrm{total}} \frac{\partial^2 w}{\partial t^2} + 2 \frac{1}{\ell \tau} p_{\mathrm{total}} \frac{\partial^2 w}{\partial x \partial t} + \frac{1}{\ell^2} 2 K_{\mathrm{total}} \frac{\partial^2 w}{\partial x^2} = 0 ,$$

or, equivalently,

$$\begin{aligned} \frac{1}{\tau^2} \left[m + m_{\mathrm{a}} \right] \frac{\partial^2 w}{\partial t^2} + 2 \frac{1}{\ell \tau} \left[m V_0 + m_{\mathrm{a}} v_{\infty} \right] \frac{\partial^2 w}{\partial x \partial t} \\ + \frac{1}{\ell^2} \left[m V_0^2 + m_{\mathrm{a}} v_{\infty}^2 \right] \frac{\partial^2 w}{\partial x^2} = 0 . \end{aligned} \tag{8.3.13}$$

This generalized added-mass type approximation is applicable for all values of V_0.

It should be pointed out that due to the forms of (8.3.8) and (8.3.13), neither approximation can qualitatively change the dynamic behavior of the model when compared to the corresponding vacuum case. These added-mass approximations only modify the coefficients of the original constant-coefficient partial differential equation. The partial differential equation itself remains identical to the vacuum case.

Thus, the dynamic stability results of the corresponding vacuum case fully qualitatively govern the stability of the approximate added-mass fluid—structure interaction model. In order to account for qualitative changes to the stability behavior, instead of the added-mass model, a more advanced fluid—structure interaction model must be used. In the context of potential flow, the analytical solution considered further above fills this role. Another option, available also for more complex flow models such as Euler or Navier—Stokes, is to numerically solve also the flow problem and to use weak coupling of the solid and fluid model components via a boundary condition at the solid surface.

8.4 Numerical Examples

In the following numerical examples, for the flow component, we will use the analytical solution of the two-dimensional potential flow problem, derived in Sect. 8.2. We state the problem, and then give the results of a linear stability analysis for representative example cases. As the axially moving structure we consider an ideal string, a linear elastic panel, and some Kelvin—Voigt panels with different degrees of viscosity. The numerical solution process is fairly standard, so we omit most of it here. The nondimensionalization we present in some detail, because it is important for interpreting the results.

Recall that the governing equation for the transverse vibrations of a tensioned, axially traveling linear elastic panel, subjected to an aerodynamic reaction q_{f} that appears as a distributed load, is

$$m\frac{\partial^2 w}{\partial t^2}+2mV_0\frac{\partial^2 w}{\partial x\partial t}+\left(mV_0^2-T_0\right)\frac{\partial^2 w}{\partial x^2}+D\frac{\partial^4 w}{\partial x^4}=q_{\mathrm{f}}\ ,\quad x\in(-\ell,\ell)\ . \tag{8.4.1}$$

Nondimensionalizing coordinates by $t=\tau t'$, $x=\ell x'$, and then omitting the primes, leads to

$$\frac{m}{\tau^2}\frac{\partial^2 w}{\partial t^2}+\frac{2mV_0}{\ell\tau}\frac{\partial^2 w}{\partial x\partial t}+\frac{1}{\ell^2}\left(mV_0^2-T_0\right)\frac{\partial^2 w}{\partial x^2}+\frac{D}{\ell^4}\frac{\partial^4 w}{\partial x^4}=q_{\mathrm{f}}\ ,\quad x\in(-1,1)\ . \tag{8.4.2}$$

In the nondimensional coordinates, the aerodynamic reaction is given by (8.2.48), the final result repeated here for convenience:

$$q_{\mathrm{f}}=-\rho_{\mathrm{f}}\frac{1}{\ell}\left(\frac{\ell}{\tau}\frac{\partial}{\partial t}+v_\infty\frac{\partial}{\partial x}\right)\int_{-1}^{1}N(\xi,x)\left(\frac{\ell}{\tau}\frac{\partial}{\partial t}+v_\infty\frac{\partial}{\partial x}\right)w(\xi,t)\,\mathrm{d}\xi\ .$$

Equation (8.4.2) is then nondimensionalized by multiplying by τ^2/m. Strictly speaking, we must nondimensionalize also the displacement w, for example by $w=hw'$, where h is the panel thickness. However, because the Eqs. (8.4.2) and (8.2.48) are linear in w, its normalization constant applies uniformly to all terms, and is therefore eliminated. Inserting q_{f} from (8.2.48) into (8.4.2) and performing the nondimensionalization, we have

$$\begin{aligned}&\frac{\partial^2 w}{\partial t^2}+2V_0\frac{\tau}{\ell}\frac{\partial^2 w}{\partial x\partial t}+\frac{\tau^2}{\ell^2}\left(V_0^2-\frac{T_0}{m}\right)\frac{\partial^2 w}{\partial x^2}+\frac{D}{m\ell^2}\frac{\tau^2}{\ell^2}\frac{\partial^4 w}{\partial x^4}\\&=-\frac{\tau^2\rho_{\mathrm{f}}}{m\ell}\left(\frac{\ell}{\tau}\frac{\partial}{\partial t}+v_\infty\frac{\partial}{\partial x}\right)\int_{-1}^{1}N(\xi,x)\left(\frac{\ell}{\tau}\frac{\partial}{\partial t}+v_\infty\frac{\partial}{\partial x}\right)w(\xi,t)\,\mathrm{d}\xi\ ,\quad x\in(-1,1)\ .\end{aligned}$$

Defining C as the critical velocity of an ideal string in absence of any surrounding fluid,

$$C\equiv\sqrt{\frac{T_0}{m}}\ , \tag{8.4.3}$$

and then choosing the characteristic time τ as

$$\tau=\frac{\ell}{C}\ , \tag{8.4.4}$$

yields

$$\begin{aligned}&\frac{\partial^2 w}{\partial t^2}+2\frac{V_0}{C}\frac{\partial^2 w}{\partial x\partial t}+\left(\frac{V_0^2}{C^2}-1\right)\frac{\partial^2 w}{\partial x^2}+\frac{D}{m\ell^2}\frac{1}{C^2}\frac{\partial^4 w}{\partial x^4}\\&=-\frac{\ell\rho_{\mathrm{f}}}{C^2 m}\left(C\frac{\partial}{\partial t}+v_\infty\frac{\partial}{\partial x}\right)\int_{-1}^{1}N(\xi,x)\left(C\frac{\partial}{\partial t}+v_\infty\frac{\partial}{\partial x}\right)w(\xi,t)\,\mathrm{d}\xi\ ,\quad x\in(-1,1)\ .\end{aligned}$$

Distributing the $1/C^2$ in the fluid terms (one $1/C$ into the differential operator in front, the other one into the integral), we have

$$\frac{\partial^2 w}{\partial t^2} + 2\frac{V_0}{C}\frac{\partial^2 w}{\partial x \partial t} + \left(\frac{V_0^2}{C^2} - 1\right)\frac{\partial^2 w}{\partial x^2} + \frac{D}{m\ell^2}\frac{1}{C^2}\frac{\partial^4 w}{\partial x^4}$$
$$= -\frac{\ell \rho_f}{m}\left(\frac{\partial}{\partial t} + \frac{v_\infty}{C}\frac{\partial}{\partial x}\right)\int_{-1}^{1} N(\xi, x)\left(\frac{\partial}{\partial t} + \frac{v_\infty}{C}\frac{\partial}{\partial x}\right) w(\xi, t)\,\mathrm{d}\xi\,, \quad x \in (-1, 1)\,.$$

Defining the nondimensional panel and free-stream flow velocities

$$c \equiv \frac{V_0}{C}\,, \quad \theta \equiv \frac{v_\infty}{C}\,, \tag{8.4.5}$$

and observing that $mC^2 = T_0$, we obtain

$$\frac{\partial^2 w}{\partial t^2} + 2c\frac{\partial^2 w}{\partial x \partial t} + \left(c^2 - 1\right)\frac{\partial^2 w}{\partial x^2} + \frac{D}{\ell^2 T_0}\frac{\partial^4 w}{\partial x^4}$$
$$= -\frac{\ell \rho_f}{m}\left(\frac{\partial}{\partial t} + \theta\frac{\partial}{\partial x}\right)\int_{-1}^{1} N(\xi, x)\left(\frac{\partial}{\partial t} + \theta\frac{\partial}{\partial x}\right) w(\xi, t)\,\mathrm{d}\xi\,, \quad x \in (-1, 1)\,.$$

Defining the nondimensional fluid reaction coefficient a and the nondimensional bending rigidity b,

$$a \equiv \frac{\ell \rho_f}{m}\,, \quad b \equiv \frac{D}{\ell^2 T_0}\,, \tag{8.4.6}$$

we have the final nondimensional governing equation for the axially traveling, tensioned, linear elastic panel subjected to an axial flow with free-stream velocity v_∞:

$$\frac{\partial^2 w}{\partial t^2} + 2c\frac{\partial^2 w}{\partial x \partial t} + \left(c^2 - 1\right)\frac{\partial^2 w}{\partial x^2} + b\frac{\partial^4 w}{\partial x^4}$$
$$= -a\left(\frac{\partial}{\partial t} + \theta\frac{\partial}{\partial x}\right)\int_{-1}^{1} N(\xi, x)\left(\frac{\partial}{\partial t} + \theta\frac{\partial}{\partial x}\right) w(\xi, t)\,\mathrm{d}\xi\,, \quad x \in (-1, 1)\,. \tag{8.4.7}$$

We use the simply supported boundary conditions

$$w(-1) = w(1) = 0\,, \quad \frac{\partial^2 w}{\partial x^2}(-1) = \frac{\partial^2 w}{\partial x^2}(1) = 0\,. \tag{8.4.8}$$

For the ideal string, $D = 0$. Then Eq. (8.4.7) reduces to a second-order partial differential equation, and the boundary conditions on $\partial^2 w/\partial x^2$ are omitted.

For the Kelvin—Voigt panel, the governing Eq. (8.4.2) is replaced by

$$m\frac{\partial^2 w}{\partial t^2} + 2mV_0\frac{\partial^2 w}{\partial x \partial t} + \left(mV_0^2 - T_0\right)\frac{\partial^2 w}{\partial x^2} + \left[D + D_{\mathrm{VE}}\left(\frac{\partial}{\partial t} + V_0\frac{\partial}{\partial x}\right)\right]\frac{\partial^4 w}{\partial x^4} = q_f\,, \quad x \in (-\ell, \ell)\,. \tag{8.4.9}$$

The only difference in form is the addition of the viscous operator to the bending term. The viscous bending rigidity D_{VE} is given by

$$D_{\text{VE}} = \tau_{\text{R}} D \,, \tag{8.4.10}$$

where τ_{R} is the retardation time (recall Chap. 5). Using (8.4.10) and nondimensional coordinates, Eq. (8.4.9) becomes

$$\frac{m}{\tau^2}\frac{\partial^2 w}{\partial t^2} + \frac{2mV_0}{\ell\tau}\frac{\partial^2 w}{\partial x \partial t} + \frac{1}{\ell^2}\left(mV_0^2 - T_0\right)\frac{\partial^2 w}{\partial x^2} + \frac{D}{\ell^4}\left[1 + \frac{\tau_{\text{R}}}{\ell}\left(\frac{\ell}{\tau}\frac{\partial}{\partial t} + V_0\frac{\partial}{\partial x}\right)\right]\frac{\partial^4 w}{\partial x^4}$$
$$= q_{\text{f}} \,, \quad x \in (-1, 1) \,. \tag{8.4.11}$$

We apply the same nondimensionalization as above. In the viscous part of the bending term, we have

$$\frac{\tau_{\text{R}}}{\ell}\left(\frac{\ell}{\tau}\frac{\partial}{\partial t} + V_0\frac{\partial}{\partial x}\right) = \frac{\tau_{\text{R}}(\ell/\tau)}{\ell}\left(\frac{\partial}{\partial t} + \frac{V_0}{\ell/\tau}\frac{\partial}{\partial x}\right) = \frac{\tau_{\text{R}}}{\tau}\left(\frac{\partial}{\partial t} + \frac{V_0}{C}\frac{\partial}{\partial x}\right) = \frac{\tau_{\text{R}}}{\tau}\left(\frac{\partial}{\partial t} + c\frac{\partial}{\partial x}\right) .$$

Defining the nondimensional retardation time

$$\gamma \equiv \frac{\tau_{\text{R}}}{\tau} \,, \tag{8.4.12}$$

we obtain the final nondimensional governing equation for the axially traveling, tensioned, Kelvin—Voigt panel subjected to an axial flow with free-stream velocity v_∞:

$$\frac{\partial^2 w}{\partial t^2} + 2c\frac{\partial^2 w}{\partial x \partial t} + \left(c^2 - 1\right)\frac{\partial^2 w}{\partial x^2} + b\left[1 + \gamma\left(\frac{\partial}{\partial t} + c\frac{\partial}{\partial x}\right)\right]\frac{\partial^4 w}{\partial x^4}$$
$$= -a\left(\frac{\partial}{\partial t} + \theta\frac{\partial}{\partial x}\right)\int_{-1}^{1} N(\xi, x)\left(\frac{\partial}{\partial t} + \theta\frac{\partial}{\partial x}\right) w(\xi, t)\, \mathrm{d}\xi \,, \quad x \in (-1, 1) \,. \tag{8.4.13}$$

The equation reduces to the linear elastic case (8.4.7) by setting $\gamma = 0$.

In the general case, because the partial differential Eq. (8.4.13) is of the fifth order in x, for the Kelvin—Voigt panel we require add one more boundary condition at the inflow boundary $x = -\ell$. In nondimensional form,

$$\frac{\partial w}{\partial x}(-1) = 0 \,. \tag{8.4.14}$$

Equations (8.4.7) and (8.4.13) are brought into a weak form in the standard manner, and then discretized in space using C^2 continuous Hermite elements (see Appendix C). Note that for any case where $v_\infty \neq 0$, the outer $\partial/\partial x$ in the aerodynamic reaction (8.2.48) must be integrated by parts to move the differentiation to the test function. This is extremely important, because the singularity of $N(\xi, x)$ is integrable but that of $\partial N/\partial x$ is not.

We then perform a linear stability analysis, see e.g. Bolotin [41]. We insert the time-harmonic trial function

$$w(x,t) \equiv \exp(st)\, W(x) \ , \tag{8.4.15}$$

thus transforming $\partial w/\partial t \to sw$. Here s is the complex-valued stability exponent (Lyapunov exponent). The appearance of $\operatorname{Re} s > 0$ indicates instability. Critical points appear where the real part of at least one eigenvalue s crosses zero. When dissipation is present, in general the critical points are distinct from bifurcation points (if any bifurcations appear). This is because dissipation in effect shifts all eigenvalues toward the half-plane where $\operatorname{Re} s < 0$.

The resulting pseudo-steady-state problem is a quadratic eigenvalue problem for nontrivial pairs (s, W). For numerical treatment, we transform it into a twice larger generalized linear eigenvalue problem via the companion form technique, see Tisseur and Meerbergen [42]. The generalized linear eigenvalue problem is then solved using a standard numerical solver.

We then perform a (quasistatic) sweep with respect to the loading parameter c, the nondimensional axial drive velocity of the moving material. Because eigenvalue solvers typically return the eigenvalues in a random order, the numerically obtained eigenvalues s_j for adjacent values of c are paired by closest-distance matching, taking care at each step to use each s_j exactly once. This reordering produces connected curves, which are convenient for visualization. We then pick $N = 4$ solution curves whose complex eigenvalues s_j have the smallest norm at $V_0 = 0$; these correspond to the lowest free vibration modes.

In the following examples, we use problem parameters typical for paper materials, see Table 8.1. In the figure headings, we display the same parameters as in Chap. 5, namely

$$\alpha = \frac{\ell}{h} \ , \tag{8.4.16}$$

$$\beta = \alpha^2 b = \frac{D}{h^2 T_0} \ , \tag{8.4.17}$$

Table 8.1 Problem parameters used in the numerical examples

ℓ [m]	h [m]	E_x [Pa]	ν	τ_R [s]	T_0 [N/m]	m [kg/m^2]	ρ_f [kg/m^3]
1	10^{-4}	10^9	0.3	0, 0.1, 1, 10 (series 1)	500	0.08	1.225
				0, 10^{-5}, 10^{-4}, 10^{-3} (series 2)			
D [Nm]			τ [s]	a α β	γ		
$9.1575 \cdot 10^{-5} \cdot \{0, 1\}$			1.2649	15.3125 10^4 18.315	$7.90569 \cdot \{0, 1, 10, 100\}$ (series 1)		
				$7.90569 \cdot \{0, 10^{-4}, 10^{-3}, 10^{-2}\}$ (series 2)			

where α is the aspect ratio of the half-span (the full span length is 2ℓ), and β is a nondimensional bending rigidity (using a normalization that yields numbers that are $O(1)$). The parameter γ is the nondimensional retardation time given by (8.4.12).

Below, we display two representative series of results, which differ in the setup of the surrounding air mass. In the first series, the whole air mass moves axially with the panel, $v_\infty = V_0$. In the second series of results, there is no axial flow, $v_\infty = 0$.

Let us first consider the first series, where $v_\infty = V_0$. This highlights that the presence of an aerodynamic reaction can qualitatively change the stability behavior also for an ideal string, which in vacuum is stable for any V_0 (recall Chap. 7). The considered cases are the ideal string, the linear elastic panel, and three Kelvin—Voigt panels with different values for the retardation time.

With the parameters in Table 8.1, when the air mass moves with the panel, the critical velocity undergoes a drastic reduction, to 39% of its vacuum value (when no surrounding flow is present). This effect was first observed by Pramila [37], in the context of another potential flow model, where the critical velocity was reduced to 25% of its vacuum value. For the model discussed here, the effect was confirmed in Jeronen [28, 29] for the setup where $v_\infty = V_0$. (This model exhibits no such effect when $v_\infty = 0$.) It is known that the corresponding reduction in the natural frequency of vibrations of the panel is a physical effect; on this, all models agree, including simple ones based on the added-mass approach. The measurement data in Pramila [37] also supports this. However, it remains unclear at the present time whether the reduction in the critical velocity is a real effect or a modeling artifact. Frondelius et al. [38] reported that when using a boundary layer model for the surrounding flow, the critical velocity is reduced to 90% of its vacuum value. For some more papers related to the question, see Niemi and Pramila [43], Pramila and Niemi [44], Pramila [40], Kulachenko et al. [45, 46].

For the considered setup where $v_\infty = V_0$, in order to see the initial postcritical behavior, as the range of c we use $c - 1 \in [-0.63, -0.36]$. The ideal string and linear elastic cases are in effect identical, so the latter is omitted. Beside the ideal string, we consider three Kelvin—Voigt panels with different viscosities. Viscosity needs to be relatively high ($\tau_\mathrm{f} = 0.1$ s) for the solution to undergo any significant change from the ideal string case. At very large viscosities ($\tau_\mathrm{R} = 10$ s), viscous behavior starts to dominate; note the qualitative similarity in the behavior of the real parts of the stability exponent to the vacuum Kelvin—Voigt results at the end of Chap. 5.

Consider Fig. 8.10 and Table 8.2. We observe behavior typical for an elastic system. In the unloaded state ($c = 0$, not shown), all eigenvalues lie on the imaginary axis. Typical to gyroscopic systems with symmetric boundary conditions, as the loading parameter c is increased quasistatically, the critical loss of stability takes the form of divergence. In Fig. 8.10 the divergence mode appears in the shape of a bubble in the real part of s, with the imaginary part simultaneously zero. When only the imaginary part of s is considered, the range of c corresponding to the first divergence mode is called *the divergence gap*, because the imaginary part of the mode $k = 1$ vanishes. We will use the term *divergence bubble* for all regions corresponding to single-mode divergence.

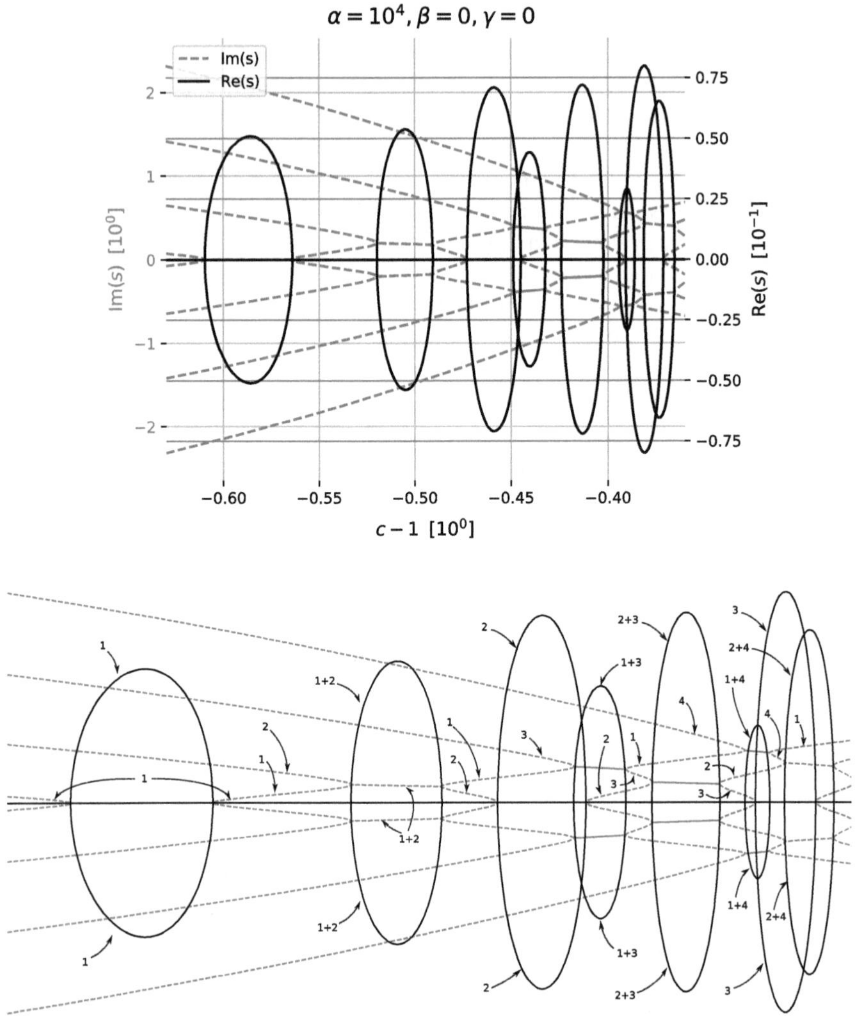

Fig. 8.10 Ideal string, with aerodynamic reaction. The whole air mass moves axially with the string, $v_\infty = V_0$. *Top*: The numerical results. Note the scale on the horizontal axis; the critical velocity is drastically reduced from the vacuum case. *Bottom*: Mode identification chart

After the first divergence bubble the system regains stability, and later loses it again via coupled-mode flutter. This is also typical. A coupled mode consists of two fundamental modes (two different values of the mode number k), and exhibits two mirror symmetries. If $s = A + Bi$ is an eigenvalue, then also $A - Bi$, $-A + Bi$ and $-a - Bi$ are. For a parametric plot in the $(\operatorname{Re} s, \operatorname{Im} s)$ plane, with respect to the parameter c, see Fig. 8.11. The range of c for this plot is the same as in Fig. 8.10.

Table 8.2 Points of interest in the range of c shown in Fig. 8.10, for an ideal string

$c-1$	k	Re(s) $[10^{-1}]$	Im(s)	Type	Note
−0.6095	1	0	0	Bifurcation; Critical point	Loss of stability (divergence)
−0.5857	1	±0.5093	0	−: min Re(s); +: max Re(s)	Extremum, 1st divergence mode
−0.5639	1	0	0	Bifurcation; Critical point	Regain of stability (divergence)
−0.5198	1 + 2	±0.2031	±0.1996	Bifurcation; Critical point	Loss of stability (flutter)
−0.5047	1 + 2	±0.5373	±0.1885	−Re: min Re(s); +Re: max Re(s)	Extremum, coupled flutter mode; all combinations of ±
−0.4906	1 + 2	0	±0.1730	Bifurcation; Critical point	Regain of stability (flutter)
−0.4729	2	0	0	Bifurcation; Critical point	Loss of stability (divergence)
−0.4587	2	±0.7093	0	−: min Re(s); +: max Re(s)	Extremum, 2nd divergence mode
−0.4488	1 + 3	0	±0.3900	Bifurcation; Critical point	Loss of stability (flutter)
−0.4450	2	0	0	Bifurcation; Critical point	Regain of stability (divergence)
−0.4403	1 + 3	±0.4417	±0.3780	−Re: min Re(s); +Re: max Re(s)	Extremum, coupled flutter mode; all combinations of ±
−0.4323	1 + 3	0	±0.3614	Bifurcation; Critical point	Regain of stability (flutter)
−0.4241	2 + 3	0	±0.2219	Bifurcation; Critical point	Loss of stability (flutter)
−0.4131	2 + 3	±0.7198	±0.2110	−Re: min Re(s); +Re: max Re(s)	Extremum, coupled flutter mode; all combinations of ±
−0.4024	2 + 3	0	±0.1994	Bifurcation; Critical point	Regain of stability (flutter)
−0.3943	1 + 4	0	±0.5685	Bifurcation; Critical point	Loss of stability (flutter)
−0.3909	3	0	0	Bifurcation; Critical point	Loss of stability (divergence)
−0.3901	1 + 4	±0.2905	±0.5558	−Re: min Re(s); +Re: max Re(s)	Extremum, coupled flutter mode; all combinations of ±
−0.3862	1 + 4	0	±0.5422	Bifurcation; Critical point	Regain of stability (flutter)
−0.3815	2 + 4	0	±0.4318	Bifurcation; Critical point	Loss of stability (flutter)
−0.3811	3	±0.7980	0	−: min Re(s); +: max Re(s)	Extremum, 3rd divergence mode
−0.3734	2 + 4	±0.6532	±0.4147	−Re: min Re(s); +Re: max Re(s)	Extremum, coupled flutter mode; all combinations of ±
−0.3714	3	0	0	Bifurcation; Critical point	Regain of stability (divergence)
−0.3656	2 + 4	0	±0.3962	Bifurcation; Critical point	Regain of stability (flutter)

Figure 8.12 and Table 8.3 similarly present the results for a Kelvin—Voigt panel with small viscosity, $\tau_R = 0.1$ s. Figure 8.13 displays close-ups of some features tabulated in Table 8.3, and Fig. 8.14 shows a parametric plot in the $(\mathrm{Re}\, s, \mathrm{Im}\, s)$ plane, with respect to the parameter c. The range of c is the same as in Fig. 8.12. Figures 8.15, 8.16, 8.17, 8.18, 8.19 and 8.20 and Tables 8.4 and 8.5 present the results for the other two Kelvin—Voigt panels similarly.

Comparing the mode identification charts in Fig. 8.10 (ideal string) and 8.12 (Kelvin—Voigt panel with $\tau_R = 0.1$ s), we see that with the introduction of small dissipation, the previously symmetric coupled flutter modes split into their fundamental single-mode components, with only one component of the pair obtaining a positive Re(s); flutter has become a single-mode phenomenon. The maximum and minimum Re(s) of the apparent flutter pair no longer necessarily lie at the same value of c (although for small dissipation, they remain near each other), nor necessarily has

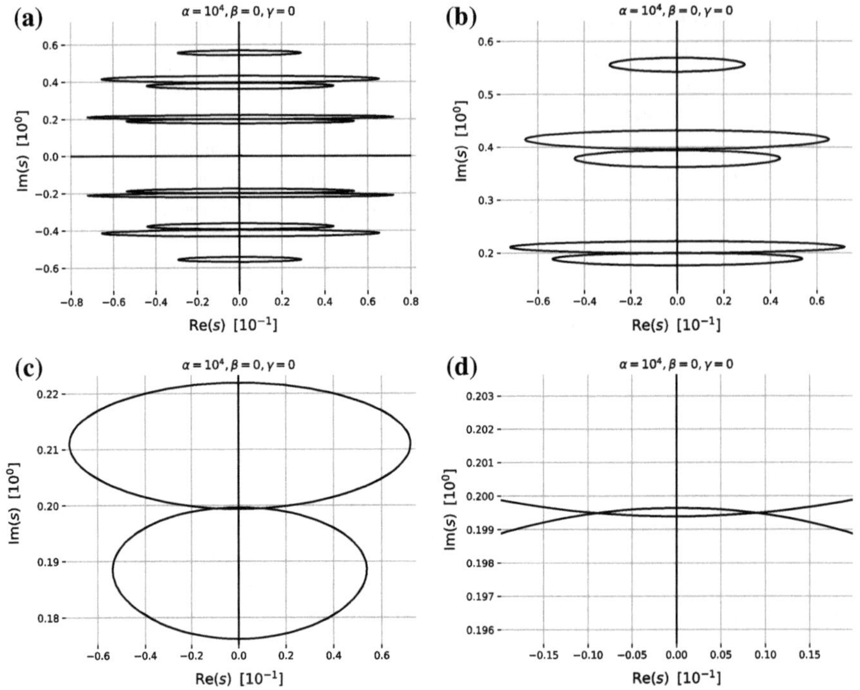

Fig. 8.11 Ideal string. Parametric plot in the $(\mathrm{Re}\, s, \mathrm{Im}\, s)$ plane, with respect to the parameter c. The range of c is the same as in Fig. 8.12. **a** Overview. **b–d** Close-ups of the upper half-plane at increasing levels of zoom

the same value for $\mathrm{Im}(s)$ for the different fundamental components (although near). We also observe that the flutter modes have lost their bifurcations; each eigenvalue curve remains distinct throughout each flutter region.

When dissipation is present, the only bifurcations occur at the beginning and at the end of a divergence region. However, because dissipation has shifted all eigenvalues toward the half-plane with negative $\mathrm{Re}(s)$, the bifurcation at the start of a divergence bubble no longer corresponds to a loss of stability. The critical point occurs at the zero crossing of $\mathrm{Re}(s)$, at a slightly higher value of c. Similarly, stability is regained already at the next zero crossing of $\mathrm{Re}(s)$, before the bifurcation point where the divergence bubble ends. This effect is best seen at high viscosities; refer to Fig. 8.20 and Table 8.5. Therefore, we conclude that a system with dissipation has much fewer bifurcation points than the corresponding ideal system, and the ones that remain are qualitatively distinct from critical points.

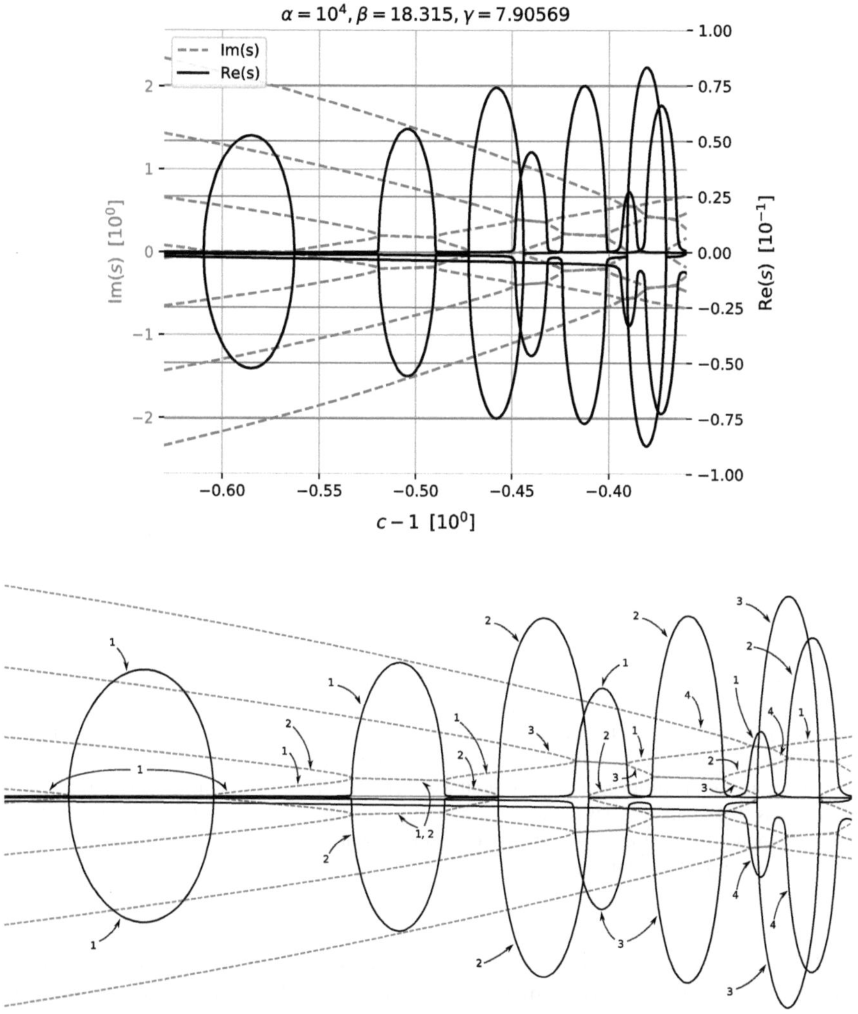

Fig. 8.12 Kelvin—Voigt panel with $\tau_R = 0.1$ s, with aerodynamic reaction. The whole air mass moves axially with the panel, $v_\infty = V_0$. Compare Fig. 8.10 for the ideal string. *Top*: The numerical results. *Bottom*: Mode identification chart

We observe also a new feature exclusive to the case with dissipation present. The eigenvalue curves exhibit local extrema of $\mathrm{Re}(s)$ not directly related to the extremum of any instability mode. In Tables 8.3, 8.4 and 8.5, these are labeled as *local extremum*. Recall that we saw this phenomenon also in the numerical results for the Kelvin—Voigt panel (in vacuum) at the end of Chap. 5.

The nondimensional critical velocity c for the panel with the highest viscosity out of those investigated ($\tau_R = 10$ s) is $c - 1 \approx -0.6094$. The value remains almost the same as for the ideal string, for which $c - 1 \approx -0.6095$. We observe that even at very

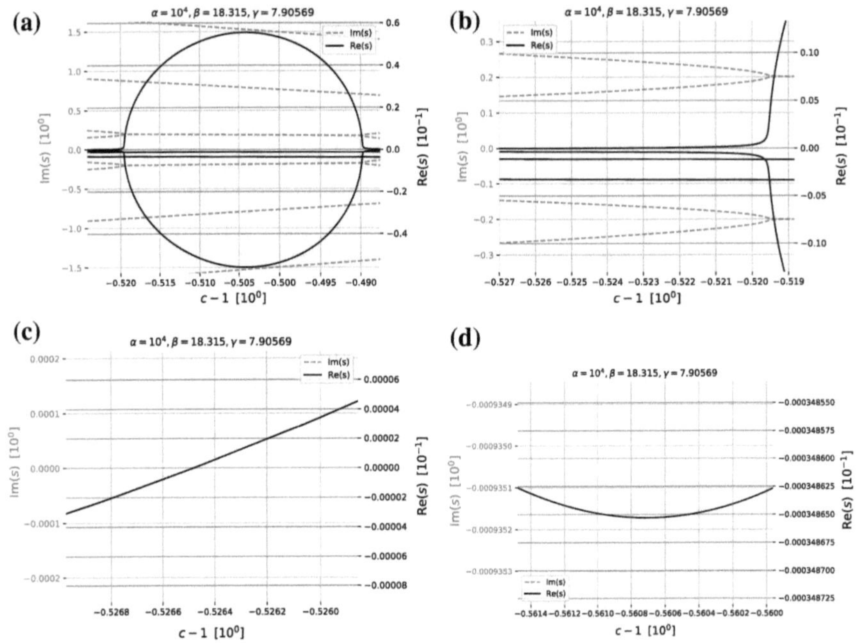

Fig. 8.13 Kelvin—Voigt panel with $\tau_R = 0.1$ s. Features corresponding to Table 8.3. **a** The $1 + 2$ flutter mode has split into its components; only the mode $k = 1$ obtains $\mathrm{Re}(s) > 0$. **b** Close-up of the beginning of the $1 + 2$ flutter mode. The imaginary parts meet at $c - 1 \approx -0.5195$. **c** Further close-up of the first flutter instability. The Loss of stability occurs at $c - 1 \approx -0.5265$, much earlier than the imaginary parts meet, at a zero crossing of $\mathrm{Re}(s)$ for mode $k = 1$ (shown) where the slope of the curve is still small. **d** A new local extremum for $\mathrm{Re}(s)$ for $k = 1$ at $c - 1 \approx -0.5607$, that does not correspond to an extremum of an instability mode. This extremum occurs after the divergence gap, but before the first flutter mode

high viscosity, even though the shapes of the eigenvalue curves change dramatically, the viscosity of the travelling material has almost no effect on the critical velocity. In the whole first series of results, the critical velocity, which is drastically reduced from the corresponding vacuum case, is effectively dictated by the flow component of the fluid—structure interaction problem.

Moving on to the second series of results, let $v_\infty = 0$. To see the initial postcritical behavior, we now use the range $c - 1 \in [-0.1, 1] \cdot 10^{-5}$. The presence of the surrounding fluid has qualitatively changed the shapes of the eigenvalue curves, but the critical velocity remains similar to the vacuum case that was considered at the end of Chap. 5. In this series of results, the presence of the surrounding fluid has no effect on the stability of the traveling ideal string, so the ideal string case is omitted.

Table 8.3 Points of interest in the range of c shown in Fig. 8.12, for a Kelvin—Voigt panel with $\tau_R = 0.1$ s

$c-1$	k	Re(s) [10^{-1}]	Im(s)	Type	Note
−0.6095	1	−0.0003058	0	Bifurcation	Start of 1st divergence bubble
−0.6095	1	0	0	Critical point	Loss of stability, after start of bubble
−0.5852	1	−0.5249, +0.5242	0	−: min Re(s); +: max Re(s)	Extremum, 1st divergence mode
−0.5630	1	0	0	Critical point	Regain of stability, before end of bubble
−0.5630	1	−0.0003488	0	Bifurcation	End of 1st divergence bubble
−0.5607	1	−0.0003487	±0.02511	min Re(s)	Local extremum
−0.5265	$1+2$	0	±0.1485 ($k=1$)	Critical point	Loss of stability (flutter)
−0.5042	1	+0.5551	±0.1896	max Re(s)	Extremum, flutter mode $1+2$
−0.5042	2	−0.5588	±0.1896	min Re(s)	Extremum, flutter mode $1+2$
−0.4783	$1+2$	0	±0.2319 ($k=1$)	Critical point	Regain of stability (flutter)
−0.4728	2	−0.003932	0	Bifurcation	Start of 2nd divergence bubble
−0.4728	2	0	0	Critical point	Loss of stability, after start of bubble
−0.4725	1	−0.00006849	±0.2741	min Re(s)	Local extremum
−0.4665	$1+3$	0	±0.2965 ($k=1$)	Critical point	Loss of stability (flutter); no regain of stability
					of curve $k=1$ before forming mode $1+4$.
−0.4581	2	−0.7484, +0.7405	0	−: min Re(s); +: max Re(s)	Extremum, 2nd divergence mode
−0.4440	2	0	0	Critical point	Regain of stability, before end of Bubble
−0.4440	2	−0.003911	0	Bifurcation	End of 2nd divergence bubble
−0.4398	1	+0.4500	±0.3800	max Re(s)	Extremum, flutter mode $1+3$
−0.4397	3	−0.4674	±0.3800	min Re(s)	Extremum, flutter mode $1+3$
−0.4278	3	−0.02333	±0.2986	max Re(s)	Local extremum
−0.4265	$2+3$	0	±0.1717 ($k=2$)	Critical point	Loss of stability (flutter)
−0.4136	1	+0.0002237	±0.4672	min Re(s)	Local extremum
−0.4125	2	+0.7495	±0.2122	max Re(s)	Extremum, flutter mode $2+3$
−0.4125	3	−0.7726	±0.2121	min Re(s)	Extremum, flutter mode $2+3$
−0.3965	$2+3$	0	±0.2756 ($k=2$)	Critical point	Regain of stability (flutter)
−0.3934	2	−0.0007304	±0.3015	min Re(s)	Local extremum
−0.3907	3	−0.02136	0	Bifurcation	Start of 3rd divergence bubble
−0.3907	3	0	0	Critical point	Loss of stability, after start of bubble
−0.3898	$2+4$	0	±0.3296 ($k=2$)	Critical point	Loss of stability (flutter)
−0.3896	1	+0.2708	±0.5582	max Re(s)	Extremum, flutter mode $1+4$
−0.3895	4	−0.3319	±0.5582	min Re(s)	Extremum, flutter mode $1+4$
−0.3837	4	−0.8305	±0.5008	max Re(s)	Local extremum
−0.3805	3	−0.8744, +0.8325	0	−: min Re(s); +: max Re(s)	Extremum, 3rd divergence mode
−0.3745	$1+4$	0	0	Critical point	Regain of stability (flutter); no separate loss of
					stability; curve $k=1$ continues from mode $1+3$.

(continued)

Table 8.3 (continued)

$c-1$	k	Re(s) $[10^{-1}]$	Im(s)	Type	Note
−0.3728	2	+0.6610	±0.4171	max Re(s)	Extremum, flutter mode 2 + 4
−0.3728	4	−0.7272	±0.4168	min Re(s)	Extremum, flutter mode 2 + 4
−0.3704	3	0	0	Critical point	Regain of stability, before end of bubble
−0.3704	3	−0.02014	0	Bifurcation	End of 3rd divergence bubble
−0.3659	1	−0.0004912	±0.6527	min Re(s)	Local extremum
−0.3614	4	−0.08682	±0.3109	max Re(s)	Local extremum

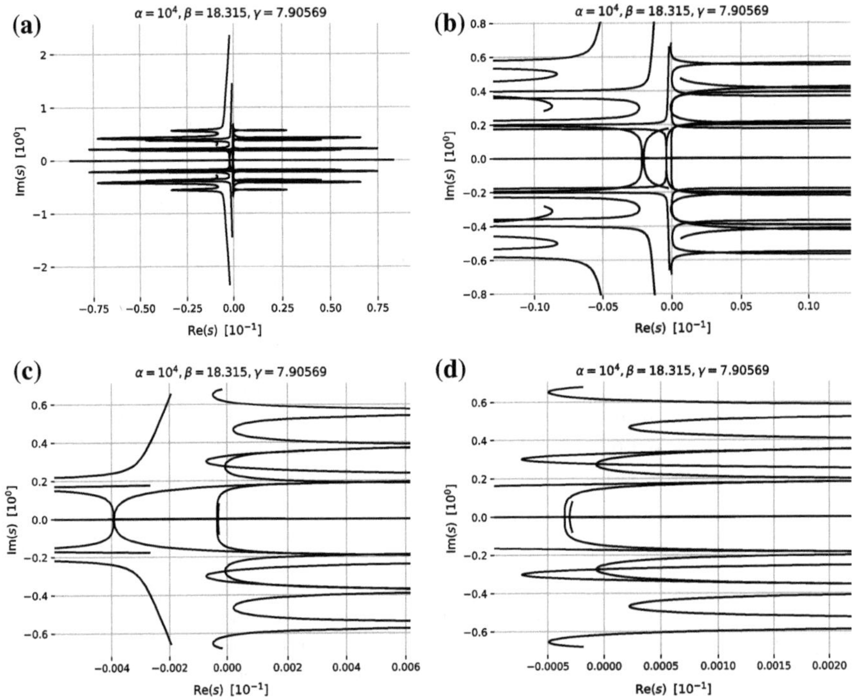

Fig. 8.14 Kelvin—Voigt panel with $\tau_R = 0.1$ s. Parametric plot in the (Re s, Im s) plane, with respect to the parameter c. The range of c is the same as in Fig. 8.12. **a** Overview. **b–d** Close-ups of the region around the origin at increasing levels of zoom

We consider the linear elastic panel and three Kelvin—Voigt panels with different values for the retardation time. The results are visualized in Figs. 8.21, 8.22, 8.23, 8.24, 8.25, 8.26, 8.27, 8.28, 8.29, 8.30, 8.31 and 8.32 and interesting points tabulated in Tables 8.6, 8.7, 8.8 and 8.9.

In the second series of results, we observe a new phenomenon: the imaginary parts of a flutter mode coalesce to zero, causing the mode to split into two divergence modes. See the mode identification charts in Figs. 8.21 and 8.24.

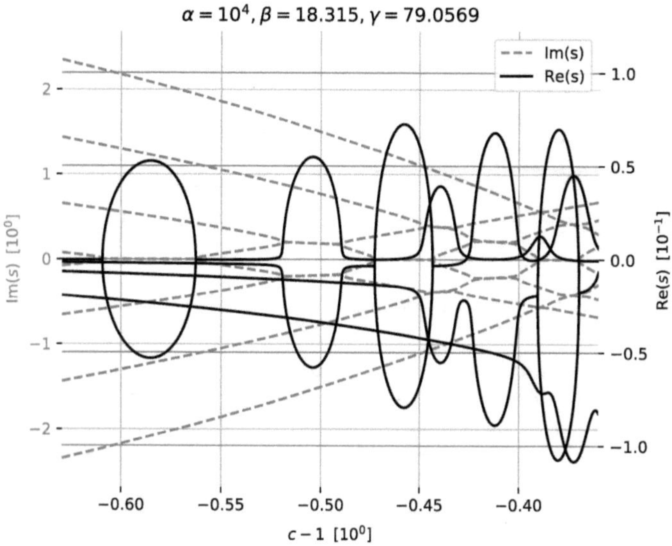

Fig. 8.15 Kelvin—Voigt panel with $\tau_R = 1$ s, with aerodynamic reaction. The whole air mass moves axially with the panel, $v_\infty = V_0$. For brevity, only the numerical results are shown. For mode identification, refer to Fig. 8.12

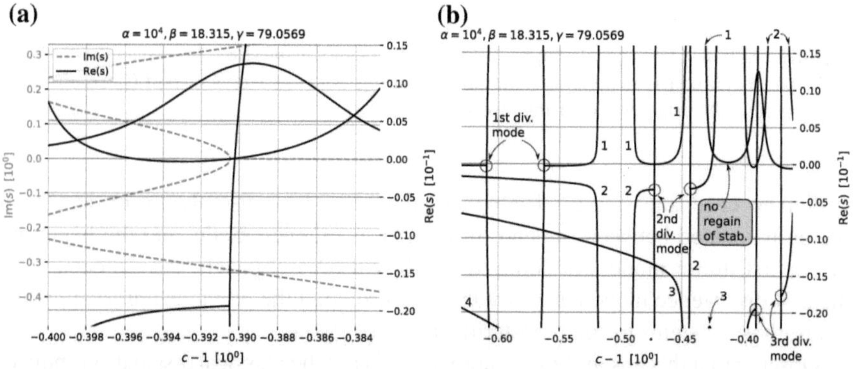

Fig. 8.16 Kelvin—Voigt panel with $\tau_R = 1$ s. Some details corresponding to Fig. 8.15 and Table 8.4. **a** The region near $c - 1 = -0.390$. Mode $k = 2$ regains stability (from flutter mode $2 + 3$) and then loses it again (in flutter mode $2 + 4$). Mode $k = 3$ undergoes divergence. Note the formation of the divergence bubble slightly before the real part becomes zero. **b** Close-up of the behavior of the real parts near zero for the same range of c as in Fig. 8.15. From here it is clear that the flutter mode $1 + 3$ never regains stability before the curve for $k = 1$ starts forming the next flutter mode $1 + 4$

When $v_\infty = 0$, after the first divergence gap, when the flutter modes begin, the next instability begins before the previous one stabilizes at least in the range of c investigated. The numerical evidence seems to suggest (but does not prove) that this

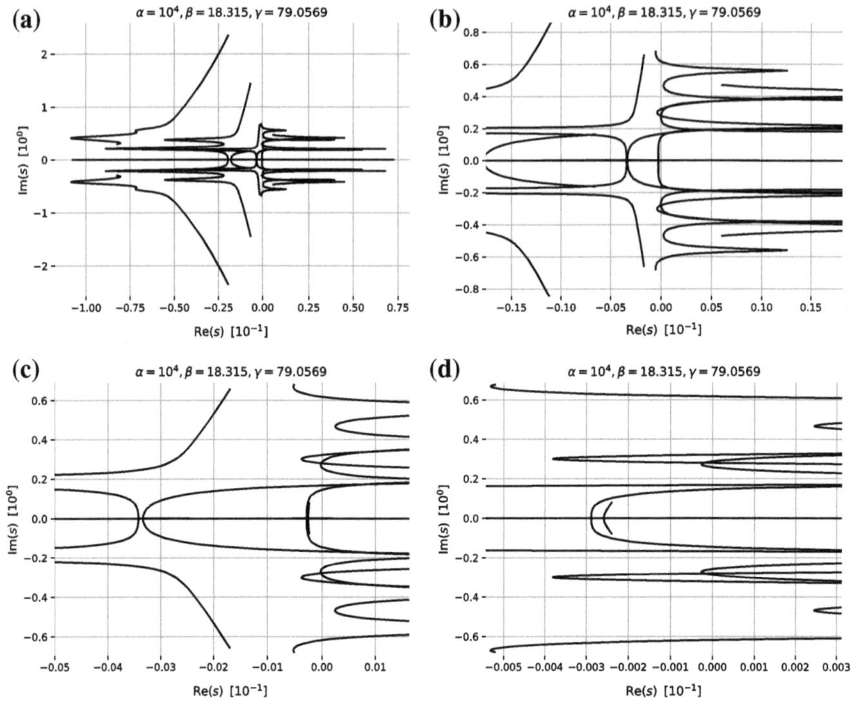

Fig. 8.17 Kelvin—Voigt panel with $\tau_R = 1$ s. Parametric plot in the $(\operatorname{Re} s, \operatorname{Im} s)$ plane, with respect to the parameter c. The range of c is the same as in Fig. 8.15. **a** Overview. **b–d** Close-ups of the region around the origin at increasing levels of zoom

system never regains stability for higher values of c. If so, its only postcritical stable region is located between the end of the first divergence mode, and the beginning of the first flutter mode. For some of the Kelvin—Voigt panels, it even occurs that the first divergence mode never restabilizes. Provided the previous observation also holds, those systems have no postcritical stable regions.

Observe that this absence of restabilization is not the classical dissipation-induced destabilization, where the first critical load drastically drops when a small dissipative term is introduced. From Tables 8.6 and 8.7 we see that the change to the critical load is minor, and that the load actually increases slightly when we introduce dissipation. We have $c - 1 \approx 0.02260 \cdot 10^{-5}$ at the first loss of stability for the linear elastic panel, versus $0.04653 \cdot 10^{-5}$ for the Kelvin—Voigt panel with $\tau_R = 10^{-5}$ s. Before we add dissipation, the *original* system—explicitly, an axially moving linear elastic panel interacting with potential flow—is conservative, and hence the condition of the Kelvin—Tait—Chataev theorem is satisfied. Thus the theorem guarantees that the introduction of dissipation into this system cannot cause a drastic decrease in the first critical load. The absence of restabilization is a different phenomenon, namely a change in the postcritical behavior.

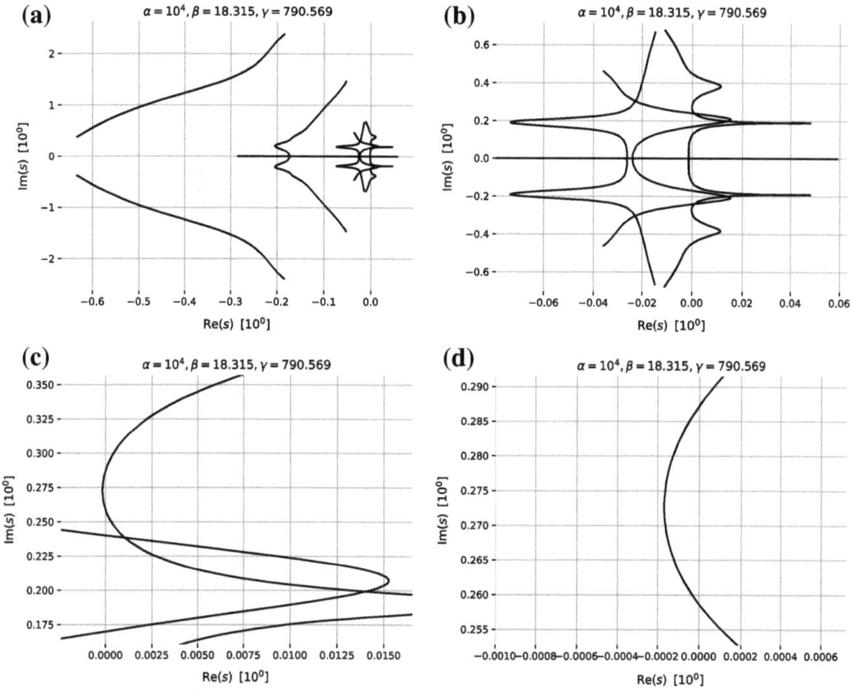

Fig. 8.18 Kelvin—Voigt panel with $\tau_R = 10$ s. Parametric plot in the $(\operatorname{Re} s, \operatorname{Im} s)$ plane, with respect to the parameter c. The range of c is the same as in Fig. 8.19. **a** Overview. **b–d** Close-ups at increasing levels of zoom

Because all real materials exhibit some viscosity, no matter how small, and air is almost always present, the practical implication is that at least for axially moving sheets, postcritical stability regions are likely a modeling artifact resulting from over-idealization, and do not physically exist.

Finally, in the analysis performed, there is one important limitation. The mode identification charts shown below have been prepared using the eigenvalues only; the eigenfunctions were not computed. To perform the identification, we have utilized two observations. First, in the initial unloaded state, each mode is a fundamental single-component mode. Secondly, because each solution curve is continuous with respect to c, bifurcations are the only points where mode identities may interact.

This is a simple information propagation analysis, which cannot tell exactly what happens at some of the more exotic bifurcation points. For example, in the second series of results for the linear elastic panel in Fig. 8.21, we see that the $1 + 2$ flutter mode splits into two divergence modes. In absence of further information, we cannot tell whether the separate modes 1 and 2 re-emerge at this bifurcation, or if the coupled mode just rearranges itself differently, yet remains coupled.

However, the present analysis does tell that the fundamental components in those modes will be different for the linear elastic and Kelvin—Voigt cases. As we observed

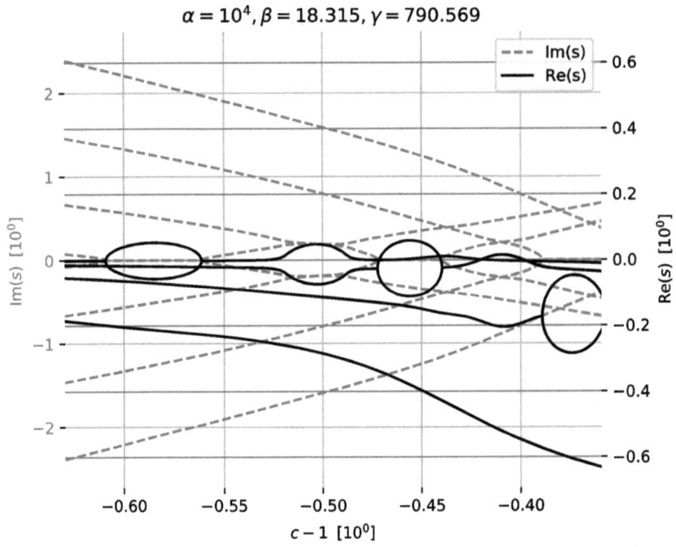

Fig. 8.19 Kelvin—Voigt panel with $\tau_R = 10$ s, with aerodynamic reaction. The whole air mass moves axially with the panel, $v_\infty = V_0$

above, the introduction of dissipation eliminates coupled-mode flutter. For the linear elastic case, the first bifurcation involving modes 1 and 2 occurs already at the start of the $1 + 2$ flutter mode, whereas in the Kelvin—Voigt case in Fig. 8.24 the eigenvalues of the modes 1 and 2 first interact at a distance, repelling each other without touching as the loading parameter c is quasistatically increased. (This is typical of systems with dissipation.) Only the fundamental mode 1 obtains a positive real part. When the corresponding bifurcation occurs, in the Kelvin—Voigt case each mode has only its own complex conjugate to interact with; the imaginary parts that coalesce to zero must belong to the same fundamental mode. The first bifurcation involving both of the first two fundamental modes occurs later, when the curves 1_a and 2_a meet. Even then the 1_b and 2_b branches remain as single-component modes until they meet in a separate bifurcation later.

For another example, in the linear elastic case in Fig. 8.21 the present analysis cannot tell which fundamental components the mode labeled $(1 + 2)_a + 3$ actually involves. To err on the side of caution, we have chosen to give a new descriptive label to each mode whose exact fundamental mode composition cannot be readily deduced from the available information.

To finish this section, we let the numerical results speak for themselves.

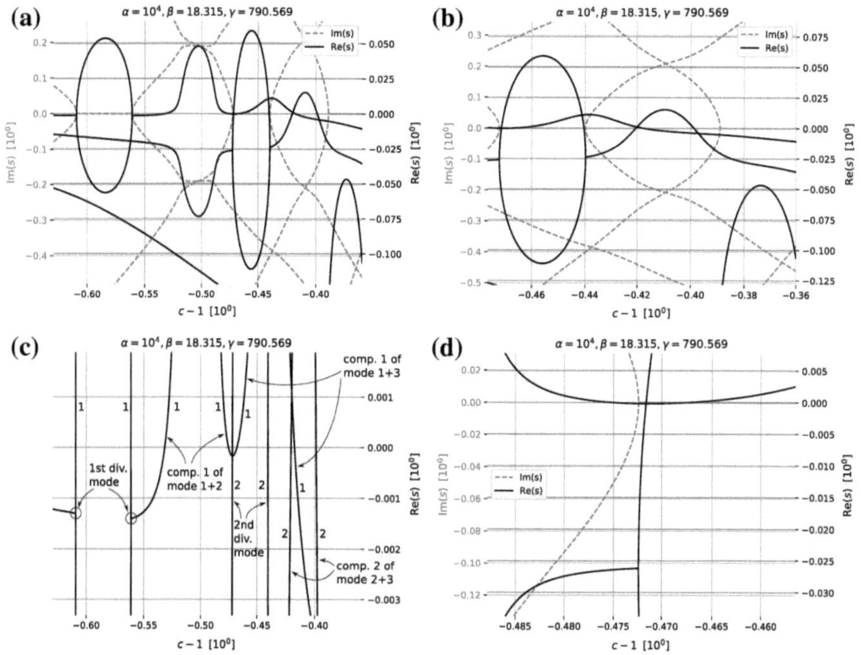

Fig. 8.20 Kelvin—Voigt panel with $\tau_R = 10\,\text{s}$. Some details corresponding to Fig. 8.19 and Table 8.5. **a** Behavior of the real parts near the origin for the full range of c. **b** Close-up starting from the 2nd divergence mode. The low bump is the flutter mode $1 + 3$ for $k = 1$. The reaction in the curve $k = 3$ is very small, see Fig. 8.19. The higher bump in the middle is $k = 2$ for the mode $2 + 3$. Finally, at the lower right, the upper branch of the 3rd divergence mode appears. This mode has become stable. **c** Close-up of the behavior of the real parts near zero for the full range of c. The mode $k = 1$ regains stability for a very small range of c between participating in the $1 + 2$ and $1 + 3$ flutter modes. **d** A detail at the start of the 2nd divergence mode, showing the real parts of modes 1 and 2 and the imaginary part of mode 2. There is a noticeable gap between the beginning of the divergence bubble (where the imaginary parts converge to zero) and the zero crossing of the real part; for a system with dissipation the bifurcation point and the critical point are distinct

8.5 Recommendations for Further Reading

To conclude this chapter, we will recommend some books and research papers that can be used to further explore various aspects related to the theoretical and numerical study of fluid—structure interaction, including fluid mechanics and finite elements. Fluid—structure interaction itself is an extensive topic, fully dealing with which would require a separate book, or possibly several.

Table 8.4 Points of interest in the range of c shown in Fig. 8.15, for a Kelvin—Voigt panel with $\tau_R = 1$ s

$c - 1$	k	Re(s) [10^{-1}]	Im(s)	Type	Note
−0.6095	1	−0.002582	0	Bifurcation	Start of 1st divergence bubble
−0.6095	1	0	0	Critical point	Loss of stability, after start of bubble
−0.5850	1	−0.5317, +0.5261	0	−: min Re(s); +: max Re(s)	Extremum, 1st divergence mode
−0.5627	1	0	0	Critical point	Regain of stability, before end of bubble
−0.5627	1	−0.002882	0	Bifurcation	End of 1st divergence bubble
−0.5272	1 + 2	0	±0.1453 ($k = 1$)	Critical point	Loss of stability (flutter)
−0.5039	1	+0.5484	±0.1901	max Re(s)	Extremum, flutter mode 1 + 2
−0.5039	2	−0.5809	±0.1900	min Re(s)	Extremum, flutter mode 1 + 2
−0.4767	1 + 2	0	±0.2573 ($k = 1$)	Critical point	Regain of stability (flutter)
−0.4729	1	−0.0002652	±0.2721	min Re(s)	Local extremum
−0.4726	2	−0.03420	0	Bifurcation	Start of 2nd divergence bubble
−0.4726	2	0	0	Critical point	Loss of stability, after start of bubble
−0.4689	1 + 3	0	±0.2870 ($k = 1$)	Critical point	Loss of stability (flutter); no regain of stability
					of curve $k = 1$ before forming mode 1 + 4.
−0.4578	2	−0.7942, +0.7257	0	−: min Re(s); +: max Re(s)	Extremum, 2nd divergence mode
−0.4434	2	0	0	Critical point	Regain of stability, before end of bubble
−0.4434	2	−0.3334	0	Bifurcation	End of 2nd divergence bubble
−0.4395	1	+0.3925	±0.3810	max Re(s)	Extremum, flutter mode 1 + 3
−0.4393	3	−0.5537	±0.3806	min Re(s)	Extremum, flutter mode 1 + 3
−0.4276	3	−0.2185	±0.3007	max Re(s)	Local extremum
−0.4267	2 + 3	0	±0.1670 ($k = 2$)	Critical point	Loss of stability (flutter)
−0.4133	1	+0.002461	±0.4768	min Re(s)	Local extremum
−0.4121	2	+0.6798	±0.2128	max Re(s)	Extremum, flutter mode 2 + 3
−0.4121	3	−0.8897	±0.2122	min Re(s)	Extremum, flutter mode 2 + 3
−0.3956	2 + 3	0	±0.2794 ($k = 2$)	Critical point	Regain of stability (flutter)
−0.3931	2	−0.003811	±0.3002	min Re(s)	Local extremum
−0.3905	3	−0.1941	0	Bifurcation	Start of 3rd divergence bubble
−0.3904	2 + 4	0	±0.3215 ($k = 2$)	Critical point	Loss of stability (flutter)
−0.3902	3	0	0	Critical point	Loss of stability, after start of bubble
−0.3893	1	+0.1256	±0.5592	max Re(s)	Extremum, flutter mode 1 + 4
−0.3878	4	−0.7175	±0.5506	min Re(s)	Extremum, flutter mode 1 + 4
−0.3851	4	−0.7112	±0.5243	max Re(s)	Local extremum
−0.3800	3	−1.076, +0.6988	0	−: min Re(s); +: max Re(s)	Extremum, 3rd divergence mode
−0.3739	1 + 4	0	±0.6178 ($k = 1$)	Critical point	Regain of stability (flutter); no separate loss of
					stability; curve $k = 1$ continues from mode 1 + 3.

(continued)

Table 8.4 (continued)

$c-1$	k	Re(s) $[10^{-1}]$	Im(s)	Type	Note
−0.3723	2	+0.4505	±0.4184	max Re(s)	Extremum, flutter mode 2 + 4
−0.3720	4	−1.085	±0.4162	min Re(s)	Extremum, flutter mode 2 + 4
−0.3698	3	0	0	Critical point	Regain of stability, before end of bubble
−0.3696	3	−0.1782	0	Bifurcation	End of 3rd divergence bubble
−0.3620	1	−0.005314	±0.6694	min Re(s)	Local extremum
−0.3614	4	−0.8051	±0.3174	max Re(s)	Local extremum

Generally speaking, simulation of fluids is challenging. As is pointed out by Gresho and Sani [47], the problem is not strictly speaking the nonlinearity of the Navier—Stokes equations, but the presence of the advection term that appears in the Eulerian formulation. Although the advection term in the Navier—Stokes equations also happens to be nonlinear, the chief difficulty is already encountered in the time-dependent advection—diffusion equation, which is a linear partial differential equation that includes an advection term. Lagrangean approaches to fluid simulation have recently been proposed; see the paper on particle FEM by Oñate et al. [48], and those on Lagrangean meshless Galerkin methods by Ponthot and Belytschko [49], Idelsohn et al. [50].

For (very) different introductory texts to the behaviour of flows around solid objects, see Acheson [10], [3], Lighthill [7], Batchelor [8] and Currie [2]. Acheson's and Anderson's books give an accessible first overview, respectively, of general fluid dynamics and aerodynamics for readers new to these topics. Lighthill's book is a systematic, easily understandable exposition of inviscid flows of both irrotational and rotational types (that is, potential and Euler flows). Batchelor introduces viscous flows immediately, and develops the exposition in that context. Currie covers a large range of topics, dividing the exposition into fundamentals (conservation laws and basic theorems), potential flows, viscous incompressible flows, and finally inviscid compressible flows.

Classic texts in fluid mechanics include Lamb [9], the volume Landau and Lifshitz [51] in their well-known series on theoretical physics, and the volume Sedov [6] in his four-volume course on continuum mechanics. The history of fluid mechanics is discussed in Tokaty [52].

For a focus specifically on rotating fluids, see Vanyo [11]. Prandtl's boundary-layer theory of flows having small viscosity is explained in the classic Schlichting [14] and its revised edition Schlichting and Gersten [13]. A recent development with regard to boundary-layer theory is the discovery that a boundary layer is not necessary for flows with small viscosity; similar flow patterns can be explained by considering a turbulent Euler flow. See the papers by Hoffman and Johnson [53, 54]. For another recent development related to this, see Verhoff [12], which suggests that

Table 8.5 Points of interest in the range of c shown in Fig. 8.19, for a Kelvin—Voigt panel with $\tau_R = 10$ s

$c - 1$	k	Re(s)	Im(s)	Type	Note
−0.6094	1	−0.001302	0	Bifurcation	Start of 1st divergence bubble
−0.6094	1	0	0	Critical point	Loss of stability, after start of bubble
−0.5841	1	+0.05377	0	max Re(s)	Extremum, 1st divergence mode
−0.5840	1	−0.05657	0	min Re(s)	Extremum, 1st divergence mode
−0.5610	1	0	0	Critical point	Regain of stability, before end of bubble
−0.5610	1	−0.001405	0	Bifurcation	End of 1st divergence bubble
−0.5317	1 + 2	0	±0.1258 (k = 1)	Critical point	Loss of stability (flutter)
−0.5029	1	+0.04820	±0.1903	max Re(s)	Extremum, flutter mode 1 + 2
−0.5026	2	−0.07345	±0.1906	min Re(s)	Extremum, flutter mode 1 + 2
−0.4751	1 + 2	0	±0.2590 (k = 1)	Critical point	Regain of stability (flutter)
−0.4724	2	−0.02622	0	Bifurcation	Start of 2nd divergence bubble
−0.4716	2	0	0	Critical point	Loss of stability, after start of bubble
−0.4715	1	−0.0001675	±0.2727	min Re(s)	Local extremum
−0.4676	1 + 3	0	±0.2871 (k = 1)	Critical point	Loss of stability (flutter)
−0.4559	2	−0.1110	0	min Re(s)	Extremum, 2nd divergence mode
−0.4558	2	+0.05921	0	max Re(s)	Extremum, 2nd divergence mode
−0.4405	2	0	0	Critical point	Regain of stability, before end of bubble
−0.4398	2	−0.02394	0	Bifurcation	End of 2nd divergence bubble
−0.4386	1	+0.01116	±0.3820	max Re(s)	Extremum, flutter mode 1 + 3; no minimum
					in pair, curve for k = 3 is decreasing.
−0.4206	2 + 3	0	±0.1699 (k = 2)	Critical point	Loss of stability (flutter)
−0.4176	1 + 3	0	±0.4486 (k = 1)	Critical point	Regain of stability (flutter)
−0.4095	2	+0.01527	±0.2072	max Re(s)	Extremum, flutter mode 2 + 3
−0.4086	3	−0.2044	±0.2058	min Re(s)	Extremum, flutter mode 2 + 3
−0.3988	2 + 3	0	±0.2400 (k = 2)	Critical point	Regain of stability (flutter)
−0.3892	3	−0.1731	0	Bifurcation	Start of 3rd divergence bubble
−0.3745	3	−0.2842	0	min Re(s)	Extremum, 3rd divergence mode (stable)
−0.3735	3	−0.04676	0	max Re(s)	Extremum, 3rd divergence mode (stable)

at least in two dimensions, potential flow theory is able to represent the initial state before transition into turbulence, if we allow introducing inviscid vortex singularities at isolated points inside the flow domain.

Concerning the related topic of free surface flows, the classical marker-and-cell (MAC) method is reviewed in McKee et al. [55]. For a computer graphics viewpoint, see Cline et al. [56]. For the more advanced volume-of-fluid (VOF) method, see, for example, Gueyffier et al. [57]. Historically, the MAC method was first published in

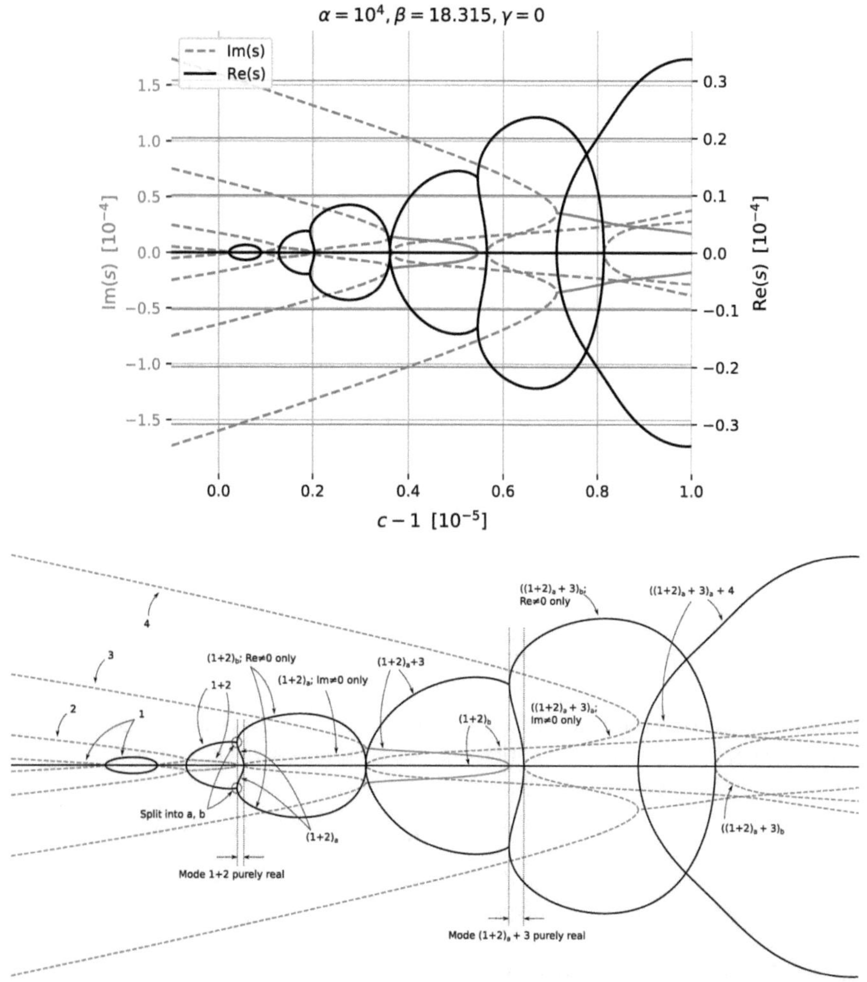

Fig. 8.21 Linear elastic panel, with aerodynamic reaction. No axial free-stream flow, $v_\infty = 0$. Compare the vacuum linear elastic results at the end of Chap. 5. *Top*: The numerical results. *Bottom*: Mode identification chart

Harlow and Welch [58]. The VOF method was introduced in the conference paper Noh and Woodward [59], the first journal publication being Hirt and Nichols [60].

Thin aerofoil (airfoil) theory is explained in, among others, Ashley and Landahl [15], Bisplinghoff and Ashley [61], Anderson [3]. Especially the first of these contains a clear presentation and uses an approach similar to the one we used in this chapter. Specifically, see the derivation of lift for the camber line of a thin aerofoil in Ashley and Landahl [15], Chap. 5–3. The second one focuses on the theory of aeroelasticity (i.e., flexible structures subjected to a flow).

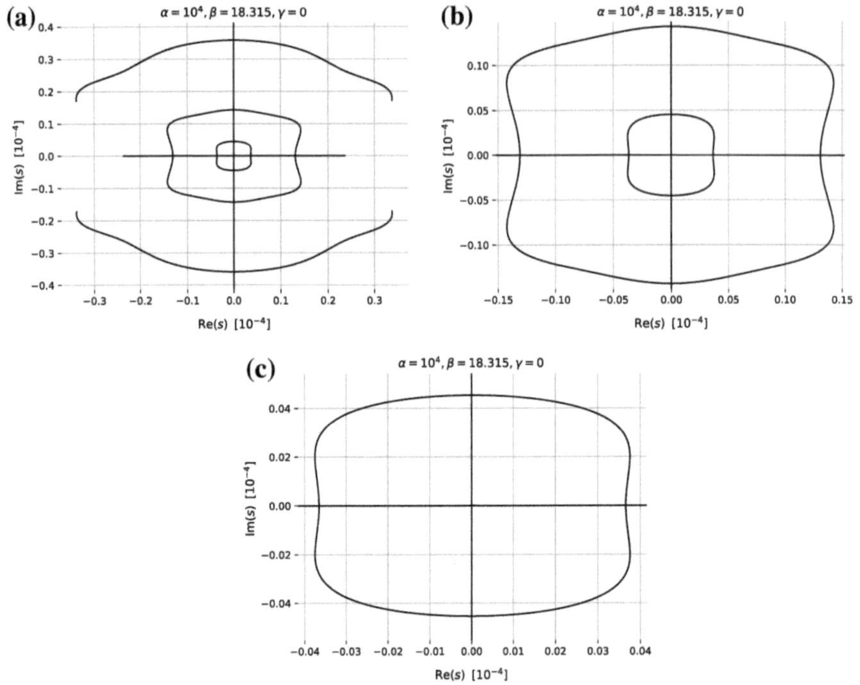

Fig. 8.22 Linear elastic panel. No axial free-stream flow, $v_\infty = 0$. Parametric plot in the (Re s, Im s) plane, with respect to the parameter c. The range of c is the same as in Fig. 8.21. **a** Overview. **b**, **c** Close-ups at increasing levels of zoom

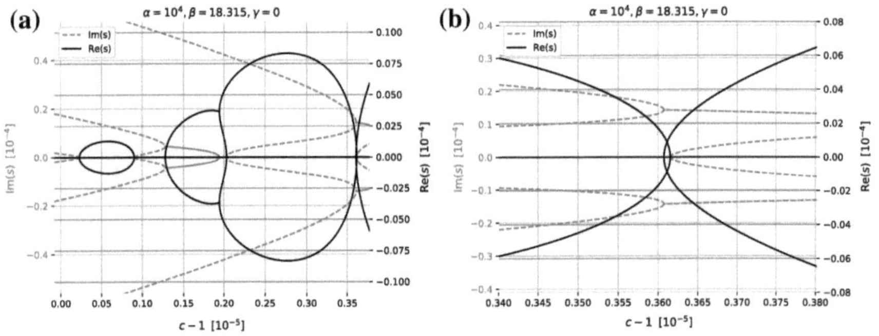

Fig. 8.23 Linear elastic panel. No axial free-stream flow, $v_\infty = 0$. Some details corresponding to Fig. 8.21 and Table 8.6. **a** Close-up of the first few instabilities. **b** Zooming at the end of the first subfigure reveals that the new instability arises first, before the previous one stabilizes

Thin aerofoil theory is of course just one possible approach, and fairly specific to the class of problems considered in this chapter. More general approaches in aerodynamics and aeroelasticity are, for example, the classical vortex panel method

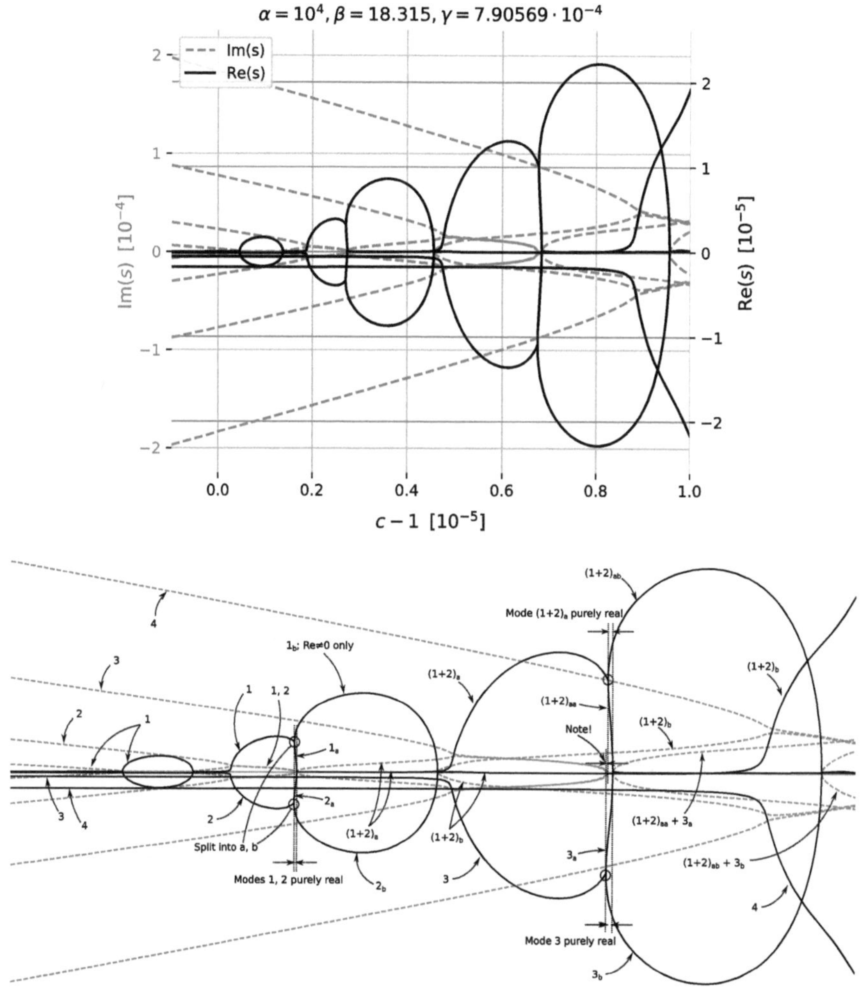

Fig. 8.24 Kelvin—Voigt panel with $\tau_R = 10^{-5}$ s, with aerodynamic reaction. No axial free-stream flow, $v_\infty = 0$. Compare the vacuum Kelvin—Voigt results at the end of Chap. 5. *Top*: The numerical results. *Bottom*: Mode identification chart

(see Anderson [3]), and fully numerical approaches using finite volumes or finite elements.

A classical simulation test case in aerodynamics is the flow around a cylinder. This is a standard topic in above-mentioned books Acheson [10], Batchelor [8], Anderson [3], Ashley and Landahl [15]. Standardized benchmarks for fluid—structure interaction in the same spirit were proposed by Turek and Hron [62], and have found their way into wide use.

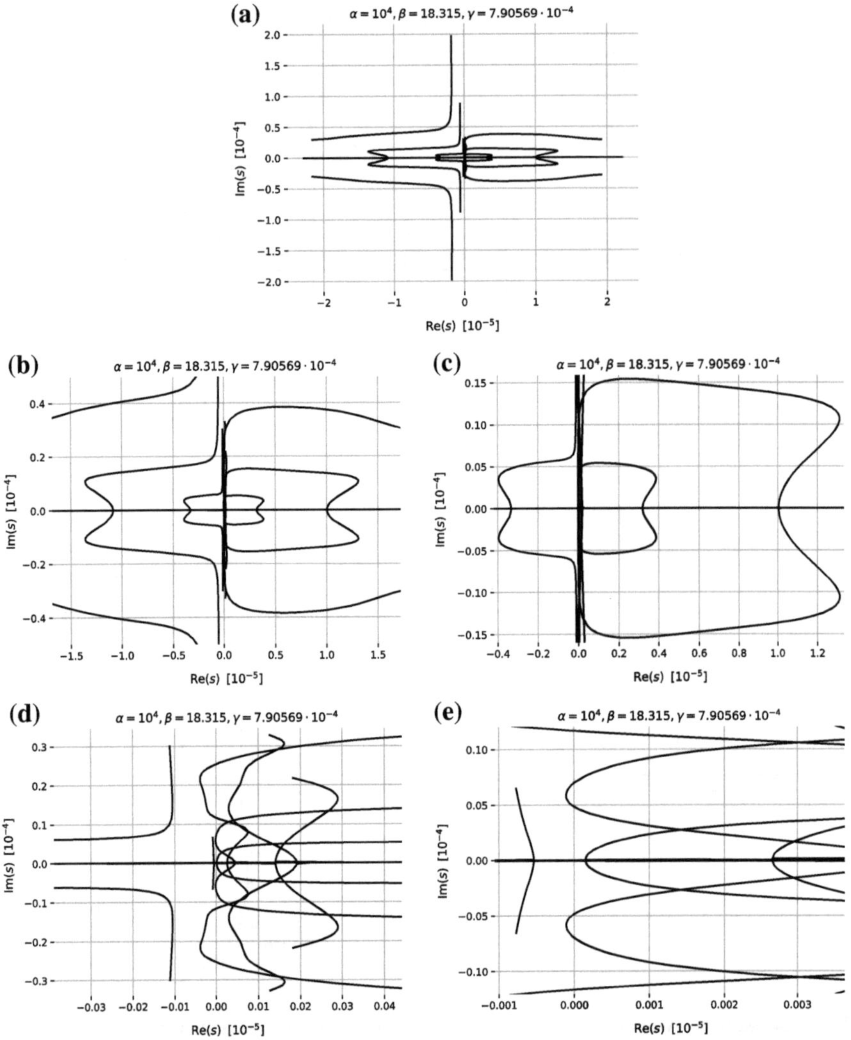

Fig. 8.25 Kelvin—Voigt panel with $\tau_R = 10^{-5}$ s. No axial free-stream flow, $v_\infty = 0$. Parametric plot in the $(\mathrm{Re}\, s, \mathrm{Im}\, s)$ plane, with respect to the parameter c. The range of c is the same as in Fig. 8.24. **a** Overview. **b, c** Close-ups at increasing levels of zoom. **d, e** Detail near the origin. Note aspect ratio of zoom

Slender bodies subjected to flow are discussed in Lighthill [7, 63] and Ashley and Landahl [15]. This topic, and other problems that share characteristics with this one, are discussed in the extensive two-volume review by Païdoussis [34, 35] and in the papers Païdoussis [39, 64].

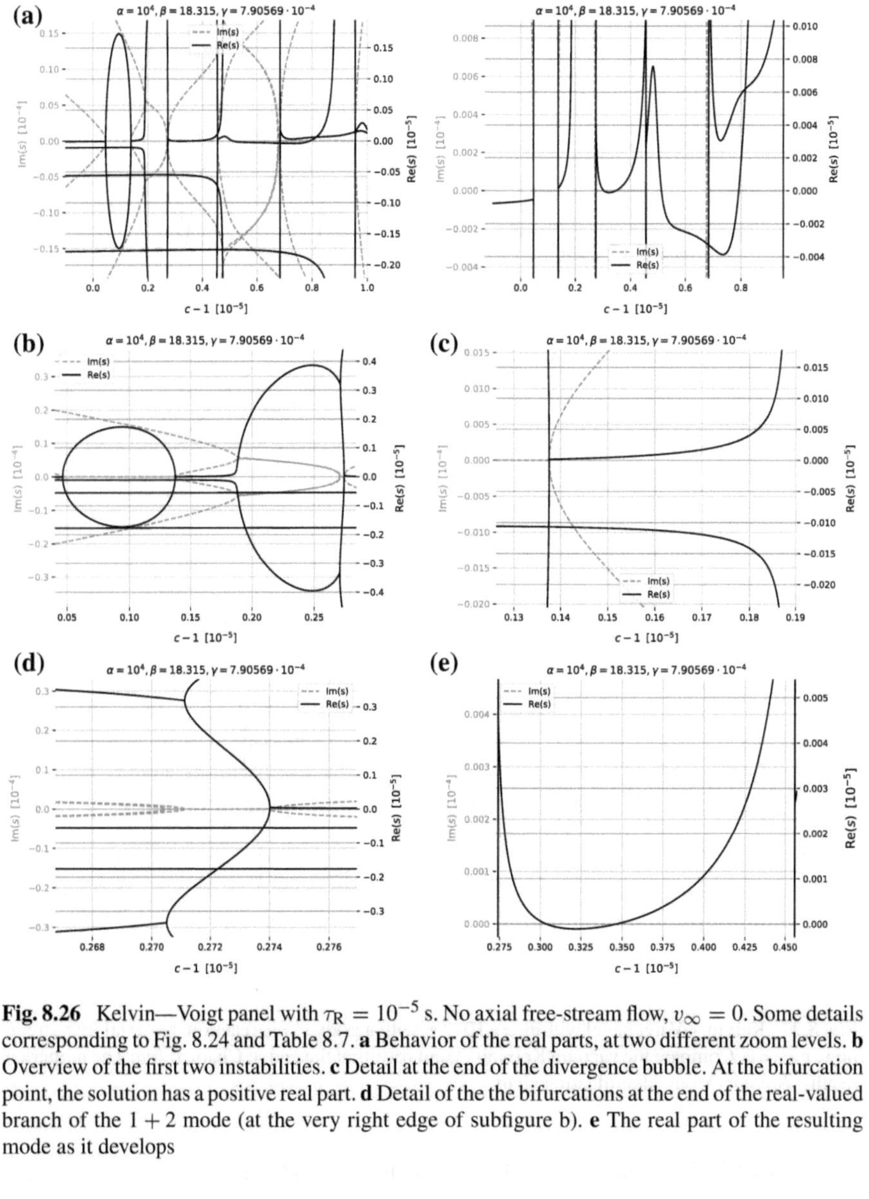

Fig. 8.26 Kelvin—Voigt panel with $\tau_R = 10^{-5}$ s. No axial free-stream flow, $v_\infty = 0$. Some details corresponding to Fig. 8.24 and Table 8.7. **a** Behavior of the real parts, at two different zoom levels. **b** Overview of the first two instabilities. **c** Detail at the end of the divergence bubble. At the bifurcation point, the solution has a positive real part. **d** Detail of the the bifurcations at the end of the real-valued branch of the $1 + 2$ mode (at the very right edge of subfigure b). **e** The real part of the resulting mode as it develops

Readers new to solid mechanics may be interested in the exposition of the Eulerian (spatial) and Lagrangean (material) formulations in Holzapfel [65], ch8Marsden and Hughes [66]. The first one is also a good reference for the more advanced reader. Specifically for axially moving materials, see the mixed Lagrangean—Eulerian formulation introduced in the paper Koivurova and Salonen [31].

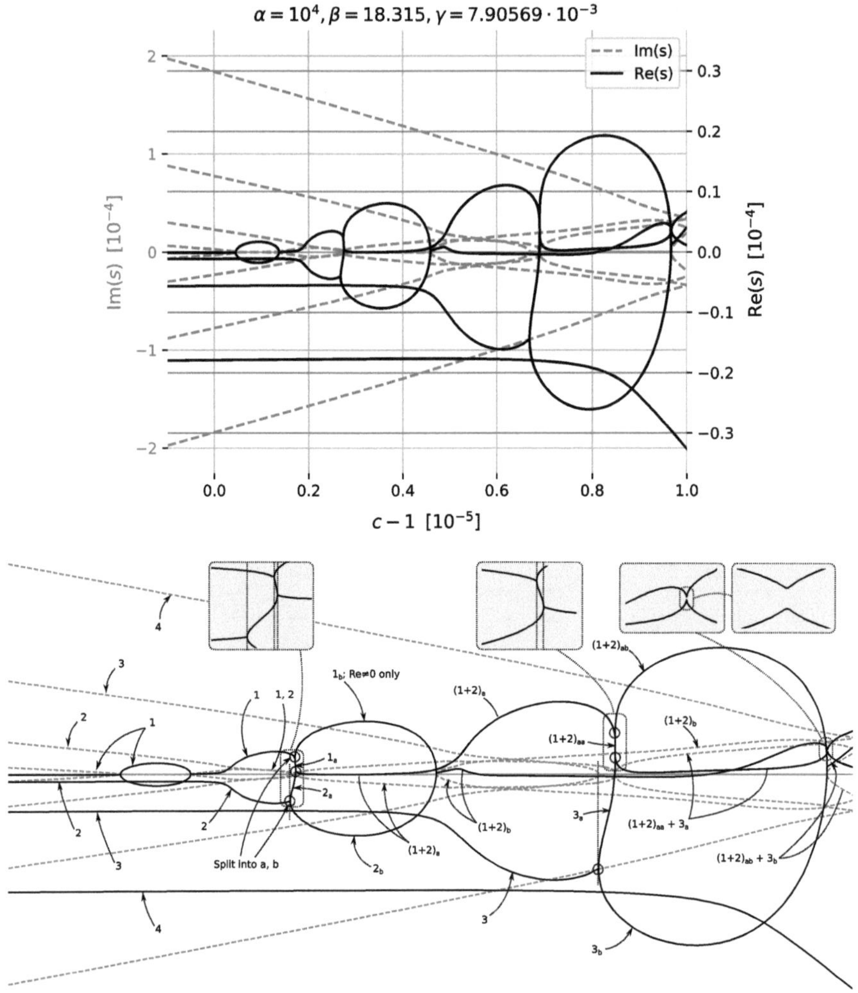

Fig. 8.27 Kelvin—Voigt panel with $\tau_R = 10^{-4}$ s, with aerodynamic reaction. No axial free-stream flow, $v_\infty = 0$. Compare the vacuum Kelvin—Voigt results at the end of Chap. 5. *Top*: The numerical results. *Bottom*: Mode identification chart

The book Allen et al. [1], besides its excellent introduction to conservation laws in continuum form, contains a chapter on finite elements. What is especially rare is that it also presents classical finite differences, and compares these two different approaches for the numerical solution of partial differential equations.

Classical texts concerning finite elements include Strang and Fix [67] and Ciarlet [68], but these may be difficult to follow. For different expositions of finite elements in general, see Johnson [69], Krizek and Neittaanmäki [70], Eriksson et al. [71], Bathe [72], Hughes [73], Belytschko et al. [74], Fish and Belytschko [75], Brenner

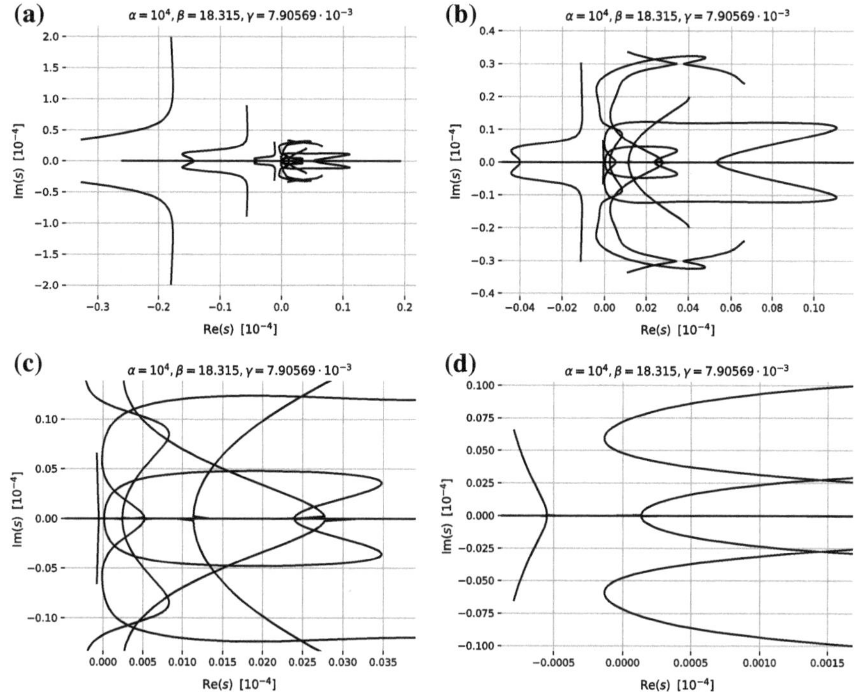

Fig. 8.28 Kelvin—Voigt panel with $\tau_{\mathrm{R}} = 10^{-4}$ s. No axial free-stream flow, $v_\infty = 0$. Parametric plot in the $(\mathrm{Re}\, s, \mathrm{Im}\, s)$ plane, with respect to the parameter c. The range of c is the same as in Fig. 8.27. **a** Overview. **b–d** Close-ups at increasing levels of zoom

and Scott [76], Zienkiewicz et al. [77–79]. Especially Hughes [73] and Eriksson et al. [71] can be recommended for their easy approachability, respectively, for readers with a background in the engineering sciences, and in mathematics. The classic reference Zienkiewicz et al. [77–79] deserves a mention for its completeness, including a section on Delaunay triangulation, which most books on finite elements omit. For the rigorous mathematical theory, see Brenner and Scott [76]. Readers new to finite elements may wish to look at Fish and Belytschko [75].

For the application of finite elements specifically to flow problems, see Gresho and Sani [47], Donea and Huerta [80]. The book by Gresho and Sani [47] explains in detail how to use fundamental Galerkin FEM effectively for flow problems. Donea and Huerta [80] takes a more standard view, and looks at approaches for numerical stabilization, including a detailed explanation of the streamline upwinding Petrov—Galerkin (SUPG) technique.

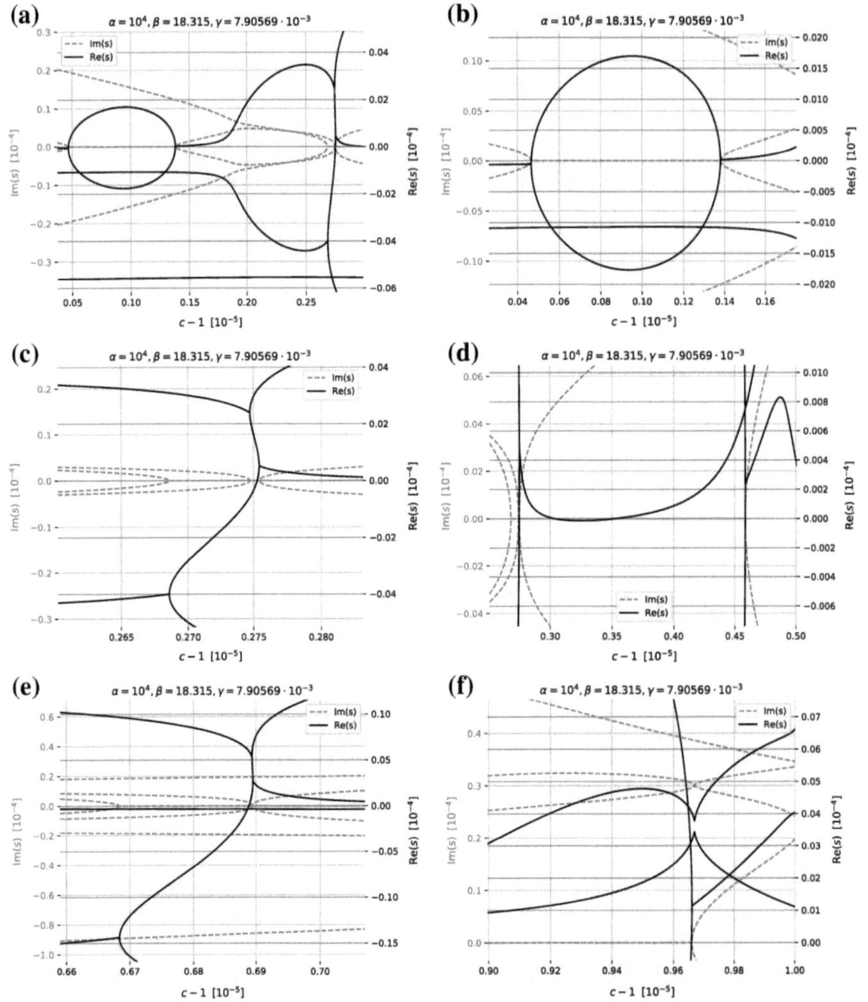

Fig. 8.29 Kelvin—Voigt panel with $\tau_R = 10^{-4}$ s. No axial free-stream flow, $v_\infty = 0$. Some details corresponding to Fig. 8.27 and Table 8.8. **a** Overview of the first two instabilities. **b** Close-up of the divergence mode. **c** Close-up of the bifurcations at the right edge of the first subfigure. **d** The real part of the resulting mode. **e** A similar set of bifurcations appears later for higher modes. **f** A pair of local extrema for Re(s); see the last two rows of Table 8.8 and subfigure b in Fig. 8.28. This is not a bifurcation; the modes never touch each other in this interaction. The bifurcation in the same view is the third last row of Table 8.8

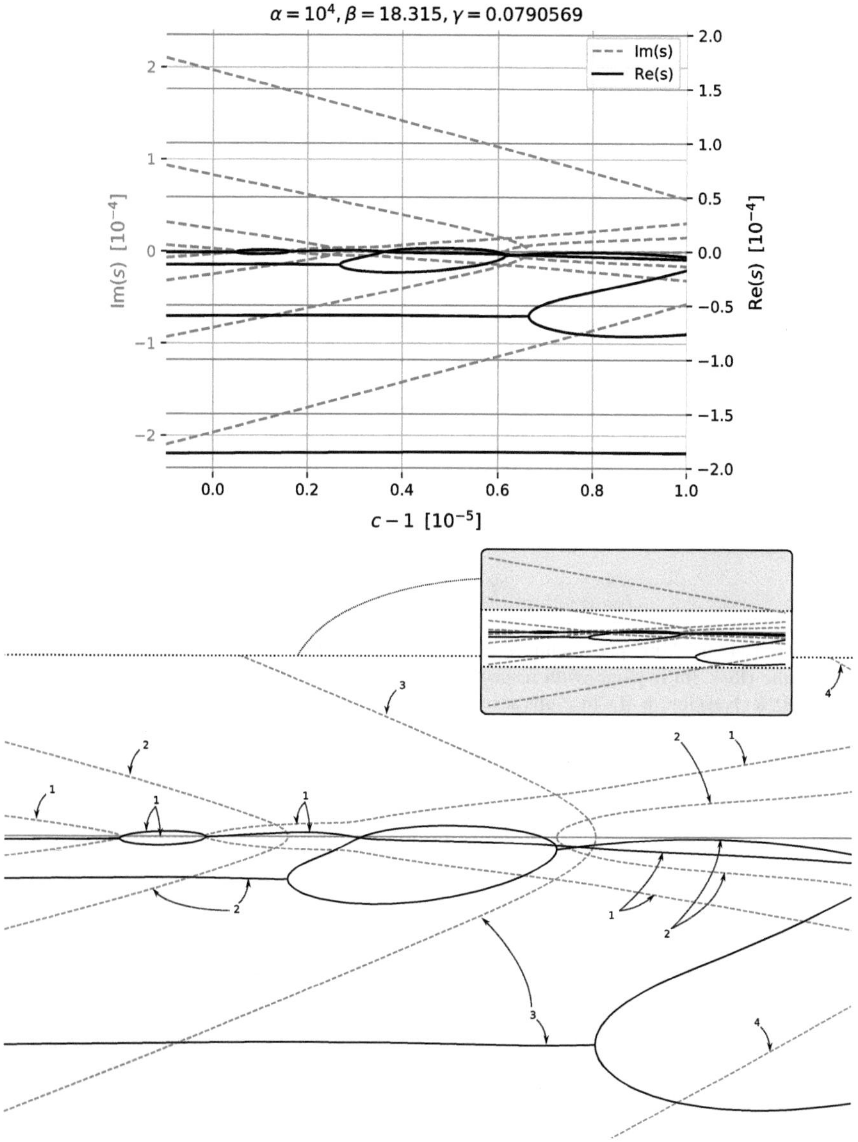

Fig. 8.30 Kelvin—Voigt panel with $\tau_R = 10^{-3}$ s, with aerodynamic reaction. No axial free-stream flow, $v_\infty = 0$. Compare the vacuum Kelvin—Voigt results at the end of Chap. 5. *Top*: The numerical results. *Bottom*: Mode identification chart

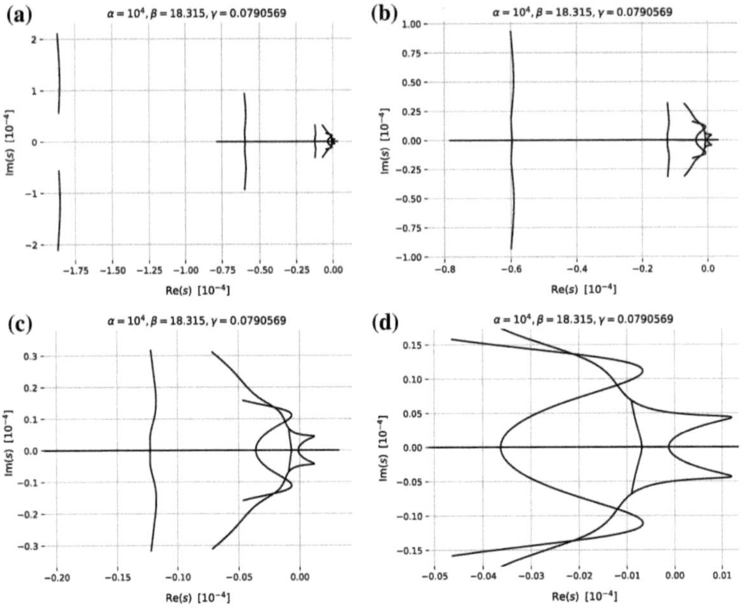

Fig. 8.31 Kelvin—Voigt panel with $\tau_R = 10^{-3}$ s. No axial free-stream flow, $v_\infty = 0$. Parametric plot in the $(\operatorname{Re} s, \operatorname{Im} s)$ plane, with respect to the parameter c. The range of c is the same as in Fig. 8.30. **a** Overview. **b–d** Close-ups at increasing levels of zoom

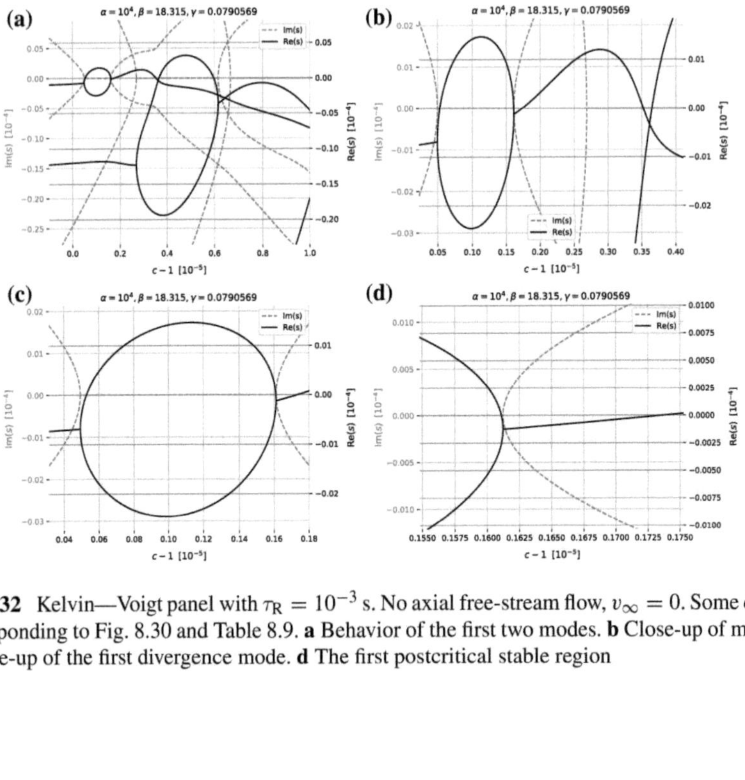

Fig. 8.32 Kelvin—Voigt panel with $\tau_R = 10^{-3}$ s. No axial free-stream flow, $v_\infty = 0$. Some details corresponding to Fig. 8.30 and Table 8.9. **a** Behavior of the first two modes. **b** Close-up of mode 1. **c** Close-up of the first divergence mode. **d** The first postcritical stable region

Table 8.6 Points of interest in the range of c shown in Fig. 8.19, for a linear elastic panel

$c-1$ [10^{-5}]	k	Re(s) [10^{-4}]	Im(s) [10^{-4}]	Type	Note
0.02260	1	0	0	Bifurcation; Critical point	1st divergence mode
0.05840	1	±0.01307	0	−: min Re(s); +: max Re(s)	Extremum, 1st divergence mode
0.09038	1	0	0	Bifurcation; Critical point	End of 1st divergence mode
0.1281	$1+2$	0	±0.04536	Bifurcation; Critical point	Loss of stability (flutter); then unstable
					For the rest of the investigated range
0.1849	$1+2$	±0.03755	±0.02001	−Re: min Re(s); +Re: max Re(s)	Extremum, coupled flutter mode
					All combinations of ±
0.1944	$1+2$	±0.03657	0	Bifurcation	$1+2$ split into a, b; changes type
					Into a divergence mode
0.2034	$(1+2)_a$	0	0	Bifurcation; Critical point	Mode $(1+2)_a$ stabilizes
0.2765	$(1+2)_b$	±0.08324	0	−: min Re(s); +: max Re(s)	Extremum, coupled divergence mode
0.3607	$(1+2)_a+3$	0	±0.1431	Bifurcation; Critical point	Mode 3 becomes coupled
0.3615	$(1+2)_b$	0	0	Bifurcation; Critical point	Mode $(1+2)_b$ stabilizes
0.5026	$(1+2)_a+3$	±0.1425	±0.08015	−Re: min Re(s); +Re: max Re(s)	Extremum, coupled flutter mode
0.5457	$(1+2)_a+3$	±0.1308	0	Bifurcation	$(1+2)_a+3$ split into a, b; changes type
					into a divergence mode
0.5649	$((1+2)_a+3)_a$	0	0	Bifurcation; Critical point	Mode $((1+2)_a+3)_a$ stabilizes
0.6706	$((1+2)_a+3)_b$	±0.2369	0	−: min Re(s); +: max Re(s)	Extremum, coupled divergence mode
0.7135	$((1+2)_a+3)_a+4$	0	±0.3584	Bifurcation; Critical point	Mode 4 becomes coupled
0.8134	$((1+2)_a+3)_b$	0	0	Bifurcation; Critical point	Mode $((1+2)_a+3)_b$ stabilizes
0.9911	$((1+2)_a+3)_a+4$	±0.3380	±0.1782	−Re: min Re(s); +Re: max Re(s)	Extremum, coupled flutter mode

Table 8.7 Points of interest in the range of c shown in Fig. 8.24, for a Kelvin—Voigt panel with $\tau_R = 10^{-5}$ s

$c-1$ [10^{-5}]	k	Re(s) [10^{-4}]	Im(s) [10^{-4}]	Type	Note
0.04652	1	−0.0005356	0	Bifurcation	Start of 1st divergence bubble
0.04653	1	0	0	Critical point	Loss of stability, after start of bubble
0.09411	1	−0.1732	0	min Re(s)	Extremum, 1st divergence mode
0.09427	1	+0.1726	0	max Re(s)	Extremum, 1st divergence mode
0.1047	2	−0.01056	±0.1507	max Re(s)	Local extremum
0.1375	1	+0.0001465	0	Bifurcation	End of 1st divergence bubble; no regain of stability
0.2484	2	−0.3956	±0.03457	min Re(s)	Extremum, flutter mode
0.2486	1	+0.3864	±0.03477	max Re(s)	Extremum, flutter mode
0.2494	3	−0.05480	±0.5081	max Re(s)	Local extremum
0.2705	2	−0.3340	0	Bifurcation	Split of mode 2 into a, b; changes type into divergence mode
0.2711	1	+0.3189	0	Bifurcation	Split of mode 1 into a, b; changes type into divergence mode
0.2740	2_a	0	0	Critical point	Zero crossing of 2_a
0.2740	$(1+2)_a$	+0.004530	0	Bifurcation	Start of $(1+2)_a$
0.3038	$(1+2)_a$	0	±0.04845	Critical point	Gain of stability
0.3222	$(1+2)_a$	−0.0001073	±0.05849	min Re(s)	Local extremum
0.3463	$(1+2)_a$	0	±0.06970	Critical point	Loss of stability
0.3591	2_b	−0.8728	0	min Re(s)	Extremum, divergence mode
0.3601	1_b	+0.8607	0	max Re(s)	Extremum, divergence mode
0.4517	4	−0.1754	±1.211	max Re(s)	Local extremum
0.4560	2_b	0	0	Critical point	Zero crossing of 2_b
0.4560	$(1+2)_b$	+0.002662	0	Bifurcation	End of divergence mode
0.4818	$(1+2)_b$	+0.007559	±0.07955	max Re(s)	Local extremum
0.6128	3	−1.361	±0.1061	min Re(s)	Extremum, flutter mode
0.6136	$(1+2)_a$	+1.310	±0.1062	max Re(s)	Extremum, flutter mode
0.6751	3	−1.085	0	Bifurcation	Split of mode 3 into a, b; changes type into divergence mode
0.6772	$(1+2)_a$	+1.005	0	Bifurcation	Split of mode $(1+2)_a$ to a, b; changes type into divergence mode
0.6834	3_a	0	0	Critical point	Zero crossing of 3_a
0.6834	$(1+2)_{aa}+3_a$	+0.01906	0	Bifurcation	Start of $(1+2)_{aa}+3_a$
0.7260	$(1+2)_{aa}+3_a$	+0.003058	±0.1403	min Re(s)	Local extremum
0.7345	$(1+2)_b$	−0.003873	±0.2160	min Re(s)	Local extremum
0.8036	3_b	−2.284	0	min Re(s)	Extremum, divergence mode
0.8068	$(1+2)_{ab}$	+2.218	0	max Re(s)	Extremum, divergence mode
0.9564	3_b	0	0	Critical point	Zero crossing of 3_b
0.9564	$(1+2)_{ab}+3_b$	+0.01406	0	Bifurcation	End of divergence mode
0.9765	$(1+2)_{aa}+3_a$	+0.01615	±0.3054	max Re(s)	Local extremum
0.9818	$(1+2)_{ab}+3_b$	+0.02893	±0.1645	max Re(s)	Local extremum

Table 8.8 Points of interest in the range of c shown in Fig. 8.27, for a Kelvin—Voigt panel with $\tau_R = 10^{-4}$ s

$c-1$ [10^{-5}]	k	Re(s) [10^{-4}]	Im(s) [10^{-4}]	Type	Note
0.04664	1	−0.0005506	0	Bifurcation	Start of 1st divergence bubble
0.04667	1	0	0	Critical point	Loss of stability, after start of bubble
0.09368	1	−0.01768	0	min Re(s)	Extremum, 1st divergence mode
0.09522	1	+0.01699	0	max Re(s)	Extremum, 1st divergence mode
0.1045	2	−0.01069	±0.1515	max Re(s)	Local extremum
0.1380	1	+0.0001346	0	Bifurcation	End of 1st divergence bubble; no regain of stability
0.2487	2	−0.04412	±0.03367	min Re(s)	Extremum, flutter mode
0.2489	3	−0.05480	±0.5081	max Re(s)	Local extremum
0.2498	1	−0.03477	±0.03578	max Re(s)	Extremum, flutter mode
0.2686	2	−0.04001	0	Bifurcation	Split of mode 2 into a, b; changes type into divergence mode
0.2747	1	+0.02397	0	Bifurcation	Split of mode 1 into a, b; changes type into divergence mode
0.2753	2_a	0	0	Critical point	Zero crossing of 2_a
0.2754	$(1+2)_a$	+0.005178	0	Bifurcation	Start of $(1+2)_a$
0.3039	$(1+2)_a$	0	±0.04796	Critical point	Gain of stability
0.3245	$(1+2)_a$	−0.0001345	±0.05919	min Re(s)	Local extremum
0.3525	$(1+2)_a$	0	±0.07198	Critical point	Loss of stability
0.3557	2_b	−0.09323	0	min Re(s)	Extremum, divergence mode
0.3663	1_b	+0.08078	0	max Re(s)	Extremum, divergence mode
0.4510	4	−0.1766	±1.217	max Re(s)	Local extremum
0.4586	2_b	0	0	Critical point	Zero crossing of 2_b
0.4587	$(1+2)_b$	+0.002402	0	Bifurcation	End of divergence mode
0.4875	$(1+2)_b$	+0.008319	±0.08482	max Re(s)	Local extremum
0.6118	3	−0.1613	±0.1060	min Re(s)	Extremum, flutter mode
0.6195	$(1+2)_a$	+0.1105	±0.1063	max Re(s)	Extremum, flutter mode
0.6683	3	−0.1437	0	Bifurcation	Split of mode 3 into a, b; changes type into divergence mode
0.6889	3_a	0	0	Critical point	Zero crossing of 3_a
0.6894	$(1+2)_a$	+0.05345	0	Bifurcation	Split of mode $(1+2)_a$ to a, b; changes type into divergence mode
0.6895	$(1+2)_{aa}+3_a$	+0.02773	0	Bifurcation	Start of $(1+2)_{aa}+3_a$
0.7332	$(1+2)_{aa}+3_a$	+0.002558	±0.1427	min Re(s)	Local extremum
0.7457	$(1+2)_b$	−0.004096	±0.2210	min Re(s)	Local extremum
0.7940	3_b	−0.2606	0	min Re(s)	Extremum, divergence mode
0.8255	$(1+2)_{ab}$	+0.1931	0	max Re(s)	Extremum, divergence mode
0.9498	$(1+2)_b$	+0.04778	±0.3193	max Re(s)	Local extremum
0.9659	3_b	0	0	Critical point	Zero crossing of 3_b
0.9661	$(1+2)_{ab}+3_b$	+0.01135	0	Bifurcation	End of divergence mode
0.9669	$(1+2)_b$	+0.03778	±0.3006	min Re(s)	Local extremum
0.9669	$(1+2)_{aa}+3_a$	+0.03439	±0.3004	max Re(s)	Local extremum

Table 8.9 Points of interest in the range of c shown in Fig. 8.30, for a Kelvin—Voigt panel with $\tau_R = 10^{-3}$ s

$c-1$ [10^{-5}]	k	Re(s) [10^{-4}]	Im(s) [10^{-4}]	Type	Note
0.04940	1	−0.006875	0	Bifurcation	Start of 1st divergence bubble
0.05293	1	0	0	Critical point	Loss of stability, after start of bubble
0.09865	1	−0.02462	0	min Re(s)	Extremum, 1st divergence mode
0.1132	1	+0.01468	0	max Re(s)	Extremum, 1st divergence mode
0.1211	2	−0.01180	±0.1552	max Re(s)	Local extremum
0.1611	1	0	0	Critical point	Regain of stability, before end of bubble
0.1612	1	−0.001254	0	Bifurcation	End of 1st divergence bubble
0.1738	1	0	±0.01347	Critical point	Loss of stability (flutter)
0.2587	2	−0.1231	±0.03026	min Re(s)	Local extremum
0.2684	2	−0.1230	±0.0008749	Bifurcation	Start of 2nd divergence bubble
0.2882	1	+0.01202	±0.04337	max Re(s)	Local extremum
0.2946	3	−0.5912	±0.5164	max Re(s)	Local extremum
0.3521	1	0	±0.04977	Critical point	Regain of stability (flutter)
0.3650	2	0	0	Critical point	Loss of stability, after start of bubble
0.3823	2	−0.1941	0	min Re(s)	Extremum, 2nd divergence mode
0.4798	2	+0.03208	0	max Re(s)	Extremum, 2nd divergence mode
0.5821	4	−1.855	±1.166	max Re(s)	Local extremum
0.5951	3	−0.5989	±0.1706	min Re(s)	Local extremum
0.5974	2	0	0	Critical point	Regain of stability, before end of bubble
0.6156	2	−0.03633	±0.0005751	Bifurcation	End of 2nd divergence bubble
0.6657	3	−0.5964	±0.001930	Bifurcation	Start of 3rd divergence bubble
0.7949	2	−0.006626	±0.1114	max Re(s)	Local extremum
0.8783	3	−0.7835	0	min Re(s)	Extremum, 3rd divergence mode

References

1. Allen MB III, Herrera I, Pinder GF (1988) Numerical modeling in science and engineering. Wiley Interscience
2. Currie IG (2002) Fundamental mechanics of fluids. CRC Press, 3rd edn, 548 p
3. Anderson JD (1985) Fundamentals of aerodynamics. McGraw-Hill
4. Sedov LI (1971) A course in continuum mechanics, vol 1. Wolters-Noordhoff Publishing, Groningen, Netherlands, English edition. ISBN 90-0179680-X
5. Kiselev SP, Vorozhtsov E, Fomin VM (2012) Foundations of fluid mechanics with applications: problem solving using mathematica. Springer, 575 p
6. Sedov LI (1972) A course in continuum mechanics, vol 3. Wolters-Noordhoff Publishing, Groningen, Netherlands, English edition. ISBN 90-0179682-6
7. Lighthill (1986) An informal introduction to theoretical fluid mechanics. Oxford Science Publications. ISBN 0-19-853630-5
8. Batchelor GK (1967) An introduction to fluid dynamics. Cambridge University Press
9. Lamb H (1932) Hydrodynamics. Dover, 6th edn, 1945
10. Acheson DJ (1990) Elementary fluid dynamics. Oxford
11. Vanyo JP (1993) Rotating fluids in engineering and science. Butterworth–Heinemann, 429 p
12. Verhoff August (2010) Two-dimensional potential flow solutions with separation. J Fluid Mech 657:238–264. https://doi.org/10.1017/S0022112010001448
13. Schlichting H, Gersten K (1999) Boundary layer theory. Springer, 8th revised and enlarged edition. ISBN 978-3-540-66270-9

14. Schlichting H (1960) Boundary layer theory. McGraw-Hill, 4th edn
15. Ashley H, Landahl M (1985) Aerodynamics of wings and bodies. Dover
16. Zeev N (1952) Conformal mapping. Dover. ISBN 0-486-61137-X
17. Driscoll TA (1996) Algorithm 756: a matlab toolbox for schwarz-christoffel mapping. ACM Trans. Math. Softw. 22(2):168–186, June 1996. ISSN 0098-3500. https://doi.org/10.1145/229473.229475. URL http://doi.acm.org/10.1145/229473.229475
18. Adams RA, Essex C (2010) Calculus: a complete course. Pearson, 7th edn
19. Min Chohong, Gibou Frédéric (2006) A second order accurate projection method for the incompressible navier-stokes equations on non-graded adaptive grids. J Comput Phys 219(2):912–929. https://doi.org/10.1016/j.jcp.2006.07.019
20. Chen Han, Min Chohong, Gibou Frédéric (2007) A supra-convergent finite difference scheme for the poisson and heat equations on irregular domains and non-graded adaptive cartesian grids. J Sci Comput 31(1–2):19–60. https://doi.org/10.1007/s10915-006-9122-8
21. Kornecki A, Dowell EH, O'Brien J (1976) On the aeroelastic instability of two-dimensional panels in uniform incompressible flow. J Sound Vib 47(2):163–178
22. Evans LC (1998) Partial differential equations. Am Math Soc. ISBN 0-8218-0772-2
23. Silverman RA (1972) Introductory complex analysis. Dover. ISBN 0-486-64686-6
24. Flanigan FJ (1972) Complex variables: harmonic and analytic functions. Dover. ISBN 0-486-61388-7
25. Sherman DI (1952) On the stress distribution in partitions, an elastic heavy medium which is weakened by elliptic holes. Izvestiya Akademii Nauk SSSR, Otdelenie Tekhnicheskikh Nauk (OTN) 7:992–1010
26. Banichuk N, Jeronen J, Neittaanmäki P, Tuovinen T (2010b) Static instability analysis for travelling membranes and plates interacting with axially moving ideal fluid. J Fluids Struct 26(2):274–291. https://doi.org/10.1016/j.jfluidstructs.2009.09.006
27. Banichuk N, Jeronen J, Neittaanmäki P, Tuovinen T (2011) Dynamic behaviour of an axially moving plate undergoing small cylindrical deformation submerged in axially flowing ideal fluid. J Fluids Struct 27(7):986–1005. ISSN 0889-9746. https://doi.org/10.1016/j.jfluidstructs.2011.07.004
28. Juha Jeronen. *On the mechanical stability and out-of-plane dynamics of a travelling panel submerged in axially flowing ideal fluid: a study into paper production in mathematical terms*. PhD thesis, Department of Mathematical Information Technology, University of Jyväskylä, 2011b. URL http://urn.fi/URN:ISBN:978-951-39-4596-1. Jyväskylä studies in computing 148. ISBN 978-951-39-4595-4 (book), ISBN 978-951-39-4596-1 (PDF)
29. Banichuk N, Jeronen J, Neittaanmäki P, Saksa T, Tuovinen T (2014) Mechanics of moving materials, volume 207 of Solid mechanics and its applications. Springer. ISBN: 978-3-319-01744-0 (print), 978-3-319-01745-7 (electronic)
30. Jeronen J, Saksa T, Tuovinen T (2016) Stability of a tensioned axially moving plate subjected to cross-direction potential flow. In Neittaanmäki P, Repin S, Tuovinen T (eds) Mathematical modeling and optimization of complex structures, dedicated to Prof. Nikolay Banichuk for his 70th anniversary, pp 105–116. Springer. ISBN 978-3-319-23563-9
31. Koivurova H, Salonen E-M (1999) Comments on non-linear formulations for travelling string and beam problems. J Sound Vib 225(5):845–856
32. Kurki M, Jeronen J, Saksa T, Tuovinen T (2012) Strain field theory for viscoelastic continuous high-speed webs with plane stress behavior. In: Eberhardsteiner J, Böhm HJ, Rammerstorfer FG (eds), CD-ROM Proceedings of the 6th European congress on computational methods in applied sciences and engineering (ECCOMAS 2012), Vienna, Austria. Vienna University of Technology. ISBN 978-3-9502481-9-7
33. Kurki M, Jeronen J, Saksa T, Tuovinen T (2016) The origin of in-plane stresses in axially moving orthotropic continua. Int J Solids Struct. https://doi.org/10.1016/j.ijsolstr.2015.10.027
34. Païdoussis MP (1998) Fluid-structure interactions: slender structures and axial flow, vol 1. Academic Press. ISBN 0-12-544360-9
35. Païdoussis MP (2004) Fluid-structure interactions: slender structures and axial flow, vol 2. Elsevier Academic Press. ISBN 0-12-544361-7

36. Eloy C, Souilliez C, Schouveiler L (2007) Flutter of a rectangular plate. J Fluids Struct 23(6):904–919
37. Pramila A (1986) Sheet flutter and the interaction between sheet and air. TAPPI J 69(7):70–74
38. Frondelius T, Koivurova H, Pramila A (2006) Interaction of an axially moving band and surrounding fluid by boundary layer theory. J Fluids Struct 22(8):1047–1056
39. Païdoussis MP (2008) The canonical problem of the fluid-conveying pipe and radiation of the knowledge gained to other dynamics problems across applied mechanics. J Sound Vib 310:462–492
40. Pramila A (1987) Natural frequencies of a submerged axially moving band. J Sound Vib 113(1):198–203
41. Bolotin VV (1963) Nonconservative problems of the theory of elastic stability. Pergamon Press, New York
42. Tisseur F, Meerbergen K (2001) The quadratic eigenvalue problem. SIAM Rev. 43:235–286
43. Niemi J, Pramila A (1986) Vibration analysis of an axially moving membrane immersed into ideal fluid by FEM. Technical report, Tampereen teknillinen korkeakoulu (Tampere University of Technology), Tampere
44. Pramila A, Niemi J (1987) FEM-analysis of transverse vibrations of an axially moving membrane immersed in ideal fluid. Int J Numer Methods Eng 24(12):2301–2313. https://doi.org/10.1002/nme.1620241205. 1-09702-07
45. Kulachenko A, Gradin P, Koivurova H (2007a) Modelling the dynamical behaviour of a paper web. Part I. Comput Struct 85:131–147. https://doi.org/10.1016/j.compstruc.2006.09.006
46. Kulachenko A, Gradin P, Koivurova H (2007b) Modelling the dynamical behaviour of a paper web. Part II. Comput Struct 85:148–157. https://doi.org/10.1016/j.compstruc.2006.09.007
47. Gresho PM, Sani RL (1999) Incompressible flow and the finite element method: advection–diffusion and isothermal laminar flow. Wiley. Reprinted with corrections. ISBN 0 471 96789 0
48. Oñate E, Idelsohn SR, del Pin F, Aubry R (2004) The particle finite element method: an overview. Int J Comput Methods 1(2):267–307. https://doi.org/10.1142/S0219876204000204
49. Ponthot J-P, Belytschko T (1998) Arbitrary Lagrangian-Eulerian formulation for element-free Galerkin method. Comput Methods Appl Mech Eng 152(1–2):19–46. https://doi.org/10.1016/S0045-7825(97)00180-1
50. Idelsohn SR, Oñate E, del Pin F (2003) A lagrangian meshless finite element method applied to fluid-structure interaction problems. Comput Struct 81(8–11):655–671. https://doi.org/10.1016/S0045-7949(02)00477-7
51. Landau LD, Lifshitz EM (1959) Fluid mechanics. English 2nd ed. published by Butterworth–Heinemann, 1987, Oxford, 2nd edn
52. Tokaty GA, History A (1994) Philosophy of fluid mechanics. Dover (1994) Republication with corrections. Original by G. T. Foulis & Co., Ltd, p 1971
53. Hoffman Johan, Johnson Claes (2010) Resolution of d'Alembert's paradox. J Math Fluid Mech 12(3):321–334. https://doi.org/10.1007/s00021-008-0290-1
54. Hoffman Johan, Johnson Claes (2008) Blow up of incompressible Euler solutions. BIT Numer Math 48(2):285–307. https://doi.org/10.1007/s10543-008-0184-x
55. McKee S, Tomé MF, Ferreira VG, Cuminato JA, Castelo A, Sousa FS, Mangiavacchi N (2008) The MAC method. Comput Fluids 37:907–930. https://doi.org/10.1016/j.compfluid.2007.10.006
56. Cline D, Cardon D, Egbert PK (2013) Fluid flow for the rest of us: tutorial of the marker and cell method in computer graphics. Technical report, Brigham Young University
57. Gueyffier D, Li J, Nadim A, Scardovelli S, Zaleski S (1999) Volume of fluid interface tracking with smoothed surface stress methods for three-dimensional flows. J Comput Phys 152:423–456. https://doi.org/10.1006/jcph.1998.6168
58. Harlow FH, Welch JE (1965) Numerical calculation of time-dependent viscous incompressible flow of fluid with free surface. Phys Fluids 8(12):2182–2189. https://doi.org/10.1063/1.1761178

59. Noh WF, Woodward P (1976) Proceedings of the fifth international conference on numerical methods in fluid dynamics June 28–July 2. Twente University, Enschede chapter SLIC (Simple Line Interface Calculation). Springer, , pp 330–340. ISBN 978-3-540-37548-7. https://doi.org/10.1007/3-540-08004-X_336
60. Hirt CW, Nichols BD (1981) Volume of fluid (VOF) method for the dynamics of free boundaries. J Comput Phys 39(1):201–225. https://doi.org/10.1016/0021-9991(81)90145-5
61. Bisplinghoff RL, Ashley H (1975) Principles of aeroelasticity. Dover Publications, Inc., New York, 1962. 2nd edn
62. Turek S, Hron J (2006) Proposal for numerical benchmarking of fluid-structure interaction between an elastic object and laminar incompressible flow. In: Bungartz H-J, Schäfer M (eds) Fluid–structure interaction—modelling, simulation, optimisation, vol 53 of Lecture Notes in Computational Science and Engineering, pp 371–385. Springer, Berlin, Heidelberg. https://doi.org/10.1007/3-540-34596-5_15. ISBN 978-3-540-34595-4
63. Lighthill MJ (1960) Note on the swimming of slender fish. J Fluid Mech 9:305–317
64. Païdoussis MP (2005) Some unresolved issues in fluid-structure interactions. J Fluids Struct 20(6):871–890
65. Holzapfel GA (2000) Nonlinear solid mechanics—a continuum approach for engineering. Wiley
66. Marsden JE, Hughes TJR (1983) Mathematical foundations of elasticity. Dover
67. Strang WG, Fix GJ (1973) An analysis of the finite element method. Wellesley Cambridge Press. ISBN 978-0961408886
68. Ciarlet PG (1978) The finite element method for elliptic problems. Studies in Mathematics and its Applications. North-Holland, Amsterdam
69. Johnson C (1987) Numerical solution of partial differential equations by the finite element method. Cambridge University Press. Reprint by Dover, 2009
70. Krizek M, Neittaanmäki P (1990) Approximation finite element, of variational problems and applications. Longman Scientific & Technical, Harlow. Copubl. Wiley, New York
71. Eriksson K, Estep D, Hansbo P, Johnson C (1996) Computational differential equations. Studentlitteratur, Lund. ISBN 91-44-49311-8
72. Klaus-Jürgen B (1996) Finite element procedures. Prentice Hall. ISBN 0-13-301458-4
73. Hughes TJR (2000) The finite element method. Linear Static and Dynamic Finite Element Analysis. Dover Publications Inc, Mineola, N.Y., USA. ISBN 0-486-41181-8
74. Belytschko T, Liu WK, Moran B (2000) Nonlinear finite elements for continua and structures. Wiley. ISBN 978-0-471-98774-1
75. Jacob F, Ted B (2007) A first course in finite elements. Wiley. ISBN 978-0-470-03580-1
76. Brenner SC, Scott LR (2010) The mathematical theory of finite element methods, vol 15 of Texts in Applied Mathematics. Springer, 3rd edn
77. Zienkiewicz OC, Taylor RL, Zhu JZ (2013a) The finite element method: its basis and fundamentals, vol 1. Butterworth–Heinemann, 7th edn
78. Zienkiewicz OC, Taylor RL, Fox DD (2013b) The finite element method for solid and structural mechanics, vol 2. Butterworth–Heinemann, 7th edn, 2013a
79. Zienkiewicz OC, Taylor RL, Nithiarasu P (2013c) The finite element method for fluid dynamics, vol 3. Butterworth–Heinemann, 7th edn
80. Jean D, Antonio H (2003) Finite element methods for flow problems. Wiley. ISBN 0-471-49666-9

Chapter 9
Optimization of Elastic Bodies Subjected to Thermal Loads

In this final chapter, we consider three thermoelastic optimization problems. Many of the arguments are developed analytically. As the first problem, we will look at the optimal thickness distribution for a beam of variable thickness, when the goal is to maximize its resistance to thermoelastic buckling, or in other words, to maximize the critical temperature at which buckling occurs. We consider the case where the ends of the beam are fixed in the axial direction, so that thermal expansion leads to a compressive reaction force in the axial direction.

In the second problem we still aim to maximize the critical temperature. The thickness of the beam is now taken to be uniform, but we allow the beam to be constructed inhomogeneously, picking from a discrete set of available materials. As the result, we will find some optimal distributions of materials that maximize the critical temperature.

The third and final problem concerns heat conduction in locally orthotropic solid bodies. By locally orthotropic, we mean a particular type of inhomogeneity, where the principal directions (axes of orthotropy) may vary as a function of the space coordinates. Using an approach based on optimization, a tensor formulation, and the calculus of variations, we derive a guaranteed double-sided estimate for energy dissipation that occurs in heat conduction in a locally orthotropic body, without assuming anything about the material orientation field. Thus we obtain guaranteed lower and upper bounds for energy dissipation that always hold regardless of how the local material orientation is distributed in the solid body. Connecting to the main theme of this book, this has applications in the analysis of heat conduction in paper materials, which is important when considering the drying process in papermaking.

9.1 Optimal Distribution of Thickness in a Thermoelastic Beam

We shall consider the static stability analysis of a heated beam. We assume that a beam having a variable thickness lies along the x axis and is simply supported at the endpoints $x = 0$ and $x = \ell$. Initially in the unheated state its length is equal to ℓ.

N. Banichuk et al., *Stability of Axially Moving Materials*, Solid Mechanics and Its Applications 259, https://doi.org/10.1007/978-3-030-23803-2_9

Before heating, there is no axial stress. Let θ denote the increase in temperature caused by heating the beam. Since the positions of the ends are fixed in the x direction, a compressive reaction force p will be generated at the supports, caused by the thermal expansion of the beam. The magnitude of p is related to the increase in temperature by

$$p = \beta\theta E\ell\left(\int_0^\ell \frac{\mathrm{d}x}{S}\right)^{-1}, \tag{9.1.1}$$

where β and E are, respectively, the coefficient of linear thermal expansion and Young's modulus, while $S = S(x)$ is a function describing the distribution of the cross-sectional area. For simplicity, and to keep our focus on the fundamentals, we will consider the material parameters such as Young's modulus as given constants, although for real materials parameters may vary as a function of temperature.

With an increase in temperature θ, the magnitude of the compressive reaction force p rises, and for some critical value of temperature θ, the compressive force reaches a value that triggers a loss of mechanical stability, leading the beam to buckle. According to Euler's theory, the magnitude of the critical load p is determined as the smallest eigenvalue for the homogeneous boundary-value problem

$$EI\frac{\mathrm{d}^2 w}{\mathrm{d}x^2} + pw = 0, \qquad w(0) = w(\ell) = 0\,, \tag{9.1.2}$$

where w is the bending moment, which for a linear-elastic beam is the second derivative of the transverse displacement.

Here, as in the preceding chapters, we consider only the design of beams for which the moment of inertia of the cross-sectional area I is a function of the form

$$EI = A_\alpha S^\alpha, \qquad \alpha = 1, 2, 3\,. \tag{9.1.3}$$

For a fixed type of cross-section and for a given function $S = S(x)$, we can solve the boundary-value problem (9.1.2) to find the value of the critical temperature θ that triggers the loss of mechanical stability.

But what if, instead, we would like to determine the optimal $S(x)$ for some given criterion? The magnitude of the reaction force p and the beam's (flexural) rigidity EI depend on the distribution of cross-sectional areas (thickness) $S(x)$, see Eqs. (9.1.1) and (9.1.3). A variation of $S(x)$ produces changes in these characteristic properties and changes the value of the critical temperature. Thus, the problem of finding the best distribution of thickness that maximizes the critical temperature is of some interest. Let us introduce a rigorous formulation of this optimization problem, following the study [1].

We need to find a function describing the distribution of cross-sectional area that satisfies the bounds

$$S_{\min} \leq S(x) \leq S_{\max} \tag{9.1.4}$$

with the volume constraint

$$\int_0^\ell S(x)\mathrm{d}x = V \;, \tag{9.1.5}$$

that maximizes the critical temperature

$$\theta_* = \max_S \theta \;, \tag{9.1.6}$$

where θ is computed from (9.1.1)–(9.1.2). $S_{\min}$, $S_{\max}$ and V are positive given constants satisfying

$$S_{\min}\ell \leq V \leq S_{\max}\ell \;.$$

In the formulation of our problem (9.1.1)–(9.1.6), the function $S(x)$ describing the distribution of the cross-sectional area plays the role of the control function, while Eq. (9.1.2) is a differential relation. We shall offer some remarks concerning the formulation of problem (9.1.1)–(9.1.6).

When considering the stability of the straight-line equilibrium configuration of a heated beam, we assume that the position of each end of the beam is fixed on the x axis. As a loss of stability occurs, the length of the beam increases, and the reaction loads at the ends decrease. The magnitude of the critical load, and of the actual compressive force that produces an equilibrium when the shape of the beam is curved but close to a straight line, may be entirely different. Therefore, this problem differs from the classical problem of buckling of an elastic beam compressed by so-called dead loads, and some additional analysis needs to be carried out with regard to the applicability of Euler's classical theory. This problem has been examined in Larichev [11], where it was shown that the difference in the beam's displacement is a quantity of higher order smallness (relative to the magnitude of a characteristic displacement), and can be ignored in computations of stability.

If in this optimization problem we do not assign a lower bound to the beam's thickness (i.e., if we substitute in the Eq. (9.1.4) that $S_{\min} = 0$), then it is possible to show that there exists a distribution of thickness for which $p = 0$ for an arbitrary value of θ and for which the beam does not buckle. Indeed, let the shape of the beam finish with a sharp spike at $x = 0$ and the function $S(x)$ obey the equation $S(x) = cx^m$, with $m \geq 1$, in the neighborhood $0 \leq x \leq \varepsilon < \ell$ of the endpoint $x = 0$. Then for an arbitrary choice of cross-sectional areas on the segment $\varepsilon \leq x \leq \ell$, which asymptotically approaches the function assigned to $0 \leq x \leq \varepsilon$, the magnitude of the reaction p (computed from (9.1.1)) approaches zero.

To find a solution that satisfies the necessary optimality condition, it is convenient to transform this problem and to eliminate the differential relation (9.1.2). To do this we note that finding the critical temperature and the corresponding functions $w(x)$ may be reduced to an equivalent variational problem of finding a minimum for the nonadditive functional

$$\theta = \min_{w} \frac{A}{\beta \ell E} \left(\int_0^\ell \frac{\mathrm{d}x}{S} \right) \left(\int_0^\ell \left(\frac{\mathrm{d}w}{\mathrm{d}x} \right)^2 \mathrm{d}x \right) \left(\int_0^\ell \frac{w^2}{S^\alpha} \mathrm{d}x \right)^{-1} , \tag{9.1.7}$$

where the minimum in (9.1.7) is sought in the class of functions satisfying the given boundary conditions. Using (9.1.7), we reduce the original problem (9.1.1)–(9.1.6) to a minimax optimization problem containing no differential relations:

$$\theta_* = \max_{S} \min_{w} \frac{A}{\beta \ell E} \left(\int_0^\ell \frac{\mathrm{d}x}{S} \right) \left(\int_0^\ell \left(\frac{\mathrm{d}w}{\mathrm{d}x} \right)^2 \mathrm{d}x \right) \left(\int_0^\ell \frac{w^2}{S^\alpha} \mathrm{d}x \right)^{-1} . \tag{9.1.8}$$

The maximum with respect to S is taken in the class of functions $S = S(x)$ that satisfy the conditions (9.1.4) and (9.1.5). The two-sided inequality (9.1.4) assigned to the design variable $S(x)$ can be automatically satisfied if we introduce an auxiliary variable φ related to S by

$$S = \mu_1 + \mu_2 \sin \varphi, \qquad \mu_1 = \frac{1}{2} \left(S_{\min} + S_{\max} \right), \qquad \mu_2 = \frac{1}{2} \left(S_{\max} - S_{\min} \right) .$$

The only constraint on the values of φ comes from inserting this representation into the volume constraint (9.1.5) and rearranging:

$$\int_0^\ell \sin \varphi \, \mathrm{d}x = \frac{V - \mu_1 \ell}{\mu_2} .$$

We can now write the formula for the first variation of the optimized functional derived from the variation of the auxiliary control function

$$\delta \theta = \int_0^\ell \Lambda \cos(\varphi) \, \delta \varphi \, \mathrm{d}x , \tag{9.1.9}$$

$$\Lambda = \frac{B_1 w^2}{(\mu_1 + \mu_2 \sin \varphi)^{(\alpha+1)}} - \frac{B_2}{(\mu_1 + \mu_2 \sin \varphi)^2} ,$$

$$B_1 = \alpha \mu_2 \left(\int_0^\ell \frac{\mathrm{d}x}{(\mu_1 + \mu_2 \sin \varphi)} \right) \left(\int_0^\ell \left(\frac{\mathrm{d}w}{\mathrm{d}x} \right)^2 \mathrm{d}x \right) \left(\int_0^\ell \frac{w^2}{(\mu_1 + \mu_2 \sin \varphi)^\alpha} \mathrm{d}x \right)^{-2}$$

$$B_2 = \mu_2 \left(\int_0^\ell \left(\frac{\mathrm{d}w}{\mathrm{d}x} \right)^2 \mathrm{d}x \right) \left(\int_0^\ell \frac{w^2}{(\mu_1 + \mu_2 \sin \varphi)^\alpha} \mathrm{d}x \right)^{-1} .$$

The isoperimetric condition assigned to all possible variations $\delta\varphi$ is given by the constraint

$$\int_0^\ell \cos(\varphi)\,\delta\varphi\,\mathrm{d}x = 0\,. \tag{9.1.10}$$

Substituting (9.1.9) and (9.1.10), the necessary optimality condition $\delta\theta = 0$ becomes

$$(\Lambda - \lambda)\cos\varphi = 0\,, \tag{9.1.11}$$

where λ is a Lagrange multiplier for the constraint (9.1.10).

For the sake of simplicity, we shall consider the case when the upper bound is inactive, and therefore, the thickness function $S(x)$ and the value of the critical reaction p are functions only of the lower bound parameter $S_{\min}$.

The optimal distribution of thickness will contain intervals on which $S(x) = S_{\min}$, and the region in which $S(x) > S_{\min}$ will change smoothly in accord with the optimality condition

$$w^2 = \gamma_1 S^{\alpha-1} + \gamma_2 S^{\alpha+1}\,, \qquad \text{where} \quad \gamma_1 = \frac{B_2}{B_1}, \quad \gamma_2 = \frac{\lambda}{B_1}\,. \tag{9.1.12}$$

For the sake of convenience, we shall use the following nondimensional variables in the subsequent discussion:

$$x' = \frac{x}{\ell}, \qquad S' = \frac{\ell S}{V}, \qquad \theta' = \frac{\beta \ell^{\alpha+1}\theta E}{A_\alpha V^{\alpha-1}}, \qquad p' = \frac{\ell^{\alpha+2} p}{A_\alpha V^\alpha}\,. \tag{9.1.13}$$

In these nondimensional variables (with primes omitted), the solution of the optimization problem will be derived on the interval $0 \le x \le 1$ for a beam with unit volume. For $\alpha = 2,\ 3$ an analytical solution was derived by Barsuk, assuming that the eigenvalues are simple. The case $\alpha = 1$ is exceptional and we shall construct a numerical solution for that case.

The optimal solution is sought in the class of symmetric distribution of thickness, that is, we assume that $S(x) = S(1-x)$. The corresponding deflection function shall be classified according to the odd or even property

$$y(x) = \pm y(1-x)\,,$$

with the plus sign corresponding to a symmetric deflection and the minus sign to an antisymmetric deflection. These relations allow us to restrict our analysis to the half-interval (0,1/2) in deriving the solution. By standard arguments, we find the following final result. The optimal distribution of the thickness is given by the quadratures

$$\int_{\varkappa}^{s} \frac{d\tau}{\Phi(\tau)} = \frac{2\sqrt{p_0}}{\sqrt{(\alpha-1)}}(x - x_*), \qquad \alpha = 2, 3 \tag{9.1.14}$$

$$\Phi(s) \equiv \frac{\sqrt{\gamma_0 + (\alpha+1)\,s}\sqrt{\gamma_0 + (\alpha-1)\,s^2}}{\left(\gamma_0 + (\alpha+1)\,s^2\right) s^{(\alpha-1)/2}}\sqrt{1-s}$$

$$\gamma_0 = \frac{\int_0^1 s(x)\mathrm{d}x}{\int_0^1 \frac{\mathrm{d}x}{s(x)}}, \qquad p = a^{\alpha} p_0, \qquad S(x) = a s(x), \qquad \varkappa = \frac{S_{\min}}{a}.$$

Here a stands for the maximum value attained by the function $S(x)$, in this context $a = S(1/2)$, and x_* is the coordinate of the point at which the graph of $S(x)$ attains its minimal value, that is, $S(x_*) = S_{\min}$.

The system of equations used to derive the values of the constants p_0, $\varkappa$ and x_* is

$$I_1(\alpha, \gamma_0, \varkappa) = \frac{2\sqrt{p_0}}{\sqrt{(\alpha-1)}}\left(\frac{1}{2} - x_*\right), \qquad \tan\left[\sqrt{\left(\frac{p_0}{\varkappa^{\alpha}}\right)}\right] x_* = \Psi(\varkappa) \tag{9.1.15}$$

$$S_{\min} I_2(\alpha, \gamma_0, \varkappa) = \frac{2\sqrt{p_0}}{\sqrt{(\alpha-1)}}\left(\frac{1}{2} - S_{\min} x_*\right)\varkappa .$$

Here we use the notation

$$I_1(\alpha, \gamma_0, \varkappa) = \int_{\varkappa}^{1} \frac{d\tau}{\Phi(\tau)} \tag{9.1.16}$$

$$I_2(\alpha, \gamma_0, \varkappa) = \int_{\varkappa}^{1} \frac{\tau d\tau}{\Phi(\tau)}$$

$$I_3(\alpha, \gamma_0, \varkappa) = \int_{\varkappa}^{1} \frac{d\tau}{\tau\Phi(\tau)}$$

$$\Psi_1(\varkappa) = \frac{\sqrt{\gamma_0 + (\alpha-1)\varkappa^2}}{\sqrt{(\alpha-1)(1-\varkappa)(\gamma_0 + (\alpha+1)\varkappa)}}.$$

The magnitude of γ_0 can be found by solving the equation

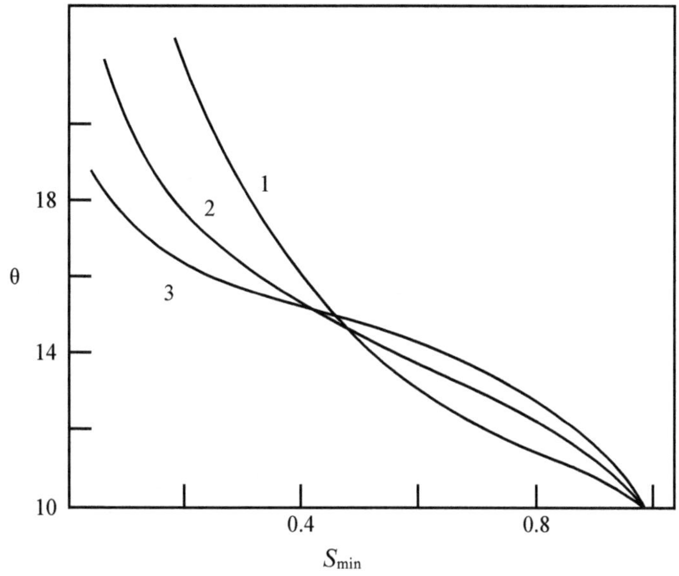

Fig. 9.1 Dependence of critical temperature on $S_{\min}$

$$\gamma_0 = \frac{\varkappa x_* + \dfrac{1}{2}\sqrt{\left(\dfrac{\alpha - 1}{p_0}\right)} I_2\,(\alpha, \gamma_0, \varkappa)}{\dfrac{x_*}{\varkappa} + \dfrac{1}{2}\sqrt{\dfrac{\alpha - 1}{p_0}} I_3(\alpha, \gamma_0, \varkappa)}\,. \tag{9.1.17}$$

The curves shown in Figs. 9.1 and 9.2 illustrate (9.1.14)–(9.1.17) for $\alpha = 2,\ 3$. For the case $\alpha = 1$, these relations were established numerically using an iterative technique.

Figure 9.1 illustrates the dependence of the critical temperature θ_* on the value of the parameter $S_{\min}$ for optimal beams. Curves 1, 2 and 3 correspond to the values of $\alpha = 1,\ 2,\ 3$, respectively. As $S_{\min}$ increases, the value of the critical temperature decreases in all cases. This may be explained by pointing out that as $S_{\min}$ increases, the amount of material that can be redistributed decreases and the available choices in the optimization are reduced.

For $S_{\min} = 1$, there is only one choice of design for the thickness function consistent with conditions (9.1.4) and (9.1.5), which is $S(x) \equiv 1$. Here all values of the optimized functional coincide.

The optimal thickness distribution functions for $\alpha = 1,\ 2,\ 3$ are shown in the Fig. 9.2. Curves 1, 2 and 3 correspond to the values $S_{\min} = 0.1,\ 0.3,\ 0.5$, respectively. Because of symmetry with respect to the point $x = 1/2$, the thickness functions $S(x)$ are drawn on the interval $1/2 < x \leq 1$ only. If all the derived solutions indicate that, for a simply supported optimal beam, most of its mass is concentrated in its middle part, and near the ends the thickness assumes the minimum permitted value. As $S_{\min}$

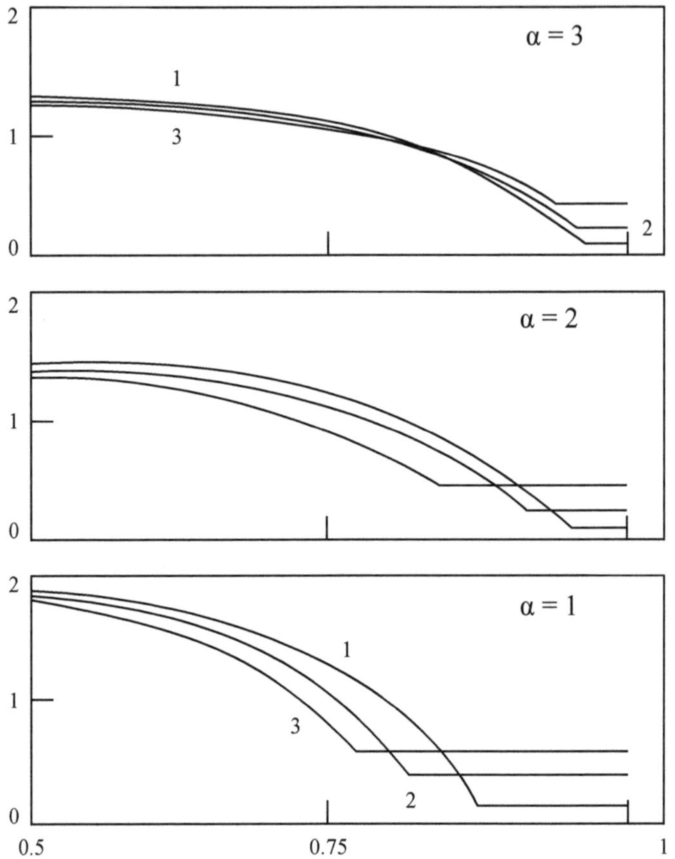

Fig. 9.2 Optimal distribution of thickness

is increased, the lengths of the segments of the optimal beam where the thickness is constant and equal to the minimal allowed value also increase.

9.2 Optimal Distribution of Materials in a Thermoelastic Beam

In this section, we considera beam composed from a discrete set of materials and subjected to compression, concentrating on the problem of maximizing the critical temperature of the loss of stability. A modified genetic algorithm is used for finding the global optimum. The computations and analysis were originally performed in the study Banichuk et al. [2].

Let an elastic beam lie along the x axis ($0 \le x \le \ell$) and be fixed at the ends $x = 0$ and $x = \ell$:

$$u(0) = u(\ell) = 0\,, \tag{9.2.1}$$

where $u(x)$ is the displacement distribution along the x axis. Let the beam consist of a discrete set of materials, distributed along the x axis and characterized by the set of parameters

$$\{E_i,\ \beta_i,\ \rho_i\}\,, \qquad i = 1, 2, \ldots, r\,, \tag{9.2.2}$$

where E_i is Young's modulus, β_i is the linear thermal expansion coefficient, ρ_i is the material density, i is the material index, and r is the number of materials. The distribution of the parameters $(E(x), \beta(x), \rho(x))$ along the beam is taken to be piecewise constant on the segment $0 \le x \le \ell$. For each point $x \in [0, \ell]$, these functions take some values from the given finite set, that is,

$$E(x) \in \{E_i\}\,, \quad \beta(x) \in \{\beta_i\}\,, \quad \rho(x) \in \{\rho_i\}\,, \tag{9.2.3}$$

where $i = 1, 2, \ldots, r$. In the following, we describe the parametrization using a piecewise constant selector function $t = t(x),\ \ x \in [0, \ell]$, taking on the values $t = t_i = i$, such that the following equalities are satisfied:

$$E\,(t(x))_{t=t_i=i} = E_i\,, \qquad \beta\,(t(x))_{t=t_i=i} = \beta_i\,, \qquad \rho\,(t(x))_{t=t_i=i} = \rho_i\,. \tag{9.2.4}$$

See Fig. 9.3.

Suppose that at the initial temperature T, the beam is in an undeformed (natural) state. The beam is heated up to the temperature $T + \Delta T$. The total strain ε is a sum of elastic ε_e and thermal ε_t strains, that is, $\varepsilon = \varepsilon_e + \varepsilon_t$, where $\varepsilon_t = \beta\ell\Delta T$. The generalized Hooke's law gives the following relation between stress σ and strain ε [12, 13, 16]:

$$\sigma = E\,(\varepsilon - \beta\Delta T)\ . \tag{9.2.5}$$

Dividing by E, rearranging, and taking into account that

$$\sigma = -\frac{P}{S}, \qquad \varepsilon = \frac{\mathrm{d}u}{\mathrm{d}x}\,, \tag{9.2.6}$$

we have

$$u(\ell) - u(0) = -P\int_0^{\ell} \frac{\mathrm{d}x}{SE} + \Delta T\int_0^{\ell} \beta \mathrm{d}x\,, \tag{9.2.7}$$

where S is the cross-sectional area of the beam and $P > 0$ is the compressive force (reaction force due to the clamped ends), acting on the inhomogeneous elastic beam. Using the clamping conditions (9.2.1) and the relation (9.2.7), we have the following

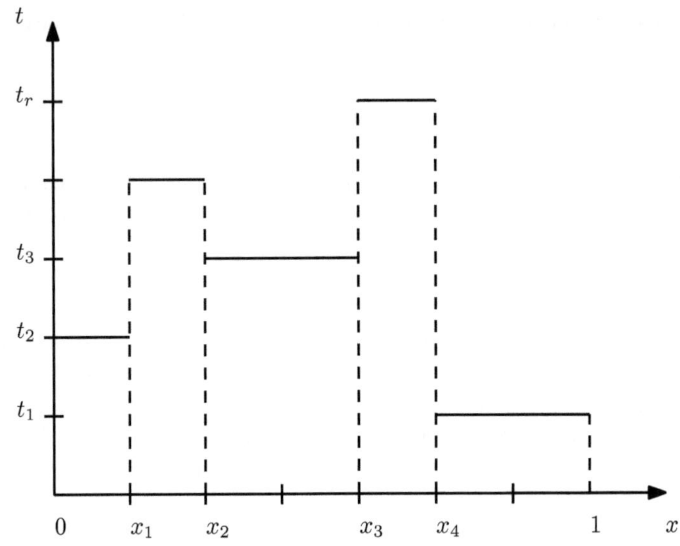

Fig. 9.3 Piecewise constant distribution of material properties along the beam

value for the compressive force:

$$P = \Delta T \frac{J_1}{J_2} \,, \tag{9.2.8}$$

where the functionals J_1 and J_2 are defined by the expressions

$$J_1 = \int_0^\ell \beta \mathrm{d}x, \qquad J_2 = \int_0^\ell \frac{\mathrm{d}x}{SE} \,. \tag{9.2.9}$$

Buckling of the compressed beam, if it occurs, is described by the following boundary value problem (see [15]):

$$EI \frac{\mathrm{d}^2 w}{\mathrm{d}x^2} + pw = 0, \qquad 0 \le x \le \ell \,, \tag{9.2.10}$$

$$w(0) = 0, \qquad w(\ell) = 0 \,. \tag{9.2.11}$$

Here $w = w(x)$ is the bending moment (the second derivative of the transverse displacement), I is the moment of inertia (second moment of area) of the beam cross-section, and p is the buckling load. Note that the problem, formulated in the form (9.2.10) and (9.2.11), corresponds to the ends being simply supported. By multiplying (9.2.10) by w, integrating by parts on the interval $[0, \ell]$, and applying the boundary conditions (9.2.11), we have the following expression for the critical buckling load

$$p = \frac{J_3}{J_4}\,, \tag{9.2.12}$$

where the functionals J_3 and J_4 are defined as

$$J_3 = \int_0^\ell \left(\frac{\mathrm{d}w}{\mathrm{d}x}\right)^2 \mathrm{d}x\,, \qquad J_4 = \int_0^\ell \frac{w^2}{EI}\mathrm{d}x\,. \tag{9.2.13}$$

Using Eqs. (9.2.8), (9.2.9), (9.2.12) and (9.2.13) for the buckling load p and the reaction force P arising from the heating process, and requiring $p = P$, we obtain the critical heating temperature

$$\Delta T = P\frac{J_2}{J_1} = \frac{J_2 J_3}{J_1 J_4}\,. \tag{9.2.14}$$

To find the buckling load P and the corresponding buckling mode $w(x)$ (integrating the result twice then gives us the transverse displacement), we can use the well-known variational principle (see, e.g., [8, 9, 17])

$$P = \min_w \frac{J_3}{J_4}\,. \tag{9.2.15}$$

The minimum in (9.2.15) is taken over the set of admissible functions satisfying the boundary conditions (9.2.11).

Suppose that the temperature increment ΔT is a given value and the beam has a cylindrical shape, that is, $S(x) = S_0$, where S_0 is a given value. Consider the problem of optimization of stability of the heated inhomogeneous beam from the discrete set of materials presented in Table 9.1.

It is required to find the design variable $t = t(x)$, determining the structure of an inhomogeneous beam from the condition of maximization of the critical temperature of the loss of stability $\Delta T = \tilde{J}$:

$$\tilde{J}_* = (\Delta T)_* = \max_t \left\{ \frac{J_2\,(E\,(t))\,J_3(w)}{J_1\,(\beta\,(t))\,J_4\,(E\,(t)\,,\,w)} \right\}\,, \tag{9.2.16}$$

under the design constraint

$$t(x) = \{t_i \mid \; x_{i-1} < x < x_i, \quad i = 1, 2, \ldots, n, \quad x_0 = 0,\; x_n = \ell\} \tag{9.2.17}$$

and the mass constraint

Table 9.1 Material parameters

No. of material	Material	Young's modulus $E \times 10^{-5}$ Mpa	Coefficient of linear expansion $\beta \times 10^6$, grad^{-1}	Specific weight $\rho g \times 10^{-4}$, N/m^3	$\mu = E\beta$
1	Molybdenum	3.3	5.6	10.2	18.48
2	Steel 1X17H2	2.0	10.3	7.75	20.6
3	Brass	1.1	19.0	8.6	20.9
4	Steel 30XTCA	1.98	11.0	7.85	21.78
5	Steel 2X13	2.2	10.1	7.75	22.22
6	Steel 20X	2.07	11.3	7.74	23.391

$$M(t) = S_0 \int_0^{\ell} \rho(t(x))\mathrm{d}x \leq M_0 \,, \tag{9.2.18}$$

where M_0 is a given upper limit for the mass of the beam, and $S = S_0$ is given a cross-sectional area.

To find the mode (shape) in which the loss of stability occurs, we use Rayleigh's principle. We have

$$\tilde{J}_* = \max_t \left\{ \frac{J_2\left(E\left(t\right)\right)}{J_1\left(\beta\left(t\right)\right)} \left[\min_w \frac{J_3(w)}{J_4\left(E\left(t\right), w\right)} \right] \right\} \,. \tag{9.2.19}$$

To solve the problem of optimization of the functional $\tilde{J}$, taking into account (9.2.18), we apply the penalty method and transform the original problem (9.2.17)–(9.2.19) to a maximization problem with respect to t of the augmented functional J^a:

$$J^a = \left\{ \frac{J_2\left(E\left(t\right)\right)}{J_1\left(\beta\left(t\right)\right)} \left[\min_w \frac{J_3(w)}{J_4\left(E\left(t\right), w\right)} \right] \right\} - \lambda\left(M - M_0\right) \,, \tag{9.2.20}$$

$$\lambda = \begin{cases} 0, & \text{if } M - M_0 \leq 0 \\ \lambda_0 & \text{if } M - M_0 > 0 \end{cases} \,, \tag{9.2.21}$$

under the constraint (9.2.17). Here λ_0 is a positive penalty multiplier.

The solution of the problem of the functional J^a, Eqs. (9.2.20) and (9.2.21), for various values of the problem parameter M_0 and material characteristics presented in Table 9.1 is performed with the help of a genetic algorithm (see [5–7]), and the algorithm of successive optimization (see [3, 4]). It is supposed that $\ell = 1$ and the interval [0, 1] of the variable x is divided by the points x_i, $i = 1, 2, \ldots, n$, into $n - 1$

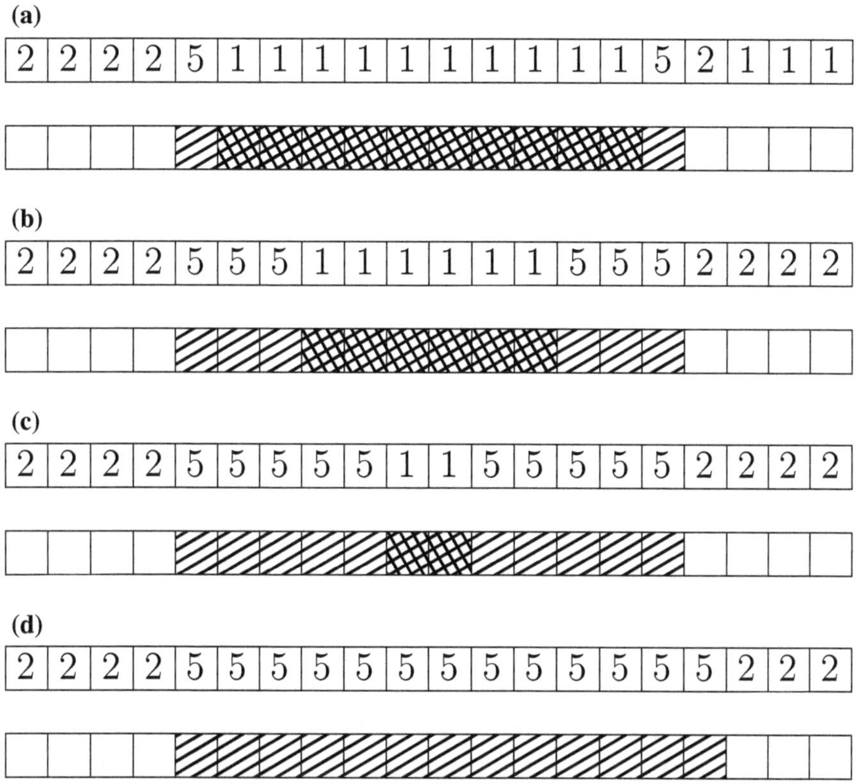

Fig. 9.4 Computed optimal distributions of the materials

subintervals, where the values β and E take on constant values corresponding to the chosen materials.

The index of the material takes values from 1 to 6. Populations under consideration consist of N individuals, each representing an admissible piecewise homogeneous beam. The number N is taken to be even, and is kept constant in the evolution process. Each j is an individual of the population, described by the set of values $t(j, i)$, representing the design variable at each node i. The optimal individual, that is, the set $t(j, i)$ maximizing the augmented functional, is sought by using the genetic algorithm.

In the first iteration, we minimize the ratio J_3/J_4 with respect to w, and find the critical force of the loss of stability for some given distribution of material $t = t(x)$, $0 \leq x \leq \ell$. This internal operation in the Eq. (9.2.20) is realized with the help of a genetic algorithm. Then, for the obtained distribution $w(i, j)$, the operation of optimization of the augmented functional is realized with the help of a genetic algorithm, and as a result, we obtain the optimal distribution $t(i, j)$ in the first approximation. Then the algorithm passes to the next iteration, which is performed analogously.

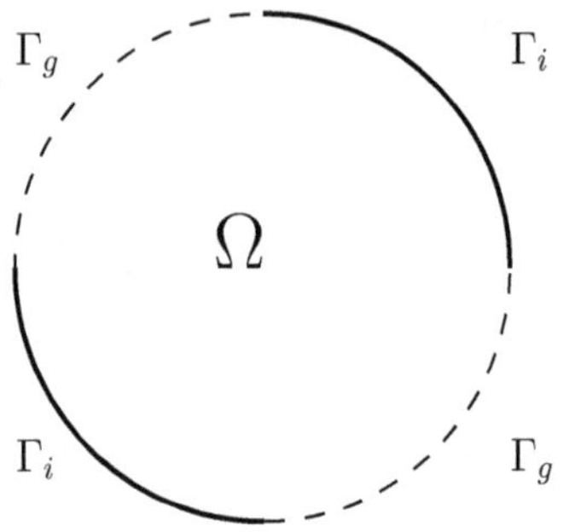

Fig. 9.5 Domain Ω with constant-temperature Γ_i (solid line) and thermally insulated Γ_g (dashed line) boundaries

Variants of material distributions a, b, c and d correspond to the values $W_0 = gM_0$ equal to 9.0, 8.5, 8.0, and 7.75. The subintervals, occupied by molybdenum ($i = 1$) are shown by crossed lines. The white subintervals correspond to the steel 1X17H2 ($i = 2$). The rising lines correspond to the steel 2X13 ($i = 5$). It is seen from Fig. 9.4 that the distribution of material is symmetric with respect to the middle of the beam. The figure also shows that placing the molybdenum at the middle part of the beam is the most effective, when the goal is to maximize the critical temperature of the loss of stability.

9.3 A Guaranteed Double-Sided Estimate for Energy Dissipation in Heat Conduction of Locally Orthotropic Solid Bodies

Let us consider the heat conduction problem for a solid body occupying the domain Ω, see Fig. 9.5, with the boundary $\Gamma = \Gamma_g + \Gamma_i$, where $\Gamma_g \cap \Gamma_i = 0$. Let the material of the body be anisotropic with respect to the heat conduction process described by the relations [10, 13]

$$\mathbf{q} = D \cdot \nabla \varphi, \qquad \varphi = \theta^{-1}, \tag{9.3.1}$$

where θ is the temperature, $\mathbf{q}$ is the heat flux (a vector field), and D is a rank-2 tensor describing the heat conduction properties of the material. If there are no heat sources in the domain Ω, the boundary conditions and the governing equation have the following form:

$$(\varphi)_{\Gamma_g} = \varphi^0, \quad (\mathbf{n} \cdot D \cdot \nabla \varphi)_{\Gamma_i} = 0, \tag{9.3.2}$$

$$\nabla \cdot (D \cdot \nabla \varphi) = 0, \quad x \in \Omega, \tag{9.3.3}$$

where φ^0 is a given function specified on Γ_g, $\mathbf{n}$ is the outwards unit normal vector of the boundary Γ_i, $(\cdot)$ denotes the scalar product, and $\nabla()$ and $\nabla \cdot ()$ denote the gradient and the divergence, respectively.

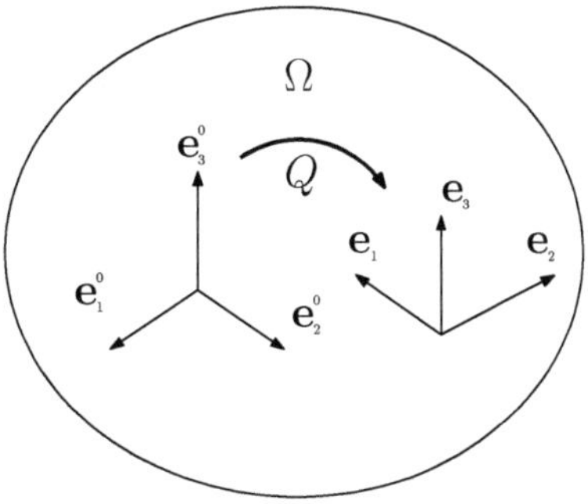

Fig. 9.6 Transformation of global unit vectors to the local material principal vectors by rotation tensor Q

Let the material be locally orthotropic, with an arbitrary orientation for the axes of orthotropicity. Let us fix the unit vectors $\mathbf{e}_1^0, \mathbf{e}_2^0, \mathbf{e}_3^0$ of a global orthogonal coordinate system x_1, x_2, x_3, see Fig. 9.6. The principal direction unit vectors $\mathbf{e}_1, \mathbf{e}_2, \mathbf{e}_3$ of the heat conduction tensor D of the orthotropic material, i.e. the axes of local symmetry, at an arbitrary point $(x_1, x_2, x_3) \in \Omega$ are related to the global coordinate vectors $\mathbf{e}_1^0, \mathbf{e}_2^0, \mathbf{e}_3^0$ via a rotation tensor $Q = Q(x)$ as

$$\mathbf{e}_i = Q * \mathbf{e}_i^0 = Q \cdot \mathbf{e}_i^0 \cdot Q^{\mathrm{T}}, \quad i = 1, 2, 3 , \tag{9.3.4}$$

$$Q^{\mathrm{T}} \cdot Q = Q \cdot Q^{\mathrm{T}} = E , \tag{9.3.5}$$

where the symbol T denotes transposition of a rank-2 tensor, and $E = \{\delta_{ij}\}$ is the rank-2 unit tensor (where δ_{ij} is a the Kronecker delta, $i, j = 1, 2, 3$), and $(*)$ is a tensor rotation. Equation (9.3.5) is an admissibility condition, which must be satisfied by the rank-2 tensor Q in order for it to describe a rotation.

Because a tensor as a whole is independent of the coordinate system it is described in, for the heat conduction tensor D we have

$$D = D_{ij}\, \mathbf{e}_i \otimes \mathbf{e}_j = Q \cdot D_{ij}^0 \mathbf{e}_i^0 \otimes \mathbf{e}_j^0 \cdot Q^{\mathrm{T}} = Q * \left(D_{ij}^0\, \mathbf{e}_i^0 \otimes \mathbf{e}_j^0\right) = Q * D^0 , \tag{9.3.6}$$

where $\otimes$ is the tensor (outer) product,

$$D^0 = D_{ij}^0 \cdot \mathbf{e}_i^0 \otimes \mathbf{e}_j^0 , \tag{9.3.7}$$

and D_{ij} (respectively D_{ij}^0) are the components of D when expressed in the material (respectively global) coordinate system. Do not be misled by the visual similarity of Eq. (9.3.6) to index notation; $\mathbf{e}_i \otimes \mathbf{e}_j$ is a rank-2 tensor. This representation is akin to writing a vector as $a_i\, \mathbf{e}_i$, or $a\,\mathbf{i} + b\,\mathbf{j} + c\,\mathbf{k}$.

We can rewrite D in the form

$$D = D_{ij}^0 Q \cdot \mathbf{e}_i^0 \otimes Q \cdot \mathbf{e}_j^0 = Q \cdot \left(D_{ij}^0 \mathbf{e}_i^0 \otimes \mathbf{e}_j^0\right) \cdot Q^{\mathrm{T}} = Q \cdot D^0 \cdot Q^{\mathrm{T}} , \tag{9.3.8}$$

where the second form is justified by manipulating the third. Both ($\otimes$) and ($\cdot$) are associative, so we may drop the parentheses; we have $\mathbf{e}_j^0 \cdot Q^{\mathrm{T}} = Q \cdot \mathbf{e}_j^0$ and D_{ij}^0 is a scalar.

If $\varkappa_i^0$ and $\mathbf{e}_i^0$ are the eigenvalues and eigenvectors of the tensor D^0,

$$D^0 \cdot \mathbf{e}_i^0 = \varkappa_i^0 \mathbf{e}_i^0 \,, \tag{9.3.9}$$

then $\varkappa_i^0$ and $\mathbf{e}_i = Q * \mathbf{e}_i^0$ are the eigenvalues and eigenvectors of the tensor $D = Q * D^0$:

$$D \cdot \mathbf{e}_i = \varkappa_i^0 \mathbf{e}_i \,. \tag{9.3.10}$$

To show this, we can use the Eqs. (9.3.4), (9.3.6) and (9.3.9):

$$D \cdot \mathbf{e}_i = \left(Q * D^0\right) \cdot \left(Q * \mathbf{e}_i^0\right) = Q * \left(D^0 \cdot \mathbf{e}_i^0\right) = \varkappa_i^0 Q * \mathbf{e}_i^0 = \varkappa_i^0 \mathbf{e}_i \,. \tag{9.3.11}$$

For a given tensor D^0 describing the heat conduction properties of the material, the value of the energy dissipation functional J,

$$J(Q, \varphi) = \int_\Omega \nabla\varphi \cdot D \cdot \nabla\varphi \, \mathrm{d}\Omega \,,$$

depends on the realization of $Q = Q(x)$, that is, the local orientation of the material axes at each point of the domain.

When $Q(x)$ is known, the energy dissipation is minimized when

$$J\left(Q, \varphi_*\right) = \min_{\varphi} J\left(Q, \varphi\right) \,, \tag{9.3.12}$$

where the minimum is taken over all functions φ satisfying the boundary conditions (9.3.2). This is the variational statement of the problem (9.3.2)–(9.3.3), Eq. (9.3.3) being its Euler equation, so the φ that minimizes dissipation will automatically be the solution of the original problem.

If there is no data concerning material orientation, that is, the tensor-valued function $Q = Q(x)$ (for $x \in \Omega$) characterizing the material distribution is unknown, then it is important to obtain lower and upper bounds for J, that is, to find guaranteed double-sided estimates $J_{\min}$ and $J_{\max}$ such that

$$J_{\min} \le J\left(Q, \varphi_*\right) \le J_{\max} \tag{9.3.13}$$

for any realization of Q satisfying the admissibility condition (9.3.5).

We apply an approach based on the solution of two optimization problems. The first problem is to find the lower bound

$$J_{\min} = \min_{Q} J\left(Q, \varphi_*\right) = \min_{Q} \min_{\varphi} J\left(Q, \varphi\right) \tag{9.3.14}$$

and the second the upper bound

$$J_{\max} = \max_{Q} J(Q, \varphi_*) = \max_{Q} \min_{\varphi} J(Q, \varphi) \ , \tag{9.3.15}$$

where min and max with respect to Q in the Eqs. (9.3.14) and (9.3.15) are taken over all Q satisfying the admissibility constraint (9.3.5). Operation min with respect to φ in the Eqs. (9.3.14) and (9.3.15) is performed taking into account the boundary conditions (9.3.2), so that the φ that is found by the minimization will be the solution of (9.3.2)–(9.3.3).

Let us now consider how to search for the extremum of J with respect to Q,

$$J \to \operatorname*{extr}_{Q} \ , \tag{9.3.16}$$

and analyze the extremum conditions.

To derive the extremum condition defining the orthogonal rotation tensor $Q = Q(x)$, characterizing the orientations of the local axes of orthotropy that lead to an extremum value of J, let us use the method of Lagrange multipliers and construct the augmented functional

$$J^L = J + J_P \tag{9.3.17}$$

$$J_P = \int_{\Omega} P \cdot\cdot \left(Q^{\mathrm{T}} \cdot Q - E\right) \mathrm{d}\Omega \tag{9.3.18}$$

$$J = \int_{\Omega} \nabla\varphi \cdot \left(Q * D^0\right) \cdot \nabla\varphi \mathrm{d}\Omega = \int_{\Omega} \nabla\varphi \cdot \left(Q \cdot D^0 \cdot Q^{\mathrm{T}}\right) \cdot \nabla\varphi \mathrm{d}\Omega \tag{9.3.19}$$

where $(\cdot\cdot)$ between tensors means the double scalar product, and the symmetric rank-2 tensor $P = P(x)$, $x \in \Omega$, is a Lagrange multiplier, defined for all of Ω and corresponding to the condition of orthogonality, Eq. (9.3.5). The dissipation energy functional J can also be written as

$$J = \int_{\Omega} B \cdot\cdot \left(Q \cdot D^0 \cdot Q^{\mathrm{T}}\right) \mathrm{d}\Omega \ , \tag{9.3.20}$$

where B is a symmetric rank-2 tensor defined by

$$B = \nabla\varphi \otimes \nabla\varphi, \quad B^{\mathrm{T}} = B \ . \tag{9.3.21}$$

Let us derive the expressions for the first variations δJ and δJ_P with respect to a variation δQ of the rotation tensor Q. We have

$$\delta J = \int_{\Omega} B \cdot\cdot \left(\delta Q \cdot D^0 \cdot Q^{\mathrm{T}} + Q \cdot D^0 \cdot \delta Q^{\mathrm{T}}\right) \mathrm{d}\Omega = 2 \int_{\Omega} \delta Q \cdot\cdot \left(D^0 \cdot Q^{\mathrm{T}} \cdot B\right) \mathrm{d}\Omega \tag{9.3.22}$$

and

$$\delta J_P = \int_{\Omega} P \cdot\cdot \left(\delta Q^{\mathrm{T}} \cdot Q + Q^{\mathrm{T}} \cdot \delta Q\right) \mathrm{d}\Omega = 2 \int_{\Omega} \delta Q \cdot\cdot \left(P \cdot Q^{\mathrm{T}}\right) \mathrm{d}\Omega \,. \tag{9.3.23}$$

Via (9.3.17)–(9.3.19), (9.3.22) and (9.3.23), we find the expression for the total variation δJ^L in terms of δQ as

$$\delta J^L = \delta J + \delta J_P = 2 \int_{\Omega} \delta Q \cdot\cdot \left(D^0 \cdot Q^{\mathrm{T}} \cdot B + P \cdot Q^{\mathrm{T}}\right) \mathrm{d}\Omega \,. \tag{9.3.24}$$

Using the extremum condition

$$\delta J^L = 0 \tag{9.3.25}$$

and the arbitrariness of δQ, we have

$$D^0 \cdot Q^{\mathrm{T}} \cdot B + P \cdot Q^{\mathrm{T}} = 0, \qquad x \in \Omega \,. \tag{9.3.26}$$

Contracting (9.3.26) from the left by Q and using (9.3.8) and (9.3.21), we find

$$D \cdot \nabla\varphi \otimes \nabla\varphi = -Q \cdot P \cdot Q^{\mathrm{T}}, \quad x \in \Omega \,. \tag{9.3.27}$$

Because the right-hand side is a symmetric rank-2 tensor, this implies the symmetry of the rank-2 tensor

$$(D \cdot \nabla\varphi) \otimes \nabla\varphi$$

that appears on the left hand side of (9.3.27), that is,

$$(D \cdot \nabla\varphi) \otimes \nabla\varphi = \nabla\varphi \otimes (D \cdot \nabla\varphi) \,. \tag{9.3.28}$$

Equation (9.3.28) is satisfied if the vectors $D \cdot \nabla\varphi$ and $\nabla\varphi$ are parallel,

$$D \cdot \nabla\varphi = \lambda \nabla\varphi \,, \tag{9.3.29}$$

where λ is some scalar value.

Next, let us obtain the double-sided estimate based on the derived extremal condition. Equation (9.3.29) represents the necessary extremum condition for the dissipation energy functional J with respect to the rotation tensor Q, or in other words, it describes a distribution of Q for which the dissipation attains an extremum. It is thus applicable to both problems (9.3.14) and (9.3.15). It also expresses the collinearity of the vectors $\nabla\varphi$ and

$$D \cdot \nabla\varphi = \left(Q \cdot D^0 \cdot Q^{\mathrm{T}}\right) \cdot \nabla\varphi\ .$$

Equation (9.3.29) is an eigenvalue problem. Consequently, the vector $\nabla\varphi$ is one of the eigenvectors of the heat conduction tensor D:

$$D \cdot \nabla\varphi = \lambda_i \nabla\varphi, \quad i = 1, 2, 3\ . \tag{9.3.30}$$

Because the eigenvalues λ_i of the tensors D and D^0 are equal, see the Eqs. (9.3.9) and (9.3.11), and we may consider these eigenvalues as given material parameters (describing heat conduction along each of the principal axes), without loss of generality we may take

$$\lambda_1 = \lambda_{\min} < \lambda_2 < \lambda_3 = \lambda_{\max}\ . \tag{9.3.31}$$

Substituting (9.3.30) into the Euler equation (9.3.3) of the functional J, we obtain the equations that determine the stationary distribution of the scalar function $\varphi = \varphi(x)$:

$$\nabla \cdot (\lambda_i \nabla\varphi) = 0, \quad i = 1, 2, 3, \quad x \in \Omega \tag{9.3.32}$$

in the case of a specified rotation tensor Q satisfying

$$\left(Q \cdot D^0 \cdot Q^{\mathrm{T}}\right) \cdot \nabla\varphi = \lambda_i \nabla\varphi\ . \tag{9.3.33}$$

The elliptic partial differential equation (9.3.32) with the boundary conditions

$$(\varphi)_{\Gamma_g} = \varphi^0, \qquad (\lambda_i \mathbf{n} \cdot \nabla\varphi)_{\Gamma_i} = 0\ , \tag{9.3.34}$$

corresponding to conditions (9.3.2) with the relations (9.3.30) constitute the conventional boundary value problem describing homogeneous or inhomogeneous isotropic processes of heat conduction. Under some known additional constraints imposed on the boundary shape $\Gamma = \Gamma_g + \Gamma_i$, where $\Gamma_g \cap \Gamma_i = 0$, we have the existence and uniqueness of the solution of (9.3.32) and (9.3.34) with given λ_i.

If the material axes of orthotropy are oriented, uniformly for the entire domain Ω, in such a way that the dissipation J is at an extremum, then λ_i is constant in Ω and the considered heat conduction process is described by the classical boundary value problem

$$\Delta\varphi = 0, \qquad x \in \Omega \tag{9.3.35}$$

$$(\varphi)_{\Gamma_g} = \varphi^0, \qquad (\mathbf{n} \cdot \varphi)_{\Gamma_i} = 0 \tag{9.3.36}$$

for a Laplace equation with (in the general case) mixed boundary conditions. Here Δ is the Laplace operator in three-dimensional space.

If the domain Ω consists of several sub-domains Ω_i such that

$$\Omega = \cup\Omega_i, \quad \Omega_i \cap \Omega_j = 0, \quad i \neq j\ , \tag{9.3.37}$$

and for each separate sub-domain Ω_i the above extremum orientation condition holds, then the isotropic heat conduction process is realized for each considered subdomain.

Let us consider the case where the orthotropic material fulfills the extremum orientation condition in the entire domain Ω. Then we will have the isotropic boundary value problem (9.3.35) and (9.3.36), and consequently the state variable φ (the reciprocal of the temperature) is independent of λ_i. As a result we obtain the following minimal and maximal values of the energy dissipation functional J:

$$\min_{Q} J = \lambda_{\min} I \ , \tag{9.3.38}$$

$$\max_{Q} J = \lambda_{\max} I \ , \tag{9.3.39}$$

where

$$I = \int_{\Omega} (\nabla\varphi)^2 \, d\Omega \, . \tag{9.3.40}$$

Thus the double-sided estimate for the energy dissipation functional can be written as

$$\lambda_{\min} \leq \frac{J}{I} \leq \lambda_{\max} \, . \tag{9.3.41}$$

We emphasize that (9.3.41) holds for any admissible material orientation field. To derive it, we leveraged the fact that the isotropic heat conduction process occurs for material orientation fields Q that minimize or maximize the energy dissipation functional J.

Let us separately consider the two-dimensional case with a plane domain Ω (see Fig. 9.7). In this case

$$\nabla\varphi = \left\{ \frac{\partial\varphi}{\partial x_1}, \frac{\partial\varphi}{\partial x_2} \right\}, \qquad x = \{x_1, x_2\} \in \Omega \, . \tag{9.3.42}$$

Then the elements of the orthogonal tensor Q are represented in the form

$$Q_{11} = Q_{22} = \cos\alpha \qquad Q_{21} = -Q_{12} = \sin\alpha \ , \tag{9.3.43}$$

where α is the angle of in-plane rotation. Using equation (9.3.33), we obtain an explicit expression relating the angle $\alpha = \alpha(x_1, x_2)$ with the function $\varphi = \varphi(x_1, x_2)$. For definiteness let us assume that the vector $\nabla\varphi$, presented in the Eq. (9.3.42), corresponds to the eigenvalue λ_i. Then the eigenvector $\mathbf{k}$, corresponding to the eigenvalue $\lambda_j, i \neq j$ is

$$\mathbf{k} = \left\{ \frac{\partial\varphi}{\partial x_2}, -\frac{\partial\varphi}{\partial x_1} \right\} , \tag{9.3.44}$$

which is orthogonal to the eigenvector $\nabla\varphi$ in Eq. (9.3.42). We form a scalar product of both sides of the vector equality (9.3.30) with the vector $\mathbf{k}$. We have

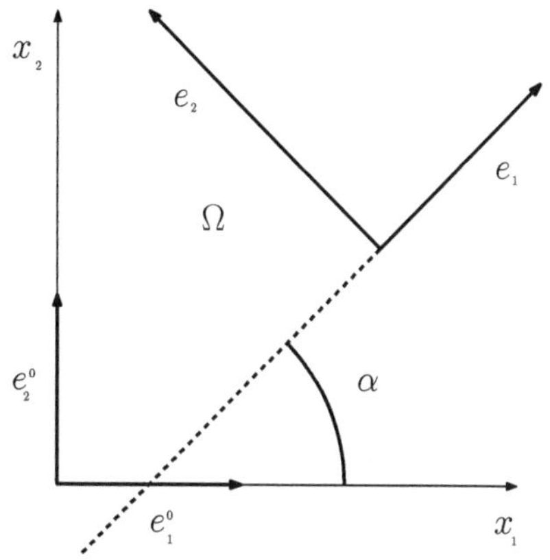

Fig. 9.7 Orientation of local orthotropicity in a two-dimensional case

$$\mathbf{k} \cdot D \cdot \nabla\varphi = 0\,. \tag{9.3.45}$$

This relation splits into two separate cases. The first is

$$\cos 2\alpha = C, \qquad \sin 2\alpha = S\,, \tag{9.3.46}$$

where

$$C = -\frac{\left(D_{11}^0 - D_{22}^0\right)\left\{\left(\dfrac{\partial\varphi}{\partial x_1}\right)^2 - \left(\dfrac{\partial\varphi}{\partial x_2}\right)^2\right\} + 4D_{12}^0\dfrac{\partial\varphi}{\partial x_1}\dfrac{\partial\varphi}{\partial x_2}}{(\nabla\varphi)^2\sqrt{\left(D_{11}^0 - D_{22}^0\right)^2 + 4\left(D_{12}^0\right)^2}} \tag{9.3.47}$$

and

$$S = \frac{2\left(D_{11}^0 - D_{22}^0\right)\dfrac{\partial\varphi}{\partial x_1}\dfrac{\partial\varphi}{\partial x_2} - 2D_{12}^0\left\{\left(\dfrac{\partial\varphi}{\partial x_1}\right)^2 - \left(\dfrac{\partial\varphi}{\partial x_2}\right)^2\right\}}{(\nabla\varphi)^2\sqrt{\left(D_{11}^0 - D_{22}^0\right)^2 + 4\left(D_{12}^0\right)^2}} \tag{9.3.48}$$

corresponds to the smaller eigenvalue λ_1, $\lambda_1 < \lambda_2$. The second case is

$$\cos 2\alpha = -C, \qquad \sin 2\alpha = -S\,, \tag{9.3.49}$$

which corresponds to the larger eigenvalue λ_2, $\lambda_2 > \lambda_1$.

Let us now consider some examples of the double-sided estimate. Suppose at first that the orthotropic material occupies the three-dimensional domain Ω situated between an internal sphere of radius r_1 and an outer sphere of radius r_2, where r_1 and r_2 are given values with $r_1 < r_2$. The temperature at the internal boundary is taken to be $\theta = \theta_1$ and at the external boundary $\theta = \theta_2$, with $\theta_1 > \theta_2$ given positive values. Thus we have the boundary conditions

$$\varphi = \varphi_1 = \frac{1}{\theta_1}, \quad r = r_1$$

$$\varphi = \varphi_2 = \frac{1}{\theta_2}, \quad r = r_2 , \tag{9.3.50}$$

where $\varphi_1 < \varphi_2$. Here we use a spherical coordinate system with the origin at $r = 0$. From the spherical symmetry of the problem geometry it follows that the extremum orientations of the axes of orthotropy with

$$\lambda_1 = \lambda_{\min} \quad \text{and} \quad \lambda_3 = \lambda_{\max} ,$$

corresponding respectively to the cases

$$J \to \min_{Q} \quad \text{and} \quad J \to \max_{Q} ,$$

are realized in the radial direction. Also $\nabla\varphi$ and the heat flux vector $\mathbf{q}$ are aligned radially at each point of the domain Ω. Due to symmetry, the heat flux $\mathbf{q}$ is absent in the circumferential directions. The following values characterize the extremal solution:

$$\mathbf{q}_{\min} = \lambda_{\min} N \mathbf{r}_0, \qquad \mathbf{q}_{\max} = \lambda_{\max} N \mathbf{r}_0 \tag{9.3.51}$$

$$\lambda_{\min} I \le J \le \lambda_{\max} I$$

$$N = \frac{\varphi_2 - \varphi_1}{r_2 - r_1}, \quad \mathbf{r}_0 = \frac{\mathbf{r}}{|\mathbf{r}|} ,$$

where

$$I = \int_{\Omega} (\nabla\varphi)^2 \, \mathrm{d}\Omega = \frac{4}{3}\pi N \left(\varphi_2 - \varphi_1\right)\left(r_1^2 + r_1 r_2 + r_2^2\right) \tag{9.3.52}$$

and $\mathbf{r}_0$ is an unit vector, oriented in the radial direction.

Next let us consider the problem of finding the double-sided estimate when a simply connected domain Ω occupied by the orthotropic material is a rectangular parallelepiped with the upper and lower faces at $x_3 = -c$ and $x_3 = c$, and side faces at $x_1 = \pm a$ and $x_2 = \pm b$. We use the Cartesian coordinate system (x_1, x_2, x_3) and take the temperature θ to be given at the lower and upper faces, with the side faces thermally insulated. The boundary conditions have the form

$$\varphi = \varphi_1 = \frac{1}{\theta_1}, \quad x_3 = -c \quad \text{and} \quad \varphi = \varphi_2 = \frac{1}{\theta_2}, \quad x_3 = c \tag{9.3.53}$$

and

$$\mathbf{q} \cdot \mathbf{n} = \mathbf{n} \cdot D \cdot \nabla\varphi = 0 \text{ at } x_1 = \pm a,\ x_2 = \pm b\,, \tag{9.3.54}$$

where $\theta_1 > 0$, $\theta_2 > 0$ and $\theta_1 > \theta_2$. The extremal solution is characterized by the existence of level surfaces where x_3 is constant, $-c < x_3 < c \in \Omega$, with a constant value for the state variable φ (constant temperature θ). The gradient of φ is parallel to the x_1 axis. Therefore the orientations with a minimal eigenvalue $\lambda = \lambda_{\min}$ (in the case $J \to \min_Q$) and with a maximal eigenvalue $\lambda = \lambda_{\max}$ (in the case of $J \to \max_Q$) are oriented in parallel with respect to the axis x_3. Such orientation provides, respectively, either the minimum or the maximum of dissipation. We have

$$\mathbf{q}_{\min} = \lambda_{\min}\Phi\mathbf{x}_3^0, \qquad \mathbf{q} = \lambda_{\max}\Phi\mathbf{x}_3^0 \tag{9.3.55}$$

$$\min_Q J = \lambda_{\min} I, \qquad \max_Q J = \lambda_{\max} I$$

$$\Phi = \frac{\varphi_2 - \varphi_1}{2c}, \qquad \nabla\varphi = \Phi\mathbf{x}_3^0, \qquad \mathbf{x}_3^0 = \frac{\mathbf{x}_3}{|\mathbf{x}_3|}\,,$$

where

$$I = \int_\Omega (\nabla\varphi)^2 \,\mathrm{d}\Omega = \frac{2ab}{c}(\varphi_2 - \varphi_1)^2 \tag{9.3.56}$$

and $\mathbf{x}_3^0$ is a unit vector of the x_3-axis, obtained when the vector $\mathbf{x}_3$ is divided by its length $|\mathbf{x}_3|$.

9.4 Conclusion

In this final chapter, we considered optimization problems in thermoelasticity. We found optimal thickness distributions for a beam to resist thermal buckling, and optimal material distributions for a beam of uniform thickness, with the same goal. Finally, we derived a guaranteed double-sided estimate that governs energy dissipation in heat conduction in a locally orthotropic solid body, which holds regardless of how the local material orientation is distributed in the solid body. Connecting to the main theme of this book, this has applications in the analysis heat conduction in paper materials, which is important when considering the drying process in papermaking.

In this book, we have covered a wide variety of topics, concentrating especially on axially moving materials. We have made brief excursions into classical solid mechanics and coupled problems involving mechanics, such as the two final chapters. Some of the topics have been given a mostly self-contained, systematic exposition, to benefit not only the specialist, but also the student or new researcher just entering the research field of axially moving materials.

It is our hope that the various solution techniques touched upon across the chapters will benefit the reader, whether in their original context or in an unexpected new application. We also hope, via detailed exposition and examples, to have made the theory of axially moving materials a more accessible topic for a new generation of researchers. Although the field of axially moving materials is already over a century old (dating back to Skutch [14]), new exciting applications await for example in printable electronics and microfluidics. This makes the field worthy of study not only from a fundamental academic viewpoint, but also for those primarily interested in applications.

Finally, many of the treatments in this book have emphasized analytical and semi-analytical approaches, sometimes with a change in perspective, or with an unconventional application of a known general result such as the implicit function theorem. Analytical tools are nowadays often overlooked in the applied sciences, but they need not be; they can serve as a basis for fundamental theoretical understanding, but also importantly, as a basis for fast numerical solvers for realtime applications, such as online prediction and control of industrial processes.

References

1. Albul AV, Banichuk NV, Barsuk AA (1980) Optimization of stability for elastic rods with thermal loads. Izvestiya Akademii Nauk SSSR, Mekhanika Tverdogo Tela (MTT), vol 3, pp 127–133 (in Russian)
2. Banichuk NV, Ivanova SY, Makeev EV, Ragnedda F (2007) Nonlocal optimization of thermoelastic rod made from discrete set of materials with application of genetic algorithm. Probl Strength Plast (69):38
3. Banichuk NV (1990) Introduction to optimization of structures. Springer, New York, 300 pages
4. Banichuk VV (1983) Problems and methods of optimal structural design. Plenum Press, New York, 313 pages
5. Goldberg DF (1989) Genetic algorithms in search, optimization and machine learning. Addison-Wesley, Reading, MA
6. Haslinger J, Mäkinen RAE (2003) Introduction to shape optimization: theory, approximation and computations. SIAM, Philadelphia, PA
7. Holland JH (1975) Adaptation in neural and artificial systems. University of Michigan Press, Ann Arbor, MI
8. Komkov V (1988) Variational principles of continuum mechanics with engineering applications. Vol 1: critical points theory. Reidel Publishing Co., Dordrecht
9. Komkov V (1988) Variational principles of continuum mechanics with engineering applications. Vol. 1: introduction to optimal design theory. Reidel Publishing Co., Dordrecht
10. Landau LD, Lifshitz EM (1970) Teoriya uprugosti (Theory of elasticity, 2nd edn). English 2nd edn. Published by Pergamon Press, Oxford, 1965
11. Larichev AD (1981) Finding a minimum volume for a beam on an elastic foundation, for a given magnitude of a critical load. In Applied and theoretical research into building structures, Moscow, pp 19–25. Kucherenko TsNIINSK (in Russian)
12. Love AEH (1944) A treatise on the mathematical theory of elasticity, 4th edn. Dover Publications, New York
13. Nowacki W (1970) Teoria sprezystosci. Panstwowe Wydawnictwo Naukowe, Warszawa
14. Skutch Rudolf (1897) Uber die Bewegung eines gespannten Fadens, weicher gezwungen ist durch zwei feste Punkte, mit einer constanten Geschwindigkeit zu gehen, und zwischen denselben in Transversal-schwingungen von gerlinger Amplitude versetzt wird. Annalen der Physik und Chemie 61:190–195

15. Timoshenko S (1956) Strength of materials, Part II: advanced theory and problems, 3rd edn. D. Van Nostrand Company
16. Timoshenko S, Goodier J (1987) Theory of elasticity, 3rd edn. McGraw-Hill
17. Washizu K (1982) Variational methods in elasticity and plasticity. Pergamon, Oxford

Appendix A
Cartesian Tensors

This Appendix reviews the basics of tensor algebra and tensor calculus in orthogonal coordinate systems. For a more thorough classical exposition of tensor calculus, refer to the books by Flügge [3] or Sokolnikoff [5].

Calculations with tensors are often written in a mix of two notations, where the tensor (or nabla) notation makes the structure apparent, while the index notation is convenient for algebraic manipulation. Below, we use both notations. For the most part, we will consider only up to rank-2 tensors. Furthermore, keep in mind that we restrict our consideration to orthogonal coordinate systems, where there is no distinction between contravariant and covariant components, and the metric tensor is the identity. Unless otherwise stated, the number of spatial dimensions (coordinate axes) is arbitrary.

We use the following notational conventions:

$\mathbf{A}$	rank-2 tensor
$\mathbf{a}$	vector, i.e. rank-1 tensor
$a,\ A$	scalar, i.e. rank-0 tensor
$\cdot$	dot product (also known as contraction)
$:$	double dot product (double contraction), with swap
$\cdot\cdot$	double dot product (double contraction), no swap
$()^{\mathrm{T}}$	transpose of rank-2 tensor
$\otimes$	outer product
$\times$	cross product, in 3 space dimensions only

Scalar multiplication is written without any multiplication symbol. In index notation, the *Einstein summation convention* is used, that is, a repeated index is summed over. In sum expressions, the repeat must occur within the same term. For example, in the expression $a_i b_i c_j$, i is considered repeated, but in $a_i b_j + a_j b_i$ it is not.

N. Banichuk et al., *Stability of Axially Moving Materials*, Solid Mechanics and Its Applications 259, https://doi.org/10.1007/978-3-030-23803-2

A.1 Tensor Algebra

The basic operations are defined as follows:

$$\mathbf{a} \cdot \mathbf{b} \equiv a_i b_i \ , \quad (\mathbf{A} \cdot \mathbf{b})_i \equiv A_{ij} b_j \ , \quad (\mathbf{a} \cdot \mathbf{B})_j \equiv a_i B_{ij} \ , \quad (\mathbf{A} \cdot \mathbf{B})_{ij} \equiv A_{ik} B_{kj}$$

$$\mathbf{A} : \mathbf{B} \equiv A_{ij} B_{ji} \ , \quad \mathbf{A} \cdot\cdot\, \mathbf{B} \equiv A_{ij} B_{ij}$$

$$\mathbf{A}^{\mathrm{T}}_{ij} \equiv A_{ji}$$

$$(\mathbf{a} \otimes \mathbf{b})_{ij} \equiv a_i b_j \ , \quad (\mathbf{A} \otimes \mathbf{b})_{ijk} \equiv A_{ij} b_k \ , \quad (\mathbf{a} \otimes \mathbf{B})_{ijk} \equiv a_i B_{jk} \ , \quad (\mathbf{A} \otimes \mathbf{B})_{ijkn} \equiv A_{ij} B_{kn}$$

$$(\mathbf{a} \times \mathbf{b})_i \equiv \varepsilon_{ijk} a_j b_k$$

Each operand of the dot and outer products must be rank-1 or higher. The double dot product is only defined for rank-2 and higher. Transpose is only defined for rank-2. The cross product is only defined in three space dimensions. It takes two vectors and produces a new vector, orthogonal to both. The symbol ε_{ijk} is the alternating unit tensor (also known as the Levi–Civita symbol). It is rank-3, and only 6 out of its $3^3 = 27$ components are nonzero:

$$\varepsilon_{ijk} = \begin{cases} +1 \ , & ijk = \{123, 231, 312\} \\ -1 \ , & ijk = \{321, 213, 132\} \\ 0 & \text{otherwise.} \end{cases} \tag{A.1.1}$$

For ε_{ijk} to be nonzero, i, j and k must be all different. For an increasing sequence (with wraparound) ε_{ijk} is positive, while for a decreasing sequence (with wraparound) it is negative. One may also consider the swapping of index pairs. Initially, ε_{ijk} has the sequence 123 and is positive. An odd number of pairwise swaps makes ε_{ijk} negative, while an even number of pairwise swaps makes it positive again. This is useful for manipulation in index notation, as swapping can be done for arbitrary index variables.

The rank of a dot product is the sum of the operands' ranks minus 2. The rank of an outer product is the sum of the operands' ranks. The double dot product produces a scalar from two rank-2 tensors. In matrix terms, $\mathbf{A} : \mathbf{B} = \mathrm{tr}(\mathbf{A} \cdot \mathbf{B})$. Note also

$$\begin{aligned} \mathbf{A} : \mathbf{B} &\equiv A_{ij} B_{ji} = A^{\mathrm{T}}_{ji} B^{\mathrm{T}}_{ij} \equiv \mathbf{A}^{\mathrm{T}} : \mathbf{B}^{\mathrm{T}} \\ \mathbf{A} \cdot\cdot\, \mathbf{B} &\equiv A_{ij} B_{ij} = A^{\mathrm{T}}_{ji} B^{\mathrm{T}}_{ji} = \mathbf{B}^{\mathrm{T}} \cdot\cdot\, \mathbf{A}^{\mathrm{T}} \end{aligned}$$

Some sources extend the double dot product for higher-rank tensors $A_{k_1 k_2 \ldots k_N}$, $B_{n_1 n_2 \ldots n_M}$, as follows:

$$\begin{aligned} (\mathbf{A} : \mathbf{B})_{k_1 k_2 \ldots k_{N-2} n_3 n_4 \ldots n_M} &\equiv A_{k_1 k_2 \ldots k_{N-2} ij} B_{jin_3 n_4 \ldots n_M} \ , \\ (\mathbf{A} \cdot\cdot\, \mathbf{B})_{k_1 k_2 \ldots k_{N-2} n_3 n_4 \ldots n_M} &\equiv A_{k_1 k_2 \ldots k_{N-2} ij} B_{ijn_3 n_4 \ldots n_M} \ . \end{aligned}$$

This is a natural generalization of the higher-rank dot product into the double-dot case. Using this definition, the rank of a double dot product is the sum of operands' ranks minus 4.

Commutativity. Binary operations up to rank 2 have the following commutation rules:

$$\begin{aligned}
\mathbf{a}\cdot\mathbf{b} &\equiv a_i b_i = b_i a_i \equiv \mathbf{b}\cdot\mathbf{a}\\
(\mathbf{A}\cdot\mathbf{b})_i &= A_{ij}b_j = b_j A_{ij} = b_j A^{\mathrm{T}}_{ji} = (\mathbf{b}\cdot\mathbf{A}^{\mathrm{T}})_i\\
(\mathbf{a}\cdot\mathbf{B})_j &= a_i B_{ij} = B_{ij}a_i = B^{\mathrm{T}}_{ji}a_i = (\mathbf{B}^{\mathrm{T}}\cdot\mathbf{a})_j\\
(\mathbf{A}\cdot\mathbf{B})_{ij} &\equiv A_{ik}B_{kj} = B_{kj}A_{ik} = B^{\mathrm{T}}_{jk}A^{\mathrm{T}}_{ki} = (\mathbf{B}^{\mathrm{T}}\cdot\mathbf{A}^{\mathrm{T}})_{ji} = (\mathbf{B}^{\mathrm{T}}\cdot\mathbf{A}^{\mathrm{T}})^{\mathrm{T}}_{ij}\\
\mathbf{A}:\mathbf{B} &\equiv A_{ij}B_{ji} = B_{ji}A_{ij} = \mathbf{B}:\mathbf{A}\\
\mathbf{A}\cdot\cdot\,\mathbf{B} &\equiv A_{ij}B_{ij} = B_{ij}A_{ij} = \mathbf{B}\cdot\cdot\,\mathbf{A}\\
(\mathbf{a}\otimes\mathbf{b})_{ij} &\equiv a_i b_j = b_j a_i = (b_i a_j)^{\mathrm{T}} \equiv (\mathbf{b}\otimes\mathbf{a})^{\mathrm{T}}_{ij}
\end{aligned}$$

Associativity. The outer product is always associative. For the dot product, this depends on the operands. For dot product triples with operands of ranks 1 and 2, the triple is associative if the middle operand is rank-2:

$$\begin{aligned}
((\mathbf{A}\cdot\mathbf{B})\cdot\mathbf{C})_{in} &= (A_{ik}B_{kj})C_{jn} = A_{ik}(B_{kj}C_{jn}) = (\mathbf{A}\cdot(\mathbf{B}\cdot\mathbf{C}))_{in}\\
((\mathbf{A}\cdot\mathbf{B})\cdot\mathbf{c})_i &= (A_{ik}B_{kj})c_j = A_{ik}(B_{kj}c_j) = (\mathbf{A}\cdot(\mathbf{B}\cdot\mathbf{c}))_i\\
((\mathbf{a}\cdot\mathbf{B})\cdot\mathbf{C})_k &= (a_i B_{ij})C_{jk} = a_i(B_{ij}C_{jk}) = (\mathbf{a}\cdot(\mathbf{B}\cdot\mathbf{C}))_k\\
(\mathbf{a}\cdot\mathbf{B})\cdot\mathbf{c} &= (a_i B_{ij})c_j = a_i(B_{ij}c_j) = \mathbf{a}\cdot(\mathbf{B}\cdot\mathbf{c})
\end{aligned}$$

If the middle operand is rank-1, the triple is non-associative:

$$\begin{aligned}
(\mathbf{A}\cdot\mathbf{b})\cdot\mathbf{c} &= A_{ij}b_j c_i\ , \quad \text{but } \mathbf{A}\cdot(\mathbf{b}\cdot\mathbf{c}) \text{ is not defined because } \mathbf{b}\cdot\mathbf{c} \text{ is a scalar}\\
\mathbf{a}\cdot(\mathbf{b}\cdot\mathbf{C}) &= a_j b_i C_{ij}\ , \quad \text{but } (\mathbf{a}\cdot\mathbf{b})\cdot\mathbf{C} \text{ is not defined because } \mathbf{a}\cdot\mathbf{b} \text{ is a scalar}\\
((\mathbf{A}\cdot\mathbf{b})\cdot\mathbf{C})_k &= A_{ij}b_j C_{ik}\ , \quad \text{but} \quad (\mathbf{A}\cdot(\mathbf{b}\cdot\mathbf{C}))_k = A_{kj}b_i C_{ij}
\end{aligned}$$

The cases marked *not defined* should instead use a scalar multiplication, producing a rank-2 tensor:

$$\begin{aligned}
(\mathbf{A}(\mathbf{b}\cdot\mathbf{c}))_{ij} &= A_{ij}(\mathbf{b}\cdot\mathbf{c}) = A_{ij}b_k c_k\\
((\mathbf{a}\cdot\mathbf{b})\mathbf{C})_{ij} &= (\mathbf{a}\cdot\mathbf{b})C_{ij} = a_k b_k C_{ij}
\end{aligned}$$

Identities for triple products.

$$
\begin{aligned}
(\mathbf{A}\cdot\mathbf{B}):\mathbf{C} &= (\mathbf{A}\cdot\mathbf{B})_{ij}C_{ji} = (A_{ik}B_{kj})C_{ji} = A_{ik}(B_{kj}C_{ji}) = A_{ik}(\mathbf{B}\cdot\mathbf{C})_{ki} = \mathbf{A}:(\mathbf{B}\cdot\mathbf{C})\\
(\mathbf{A}\cdot\mathbf{b})\cdot\mathbf{c} &= (A_{ij}b_j)c_i = A_{ij}(b_jc_i) = A_{ij}(\mathbf{b}\otimes\mathbf{c})_{ji} = \mathbf{A}:(\mathbf{b}\otimes\mathbf{c})\\
\mathbf{a}\cdot(\mathbf{b}\cdot\mathbf{C}) &= a_j(b_iC_{ij}) = (a_jb_i)C_{ij} = (\mathbf{a}\otimes\mathbf{b})_{ji}C_{ij} = (\mathbf{a}\otimes\mathbf{b}):\mathbf{C}\\
(\mathbf{a}\otimes\mathbf{b}):\mathbf{C} &= \mathbf{a}\cdot(\mathbf{b}\cdot\mathbf{C})\\
\mathbf{A}:(\mathbf{b}\otimes\mathbf{c}) &= (\mathbf{A}\cdot\mathbf{b})\cdot\mathbf{c}\\
((\mathbf{A}\cdot\mathbf{b})\cdot\mathbf{C})_k &= A_{ij}b_jC_{ik} = C^{\mathrm{T}}_{ki}A_{ij}b_j = (\mathbf{C}^{\mathrm{T}}\cdot\mathbf{A}\cdot\mathbf{b})_k\\
(\mathbf{A}\cdot(\mathbf{b}\cdot\mathbf{C}))_k &= A_{kj}b_iC_{ij} = b_iC_{ij}A^{\mathrm{T}}_{jk} = (\mathbf{b}\cdot\mathbf{C}\cdot\mathbf{A}^{\mathrm{T}})_k\\
(\mathbf{A}^{\mathrm{T}}\cdot\mathbf{B}\cdot\mathbf{c})_k &= ((\mathbf{B}\cdot\mathbf{c})\cdot\mathbf{A})_k\\
(\mathbf{a}\cdot\mathbf{B}\cdot\mathbf{C}^{\mathrm{T}})_k &= (\mathbf{C}\cdot(\mathbf{a}\cdot\mathbf{B}))_k\\
(\mathbf{a}\cdot(\mathbf{b}\otimes\mathbf{c}))_j &= a_i(b_ic_j) = (c_jb_i)a_i = ((\mathbf{c}\otimes\mathbf{b})\cdot\mathbf{a})_j\\
((\mathbf{a}\otimes\mathbf{b})\cdot\mathbf{c})_j &= (\mathbf{c}\cdot(\mathbf{b}\otimes\mathbf{a}))_j
\end{aligned}
$$

In three space dimensions, we also have the triple product from vector algebra:

$$(\mathbf{a}\times\mathbf{b})\cdot\mathbf{c} = (\varepsilon_{ijk}a_jb_k)c_i = a_j(\varepsilon_{ijk}b_kc_i) = a_j(-\varepsilon_{jik}b_kc_i) = a_j(\varepsilon_{jki}b_kc_i) = \mathbf{a}\cdot(\mathbf{b}\times\mathbf{c})$$

Identities involving four vectors.

$$((\mathbf{a}\otimes\mathbf{b})\cdot(\mathbf{c}\otimes\mathbf{d}))_{ij} = (a_ib_k)(c_kd_j) = (a_ic_k)(b_kd_j) = ((\mathbf{a}\otimes\mathbf{c})\cdot(\mathbf{b}\otimes\mathbf{d}))_{ij}$$

$$
\begin{aligned}
((\mathbf{a}\otimes\mathbf{b})\cdot(\mathbf{c}\otimes\mathbf{d}))_{ij} &= ((\mathbf{a}\otimes\mathbf{c})\cdot(\mathbf{b}\otimes\mathbf{d}))_{ij} = (a_ic_k)(b_kd_j) = (d_jc_k)(b_ka_i)\\
&= ((\mathbf{d}\otimes\mathbf{c})\cdot(\mathbf{b}\otimes\mathbf{a}))_{ji} = ((\mathbf{d}\otimes\mathbf{c})\cdot(\mathbf{b}\otimes\mathbf{a}))^{\mathrm{T}}_{ij}
\end{aligned}
$$

$$((\mathbf{a}\otimes\mathbf{b})\cdot(\mathbf{c}\otimes\mathbf{d}))_{ij} = (a_ib_k)(c_kd_j) = (a_id_j)(b_kc_k) = (\mathbf{a}\otimes\mathbf{d})_{ij}(\mathbf{b}\cdot\mathbf{c})$$

$$(\mathbf{a}\otimes\mathbf{b})_{ij}(\mathbf{c}\cdot\mathbf{d}) = ((\mathbf{a}\otimes\mathbf{c})\cdot(\mathbf{d}\otimes\mathbf{b}))_{ij}$$

$$(\mathbf{a}\cdot\mathbf{b})(\mathbf{c}\otimes\mathbf{d})_{ij} = ((\mathbf{c}\otimes\mathbf{a})\cdot(\mathbf{b}\otimes\mathbf{d}))_{ij}$$

$$(\mathbf{a}\otimes\mathbf{b}):(\mathbf{c}\otimes\mathbf{d}) = (a_ib_j)(c_jd_i) = (a_id_i)(b_jc_j) = (\mathbf{a}\cdot\mathbf{d})(\mathbf{b}\cdot\mathbf{c})$$

$$
\begin{aligned}
(\mathbf{a}\otimes\mathbf{b}):(\mathbf{c}\otimes\mathbf{d}) &= (\mathbf{a}\cdot\mathbf{d})(\mathbf{b}\cdot\mathbf{c}) = (a_id_i)(b_jc_j) = (d_ia_i)(c_jb_j)\\
&= (\mathbf{d}\cdot\mathbf{a})(\mathbf{c}\cdot\mathbf{b}) = (\mathbf{d}\otimes\mathbf{c}):(\mathbf{b}\otimes\mathbf{a})
\end{aligned}
$$

$$\mathbf{a}\cdot(\mathbf{b}\cdot(\mathbf{c}\otimes\mathbf{d})) = a_j(b_i(c_id_j)) = (a_jd_j)(b_ic_i) = (\mathbf{a}\cdot\mathbf{d})(\mathbf{b}\cdot\mathbf{c})$$

$$((\mathbf{a}\otimes\mathbf{b})\cdot\mathbf{c})\cdot\mathbf{d} = ((a_ib_j)c_j)d_i = (a_id_i)(b_jc_j) = (\mathbf{a}\cdot\mathbf{d})(\mathbf{b}\cdot\mathbf{c})$$

$$(\mathbf{a}\cdot\mathbf{b})(\mathbf{c}\cdot\mathbf{d}) = (\mathbf{a}\otimes\mathbf{c}) : (\mathbf{d}\otimes\mathbf{b}) = \mathbf{a}\cdot(\mathbf{c}\cdot(\mathbf{d}\otimes\mathbf{b})) = ((\mathbf{a}\otimes\mathbf{c})\cdot\mathbf{d})\cdot\mathbf{b}$$

$$\mathbf{a}\cdot(\mathbf{b}\otimes\mathbf{c})\cdot\mathbf{d} = a_i(b_ic_j)d_j = (a_ib_i)(c_jd_j) = (\mathbf{a}\cdot\mathbf{b})(\mathbf{c}\cdot\mathbf{d})$$

A.2 Tensor Calculus

The difference between the algebraic formulas and formulas involving nabla is that the derivative operator distributes into products (by the rule of the derivative of a product), and that (being an operator) it does not commute even for scalars. We will restrict our consideration to relatively simple cases, encountered often in applications.

There are two notational conventions for the gradient of a vector field. The first one is

$$(\nabla\mathbf{a})_{ij} \equiv \frac{\partial}{\partial x_i}a_j \equiv \partial_i a_j \ . \tag{A.2.1}$$

This convention works well with the prefix notation for derivatives, where ∂_i is shorthand for $\partial/\partial x_i$. Note that (A.2.1) defines the gradient of a vector field as the transpose of the standard Jacobian.

According to the convention (A.2.1), the directional derivative $(\mathbf{a}\cdot\nabla)()$ of a vector field $\mathbf{b}$ is

$$(\mathbf{a}\cdot\nabla\mathbf{b})_j \equiv a_i(\partial_i b_j) \ .$$

In order to avoid special-casing the rules of tensor algebra, the direction vector multiplies from the left.

The other notational convention for the gradient of a vector field is

$$(\nabla\mathbf{a})_{ij} \equiv (\mathbf{J}(\mathbf{a}))_{ij} \equiv \frac{\partial a_i}{\partial x_j} \equiv a_{i,j} \ , \tag{A.2.2}$$

where $\mathbf{J}()$ denotes the standard Jacobian. This convention is well suited for the comma-suffix notation for derivatives. With the prefix notation, it requires a transpose:

$$(\nabla\mathbf{a})_{ij} = \partial_j a_i = (\partial_i a_j)^{\mathrm{T}} \ .$$

According to the convention (A.2.2), the directional derivative $(\mathbf{a}\cdot\nabla)()$ of a vector field $\mathbf{b}$ is

$$((\nabla\mathbf{b})\cdot\mathbf{a})_i = (\mathbf{J}(\mathbf{b})\cdot\mathbf{a})_i = b_{i,j}a_j \ .$$

In order to avoid special-casing the rules, the direction vector must multiply from the right. Be aware that many authors that use the convention (A.2.2) still write the directional derivative in nabla notation as $\mathbf{a}\cdot\nabla\mathbf{b}$. Then it must be kept in mind that in this particular construct, the direction vector $\mathbf{a}$ multiplies *from the right*.

In either convention, in the directional derivative, the indices on the direction vector and the derivative are the same. For the rest of this Appendix, we will use the *transpose Jacobian* convention (A.2.1) with prefix notation for derivatives.

Basic definitions. The gradient of a scalar field is

$$(\nabla a)_i = \partial_i a \ .$$

The gradient of a vector field, using convention (A.2.1), is

$$(\nabla \mathbf{a})_{ij} = \partial_i a_j \ .$$

In convention (A.2.1), the gradient is formally the outer product of nabla with its operand: $\nabla(\dots) \equiv (\nabla) \otimes (\dots)$.

The directional derivative of a scalar field b in the direction of a vector $\mathbf{a}$ is

$$\mathbf{a} \cdot \nabla b \equiv a_i \partial_i b \ .$$

The directional derivative of a vector field, consistent with convention (A.2.1), is

$$(\mathbf{a} \cdot \nabla \mathbf{b})_j \equiv a_i \partial_i b_j \ .$$

The jth component of the result is the directional derivative of the component b_j; the directional derivative is effectively a scalar operator. As per convention (A.2.1), the direction vector $\mathbf{a}$ multiplies from the left. Note that

$$(\mathbf{a} \cdot \nabla)\mathbf{b} = (a_i \partial_i) b_j = a_i (\partial_i b_j) = \mathbf{a} \cdot (\nabla \mathbf{b}) \ ,$$

so we may write $\mathbf{a} \cdot \nabla \mathbf{b}$ without parentheses. This is often used when manipulating directional derivatives, and holds also for operands of any rank in place of $\mathbf{b}$.

Generally, the directional derivative for an operand of any rank is

$$(\mathbf{a} \cdot \nabla)(\dots) \equiv a_i \partial_i (\dots) \ .$$

The divergence of a vector is

$$\nabla \cdot \mathbf{a} \equiv \partial_i a_i$$

and that of a rank-2 tensor is

$$(\nabla \cdot \mathbf{A})_j \equiv \partial_i A_{ij} \ .$$

In convention (A.2.1), the divergence is formally the dot product of nabla with its operand. In three space dimensions, the curl of a vector field is

$$(\nabla \times \mathbf{a})_i = \varepsilon_{ijk} \partial_j a_k \ ,$$

where ε_{ijk} is the alternating unit tensor (A.1.1). The curl is formally the cross product of nabla with its operand.

Identities involving the divergence operator. Identities such as the following are needed for creating multidimensional integration-by-parts formulas when deriving weak forms for partial differential equations. The first one appears in the mass balance in continuum form (already in the strong form).

$$\nabla \cdot (\mathbf{a}b) = \partial_i(a_i b) = (\partial_i a_i)b + a_i(\partial_i b) = (\nabla \cdot \mathbf{a})b + \mathbf{a} \cdot \nabla b$$

$$\begin{aligned}(\nabla \cdot (\mathbf{A}b))_j &= \partial_i(A_{ij}b) = (\partial_i A_{ij})b + A_{ij}(\partial_i b) = (\partial_i A_{ij})b + A^{\mathrm{T}}_{ji}(\partial_i b) \\ &= ((\nabla \cdot \mathbf{A})b + \mathbf{A}^{\mathrm{T}} \cdot \nabla b)_j\end{aligned}$$

$$(\nabla \cdot (\mathbf{a} \otimes \mathbf{b}))_j = \partial_i(a_i b_j) = (\partial_i a_i)b_j + a_i(\partial_i b_j) = ((\nabla \cdot \mathbf{a})\mathbf{b} + \mathbf{a} \cdot \nabla \mathbf{b})_j$$

$$\begin{aligned}\nabla \cdot (\mathbf{A} \cdot \mathbf{b}) &= \partial_i(A_{ij}b_j) = (\partial_i A_{ij})b_j + A_{ij}(\partial_i b_j) = (\nabla \cdot \mathbf{A}) \cdot \mathbf{b} + \mathbf{A} : (\nabla \mathbf{b})^{\mathrm{T}} \\ &= (\nabla \cdot \mathbf{A}) \cdot \mathbf{b} + \mathbf{A}^{\mathrm{T}} : \nabla \mathbf{b}\end{aligned}$$

$$\nabla \cdot (\mathbf{a} \cdot \mathbf{B}) = \nabla \cdot (\mathbf{B}^{\mathrm{T}} \cdot \mathbf{a}) = (\nabla \cdot \mathbf{B}^{\mathrm{T}}) \cdot \mathbf{a} + \mathbf{B} : \nabla \mathbf{a}$$

$$\begin{aligned}(\nabla \cdot (\mathbf{A} \cdot \mathbf{B}))_j &= \partial_i(A_{ik}B_{kj}) = (\partial_i A_{ik})B_{kj} + A_{ik}(\partial_i B_{kj}) = (\partial_i A_{ik})B_{kj} + A^{\mathrm{T}}_{ki}(\partial_i B_{kj}) \\ &= ((\nabla \cdot \mathbf{A}) \cdot \mathbf{B} + \mathbf{A}^{\mathrm{T}} : \nabla \mathbf{B})_j\end{aligned}$$

Identities involving divergence and three vectors. These are also occasionally needed when deriving weak forms.

$$\nabla \cdot (\mathbf{a} \cdot (\mathbf{b} \otimes \mathbf{c})) = \mathbf{b} \cdot (\mathbf{c} \cdot \nabla \mathbf{a}) + \mathbf{a} \cdot (\nabla \cdot (\mathbf{c} \otimes \mathbf{b}))$$

$$\nabla \cdot ((\mathbf{a} \otimes \mathbf{b}) \cdot \mathbf{c}) = \mathbf{b} \cdot (\mathbf{a} \cdot \nabla \mathbf{c}) + \mathbf{c} \cdot (\nabla \cdot (\mathbf{a} \otimes \mathbf{b}))$$

As an example, let us derive these two identities. The first one:

$$\begin{aligned}\nabla \cdot (\mathbf{a} \cdot (\mathbf{b} \otimes \mathbf{c})) &= \partial_j(a_i b_i c_j) \\ &= (\partial_j a_i)b_i c_j + a_i(\partial_j b_i)c_j + a_i b_i(\partial_j c_j) \\ &= c_j(\partial_j a_i)b_i + \left[c_j(\partial_j b_i) + (\partial_j c_j)b_i\right] a_i \\ &= \mathbf{c} \cdot (\nabla \mathbf{a}) \cdot \mathbf{b} + [\mathbf{c} \cdot (\nabla \mathbf{b}) + (\nabla \cdot \mathbf{c})\mathbf{b}] \cdot \mathbf{a} \\ &= (\mathbf{c} \cdot \nabla \mathbf{a}) \cdot \mathbf{b} + [\nabla \cdot (\mathbf{c} \otimes \mathbf{b})] \cdot \mathbf{a} \\ &= \mathbf{b} \cdot (\mathbf{c} \cdot \nabla \mathbf{a}) + \mathbf{a} \cdot [\nabla \cdot (\mathbf{c} \otimes \mathbf{b})] \ .\end{aligned}$$

The second one follows by applying the triple-product identity

$$(\mathbf{a} \otimes \mathbf{b}) \cdot \mathbf{c} = \mathbf{c} \cdot (\mathbf{b} \otimes \mathbf{a})$$

on the left-hand side of the first one, and then renaming the symbols.

An identity for flow problems. As the final example for this section, let us derive the identity we used in Chap. 8, when dealing with the equations of fluid flow:

$$\mathbf{v}\cdot\nabla\mathbf{v} = \frac{1}{2}\nabla(v^2) - \mathbf{v}\times(\nabla\times\mathbf{v})\,, \tag{A.2.3}$$

where $v^2 \equiv \mathbf{v}\cdot\mathbf{v}$, and we are working in three space dimensions. To show (A.2.3), note the right-hand side is a vector; let us call it $\mathbf{V}$. In index notation, we have

$$V_j = \frac{1}{2}\partial_j(v_i v_i) - \varepsilon_{jkn} v_k(\varepsilon_{nop}\partial_o v_p)\,. \tag{A.2.4}$$

The first term of (A.2.4) can be written as

$$\frac{1}{2}\partial_j(v_i v_i) = \frac{1}{2}\big((\partial_j v_i)v_i + v_i(\partial_j v_i)\big) = v_i\partial_j v_i\,.$$

Therefore

$$V_j = v_i\partial_j v_i - \varepsilon_{jkn}\varepsilon_{nop} v_k \partial_o v_p\,.$$

It is convenient to place the same index first in both Levi–Civita symbols. We perform two pairwise swaps, $jkn \to nkj \to njk$, noting that an even number of swaps does not change the sign. We have

$$V_j = v_i\partial_j v_i - \varepsilon_{njk}\varepsilon_{nop} v_k \partial_o v_p\,.$$

Now we may use one of the basic properties of the Levi–Civita symbol (given here for general tensors, where co- and contravariance matters):

$$\varepsilon_{ijk}\varepsilon^{imn} = \delta^m_j\delta^n_k - \delta^n_j\delta^m_k\,. \tag{A.2.5}$$

Working with cartesian tensors, we may trivially lower all indices. We have

$$V_j = v_i\partial_j v_i - (\delta_{jo}\delta_{kp} - \delta_{jp}\delta_{ko})v_k\partial_o v_p\,,$$

and expanding the parentheses,

$$V_j = v_i\partial_j v_i - \delta_{jo}\delta_{kp}v_k\partial_o v_p + \delta_{jp}\delta_{ko}v_k\partial_o v_p\,.$$

Applying the Kronecker deltas eliminates the indices o and p, leaving

$$V_j = v_i\partial_j v_i - v_k\partial_j v_k + v_k\partial_k v_j\,.$$

Since the name of the symbol used as a summation index does not matter, the first two terms cancel, and we have

$$V_j = v_k\partial_k v_j\,.$$

On the other hand, the left-hand side of (A.2.3) is

$$(\mathbf{v} \cdot \nabla \mathbf{v})_j = v_i \partial_i v_j \ ,$$

which is exactly V_j, thus establishing the identity (A.2.3).

A.3 Integration by Parts in Multiple Dimensions

When deriving weak forms of partial differential equations, from a practical viewpoint the aim is to integrate by parts in order to reduce the continuity requirements on the unknown field. There are (at least) two theorems that can be used to achieve this in multiple spatial dimensions: the Gauss–Green–Ostrogradsky divergence theorem for the divergence operator, and the classical Stokes' theorem for the curl operator. In this Appendix, we will consider the divergence operator only, as it appears more often in mechanics.

The Gauss–Green–Ostrogradsky divergence theorem states that

$$\int_\Omega \nabla \cdot \mathbf{a} \, \mathrm{d}x = \int_\Gamma \mathbf{n} \cdot \mathbf{a} \, \mathrm{d}s \ , \tag{A.3.1}$$

where Ω is a region enclosed by the surface (in two dimensions, boundary) Γ, and $\mathbf{n}$ is the unit outer normal vector of the surface. For a proof of the theorem, refer to calculus textbooks such as Adams and Essex [1].

In the special case of one space dimension, we have

$$\int_b^c \partial_1 a_1 \, \mathrm{d}x_1 = a_1|_{x_1=c} - a_1|_{x_1=b} \ ,$$

which is the fundamental theorem of calculus.

A general way to integrate by parts in multiple space dimensions in many problems in mechanics is to manipulate the expression into the divergence of some expression, minus another expression, and then apply the Gauss–Green–Ostrogradsky theorem to the divergence term. Let us look at some examples.

General strategy. Let us illustrate the general strategy using a one-dimensional example. Consider an expression of the form

$$(\partial_1 u) v \ .$$

To integrate by parts, the goal is to rewrite this in the form $(\partial_1(\ldots)) - (\ldots)$. The general pattern is to take the derivative operator off one of the factors in the product (the one that we aim to integrate by parts), apply the operator to the whole expression instead, and evaluate the result. In this case, we find

$$\partial_1(uv) = (\partial_1 u)v + u(\partial_1 v) \ .$$

Note this has generated the original expression as one of the terms on the right-hand side. Rearranging, we have

$$(\partial_1 u)v = \partial_1(uv) - u(\partial_1 v) \ .$$

Integrating on both sides over the domain (a, b), we obtain

$$\int_a^b (\partial_1 u)v \, \mathrm{d}x_1 = \int_a^b \partial_1(uv) \, \mathrm{d}x_1 - \int_a^b u(\partial_1 v) \, \mathrm{d}x_1 \ .$$

Now we can apply the fundamental theorem of calculus (that is, the one-dimensional analogue of the Gauss–Green–Ostrogradsky theorem) to the first term on the right-hand side. We obtain the result

$$\int_a^b (\partial_1 u)v \, \mathrm{d}x_1 = \big((uv)|_{x_1=b} - (uv)|_{x_1=a}\big) - \int_a^b u(\partial_1 v) \, \mathrm{d}x_1 \ ,$$

which is just the formula for one-dimensional integration by parts.

In multiple spatial dimensions, there are tensors of various ranks, and various different products that can be applied to them. The details of the procedure will thus differ depending on the form of the equation being worked on; there is no one formula for integration by parts in multiple dimensions. The Gauss–Green–Ostrogradsky theorem and the Stokes' theorem are appropriate tools when working with multiple dimensions. Let us now look at some examples where we may use the divergence theorem.

Green's first integral identity. As our first multidimensional example, let us consider Green's first integral identity. This identity is often used with the Poisson and heat equations. It follows from the Gauss–Green–Ostrogradsky theorem. Consider the expression

$$\nabla \cdot (\phi \nabla u) = \nabla \phi \cdot \nabla u + \phi(\nabla \cdot \nabla u) \equiv \nabla \phi \cdot \nabla u + \phi(\Delta u) \ ,$$

which yields, by rearranging,

$$\phi(\Delta u) = \nabla \cdot (\phi \nabla u) - \nabla \phi \cdot \nabla u \ .$$

Integrating on both sides, we obtain

$$\int_\Omega \phi(\Delta u) \, \mathrm{d}x = \int_\Omega \nabla \cdot (\phi \nabla u) \, \mathrm{d}x - \int_\Omega \nabla \phi \cdot \nabla u \, \mathrm{d}x \ .$$

Applying (A.3.1) to the divergence term on the right-hand side produces

$$\int_{\Omega} \phi(\Delta u)\, \mathrm{d}x = \int_{\Gamma} \mathbf{n} \cdot (\phi \nabla u)\, \mathrm{d}s - \int_{\Omega} \nabla \phi \cdot \nabla u\, \mathrm{d}x\ .$$

Rearranging the products in each integrand, we have

$$\int_{\Omega} (\Delta u)\phi\, \mathrm{d}x = \int_{\Gamma} (\mathbf{n} \cdot \nabla u)\phi\, \mathrm{d}s - \int_{\Omega} \nabla u \cdot \nabla \phi\, \mathrm{d}x\ ,$$

which is Green's first integral identity.

Balance of linear momentum. In mechanics, the dynamic balance of linear momentum in continuum form is described by (see, for example, Allen et al. [2])

$$\rho \frac{\mathrm{d}^2 \mathbf{u}}{\mathrm{d}t^2} - \nabla \cdot \boldsymbol{\sigma}^{\mathrm{T}} = \mathbf{f}\ , \qquad \text{(A.3.2)}$$

where ρ is the density of the material, $\mathbf{u}$ is the displacement, $\boldsymbol{\sigma}$ is the stress tensor (rank 2), and the vector $\mathbf{f}$ represents external body forces. When deriving the weak form of (A.3.2), we would like to integrate by parts in the stress term. For clarity of presentation, in the following, we will consider the stress term only; the inertial and body force terms do not require integration by parts.

Above, we gave the identity

$$\nabla \cdot (\mathbf{a} \cdot \mathbf{B}) = \nabla \cdot (\mathbf{B}^{\mathrm{T}} \cdot \mathbf{a}) = (\nabla \cdot \mathbf{B}^{\mathrm{T}}) \cdot \mathbf{a} + \mathbf{B} : \nabla \mathbf{a}\ .$$

Let $\mathbf{a} = \boldsymbol{\varphi}$ and $\mathbf{B} = \boldsymbol{\sigma}$, where the symbol $\boldsymbol{\varphi}$ denotes a vector-valued test function. We have

$$\nabla \cdot (\boldsymbol{\varphi} \cdot \boldsymbol{\sigma}) = (\nabla \cdot \boldsymbol{\sigma}^{\mathrm{T}}) \cdot \boldsymbol{\varphi} + \boldsymbol{\sigma} : \nabla \boldsymbol{\varphi}\ .$$

Integrating on both sides over the domain Ω, we obtain

$$\int_{\Omega} \nabla \cdot (\boldsymbol{\varphi} \cdot \boldsymbol{\sigma})\, \mathrm{d}x = \int_{\Omega} (\nabla \cdot \boldsymbol{\sigma}^{\mathrm{T}}) \cdot \boldsymbol{\varphi}\, \mathrm{d}x + \int_{\Omega} \boldsymbol{\sigma} : \nabla \boldsymbol{\varphi}\, \mathrm{d}x\ .$$

Rearranging yields

$$\int_{\Omega} (\nabla \cdot \boldsymbol{\sigma}^{\mathrm{T}}) \cdot \boldsymbol{\varphi}\, \mathrm{d}x = \int_{\Omega} \nabla \cdot (\boldsymbol{\varphi} \cdot \boldsymbol{\sigma})\, \mathrm{d}x - \int_{\Omega} \boldsymbol{\sigma} : \nabla \boldsymbol{\varphi}\, \mathrm{d}x\ .$$

Applying the Gauss–Ostrogradsky theorem (A.3.1) in the divergence term on the right-hand side, we obtain the following formula for integration by parts in the stress term:

$$\int_{\Omega} (\nabla \cdot \boldsymbol{\sigma}^{\mathrm{T}}) \cdot \boldsymbol{\varphi}\, \mathrm{d}x = \int_{\Gamma} \mathbf{n} \cdot (\boldsymbol{\varphi} \cdot \boldsymbol{\sigma})\, \mathrm{d}s - \int_{\Omega} \boldsymbol{\sigma} : \nabla \boldsymbol{\varphi}\, \mathrm{d}x\ . \qquad \text{(A.3.3)}$$

An interpretation of (A.3.3) common in the engineering sciences is to consider the test function $\boldsymbol{\varphi}$ as a virtual displacement, $\boldsymbol{\varphi} = \delta \mathbf{u}$. The terms in the linear momentum balance Eq. (A.3.2) represent forces. The test function is the dual variable of force, hence it can be considered as a displacement. The dot-product of a force and a displacement is the work done by the force; thus (A.3.3) (plus the weak forms of the inertial and body force terms) represents the virtual work.

In linear elasticity, the strain is defined as

$$\boldsymbol{\varepsilon} \equiv \operatorname{sym} \nabla \mathbf{u} \,, \tag{A.3.4}$$

where the symmetric gradient of a differentiable vector field $\mathbf{a}$ is defined as the following rank-2 tensor:

$$(\operatorname{sym} \nabla \mathbf{a})_{ij} \equiv \frac{1}{2}(\partial_i a_j + \partial_j a_i) = \frac{1}{2}\left(\nabla \mathbf{a} + (\nabla \mathbf{a})^{\mathrm{T}}\right)_{ij} \,.$$

On the right-hand side of the integration-by-parts formula (A.3.3), some authors use a virtual strain $\delta\boldsymbol{\varepsilon}$ instead of $\nabla\boldsymbol{\varphi} = \nabla(\delta\mathbf{u})$. This is possible due to the symmetry of the stress tensor $\boldsymbol{\sigma}$. Because $\boldsymbol{\sigma}$ is symmetric, for any differentiable vector-valued function $\boldsymbol{\varphi}$ it holds that

$$\begin{aligned}
\operatorname{sym} \nabla\boldsymbol{\varphi} : \boldsymbol{\sigma} &= \frac{1}{2}\left(\partial_i \varphi_j + \partial_j \varphi_i\right)\sigma_{ji} = \frac{1}{2}(\partial_i \varphi_j)\sigma_{ji} + \frac{1}{2}(\partial_j \varphi_i)\sigma_{ji} \\
&= \frac{1}{2}(\partial_i \varphi_j)\sigma_{ji} + \frac{1}{2}(\partial_j \varphi_i)\sigma_{ij}^{\mathrm{T}} = \frac{1}{2}\nabla\boldsymbol{\varphi} : \boldsymbol{\sigma} + \frac{1}{2}\nabla\boldsymbol{\varphi} : \boldsymbol{\sigma}^{\mathrm{T}} \\
&= \frac{1}{2}\nabla\boldsymbol{\varphi} : \boldsymbol{\sigma} + \frac{1}{2}\nabla\boldsymbol{\varphi} : \boldsymbol{\sigma} = \nabla\boldsymbol{\varphi} : \boldsymbol{\sigma} \,.
\end{aligned}$$

Thus, instead of the gradient of the virtual displacement, $\nabla\boldsymbol{\varphi} = \nabla(\delta\mathbf{u})$, one can use the virtual strain $\delta\boldsymbol{\varepsilon} = \operatorname{sym} \nabla(\delta\mathbf{u})$ as the test function in the last term of (A.3.3):

$$\int_{\Omega} \boldsymbol{\sigma} : \nabla(\delta\mathbf{u}) \, \mathrm{d}x = \int_{\Omega} \boldsymbol{\sigma} : (\operatorname{sym} \nabla(\delta\mathbf{u})) \, \mathrm{d}x = \int_{\Omega} \boldsymbol{\sigma} : \delta\boldsymbol{\varepsilon} \, \mathrm{d}x \,.$$

The constitutive stress-strain relation from a material model can be inserted into (A.3.3) to produce weak form linear momentum balance equations for specific classes of materials. For example, in linear elasticity one has the stress-strain relation

$$\boldsymbol{\sigma} = \mathbf{C} : \boldsymbol{\varepsilon} \,,$$

where $\boldsymbol{\varepsilon}$ is the strain tensor (rank 2), and the rank-4 symmetric tensor $\mathbf{C}$ contains the elastic coefficients (compliances) of the material. If one wishes to obtain an equation written directly in terms of the displacement $\mathbf{u}$, the final step is to insert the definition of the strain, for example (A.3.4).

Directional derivative of a vector field. Let us conclude this Appendix with an example that arises, at least almost exclusively, in problems involving axially moving materials. In the transformation from axially co-moving coordinates to laboratory coordinates, in the Eq. (A.3.2) the second time derivative becomes

$$\frac{\mathrm{d}^2\mathbf{u}}{\mathrm{d}t^2} = \frac{\mathrm{d}}{\mathrm{d}t}\left(\frac{\partial \mathbf{u}}{\partial t} + \mathbf{U}\cdot\nabla\mathbf{u}\right) = \frac{\partial^2 \mathbf{u}}{\partial t^2} + 2\mathbf{U}\cdot\nabla\left(\frac{\partial \mathbf{u}}{\partial t}\right) + (\mathbf{U}\cdot\nabla)(\mathbf{U}\cdot\nabla\mathbf{u})\,.$$

The steady-state part of the inertia operator has produced a second directional derivative along the direction of the velocity field $\mathbf{U}$ that describes the axial motion of the co-moving coordinate system.

Above, we derived the following three-vector identity involving the divergence operator:

$$\nabla\cdot(\mathbf{a}\cdot(\mathbf{b}\otimes\mathbf{c})) = \mathbf{b}\cdot(\mathbf{c}\cdot\nabla\mathbf{a}) + \mathbf{a}\cdot(\nabla\cdot(\mathbf{c}\otimes\mathbf{b}))\,.$$

Let $\mathbf{a} = (\mathbf{U}\cdot\nabla\mathbf{u})$, $\mathbf{b} = \boldsymbol{\varphi}$ and $\mathbf{c} = \mathbf{U}$. The interpretation is that $\mathbf{u}$ is the vector quantity being studied, $\mathbf{U}$ is the velocity field, and $\boldsymbol{\varphi}$ is a vector-valued test function. We have

$$\nabla\cdot((\mathbf{U}\cdot\nabla\mathbf{u})\cdot(\boldsymbol{\varphi}\otimes\mathbf{U})) = \boldsymbol{\varphi}\cdot(\mathbf{U}\cdot\nabla(\mathbf{U}\cdot\nabla\mathbf{u})) + (\mathbf{U}\cdot\nabla\mathbf{u})\cdot(\nabla\cdot(\mathbf{U}\otimes\boldsymbol{\varphi}))$$

Note the appearance of $(\mathbf{U}\cdot\nabla)^2\mathbf{u}$ in the first term on the right-hand side. Rearranging yields

$$\boldsymbol{\varphi}\cdot(\mathbf{U}\cdot\nabla(\mathbf{U}\cdot\nabla\mathbf{u})) = \nabla\cdot((\mathbf{U}\cdot\nabla\mathbf{u})\cdot(\boldsymbol{\varphi}\otimes\mathbf{U})) - (\mathbf{U}\cdot\nabla\mathbf{u})\cdot(\nabla\cdot(\mathbf{U}\otimes\boldsymbol{\varphi}))\,.$$

Integrating over the domain Ω, we obtain

$$\int_\Omega \boldsymbol{\varphi}\cdot(\mathbf{U}\cdot\nabla(\mathbf{U}\cdot\nabla\mathbf{u}))\,\mathrm{d}x = \int_\Omega \nabla\cdot((\mathbf{U}\cdot\nabla\mathbf{u})\cdot(\boldsymbol{\varphi}\otimes\mathbf{U}))\,\mathrm{d}x - \int_\Omega (\mathbf{U}\cdot\nabla\mathbf{u})\cdot(\nabla\cdot(\mathbf{U}\otimes\boldsymbol{\varphi}))\,\mathrm{d}x\,.$$

We may now apply the Gauss–Green–Ostrogradsky theorem (A.3.1) to the divergence term on the right-hand side. This results in the following integration-by-parts formula for the second directional derivative:

$$\int_\Omega \boldsymbol{\varphi}\cdot(\mathbf{U}\cdot\nabla(\mathbf{U}\cdot\nabla\mathbf{u}))\,\mathrm{d}x = \int_\Gamma \mathbf{n}\cdot((\mathbf{U}\cdot\nabla\mathbf{u})\cdot(\boldsymbol{\varphi}\otimes\mathbf{U}))\,\mathrm{d}s - \int_\Omega (\mathbf{U}\cdot\nabla\mathbf{u})\cdot(\nabla\cdot(\mathbf{U}\otimes\boldsymbol{\varphi}))\,\mathrm{d}x\,. \tag{A.3.5}$$

This formula can be used to treat the inertial term in problems of axially moving materials. For a practical example of the use of this result in the analysis of in-plane deformation of a travelling viscoelastic sheet, see Kurki et al. [4].

Generalizing slightly, in the three-vector identity above, let $\mathbf{a}$ be any differentiable vector field, and $\mathbf{b} = \boldsymbol{\varphi}$, $\mathbf{c} = \mathbf{U}$. Here $\boldsymbol{\varphi}$ is still a vector-valued test function, but $\mathbf{U}$ represents an arbitrary direction field. We have

$$\nabla\cdot(\mathbf{a}\cdot(\boldsymbol{\varphi}\otimes\mathbf{U})) = \boldsymbol{\varphi}\cdot(\mathbf{U}\cdot\nabla\mathbf{a}) + \mathbf{a}\cdot(\nabla\cdot(\mathbf{U}\otimes\boldsymbol{\varphi}))\,.$$

Following the same steps as above, we obtain

$$\int_\Omega \boldsymbol{\varphi}\cdot(\mathbf{U}\cdot\nabla\mathbf{a})\,\mathrm{d}x = \int_\Gamma \mathbf{n}\cdot(\mathbf{a}\cdot(\boldsymbol{\varphi}\otimes\mathbf{U}))\,\mathrm{d}s - \int_\Omega \mathbf{a}\cdot(\nabla\cdot(\mathbf{U}\otimes\boldsymbol{\varphi}))\,\mathrm{d}x\,. \tag{A.3.6}$$

This result allows to integrate by parts in any directional derivative of a vector field.

References

1. Adams RA, Essex C (2010) Calculus: a complete course, 7th edn. Pearson
2. Allen III MB, Herrera I, Pinder GF (1988) Numerical modeling in science and engineering. Wiley Interscience
3. Flügge W (1972) Tensor analysis and continuum mechanics. Springer
4. Kurki M, Jeronen J, Saksa T, Tuovinen T (2016) The origin of in-plane stresses in axially moving orthotropic continua. Int J Solids Struct. https://doi.org/10.1016/j.ijsolstr.2015.10.027
5. Sokolnikoff IS (1951) Tensor analysis: theory and applications. Wiley

Appendix B
Numerical Integration of ODEs and Semidiscrete PDEs

Ordinary differential equations (ODEs) are often numerically treated (possibly after transformation) in the form of the first-order initial value problem

$$\frac{\partial u}{\partial t} = f(\,u(t),\, t\,)\,, \tag{B.0.1}$$

$$u(0) = u_0\,. \tag{B.0.2}$$

In this Appendix we will summarize and comment on some methods for the numerical solution of problem (B.0.1)–(B.0.2), concentrating especially on nonlinear problems, where we make no assumption about the structure of f beyond the minimal necessary continuity for each considered method. Beyond the very basics, we build a brief theoretical development of iterative implicit methods based on the Banach fixed point theorem, which ties in to our discussion of classical implicit methods, and finally provide an exposition of the time-discontinuous Galerkin method (dG).

A close relative of (B.0.1)–(B.0.2) is the initial value problem of a semilinear system of ordinary differential equations

$$\mathbf{M}\frac{\partial \mathbf{u}}{\partial t} = \mathbf{f}(\,\mathbf{u}(t),\, t\,)\,, \tag{B.0.3}$$

$$\mathbf{u}(0) = \mathbf{u}_0\,, \tag{B.0.4}$$

where $\mathbf{M}$ is a matrix (usually called the mass matrix), and $\mathbf{u}$ and $\mathbf{f}$ are vector-valued. This form is commonly encountered after space discretization of partial differential equations (PDEs) involving only first time derivatives, such as the heat equation, the first-order transport equation, and the Euler and Navier–Stokes fluid flow equations. PDEs that are of second order in time, such as the wave equation and equations describing mechanical vibrations in solids, can be also reduced to this form by defining $\mathbf{v} = \partial\mathbf{u}/\partial t$, and then writing (B.0.3)–(B.0.4) for $\mathbf{w} = [\mathbf{u}, \mathbf{v}]^{\mathrm{T}}$.

Nontrivial matrices $\mathbf{M}$ arise from space discretization by the finite element method, when the basis is not L^2-orthogonal, leading to connections between the

N. Banichuk et al., *Stability of Axially Moving Materials*, Solid Mechanics and Its Applications 259, https://doi.org/10.1007/978-3-030-23803-2

degrees of freedom in the mass matrix $\mathbf{M}$. Finite difference methods and the moving least squares meshless method (MLS) lead to $\mathbf{M}$ being the unit matrix, as do Galerkin methods when the basis is orthonormal. Orthogonal bases lead to diagonal mass matrices.

In (B.0.3)–(B.0.4), for each component equation i, the load component f_i depends on all components of $\mathbf{u}$, making the treatment of (B.0.3)–(B.0.4) slightly different from that of the one degree of freedom model problem (B.0.1)–(B.0.2).

Formally, provided that $\mathbf{M}$ is invertible, which is usually true in problems with a physical background, we may multiply (B.0.3) from the left by the inverse matrix $\mathbf{M}^{-1}$, obtaining

$$\frac{\partial \mathbf{u}}{\partial t} = \mathbf{M}^{-1}\mathbf{f}(\,\mathbf{u}(t),\, t\,)\ . \tag{B.0.5}$$

In practice this is not possible beyond very small systems due to the prohibitive cost of matrix inversion. In practical numerics, we proceed by recognizing that we have explicit access to $\mathbf{f}$ (given a candidate $\mathbf{u}$; in nonlinear problems some kind of iteration is often needed), we also have $\mathbf{M}$, and we need $\mathbf{M}^{-1}\mathbf{f}$. Defining the effective load

$$\mathbf{g}(\,\mathbf{u}(t),\, t\,) = \mathbf{M}^{-1}\mathbf{f}(\,\mathbf{u}(t),\, t\,)\ , \tag{B.0.6}$$

equation (B.0.5) becomes

$$\frac{\partial \mathbf{u}}{\partial t} = \mathbf{g}(\,\mathbf{u}(t),\, t\,)\ , \tag{B.0.7}$$

making it possible, in principle, to integrate each component equation separately, using methods designed for the problem (B.0.1)–(B.0.2). Multiplying (B.0.6) from the left by $\mathbf{M}$, we obtain

$$\mathbf{M}\,\mathbf{g} = \mathbf{f}\ . \tag{B.0.8}$$

By solving the linear equation system (B.0.8) for the unknown vector $\mathbf{g}$, we obtain the value of $\mathbf{M}^{-1}\mathbf{f}$ whenever it is needed. This turns $\mathbf{M}^{-1}(\ldots)$ into a linear operator, which can be evaluated in practice also for large systems.

If $\mathbf{M}$ is independent of $\mathbf{u}$ and t, it may be possible to speed up the solution of (B.0.8). For example, in moderately sized problems, one may precompute the LU decomposition of $\mathbf{M}$, and then reuse it for each new $\mathbf{f}$.

For nonlinear problems, solving (B.0.8) may be expensive, because (B.0.8) may need to be solved several times for each iteration of the fixed point loop. Equation (B.0.5) is typically converted into an integral form; the right-hand side must be evaluated at the quadrature points when the integral is approximated numerically.

When solving initial-boundary value problems, one must also make sure the boundary conditions hold. If the representation of the unknown field is chosen such that the boundary conditions are automatically enforced, there is no need to do anything. For example, in finite element methods, the degrees of freedom determined by Dirichlet boundary conditions are simply eliminated from the discrete equation system (see, e.g., Zienkiewicz et al. [3]), and the Neumann boundary conditions are weakly enforced by adding the appropriate contributions to the discrete equation system. This is essentially a clever choice of representation, which encodes the boundary condition information from the PDE problem into the ODE system (B.0.3).

However, finite difference (and other collocation) methods may require manual enforcement of boundary conditions, especially when prescribing derivatives at a boundary, since the collocation representation does not encode such information in a natural fashion. The ODE system (B.0.3) may predict values for $\mathbf{u}$ at the boundary degrees of freedom that conflict with the boundary conditions. Then special measures must be taken to enforce the BCs. In that case, when using methods that compute intermediate results, such as the explicit Runge–Kutta methods summarized in the next section, it may improve the quality of the results in practice if the boundary conditions are enforced not only for the final result from the timestep, but also for the intermediate field values. Also, it may be useful to look at what the boundary conditions imply for $\partial \mathbf{u}/\partial t$ at the boundary, and enforce this in the $\mathbf{k}_j$, since they represent approximations to the time derivative.

B.1 Explicit Runge–Kutta Methods

Explicit methods are particularly convenient, since they require access only to already available values. Some explicit Runge–Kutta (RK) methods have been collected below; note these cannot be unconditionally stable. The methods are presented in an algorithmic form. For $\mathbf{u}$ and t, the subscripts n and $n+1$ refer to timestep numbers, such as $\mathbf{u}_n \equiv \mathbf{u}(t = t_n)$. The subscripts for the $\mathbf{k}_j$ simply label the quantities.

When $\mathbf{M}$ is not the identity matrix, observe that by (B.0.1) the $\mathbf{k}_j = \mathbf{f}(\ldots)$ represent approximations to the derivative. Hence if the ODE system has a nontrivial mass matrix $\mathbf{M}$, evaluating $\mathbf{f}$ actually gives

$$\mathbf{M}\,\mathbf{k}_j = \mathbf{f}(\ldots)$$

and $\mathbf{k}_j$ can be obtained by solving this linear equation system. This solution process must be repeated for each j, so in this setting, for example RK4 requires solving four linear equation systems per timestep.

The first-order explicit RK method is the forward Euler method, which is extremely unstable, and omitted here.

The **second-order explicit Runge–Kutta method (RK2)** is given in parametric form by

$$\mathbf{k}_1 = \mathbf{f}(\mathbf{u}_n,\ t_n)\ , \tag{B.1.1}$$

$$\mathbf{k}_2 = \mathbf{f}(\mathbf{u}_n + \beta\,\Delta t\,\mathbf{k}_1,\ t_n + \beta\,\Delta t)\ , \tag{B.1.2}$$

$$\mathbf{u}_{n+1} = \mathbf{u}_n + \Delta t\left[(1 - \frac{1}{2\beta})\mathbf{k}_1 + \frac{1}{2\beta}\mathbf{k}_2\right]\ . \tag{B.1.3}$$

Classical choices for the parameter β are $1/2$ (explicit midpoint method), $2/3$ (Ralston's method), and 1 (Heun's method, also known as the explicit trapezoid rule). **Kutta's third-order method (RK3)** is

$$\mathbf{k}_1 = \mathbf{f}(\mathbf{u}_n,\ t_n)\ , \tag{B.1.4}$$

$$\mathbf{k}_2 = \mathbf{f}(\mathbf{u}_n + \frac{\Delta t}{2}\mathbf{k}_1,\ t_n + \frac{\Delta t}{2})\ , \tag{B.1.5}$$

$$\mathbf{k}_3 = \mathbf{f}(\mathbf{u}_n - \Delta t\,\mathbf{k}_1 + 2\Delta t\,\mathbf{k}_2,\ t_n + \Delta t)\ , \tag{B.1.6}$$

$$\mathbf{u}_{n+1} = \mathbf{u}_n + \frac{\Delta t}{6}(\mathbf{k}_1 + 4\mathbf{k}_2 + \mathbf{k}_3)\ . \tag{B.1.7}$$

The classical **fourth-order Runge–Kutta (RK4)**, which is the most popular of the RK methods, is

$$\mathbf{k}_1 = \mathbf{f}(\mathbf{u}_n,\ t_n)\ , \tag{B.1.8}$$

$$\mathbf{k}_2 = \mathbf{f}(\mathbf{u}_n + \frac{\Delta t}{2}\mathbf{k}_1,\ t_n + \frac{\Delta t}{2})\ , \tag{B.1.9}$$

$$\mathbf{k}_3 = \mathbf{f}(\mathbf{u}_n + \frac{\Delta t}{2}\mathbf{k}_2,\ t_n + \frac{\Delta t}{2})\ , \tag{B.1.10}$$

$$\mathbf{k}_4 = \mathbf{f}(\mathbf{u}_n + \Delta t\,\mathbf{k}_3,\ t_n + \Delta t)\ , \tag{B.1.11}$$

$$\mathbf{u}_{n+1} = \mathbf{u}_n + \frac{\Delta t}{6}(\mathbf{k}_1 + 2\mathbf{k}_2 + 2\mathbf{k}_3 + \mathbf{k}_4)\ . \tag{B.1.12}$$

Kutta also proposed another fourth-order explicit RK method, known as **the 3/8 rule**:

$$\mathbf{k}_1 = \mathbf{f}(\mathbf{u}_n,\ t_n)\ , \tag{B.1.13}$$

$$\mathbf{k}_2 = \mathbf{f}(\mathbf{u}_n + \frac{1}{3}\Delta t\,\mathbf{k}_1,\ t_n + \frac{1}{3}\Delta t)\ , \tag{B.1.14}$$

$$\mathbf{k}_3 = \mathbf{f}(\mathbf{u}_n - \frac{1}{3}\Delta t\,\mathbf{k}_1 + \Delta t\,\mathbf{k}_2,\ t_n + \frac{2}{3}\Delta t)\ , \tag{B.1.15}$$

$$\mathbf{k}_4 = \mathbf{f}(\mathbf{u}_n + \Delta t\,\mathbf{k}_1 - \Delta t\,\mathbf{k}_2 + \Delta t\,\mathbf{k}_3,\ t_n + \Delta t)\ , \tag{B.1.16}$$

$$\mathbf{u}_{n+1} = \mathbf{u}_n + \frac{\Delta t}{8}(\mathbf{k}_1 + 3\mathbf{k}_2 + 3\mathbf{k}_3 + \mathbf{k}_4)\ . \tag{B.1.17}$$

B.2 Classical Implicit Methods

In principle, the classical **implicit midpoint rule (IMR)** works as follows:

$$\mathbf{k} = \mathbf{f}(\mathbf{u}_{n+1/2},\ t_{n+1/2})\ , \tag{B.2.1}$$

$$\mathbf{u}_{n+1} = \mathbf{u}_n + \Delta t\,\mathbf{k}\ , \tag{B.2.2}$$

where the half-timestep (midpoint) value of the state vector is taken as the linear interpolant

$$\mathbf{u}_{n+1/2} \approx \frac{1}{2}(\mathbf{u}_n + \mathbf{u}_{n+1}) \,. \tag{B.2.3}$$

The accuracy of the method is $O(\,(\Delta t)^2\,)$; furthermore, the error tends to oscillate around zero. The method is symplectic and approximately conserves energy, which are useful properties especially for problems in Hamiltonian mechanics. The method is not unconditionally stable; a problem-specific maximum timestep size exists above which numerical stability is lost.

As all implicit methods, IMR requires access to $\mathbf{u}_{n+1}$, which is unknown, so the interpolant (B.2.3) is not directly computable. For linear problems, the standard approach is to set up a linear equation system, inserting (B.2.3) into (B.2.1), expanding $\mathbf{f}$ for each specific problem, and then moving terms with $\mathbf{u}_{n+1}$ to the left-hand side and $\mathbf{u}_n$ to the right-hand side. The resulting linear equation system, when numerically solved, yields $\mathbf{u}_{n+1}$.

For nonlinear problems in general, no standard form is possible, since each nonlinear problem class is different. A general approach however exists. We replace $\mathbf{u}_{n+1}$ in (B.2.3) with some computable approximation $\mathbf{u}^*$:

$$\mathbf{u}_{n+1/2} \approx \frac{1}{2}(\mathbf{u}_n + \mathbf{u}^*) \,. \tag{B.2.4}$$

The result is then refined by iteration. Given an initial guess for $\mathbf{u}^*$, evaluate the approximate $\mathbf{u}_{n+1/2}$ from (B.2.4), insert it into (B.2.1), and evaluate (B.2.2), interpreting the left-hand side as the new value for $\mathbf{u}^*$. Then repeat. Continue iteration until $\mathbf{u}^* \approx \mathbf{u}_{n+1}$ has converged, or a prescribed maximum number of iterations is reached.

An easy choice for the initial guess for $\mathbf{u}^*$ is to use $\mathbf{u}_n$. Then, initially $\mathbf{u}_{n+1/2} = \mathbf{u}_n$, and we see from (B.2.1)–(B.2.2) that the first iteration of IMR will compute a forward Euler prediction. Thus, the do-nothing initial guess is effectively after one iteration equivalent with using a forward Euler initial guess.

The requirement of contractivity of fixed-point iteration is the theoretical reason behind the upper limit for timestep size in iterative IMR for nonlinear problems; unlike in linear IMR, where the maximum timestep size arises from a von Neumann stability analysis of the linear system.

Sometimes unconditional stability and high numerical dissipation are desirable properties. Then the classical **backward Euler method (BE)** is useful:

$$\mathbf{k} = \mathbf{f}(\mathbf{u}_{n+1},\ t_{n+1}) \tag{B.2.5}$$

$$\mathbf{u}_{n+1} = \mathbf{u}_n + \Delta t\, \mathbf{k} \,, \tag{B.2.6}$$

The iterative approximation is simply

$$\mathbf{u}_{n+1} \approx \mathbf{u}^* \,, \tag{B.2.7}$$

and the iteration proceeds using the same procedure as for IMR. The accuracy of BE is $O(\Delta t)$.

Note unconditional stability is only guaranteed for linear problems, using the approach of reduction to a linear equation system (as outlined for IMR above). In case of iterative solution of nonlinear problems, the contractivity of the iterative procedure (B.2.7), (B.2.5), (B.2.6) usually fails above some problem-specific critical timestep size.

B.3 The Theoretical Basis of Iterative Implicit Methods

Let us look at why a maximum timestep size arises in implicit methods for nonlinear problems, when the method is cast as an iterative procedure. Using IMR as an example, mathematically the iteration procedure (B.2.4), (B.2.1), (B.2.2) is based on contractive self-maps on metric spaces. Each iteration maps the metric space to itself. Recall that a metric space is an ordered pair (X, d), where X is a set, and d is a metric, which defines the distance between any two points in X.

The **Banach fixed point theorem** (see [2], p. 170 ff.) states that on a metric space, if a self-map is contractive, then a fixed point exists for the map, and moreover, the fixed point is unique. In other words, let (X, d) be a metric space. For a contractive self-map $T : X \to X$, there exists a unique point x^* such that $T(x^*) = x^*$. The mapping T is contractive if it satisfies the Lipschitz condition $d(\,T(x_1),\ T(x_2)\,) \leq q\,d(x_1, x_2)$, where $q \in [0, 1)$ is the Lipschitz constant of the mapping T, and $x_1, x_2 \in X$ are arbitrary.

Lipschitz continuity is a fairly stringent requirement, and it may also depend on the domain of the map. For example, in one dimension, the map $x \mapsto x^2$ is Lipschitz on any finite interval, but on the whole of $\mathbb{R}$ it is not. Note also that this particular example is contractive only in the set $|x| \leq 1$.

The Banach fixed point theorem is closely related to the Picard–Lindelöf theorem (see [2], p. 177 ff.), which is concerned with the existence and uniqueness of solutions to the first-order initial value problem. Indeed, the first is invoked in the proof of the latter.

The iterative IMR procedure presented above is essentially a discrete application of the Picard–Lindelöf idea of recasting the first-order initial value problem (B.0.1)–(B.0.2) as a fixed-point problem for an integral operator. Observe that the Eq. (B.2.2) is a discrete approximation of the integral

$$\mathbf{u}_{n+1} = \mathbf{u}_n + \int_{t_n}^{t_{n+1}} \mathbf{f}(\,\mathbf{u}(t),\ t\,)\ \mathrm{d}t\ , \tag{B.3.1}$$

which comes from the fundamental theorem of calculus, after using the standard form of the problem, Eq. (B.0.1), to represent the derivative. The same observation holds for the formula for $\mathbf{u}_{n+1}$ in the backward Euler method, Eq. (B.2.6).

Equation (B.3.1) is the integral that is considered in the Picard–Lindelöf theorem. The initial value $\mathbf{u}_n$ in (B.3.1) is effectively a constant, but the $\mathbf{u}(t)$ in the integrand may need to be evaluated anywhere in the interval $[t_n, t_{n+1}]$, depending on the details of the numerical method.

The observation also holds for the formulas for $\mathbf{u}_{n+1}$ in the explicit RK methods, namely Eqs. (B.1.3), (B.1.7), (B.1.12) and (B.1.17). Being explicit methods, however, the approximation to the integral is there constructed in such a way that no future data is needed. For explicit RK methods, although they too can be thought of as being based on rewriting the problem in the integral form (B.3.1), there is no need to invoke the fixed point theorem, since all needed quantities can be computed by an explicit sequence of operations.

To produce a practical numerical method for (B.3.1), two things must be specified. First, because the $\mathbf{u}(t)$ in the integrand is unknown, it must be modeled. An algorithm is specified to construct an approximate $\widetilde{\mathbf{u}}(t) \approx \mathbf{u}(t)$, given some discrete set of data, for example, point values at a set of points in the interval $[t_n, t_{n+1}]$, or alternatively, coefficients for a Galerkin representation. At each timestep, we seek to solve the approximate problem

$$\mathbf{u}_{n+1} \approx \mathbf{u}_n + \int_{t_n}^{t_{n+1}} \mathbf{f}(\widetilde{\mathbf{u}}(t),\, t)\, \mathrm{d}t\ . \tag{B.3.2}$$

At the beginning of the timestep, $\widetilde{\mathbf{u}}(t_n) = \mathbf{u}_n$. In an implicit method, $\widetilde{\mathbf{u}}(t_{n+1}) = \mathbf{u}_{n+1}$.

In this view, the choice of the model $\widetilde{\mathbf{u}}(t)$ is the difference between explicit and implicit methods. In explicit methods, the model is constructed such that for any fixed $\tau \in (t_n, t_{n+1}]$, the value $\widetilde{\mathbf{u}}(\tau)$ is based on information at $t < \tau$ only. Implicit methods lift this restriction, allowing the use of information from anywhere in $[t_n, t_{n+1}]$.

Second, an algorithm to evaluate the integral must be specified. A typical choice is a quadrature formula, approximating the integral as a sum of weighted values of the integrand at a preselected set of points.

To bring (B.3.2) into a form where fixed point theorems are applicable, we define the self-map (on the space where the instantaneous field values $\mathbf{u}_{n+1}$ live)

$$\mathbf{u}_{n+1}^{(i+1)} = \mathbf{u}_n + \int_{t_n}^{t_{n+1}} \mathbf{f}(\widetilde{\mathbf{u}}^{(i)}(t),\, t)\, \mathrm{d}t\ , \tag{B.3.3}$$

where the superscript in parentheses indexes the sequence of iterates. Considering the class of methods which approximate the integral as a quadrature, such as IMR and BE, we further approximate

$$\mathbf{u}_{n+1}^{(i+1)} \approx \mathbf{u}_n + \Delta t \sum_{k=1}^{N} q_k\, \mathbf{f}(\widetilde{\mathbf{u}}^{(i)}(\tau_k),\, \tau_k)\ , \quad \tau_k \equiv t_n + p_k \Delta t\ , \tag{B.3.4}$$

where N is the number of quadrature points, q_k are their weights and $p_k \in [0, 1]$ their positions on the standard unit interval.

In collocation methods, the model iterate $\widetilde{\mathbf{u}}^{(i)}(t)$ depends on $\mathbf{u}_n$, $\mathbf{u}_{n+1}^{(i)}$, and possibly a set of in-between point values that must be determined internally. In the case of IMR and BE, only the endpoint values are used. For these methods, we choose $\widetilde{\mathbf{u}}^{(i)}(t)$ as the linear interpolant expressed on the standard unit interval

$$\widetilde{\mathbf{u}}^{(i)}(p) = (1-p)\,\mathbf{u}_n + p\,\mathbf{u}_{n+1}^{(i)}\,, \quad p \in [0,1]\,, \tag{B.3.5}$$

transforming the quadrature (B.3.4) into

$$\mathbf{u}_{n+1}^{(i+1)} = \mathbf{u}_n + \Delta t \sum_{k=1}^{N} q_k\, \mathbf{f}(\,(1-p_k)\,\mathbf{u}_n + p_k\,\mathbf{u}_{n+1}^{(i)}\,,\ \tau_k\,)\,, \quad \tau_k \equiv t_n + p_k \Delta t\,. \tag{B.3.6}$$

In IMR, the quadrature is simply the midpoint rule $N = 1$, $p_1 = 1/2$, $q_1 = 1$, leading to

$$\mathbf{u}_{n+1}^{(i+1)} = \mathbf{u}_n + \Delta t\, \mathbf{f}(\,\frac{1}{2}\mathbf{u}_n + \frac{1}{2}\mathbf{u}_{n+1}^{(i)}\,,\ t_n + \frac{1}{2}\Delta t\,)\,, \tag{B.3.7}$$

which matches (B.2.1), (B.2.2) and (B.2.4).

On the left-hand side of (B.3.6) only the iterate $i+1$ appears, and on the right-hand side only i. Hence (B.3.6) can be explicitly iterated to produce a sequence of values $\mathbf{u}_{n+1}^{(i+1)}$, $i = 0, 1, 2, \ldots$; it is the mapping $T:\ \mathbf{u}_{n+1}^{(i)} \mapsto \mathbf{u}_{n+1}^{(i+1)}$ for which a fixed point is being sought. By the Banach fixed point theorem, if T is contractive (i.e., the mapped points $T(\mathbf{u}_{n+1}^{(i_1)})$ and $T(\mathbf{u}_{n+1}^{(i_2)})$ are closer together than the original points $\mathbf{u}_{n+1}^{(i_1)}$ and $\mathbf{u}_{n+1}^{(i_2)}$, for any choice of original points), it has a unique fixed point. Moreover, the fixed point is the solution $\mathbf{u}_{n+1}$. Note that in (B.3.6), $\mathbf{f}$ and $\mathbf{u}_n$ can be considered fixed, whereas $\mathbf{u}_{n+1}^{(0)}$ and Δt are free.

As a final remark on Eq. (B.3.6), the technique of relaxation is sometimes offered as a convergence aid for fixed point iteration methods. For example, for IMR, we replace the update procedure (B.3.7) with the following modified version:

$$\widehat{\mathbf{u}}_{n+1}^{(i+1)} = \mathbf{u}_n + \Delta t\, \mathbf{f}(\,\frac{1}{2}\mathbf{u}_n + \frac{1}{2}\mathbf{u}_{n+1}^{(i)}\,,\ t_n + \frac{1}{2}\Delta t\,)\,, \tag{B.3.8}$$

$$\mathbf{u}_{n+1}^{(i+1)} = (1-\alpha)\mathbf{u}_{n+1}^{(i)} + \alpha\widehat{\mathbf{u}}_{n+1}^{(i+1)}\,, \quad \alpha \in (0,1) \text{ given.} \tag{B.3.9}$$

The first equation is the same as (B.3.7); we have simply renamed the left-hand side. Subtracting $\mathbf{u}_{n+1}^{(i)}$ from (B.3.9), we see that

$$\mathbf{u}_{n+1}^{(i+1)} - \mathbf{u}_{n+1}^{(i)} = \alpha(\widehat{\mathbf{u}}_{n+1}^{(i+1)} - \mathbf{u}_{n+1}^{(i)})\,. \tag{B.3.10}$$

The application of the classical relaxation technique thus makes the distance between successive iterates smaller. Compared to the version without relaxation, this has two effects: contractivity is improved, but on the other hand, convergence requires more iterations, because the steps are smaller. Despite the slower convergence, this may be useful in cases where contractivity would otherwise fail. The same remark applies to any iterative implicit method, replacing (B.3.8) with the appropriate update formula.

In this connection we must mention that based on the Picard–Lindelöf theorem and Ostrowski estimates, a very promising implicit method for reliable integration of the first-order initial value problem has been recently developed by [1]. The method provides error bounds that are both computable and guaranteed, and is the first known method that can integrate, down to a prescribed error tolerance, the first-order initial value problem $\partial u/\partial t = f(u(t), t), u(0) = u_0$. Methods to do this for known functions have been available for a long time; practically all numerical quadrature codes provide some form of guaranteed error control. What is new is the ability to do this for unknown functions, of which we know only that they are solutions of the first-order initial value problem.

Both the classical iterative solution procedure for implicit $\mathbf{u}_{n+1}$ for arbitrary nonlinear problems, and the new method by Matculevich et al. [1], can be considered to have a common theoretical origin in the Picard–Lindelöf problem recasting approach.

This is hardly the standard way to look at the classical implicit midpoint rule. Viewing IMR primarily in the context of linear problems leads into a completely different direction, even giving the method a completely different theoretical origin, in finite differences. Although (B.3.7) can be re-interpreted as a finite difference expression, keep in mind that Eq. (B.3.1), or indeed the entire derivation of (B.3.7), made no use of finite differences.

Specialization of IMR to linear problems enables additional formal manipulations to be performed, leading to linear equation systems and von Neumann stability analysis. The linear and iterative nonlinear versions of IMR do not have much in common, beside the core idea of seeking an implicit approximation at the midpoint—which is formally encoded into Eqs. (B.2.1)–(B.2.3).

We note that the rate of convergence of the sequence of fixed point iterates x_n is characterized by the following equivalent expressions:

$$d(x^*, x_n) \le \frac{q^n}{1-q} d(x_1, x_0) , \tag{B.3.11}$$

$$d(x^*, x_{n+1}) \le \frac{q}{1-q} d(x_{n+1}, x_n) , \tag{B.3.12}$$

$$d(x^*, x_{n+1}) \le q\, d(x^*, x_n) . \tag{B.3.13}$$

Any value of $q \in [0, 1)$ satisfying these relations for a given contractive mapping T is a Lipschitz constant of T; the smallest possible value is the best Lipschitz constant of T.

Relations (B.3.11)–(B.3.13) suggest that the points x_n form a geometric sequence converging toward the limit point x^*. This also suggests a linear convergence rate, as is well known for fixed point procedures; the geometric nature of the sequence, roughly speaking, adds a constant number of correct digits in each iteration.

If the sequence is at least approximately geometric, q can be approximated in a simple manner as

$$q \approx \frac{d(x_{n+1}, x_n)}{d(x_n, x_{n-1})} \,. \tag{B.3.14}$$

The expression (B.3.14) should stay approximately constant throughout the iteration. One can then use (B.3.11) or (B.3.12) to estimate the remaining distance to the (unknown) fixed point, providing a computable *error indicator* that can be used for error control. The test (B.3.14) also tells us when contractivity has failed and hence, when the sequence x_n cannot be guaranteed to converge.

Contractivity may be local. Typically, for fixed-point procedures arising in numerical integration of differential equations, contractivity occurs if the timestep Δt is small enough and/or the initial guess for $\mathbf{u}^*$ (iterate x_0) is close enough to the solution. In this case, the set X must be chosen appropriately such that T remains contractive within it; then X can be considered the basin of attraction of the fixed point. What typically happens outside the basin of attraction is that the contractivity of the iterative procedure is violated, and the sequence fails to converge.

The implied comparison with dynamical systems, which typically have several basins of attraction, may however be a bit misleading, so the following must be emphasized. The **Picard–Lindelöf theorem** states that the solution of the first-order initial value problem $\partial u / \partial t = f(u(t), t)$, $u(t_0) = u_0$ exists *and is unique* on an interval $t \in [t_0 - \varepsilon, t_0 + \varepsilon]$ for some $\varepsilon > 0$, if the load function f is uniformly Lipschitz continuous in u, and continuous in t. Concerning the design of numerical methods, it may be of interest to note that in the proof, the Banach fixed point theorem is invoked with the initial iterate defined as $u(t) = u_0$ on the whole interval.

Obviously, different initial values $\mathbf{u}_n$ will typically lead to different solutions $\mathbf{u}_{n+1}$. What the Banach and PL theorems say is only that *for a given initial value* $\mathbf{u}_n$, the choice of the initial iterate within the set X does not matter; if contractivity holds, then the unique solution $\mathbf{u}_{n+1}$ will be found.

Finally, this is the theoretical origin of the upper limit on timestep size for IMR when cast as an iterative procedure for nonlinear problems. Contrast this with IMR specialized for linear problems, where the upper limit arises via von Neumann stability analysis. There is no requirement for these maximum timestep sizes to be the same.

If we use an explicit integration method to generate the initial guess for $\mathbf{u}^*$, the above basin-of-attraction consideration is separate from the usual stability limit of that method, since that method will not be used for the actual integration. This is true regardless of how the stability limit was derived; the stability limit only tells us whether integration *using that method* would remain stable. But here, the relevant question is entirely different: whether our iterative procedure, with the given initial guess, will converge to the fixed point $\mathbf{u}_{n+1}$ or not.

Is there only one basin of attraction? By the Picard–Lindelöf theorem, if the timestep is small enough, existence and uniqueness hold for the solution of the first-order initial value problem. Thus, it follows that, with Δt small enough, and with $\mathbf{u}_n$ and $\mathbf{f}$ kept fixed, there can be no competing basins of attraction leading to other solutions. However, this is assuming that the approximations made when constructing the numerical integration method have not changed the qualitative behavior.

Finally, what about global existence and the uniqueness of the solution on $t \in [0, t_{\mathrm{f}}]$, where t_{f} is the end time of the simulation? The key observation is we may treat each timestep as a separate initial value problem. On the interval $t \in [t_n, t_{n+1}]$, we use the initial condition $\mathbf{u}(t_n) = \mathbf{u}_n$, and the duration of integration, considering only this subproblem, is only Δt. Then we obtain the final value $\mathbf{u}(t_{n+1}) = \mathbf{u}_{n+1}$, which becomes the initial condition for the next timestep. Thus, even if the ε in the Picard–Lindelöf theorem turns out to be small, this does not pose a problem for the global existence and uniqueness of the solution.

B.4 Time-Discontinuous Galerkin (dG)

Traditionally, ODE integration has been based on difference methods, even though the space discretization of PDEs has long been routinely treated in weak form using Galerkin methods such as finite elements. Taking a modern approach, the family of time-discontinuous Galerkin methods seeks a weak solution of the initial value problem (B.0.1)–(B.0.2).

To cast our problem into a weak form, we begin by multiplying (B.0.1) by a test function $w(t)$, and integrate in time from 0 to the simulation end time t_{f}:

$$\int_0^{t_{\mathrm{f}}} \frac{\partial u}{\partial t} w \, \mathrm{d}t = \int_0^{t_{\mathrm{f}}} f(\, u(t),\, t\,)\, w \, \mathrm{d}t \;, \quad \forall w \,. \tag{B.4.1}$$

The quantifier is taken over admissible test functions w.

Unlike in finite elements, integration by parts would not do us much good here, since (B.4.1) only needs the first derivative. In a pure Galerkin method, the same set of functions is used as both the basis functions of u, and as the test functions w. In such a setting, moving the differentiation from u to w would not gain us anything.

Instead, the idea is to relax the requirements on u and w: we seek a solution in C^{-1}, in other words, allow u and w to have finite discontinuities across the element boundaries. This complicates the method slightly because we must account for derivatives of finite discontinuities, that is, Dirac deltas.

Let us begin with some background needed to develop the argument. We will start with piecewise continuous functions and jumps. Let $v(t)$ be a left-continuous or right-continuous function with a jump at t_0,

$$v(t) = \begin{cases} v_-(t) \;, & t \leq t_0 \text{ (if L.C.)} \;, \quad t < t_0 \text{ (if R.C.)} \;, \\ v_+(t) \;, & t > t_0 \text{ (if L.C.)} \;, \quad t \geq t_0 \text{ (if R.C.)} \;, \end{cases} \tag{B.4.2}$$

where $v_-(t)$ and $v_+(t)$ are continuous functions. According to the theory of distributions, the derivative of $v(t)$ can be written as

$$\frac{\partial v}{\partial t}(t) = \begin{cases} \dfrac{\partial v_-}{\partial t}(t) \,, & t < t_0 \,, \\ \lim\limits_{\varepsilon \to 0^+} \left(v_+(t_0+\varepsilon) - v_-(t_0-\varepsilon)\right) \delta(t-t_0) \,, & t = t_0 \,, \\ \dfrac{\partial v_+}{\partial t}(t) \,, & t > t_0 \,, \end{cases} \tag{B.4.3}$$

where $\delta(\ldots)$ is the Dirac delta distribution. Equations (B.4.2)–(B.4.3) naturally generalize to any finite number of discontinuities.

To shorten the notation, we abbreviate the limit expression in (B.4.3) by defining the *jump operator* $[\ldots]$, which maps functions to functions:

$$[F](t) := \lim_{\varepsilon \to 0^+} \left(F(t+\varepsilon) - F(t-\varepsilon)\right) \,. \tag{B.4.4}$$

Note that if F is continuous at a point t, then at that point $[F](t) = F(t)$; for continuous functions $[\ldots]$ is the identity operator. In general, we consider the case where F is sufficiently mildly behaved so that the left and right limits exist at the point t so that the definition (B.4.4) is admissible.

Inserting (B.4.4) in (B.4.3), we have

$$\frac{\partial v}{\partial t}(t) = \begin{cases} \dfrac{\partial v_-}{\partial t}(t) \,, & t < t_0 \,, \\ [v](t_0)\, \delta(t-t_0) \,, & t = t_0 \,, \\ \dfrac{\partial v_+}{\partial t}(t) \,, & t > t_0 \,, \end{cases} . \tag{B.4.5}$$

Observe that $[v](t_0)$ can be treated simply as a number, which is in principle explicitly obtainable from the definitions (B.4.4) and (B.4.2). Thus, as well as the jump operator $[\ldots]$, we may speak of *the jump* $[v](t_0)$, meaning the result when the jump operator is applied to a given function at a given point.

Formally, the Dirac delta is given by

$$\delta(\tau) = \begin{cases} +\infty \,, & \tau = 0 \,, \\ 0 \,, & \tau \neq 0 \,, \end{cases} \tag{B.4.6}$$

with the constraint

$$\int_{-\infty}^{+\infty} \delta(\tau)\, \mathrm{d}\tau = 1 \,, \tag{B.4.7}$$

which defines the singularity to have unit mass when the delta distribution is interpreted as the density function of an ideal point mass. The Dirac delta distribution satisfies the property

$$\int_{-\infty}^{+\infty} f(\tau)\, \delta(\tau)\, \mathrm{d}\tau = f(0) \,. \tag{B.4.8}$$

Importantly, the Dirac delta distribution is not square integrable ($\delta \notin L^2(\mathbb{R})$).

Strictly speaking, (B.4.8) is an abuse of notation, because no function satisfying (B.4.8) exists. What Eq. (B.4.8) actually means must be defined in some mathematically rigorous fashion. One option is to define δ as a measure. When given a subset A of the real line $\mathbb{R}$, we define $\delta(A) = 1$ if $0 \in A$, and $\delta(A) = 0$ otherwise. Then we take the left-hand side of (B.4.8) to mean the integral of f against this measure, $\int_{-\infty}^{+\infty} f(\tau)\delta\{\mathrm{d}\tau\}$. Alternatively, one may use the theory of distributions, where distributions are integral functionals; then $\delta(\tau)$ does not need to exist as an independent object outside the context of the integral (B.4.8).

The Dirac delta can be understood as a limit of a sequence of functions. For example, consider the following sequence of piecewise constant functions with two jumps:

$$d_n(\tau) := \begin{cases} 0\,, & \tau < -1/2n\,, \\ n\,, & -1/2n < \tau < +1/2n\,, \\ 0\,, & \tau > +1/2n\,, \end{cases} \tag{B.4.9}$$

where $n = 1, 2, \ldots$ Left- or right-continuity does not matter for this example, so we simply leave $d_n(\tau)$ undefined at the discontinuities. Geometrically, (B.4.9) describes a sequence of rectangles, which become narrower and higher as n increases. The width and height are chosen such that the area remains constant; each $d_n(\tau)$ satisfies the normalization (B.4.7). At $n \to \infty$, the limit of the sequence (B.4.9) is the Dirac delta. Alternatively, one may also use a sequence of zero-centered normal distributions

$$\rho_a(\tau) := \frac{1}{a\sqrt{\pi}} \exp(-x^2/a^2)\,,$$

and take the limit $a \to 0$.

A close relative of the Dirac delta distribution is the Heaviside step function:

$$H(t) = \begin{cases} 0\,, & t \le t_0 \text{ (if L.C.)}\,, \quad t < t_0 \text{ (if R.C.)}\,, \\ 1\,, & t > t_0 \text{ (if L.C.)}\,, \quad t \ge t_0 \text{ (if R.C.)}\,, \end{cases} \tag{B.4.10}$$

These are related as follows. Observe that if $b < 0$, then

$$\int_{-\infty}^{b} \delta(\tau)\,\mathrm{d}\tau = 0\,, \quad b < 0\,, \tag{B.4.11}$$

because $\delta(\tau) = 0$ for all $\tau < 0$. On the other hand, for any $b \neq 0$, by (B.4.8) we have

$$1 = \int_{-\infty}^{\infty} \delta(\tau)\,\mathrm{d}\tau = \int_{-\infty}^{b} \delta(\tau)\,\mathrm{d}\tau + \int_{b}^{+\infty} \delta(\tau)\,\mathrm{d}\tau\,. \tag{B.4.12}$$

Consider the case $b > 0$. The second term vanishes because $\delta(\tau) = 0$ for all $\tau > 0$. Only the first term remains, with the result

$$\int_{-\infty}^{b} \delta(\tau)\,\mathrm{d}\tau = 1\,, \quad b > 0\,. \tag{B.4.13}$$

Comparing (B.4.11), (B.4.13) and (B.4.10), we see that for any $b \neq 0$, the Heaviside step function is the cumulative distribution function of the Dirac delta distribution:

$$\int_{-\infty}^{b} \delta(\tau)\,\mathrm{d}\tau = H(b)\,, \quad b \neq 0\,. \tag{B.4.14}$$

To cover the case $b = 0$, as always when dealing with infinities, one must be careful. The singularity must be counted only once. One way to do this is to modify (B.4.12) to read

$$1 = \int_{-\infty}^{\infty} \delta(\tau)\,\mathrm{d}\tau = \lim_{\varepsilon\to 0^+} \left(\int_{-\infty}^{\varepsilon} \delta(\tau)\,\mathrm{d}\tau + \int_{\varepsilon}^{+\infty} \delta(\tau)\,\mathrm{d}\tau \right)\,, \tag{B.4.15}$$

which places the split unambiguously on one side of the origin. Then we evaluate the integrals and take the limit, in that order. The first term remains, while the second one vanishes. Considering that the result is similar in form to (B.4.14), but now with $b = 0$, we conclude that $H(0) = 1$. Thus placing the split on the positive side of the origin, as in (B.4.15), corresponds to choosing H to be right-continuous.

If, on the other hand, the split is placed on the negative side of the origin,

$$1 = \int_{-\infty}^{\infty} \delta(\tau)\,\mathrm{d}\tau = \lim_{\varepsilon\to 0^+} \left(\int_{-\infty}^{-\varepsilon} \delta(\tau)\,\mathrm{d}\tau + \int_{-\varepsilon}^{+\infty} \delta(\tau)\,\mathrm{d}\tau \right)\,, \tag{B.4.16}$$

then the first term, which tentatively defines $H(0)$, is zero. Hence this choice corresponds to choosing H to be left-continuous.

We conclude that

$$\int_{-\infty}^{b} \delta(\tau)\,\mathrm{d}\tau = H(b) \tag{B.4.17}$$

for any b, with the case $b = 0$ requiring special interpretation as explained above. (If H is L.C., then also the LHS is zero; if H is R.C., then the LHS is 1.)

Finally, considering b as a variable and formally differentiating (B.4.17) with respect to b, it is seen that

$$\delta(b) = \frac{\partial H}{\partial b}(b)\,. \tag{B.4.18}$$

Rigorously showing (B.4.18) requires further development beyond the scope of this Appendix.

This machinery allows us to define the *continuous part* of $v(t)$, here denoted by $\widehat{v}(t)$:

$$\widehat{v}(t) := v(t) - [v](t_0)\,H(t - t_0)\,, \tag{B.4.19}$$

where $H(\dots)$ is the Heaviside step function. The jump $[v](t_0)$ is simply a number that can be obtained from $v(t)$. To justify the name, it must be shown that (B.4.19) is indeed continuous. The continuity follows from (B.4.2). At any fixed $t \neq t_0$, (B.4.19) is continuous, because v is, and the Heaviside function is then constant in a small neighborhood of t.

Thus we only need to show continuity at the point $t = t_0$. The limit from the left is

$$\begin{aligned} \lim_{\varepsilon \to 0^+} \widehat{v}(t_0 - \varepsilon) &= v_-(t_0) - [v](t_0) \lim_{\varepsilon \to 0^+} H(-\varepsilon) \\ &= v_-(t_0) , \end{aligned} \tag{B.4.20}$$

where we have used the fact that $H(-t)$ is zero for all $t > 0$. In order to have continuity, (B.4.20) must match the limit from the right,

$$\begin{aligned} \lim_{\varepsilon \to 0^+} \widehat{v}(t_0 + \varepsilon) &= v_+(t_0) - [v](t_0) \lim_{\varepsilon \to 0^+} H(+\varepsilon) \\ &= v_+(t_0) - [v](t_0) \\ &= v_+(t_0) - (v_+(t_0) - v_-(t_0)) \\ &= v_-(t_0) , \end{aligned} \tag{B.4.21}$$

where we have used $H(t) = 1$ for all $t > 0$. Thus continuity holds also at t_0. Here it does not matter whether H is taken to be left- or right-continuous, because it is never evaluated at $\varepsilon = 0$; only the one-sided limits are needed.

We may reinterpret the definition of $\widehat{v}(t)$, Eq. (B.4.19), as decomposing $v(t)$ into a continuous function and a Heaviside step function times the jump:

$$v(t) = \widehat{v}(t) + [v](t_0)\, H(t - t_0) . \tag{B.4.22}$$

For this interpretation to make sense at $t = t_0$, the left- or right-continuity of H must be chosen to match that of v.

Next, we must consider the differentiation and definite integration of piecewise continuous functions. The decomposition (B.4.22) allows us to rewrite (B.4.5) as

$$\frac{\partial v}{\partial t}(t) = \frac{\partial \widehat{v}}{\partial t}(t) + [v](t_0)\, \delta(t - t_0) , \tag{B.4.23}$$

using the relation between the Heaviside step function and the Dirac delta distribution. Equation (B.4.23) gives us the means to formally treat the integration of $v'(t)$ over an interval (a, b) containing the singularity at t_0, that is, $a < t_0 < b$. We have

$$
\begin{aligned}
\int_a^b v'(t)\,\mathrm{d}t &= \int_a^b \widehat{v}'(t)\,\mathrm{d}t + \int_a^b [v](t_0)\,\delta(t-t_0)\,\mathrm{d}t \\
&= \int_a^b \widehat{v}'(t)\,\mathrm{d}t + [v](t_0)\int_a^b \delta(t-t_0)\,\mathrm{d}t \\
&= \int_a^b \widehat{v}'(t)\,\mathrm{d}t + [v](t_0) \\
&= \widehat{v}(b) - \widehat{v}(a) + [v](t_0)\,.
\end{aligned}
$$

The last integral on the second line evaluates to 1, because we are considering the case where $t_0 \in (a, b)$.

Finally, consider terms of the same form as the first term in the weak form, equation (B.4.1), and let both u and w have a jump at $t_0 \in (a, b)$. This choice is motivated by our aim to construct a classical Galerkin method, where the basis and test functions are taken to be the same. By (B.4.22) and (B.4.23), we have

$$
\begin{aligned}
\int_a^b u'(t)w(t)\,\mathrm{d}t &= \int_a^b \Big(\widehat{u}'(t) + [u](t_0)\,\delta(t-t_0)\Big)\Big(\widehat{w}(t) + [w](t_0)\,H(t-t_0)\Big)\,\mathrm{d}t \\
&= \int_a^b \widehat{u}'(t)\widehat{w}(t)\,\mathrm{d}t \\
&\quad + [w](t_0)\int_a^b \widehat{u}'(t)H(t-t_0)\,\mathrm{d}t \\
&\quad + [u](t_0)\int_a^b \widehat{w}(t)\,\delta(t-t_0)\,\mathrm{d}t \\
&\quad + [u](t_0)[w](t_0)\int_a^b H(t-t_0)\delta(t-t_0)\,\mathrm{d}t \\
&= \int_a^b \widehat{u}'(t)\widehat{w}(t)\,\mathrm{d}t + [w](t_0)\int_{t_0}^b \widehat{u}'(t)\,\mathrm{d}t + [u](t_0)\widehat{w}(t_0) + [u](t_0)[w](t_0)H(0) \\
&= \int_a^b \widehat{u}'(t)\widehat{w}(t)\,\mathrm{d}t + [w](t_0)(\widehat{u}(b) - \widehat{u}(t_0)) + [u](t_0)(\widehat{w}(t_0) + [w](t_0)H(0)) \\
&= \int_a^b \widehat{u}'(t)\widehat{w}(t)\,\mathrm{d}t + [w](t_0)(\widehat{u}(b) - \widehat{u}(t_0)) + [u](t_0)\,w(t_0)\,, \qquad \text{(B.4.24)}
\end{aligned}
$$

where in the last step we used (B.4.22). Here it does not matter whether u and w are left- or right-continuous.

To finish the preliminaries, let us consider how to treat functions with several discontinuities. Let $u(t)$ be a left-continuous function with $N \geq 2$ jumps, located at points t_k, $k = 1, 2, \ldots, N$:

$$
u(t) = \widehat{u}(t) + \sum_{k=1}^{N} [u](t_k)\,H(t-t_k)\,. \qquad \text{(B.4.25)}
$$

Because $H(\tau) = 1$ for all $\tau > 0$, we may write for given κ and $t > t_\kappa$ that

$$u_\kappa(t) = \widehat{u}(t) + c_\kappa + \sum_{k=\kappa+1}^{N} [u](t_k)\, H(t - t_k) \,, \quad 1 \le \kappa \le N,\ t > t_\kappa \,, \qquad \text{(B.4.26)}$$

where

$$c_\kappa := \sum_{k=1}^{\kappa} [u](t_k) \,. \qquad \text{(B.4.27)}$$

By convention, if $\kappa = N$, we ignore the sum in (B.4.26) because then its start index is greater than its end index, and similarly for the case $\kappa = 0$ in (B.4.27).

If we further restrict our consideration to the interval $t \in (t_\kappa, t_{\kappa+2})$, then

$$u_\kappa(t) = \widehat{u}(t) + c_\kappa + [u](t_{\kappa+1})\, H(t - t_{\kappa+1}) \,, \quad 1 \le \kappa \le N,\ t \in (t_\kappa, t_{\kappa+2}) \,, \qquad \text{(B.4.28)}$$

locally returning to a form with only one discontinuity, allowing us to use the equations already developed. For the purposes of (B.4.28), we use the convention that for any $k > N$, $t_k = +\infty$.

To obtain a version of (B.4.24) for the present case, it is convenient to split the interval of integration so that each subinterval contains up to one discontinuity, allowing us to apply (B.4.24) separately to each, and for this, use the form (B.4.28) to represent u and w.

Considering that the aim is to develop a timestepping method on an interval $t \in [a, b]$, we would like to choose the limits of integration as $I_k = (t_k, t_{k+1})$ for $k = 0, 1, \ldots, N$, setting $t_0 = a$ and $t_{N+1} = b$. However, to avoid ambiguities in handling the delta distribution, we must actually use $I_k \equiv \lim_{\varepsilon \to 0^+} I_k^\varepsilon$, where $I_k^\varepsilon = (t_k - \varepsilon, t_{k+1} - \varepsilon)$.

Planning to treat our solution as left-continuous, to keep things in line with the intuitive picture, we require $\varepsilon < \frac{1}{2} \min\{t_{k+1} - t_k : k = 0, 1, \ldots, N\}$, guaranteeing that each interval covers only one discontinuity. The epsilon in the lower limit makes the local discontinuity fall strictly inside each I_k^ε, and the one in the upper limit retains the lengths and the nonoverlapping property of the intervals.

Thus, we have

$$\int_a^b u'(t) w(t)\, \mathrm{d}t = \lim_{\varepsilon \to 0^+} \sum_{k=0}^{N} \int_{t_k - \varepsilon}^{t_{k+1} - \varepsilon} u'(t) w(t)\, \mathrm{d}t \,. \qquad \text{(B.4.29)}$$

It will however be simpler in practice to consider the case with just one timestep, sidestepping the need for (B.4.29), and also avoiding the need to keep track of the constants c_κ for the representation (B.4.28). This is possible for two reasons.

First, we observe that each jump connects only adjacent intervals. Due to the form of our problem (B.0.1)–(B.0.2), we may split it into a sequence of subproblems on I_k, with the final value $u(t_{k+1})$ from the current timestep fed into the next one as the initial condition u_0.

Secondly, without loss of generality, in the Galerkin setup to follow later, we may choose the global basis functions, and in a pure Galerkin method, hence also the global test functions, to have support only on one element each. Note that this differs from classical nodal finite element methods, where the global basis functions are attached to nodes, and have support on the patch of elements surrounding the node. There the global basis functions are pieced together in a matching way from local basis functions on each element belonging to the patch. Here, because the dG method is constructed to handle discontinuities of the basis across element boundaries, there is no need to perform this patching.

If we have a function, continuous across element boundaries, having support on several elements, that we would like to use in the basis, then, because this is a discontinuous method, we may always split it into parts restricted to individual elements, by multiplying with a suitable indicator function, and use those parts instead of using the original function. If the parts are used separately, this allows more freedom than the original continuous function, since the pieces may then have different values for their Galerkin coefficients. If used as a sum, this reduces to using the original continuous function.

From these two properties we conclude that the method can be designed to work locally, and hence it is sufficient to develop an algorithm to treat a single interval (t_k, t_{k+1}) at a time.

Now we have all the tools required for developing the weak form. Let us split the domain $[0, t_\mathrm{f}]$ into N intervals $I_k = [t_k, t_{k+1}]$, with $t_1 = 0$ and $t_{N+1} = t_\mathrm{f}$. Let $u(t) = u_0$ for all $t < 0$.

Let u be left-continuous and w right-continuous. This complementary choice of the continuity types is needed so that the test at I_k will see the final value of u on I_{k-1}. At the start of the kth timestep, $u(t_k)$ is the final value of the solution from the previous timestep, while $w(t_k)$ is the starting value of the test function for the current timestep.

Keep in mind that the weak form (B.4.1) must hold separately for each choice of w. Thus, each of the test functions must be defined on the whole interval $[0, t_\mathrm{f}]$. In practice, we will choose them to be zero in most of the domain, nonzero on one element, and having up to two jumps, one at each endpoint of its support.

Denote $u(t_k) = u_0$, the initial condition for this timestep. Let ε be arbitrary, with $0 < \varepsilon < \frac{1}{2}\min\{t_{k+1} - t_k : k = 0, 1, \ldots, N\}$. Consider a right-continuous test function w having support only on $t \in [t_k - \varepsilon, t_{k+1} - \varepsilon)$, and having a jump at t_k. Equation (B.4.1), that is,

$$\int_0^{t_\mathrm{f}} \frac{\partial u}{\partial t} w \, \mathrm{d}t = \int_0^{t_\mathrm{f}} f(u(t), t)\, w \, \mathrm{d}t \ , \quad \forall w \ ,$$

for such w becomes

$$\lim_{\varepsilon \to 0^+} \int_{t_k - \varepsilon}^{t_{k+1} - \varepsilon} \frac{\partial u}{\partial t} w \, \mathrm{d}t = \lim_{\varepsilon \to 0^+} \int_{t_k - \varepsilon}^{t_{k+1} - \varepsilon} f(u(t), t)\, w \, \mathrm{d}t \ , \tag{B.4.30}$$

because elsewhere $w = 0$. As was explained above, we must use limits in order to place the singularity in $u'(t)$ strictly inside the domain of integration. This is also needed to count only the singularity at t_k. Because u is left-continuous and we approach t_{k+1} from the left, *within this timestep* $u'(t)$ has no singularity at t_{k+1}, in the sense of the limit in (B.4.30).

On the right-hand side of (B.4.30), by the problem definition (B.0.1)–(B.0.2), the load function f involves no time differentiations of u or w. These functions are C^{-1} continuous. Thus, if f itself is at least C^{-1} continuous in u and t, then the integrand will not be worse than C^{-1}, and hence does not need any special consideration. This allows us to immediately drop the limit from the right-hand side:

$$\lim_{\varepsilon \to 0^+} \int_{t_k - \varepsilon}^{t_{k+1} - \varepsilon} \frac{\partial u}{\partial t} w \, \mathrm{d}t = \int_{t_k}^{t_{k+1}} f(\, u(t),\, t\,)\, w \, \mathrm{d}t \;.$$

The left-hand side now has exactly one singularity, which is strictly inside the domain of integration, so we may apply (B.4.24), obtaining

$$\lim_{\varepsilon \to 0^+} \left\{ \int_{t_k - \varepsilon}^{t_{k+1} - \varepsilon} \frac{\partial \widehat{u}}{\partial t} \widehat{w} \, \mathrm{d}t + [w](t_k)(\, \widehat{u}(t_{k+1} - \varepsilon) - \widehat{u}(t_k)\,) + [u](t_k)\, w(t_k) \right\}$$
$$= \int_{t_k}^{t_{k+1}} f(\, u(t),\, t\,)\, w \, \mathrm{d}t \;.$$

Evaluating the limit on the left-hand side yields

$$\int_{t_k}^{t_{k+1}} \frac{\partial \widehat{u}}{\partial t} \widehat{w} \, \mathrm{d}t + [w](t_k)(\, \widehat{u}(t_{k+1}) - \widehat{u}(t_k)\,) + [u](t_k)\, w(t_k) = \int_{t_k}^{t_{k+1}} f(\, u(t),\, t\,)\, w \, \mathrm{d}t \;. \tag{B.4.31}$$

Note that w, which was defined to have support on $t \in [t_k - \varepsilon, t_{k+1} - \varepsilon)$, because $\varepsilon \to 0$, has now converged into having support on $t \in [t_k, t_{k+1})$. We could denote the original by w_ε and its limit by w, but the presentation looks clearer without the extra subscript.

Equation (B.4.31) still contains a mix of u, $\widehat{u}$, w and $\widehat{w}$. Observe that

$$u(t_k) = u_0 \qquad \text{(initial condition for this timestep)}\;, \tag{B.4.32}$$
$$\widehat{u}(t_k) = u(t_k) \qquad (u \text{ L.C.})\;, \tag{B.4.33}$$
$$\widehat{u}(t) = u(t) - [u](t_k)\;, \quad t \in (t_k, t_{k+1}] \qquad \text{(Eq. (B.4.19), } u \text{ L.C.)}\;, \tag{B.4.34}$$
$$\widehat{w}(t) = w(t) - [w](t_k)\;, \quad t \in [t_k, t_{k+1}) \qquad \text{(Eq. (B.4.19), } w \text{ R.C.)}\;. \tag{B.4.35}$$

It also holds that

$$[w](t_k) = w(t_k) \qquad \text{(because } \lim_{\epsilon \to 0^+} w(t_k - \epsilon) = 0)\;, \tag{B.4.36}$$

which may be useful in a numerical implementation because it allows us to process each element locally.

The expression $\widehat{u}(t_{k+1}) - \widehat{u}(t_k)$ may be treated using one-sided limits of u from inside the timestep:

$$\begin{aligned}\widehat{u}(t_{k+1}) - \widehat{u}(t_k) &= (u(t_{k+1}) - [u](t_k)) - u(t_k) \\ &= u(t_{k+1}) - (u(t_k) + [u](t_k)) \\ &= \lim_{\varepsilon \to 0^+} u(t_{k+1} - \varepsilon) - \lim_{\varepsilon \to 0^+} u(t_k + \varepsilon) \,,\end{aligned} \tag{B.4.37}$$

where in the last form we have used the left-continuity of u. This holds very generally; in the context of multiple timesteps, if there is any offset between u and $\widehat{u}$ due to the history of jumps at earlier timesteps, in (B.4.37) this offset is canceled by the subtraction; what matters is only how much the continuous part $\widehat{u}$, or indeed u itself, changes across the timestep.

Rewriting (B.4.31), that is,

$$\int_{t_k}^{t_{k+1}} \frac{\partial \widehat{u}}{\partial t} \widehat{w} \, \mathrm{d}t + [w](t_k)(\widehat{u}(t_{k+1}) - \widehat{u}(t_k)) + [u](t_k)\, w(t_k) = \int_{t_k}^{t_{k+1}} f(u(t), t)\, w \, \mathrm{d}t \,,$$

with the help of (B.4.34), (B.4.35) and (B.4.37) to eliminate $\widehat{u}$ and $\widehat{w}$, we have

$$\begin{aligned}&\int_{t_k}^{t_{k+1}} \frac{\partial u(t)}{\partial t} \big(w(t) - [w](t_k)\big) \, \mathrm{d}t + [w](t_k)\big(\lim_{\varepsilon \to 0^+} u(t_{k+1} - \varepsilon) - \lim_{\varepsilon \to 0^+} u(t_k + \varepsilon)\big) \\ &\quad + [u](t_k)\, w(t_k) = \int_{t_k}^{t_{k+1}} f(u(t), t)\, w \, \mathrm{d}t \,.\end{aligned} \tag{B.4.38}$$

When manipulating the equations, it is important to keep in mind that u is left-continuous, whereas w is right-continuous. Here we have used the fact that $\partial \widehat{u}/\partial t = \partial u/\partial t$ on $t \in (t_k, t_{k+1}]$, because the jump is a constant.

Noting further that $[w](t_k)$ is a constant, and u is continuous on $t \in (t_k, t_{k+1}]$, we may evaluate the second term in the first integral, obtaining

$$\begin{aligned}&\int_{t_k}^{t_{k+1}} \frac{\partial u(t)}{\partial t} w(t) \, \mathrm{d}t - [w](t_k)\big(u(t_{k+1}) - \lim_{\varepsilon \to 0^+} u(t_k + \varepsilon)\big) \\ &\quad + [w](t_k)\big(u(t_{k+1}) - \lim_{\varepsilon \to 0^+} u(t_k + \varepsilon)\big) + [u](t_k)\, w(t_k) = \int_{t_k}^{t_{k+1}} f(u(t), t)\, w \, \mathrm{d}t \,.\end{aligned}$$

Now the terms involving $[w](t_k)$ cancel, yielding the weak form for piecewise continuous u and w:

$$\int_{t_k}^{t_{k+1}} \frac{\partial u(t)}{\partial t} w(t) \, \mathrm{d}t + [u](t_k)\, w(t_k) = \int_{t_k}^{t_{k+1}} f(u(t), t)\, w \, \mathrm{d}t \,. \tag{B.4.39}$$

Keep in mind that when deriving (B.4.39), we have taken the solution u as left-continuous and the test function w as right-continuous, with jumps occurring at the timestep boundaries.

So far we have treated the weak initial value problem, first in the context of general piecewise continuous functions having finite discontinuities at timestep boundaries, and then with test functions having their support restricted to $[t_k, t_{k+1})$. Now we are in a position to introduce the Galerkin component of the method.

First, let $\phi_n(t), n = 1, 2, \ldots$ be a set of continuous functions defined on the closed interval $[t_k, t_{k+1}]$. Define the half-open intervals $L_k = (t_k, t_{k+1}]$ and $R_k = [t_k, t_{k+1})$. Let $\chi_A(t)$ denote the indicator function of a set A:

$$\chi_A(t) = \begin{cases} 1\,, & t \in A\,, \\ 0\,, & t \notin A\,. \end{cases} \tag{B.4.40}$$

Define

$$\varphi_n(t) := \chi_{L_k}(t)\phi_n(t)\,, \tag{B.4.41}$$

$$\psi_n(t) := \chi_{R_k}(t)\phi_n(t)\,, \tag{B.4.42}$$

making φ_n (respectively ψ_n) suitable as a basis for left-continuous (resp. right-continuous) functions. The only effect the multiplication by χ has is that it removes the inappropriate endpoint from the support. In the Galerkin spirit, in all other respects the bases are the same.

As for the actual choice of the basis functions, we only make the following short remark: the Lobatto basis (also known as the hierarchical basis) is particularly useful. The reference element is $t \in [0, 1]$. The first two basis functions are $\phi_0(t) = 1 - t$, $\phi_1(t) = t$. All the others are bubbles, obtained as the integrals of Legendre polynomials. This allows for high-order dG(q) methods if needed, although in practice dG(1) (linear, no bubbles) and dG(2) (one bubble) are perhaps the most useful.

We now represent u as a Galerkin series,

$$u(t) := \sum_{m=1}^{\infty} u_m \varphi_m(t)\,, \tag{B.4.43}$$

where u_m are the Galerkin coefficients, and $\varphi_m(t)$ are the global basis functions. We choose the test functions w_i as

$$w_i(t) := \psi_i(t)\,, \quad i = 1, 2, \ldots\,, \tag{B.4.44}$$

that is, otherwise the same as the basis for u, but with right-continuity.

Differing from continuous Galerkin methods, because now there is no requirement of continuity across the timestep boundaries, we may actually take the local basis functions—appropriately positioned and scaled on the t axis separately for each element, as the time integration proceeds—as the global basis functions.

Inserting (B.4.43)–(B.4.44) into (B.4.39) obtains

$$\int_{t_k}^{t_{k+1}} \frac{\partial}{\partial t}\left(\sum_{m=1}^{\infty} u_m \varphi_m(t)\right) \psi_i(t)\,\mathrm{d}t + \left(\left(\sum_{m=1}^{\infty} u_m \lim_{\varepsilon\to 0^+} \varphi_m(t_k+\varepsilon)\right) - u_0\right)\psi_i(t_k)$$
$$= \int_{t_k}^{t_{k+1}} f\Big(\sum_{n=1}^{\infty} u_m \varphi_m(t),\, t\Big)\, \psi_i(t)\,\mathrm{d}t\ , \quad \forall i = 1, 2, \ldots\ , \tag{B.4.45}$$

where u_0 is the initial condition for this timestep.

Because there are no discontinuities remaining in the first term, we may rearrange it as

$$\int_{t_k}^{t_{k+1}} \frac{\partial}{\partial t}\left(\sum_{m=1}^{\infty} u_m \varphi_m(t)\right) \psi_i\,\mathrm{d}t = \int_{t_k}^{t_{k+1}} \psi_i \sum_{m=1}^{\infty} u_m \frac{\partial \varphi_m}{\partial t}(t)\,\mathrm{d}t$$
$$= \sum_{m=1}^{\infty} u_m \int_{t_k}^{t_{k+1}} \psi_i \frac{\partial \varphi_m}{\partial t}(t)\,\mathrm{d}t\ , \tag{B.4.46}$$

first using the fact that every continuous function on [0, 1], which is our reference element, is a uniform limit of polynomials, so that we may exchange differentiation and infinite summation, and then using Tonelli's and Fubini's theorems to justify the exchange of integration and infinite summation.

The final step of the derivation of the dG procedure is to apply (B.4.46) and (B.4.45). In practical numerics, we then formally truncate the Galerkin series, obtaining

$$\sum_{m=1}^{M} u_m \int_{t_k}^{t_{k+1}} \psi_i \frac{\partial \varphi_m}{\partial t}(t)\,\mathrm{d}t + \left(\left(\sum_{m=1}^{M} u_m \lim_{\varepsilon\to 0^+} \varphi_m(t_k+\varepsilon)\right) - u_0\right)\psi_i(t_k)$$
$$= \int_{t_k}^{t_{k+1}} f\Big(\sum_{m=1}^{M} u_m \varphi_m(t),\, t\Big)\, \psi_i(t)\,\mathrm{d}t\ , \quad \forall i = 1, 2, \ldots, M\ , \tag{B.4.47}$$

where M is the total number of basis functions ϕ_n in the discrete basis. Transferring terms and recognizing that this can be formally written as a linear equation system, we have

$$\mathbf{A}\mathbf{u} = \mathbf{b}(\mathbf{u})\ , \tag{B.4.48}$$

where $\mathbf{u}$ is an M-element vector consisting of the coefficients u_m, and

$$A_{im} = \int_{t_k}^{t_{k+1}} \psi_i(t) \, \frac{\partial \varphi_m}{\partial t}(t) \, \mathrm{d}t + \psi_i(t_k) \lim_{\varepsilon \to 0^+} \varphi_m(t_k + \varepsilon) \, , \tag{B.4.49}$$

$$b_i(\mathbf{u}) = \int_{t_k}^{t_{k+1}} f\big(\sum_{m=1}^{M} u_m \varphi_m(t), \, t\big) \, \psi_i(t) \, \mathrm{d}t + \psi_i(t_k) u_0 \, . \tag{B.4.50}$$

Treating the one degree of freedom model problem (B.0.1)–(B.0.2) has led to an equation system with M equations, because there are M basis functions modeling the behavior of this degree of freedom across a single timestep.

Equations (B.4.48)–(B.4.50), in principle, allow us to solve for the Galerkin coefficients u_m for this timestep. Together with relation (B.4.43), we obtain the behavior of $u(t)$ for all $t \in (t_k, t_{k+1}]$. Especially, evaluating $u(t_{k+1})$ from (B.4.43) then gives us the initial condition for the next timestep. The limit expression in (B.4.49) is easy to evaluate; it is simply $\phi_m(t_k)$.

For nonlinear problems in general, $\mathbf{b}$ depends on $\mathbf{u}$; this is why we have presented the method in the above form. Fixed point iteration, for the Galerkin coefficients $\mathbf{u}$, is thus needed to evaluate the load vector $\mathbf{b}$. If f is linear, its linear dependence on $\mathbf{u}$ could of course be extracted and transferred into $\mathbf{A}$, but here we concentrate on the general nonlinear case.

The final step in practical numerics is to rewrite the integrals in (B.4.49)–(B.4.50) using quadratures on the reference element $[0, 1]$. The standard tools from finite elements methods can be used for this.

To conclude our presentation of dG, consider the case with a nontrivial mass matrix $\mathbf{M}$. The weak form of the problem, Eq. (B.4.1), does not require many changes. Defining the act of testing a vector-valued quantity by a vector-valued test function as a projection into the direction of the test function, we write

$$\int_0^{t_f} \frac{\partial \mathbf{u}}{\partial t} \cdot \mathbf{w} \, \mathrm{d}t = \int_0^{t_f} \big(\mathbf{M}^{-1} \mathbf{f}(\mathbf{u}(t), \, t)\big) \cdot \mathbf{w} \, \mathrm{d}t \, , \quad \forall \mathbf{w} \, , \tag{B.4.51}$$

where $\mathbf{w}$ is a vector-valued test function, and the dot denotes the usual Euclidean inner product. By defining the effective load

$$\mathbf{g}(\mathbf{u}(t), \, t) = \mathbf{M}^{-1} \mathbf{f}(\mathbf{u}(t), \, t) \tag{B.4.52}$$

we have

$$\int_0^{t_f} \frac{\partial \mathbf{u}}{\partial t} \cdot \mathbf{w} \, \mathrm{d}t = \int_0^{t_f} \big(\mathbf{g}(\mathbf{u}(t), \, t)\big) \cdot \mathbf{w} \, \mathrm{d}t \, , \quad \forall \mathbf{w} \, . \tag{B.4.53}$$

Looking at the contribution of a single component of the test, w_j, we have

$$\int_0^{t_f} \frac{\partial u_j}{\partial t} w_j \, \mathrm{d}t = \int_0^{t_f} \big(g_j(u_1(t), u_2(t), \ldots, \, t)\big) w_j \, \mathrm{d}t \, , \quad \forall w_j \, . \tag{B.4.54}$$

The machinery presented above applies directly to each of the Eqs. (B.4.54), producing equations of the form (B.4.39).

After summing the results over j, and finally using (B.4.52) to replace $\mathbf{g}$ with its definition, we have the result

$$\int_{t_k}^{t_{k+1}} \frac{\partial \mathbf{u}(t)}{\partial t} \cdot \mathbf{w}(t)\, \mathrm{d}t + [\mathbf{u}](t_k) \cdot \mathbf{w}(t_k) = \int_{t_k}^{t_{k+1}} \mathbf{M}^{-1}\mathbf{f}(\mathbf{u}(t),\, t) \cdot \mathbf{w}\, \mathrm{d}t\ , \quad \forall \mathbf{w}. \tag{B.4.55}$$

From here we may proceed by treating $\mathbf{M}^{-1}\mathbf{f}$ as an effective load, and solving the linear equation system $\mathbf{M}\,\mathbf{g} = \mathbf{f}$ for $\mathbf{g}$ whenever its numerical value is needed.

Because the degrees of freedom have been decoupled on the left-hand side, it is convenient to test them individually, taking the test functions $\mathbf{w}$ as w_{ji}, $j = 1, 2, \ldots, N$ (indexing the degrees of freedom), $i = 1, 2, \ldots, M$ (indexing the set of test functions in a timestep):

$$w_{1i}(t) = (\, w_i(t),\, 0,\, 0, \ldots,\, 0\,),$$

$$w_{2i}(t) = (\,0,\, w_i(t),\, 0,\, 0, \ldots,\, 0\,),$$

$$\ldots$$

$$w_{Ni}(t) = (\,0,\, 0,\, 0, \ldots,\, 0,\, w_i(t)\,)\,,$$

where $w_i(t)$, $i = 1, 2, \ldots, M$, are the test functions $\psi_i(t)$ for a scalar quantity on the timestep. This produces NM equations in total.

Each evaluation of $\mathbf{g}$ will require the solution of a linear equation system. Furthermore, because $\mathbf{u}(t)$ is unknown at any point $t > t_k$, for nonlinear problems this expensive solution of multiple linear equation systems for $\mathbf{g}$ must occur for each iteration inside the fixed point iteration loop. Thus, for a general semilinear system of ODEs, dG can be expensive.

Finally, if the space discretization is such as to make $\mathbf{M}$ the unit matrix, then $\mathbf{g} = \mathbf{f}$ and (B.4.55) becomes cheaper to solve. One particularly simple implementation for nonlinear problems of this form is to just use (B.4.48)–(B.4.50) (in principle) independently for each space degree of freedom, with fixed point iteration to make it possible to explicitly evaluate the right-hand side. The only required modification is that the last available iterate from all space degrees of freedom is needed to evaluate f. Each space degree of freedom $u_j(t)$ has its own Galerkin coefficients in time, $(u_j)_m$, so we replace the original f with

$$f(\,u_1(t),\, u_2(t),\, \ldots,\, t\,) = f\Big(\sum_{m=1}^{M}(u_1)_m\varphi_m(t),\, \sum_{m=1}^{M}(u_2)_m\varphi_m(t),\, \ldots,\, t\Big).$$

If the ODE system originates in a PDE problem and $u(x,t)$ is discretized using finite elements, we have

$$u(x,t) = \sum_{n=1}^{N} u_n(t)\widetilde{\varphi}_n(x) , \tag{B.4.56}$$

where $\widetilde{\varphi}_n(x)$ are the (global) space basis functions, and

$$u_n(t) = \sum_{m=1}^{M} (u_n)_m \varphi_m(t) , \tag{B.4.57}$$

where $\varphi_m(t)$ are the time basis functions. This gives the truncated Galerkin representation of u at an arbitrary point $(x,t) \in \Omega \times (t_k, t_{k+1}]$ as

$$u(x,t) = \sum_{n=1}^{N} \widetilde{\varphi}_n(x) \sum_{m=1}^{M} (u_n)_m \varphi_m(t) \tag{B.4.58}$$

in terms of the Galerkin coefficients. Using (B.4.57) in (B.4.55), we may write out the following system of NM equations, from which the Galerkin coefficients $(u_n)_m$ may be solved:

$$\begin{aligned}&\sum_{m=1}^{M} (u_n)_m \int_{t_k}^{t_{k+1}} \psi_i \frac{\partial \varphi_m}{\partial t}(t)\, \mathrm{d}t + \left(\left(\sum_{m=1}^{M} (u_n)_m \lim_{\varepsilon \to 0^+} \varphi_m(t_k + \varepsilon) \right) - u_0 \right) \psi_i(t_k) \\ &= \int_{t_k}^{t_{k+1}} \left(g\left(\sum_{m=1}^{M} (u_1)_m \varphi_m(t), \sum_{m=1}^{M} (u_2)_m \varphi_m(t), \ldots, t \right) \right)_n \psi_i(t)\, \mathrm{d}t , \quad \forall i = 1, 2, \ldots, M , \quad \forall n = 1, 2, \ldots N .\end{aligned} \tag{B.4.59}$$

The meaning of the indices is as follows. The index k denotes a timestep. Effectively it is a placeholder, since we consider integration over a single timestep only. The index n denotes the space degrees of freedom, $u_n(t)$ in (B.4.56)–(B.4.57). The index m refers to the time degrees of freedom, $(u_n)_m$ in (B.4.57)–(B.4.58). In (B.4.59), m only appears as a summation index in the truncated Galerkin series that represents $u_n(t)$. Finally, i refers to the time test functions $\psi_i(t)$.

Instead of $\psi_i(t)$, we ought to actually speak of $w_{ji}(x,t)$, where j refers to the space test functions $\widetilde{\psi}_j(x)$. However, choosing the test functions in the manner explained above, and noting that the matrix $\mathbf{M}$ (hidden inside g) encodes the connections between the space degrees of freedom, we are left with only $\psi_i(t)$ as the test in (B.4.59).

There is no summation over n, although n is present in the Galerkin coefficient $(u_n)_m$; instead, we have left n free. This is possible due to the decoupling; the standard alternative would be to sum over n and leave j free.

Spelling out g_j in component form, with summation over repeated indices implied, we have

$$\big(g(u_1(t), u_2(t), \ldots, t)\big)_n = (M^{-1})_{nq}\big(f(u_1(t), u_2(t), \ldots, t)\big)_q , \qquad \text{(B.4.60)}$$

where q indexes the components of f. There is one component of g_n directly corresponding to each space degree of freedom $u_n(t)$. This utilizes the fact that the multiplication by $\mathbf{M}^{-1}$ has decoupled the space degrees of freedom; the components f_q could not be interpreted similarly.

To simplify the notation, we may effectively consider

$$\widehat{\varphi}_{nm}(x,t) = \widetilde{\varphi}_n(x)\varphi_m(t) \qquad \text{(B.4.61)}$$

as a spacetime basis on $C^0(\Omega) \times C^{-1}([t_k, t_{k+1}])$. The index pair nm can be replaced, if desired, with a single linear index, by a one-to-one mapping such as $p = (n-1)M + m$. Then, we have

$$u(x,t) = \sum_{p=1}^{P} u_p \widehat{\varphi}_p(x,t) ,$$

where $P = NM$, and u_p are the Galerkin coefficients to be solved. Now the test functions in the Galerkin method become

$$w_{ji}(x,t) = \widehat{\varphi}_{ji}(x,t) , \quad j = 1, 2, \ldots, N , \quad i = 1, 2, \ldots M ,$$

or using the linear index p,

$$w_p(x,t) = \widehat{\varphi}_p(x,t) , \quad p = 1, 2, \ldots, P .$$

The total number of equations will still be NM; this only changes the notation.

References

1. Matculevich S, Neittaanmäki P, Repin S (2013) Guaranteed error bounds for a class of Picard–Lindelöf iteration methods. In: Repin S, Tiihonen T, Tuovinen T, (eds) Numerical methods for differential equations, optimization, and technological problems. Dedicated to Professor P. Neittaanmäki on his 60th Birthday, vol 27. Computational methods in applied sciences. Springer Netherlands, pp 175–189 ISBN: 978-94-007-5287-0 (Print), 978-94-007-5288-7 (Online)
2. Rosenlicht M (1968) Introduction to analysis. Dover, 1985. ISBN 978-0-486-65038-3. Republication of the edition published by Scott Foresman & Co
3. Zienkiewicz OC, Taylor RL, Zhu JZ (2013) The finite element method: its basis and fundamentals, vol 1, 7th edn. Butterworth–Heinemann

Appendix C
Finite Elements of the Hermite Type

The axially travelling Kelvin–Voigt beam (or panel) that was considered in Chap. 5 is nonstandard from the viewpoint of literature on partial differential equations, as the strong form of the equation is of the fifth order in x. In the weak form, up to third-order derivatives appear (due to the bending moment $M(x)$), requiring C^2 continuity across element boundaries in order to enforce integrability. In one space dimension, Hermite elements can provide the required continuity.

In this Appendix, we give a brief exposition of finite elements of the Hermite type, which include as degrees of freedom not only the values, but also the derivatives of the unknown function. We only provide material that is specific to the use of Hermite elements. We assume familiarity with linear finite elements and the use of a reference element. If needed, a general introduction to the finite element method can be found in textbooks such as Hughes [1], Jacob and Ted [2], Zienkiewicz et al. [3].

Let us consider the second-order Hermite element, which is C^2 continuous, continuously interpolating w, w' and w''. If w'' is C^0 continuous across element boundaries, then w''' is finitely discontinuous across element boundaries. Hence, we may have up to third order derivatives in the weak form before losing integrability.

Let us walk through the derivation of Hermite elements. The procedure easily generalizes to any order of the highest derivative that needs to be interpolated. The standard C^1-continuous beam element, which is the first-order Hermite element, is derived in the same way.

We start by imposing the following *interpolation conditions* in element-local coordinates $x \in [0, 1]$:

$$w(0) = w_0 \,, \quad w'(0) = w'_0 \,, \quad w''(0) = w''_0 \,, \tag{C.0.1}$$

$$w(1) = w_1 \,, \quad w'(1) = w'_1 \,, \quad w''(1) = w''_1 \,, \tag{C.0.2}$$

where the right-hand sides are prescribed constants, that is, the values we would like to interpolate. We then look for a polynomial that fulfills (C.0.1)–(C.0.2) on the reference element (local element). We write

N. Banichuk et al., *Stability of Axially Moving Materials*, Solid Mechanics and Its Applications 259, https://doi.org/10.1007/978-3-030-23803-2

$$w(x) \equiv \sum_{j=0}^{5} a_j x^j \equiv a_0 + a_1 x + a_2 x^2 + a_3 x^3 + a_4 x^4 + a_5 x^5 \,, \tag{C.0.3}$$

specifically choosing a fifth-degree polynomial, because it has six coefficients, matching the number of conditions in (C.0.1)–(C.0.2). Inserting (C.0.3) to the left-hand sides of (C.0.1)–(C.0.2), we obtain a linear equation system for the a_j:

$$w(0) = a_0 = w_0 \,, \tag{C.0.4}$$

$$w'(0) = a_1 = w_0' \,, \tag{C.0.5}$$

$$w''(0) = a_2 = w_0'' \,, \tag{C.0.6}$$

$$w(1) = \sum_{j=0}^{5} a_j = w_1 \,, \tag{C.0.7}$$

$$w'(1) = \sum_{j=1}^{5} j a_j = w_1' \,, \tag{C.0.8}$$

$$w''(1) = \sum_{j=2}^{5} j(j-1) a_j = w_1'' \,. \tag{C.0.9}$$

In matrix form, Eqs. (C.0.4)–(C.0.9) become

$$\begin{bmatrix} 1 & & & & & \\ & 1 & & & & \\ & & 1 & & & \\ 1 & 1 & 1 & 1 & 1 & 1 \\ & 1 & 2 & 3 & 4 & 5 \\ & & 2 & 6 & 12 & 20 \end{bmatrix} \begin{bmatrix} a_0 \\ a_1 \\ a_2 \\ a_3 \\ a_4 \\ a_5 \end{bmatrix} = \begin{bmatrix} w_0 \\ w_0' \\ w_0'' \\ w_1 \\ w_1' \\ w_1'' \end{bmatrix} . \tag{C.0.10}$$

Solving (C.0.10) yields expressions for the a_j as linear combinations of the known quantities on the right-hand side.

We insert the resulting expressions for the a_j into (C.0.3), and then group the result into subexpressions of the form $w_0 N_1(x)$, $w_0' N_2(x)$, …, $w_1'' N_6(x)$, where each $N_j(x)$ is a polynomial of x with explicit numerical coefficients that do not involve the quantities w_0, w_0', w_0'', w_1, w_1' and w_1''. The representation (C.0.3) becomes

$$w(x) = w_0 N_1(x) + w_0' N_2(x) + w_0'' N_3(x) + w_1 N_4(x) + w_1' N_5(x) + w_1'' N_6(x) \,. \tag{C.0.11}$$

The expressions $N_1(x), N_2(x) \ldots, N_6(x)$ are the element-local basis functions, which satisfy

$$
\begin{aligned}
N_1(0) &= 1\,, \quad N_1'(0) = 0\,, \quad N_1''(0) = 0\,,\\
N_2(0) &= 0\,, \quad N_2'(0) = 1\,, \quad N_2''(0) = 0\,,\\
N_3(0) &= 0\,, \quad N_3'(0) = 0\,, \quad N_3''(0) = 1\,,\\
N_4(1) &= 1\,, \quad N_4'(1) = 0\,, \quad N_4''(1) = 0\,,\\
N_5(1) &= 0\,, \quad N_5'(1) = 1\,, \quad N_5''(1) = 0\,,\\
N_6(1) &= 0\,, \quad N_6'(1) = 0\,, \quad N_6''(1) = 1\,.
\end{aligned}
$$

Thus in practice, the interpolation conditions (C.0.1)–(C.0.2) are fulfilled by simply inserting the desired values into the representation (C.0.11). The degrees of freedom are the values at the nodes of precisely the quantities that we wish to interpolate.

Explicitly, the C^2 Hermite basis functions are

$$N_1(x) = -6x^5 + 15x^4 - 10x^3 + 1\,, \tag{C.0.12}$$

$$N_2(x) = -3x^5 + 8x^4 - 6x^3 + x\,, \tag{C.0.13}$$

$$N_3(x) = -\frac{1}{2}x^5 + \frac{3}{2}x^4 - \frac{3}{2}x^3 + \frac{1}{2}x^2\,, \tag{C.0.14}$$

$$N_4(x) = +6x^5 - 15x^4 + 10x^3\,, \tag{C.0.15}$$

$$N_5(x) = -3x^5 + 7x^4 - 4x^3\,, \tag{C.0.16}$$

$$N_6(x) = +\frac{1}{2}x^5 - x^4 + \frac{1}{2}x^3\,. \tag{C.0.17}$$

As was mentioned, the classical beam element, which is the C^1 Hermite element, is derived using the exact same procedure. We require interpolation for w and w' at the points $x = 0$ and $x = 1$ in element-local coordinates, $x \in [0, 1]$, and seek a third-order polynomial. Applying the same steps as above, the result is

$$w(x) = w_0 B_1(x) + w_0' B_2(x) + w_1 B_3(x) + w_1' B_4(x)\,, \tag{C.0.18}$$

where

$$B_1(x) = 2x^3 - 3x^2 + 1\,, \tag{C.0.19}$$

$$B_2(x) = x^3 - 2x^2 + x\,, \tag{C.0.20}$$

$$B_3(x) = -2x^3 + 3x^2\,, \tag{C.0.21}$$

$$B_4(x) = x^3 - x^2\,. \tag{C.0.22}$$

Beam elements can accommodate up to second-order derivatives in the weak form before losing integrability. They are useful for the transverse component w of a linear elastic beam (whether axially travelling or stationary), and for the axial component u of an axially travelling Kelvin–Voigt beam or panel, where $\partial^2 u/\partial x^2$ appears in the expression of the resultant axial force $N(x)$.

C.1 Coordinate Mapping and the Derivative Degrees of Freedom

For mapping from the reference element $x \in [0, 1]$ to the global $X \in \mathbb{R}$ for each element in the mesh, an affine coordinate transformation is particularly convenient:

$$X_e(x) = a_0 + a_1 x \ . \tag{C.1.1}$$

The subscript e indicates that each element e in the mesh has its own values for the coefficients $a_0 \equiv a_0(\mathrm{e})$ and $a_1 \equiv a_1(\mathrm{e})$. Using the change of variable in an integral and the coordinate mapping (C.1.1), for each element we may write

$$\begin{aligned} \int_{X_1}^{X_2} f(X)\,\mathrm{d}X &= \int_{x_1}^{x_2} f(x)\,|\det J(\,X_e(x)\,)|\,\mathrm{d}x \\ &= \int_{x_1}^{x_2} f(x) \left| \frac{\partial X_e}{\partial x}(x) \right| \mathrm{d}x = a_1 \int_0^1 f(x)\,\mathrm{d}x \ , \end{aligned} \tag{C.1.2}$$

that is, the Jacobian of the affine mapping is an elementwise constant. Here the function $f(X)$ is defined in global coordinates, while $f(x) \equiv f(\,X_e(x)\,)$ is the same function as expressed in the element-local coordinates for element e.

By the chain rule, differentiation with respect to the global X transforms as

$$\frac{\partial f}{\partial X} = \frac{\partial x}{\partial X}\frac{\partial f}{\partial x} = \frac{1}{a_1}\frac{\partial f}{\partial x} \ , \tag{C.1.3}$$

$$\frac{\partial^2 f}{\partial X^2} = \frac{\partial}{\partial X}\left(\frac{\partial f}{\partial X}\right) = \frac{\partial}{\partial X}\left(\frac{1}{a_1}\frac{\partial f}{\partial x}\right) = \frac{1}{a_1}\frac{\partial}{\partial X}\left(\frac{\partial f}{\partial x}\right) = \frac{1}{a_1}\frac{\partial x}{\partial X}\frac{\partial}{\partial x}\left(\frac{\partial f}{\partial x}\right) = \frac{1}{a_1^2}\frac{\partial^2 f}{\partial x^2} \ , \tag{C.1.4}$$

and likewise for higher derivatives. The constant a_1 depends only on the length of each element.

When using Hermite elements, one must be careful with the derivative degrees of freedom. For example, for the C^2 Hermite elements, the Galerkin representation of w is (C.0.11), namely

$$w(x) = w_0 N_1(x) + w_0' N_2(x) + w_0'' N_3(x) + w_1 N_4(x) + w_1' N_5(x) + w_1'' N_6(x) \ .$$

As it stands, this representation is written using element-local degrees of freedom $w_0, \ldots, w_1''$ corresponding to element-local $x \in [0, 1]$. Explicitly, we may write

$$\begin{aligned} w(x) = {} & w_0|_{\text{local}}\, N_1(x) + w_0'|_{\text{local}}\, N_2(x) + w_0''|_{\text{local}}\, N_3(x) + w_1|_{\text{local}}\, N_4(x) \\ & + w_1'|_{\text{local}}\, N_5(x) + w_1''|_{\text{local}}\, N_6(x) \ . \end{aligned} \tag{C.1.5}$$

Since we would like to represent a function for which $\partial w/\partial X$ is continuous on the global mesh, using the global X coordinate, this implies that if the lengths of adjacent elements differ, the values of the local derivative degrees of freedom (DOFs) $w_0'|_{\text{local}}, \ldots, w_1''|_{\text{local}}$ must be different on each side of a mesh node where elements with different lengths meet.

However, we would like to have just one set of global degrees of freedom per global mesh node. Thus, as the global DOFs, we use the *global* derivatives $\partial w/\partial X$ and $\partial^2 w/\partial X^2$, which at a given mesh node agree on both sides of the node, regardless of the element sizes.

Thus, we must convert our global derivative degrees of freedom to their corresponding local values by applying equations (C.1.3)–(C.1.4) in reverse. At any given element e,

$$
\begin{aligned}
w_0'|_{\text{local}} &\equiv \frac{\partial w}{\partial x}|_{x=0} = a_1 \frac{\partial w}{\partial X}|_{X_e(x=0)} \equiv a_1 w_0'|_{\text{global}} \,, \\
w_1'|_{\text{local}} &\equiv \frac{\partial w}{\partial x}|_{x=1} = a_1 \frac{\partial w}{\partial X}|_{X_e(x=1)} \equiv a_1 w_1'|_{\text{global}} \,, \\
w_0''|_{\text{local}} &\equiv \frac{\partial^2 w}{\partial x^2}|_{x=0} = a_1^2 \frac{\partial^2 w}{\partial X^2}|_{X_e(x=0)} \equiv a_1^2 w_0''|_{\text{global}} \,, \\
w_1''|_{\text{local}} &\equiv \frac{\partial^2 w}{\partial x^2}|_{x=1} = a_1^2 \frac{\partial^2 w}{\partial X^2}|_{X_e(x=1)} \equiv a_1^2 w_1''|_{\text{global}} \,.
\end{aligned}
$$

Therefore, at element e, we have the representation

$$
\begin{aligned}
w_e(x) \equiv w(X_e(x)) = {} & w_0 N_1(x) + a_1 w_0' N_2(x) + a_1^2 w_0'' N_3(x) + w_1 N_4(x) \\
& + a_1 w_1' N_5(x) + a_1^2 w_1'' N_6(x) \,, \qquad \text{(C.1.6)}
\end{aligned}
$$

where $x \in [0, 1]$, the factors $w_0, \ldots, w_1''$ now refer to the global degrees of freedom $w_0|_{\text{global}}, \ldots, w_1''|_{\text{global}}$, and a_1 is evaluated for the element e. Similarly, for beam elements, we may write

$$
w_e(x) \equiv w(X_e(x)) = w_0 B_1(x) + a_1 w_0' B_2(x) + w_1 B_3(x) + a_1 w_1' B_4(x) \,. \qquad \text{(C.1.7)}
$$

We must use (C.1.6)–(C.1.7) not only when assembling the solution, but also when substituting the Galerkin representation of w into the weak form to generate the problem matrices—because the problem is assembled in terms of the global DOFs, that is, the DOF vector containing the discrete unknowns refers to the global DOFs. This substitution is just a particular application of the Galerkin assembly procedure.

Derivatives of w in the weak form are handled in the usual way, using (C.1.6)–(C.1.7) and the appropriate conversions. For example, if a $\partial w/\partial X$ appears in the weak form and we use C^2 Hermite elements with the affine coordinate mapping (C.1.1), we apply the equations (C.1.3) and (C.1.6), obtaining

$$\begin{aligned}\frac{\partial w(X)}{\partial X} &= \frac{\partial x}{\partial X}\frac{\partial w_{\mathrm{e}}(x)}{\partial x} \\ &= \frac{1}{a_1}\frac{\partial w_{\mathrm{e}}}{\partial x} \\ &= \frac{w_0}{a_1}\frac{\partial N_1}{\partial x}(x) + w_0'\frac{\partial N_2}{\partial x}(x) + a_1 w_0''\frac{\partial N_3}{\partial x}(x) + \frac{w_1}{a_1}\frac{\partial N_4}{\partial x}(x) + w_1'\frac{\partial N_5}{\partial x}(x) + a_1 w_1''\frac{\partial N_6}{\partial x}(x)\;. \end{aligned} \tag{C.1.8}$$

Finally, note that in the global DOFs, the pairs $w_1|_{\mathrm{e}-1} = w_0|_{\mathrm{e}}$, $w_1'|_{\mathrm{e}-1} = w_0'|_{\mathrm{e}}$ and $w_1''|_{\mathrm{e}-1} = w_0''|_{\mathrm{e}}$ refer to the same global degree of freedom (respectively). This is taken into account in matrix assembly as usual when dealing with finite elements.

In conclusion, let us summarize the key points of the procedure when applying Hermite elements to a weak form. As usual with finite elements, we first rewrite the integral over the one-dimensional domain $\Omega = \{\, X : X_1 < X < X_2 \,\}$ as a sum of integrals over the elements

$$\int_{\Omega} \ldots \, \mathrm{d}X = \sum_{\mathrm{e}} \int_{\Omega_{\mathrm{e}}} \ldots \, \mathrm{d}X \;.$$

Since all the integrands are, at most, finitely discontinuous, there are no Dirac deltas in the integrands anywhere in the domain. This applies especially at the boundaries between the elements. Thus the element boundaries generate no extra terms.

Next, we represent each integral over an element e as an integral over the reference element [0, 1]:

$$\int_{\Omega_{\mathrm{e}}} \ldots \, \mathrm{d}X = a_1 \int_0^1 \ldots \, \mathrm{d}x \;,$$

where a_1 is evaluated at the element e. This step is the change of variable in the integral. Then, in each integrand, we represent each global derivative in terms of its reference element equivalent, via the chain rule:

$$\frac{\partial}{\partial X^k}(\ldots) = \frac{1}{a_1^k}\frac{\partial}{\partial x^k}(\ldots)\;,$$

where a_1 is evaluated at the element e.

Finally, we insert the Galerkin representation of w, as expressed on the reference element, *but in terms of the global DOFs*. Refer to the Eq. (C.1.6) for C^2 Hermite elements and (C.1.7) for beam elements.

The first three steps are the same as with standard C^0 continuous finite elements. The conversion in the last step is the main difference between elements having derivative DOFs, such as Hermite elements, and elements having only function values as their degrees of freedom.

In principle, all four steps must be performed with any element type, but for continuously interpolating elements with no derivative DOFs (such as linear elements, the classical P^k Lagrange elements, or the modern hierarchical elements), the element-local and the global DOFs coincide. In practice, the conversion is in that case omitted, and one uses the element-local representation directly.

References

1. Hughes TJR (2000) The finite element method. Linear static and dynamic finite element analysis. Dover Publications, Inc., Mineola, N.Y., USA. ISBN 0-486-41181-8
2. Jacob F, Ted B (2007) A first course in finite elements. Wiley. ISBN 978-0-470-03580-1
3. Zienkiewicz OC, Taylor RL, Zhu JZ (2013) The finite element method: its basis and fundamentals, vol 1,7th edn. Butterworth–Heinemann

Index

N. Banichuk et al., *Stability of Axially Moving Materials*, Solid Mechanics and Its Applications 259, https://doi.org/10.1007/978-3-030-23803-2

MIX
Papier aus verantwortungsvollen Quellen
Paper from responsible sources
FSC® C105338

If you have any concerns about our products,
you can contact us on
ProductSafety@springernature.com

In case Publisher is established outside the EU,
the EU authorized representative is:
Springer Nature Customer Service Center GmbH
Europaplatz 3, 69115 Heidelberg, Germany

Printed by Libri Plureos GmbH
in Hamburg, Germany